VOLUME ONE HUNDRED AND SEVENTEEN

PROGRESS IN MOLECULAR BIOLOGY AND TRANSLATIONAL SCIENCE

Oligomerization in Health and Disease

VOLUME ONE HUNDRED AND SEVENTEEN

PROGRESS IN MOLECULAR BIOLOGY AND TRANSLATIONAL SCIENCE

Oligomerization in Health and Disease

Edited by

JESÚS GIRALDO

Institut de Neurociències and Unitat de Bioestadística
Facultat de Medicina
Universitat Autònoma de Barcelona
08193 Bellaterra, Spain

FRANCISCO CIRUELA

Unitat de Farmacologia
Departament de Patologia i Terapèutica Experimental
Facultat de Medicina
IDIBELL-Universitat de Barcelona
08907 Barcelona, Spain

ELSEVIER

AMSTERDAM • BOSTON • HEIDELBERG • LONDON
NEW YORK • OXFORD • PARIS • SAN DIEGO
SAN FRANCISCO • SINGAPORE • SYDNEY • TOKYO
Academic Press is an imprint of Elsevier

Academic Press is an imprint of Elsevier
The Boulevard, Langford Lane, Kidlington, Oxford, OX5 1GB, UK
32 Jamestown Road, London NW1 7BY, UK
Radarweg 29, PO Box 211, 1000 AE Amsterdam, The Netherlands
225 Wyman Street, Waltham, MA 02451, USA
525 B Street, Suite 1800, San Diego, CA 92101-4495, USA

First edition 2013

Library of Congress Cataloging-in-Publication Data
A catalog record for this book is available from the Library of Congress

British Library Cataloguing in Publication Data
A catalogue record for this book is available from the British Library

ISBN: 978-0-12-386931-9
ISSN: 1877-1173

Printed and bound by CPI Group (UK) Ltd, Croydon, CR0 4YY
Transferred to digital print 2012

CONTENTS

CONTRIBUTORS

Md. Faiz Ahmad
Department of Pharmacology, School of Medicine, Case Western Reserve University, Cleveland, Ohio, USA

Benedetta Bolognesi
Centre for Genomic Regulation, CRG and UPF, Barcelona, Spain

Lioubov I. Brueggemann
Department of Molecular Pharmacology and Therapeutics, Loyola University Chicago, Maywood, Illinois, USA

Kenneth L. Byron
Department of Molecular Pharmacology and Therapeutics, Loyola University Chicago, Maywood, Illinois, USA

Daniele Casciari
Department of Life Sciences, Dulbecco Telethon Institute, University of Modena and Reggio Emilia, Modena, Italy

Moraje V. Chandrakala
Department of Pharmacology and Systems Therapeutics, Mount Sinai School of Medicine, New York, USA

Eric Chevet
Inserm U1053, Université Bordeaux Segalen, Bordeaux, France

Oliver Daumke
Max-Delbrück-Centrum for Molecular Medicine, Crystallography, Berlin, Germany

Chris G. Dealwis
Department of Pharmacology, School of Medicine, and Center for Proteomics and Bioinformatics, Case Western Reserve University, Cleveland, Ohio, USA

Thomas E. DeCoursey
Department of Molecular Biophysics and Physiology, Rush University, Chicago Illinois, USA

Lakshmi A. Devi
Department of Pharmacology and Systems Therapeutics, and Department of Neuroscience and Friedman Brain Institute, Mount Sinai School of Medicine, New York, USA

Katja Faelber
Max-Delbrück-Centrum for Molecular Medicine, Crystallography, Berlin, Germany

Francesca Fanelli
Department of Life Sciences, Dulbecco Telethon Institute, University of Modena and Reggio Emilia, Modena, Italy

Angelo Felline
Department of Life Sciences, Dulbecco Telethon Institute, University of Modena and Reggio Emilia, Modena, Italy

Rafael Franco
Applied Medical Research Center (CIMA), University of Navarra, Pamplona, Spain, and Department of Biochemistry and Molecular Biology, University of Barcelona, Barcelona, Spain

Wakako Fujita
Department of Pharmacology and Systems Therapeutics, Mount Sinai School of Medicine, New York, USA

Kenji Funami
Department of Microbiology and Immunology, Hokkaido University Graduate School of Medicine, Sapporo, Japan

Song Gao
Max-Delbrück-Centrum for Molecular Medicine, Crystallography, Berlin, Germany, and State Key Laboratory of Oncology in South China, Sun Yat-sen University Cancer Center, Guangzhou, PR China

Saverio Gentile
Department of Molecular Pharmacology and Therapeutics, Loyola University Chicago, Maywood, Illinois, USA

Ivone Gomes
Department of Pharmacology and Systems Therapeutics, Mount Sinai School of Medicine, New York, USA

Javier González-Maeso
Department of Psychiatry; Department of Neurology, and Friedman Brain Institute, Mount Sinai School of Medicine, One Gustave L. Levy Place, New York, USA

Ingo H. Greger
Neurobiology Division, MRC Laboratory of Molecular Biology, Cambridge, United Kingdom

Lynn Gregory-Pauron
Unité de Recherche Animal et Fonctionnalités des Produits Animaux, Université de Lorraine, INRA, Vandoeuvre-lès-Nancy, France

Tracy M. Handel
Skaggs School of Pharmacy & Pharmaceutical Science, University of California, La Jolla, San Diego, California, USA

Aylin Carla Hanyaloglu
Department of Surgery and Cancer, Institute of Reproductive and Developmental Biology, Hammersmith Campus, Imperial College London, London, United Kingdom

Kosuke Hashimoto
Omics Science Center, RIKEN, Yokohama, Japan

Volker Haucke
Leibniz-Institute for Molecular Pharmacology (FMP) & Freie Universität Berlin, Berlin, Germany

Martin Held
Institute for Mathematics, Freie Universität Berlin, Berlin, Germany

Beatriz Herguedas
Neurobiology Division, MRC Laboratory of Molecular Biology, Cambridge, United Kingdom

Terrell Holloway
Department of Psychiatry, Mount Sinai School of Medicine, One Gustave L. Levy Place, New York, USA

Ilpo Huhtaniemi
Department of Physiology, Institute for Biomedicine, University of Turku, Turku, Finland, and Department of Surgery and Cancer, Institute of Reproductive and Developmental Biology, Hammersmith Campus, Imperial College London, London, United Kingdom

Albert Jeltsch
Institute of Biochemistry, Stuttgart University, Stuttgart, Germany

Kim Carol Jonas
Department of Surgery and Cancer, Institute of Reproductive and Developmental Biology, Hammersmith Campus, Imperial College London, London, United Kingdom

Renata Z. Jurkowska
Institute of Biochemistry, Stuttgart University, Stuttgart, Germany

Nurit Keinan
Department of Life Sciences, and The National Institute for Biotechnology in the Negev, Ben-Gurion University of the Negev, Beer-Sheva, Israel

Amir R. Khan
School of Biochemistry and Immunology, Trinity Biomedical Sciences Institute, Trinity College Dublin, Dublin 2, Ireland

James Krieger
Neurobiology Division, MRC Laboratory of Molecular Biology, Cambridge, United Kingdom

Emmanuel D. Levy
Department of Structural Biology, Weizmann Institute of Science, Rehovot, Israel

Thomas Madej
National Center for Biotechnology Information, National Library of Medicine, National Institutes of Health, Bethesda, Maryland, USA

Eva Martínez-Pinilla
Applied Medical Research Center (CIMA), University of Navarra, Pamplona, Spain

Misako Matsumoto
Department of Microbiology and Immunology, Hokkaido University Graduate School of Medicine, Sapporo, Japan

Peter J. McCormick
Department of Biochemistry and Molecular Biology, University of Barcelona, Barcelona; Centro de Investigación Biomédica en Red de Enfermedades Neurodegenerativas (CIBERNED), Madrid, and Institute of Biomedicine of the University of Barcelona (IBUB), Barcelona, Spain

Andreas Merdes
Centre de Biologie du Développement, Université Paul Sabatier, Toulouse, France

Dario Mizrachi
Department of Life Sciences, and The National Institute for Biotechnology in the Negev, Ben-Gurion University of the Negev, Beer-Sheva, Israel

José L. Moreno
Department of Psychiatry, Mount Sinai School of Medicine, One Gustave L. Levy Place, New York, USA

Hafumi Nishi
National Center for Biotechnology Information, National Library of Medicine, National Institutes of Health, Bethesda, Maryland, USA, and Current address: Yokohama City University, Yokohama, Japan

Frank Noé
Institute for Mathematics, Freie Universität Berlin, Berlin, Germany

Hiroyuki Oshiumi
Department of Microbiology and Immunology, Hokkaido University Graduate School of Medicine, Sapporo, Japan

Anna R. Panchenko
National Center for Biotechnology Information, National Library of Medicine, National Institutes of Health, Bethesda, Maryland, USA

York Posor
Leibniz-Institute for Molecular Pharmacology (FMP) & Freie Universität Berlin, Berlin, Germany

James H. Prestegard
Complex Carbohydrate Research Center, University of Georgia, Athens, Georgia, USA

Francesco Raimondi
Department of Life Sciences, Dulbecco Telethon Institute, University of Modena and Reggio Emilia, Modena, Italy

Marie-Hélène Remy
Centre de Biologie du Développement, Université Paul Sabatier, Toulouse, France

Ana Ricobaraza
Applied Medical Research Center (CIMA), University of Navarra, Pamplona, Spain

Adolfo Rivero-Müller
Department of Physiology, Institute for Biomedicine, University of Turku, Turku, Finland

Xavier Salvatella
ICREA, Barcelona, Spain, and Joint BSC-IRB Research Programme in Computational Biology, IRB Barcelona, Barcelona, Spain

Michele Seeber
Department of Life Sciences, Dulbecco Telethon Institute, University of Modena and Reggio Emilia, Modena, Italy

Tsukasa Seya
Department of Microbiology and Immunology, Hokkaido University Graduate School of Medicine, Sapporo, Japan

Joshua S. Sharp
Complex Carbohydrate Research Center, University of Georgia, Athens, Georgia, USA

Varda Shoshan-Barmatz
Department of Life Sciences, and The National Institute for Biotechnology in the Negev, Ben-Gurion University of the Negev, Beer-Sheva, Israel

Susan M.E. Smith
Department of Pathology and Laboratory Medicine, Emory School of Medicine, Atlanta Georgia, USA

Saïd Taouji
Inserm U1053, Université Bordeaux Segalen, Bordeaux, France

Gian Gaetano Tartaglia
Centre for Genomic Regulation, CRG and UPF, Barcelona, Spain

Sarah A. Teichmann
MRC Laboratory of Molecular Biology, and Wellcome Trust Genome Campus, Cambridge, United Kingdom

Xu Wang
Department of Chemistry and Biochemistry, Arizona State University, Tempe, Arizona, USA

Sebastian Wolf
Inserm U1053, Université Bordeaux Segalen, Bordeaux, France

PREFACE

In cells, nearly all functions are mediated by multiprotein complexes, and also it is believed that approximately one-third of cellular proteins are oligomeric, thus showing a quaternary association. Therefore, it seems reasonable that oligomerization of proteins in cellular systems constitutes a widespread biological phenomenon. Indeed, two or more polypeptide chains often associate forming an oligomer. Thus, it can be accepted that protein oligomerization is prevalent in nature since this phenomenon, which appeared early in protein development, has been preserved through protein evolution. Certainly, many primitive proteins, mostly enzymes and soluble proteins, assembled into homo- or hetero-oligomeric structures constituting an entrenched feature of these proteins. In such cases, oligomerization has been shown to boost protein stability, to allow protein–protein allosterism, to let the compartmentalization of reactions, and also to make possible the interaction between multitask sites within oligomeric soluble proteins. On the other hand, the fundamental rules driving membrane protein oligomerization are relatively constraint and thus yet to be completely understood, although a large number of investigations have been performed in this direction. Overall, in this book we aim to bring together a number of leading experts in the field of protein oligomerization in order to summarize the main aspects revolving around oligomeric proteins functioning both in normal and pathological conditions.

The book is structured in five sections. Beginning with the topic of structural biology, Section I of the book embraces all these structural and physicochemical aspects related to protein oligomerization placed in an evolutionary context. Chapter 1, by Hafumi Nishi, Kosuke Hashimoto, Thomas Madej, and Anna R. Panchenko, provides a detailed analysis of the evolutionary mechanisms responsible for protein oligomerization remarking particularly the relation that the dynamic equilibrium between different oligomerization states may have to physiological conditions both in health and disease. Oligomer structure and evolutionary mechanisms are further discussed in Chapter 2, in which Emmanuel D. Levy and Sarah A. Teichmann review the existing methods for quaternary structure analysis and examine interface properties. In Chapter 3, by Benedetta Bolognesi and Gian Gaetano Tartaglia, the physicochemical determinants of protein oligomerization are comprehensively presented and their link with aggregation rates examined. The analysis reveals interesting features in the context of evolutionary pressure, which may provide useful predictors for protein misfolding and aggregation. Section I ends with Chapter 4 in which Xavier Salvatella reviews the structure of amyloid fibrils, highly organized protein aggregates which have been associated with some neurological diseases.

Next, the book moves on to the study of membrane proteins oligomerization. First, five chapters in Section II provide a state-of-the-art flavor of seven-transmembrane receptor or G protein-coupled receptor (GPCR) oligomerization. In Chapter 5, by Francesca Fanelli, Michele Seeber, Angelo Felline, Daniele Casciari, and Francesco Raimondi, a structural view of GPCR oligomerization is provided with special emphasis on the FiPD-based approach. Interestingly, this approach has been proved to be effective in predicting the supramolecular organization of membrane proteins including GPCRs. Furthermore, the GPCR oligomer-based drug discovery is discussed from a technical point of view by Rafael Franco, Eva Martínez-Pinilla, Ana Ricobaraza, and Peter J. McCormick (Chapter 6). Chapter 7, by Adolfo Rivero-Müller, Kim Carol Jonas, Aylin Carla Hanyaloglu, and Ilpo Huhtaniemi, summarizes the evidence for GPCR di/oligomerization in living organisms with a particular focus of their involvement in health and disease. José L. Moreno, Terrell Holloway, and Javier González-Maeso in Chapter 8 review current data on the role of GPCR oligomerization in receptor signaling as well as its potential implication in neuropsychiatric disorders such as schizophrenia, depression, and Parkinson's disease. Finally, Section II dealing with GPCR oligomerization ends with Chapter 9 by Ivone Gomes, Wakako Fujita, Moraje V. Chandrakala, and Lakshmi A. Devi which describes the evidences so far for opioid, cannabinoid, and dopamine receptor oligomerization and the potential role of these receptor oligomers in pathophysiological conditions associated with drugs of abuse.

Subsequently, in Section III, four chapters deal with several aspects revolving around the relationship between ion channel functioning and oligomerization. Therefore, Chapter 10, by Lioubov I. Brueggemann, Saverio Gentile, and Kenneth L. Byron, describes the types of protein–protein interactions among pore-forming voltage-gated potassium channel subunits and their auxiliary protein partners. Next, the oligomerization of the mitochondrial protein voltage-dependent anion channel (VDAC1) is analyzed in Chapter 11; thus, Varda Shoshan-Barmatz, Dario Mizrachi, and Nurit Keinan provide insight into the key role of VDAC1 in mammalian cell life and death and emphasize VDAC1 function in apoptosis, focusing on VDAC1 oligomerization as a key step in mitochondria-mediated apoptosis. In Chapter 12, Susan M.E. Smith and Thomas E. DeCoursey analyzed the consequences of voltage-gated proton channel (H_V1) dimerization which provides the chance for a cooperative gating mechanism that gives rise to a steeply voltage-dependent conductance, important feature for optimum compensation for Nox activity and phagocyte function. Finally, this section ends with Chapter 13 where Beatriz Herguedas, James Krieger, and Ingo H. Greger discuss the principles of oligomeric ion channel assembly using neuronal ionotropic glutamate receptors as a prototypic example.

Once the oligomerization of membrane proteins was analyzed, the volume moves on to the study of cytosolic/soluble protein oligomerization. Accordingly, two sections formed by four chapters each tackle several aspects surrounding the oligomerization of enzymes (Section IV) and regulatory proteins (Section V). Therefore, Chapter 14, by Md. Faiz Ahmad and Chris G. Dealwis, discusses how subunit oligomerization impinges into allosteric regulation and substrate recognition of ribonucleotide reductase, the enzyme that converts ribonucleosides to deoxyribonucleosides. Katja Faelber, Song Gao, Martin Held, York Posor, Volker Haucke, Frank Noé, and Oliver Daumke in Chapter 15 describe the molecular mechanisms of how dynamin oligomerizes. Interestingly, they introduce a model describing how dynamin oligomerization is linked to membrane fission. In addition, in this chapter, the molecular mechanisms of how mutations in dynamin could lead to the congenital diseases, centronuclear myopathy and Charcot-Marie-Tooth disease, are discussed. In Chapter 16, by Albert Jeltsch and Renata Z. Jurkowska, the complex oligomerization of the Dnmt3a DNA methyltransferase and its functional implications are analyzed. Section IV ends with Chapter 17 by Saïd Taouji, Sebastian Wolf, and Eric Chevet which deals with the oligomerization of ER stress transducers and in particular inositol-requiring enzyme (IRE1). Thus, they describe the oligomerization-dependent mechanisms for modulating unfolded protein response signals on and off.

The volume ends with Section V which focuses on the role of oligomerization in regulatory protein function in health and disease. Therefore, Chapter 18, by Misako Matsumoto, Kenji Funami, Hiroyuki Oshiumi, and Tsukasa Seya, deals with the oligomerization of Toll-IL-1-receptor-containing adaptor molecule-1 with downstream signaling molecules, thus forming a speckle-like signalosome in the cytosol. Interestingly, this oligomerization phenomenon plays a key role in virus infection and antitumor immunity. In Chapter 19, by Marie-Hélène Remy, Andreas Merdes, and

Lynn Gregory-Pauron, the potential of the gamma-tubulin complex as a pharmacological target (i.e., to control proliferation of cells) is highlighted. Next, Xu Wang, Joshua S. Sharp, Tracy M. Handel, and James H. Prestegard in Chapter 20 summarize the existing knowledge regarding to the structure and function of chemokine oligomers. In addition, in this chapter, the importance of chemokine interactions with cell-surface glycosaminoglycans is analyzed in view that this interaction appears to be chemokine oligomer dependent. Finally, Section V ends with Chapter 21 by Amir R. Khan which deals with the oligomerization of Rab/effector complexes in the regulation of vesicle trafficking.

Collectively, from the information reported in this book, it is concluded that protein oligomerization constitutes a superior device that provides diversity, complexity, and regulation to nearly all cellular processes. Indeed, within the benefits of protein oligomerization to cellular functioning, we can remark the increase in the efficiency of processes involving multiple proteins (e.g., signal transduction and enzymatic reactions, within others), the role in stabilizing and protecting proteins against proteolytic processes (e.g., chaperone activity), and the allosteric regulation of protein functionality by protein–protein interactions (e.g., *trans*-inhibition and *trans*-activation phenomena of receptors and enzymes). It is important to mention here that communication across oligomeric proteins is fundamental for proper oligomer biological activity; thus, the knowledge of the interfaces between protomers forming a named oligomer seems critical to accurately understand oligomer's functionality. In conclusion, protein oligomerization may often be an advantageous feature; however, oligomerization can have eventually deleterious consequences when nonnative oligomers associate with pathogenic states, and in such cases the knowledge of the oligomeric protein biology seems critical to allow potential pharmacological intervention in these pathological conditions associated with oligomeric protein misfunction.

Finally, we would like to express our best thanks to all the contributing authors and co-authors of this book for their commitment, time, experience, and patience. Also, our special thanks to the anonymous reviewers for their excellent contributions to the peer-review process. In addition, we want to express our special gratitude to Professor P. Michael Conn, editor-in-chief of *Progress in Molecular Biology and Translational Science*, for giving us this opportunity, and to the staff at Elsevier for their assistance and cooperation.

Jesús Giraldo
Francisco Ciruela
Barcelona 2013

SECTION I

Structural, Physicochemical, and Evolutionary Aspects of Protein Oligomerization

CHAPTER ONE

Evolutionary, Physicochemical, and Functional Mechanisms of Protein Homooligomerization

Hafumi Nishi[*,1], **Kosuke Hashimoto**[†], **Thomas Madej**[*], **Anna R. Panchenko**[*]

[*]National Center for Biotechnology Information, National Library of Medicine, National Institutes of Health, Bethesda, Maryland, USA
[†]Omics Science Center, RIKEN, Yokohama, Japan
[1]Current address: Yokohama City University, Yokohama, Japan

Contents

Abstract

Protein homooligomers afford several important benefits for the cell; they mediate and regulate gene expression, activity of many enzymes, ion channels, receptors, and cell–cell adhesion processes. The evolutionary and physical mechanisms of oligomer formation are very diverse and are not well understood. Certain homooligomeric states may be conserved within protein subfamilies and between different subfamilies, therefore providing the specificity to particular substrates while minimizing interactions with unwanted partners. In addition, transitions between different oligomeric states may regulate protein activity and support the switch between different pathways. In this chapter, we summarize the biological importance of homooligomeric assemblies, physicochemical properties of their interfaces, experimental methods for their identification, their evolution, and role in human diseases.

Progress in Molecular Biology and Translational Science, Volume 117
ISSN 1877-1173
http://dx.doi.org/10.1016/B978-0-12-386931-9.00001-5

1. FUNCTIONAL ROLES OF HOMOOLIGOMERS IN A CELL

Only a small fraction of proteins function in isolation and many soluble and membrane-bound proteins form oligomeric complexes defined as those having identical or very similar homologous chains (called "homooligomers" hereafter).[1–3] It is difficult to overestimate the functional importance of protein homooligomers which provide the diversity and specificity of many pathways, regulate their cross talk, and may mediate and regulate gene expression, activity of enzymes, ion channels, receptors, and cell–cell adhesion processes.[4] Moreover, large assemblies consisting of many identical subunits have advantageous regulatory properties as they can undergo sensitive phase transitions.[3,5] The molecular structure of homooligomers can also provide sites for allosteric regulation, generate new binding sites at interfaces to increase specificity, and diversity.[6] In addition, oligomerization allows proteins to form large assemblies without increasing genome size and may enhance stability, while the reduced surface area of the components in a complex can offer protection against denaturation.[2,5,7]

There are certain characteristic protein regions (discussed in detail in the following sections) that might be important for providing a given oligomeric state. These features can function as specificity determinants and be responsible for high binding specificity to certain interacting partners while minimizing interactions with other unwanted partners. Some protein regions modulating oligomerization may be preserved in evolution only within a specific protein subfamily which can be characterized by a well-defined oligomeric state. Such an evolutionary mechanism would be essential for the separation of functional pathways of close paralogs, preventing the possible usage of similar surface regions interacting with the same or very similar partners or facilitating through specific features the interactions with the novel partners. As shown in Table 1.1, proteins in different oligomeric states may have various binding affinities and functional activities. Figure 1.1 and the two following examples illustrate the development of new protein specificities through homooligomerization in different organisms or in paralogs from the same organism.

The first example involves proteins from the human p53 C-terminal domain family which function as homotetramers. Two of these family members (p63 and p73) can also form mixed heterotetramers; however, p53 protein cannot associate with either p63 or p73.[18] Interestingly, a recent study showed that p63 and p73 are different from p53 in that the former have

Table 1.1 Experimental examples of proteins regulated through transitions between different oligomeric states

Protein name	Regulation of transition	Function of one oligomeric state	Function of the other oligomeric state	Reference
Pyruvate kinase M2	Tetramer–dimer ratio is regulated by fructose 1,6-P2, serine, and phosphorylation	Inactive dimer leads to accumulation of phosphometabolites in tumor cells	Active tetramer associates with glycolytic enzyme complex to perform its functions in normal cell	8
Rad53, DNA damage checkpoint kinase in yeast	Assembly/disassembly is regulated through phosphorylation		Rad53 forms oligomers upon DNA damage. Rad53/Chk2 homooligomerization is a mechanism to activate signal in DNA damage responses	9
Human thymidylate synthase		Conformational changes accompanying dimer/monomer transition lead to exposure of Cys residues of monomer and binding to TS mRNA and autoregulation of translation by its protein product	Dimeric form is an active thymidylate synthase	10
Estrogen receptor	Phosphorylation of Tyr547 by Src tyrosine kinases controls receptor dimerization	Monomer probably does not bind DNA	Dimerization induces DNA binding	11
Hexokinase PII	Reversible phosphorylation is carbon source dependent. Glucose promotes dephosphorylation	When monomer is phosphorylated, cells cannot provide glucose repression of invertase	Dimer is dephosphorylated	12

Continued

Table 1.1 Experimental examples of proteins regulated through transitions between different oligomeric states—cont'd

Protein name	Regulation of transition	Function of one oligomeric state	Function of the other oligomeric state	Reference
Mammalian STE20-like kinase (MST)	Okadaic acid induces phosphorylation in the activation loop and monomerization	Phosphorylated active monomeric enzyme is translocated into nucleus	Dimerization contributes to the cytoplasmic retention of MST	13
Bacterial enhancer-binding protein nitrogen regulatory protein C (NtrC)	Phosphorylation induces oligomerization and changes interdomain interactions		Oligomerization activates the ATPase activity	14
C-Terminal Src kinase (Csk)			SH3 domains form homodimers and bury recognition surface for SH3 interactors including tyrosine phosphatase	15
D-Hydantoinase	Removal of C-terminal Arg induces monomer formation	Monomer shows 40% activity reduction	Dimer and tetramers are active	16
14-3-3 Proteins	Phosphorylation of Ser58 disrupts dimers	Monomer–dimer transition modulates target protein activity	Monomer–dimer transition modulates target protein activity	17

Adapted from Ref. 3.

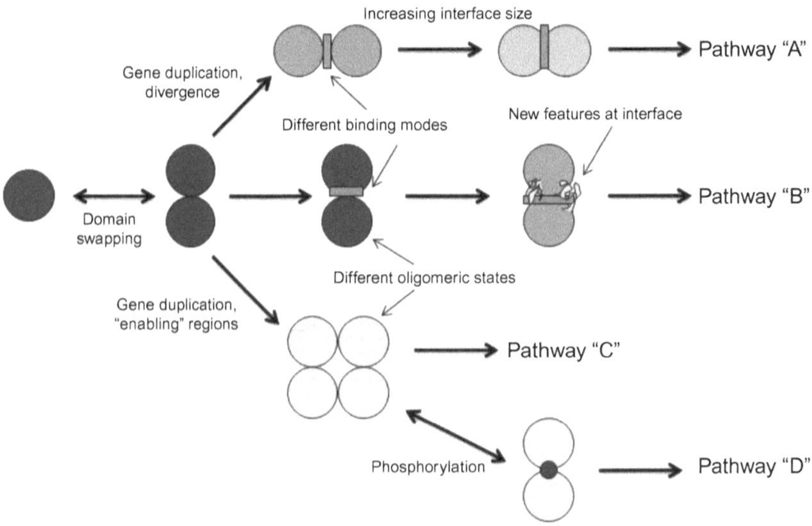

Figure 1.1 Illustration for the development of novel protein specificities and regulation of protein activity through homooligomerization. *Adapted from Ref. 3.* (For color version of this figure, the reader is referred to the online version of this chapter.)

an additional alpha-helix which stabilizes the tetramer.[19] The absence of this helix in p53 explains its inability to oligomerize with p63/p73, resulting in the separation of the p53 pathway from the paralogous p63/p73 pathways. Another example is the LIM-domain-binding protein Ldb1, a nuclear adaptor protein which interacts with diverse proteins containing LIM domains and plays essential roles in development and cellular differentiation.[20,21] Humans have two close paralogs Ldb1 and Ldb2. Since the loss of Ldb1 causes severe developmental defects in embryos that are not compensated for by Ldb2, Ldb1 and Lbd2 are likely to participate in different pathways. It is known that Ldb1 forms a trimer while Ldb2 exists in a monomer–tetramer–octamer equilibrium, suggesting that the oligomeric differences might enable Ldb1 and Ldb2 to interact with different partners in different pathways.[22]

2. EXPERIMENTAL CHARACTERIZATION AND COMPUTATIONAL PREDICTION

Experimental characterization of homooligomeric structures, their dynamic equilibrium between different oligomeric states, and computational inference of the biologically relevant oligomeric form from the crystalline state have always been very challenging. Various experimental and computational methods have been devised over the years to meet these challenges. They aim to accurately identify biological interactions and distinguish them from nonbiological crystal packing or nonspecific aggregation. The main experimental techniques include X-ray and neutron scattering, mass spectrometry, gel filtration, dynamic light scattering, analytical ultracentrifugation, and fluorescence resonance energy transfer (FRET)[23] (see also references for Table 1.1). For instance, analytical centrifugation and gel filtration chromatography provide data on molecular mass distribution, the subunit stoichiometry of the complexes, and equilibrium constants. FRET characterizes the kinetics and dynamics of complex formation, monitoring the extent of energy transfer between donor and acceptor, whereas X-ray and neutron scattering offer the atomic details of interaction interfaces. Nowadays, proteins are being crystallized using high-throughput techniques and very often without the extensive biochemical or biophysical characterization of their oligomeric states. This raises the importance of computational methods to reconstruct biological assemblies from crystalline states.[24–26] These algorithms apply crystallographic symmetry operations and then attempt to differentiate the biological from the crystal packing interfaces by computational criteria. For example, the PISA algorithm applies graph theory to find the set of stable assemblies, which fill all the crystal space in a regular manner, with nodes and edges in the graph corresponding to protein monomers and interfaces between them.[25] To distinguish "biologically relevant" from crystal packing interactions, one can also use *ad-hoc* scoring schemes which are determined by interface area, amino acid composition, number of contacts, topological complementarity, hydrogen bonding, and other characteristics.[27–31]

3. PHYSICOCHEMICAL PROPERTIES OF HOMOOLIGOMERIC INTERFACES

The amino acid composition of homooligomeric interfaces differs from those of crystal packing interfaces, heterooligomeric interfaces, and solvent-exposed surfaces and largely depends on the type of homooligomeric complexes.[27–30,32,33] Moreover, besides interface composition, allosteric mutations may also affect intersubunit geometry and contribute to the evolution of oligomeric states.[34] Obligate homooligomers (complexes where monomers are unstable and are prone to unfold upon isolation) are characterized by the large fraction of hydrophobic and, to a lesser extent, aromatic residues on their interfaces, while nonobligate (transient) complexes include more polar and charged residues.[28,32,35] Interfaces of obligate homooligomers are usually larger[28] but contain fewer hydrogen bonds per residue compared to heterooligomers.[33] The analysis of kinetic and equilibrium data on dimeric proteins shows that per-residue interface and surface areas of "three-state dimers" (monomers are stable on their own) are significantly smaller than that of "two-state dimers" (monomers are not stable when separated from the complex).[36]

As can be seen in Fig. 1.2, homooligomers in the Protein Databank (PDB) mostly form dimers (37%), then trimers (5%) and tetramers (7%) with predominantly cyclic or dihedral symmetries. Moreover, eukaryotes have a considerably smaller fraction of homooligomers (40%) compared to eubacteria and archaea (60%). This might be explained by previous observations that eukaryotic proteins evolved through the more extensive domain fusion and shuffling events and contain more heterooligomers and multidomain proteins.[37] The self-attraction and symmetry in homooligomers might occur due to stability, foldability, and evolutionary optimization.[38,39] The physical effect of a statistically enhanced self-attraction was recently modeled to show that interactions between identical random surfaces are stronger than attractive interactions between different random surfaces of the same size.[38] Furthermore, it was demonstrated that the efficiency of co-aggregation between different monomers and protein domains decreases with decreasing sequence identity.[40] Binding arrangements involving isologous homooligomeric interfaces with a two-fold symmetry axis seemed to be more frequently conserved in evolution compared to nonisologous interfaces.[41] In addition, symmetrical isologous dimers were shown to

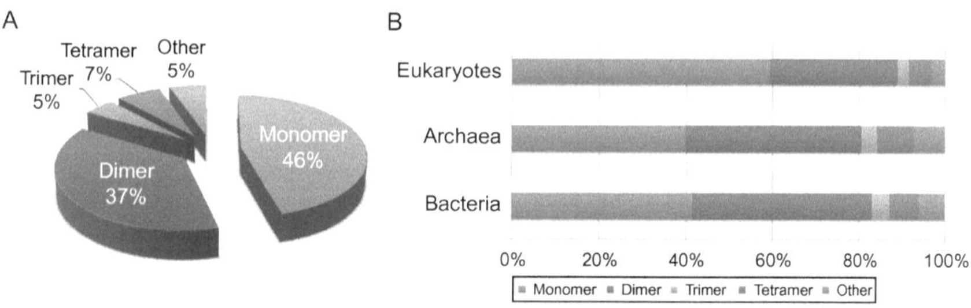

Figure 1.2 Distribution of different homooligomeric states in a nonredundant set of Protein Data Bank (PDB) structures (A) and in Eukaryotes, Archaea, and Eubacteria (B). The nonredundant set of structures was obtained using the criteria of BLAST p value $<10^{-7}$ on all PDB chains and oligomeric state annotations were taken from PDB. (For color version of this figure, the reader is referred to the online version of this chapter.)

contain more residues in disordered regions than heterodimers or non-symmetrical dimer interfaces,[42] which might modulate the high specificity of interactions between the dimer complexes and their interacting partners and play an important role in allosteric regulation.[43]

4. EVOLUTIONARY MECHANISMS TO FORM HOMOOLIGOMERS

A number of evolutionary mechanisms of homooligomerization have been proposed. According to some of them, evolutionary pathways may follow kinetic pathways of two-state or three-state folding.[44,45] At the same time, assembly pathways of oligomers may mimic evolutionary pathways, and it was suggested and verified experimentally that homomeric proteins with dihedral symmetry may evolve and assemble through their cyclic intermediates.[46,47] One of the major mechanisms of oligomerization is gene duplication with subsequent diversification where one copy of the gene retains its original function whereas the other gene copy is under relaxed evolutionary constraints, allowing it to develop new functional specificities. This mechanism may lead to the formation of oligomeric paralogs and may create protein complexes with novel specificities.[41,48–51] While in *S. cerevisiae* up to 20% of complexes evolved by stepwise partial duplications,[48] this mechanism was found to be less prevalent in *Escherichia coli*.[49]

Similarity in protein sequences, folds, and functions between two orthologous proteins does not necessarily imply that they will have the same interacting partners.[52] Although homooligomeric states and binding interfaces are usually conserved within the clades on phylogenetic trees, they can only be reliably transferred from very close homologs (sharing more than 30% sequence identity for oligomeric states and sharing more than 50–70% identity for binding modes inference).[41,46] In some cases, the oligomeric state can be evolutionarily conserved while the binding arrangement can be markedly different. This points to the possibility that interactions between monomers are not necessarily inherited from the ancestral homo-oligomer but rather can develop anew in evolution. For example, proteins from the glycosyltransferase family may function as monomers, dimers, or tetramers in different organisms. Although the dimeric and tetrameric proteins appear to be as ancient as the monomers, the binding modes can differ even between very similar proteins, for instance, engaging completely opposite sites on the molecule.[50] Despite this diversity, there are certain characteristics of evolutionarily conserved homooligomeric binding modes.

For example, binding modes with larger interfaces are more frequently conserved, while smaller interfaces are acquired more recently in evolution.[41,46]

In the next few sections, we will consider several evolutionary mechanisms that play key roles in homooligomerization: domain swapping, formation of Leucine zippers, amino acid substitutions, and insertions/deletions.

4.1. Domain swapping

Domain swapping includes opening up of the monomeric conformation and exchanging identical regions between two monomers. Although the term "domain swapping" was introduced by Eisenberg and coworkers in 1994,[53] the first example of this type of oligomer, bovine pancreatic ribonuclease (RNase A, Fig. 1.3A), was found by Crestfield *et al.* in 1962.[54] Later on, it was shown that RNase A exchanges N-termini upon dimerization. Originally, domain swapping was applied only in the context of homooligomers; however, the usage of the term has been extended lately and it now encompasses different cases including those where swapping may occur between regions of closely homologous but nonidentical monomers. Overall, about 300 examples of domain swapping are reported in protein structures from the Protein Data Bank.[55,56] The length of the swapped region may vary considerably, from a single secondary structure element to regions of over 100 residues. Although swapping of terminal regions seems to be more common, the regions can be located anywhere in the structure, and on occasion, two or more distinct regions may be swapped.[55,57] For example, RNase A has two swapped regions at the N- and C- termini and can form a dimer, trimer, or higher-order oligomers (Fig. 1.3A).[58]

Interdomain linker regions have been studied in order to identify regions responsible for domain swapping.[59–61] The importance of proline residues in linker regions was recognized by several experimental studies,[59,60] and recently, the amino acid proline has been shown to have the highest propensity in hinge regions, similar to loop structures.[62] However, the roles and number of these proline residues varied between the proteins. Cell cycle regulatory protein, p13suc1, for example, has two prolines in the hinge loop which control its monomer–dimer equilibrium.[59] Interestingly, the replacement of the first proline by alanine results in the shift of monomer–dimer equilibrium toward the monomer, whereas the replacement of the second proline shifts the equilibrium toward the dimer. In another example, in the hinge region of cyanovirin-N, the substitution of proline by glycine

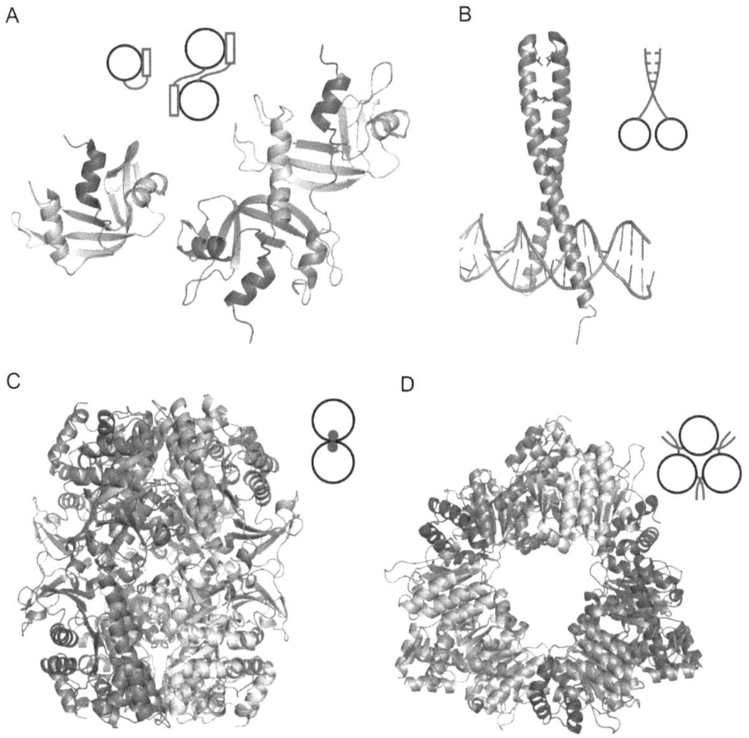

Figure 1.3 Different mechanisms of homooligomerization. (A) Domain swapping of RNase A: monomer (1RTB) and domain-swapped dimer (1A2W) of RNase A. The swapped regions and domain linker regions are shown as red and yellow, respectively. (B) Leu-zipper of GCN4 (1YSA). Leu at position "d" is shown as stick model and colored in red. (C) Amino acid substitutions for designed homooctamer of L-rhamnulose-1-phosphatase aldolase (2UYU). The original protein exists as a tetramer which is shifted to the octamer upon a single substitution (A88F, shown in red). (D) Insertions of *N*-acetyl-L-glutamate kinase hexamer (2BUF). Inserted N-terminal helix shown in red enables the hexamer formation. *Adapted from Ref. 3.* (See Color Insert.)

stabilizes both the monomeric and dimeric states, while the replacement of serine by proline results in an exclusively dimeric protein.[60] The length of the domain linker itself may also affect the monomer–dimer equilibrium. For instance, interleukin-5 (IL-5) belongs to the short-chain helical cytokine subfamily and forms a domain-swapped homodimer although other members in the family form monomers. The loop between helices C and D is shorter in IL-5 compared to other family members and the engineered lengthening of the loop enabled IL-5 to form an active monomer.[63]

There are computational and theoretical studies which aim to examine the mechanism of domain swapping. Using a model which incorporates only the monomeric conformation in an energy function, it has been shown that the intermediate states represent open-ended, domain-swapped dimers, and it also appears that overall monomer topology is a main determinant of the structure of domain-swapped dimers.[64] According to another mechanism for domain swapping, the swapping starts from the C-terminus and progresses by exchanging an increasing portion of the chains until a stable conformational state is reached. In contrast to the previous study, this process does not include complete monomer unfolding.[65]

4.2. Structural oligomerization motifs

There are several known structural motifs used by proteins for oligomerization, the alpha-helical coiled-coil being the most common one among them. The coiled-coil motif appears in proteins with different functions, for example, alpha-keratin and vimentin (intermediate filaments of the cytoskeleton), myosin and kinesin (motor proteins), SNARE proteins (mediation of vesicle fusion), and the b-Zip family (transcription factor).[66] For instance, the b-Zip family uses the Leu-zipper motif for DNA binding (Fig. 1.3B). The coiled-coil motif is observed as a series of continuous heptad repeats $(abcdefg)_n$ in protein sequences.[67] When the motif forms an alpha-helical structure, hydrophobic residues at the "a" and "d" positions interact with each other to form a helical bundle. The "Leu-zipper" motif is a type of coiled-coil structure where leucine is frequently observed at the "d" position (Fig. 1.3B).[68] The coiled-coil motifs may form different oligomeric states depending on the particular amino acids in the heptad repeat. For example, the yeast transcription factor GCN4 forms a dimeric coiled coil; however, the replacements of Ile by Leu at the "a" and Leu by Ile at the "d" positions enables GCN4 to form a tetramer.[69] Other combinations of Leu, Ile, and Val at the "a" and "d" positions may also cause trimerization,[69,70] whereas replacements in positions "e" or "g" can lead to higher-order oligomers such as heptamers.[71] Moutevelis and Woolfson organized coiled-coil structures in a "periodic table"[72] where coiled-coil structures were arranged in columns with increasing numbers of alpha-helices. According to the authors' results, 74% of coiled-coil structures were dimers and 12% formed trimers.

4.3. Amino acid substitutions

Amino acid substitutions introduced on the protein surface/interface may cause association or dissociation of homooligomers. A simple rule for amino acid substitutions that mediate the oligomeric states was proposed recently:

the replacement of solvent-exposed residues by more hydrophobic and/or larger protruding residues tends to shift the equilibrium toward the formation of oligomers.[73] Another study found that as few as two substitutions could shift the average composition of a surface patch to that of an equivalent interface.[74] The artificial design of new oligomers by amino acid substitutions was performed on four protein assemblies[75] (Fig. 1.3C). Some of these proteins needed only a single amino acid replacement to associate to a higher-order complex, and it was discovered that introducing large nonpolar side chains, such as phenylalanines or tryptophans, facilitated the complex formation.

Residues or sites with a large contribution to binding energy are sometimes called "binding hotspots." Originally, hotspots were recognized by alanine scanning of the interface region between human growth hormone and the extracellular domain of its receptor as those few residues which caused the largest reduction in the binding affinity.[76] Bogan and Thorn compiled a database of alanine mutants and described some important properties of hotspot residues.[77] It was shown that hotspots were located near the center of the interface completely occluded from bulk solvent, with an abundance of arginine, tryptophan, and tyrosine.[77,78] Additionally, it was suggested that some tightly packed binding hot spots could form a cluster on an interface.[79]

Amino acid substitutions on homooligomer interfaces may change the functional activity of a protein and have been implicated in several diseases. For example, mutations leading to the disease fructose intolerance are shown to destabilize the tetramer of the enzyme fructose-1,6-biphosphate aldolase A and thereby decrease its activity,[80] while certain substitutions in glutamate receptors may stabilize their dimer interface and reduce desensitization.[81]

4.4. Insertions and deletions

Insertions and deletions at oligomer interfaces provide another important mechanism to modulate different functional specificities and protein oligomeric states in evolution. The protein *N*-acetyl-L-glutamate kinase (NAGK) from *Pseudomonas aeruginosa* is such an example; it catalyzes the second step in arginine biosynthesis. *P. aeruginosa* NAGK has an extra alpha-helix on its N-terminus and forms a homohexamer (trimers of dimers), although the same kinase from *E. coli* does not have this helix and exists as a homodimer (Fig. 1.3D). The homohexameric NAGK has a built-in negative feedback mechanism; binding of arginine causes a conformational change which inhibits the enzyme function by stabilizing an enlarged active center conformation.

The arginine binding site is formed by the N-terminal helix and C-terminal lobe, flanking the junction between dimers, thus, hexamerization mediated by the N-terminal helix would be essential for this inhibition.[82]

The inspection of inserted and deleted regions among homologous proteins existing in different oligomeric states has revealed that about a quarter of them are located on interaction interfaces[83] and many of them are responsible for enabling or disabling the formation of oligomers.[50,84,85] Indeed, removal of enabling regions from protein structures may result in the complete or partial loss of oligomer stability.[83,85] According to these investigations, the insertions and deletions are located preferentially in loop regions and to a lesser extent on alpha-helices and beta-strands. Furthermore, the insertions and deletions modulating oligomerization usually have a lower aggregation propensity and contain a larger fraction of polar and charged residues compared to conventional interfaces and protein surfaces.[83] The residue composition of enabling regions depends on the size of the interface area; large insertions (>500 Å^2 on interface area) tend to be hydrophobic, similar to conventional interfaces, but small insertions have polar or small residues such as Asn, Ser, Gly, or Pro. This tendency can be explained from the viewpoint of structural restraints for these insertions; the small insertions mainly form turns in the protein secondary structure, thus residues with high turn propensities are required to maintain their local structures.[85] Interestingly, different oligomerization mechanisms can be employed within different proteins from the same family. For example, in the dihydrofolate reductase family, the enabling loop of bacteriophage T4 leads to the formation of a homodimer,[84] while dimerization of the same enzyme from *Thermotoga maritima* is achieved by amino acid substitutions.[73]

5. REGULATION OF PROTEIN ACTIVITY THROUGH OLIGOMERIZATION

Cellular processes are extremely complex, requiring many factors to provide desired outcomes and prevent inefficiency. This control might be applied at the level of gene expression or protein–protein interactions, by changing protein activity, posttranslational modifications, or other means. Moreover, proteins can exist in dynamic equilibrium between different oligomeric states which can be controlled by physiological conditions (pH, certain ionic strength, temperature), ligands, and posttranslational modifications.[86] Protein activity may be also regulated through the dynamic transitions between oligomeric states with differential activity. It has also been shown that certain mutations may induce changes in oligomeric state and activity but do not compromise stability.[16] In addition, reversible transitions between discrete conformations and oligomeric states might account for protein cooperative binding properties and allosteric mechanisms in signal transduction.[6]

Here we present different scenarios which can explain how shifting the equilibrium between different oligomeric states might serve as a regulatory mechanism. These scenarios include: -homooligomerization may be important for protein self-activation when the active or binding site occupies the oligomeric binding interface; -conformational changes accompanying oligomerization could lead to exposure/suppression of the active or protein binding sites; -the formation of an oligomer may inhibit binding of a monomer to its substrate through competitive binding even without conformational changes; -the posttranslational modifications or binding of small molecules at or near the oligomeric interface can shift the equilibrium between different oligomeric states. In addition to examples from the previous study,[86] we manually compiled a list of experimentally verified examples of the above-mentioned mechanisms in Table 1.1. As can be seen from this table, phosphorylation is a very widespread mechanism used by the cell to control protein activity through oligomerization. Indeed, recent work has shown that phosphorylation sites tend to be located on binding interfaces in heterooligomeric and weak transient homooligomeric complexes, and phosphorylation may potentially mediate or in many cases disrupt the complex formation.[87]

We will describe a specific example, which shows how the transition between the dimeric and tetrameric forms of pyruvate kinase can be implicated in tumor formation. Pyruvate kinase is a key glycolytic enzyme and its activity is consistently altered during tumorigenesis. It has been noted that, during tumor formation, the M2 isoform of pyruvate kinase is overexpressed.[8] The active tetrameric form of this enzyme from normal cells has high affinity to its substrate phosphoenolpyruvate (PEP), associates with the glycolytic enzyme complex, and produces high levels of ATP. The dimeric form has low affinity to PEP, does not associate with the glycolytic enzyme complex, and is accompanied by low levels of ATP. Having the inactive dimeric form is advantageous for tumor cells as phosphometabolites above pyruvate kinases in the glycolytic cycle accumulate in tumor cells and are then available as precursors for synthesis required by tumorigenesis. The tetramer–dimer ratio is regulated by different factors, for example, by fructose 1,6-P2 and serine concentrations. Moreover, Rous sarcoma virus may also phosphorylate M2-PK and lead to its dimerization and disassociation from the glycolytic enzyme complex.[88]

6. OLIGOMERIZATION, PROTEIN AGGREGATION, AND RELATED DISEASES

Protein aggregation and amyloid fibril formation are associated with a number of devastating human diseases, including Alzheimer's disease, Parkinson's disease, and Huntington's disease. Amyloid fibril is an insoluble protein fibril, which usually represents a straight and unbranched polymer of several thousand angstroms in length and is deposited mainly in extracellular regions.[89] The major difference from other biological polymers is that amyloid fibers have very high stability, are highly insoluble in detergent, and have a considerable amount of β-sheet structure.[90] The mechanism of amyloidogenesis has been extensively studied, and several models have been proposed. The prevalent model to explain the self-assembly process is nucleation-dependent polymerization.[91] According to this model, a nucleus is required to initiate the fibrillation process. After the formation of a nucleus, the elongation of fiber becomes thermodynamically favorable and a large percentage of the starting protein material is assembled into fibrils. Another mechanism is double-concerted fibrillation.[92] In this model, the amyloid fibril formation is achieved via two consecutive processes. The first process involves concerted associations of monomers, forming oligomeric granular species. In the second process, the oligomeric species act as a growing unit and form fibrils. In addition, a number of theoretical and computational approaches have been developed to elucidate the detailed process of amyloid fibril formation.[93] An example of this is a molecular dynamics simulation showing the aggregation process of eight SH3 domains, where the initial step of aggregation is the dimerization, followed by the formation of one aggregate that consists of all eight proteins.[94]

An important question is why disease-related amyloids are toxic. In the case of Alzheimer's disease, which is one of the most common neurodegenerative diseases, many studies suggest that a high concentration of amyloid-β (Aβ) proteins causes neuronal alterations.[95] Interestingly, not only assembled amyloid fibrils but also prefibrillar aggregates including dimers, trimers, and relatively short oligomers show cytotoxicity.[96] In particular, the Aβ dimers extracted directly from brains of Alzheimer's disease patients inhibited long-term potentiation, enhanced long-term depression, and reduced dendritic spine density.[97] In contrast, the same study demonstrated that amyloid plaque cores did not impair long-term potentiation unless Aβ dimers were released from the amyloid. This suggests that the amyloid core is not toxic by itself but the dynamic release of oligomers from the amyloid is toxic.

It remains unclear why natively soluble proteins aggregate under some specific conditions. Certainly, one of the factors is mutation, which could trigger diseases or increase the risk of diseases. For example, myofibrillar myopathies are a group of neuromuscular disorders, characterized by the cytoplasmic aggregation of multiple proteins, resulting in slowly progressive weakening of limb muscles. Desmin, one of the aggregated proteins in this disorder, is a muscle-specific filament protein, which plays an important role as a structural component of the muscle cytoarchitecture. A genetic study identified three different missense mutations in the desmin protein, namely R350P, E413K, and R454W.[98] These mutations have a profound effect on the stability of the dimer and tetramer, which impairs the proper filament assembly. A nonsense mutation was found in the gene encoding another aggregated protein, FLNC. The mutation causes the truncation of a part of the dimerization domain, leading to the inability to form a proper dimer.[99] Overall, more than 20 missense mutations that cause Alzheimer's disease have been identified in or near the coding region of Aβ protein,[96] which enhance the aggregation of the peptides.

On the other hand, the ability to form amyloid fibril is not limited to a small group of disease-related proteins.[100,101] There is a growing list of functional amyloids, where proteins natively form filamentous aggregates and fulfill specific biological functions.[90] A recent study demonstrated that more than 30 human peptide hormones can be spontaneously assembled and form amyloids, which function as a storage of hormones.[102] This study also showed that the amyloid fibrils formed by the peptide hormones were able to release active monomeric hormones despite the high stability of the amyloids. Although the mechanism of the isolation of the monomers from the amyloid fibrils is still unknown, the secretion of peptide hormones is a highly regulated process. Other examples of functional amyloids include the Pmel17 protein forming intramelanosome fibrils through normal biological proteolytic processing associated with skin pigmentation[103] and the curli fiber of *E. coli*, which are involved in adhesion to surfaces and biofilm formation.[104] These functional amyloids are regulated and contribute to cellular biology without causing significant cytotoxicity. In addition, oligomerization is used in nature to counterbalance the destabilizing effects of mutations or higher temperatures and protect against nonspecific aggregation.[105,106]

7. CONCLUSIONS

Analysis of high-throughput protein–protein interaction networks shows that there are significantly more self-interacting proteins than expected by chance.[107] Despite the importance and abundance of homooligomers in a cell, they are usually neglected in studies of protein–protein interaction networks. An explanation comes from the ambiguity of the experimental characterization of homooligomers and the difficulty of their computational prediction. In this review, we attempted to summarize the biological importance of homooligomeric assemblies, their evolution, and physicochemical properties. We also outlined the regulatory properties of homooligomers which might exist in dynamic equilibrium between different oligomeric states and can be controlled by physiological conditions, ligands, and post-translational modifications. The disruption of their regulatory functions may lead to many human diseases, namely certain mutations may induce changes in oligomeric state and protein activity and in some cases trigger non-specific protein aggregation and amyloid fibril formation.

ACKNOWLEDGMENT

This work was supported by the Intramural Research Program of the National Library of Medicine at the United States.

REFERENCES

1. Cornish-Bowden AJ, Koshland Jr DE. The quaternary structure of proteins composed of identical subunits. *J Biol Chem* 1971;**246**:3092–102.
2. Jones S, Thornton JM. Protein-protein interactions: a review of protein dimer structures. *Prog Biophys Mol Biol* 1995;**63**:31–65.
3. Hashimoto K, Nishi H, Bryant S, Panchenko AR. Caught in self-interaction: evolutionary and functional mechanisms of protein homooligomerization. *Phys Biol* 2011;**8**:035007.
4. Ali MH, Imperiali B. Protein oligomerization: how and why. *Bioorg Med Chem* 2005;**13**:5013–20.
5. Goodsell DS, Olson AJ. Structural symmetry and protein function. *Annu Rev Biophys Biomol Struct* 2000;**29**:105–53.
6. Changeux JP, Edelstein SJ. Allosteric mechanisms of signal transduction. *Science* 2005;**308**:1424–8.
7. Miller S, Lesk AM, Janin J, Chothia C. The accessible surface area and stability of oligomeric proteins. *Nature* 1987;**328**:834–6.
8. Mazurek S, Boschek CB, Hugo F, Eigenbrodt E. Pyruvate kinase type M2 and its role in tumor growth and spreading. *Semin Cancer Biol* 2005;**15**:300–8.
9. Jia-Lin Ma N, Stern DF. Regulation of the Rad53 protein kinase in signal amplification by oligomer assembly and disassembly. *Cell Cycle* 2008;**7**:808–17.

10. Lin X, Liu J, Maley F, Chu E. Role of cysteine amino acid residues on the RNA binding activity of human thymidylate synthase. *Nucleic Acids Res* 2003;**31**:4882–7.
11. Arnold SF, Vorojeikina DP, Notides AC. Phosphorylation of tyrosine 537 on the human estrogen receptor is required for binding to an estrogen response element. *J Biol Chem* 1995;**270**:30205–12.
12. Randez-Gil F, Sanz P, Entian KD, Prieto JA. Carbon source-dependent phosphorylation of hexokinase PII and its role in the glucose-signaling response in yeast. *Mol Cell Biol* 1998;**18**:2940–8.
13. Lee KK, Yonehara S. Phosphorylation and dimerization regulate nucleocytoplasmic shuttling of mammalian STE20-like kinase (MST). *J Biol Chem* 2002;**277**:12351–8.
14. Hwang I, Thorgeirsson T, Lee J, Kustu S, Shin YK. Physical evidence for a phosphorylation-dependent conformational change in the enhancer-binding protein NtrC. *Proc Natl Acad Sci USA* 1999;**96**:4880–5.
15. Levinson NM, Visperas PR, Kuriyan J. The tyrosine kinase Csk dimerizes through its SH3 domain. *PLoS One* 2009;**4**:e7683.
16. Devenish SR, Gerrard JA. The role of quaternary structure in (beta/alpha)(8)-barrel proteins: evolutionary happenstance or a higher level of structure-function relationships? *Org Biomol Chem* 2009;**7**:833–9.
17. Woodcock JM, Murphy J, Stomski FC, Berndt MC, Lopez AF. The dimeric versus monomeric status of 14-3-3zeta is controlled by phosphorylation of Ser58 at the dimer interface. *J Biol Chem* 2003;**278**:36323–7.
18. Davison TS, Vagner C, Kaghad M, Ayed A, Caput D, Arrowsmith CH. p73 and p63 are homotetramers capable of weak heterotypic interactions with each other but not with p53. *J Biol Chem* 1999;**274**:18709–14.
19. Joerger AC, Rajagopalan S, Natan E, Veprintsev DB, Robinson CV, Fersht AR. Structural evolution of p53, p63, and p73: implication for heterotetramer formation. *Proc Natl Acad Sci USA* 2009;**106**:17705–10.
20. Matthews JM, Visvader JE. LIM-domain-binding protein 1: a multifunctional cofactor that interacts with diverse proteins. *EMBO Rep* 2003;**4**:1132–7.
21. Matthews JM, Bhati M, Craig VJ, Deane JE, Jeffries C, Lee C, et al. Competition between LIM-binding domains. *Biochem Soc Trans* 2008;**36**:1393–7.
22. Cross AJ, Jeffries CM, Trewhella J, Matthews JM. LIM domain binding proteins 1 and 2 have different oligomeric states. *J Mol Biol* 2010;**399**:133–44.
23. Dafforn TR. So how do you know you have a macromolecular complex? *Acta Crystallogr D Biol Crystallogr* 2007;**63**:17–25.
24. Henrick K, Thornton JM. PQS: a protein quaternary structure file server. *Trends Biochem Sci* 1998;**23**:358–61.
25. Krissinel E, Henrick K. Inference of macromolecular assemblies from crystalline state. *J Mol Biol* 2007;**372**:774–97.
26. Zhu H, Domingues FS, Sommer I, Lengauer T. NOXclass: prediction of protein-protein interaction types. *BMC Bioinformatics* 2006;**7**:27.
27. Ponstingl H, Henrick K, Thornton JM. Discriminating between homodimeric and monomeric proteins in the crystalline state. *Proteins* 2000;**41**:47–57.
28. Bahadur RP, Chakrabarti P, Rodier F, Janin J. Dissecting subunit interfaces in homodimeric proteins. *Proteins* 2003;**53**:708–19.
29. Janin J, Rodier F. Protein-protein interaction at crystal contacts. *Proteins* 1995;**23**: 580–7.
30. Dasgupta S, Iyer GH, Bryant SH, Lawrence CE, Bell JA. Extent and nature of contacts between protein molecules in crystal lattices and between subunits of protein oligomers. *Proteins* 1997;**28**:494–514.
31. Shoemaker BA, Panchenko AR, Bryant SH. Finding biologically relevant protein domain interactions: conserved binding mode analysis. *Protein Sci* 2006;**15**:352–61.

32. Ofran Y, Rost B. Analysing six types of protein-protein interfaces. *J Mol Biol* 2003;**325**:377–87.
33. Zhanhua C, Gan JG, Lei L, Sakharkar MK, Kangueane P. Protein subunit interfaces: heterodimers versus homodimers. *Bioinformation* 2005;**1**:28–39.
34. Perica T, Chothia C, Teichmann SA. Evolution of oligomeric state through geometric coupling of protein interfaces. *Proc Natl Acad Sci USA* 2012;**109**:8127–32.
35. Dey S, Pal A, Chakrabarti P, Janin J. The subunit interfaces of weakly associated homodimeric proteins. *J Mol Biol* 2010;**398**:146–60.
36. Gunasekaran K, Tsai CJ, Nussinov R. Analysis of ordered and disordered protein complexes reveals structural features discriminating between stable and unstable monomers. *J Mol Biol* 2004;**341**:1327–41.
37. Lynch M. The evolution of multimeric protein assemblages. *Mol Biol Evol* 2012;**29**: 1353–66.
38. Lukatsky DB, Shakhnovich BE, Mintseris J, Shakhnovich EI. Structural similarity enhances interaction propensity of proteins. *J Mol Biol* 2007;**365**:1596–606.
39. Andre I, Strauss CE, Kaplan DB, Bradley P, Baker D. Emergence of symmetry in homooligomeric biological assemblies. *Proc Natl Acad Sci USA* 2008;**105**:16148–52.
40. Wright CF, Teichmann SA, Clarke J, Dobson CM. The importance of sequence diversity in the aggregation and evolution of proteins. *Nature* 2005;**438**:878–81.
41. Dayhoff JE, Shoemaker BA, Bryant SH, Panchenko AR. Evolution of protein binding modes in homooligomers. *J Mol Biol* 2010;**395**:860–70.
42. Fong JH, Shoemaker BA, Garbuzynskiy SO, Lobanov MY, Galzitskaya OV, Panchenko AR. Intrinsic disorder in protein interactions: insights from a comprehensive structural analysis. *PLoS Comput Biol* 2009;**5**:e1000316.
43. Simon SM, Sousa FJ, Mohana-Borges R, Walker GC. Regulation of *Escherichia coli* SOS mutagenesis by dimeric intrinsically disordered umuD gene products. *Proc Natl Acad Sci USA* 2008;**105**:1152–7.
44. Xu D, Tsai CJ, Nussinov R. Mechanism and evolution of protein dimerization. *Protein Sci* 1998;**7**:533–44.
45. Tiana G, Broglia RA. Design and folding of dimeric proteins. *Proteins* 2002;**49**:82–94.
46. Levy ED, Boeri Erba E, Robinson CV, Teichmann SA. Assembly reflects evolution of protein complexes. *Nature* 2008;**453**:1262–5.
47. Villar G, Wilber AW, Williamson AJ, Thiara P, Doye JP, Louis AA, et al. Self-assembly and evolution of homomeric protein complexes. *Phys Rev Lett* 2009;**102**:118106.
48. Pereira-Leal JB, Teichmann SA. Novel specificities emerge by stepwise duplication of functional modules. *Genome Res* 2005;**15**:552–9.
49. Reid AJ, Ranea JA, Orengo CA. Comparative evolutionary analysis of protein complexes in *E. coli* and yeast. *BMC Genomics* 2010;**11**:79.
50. Hashimoto K, Madej T, Bryant SH, Panchenko AR. Functional states of homooligomers: insights from the evolution of glycosyltransferases. *J Mol Biol* 2010; **399**:196–206.
51. Pereira-Leal JB, Levy ED, Kamp C, Teichmann SA. Evolution of protein complexes by duplication of homomeric interactions. *Genome Biol* 2007;**8**:R51.
52. Yu H, Luscombe NM, Lu HX, Zhu X, Xia Y, Han JD, et al. Annotation transfer between genomes: protein-protein interologs and protein-DNA regulogs. *Genome Res* 2004;**14**:1107–18.
53. Bennett MJ, Choe S, Eisenberg D. Domain swapping: entangling alliances between proteins. *Proc Natl Acad Sci USA* 1994;**91**:3127–31.
54. Crestfield AM, Stein WH, Moore S. On the aggregation of bovine pancreatic ribonuclease. *Arch Biochem Biophys Suppl* 1962;**1**:217–22.
55. Gronenborn AM. Protein acrobatics in pairs: dimerization via domain swapping. *Curr Opin Struct Biol* 2009;**19**:39–49.

56. Shameer K, Shingate PN, Manjunath SC, Karthika M, Pugalenthi G, Sowdhamini R. 3DSwap: curated knowledgebase of proteins involved in 3D domain swapping. *Database (Oxford)* 2011;**2011**:bar042.
57. Liu Y, Eisenberg D. 3D domain swapping: as domains continue to swap. *Protein Sci* 2002;**11**:1285–99.
58. Liu Y, Gotte G, Libonati M, Eisenberg D. A domain-swapped RNase A dimer with implications for amyloid formation. *Nat Struct Biol* 2001;**8**:211–4.
59. Rousseau F, Schymkowitz JW, Wilkinson HR, Itzhaki LS. Three-dimensional domain swapping in p13suc1 occurs in the unfolded state and is controlled by conserved proline residues. *Proc Natl Acad Sci USA* 2001;**98**:5596–601.
60. Barrientos LG, Louis JM, Botos I, Mori T, Han Z, O'Keefe BR, et al. The domain-swapped dimer of cyanovirin-N is in a metastable folded state: reconciliation of X-ray and NMR structures. *Structure* 2002;**10**:673–86.
61. Orlikowska M, Jankowska E, Kolodziejczyk R, Jaskolski M, Szymanska A. Hinge-loop mutation can be used to control 3D domain swapping and amyloidogenesis of human cystatin C. *J Struct Biol* 2011;**173**:406–13.
62. Shingate P, Sowdhamini R. Analysis of domain-swapped oligomers reveals local sequence preferences and structural imprints at the linker regions and swapped interfaces. *PLoS One* 2012;**7**:e39305.
63. Dickason RR, Huston DP. Creation of a biologically active interleukin-5 monomer. *Nature* 1996;**379**:652–5.
64. Yang S, Cho SS, Levy Y, Cheung MS, Levine H, Wolynes PG, et al. Domain swapping is a consequence of minimal frustration. *Proc Natl Acad Sci USA* 2004;**101**:13786–91.
65. Malevanets A, Sirota FL, Wodak SJ. Mechanism and energy landscape of domain swapping in the B1 domain of protein G. *J Mol Biol* 2008;**382**:223–35.
66. Burkhard P, Stetefeld J, Strelkov SV. Coiled coils: a highly versatile protein folding motif. *Trends Cell Biol* 2001;**11**:82–8.
67. Mason JM, Arndt KM. Coiled coil domains: stability, specificity, and biological implications. *Chembiochem* 2004;**5**:170–6.
68. O'Shea EK, Rutkowski R, Kim PS. Evidence that the leucine zipper is a coiled coil. *Science* 1989;**243**:538–42.
69. Harbury PB, Zhang T, Kim PS, Alber T. A switch between two-, three-, and four-stranded coiled coils in GCN4 leucine zipper mutants. *Science* 1993;**262**:1401–7.
70. Harbury PB, Kim PS, Alber T. Crystal structure of an isoleucine-zipper trimer. *Nature* 1994;**371**:80–3.
71. Liu J, Zheng Q, Deng Y, Cheng CS, Kallenbach NR, Lu M. A seven-helix coiled coil. *Proc Natl Acad Sci USA* 2006;**103**:15457–62.
72. Moutevelis E, Woolfson DN. A periodic table of coiled-coil protein structures. *J Mol Biol* 2009;**385**:726–32.
73. Nishi H, Ota M. Amino acid substitutions at protein-protein interfaces that modulate the oligomeric state. *Proteins* 2010;**78**:1563–74.
74. Levy ED. A simple definition of structural regions in proteins and its use in analyzing interface evolution. *J Mol Biol* 2010;**403**:660–70.
75. Grueninger D, Treiber N, Ziegler MO, Koetter JW, Schulze MS, Schulz GE. Designed protein-protein association. *Science* 2008;**319**:206–9.
76. Clackson T, Wells JA. A hot spot of binding energy in a hormone-receptor interface. *Science* 1995;**267**:383–6.
77. Bogan AA, Thorn KS. Anatomy of hot spots in protein interfaces. *J Mol Biol* 1998;**280**:1–9.
78. Chakrabarti P, Janin J. Dissecting protein-protein recognition sites. *Proteins* 2002;**47**:334–43.

79. Keskin O, Ma B, Nussinov R. Hot regions in protein-protein interactions: the organization and contribution of structurally conserved hot spot residues. *J Mol Biol* 2005;**345**:1281–94.
80. Malay AD, Allen KN, Tolan DR. Structure of the thermolabile mutant aldolase B, A149P: molecular basis of hereditary fructose intolerance. *J Mol Biol* 2005;**347**:135–44.
81. Sun Y, Olson R, Horning M, Armstrong N, Mayer M, Gouaux E. Mechanism of glutamate receptor desensitization. *Nature* 2002;**417**:245–53.
82. Ramon-Maiques S, Fernandez-Murga ML, Gil-Ortiz F, Vagin A, Fita I, Rubio V. Structural bases of feed-back control of arginine biosynthesis, revealed by the structures of two hexameric N-acetylglutamate kinases, from *Thermotoga maritima* and *Pseudomonas aeruginosa*. *J Mol Biol* 2006;**356**:695–713.
83. Hashimoto K, Panchenko AR. Mechanisms of protein oligomerization, the critical role of insertions and deletions in maintaining different oligomeric states. *Proc Natl Acad Sci USA* 2010;**107**:20352–7.
84. Akiva E, Itzhaki Z, Margalit H. Built-in loops allow versatility in domain-domain interactions: lessons from self-interacting domains. *Proc Natl Acad Sci USA* 2008;**105**:13292–7.
85. Nishi H, Koike R, Ota M. Cover and spacer insertions: small nonhydrophobic accessories that assist protein oligomerization. *Proteins* 2011;**79**:2372–9.
86. Nooren IM, Thornton JM. Structural characterisation and functional significance of transient protein-protein interactions. *J Mol Biol* 2003;**325**:991–1018.
87. Nishi H, Hashimoto K, Panchenko AR. Phosphorylation in protein-protein binding: effect on stability and function. *Structure* 2011;**19**:1807–15.
88. Presek P, Reinacher M, Eigenbrodt E. Pyruvate kinase type M2 is phosphorylated at tyrosine residues in cells transformed by Rous sarcoma virus. *FEBS Lett* 1988;**242**:194–8.
89. Sipe JD, Benson MD, Buxbaum JN, Ikeda S, Merlini G, Saraiva MJ, et al. Amyloid fibril protein nomenclature: 2010 recommendations from the nomenclature committee of the International Society of Amyloidosis. *Amyloid* 2010;**17**:101–4.
90. Hammer ND, Wang X, McGuffie BA, Chapman MR. Amyloids: friend or foe? *J Alzheimers Dis* 2008;**13**:407–19.
91. Xue WF, Homans SW, Radford SE. Systematic analysis of nucleation-dependent polymerization reveals new insights into the mechanism of amyloid self-assembly. *Proc Natl Acad Sci USA* 2008;**105**:8926–31.
92. Bhak G, Choe YJ, Paik SR. Mechanism of amyloidogenesis: nucleation-dependent fibrillation versus double-concerted fibrillation. *BMB Rep* 2009;**42**:541–51.
93. Straub JE, Thirumalai D. Toward a molecular theory of early and late events in monomer to amyloid fibril formation. *Annu Rev Phys Chem* 2011;**62**:437–63.
94. Ding F, Dokholyan NV, Buldyrev SV, Stanley HE, Shakhnovich EI. Molecular dynamics simulation of the SH3 domain aggregation suggests a generic amyloidogenesis mechanism. *J Mol Biol* 2002;**324**:851–7.
95. Mucke L, Masliah E, Yu GQ, Mallory M, Rockenstein EM, Tatsuno G, et al. High-level neuronal expression of abeta 1–42 in wild-type human amyloid protein precursor transgenic mice: synaptotoxicity without plaque formation. *J Neurosci* 2000;**20**:4050–8.
96. Selkoe DJ. Alzheimer's disease. *Cold Spring Harb Perspect Biol* 2011;**3**(7). Pii: a004457. http://dx.doi.org/10.1101/cshperspect.a004457.
97. Shankar GM, Li S, Mehta TH, Garcia-Munoz A, Shepardson NE, Smith I, et al. Amyloid-beta protein dimers isolated directly from Alzheimer's brains impair synaptic plasticity and memory. *Nat Med* 2008;**14**:837–42.
98. Levin J, Bulst S, Thirion C, Schmidt F, Botzel K, Krause S, et al. Divergent molecular effects of desmin mutations on protein assembly in myofibrillar myopathy. *J Neuropathol Exp Neurol* 2010;**69**:415–24.

99. Vorgerd M, van der Ven PF, Bruchertseifer V, Lowe T, Kley RA, Schroder R, et al. A mutation in the dimerization domain of filamin c causes a novel type of autosomal dominant myofibrillar myopathy. *Am J Hum Genet* 2005;**77**:297–304.
100. Dobson CM. Protein misfolding, evolution and disease. *Trends Biochem Sci* 1999;**24**:329–32.
101. Shewmaker F, McGlinchey RP, Wickner RB. Structural insights into functional and pathological amyloid. *J Biol Chem* 2011;**286**:16533–40.
102. Maji SK, Perrin MH, Sawaya MR, Jessberger S, Vadodaria K, Rissman RA, et al. Functional amyloids as natural storage of peptide hormones in pituitary secretory granules. *Science* 2009;**325**:328–32.
103. Berson JF, Theos AC, Harper DC, Tenza D, Raposo G, Marks MS. Proprotein convertase cleavage liberates a fibrillogenic fragment of a resident glycoprotein to initiate melanosome biogenesis. *J Cell Biol* 2003;**161**:521–33.
104. Barnhart MM, Chapman MR. Curli biogenesis and function. *Annu Rev Microbiol* 2006;**60**:131–47.
105. Ma BG, Goncearenco A, Berezovsky IN. Thermophilic adaptation of protein complexes inferred from proteomic homology modeling. *Structure* 2010;**18**:819–28.
106. Bershtein S, Mu W, Shakhnovich EI. Soluble oligomerization provides a beneficial fitness effect on destabilizing mutations. *Proc Natl Acad Sci USA* 2012;**109**:4857–62.
107. Ispolatov I, Yuryev A, Mazo I, Maslov S. Binding properties and evolution of homodimers in protein-protein interaction networks. *Nucleic Acids Res* 2005;**33**:3629–35.

CHAPTER TWO

Structural, Evolutionary, and Assembly Principles of Protein Oligomerization

Emmanuel D. Levy[*], **Sarah A. Teichmann**[†,‡]
[*]Department of Structural Biology, Weizmann Institute of Science, Rehovot, Israel
[†]MRC Laboratory of Molecular Biology, Cambridge, United Kingdom
[‡]Wellcome Trust Genome Campus, Cambridge, United Kingdom

Contents

Progress in Molecular Biology and Translational Science, Volume 117
ISSN 1877-1173
http://dx.doi.org/10.1016/B978-0-12-386931-9.00002-7

Abstract

In the protein universe, 30–50% of proteins self-assemble to form symmetrical complexes consisting of multiple copies of themselves, called homomers. The prevalence of homomers motivates us to review many of their properties. In Section 1, we describe the methods and challenges associated with quaternary structure inference—these methods are indeed at the basis of any analysis on homomers. In Section 2, we describe the morphological properties of homomers, as well as the database 3DComplex, which provides a taxonomy for both homomeric and heteromeric protein complexes. In Section 3, we review interface properties of homomeric complexes. In Section 4, we then present recent findings on the evolution of homomer interfaces, which we link in Section 5 to the evolution of homomers as entire entities. In Section 6, we discuss mechanisms involved in their assembly and how these mechanisms can be linked to evolution.

1. INTRODUCTION: FROM A CRYSTAL TO A QUATERNARY STRUCTURE

1.1. Increase in structural data over time

A homomer is formed by self-interacting copies of a protein unit. As noted by Janin *et al.*,[1] the characterization of many homomers began long before protein structures could be solved by X-ray crystallography. At the time, the method of choice was analytical centrifugation. By 1975, this led to the characterization of the quaternary state for hundreds of oligomeric assemblies.[2] At the same time, there were less than 13 structures deposited in the Protein Data Bank. Structural data have, however, grown steadily since then,[3] particularly over the past two decades. For example, 829 structures were recorded in 1992, 16,322 structures were recorded in 2002, and this number rose to 74,937 as of October 2012. Note that these numbers reflect only protein structures solved by X-ray crystallography, which is the most commonly used method for solving the structure of proteins.

1.2. The crystalline lattice and the problem of nonbiological contacts

This abundance of structural data means that today, the largest body of information available on protein quaternary structure comes from X-ray crystallography. Importantly, however, a prerequisite to solving a protein's structure by X-ray crystallography is to obtain a crystal, that is, to induce the protein (or protein complex) into self-assembling with other copies of itself in a regular fashion to form a lattice. The structure solved by crystallographers

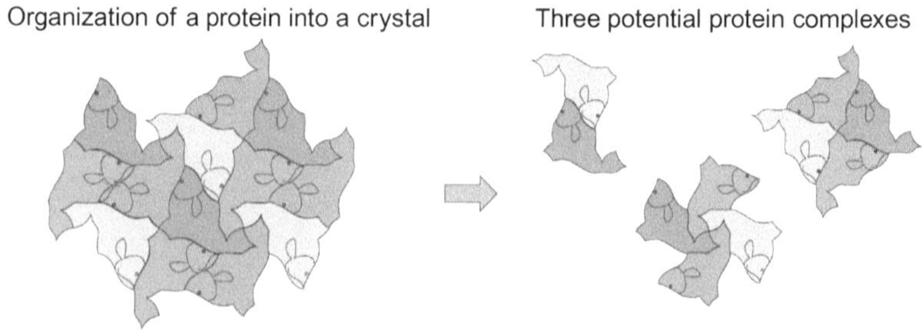

Figure 2.1 A schematic protein lattice and potential complexes that can be inferred from it. In order to solve the structure of a protein by X-ray crystallography, a necessary step is to induce the protein into forming a crystal. Within a crystal, proteins are organized into a lattice as illustrated in the figure, where each carp represents a protein. Proteins contact each other in different ways in the lattice (left), and different contacts correspond to different potential protein complexes (right). Importantly, however, most protein–protein interfaces in the crystal are artifacts (due to crystal formation). A key step in dealing with structural data is therefore to identify the biologically relevant interfaces. There is a large body of literature on this topic, as discussed in the main text. Note that crystals used for X-ray crystallography are three dimensional, while the illustration is two dimensional. The mosaic image is a reproduction of "carps" from Escher. (For color version of this figure, the reader is referred to the online version of this chapter.)

corresponds to the asymmetric unit (ASU), which is the minimal unit from which the whole crystal can be reconstructed using symmetry operations. Thus, a protein crystal can be seen as a mosaic, whose repeating unit is the ASU. This is illustrated in Fig. 2.1, with a "fish" being an ASU.

Importantly, the ASU does not necessarily reflect the biological state of a protein. For example, an ASU may contain a dimer of a protein that exists as a monomer *in vivo*. Conversely, the ASU may contain a single copy of a protein that actually forms a dimer or a tetramer *in vivo*, as illustrated in Fig. 2.1.

In sum, due to the nature of X-ray crystallography, artificial crystal contacts must be formed between proteins, resulting in many assemblies that can potentially be considered as biological. A task of fundamental importance is therefore to distinguish biological protein interfaces from nonbiological interfaces in the crystal.

1.3. Distinguishing biological from nonbiological protein contacts

We all know that strong friendships require time to develop and that people rarely become "best friends" in a day. Similarly, a pair of proteins is unlikely to exhibit a large interaction interface due to chance. In other words, a large interface is likely to be the result of an evolutionary process. Consistent with this idea, the buried surface area of crystal contacts, which occur by chance and are not selected for, is on average significantly smaller than that of biological interfaces.[4–6] Typically, biological interfaces bury more than 400 Å^2 per subunit (800 Å^2 total), while crystal contacts bury on average less than 400 Å^2 per subunit. It can happen, however, that crystal contacts bury 400–1000 Å^2, and in some instances even more.[a] Thus, a number of properties must be considered to discriminate between biological and nonbiological interfaces as accurately as possible. These properties can be broadly classified into one of two types: (i) structural properties, deduced directly from the protein structure or from the crystal lattice, and (ii) comparative properties, which make use of external data not present in the structure itself.

Among structural properties, interface amino acid composition has probably been the most frequently described, showing that crystal contacts are more polar on average than biological interfaces. Another property described by Pal *et al.* is the "number of polypeptide segments" at interfaces, with biological interfaces involving, on average, a smaller number of longer segments.[7] Other promising descriptors are the fraction of atoms that are fully buried at the interface and the associated packing density at the interface, indicating that biological interfaces tend to exhibit better packing and involve a larger fraction of fully buried atoms.[8] An interesting observation is also that amino acids at crystal contacts have on average lower conformational entropies than amino acids found at the surface.[9,10] This may prove a useful feature to consider in classifiers developed in the future. Such existing classifiers include NOXclass[11] and IPAC,[12] which are based on a combination of structural parameters coupled to a machine-learning framework.

In terms of comparative properties, conservation has been the most widely used descriptor. A premise in using conservation information is that functional elements are on average more conserved than less functional elements. Amino acids at biological interfaces, which have a specific function (to maintain or regulate the interaction), are thus expected to be more conserved than surface ones, which may or may not be involved in specific functions. A number of studies have indeed confirmed that interface amino acids are more evolutionarily conserved than surface ones.[13–25] Rather than considering evolutionary conservation, Xu and Dunbrack have assessed the conservation of interfaces in different crystal forms.[26,27] Since proteins are

[a] Methods used to measure protein surfaces and volumes are discussed in Section 3.

frequently crystallized in multiple conditions (e.g., with different ligands), it is possible to discriminate contacts that are recurrent across crystal forms from those that are unique to a particular crystallization condition. As expected, recurrent contacts tend to be biological.

1.4. Inferring biological assemblies

Historically, the ASU was the default information downloadable from the Protein Data Bank. This prompted the development of protein quaternary structure (PQS) (1998), which was the first automatic approach to infer quaternary structures from crystal lattices.[28] Later, the PDB started to provide information about biological quaternary structures, called "Biological Units" at the time and "Biological Assemblies" nowadays. The PDB relies on three sources for inferring Biological Assemblies: the authors of the structure; the PQS server[28]; and the protein interfaces, surfaces, and assemblies (PISA) server.[29] Nevertheless, errors persist due to the difficult nature of the problem. The database ProtBuD in fact provides a side-by-side comparison between the biological units provided by PQS and PDB.[30] PQS, which is hosted at the European Bioinformatics Institute, is however going to be discontinued soon, as another server, PISA,[29] replaces it. Interestingly, PISA uses a different approach from those described above. It aims at computing the ΔG of association between proteins, which in theory should integrate all the structural features in the best possible way. PISA provides accurate results in over 80% of cases. Yet, another approach is taken with our PiQSi server, which makes quaternary structure annotation community based.[31] Anyone can register with PiQSi and annotate quaternary structures with the help of the interactive interface to explore, for example, homologous complexes. Annotations are then reviewed internally before being added to the database.

2. HOMOMER MORPHOLOGY

2.1. Size of homomers

The most basic property that can be used to describe a homomer is its number of subunits. In a thorough review on structural symmetry and protein function, Goodsell and Olson record the occurrence of different oligomeric states for proteins found in *Escherichia coli*.[32] Not considering heterocomplexes, they observe 23% of monomers and 77% of homomers. Certainly, these numbers should be critically assessed as they may be biased to some extent. For example, a scientist may be more likely to report an oligomeric state when a dimer is observed than when a monomer is observed. In other words, oligomers might be more "noticeable" compared to monomers. This prompts us to assess their prevalence in crystal structures, which represents another independent type of data.

In Fig. 2.2, we show the distribution of the number of subunits in homomers as obtained from structural data. Importantly, these data may not suffer from the same "reporting" bias but are likely to suffer from other biases related to, for example, the probability of crystallization of different types of proteins. Yet, it also shows that homomers are widespread. Similar to the data from Goodsell and Olson, we observe only 35% of monomers in *E. coli*. In *Homo sapiens*, this number goes up to 54% (whether this difference is biological or results from different target choices by crystallographers would need to be investigated further—we also note an important difference between both species: *E. coli* structures often span full length proteins, while *H. sapiens* structures often correspond to protein fragments). Taken together, these data suggest that among proteins that are not part of a stable heterocomplex (e.g., such as the ribosome or the proteasome), homomers are formed at frequencies ranging from ~50% to ~70%.

Considering the frequencies for the different types of homomers, we note a general decrease as the number of subunits increases in Fig. 2.2. However, it is also clear that homomers with even numbers of subunits are favored over those with odd numbers. For example, in *E. coli*, there are 4.0% homomers with three subunits and 8.6% with four. The reason for this effect is that two types of symmetry (discussed below) can make up

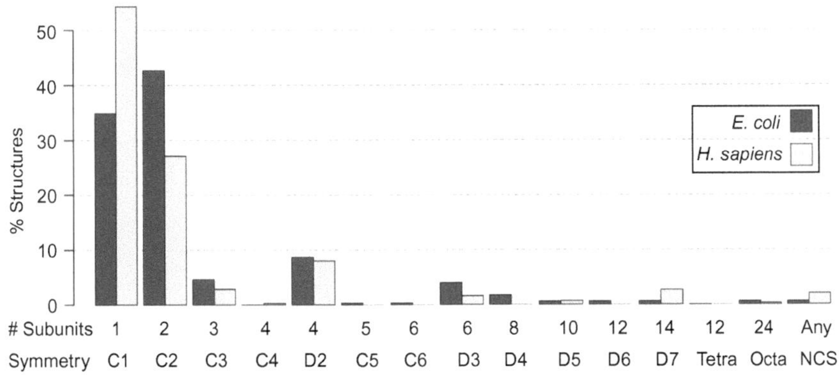

Figure 2.2 Number of quaternary structures per symmetry type. The frequency of different symmetry types obtained using the *E. coli* and the *H. sapiens* data set from Ref. 33 available at http://www.weizmann.ac.il/Structural_Biology/faculty_pages/ELevy/intDef/interface_def.html.

even-numbered homomers, while a single type can lead to odd-numbered ones.[34] This is also illustrated in Fig. 2.2, where the corresponding symmetry type is indicated under each number of subunits. There is indeed not a single "C4" tetramer in the *E. coli* data set, but there are 8.6% of "D2" tetramers. The decrease observed with increasing numbers of subunits should thus be considered independently for both symmetry types (C type or *cyclic*, and D type or *dihedral*).

2.2. Symmetry types

Cyclic and dihedral symmetries are the two main types encountered in homomers. In a cyclic symmetry, the subunits are related by a single *n*-fold symmetry axis, where *n* denotes the symmetry type (i.e., Cn). In a C2 dimer, for example, the subunits are related by a twofold symmetry axis, meaning that rotating the complex by 180° around the axis results in the subunits being mutually superposed. In a C3 trimer, the same occurs after a 120° rotation (Fig. 2.3).

Importantly, among all cyclic symmetries, C2 is the only one where the contact between subunits is mediated by the same surface patch on both subunits—Monod termed this type of contact "*iso*logous," where subunits contact each other in a "face-to-face" fashion. In contrast, when considering a higher-order symmetry like C4, the subunits contact each other in a "face-to-back" fashion, so that contacts are now "*hete*rologous". In other words,

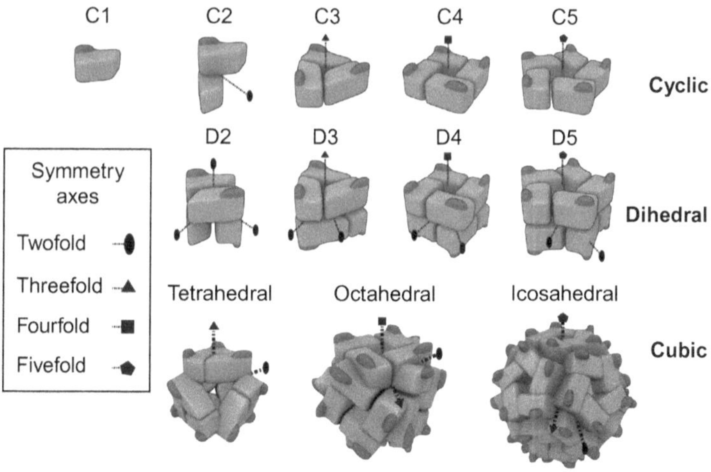

Figure 2.3 Main symmetry types that homomers adopt. (For color version of this figure, the reader is referred to the online version of this chapter.)

interfaces associated with a twofold symmetry are symmetric themselves (the axis of symmetry passes through the center of mass of the interface), while interfaces associated with higher-order symmetries are not symmetric (and the axis of symmetry does not coincide with the center of the interface).

Dihedral symmetries (D2–D5 in Fig. 2.3) can be seen as two cyclic structures stacked onto each other and related by twofold symmetry axes. The resulting geometry has important functional implications in terms of, for example, directionality.[32] In contrast to cyclic complexes, dihedral complexes are nondirectional and are not found in membranes (a notable exception is the nuclear pore complex, which exhibits dihedral pseudosymmetry but spans two lipid bilayers[35]).

A cyclic complex can therefore acquire a new isologous interface to become dihedral. A dihedral complex, however, cannot acquire a new isologous interface as this would lead to the formation of an infinite polymer. This is illustrated in Fig. 2.4, with the example of hemoglobin. An allelic variant of the β chain (an aspartic acid mutated to a valine) induces the formation of a new protein–protein interaction interface. Because of the symmetry of a hemoglobin tetramer, this interface appears twice per tetramer (once on each copy of the β chain), so that one tetramer interacts with another, which interacts with another, etc. This allelic variant thus leads to the formation of fibrils in red blood cell, conferring on them an elongated

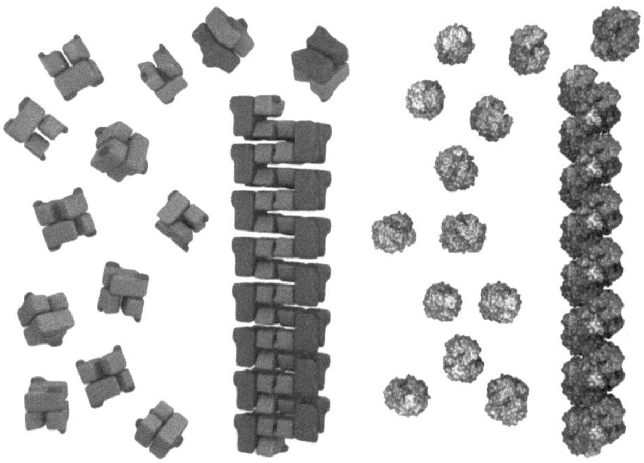

Figure 2.4 Open symmetries can be triggered by single point mutations as in sickle cell anemia. The left side of the image shows a schematic representation of hemoglobin, while the right side shows a molecular surface. Each representation shows free (wild-type) hemoglobin and an allelic variant associated to sickle cell anemia, which forms fibrils. Note that the fibrils have an illustrative purpose only and may not be accurate with respect to the orientation of the subunits. (See Color Insert.)

shape characteristic of the sickle cell disease. Such "open" symmetries, which appear dangerous due to their potential to form polymers, are uncommon. Their prevalence is shown in the histogram Fig. 2.2 in the category "NCS" (no closed symmetry).

2.3. Emergence of symmetric structures

Whether the axis of symmetry is within or outside of the interface introduces a fundamental difference in terms of the energy of interaction. If inside, as put by Monod *et al.* "mutation of one residue, conferring upon it the capacity to form a bond with its partner, would result in two new bonds being achieved in the dimer."[37] The fact that one mutation has double the effects in a C2 association indeed has important energetic consequences. Let us consider a very simple model where a random mutation *i* favors or disfavors the creation of a new interaction by ΔG_i, and let us assume that ΔG_i is normally distributed. One can visualize intuitively that considering the "double effect" at C2 interfaces results in a larger variance for the distribution of ΔG_i.[36] This suggests that the impact of mutations should have more often extreme energetic effects (both positive and negative) on isologous interfaces when compared to heterologous interfaces. Several studies have investigated these notions in detail—from the statistical energetic effects associated with self-association,[38] to their implication in proteins using a coarse-grained model,[39] an atom-based model[40] or an evolutionary-oriented model.[41] These studies all point to the fact that isologous (C2) interfaces are energetically more favorable than heterologous interfaces.

Considering the above arguments helps understand why cyclic symmetries with three and more subunits are rare (as these do not involve isologous but heterologous interfaces). On top of this, geometry can also help explain their scarcity. Considering a monomer, a new isologous interface naturally yields a *closed* dimer. However, when considering a new heterologous interface, the two surface patches on the monomer must be precisely oriented not to yield an *open* screw-type symmetry (i.e., involving rotations plus translations). In sum, the evolution of new cyclic complexes is limited by two factors: the type and the orientation of the interfaces that need to be created.

2.4. Classification of homomers to facilitate their study

Hemoglobin was the first protein oligomer whose structure was solved.[42] Since then, many variants in different forms (e.g., with or without oxygen bound to the heme) or from different species have been crystallized, so that

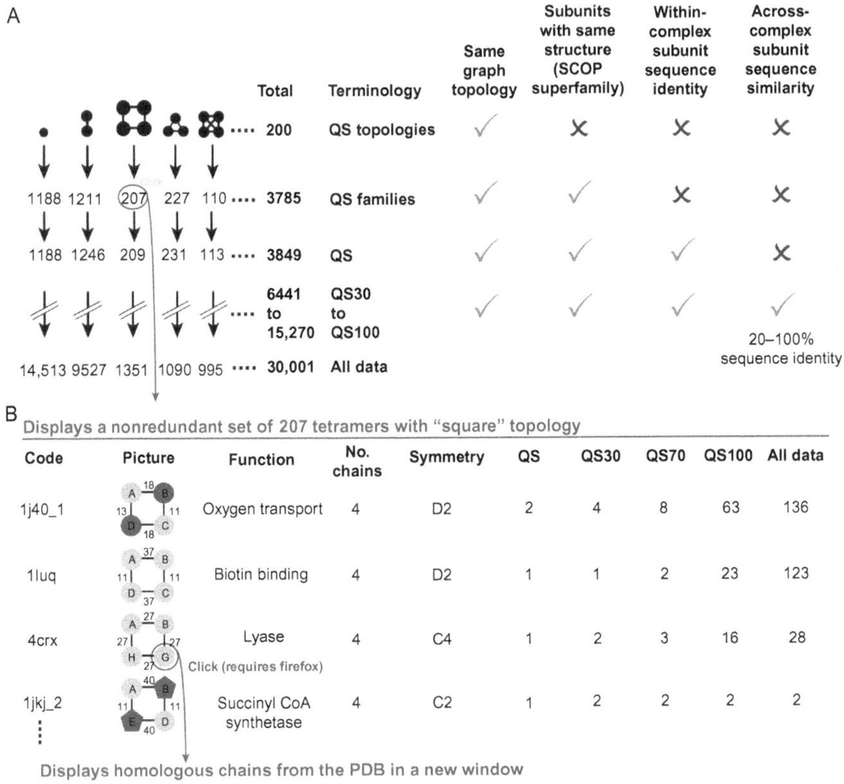

Figure 2.5 The 3DComplex hierarchical classification. QS stands for "quaternary structure," which refers to the spatial arrangement of subunits in the complex. The names of different levels of the hierarchy are indicated on the left, and the corresponding number of groups at each level is given on the right. The classification is hierarchical with one root at the top and 21,037 leaves at the bottom. At the top most level, lenient similarity is sufficient to group complexes, and these groups are split into subgroups toward the bottom of the hierarchy, where similarity of various detailed features is required. This scheme is similar to the taxonomic classification of species, where human beings are in the same group as baker's yeast at the top of the hierarchy, but are separated at lower levels. (For color version of this figure, the reader is referred to the online version of this chapter.)

there are over 200 hemoglobin entries archived in the PDB.[43] Importantly, however, for any analysis relating to quaternary structures, it is usually necessary to remove redundant structures (e.g., to estimate the frequency of different symmetries). This was a driving force for the development of 3DComplex, which is a classification of protein complexes of known structure. 3DComplex organizes protein complexes according to many structural features, the three main of which are (i) the three-dimensional structure of the constituent subunits in terms of domains and protein families, (ii) the pattern of binary interactions between the subunits, and (iii) the symmetry (if any) adopted by the subunits.

Figure 2.5 depicts the data accessible from the Web site www.3Dcomplex.org, where complexes are organized into a hierarchy. At the top most level, the similarity of simple features is sufficient to group complexes, and these groups are split into subgroups toward the bottom of the hierarchy, where similarity of more detailed features (such as minimal sequence similarity of subunits) is required. The classification can be used to subselect specific groups of complexes such as: "tetramers where each subunit contacts two others, and where all subunits have a globin fold and exhibit D2 symmetry."

3. INTERFACES IN HOMOMERS

3.1. Defining the protein surface and the interface

In a homomer, protein subunits are held together by interaction interfaces. To study these regions of contact between subunits, we first need to define them (i.e., their boundaries). There are several approaches for this purpose, and a first group of approaches involves the calculation of a "protein surface," while a second is based on amino acid contacts (described in the last paragraph of this section). In terms of the first approach, several methodologies exist to calculate protein surfaces, a thorough description of which can be found at http://www.netsci.org/Science/Compchem/feature14.html. Below we briefly outline three methodologies.

One of the first methods to calculate protein surfaces was that of Lee and Richards.[44] It consists of "rolling a ball" of arbitrary radius (a typical value is the radius of a water molecule, 1.4 Å) over the protein of interest. The surface accessible to the "rolling ball" is then defined as the solvent-accessible surface area (ASA). An implementation of this algorithm is available in the programs *Naccess*,[b] MSMS,[c,45] as well as in the program SURFACE that is part of the CCP4 package.[d,46] Another implementation from Michael Connolly[47] is also available.[e]

A conceptually similar method consists in placing equidistant points on a sphere around each atom and then considering only those points that are not

[b] http://www.bioinf.manchester.ac.uk/naccess/.

[c] http://mgltools.scripps.edu/packages/MSMS.

[d] http://www.ccp4.ac.uk.

[e] http://biohedron.drupalgardens.com/content/molecular-surface-package.

included in another sphere. This method was described in Ref. 48 and is implemented in the program AREAIMOL, also part of the CCP4 package,[46] in the program DSSP,[f] where the area is calculated by "geodesic sphere integration,"[49] as well as in SurfRace.[50]

Another method is based on Voronoi and Voronoi-related tessellations. Such methods have been thoroughly reviewed in Ref. 51. An advantage of Voronoi-related tessellations is that they provide access to a variety of descriptors of the geometry of the surface, as well as of amino acid contacts at the interface. Bernauer *et al.* used such descriptors in combination with support vector machines to distinguish crystal contacts from biological interfaces. A recent software called "Intervor" enables the computation of such surfaces.[52]

Considering the first two methods, calculating the solvent-accessible areas of a complex does not directly yield information on the interaction interface. In order to obtain this information, one needs to measure the ASA of the complex ($ASA_{complex}$) and that of each subunit individually (ASA_{sub1}, ASA_{sub2}). Thus, the surface buried at the interface can be obtained by taking $ASA_{buried} = (ASA_{sub1} + ASA_{sub2}) - ASA_{complex}$. In order to measure the involvement of a residue at the interface, the same principle can be applied, that is, measure the ASA of the residue in the subunit alone and in complex. Depending on the goal of the study, it can be important to consider the relative ASA rather than the absolute ASA, as not all amino acids have the same size (i.e., a fully buried glycine is likely to bury less surface than a partially buried tryptophan).

Finally, protein–protein interfaces can be defined based on contacts between subunits rather than based on ASA. The definition can be loose, for instance, any pair of residues is considered in contact when their alpha or beta carbons are closer than a given cut-off value. It can be more detailed by taking into account side chains (e.g., van der Waals radii[53] plus 0.5 Å), and even more detailed by taking into account salt bridges, hydrogen bonds, solvent molecules, and ions.[54]

3.2. Interface properties: Focusing on the amino acid composition

The literature on the general properties of protein–protein interfaces is so extensive that it cannot be thoroughly reviewed here. We thus refer the reader to the following books[55–58] and reviews.[1,59–68] In this section, we focus specifically on the amino acid composition of interfaces and how it

[f] http://swift.cmbi.ru.nl/gv/dssp/.

varies across different regions within the interface. In particular, we will try to measure the extent to which interface and surface differ.

Protein–protein interfaces, in general, tend to be more hydrophobic than protein surfaces,[60,69–71] and homomers have frequently been described as having more hydrophobic interfaces than hetero-oligomers.[72,73] The same is true of obligate homomers, the interfaces of which appear more hydrophobic than those of transient homomers.[72–74] To some extent, this is linked to differences in interface sizes, because polar amino acids tend to be exposed to the solvent, while hydrophobic amino acids tend to be buried. Considering a simplified elliptic interface, we note that the ratio "perimeter divided by surface" decreases as the surface increases. In other words, the perimeter, which is exposed to the solvent and therefore hydrophilic, represents a smaller fraction of larger interfaces. This "perimeter/surface" dichotomy is reflected in the core–rim model of protein–protein interfaces,[75–77] where the rim is composed of residues having all of their atoms accessible to the solvent and the core comprises residues with at least one fully buried atom.[76]

The core–rim model was extended with a third category, called the support. The support, rim, and core are illustrated in Fig. 2.6, which also depicts how they are defined.[33] This definition relies first on the dichotomy of structures into two classes: the "surface" (relative ASA > 25%) and the "interior" (relative ASA < 25%). The value of 25% is chosen because it maximizes the difference in composition between these two regions. Based on this first dichotomy

- a rim residue is a surface residue in the structure alone AND in the complex
- a support residue is an interior residue in the structure alone AND in the complex
- a core residue is a surface residue in the structure alone and it becomes an interior residue in the complex.

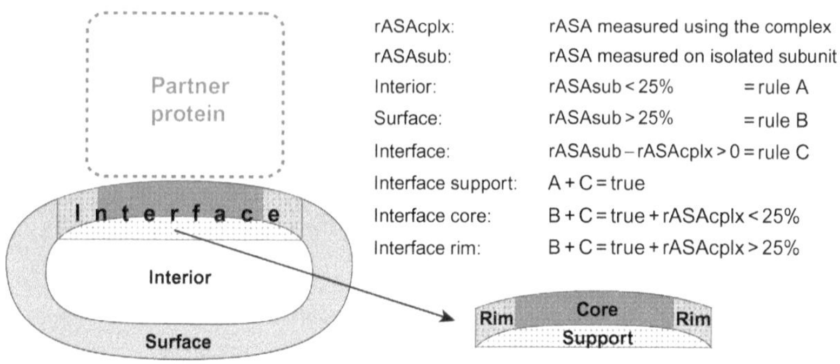

Figure 2.6 A definition of structural regions for residues at interfaces.

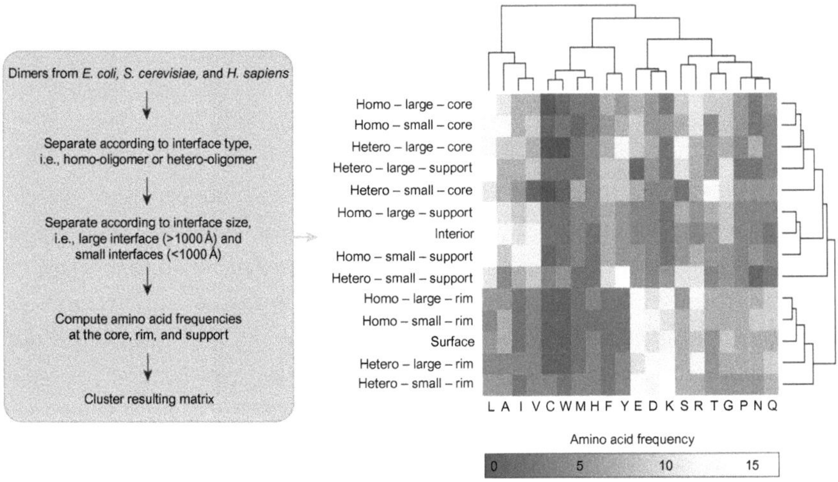

Figure 2.7 It is important to consider the above structural regions when investigating and comparing the quaternary structure of homo-oligomers with each other and with other types of complexes. *Figure partially reproduced from Ref. 33.* (For color version of this figure, the reader is referred to the online version of this chapter.)

An interesting property of this definition is that it allows one to compare structurally equivalent regions. This can be useful within the structure, for example, to compare the evolutionary conservation of the support with that of the interior, and the evolutionary conservation of the rim or core with that of the surface. It can also be useful across structures, because the definition yields regions that are stable in their composition even across interfaces of different sizes and of different types (Fig. 2.7). In other words, while we know that the global composition of interfaces changes with size, it seems that the breakdown into support, core, and rim makes individual regions more comparable to each other, even across different types of complexes.

4. CREATING INTERFACES IN HOMOMERS

4.1. Are point mutations likely to create new interfaces?

The dichotomy of interfaces into the regions described above also enables us to ask the following question: from a simple amino acid composition perspective, "how many mutations are required to turn an average $1000\ \text{Å}^2$ surface patch into an average $1000\ \text{Å}^2$ interface patch?" Surprisingly, it was found that as little as two amino acid changes are sufficient. This illustrates that protein surfaces are in fact quite close to interfaces in terms of their chemical composition. This is also reflected in the fact that it is difficult to distinguish interface from surface based on amino acid composition alone.[78,79] In other words, protein surfaces may be more "sticky" than commonly assumed.[33,80]

This character suggests that it may be easier than commonly thought to create new protein interfaces. Surely, a number of factors such as shape and physicochemical complementarity must be considered to infer if two specific proteins are likely to interact, but the above result tells us about *average* stickiness properties. Interestingly, we saw in Section 2.2 that symmetric (isologous) interfaces are expected to be more frequently favorable energetically than heterologous interfaces. Consistent with this notion, there are several examples where a single or a couple of point mutations are associated with a dimerization event,[81–84] but to our knowledge, there is no such example for heterologous interfaces.

The chemical proximity between surface and interface also suggests that functional, nonfunctional, and deleterious interactions are likely explored during the random course of evolution. One such example is illustrated by sickle cell anemia, where a single amino acid substitution is responsible for the formation of fibrils.[84] Other examples are reflected in homologous proteins forming different types of dimers, that is, where the dimeric interface is situated in different regions of the structure on both homologues, as in, for example, sulfhydryl oxidases.[85]

4.2. Molecular mechanisms associated to changes in oligomeric state

Analyses of homologous proteins forming different quaternary structures can reveal molecular mechanisms associated with evolutionary changes in homomeric state. Several molecular mechanisms have been observed, and we describe them below starting from the most indirect effects (i.e., where a mutation is away from the interface) to the most direct effects (i.e., where the mutations take place at the interface).

Perica *et al.* recently estimated that up to 50% of changes in oligomeric state might originate from allosteric effects, that is, they result from changes in sequence away from the interface.[86] The PyrR family illustrates such a case, where a rotation of the subunits at the dimeric interface creates a geometric incompatibility at the tetrameric interface.[86]

A distinct mechanism by which a mutation may induce a change in oligomeric state is through domain swapping.[87] In this case, two or more proteins mutually exchange a part of their structure, usually at the N- or

C-terminus, with each other. An interesting case was recently described where a domain-swapping event rescued the stability of a monomeric protein. This suggests that homomers may sometime appear in evolution simply as a mechanism to rescue stability, and may not necessarily be driven by function *per se*.[83]

Several studies[88–91] have carried out comprehensive comparisons of homologous protein pairs forming monomers and dimers. They focused on the interface region in the dimer and a structurally equivalent region in the monomer to identify differences that may explain changes in oligomeric state. Protein loops were found to have an important impact, both in promoting (enabling loops) and avoiding (disabling loops) dimerization.[89] These loop regions have been characterized in terms of several properties including their length (mostly less than 10 amino acids) and their structure (mostly devoid of secondary structure).[88] Another study focusing on the family of glycosyltransferases arrived at similar conclusions.[92]

5. EVOLUTION OF HOMOMER GEOMETRY AND DIVERSITY THROUGH GENE DUPLICATION

5.1. Probable oligomeric state transitions

In the previous section, we discussed molecular mechanisms that can explain evolutionary changes in oligomeric state. Below we ask first, how frequently do such changes take place during the course of evolution? And when they do, which oligomeric transitions are most common?

In recent years, a significant body of work has focused on "interface" conservation and "interface geometry" conservation,[93–96] as exemplified by the many databases making this information available.[27,97–99] These allow one to predict whether two proteins are likely to interact with each other and if they do, which geometry is adopted. Such information is very important for the structural modeling of protein interactions and to analyze the evolutionary conservation of individual interfaces. Importantly, however, information on binary interactions between domains is not sufficient to assess the conservation of homomers, which may involve two or more distinct interfaces. An estimate of the conservation of quaternary structure during evolution has been made based on the symmetry of homomers. This shows that at sequence identity levels of 90%, 50%, and 30%, there are 93%, 81%, and 70% of proteins that exhibit identical symmetry types, respectively.[100] However, this estimate does not take domain geometry into account, so that two homologous proteins forming two C2 dimers may be counted as having an identical quaternary structure although they exhibit distinct interfaces. In the future, it will therefore be important to systematically combine symmetry information with domain-interaction-geometry information in order to refine these numbers.

Nevertheless, these numbers are informative as they represent upper boundaries. This means that two sequences sharing 50% identity have similar quaternary structures in at most 81% of cases. What then happens in the remaining cases? As we have seen in Section 2.2, some symmetry types are geometrically closer to each other than others. In other words, while the transition from a C2 homomer to a D2 homomer can be achieved by creation of a single interface, the transition between a C2 and a C4 homomer would require many more events—namely, breaking the C2 interface and creating two complementary interfaces yielding a C4 symmetry. As a result, and as expected, we observe that geometrically favorable transitions occur more frequently than nonfavorable ones.[100] One example is illustrated by the enzyme phosphoglycerate mutase, which exists as a dimer in *H. sapiens* and as a tetramer in *S. cerevisiae* (Fig. 2.8). We can see, however, that the *S. cerevisiae* tetramer is made of two dimers that are structurally almost identical to that found in *H. sapiens*. This notion is summarized in Fig. 2.9, where we highlight the geometrically most favorable transitions.

5.2. Evolution through gene duplication

Interestingly, changing oligomeric state is not the only option a homomer has to evolve. Another route involving gene duplication is frequently taken. Such an event turns a homomer into a "pseudohetero" protein complex, made of homologous proteins. This mechanism has been observed to occur frequently in protein networks, whereby homologous proteins tend to interact more frequently than would be expected by chance.[101,102]

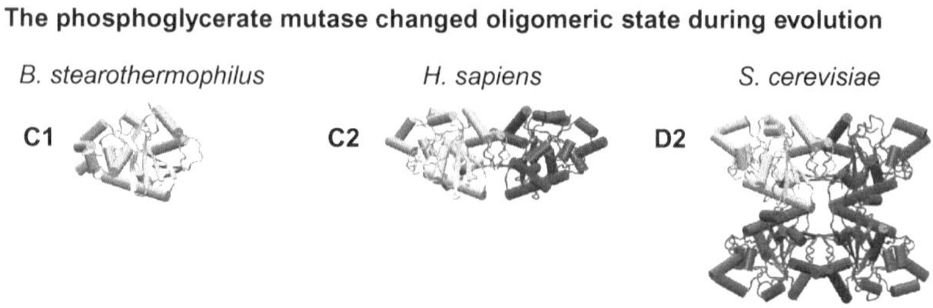

Figure 2.8 Changes in oligomeric state among homologues are often related geometrically. (For color version of this figure, the reader is referred to the online version of this chapter.)

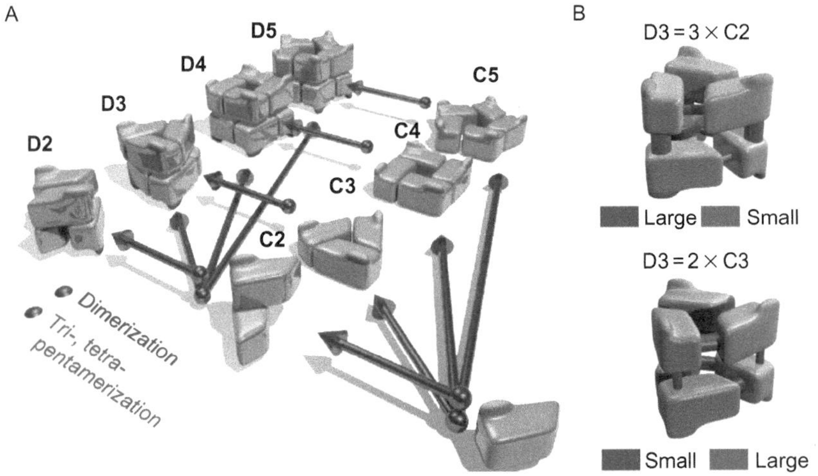

Figure 2.9 Model of quaternary structure evolution in homomers, as proposed in Ref. 100. (See Color Insert.)

As a matter of fact, many protein complexes are the product of such a scenario. Well-known examples are hemoglobin, RNA polymerases, the photosystem I, the CCT complex, the 20S proteasome, and several transcription factor families such as the Myc/Max factors or NF-κB family which form heterodimers of homologous subunits. All these examples illustrate that symmetry was present at the origin of very fundamental molecular machines and led to "pseudosymmetry," after gene duplication occurred. Symmetry breaking by gene duplication enables each subunit to evolve independently, to regulate the system and be regulated in an independent fashion. The photosystem I complex illustrates this well, where each duplicate of the "ancestral homodimer" binds different sets of proteins.

We summarize this section with a hypothetical pathway of how the proteasome may have evolved from a single domain into a large and complex cellular machine (Fig. 2.10). All subunits of the 20S proteasome share a similar structure, which belongs to the "N-terminal nucleophile aminohydrolase domain" family, suggesting that they descend from a common ancestral protein. The scenario starts with a single domain (pdb: 1xfg, glutaminase domain[103]), which may have formed a dimer or a hexameric ring (no such intermediates are known to our knowledge). Six copies of the dimer (6 × C2) or two copies of the hexamer (2 × C6) could have led to a structure similar to the Heat Shock Locus V protein complex (D6, pdb: 1ned[104]). A gene duplication event would have allowed two D6 complexes to be stacked on top of each other to form a "D6 dimer." (Note that

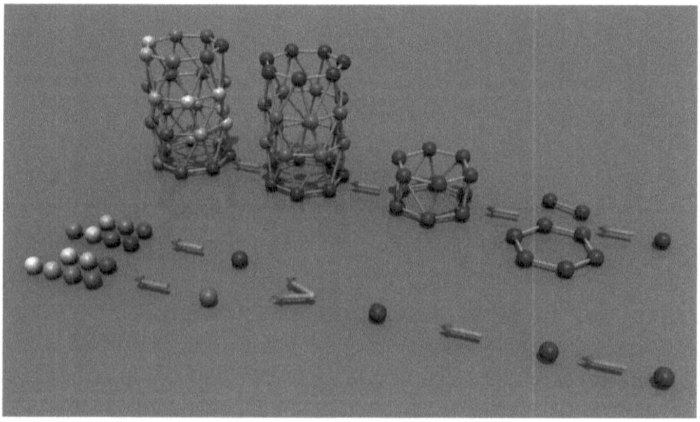

Figure 2.10 Inferring a possible evolutionary scenario from a single domain to a large molecular machine. (See Color Insert.)

without gene duplication, this would not be possible as it would yield a fibril.) Such a dimerization event and an "opening" of the ring to accommodate one additional subunit yield a structure similar to the archaeal proteasome (pdb: 1pma[105]). Finally, successive gene duplications yield the eukaryotic 20S proteasome characterized in yeast (pdb: 1ryp[106]).

6. ASSEMBLY OF HOMOMERS

6.1. Assembling into dimers

After being synthesized by the ribosome, polypeptide chains must find each other and assemble in order to form a homomer. To describe this process, we will distinguish two types of assembly events: (i) "binding steps," which we use to refer to two subunits or groups of subunits binding each other, and (ii) "assembly pathway," which we use to describe the sequence of binding steps followed in the assembly of a homomer. Considering a dimer, a single binding step (and therefore a single pathway) can be followed. The binding step can nevertheless take different forms. Schematically, the polypeptide chains can undergo folding and binding either simultaneously (two-state binding) or sequentially (three-state binding). Such a dichotomy is also reflected in the distinction frequently found in the literature between "obligate interactions" (where subunits are not stable on their own) and "non-obligate interactions" (where subunits can exist in their folded state in the absence of their partner).[63] Of course, many cases are situated in between these two extreme scenarios.

Interestingly, the native-state topology of the protein can in fact be used to infer the extent to which folding and binding are likely to occur simultaneously.[107–110] Simply put, the more the interface contributes to the stability of the native state, the more it is likely to involve simultaneous folding and binding. In other words, the larger the area buried at the interface relative to the area buried within each subunit, the more likely it is that the binding is coupled to folding.[111]

It is tempting to link these assembly mechanisms to evolution. One could indeed expect that a stable monomer evolving and forming a dimer is likely to assemble through a "three-state" mechanism where the folded intermediate corresponds to the ancestral monomer.[112,113] In contrast, dimers burying large interfaces and assembling through a two-state binding may be less likely to exist (or have existed) as monomers in other species. Little is known about this notion, which will be worth investigating in the future.

6.2. Assembling into higher-order oligomers

In 1979, the assembly pathway for many homomers and hetero-oligomers had already been characterized,[114] and in the following two decades, many more were characterized.[115,116] Today, the assembly pathway best characterized might be that of the tetramerization domain of p53.[117,118] It takes place in two binding steps: the first is a two-state binding of two monomers into a dimer, and the second is the binding of two folded dimers to form a tetramer.

This assembly pathway is frequently observed among tetramers, that is, where a dimer forms first and two dimers subsequently assemble into a tetramer. This can be expected among D2 tetramers, which often involve two interfaces of different strengths. The assembly pathway observed for p53 results from a "hierarchy" between interfaces, with a strong and a weak interface. Such a hierarchy in the affinity between the two possible dimers has the advantage of favoring a single pathway and minimizing complexes trapped in non-native states.[119] Assembly through a well-defined pathway becomes even more essential if considering large homomers such as virus capsids. For example, considering the assembly of an icosahedron formed by 60 subunits, the number of possible assembly pathways reduces drastically if considering solely pre-assembled C5 pentamers, 12 of which may form a complete capsid.[120] Although this is a simplification of a highly complex process,[121] it illustrates that introducing a hierarchy of interface strength can favor a specific assembly pathway.

6.3. When assembly recapitulates evolution

We have seen in Section 6.1 how the assembly pathway of a dimer can be reflected in its evolution. Regarding higher-order homomers, a similarity between "ontogeny" and "phylogeny" can also be inferred, although based on a different rationale.

The rationale is that a hierarchy of interface sizes is naturally expected. The expected area buried within an oligomer follows a simple power law relationship: $ASA = 5.3 \cdot M^{0.76}$,[122] where ASA is the accessible surface area in Å^2 and M the molecular weight in daltons. The expected ASA of a 30-kD protein is therefore ~13,400 Å^2. If that protein forms dimers and tetramers, their expected ASA will be 22,700 and 38,400 Å^2, respectively. We can thus calculate the expected area buried in the formation of a dimer ($2 \times 13{,}400 - 22{,}700 = 4100$ Å^2) and a tetramer ($2 \times 22{,}700 - 38{,}400 = 6950$ A^2). Importantly, in the formation of the tetramer, two interfaces (and up to four) are formed, so that the size obtained must be divided by two (or four). In sum, a 30-kD monomer that evolves into a dimer is expected to bury 4100 Å^2 at the dimeric interface, and the same dimer that evolves to a tetramer is expected to bury from 1740 to 3745 Å^2 at the tetrameric interface(s). This illustrates why, on average, interfaces associated with higher oligomeric transitions (Cn ↔ Dn) are expected to be smaller than those associated to lower oligomeric transitions (C1 ↔ Cn)—as depicted in Fig. 2.9B.

An implication of this observation, which was confirmed in ~80% of homomers with different symmetry,[100] is that large interfaces in dihedral homomers are likely to be conserved in cyclic homologues. In addition, considering that interface size is indicative of "affinity," it suggests that the predicted evolutionary intermediate in a complex is also likely to represent an assembly intermediate.[100,123]

7. CONCLUSION

We have seen that homomers are frequent, with up to 50% of proteins forming such structures. Importantly, however, most proteins do not act alone in cells. In other words, most of the homomers observed in structural data are expected to form higher-order complexes in cells. This is illustrated in a recent study of protein complexes in *Mycoplasma pneumonia*,[124] where it was estimated that up to 47% of soluble proteins have stoichiometries larger than one, that is, more than one copy of the protein is observed in the complex they form. This suggests that in future studies, it will be particularly interesting to consider the properties of homomers in the broader context of complexes made up of hetero-interactions as well. This will likely reveal new functional[32,125,126] and biophysical[127] properties associated with the cellular machinery.

ACKNOWLEDGMENTS

We thank Joël Janin, Peter Rose, Tina Terica, Joseph Marsh, Eviatar Natan, and Ezra Peisach for their comments on this chapter.

REFERENCES

1. Janin J, Bahadur RP, Chakrabarti P. Protein-protein interaction and quaternary structure. *Q Rev Biophys* 2008;**41**(2):133–80.
2. Darnall DW, Klotz IM. Subunit constitution of proteins: a table. *Arch Biochem Biophys* 1975;**166**(2):651–82.
3. Levitt M. Growth of novel protein structural data. *Proc Natl Acad Sci USA* 2007;**104**(9): 3183–8.
4. Janin J. Specific versus non-specific contacts in protein crystals. *Nat Struct Biol* 1997; **4**(12):973–4.
5. Dasgupta S, Iyer GH, Bryant SH, Lawrence CE, Bell JA. Extent and nature of contacts between protein molecules in crystal lattices and between subunits of protein oligomers. *Proteins* 1997;**28**(4):494–514.
6. Carugo O, Argos P. Protein-protein crystal-packing contacts. *Protein Sci* 1997; **6**(10):2261–3.
7. Pal A, Chakrabarti P, Bahadur R, Rodier F, Janin J. Peptide segments in protein-protein interfaces. *J Biosci* 2007;**32**(1):101–11.
8. Bahadur RP, Chakrabarti P, Rodier F, Janin J. A dissection of specific and non-specific protein-protein interfaces. *J Mol Biol* 2004;**336**(4):943–55.
9. Cieslik M, Derewenda ZS. The role of entropy and polarity in intermolecular contacts in protein crystals. *Acta Crystallogr D Biol Crystallogr* 2009;**65**(Pt 5):500–9.
10. Price 2nd. WN, Chen Y, Handelman SK, Neely H, Manor P, Karlin R, et al. Understanding the physical properties that control protein crystallization by analysis of large-scale experimental data. *Nat Biotechnol* 2009;**27**(1):51–7.
11. Zhu H, Domingues FS, Sommer I, Lengauer T. NOXclass: prediction of protein-protein interaction types. *BMC Bioinformatics* 2006;**7**:27.
12. Mitra P, Pal D. Combining Bayes classification and point group symmetry under Boolean framework for enhanced protein quaternary structure inference. *Structure* 2011;**19**(3):304–12.
13. Valdar WS, Thornton JM. Conservation helps to identify biologically relevant crystal contacts. *J Mol Biol* 2001;**313**(2):399–416.
14. Elcock AH, McCammon JA. Identification of protein oligomerization states by analysis of interface conservation. *Proc Natl Acad Sci USA* 2001;**98**(6):2990–4.
15. Valdar WS, Thornton JM. Protein-protein interfaces: analysis of amino acid conservation in homodimers. *Proteins* 2001;**42**(1):108–24.
16. Guharoy M, Chakrabarti P. Conservation and relative importance of residues across protein-protein interfaces. *Proc Natl Acad Sci USA* 2005;**102**(43):15447–52.
17. Guharoy M, Chakrabarti P. Conserved residue clusters at protein-protein interfaces and their use in binding site identification. *BMC Bioinformatics* 2010;**11**:286.

18. Armon A, Graur D, Ben-Tal N. ConSurf: an algorithmic tool for the identification of functional regions in proteins by surface mapping of phylogenetic information. *J Mol Biol* 2001;**307**(1):447–63.
19. Ma B, Elkayam T, Wolfson H, Nussinov R. Protein-protein interactions: structurally conserved residues distinguish between binding sites and exposed protein surfaces. *Proc Natl Acad Sci USA* 2003;**100**(10):5772–7.
20. Eames M, Kortemme T. Structural mapping of protein interactions reveals differences in evolutionary pressures correlated to mRNA level and protein abundance. *Structure* 2007;**15**(11):1442–51.
21. Rahat O, Yitzhaky A, Schreiber G. Cluster conservation as a novel tool for studying protein-protein interactions evolution. *Proteins* 2008;**71**(2):621–30.
22. Choi YS, Yang JS, Choi Y, Ryu SH, Kim S. Evolutionary conservation in multiple faces of protein interaction. *Proteins* 2009;**77**(1):14–25.
23. Franzosa EA, Xia Y. Structural determinants of protein evolution are context-sensitive at the residue level. *Mol Biol Evol* 2009;**26**(10):2387–95.
24. Scharer MA, Grutter MG, Capitani G. CRK: an evolutionary approach for distinguishing biologically relevant interfaces from crystal contacts. *Proteins* 2010;**78**(12):2707–13.
25. Mintseris J, Weng Z. Structure, function, and evolution of transient and obligate protein-protein interactions. *Proc Natl Acad Sci USA* 2005;**102**(31):10930–5.
26. Xu Q, Canutescu AA, Wang G, Shapovalov M, Obradovic Z, Dunbrack Jr RL. Statistical analysis of interface similarity in crystals of homologous proteins. *J Mol Biol* 2008;**381**(2):487–507.
27. Xu Q, Dunbrack Jr RL. The protein common interface database (ProtCID)—a comprehensive database of interactions of homologous proteins in multiple crystal forms. *Nucleic Acids Res* 2011;**39**:D761–D770 (Database issue).
28. Henrick K, Thornton JM. PQS: a protein quaternary structure file server. *Trends Biochem Sci* 1998;**23**(9):358–61.
29. Krissinel E, Henrick K. Inference of macromolecular assemblies from crystalline state. *J Mol Biol* 2007;**372**(3):774–97.
30. Xu Q, Canutescu A, Obradovic Z, Dunbrack Jr RL. ProtBuD: a database of biological unit structures of protein families and superfamilies. *Bioinformatics (Oxford, England)* 2006;**22**(23):2876–82.
31. Levy ED. PiQSi: Protein Quaternary Structure Investigation. *Structure* 2007;**15**(11):4.
32. Goodsell DS, Olson AJ. Structural symmetry and protein function. *Annu Rev Biophys Biomol Struct* 2000;**29**:105–53.
33. Levy ED. A simple definition of structural regions in proteins and its use in analyzing interface evolution. *J Mol Biol* 2010;**403**(4):660–70.
34. Claverie P, Hofnung M, Monod J. Sur certaines implications de l'hypothèse d'équivalence stricte entre les protomères des protéines oligomériques. *C R Séances Acad Sci* 1968;**266**:1616–8.
35. Alber F, et al. The molecular architecture of the nuclear pore complex. *Nature* 2007;**450**(7170):695–701.
36. Wales DJ. Symmetry, near-symmetry and energetics. *Chem Phys Lett* 1998;**285**(5–6):330–6.
37. Monod J, Wyman J, Changeux JP. On the Nature of Allosteric Transitions: a Plausible Model. *J Mol Biol* 1965;**12**:88–118.
38. Lukatsky DB, Zeldovich KB, Shakhnovich EI. Statistically enhanced self-attraction of random patterns. *Phys Rev Lett* 2006;**97**(17):178101.
39. Lukatsky DB, Shakhnovich BE, Mintseris J, Shakhnovich EI. Structural similarity enhances interaction propensity of proteins. *J Mol Biol* 2007;**365**(5):1596–606.

40. Andre I, Strauss CE, Kaplan DB, Bradley P, Baker D. Emergence of symmetry in homooligomeric biological assemblies. *Proc Natl Acad Sci USA* 2008;**105**(42):16148–52.
41. Schulz GE. The dominance of symmetry in the evolution of homo-oligomeric proteins. *J Mol Biol* 2010;**395**(4):834–43.
42. Perutz MF. Structure of hemoglobin. *Brookhaven Symp Biol* 1960;**13**:165–83.
43. Dey S, Chakrabarti P, Janin J. A survey of hemoglobin quaternary structures. *Proteins* 2011;**79**(10):2861–70.
44. Lee B, Richards FM. The interpretation of protein structures: estimation of static accessibility. *J Mol Biol* 1971;**55**(3):379–400.
45. Sanner MF, Olson AJ, Spehner J-C. Reduced surface: an efficient way to compute molecular surfaces. *Biopolymers* 1996;**38**(3):305–20.
46. Collaborative Computational Project . The CCP4 suite: programs for protein crystallography. *Acta Crystallogr D Biol Crystallogr* 1994;**50**(Pt 5):760–3.
47. Connolly ML. The molecular surface package. *J Mol Graph* 1993;**11**(2):139–41.
48. Shrake A, Rupley JA. Environment and exposure to solvent of protein atoms. Lysozyme and insulin. *J Mol Biol* 1973;**79**(2):351–71.
49. Kabsch W, Sander C. Dictionary of protein secondary structure: pattern recognition of hydrogen-bonded and geometrical features. *Biopolymers* 1983;**22**(12):2577–637.
50. Tsodikov OV, Record Jr MT, Sergeev YV. Novel computer program for fast exact calculation of accessible and molecular surface areas and average surface curvature. *J Comput Chem* 2002;**23**(6):600–9.
51. Poupon A. Voronoi and Voronoi-related tessellations in studies of protein structure and interaction. *Curr Opin Struct Biol* 2004;**14**(2):233–41.
52. Loriot S, Cazals F. Modeling macro-molecular interfaces with Intervor. *Bioinformatics (Oxford, England)* 2010;**26**(7):964–5.
53. Tsai J, Taylor R, Chothia C, Gerstein M. The packing density in proteins: standard radii and volumes. *J Mol Biol* 1999;**290**(1):253–66.
54. Andreani J, Faure G, Guerois R. Versatility and invariance in the evolution of homologous heteromeric interfaces. *PLoS Comput Biol* 2012;**8**(8):e1002677.
55. Nussinov R, Schreiber G, editors. *Computational protein-protein interactions*. Boca Raton: CRC Press; 2009.
56. Kleanthous C, editor. *Protein-protein recognition: frontiers in molecular biology*. New York: Oxford University Press; 2000, p. 314.
57. Janin J, Wodak SJ, editors. *Protein modules and protein-protein interaction. Advances in protein chemistry,* **61**:San Diego: London Academic Press; 2003, p. 333.
58. Golemis E, Adams PD. *Protein-protein interactions: a molecular cloning manual*. New York: Cold Spring Harbor Laboratory Press; 2005, p. 938.
59. Aloy P, Russell RB. Structural systems biology: modelling protein interactions. *Nat Rev Mol Cell Biol* 2006;**7**(3):188–97.
60. Chothia C, Janin J. Principles of protein-protein recognition. *Nature* 1975;**256**(5520):705–8.
61. Deremble C, Lavery R. Macromolecular recognition. *Curr Opin Struct Biol* 2005;**15**(2):171–5.
62. Han JH, Batey S, Nickson AA, Teichmann SA, Clarke J. The folding and evolution of multidomain proteins. *Nat Rev Mol Cell Biol* 2007;**8**(4):319–30.
63. Jones S, Thornton JM. Principles of protein-protein interactions. *Proc Natl Acad Sci USA* 1996;**93**(1):13–20.
64. Keskin O, Gursoy A, Ma B, Nussinov R. Principles of protein-protein interactions: what are the preferred ways for proteins to interact? *Chem Rev* 2008;**108**(4):1225–44.
65. Levy ED, Pereira-Leal JB. Evolution and dynamics of protein interactions and networks. *Curr Opin Struct Biol* 2008;**18**(3):349–57.

66. Nooren IM, Thornton JM. Diversity of protein-protein interactions. *EMBO J* 2003;**22**(14):3486–92.
67. Ponstingl H, Kabir T, Gorse D, Thornton JM. Morphological aspects of oligomeric protein structures. *Prog Biophys Mol Biol* 2005;**89**(1):9–35.
68. Reichmann D, Rahat O, Cohen M, Neuvirth H, Schreiber G. The molecular architecture of protein-protein binding sites. *Curr Opin Struct Biol* 2007;**17**(1):67–76.
69. Young L, Jernigan RL, Covell DG. A role for surface hydrophobicity in protein-protein recognition. *Protein Sci* 1994;**3**(5):717–29.
70. Tsai CJ, Lin SL, Wolfson HJ, Nussinov R. Studies of protein-protein interfaces: a statistical analysis of the hydrophobic effect. *Protein Sci* 1997;**6**(1):53–64.
71. Bahadur RP, Chakrabarti P, Rodier F, Janin J. Dissecting subunit interfaces in homodimeric proteins. *Proteins* 2003;**53**(3):708–19.
72. Nooren IM, Thornton JM. Structural characterisation and functional significance of transient protein-protein interactions. *J Mol Biol* 2003;**325**(5):991–1018.
73. Ofran Y, Rost B. Analysing six types of protein-protein interfaces. *J Mol Biol* 2003;**325**(2):377–87.
74. Dey S, Pal A, Chakrabarti P, Janin J. The subunit interfaces of weakly associated homodimeric proteins. *J Mol Biol* 2010;**398**(1):146–60.
75. Lo Conte L, Chothia C, Janin J. The atomic structure of protein-protein recognition sites. *J Mol Biol* 1999;**285**(5):2177–98.
76. Chakrabarti P, Janin J. Dissecting protein-protein recognition sites. *Proteins* 2002;**47**(3):334–43.
77. Larsen TA, Olson AJ, Goodsell DS. Morphology of protein-protein interfaces. *Structure* 1998;**6**(4):421–7.
78. Neuvirth H, Raz R, Schreiber G. ProMate: a structure based prediction program to identify the location of protein-protein binding sites. *J Mol Biol* 2004;**338**(1):181–99.
79. de Vries SJ, Bonvin AM. How proteins get in touch: interface prediction in the study of biomolecular complexes. *Curr Protein Pept Sci* 2008;**9**(4):394–406.
80. Levy ED, De S, Teichmann SA. Cellular crowding imposes global constraints on the chemistry and evolution of proteomes. *Proc Natl Acad Sci USA* 2012;**109**(50):20461–6.
81. Rollins CT, Rivera VM, Woolfson DN, Keenan T, Hatada M, Adams SE, et al. Aligand-reversible dimerization system for controlling protein-protein interactions. *Proc Natl Acad Sci USA* 2000;**97**(13):7096–101.
82. Grueninger D, Treiber N, Ziegler MO, Koetter JW, Schulze MS, Schulz GE. Designed protein-protein association. *Science* 2008;**319**(5860):206–9.
83. Bershtein S, Mu W, Shakhnovich EI. Soluble oligomerization provides a beneficial fitness effect on destabilizing mutations. *Proc Natl Acad Sci USA* 2012;**109**(13):4857–62.
84. Pauling L, Itano HA. Sickle cell anemia, a molecular disease. *Science* 1949;**109**(2835):443.
85. Hakim M, Fass D. Dimer interface migration in a viral sulfhydryl oxidase. *J Mol Biol* 2009;**391**(4):758–68.
86. Perica T, Chothia C, Teichmann SA. Evolution of oligomeric state through geometric coupling of protein interfaces. *Proc Natl Acad Sci USA* 2012;**109**(21):8127–32.
87. Bennett MJ, Schlunegger MP, Eisenberg D. 3D domain swapping: a mechanism for oligomer assembly. *Protein Sci* 1995;**4**(12):2455–68.
88. Hashimoto K, Panchenko AR. Mechanisms of protein oligomerization, the critical role of insertions and deletions in maintaining different oligomeric states. *Proc Natl Acad Sci USA* 2010;**107**(47):20352–7.
89. Akiva E, Itzhaki Z, Margalit H. Built-in loops allow versatility in domain-domain interactions: lessons from self-interacting domains. *Proc Natl Acad Sci USA* 2008;**105**(36):13292–7.

90. Nishi H, Koike R, Ota M. Cover and spacer insertions: small nonhydrophobic accessories that assist protein oligomerization. *Proteins* 2011;**79**(8):2372–9.
91. Nishi H, Ota M. Amino acid substitutions at protein-protein interfaces that modulate the oligomeric state. *Proteins* 2010;**78**(6):1563–74.
92. Hashimoto K, Madej T, Bryant SH, Panchenko AR. Functional states of homooligomers: insights from the evolution of glycosyltransferases. *J Mol Biol* 2010;**399**(1):196–206.
93. Han JH, Kerrison N, Chothia C, Teichmann SA. Divergence of interdomain geometry in two-domain proteins. *Structure* 2006;**14**(5):935–45.
94. Dayhoff JE, Shoemaker BA, Bryant SH, Panchenko AR. Evolution of protein binding modes in homooligomers. *J Mol Biol* 2010;**395**(4):860–70.
95. Zhang QC, Petrey D, Norel R, Honig BH. Protein interface conservation across structure space. *Proc Natl Acad Sci USA* 2010;**107**(24):10896–901.
96. Aloy P, Ceulemans H, Stark A, Russell RB. The relationship between sequence and interaction divergence in proteins. *J Mol Biol* 2003;**332**(5):989–98.
97. Faure G, Andreani J, Guerois R. InterEvol database: exploring the structure and evolution of protein complex interfaces. *Nucleic Acids Res* 2012;**40**:D847–D856 (Database issue).
98. Stein A, Ceol A, Aloy P. 3did: identification and classification of domain-based interactions of known three-dimensional structure. *Nucleic Acids Res* 2011;**39**:D718–D723 (Database issue).
99. Winter C, Henschel A, Kim WK, Schroeder M. SCOPPI: a structural classification of protein-protein interfaces. *Nucleic Acids Res* 2006;**34**:D310–D314 (Database issue).
100. Levy ED, Boeri Erba E, Robinson CV, Teichmann SA. Assembly reflects evolution of protein complexes. *Nature* 2008;**453**(7199):1262–5.
101. Ispolatov I, Yuryev A, Mazo I, Maslov S. Binding properties and evolution of homodimers in protein-protein interaction networks. *Nucleic Acids Res* 2005;**33**(11):3629–35.
102. Pereira-Leal JB, Levy ED, Kamp C, Teichmann SA. Evolution of protein complexes by duplication of homomeric interactions. *Genome Biol* 2007;**8**(4):R51.
103. Isupov MN, Obmolova G, Butterworth S, Badet-Denisot M-A, Badet B, Polikarpov I, et al. Substrate binding is required for assembly of the active conformation of the catalytic site in Ntn amidotransferases: evidence from the 1.8 A crystal structure of the glutaminase domain of glucosamine 6-phosphate synthase. *Structure* 1996;**4**(7):801–10.
104. Bochtler M, Ditzel L, Groll M, Huber R. Crystal structure of heat shock locus V (HslV) from Escherichia coli. *Proc Natl Acad Sci USA* 1997;**94**(12):6070–4.
105. Seemuller E, Lupas A, Stock D, Löwe J, Huber R, Baumeister W. Proteasome from Thermoplasma acidophilum: a threonine protease. *Science* 1995;**268**(5210):579–82.
106. Groll M, Ditzel L, Löwe J, Stock D, Bochtler M, Bartunik HD, et al. Structure of 20S proteasome from yeast at 2.4 A resolution. *Nature* 1997;**386**(6624):463–71.
107. Tsai CJ, Xu D, Nussinov R. Structural motifs at protein-protein interfaces: protein cores versus two-state and three-state model complexes. *Protein Sci* 1997;**6**(9):1793–805.
108. Tsai CJ, Xu D, Nussinov R. Protein folding via binding and vice versa. *Fold Des* 1998;**3**(4):R71–R80.
109. Levy Y, Wolynes PG, Onuchic JN. Protein topology determines binding mechanism. *Proc Natl Acad Sci USA* 2004;**101**(2):511–6.
110. Karthikraja V, Suresh A, Lulu S, Kangueane U, Kangueane P. Types of interfaces for homodimer folding and binding. *Bioinformation* 2009;**4**(3):101–11.
111. Janin J, Miller S, Chothia C. Surface, subunit interfaces and interior of oligomeric proteins. *J Mol Biol* 1988;**204**(1):155–64.

112. Xu D, Tsai CJ, Nussinov R. Mechanism and evolution of protein dimerization. *Protein Sci* 1998;**7**(3):533–44.
113. D'Alessio G. The evolutionary transition from monomeric to oligomeric proteins: tools, the environment, hypotheses. *Prog Biophys Mol Biol* 1999;**72**(3):271–98.
114. Friedman FK, Beychok S. Probes of subunit assembly and reconstitution pathways in multisubunit proteins. *Annu Rev Biochem* 1979;**48**:217–50.
115. Jaenicke R, Lilie H. Folding and association of oligomeric and multimeric proteins. *Adv Protein Chem* 2000;**53**:329–401.
116. Jaenicke R. Folding and association of proteins. *Prog Biophys Mol Biol* 1987;**49** (2–3):117–237.
117. Mateu MG, Sanchez Del Pino MM, Fersht AR. Mechanism of folding and assembly of a small tetrameric protein domain from tumor suppressor p53. *Nat Struct Biol* 1999;**6** (2):191–8.
118. Shakhnovich EI. Folding by association. *Nat Struct Biol* 1999;**6**(2):99–102.
119. Powers ET, Powers DL. A perspective on mechanisms of protein tetramer formation. *Biophys J* 2003;**85**(6):3587–99.
120. Bahadur RP, Rodier F, Janin J. A dissection of the protein-protein interfaces in icosahedral virus capsids. *J Mol Biol* 2007;**367**(2):574–90.
121. Katen S, Zlotnick A. The thermodynamics of virus capsid assembly. *Methods Enzymol* 2009;**455**:395–417.
122. Miller S, Lesk AM, Janin J, Chothia C. The accessible surface area and stability of oligomeric proteins. *Nature* 1987;**328**(6133):834–6.
123. Villar G, Wilber AW, Williamson AJ, Thiara P, Doye JPK, Louis AA, et al. Self-assembly and evolution of homomeric protein complexes. *Phys Rev Lett* 2009;**102** (11):118106.
124. Kuhner S, van Noort V, Betts MJ, Leo-Macias A, Batisse C, Rode M, et al. Proteome organization in a genome-reduced bacterium. *Science* 2009;**326**(5957):1235–40.
125. Marianayagam NJ, Sunde M, Matthews JM. The power of two: protein dimerization in biology. *Trends Biochem Sci* 2004;**29**(11):618–25.
126. Griffin MD, Gerrard JA. The relationship between oligomeric state and protein function. *Adv Exp Med Biol* 2012;**747**:74–90.
127. Li P, Banjade S, Cheng HC, Kim S, Chen B, Guo L, et al. Phase transitions in the assembly of multivalent signalling proteins. *Nature* 2012;**483**(7389):336–40.

CHAPTER THREE

Physicochemical Principles of Protein Aggregation

Benedetta Bolognesi, Gian Gaetano Tartaglia
Centre for Genomic Regulation, CRG and UPF, Barcelona, Spain

Contents

Abstract

This chapter provides a theoretical framework on the quantitative description of protein aggregation. The reader is provided with an overview of the fundamental theory of linear and helical polymers, as well as an introduction on the parameters governing evolution of aggregates over time. The models presented for the interpretation of the protein aggregation process take into account the contributions of different physicochemical parameters such as charge, hydrophobicity, and secondary structure propensity. Finally, we discuss our current understanding of how prediction of aggregation rates and identification of aggregation-prone protein regions are predicted from the information contained in the primary amino acid sequence.

1. INTRODUCTION

β-Sheet structures are, together with α helices, the most common regular motifs in natively folded proteins. Partial or complete disruption of the native fold is observed when a protein is subjected to stress originating from

Progress in Molecular Biology and Translational Science, Volume 117
ISSN 1877-1173
http://dx.doi.org/10.1016/B978-0-12-386931-9.00003-9

unsuitable chemical (e.g., high or low pH, high salt concentrations, hydrophobic environment) or physical (e.g., high temperature, high pressure) agents.[1,2] Denatured proteins have no defined secondary and tertiary structure and, especially at high concentrations, tend to aggregate into insoluble deposits. Many aggregates, known as amyloid fibrils, share a common morphology and can be recognized under the electron microscope as regular rope-like structures that measure micrometers in length and a few nanometers in diameter. As shown by biophysical techniques, such as circular dichroism and Fourier transform infrared spectroscopy, amyloid fibrils have a high content of β structure, whichever the structure of the monomeric molecule in the native state. X-Ray diffraction analysis of fibrils yields a typical cross-β diffraction pattern, signature of an intermolecular β-sheet structure, where the hydrogen bonding among β strands runs parallel to the main fibril axis generating a pleated β-sheet structure. Investigations using electron and atomic force microscopy show that amyloid fibrils consist of a variable number of protofilaments, each of a diameter of approximately two nanometers and twisted around each other to form supercoiled rope-like structure.[3,4] Packing of the filaments is dependent on the protein system, but a single system can also display multiple fibril morphologies.[5]

One of the main causes of incorrect protein folding *in vivo* is cell stress, which can be caused by heat shock, nutrient depletion, or other stimuli.[6,7] Production of inactive proteins not only represents an energetic drain and a metabolic load for the cell but also may result in accumulation of the unfolded proteins within inclusion bodies that are responsible for cell damage. Indeed, misfolded proteins that escape the quality control mechanisms of the cell may lead to the impairment of relevant biological processes and affect the viability of the organism. Up to now, protein aggregation has been associated with more than 30 diseases and in particular amyloid fibrils have been found involved in a number of debilitating pathologies including Alzheimer's, Parkinson's, Huntington's, prion disease, and type II diabetes.[8,9] The propensity of different proteins to form amyloid fibrils can vary widely depending on the physicochemical properties of the specific amino acid sequence involved.[10,11]

Misfolding and aggregation of proteins *in vivo* differ from similar processes taking place under *in vitro* experimental conditions as they occur in complex cellular environments containing a host of factors that are known to modulate protein aggregation and attempt to protect against any subsequent toxicity.[12] In fact, efficient folding of many newly synthesized proteins depends on assistance from molecular chaperones, which prevent protein misfolding and aggregation in the crowded environment of the cell.[13] It is estimated that more than 30% of the newly synthesized proteins are degraded by proteasome due to translation errors or improper folding.[14] Nevertheless, it is important to mention that the same physicochemical principles that lead to aggregation *in vitro* are present in cellular environments, and once the physicochemical determinants for aggregation are identified, they can be conveyed almost straightforwardly into a mathematical model.

In this chapter, we discuss physicochemical principles that lie behind protein polymerization, including kinetic models for oligomerization and evolutionary pressures arising against toxic aggregation.

2. LINEAR POLYMERS

Let us consider a solution of macromolecules with the ability to form aggregates by end-to-end association (Fig. 3.1A). In equilibrium state, the solution contains dispersed monomers and linear polymers of various lengths. The following mass action law[15] exists between the concentration of monomers μ_1 and dimers μ_2

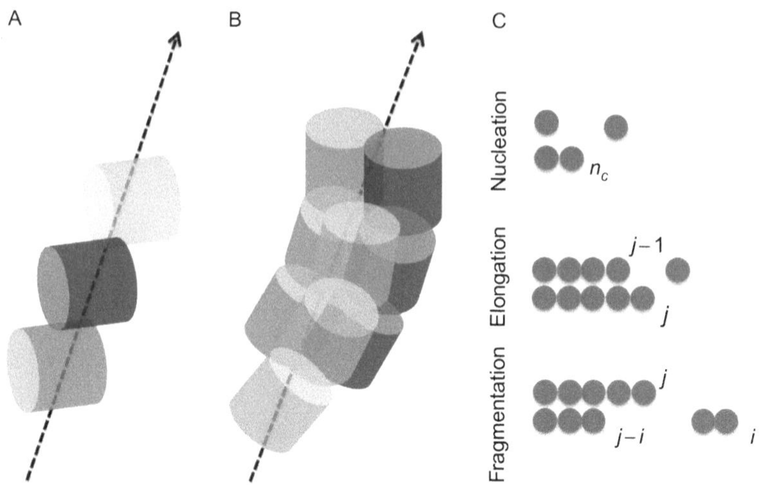

Figure 3.1 Protein oligomerization. Models of protein aggregation: (A) linear polymers and (B) helical polymers. In amyloid fibrils, each building block will acquire a β structure once incorporated in the polymer. (C) Growth trough nucleation, elongation, and fragmentation. (For color version of this figure, the reader is referred to the online version of this chapter.)

$$\mu_2 = z_1 \mu_1^2 \qquad [3.1]$$

where z_1 is the equilibrium constant of dimerization. Similarly for the transition from $(i+1)$-mer to (i)-mer

$$\mu_{i-1} = z_i \mu_i \mu_1 \qquad [3.2]$$

When z_1 is independent of i or when the binding free energy of monomer to i-mer is independent of i, the equilibrium concentration of the i-mer is given by

$$\mu_i = z^{-1}(z\mu)^i \qquad [3.3]$$

In Eq. (3.3), we have a function of the equilibrium constant $z(=z_i)$ and monomer concentration $\mu(=\mu_i)$.

The total mass m can be expressed as

$$m = \sum_{i=1} i\mu_i \qquad [3.4]$$

If z_i is independent of i, we have

$$m = \sum iz^{-1}(z\mu)^i = z^{-1}\frac{z\mu}{(1-z\mu)^2} \qquad [3.5]$$

Using Eq. (3.5), we can determine the concentration of monomer at given m, and from μ we can calculate the concentration of i-mers.

In general, the following scheme can be derived when the aggregates are in equilibrium with soluble monomeric material $M_S = \mu_1$

$$M_S + \mu_{i-1} \leftrightarrow \mu_i \qquad [3.6]$$

where the equilibrium constant is $Z_{eq} = \mu_i / M_S \mu_{i-1}$.

In this case, the total mass concentration $m = \sum_{i=1} i\mu_i = M_S/(1 - Z_{eq}M_S)^2$ can be rearranged as

$$Z_{eq} = \frac{1}{M_S} - \frac{1}{\sqrt{M_S m}} \qquad [3.7]$$

When $M_S \gg m$, the equilibrium expression reduces to the situation where fibril ends E are in equilibrium with monomers

$$M_S + E \leftrightarrow E \qquad [3.8]$$

where the equilibrium constant is $Z_{eq} = 1/M_S$.

3. HELICAL POLYMERS

In the helix formed by a linear polypeptide, each amino acid is bound to ~4 amino acids by two kinds of bonds, one being the primary bond with neighboring amino acids in the polypeptide chain and the other the hydrogen bond with the third preceding amino acid along the chain. Similarly, aggregates can be formed by monomer macromolecules with two bonds with other monomers. In the linear polymer, each monomer is bound to two other monomers and usually all bonds are of the same chemical nature (Fig. 3.1A). In the helical polymer having three monomers per turn, each monomer is bound with neighboring monomers along the linear chain and simultaneously with the third preceding and succeeding monomers (Fig. 3.1B). If we consider an equilibrium solution containing both linear polymers and helical polymers,[15] the concentration of the helical trimer $\mu_{3\mathrm{h}}$ can be related to that of the linear trimer $\mu_{3\mathrm{l}}$ by

$$\mu_{3\mathrm{h}} = \sigma\mu_{3\mathrm{l}} \qquad [3.9]$$

where $\sigma = \exp(-\partial f/KT)$ and ∂f is the free energy increment for helical trimers.

If we introduce the chemical constant z_h of equilibrium between the fourth monomer and the helical trimer, the concentration of the shortest helical polymer $\mu_{4\mathrm{h}}$ will follow

$$\mu_{4\mathrm{h}}/\sigma = z_\mathrm{h}\mu_{3\mathrm{h}}\mu_1/\sigma = \mu z^{-1} z_\mathrm{h}(z\mu)^3 = z^{-1}(z/z_\mathrm{h})^3(z_\mathrm{h}\mu)^4 \qquad [3.10]$$

Since the fourth monomer in the helix can bind two monomers, z_h is usually larger than z. For the further growth of the helical polymer by attaching monomers to the helical nucleus, we can assume the same chemical constant z_h and obtain

$$\mu_{i\mathrm{h}} = z^{-1}(z/z_\mathrm{h})^3(z_\mathrm{h}\mu)^i\sigma = z^{-1}\nu(z_\mathrm{h}\mu)^i \qquad [3.11]$$

where $\nu = \sigma(z/z_\mathrm{h})^3$.

In the solution containing monomers and linear and helical polymers, we have

$$m = \mu + m_\mathrm{l} + m_\mathrm{h} \qquad [3.12]$$

where $m_\mathrm{l} = \sum_{i=2} iz^{-1}(z\mu)^i$ and $m_2 = \sum_{i=3} iz^{-1}\nu(z_\mathrm{h}\mu)^i$.

For very small values of m, monomer concentration μ increases proportionally to m and a small number of linear polymers (dimers, trimers, etc.)

will appear in solution. When μ approaches z^{-1}, helical polymers will start to appear (if z_h is larger than z) and the total concentration approaches

$$m_c = \frac{1/z_h}{(1 - z/z_h)^2} \qquad [3.13]$$

If $m \gg m_c$, $m \sim m_h$ and helical polymers prevail in solution.[15]

4. TIME EVOLUTION OF LINEAR AND HELICAL POLYMERS

Let us denote the concentration of monomers, linear (i)-mers, and helical (i)-mers at time t as $\mu(t)$, $\mu_{il}(t)$, and $\mu_{ih}(t)$, respectively.[15] The growth rate of helical (i)-mers to helical ($i+1$)-mers can be expressed as $k_+\mu(t)\mu_{ih}(t)$ and the rate of detachment from (i)-mers will be $k_-\mu_{ih}(t)$. Similarly, the rate of transformation from linear to helical trimers can be expressed as $k'_+\mu_{3l}(t)$ and the reverse reaction will be driven by $k'_-\mu_{3h}(t)$. Hence, we have for the total number concentration[16]

$$\frac{dp}{dt} = \frac{d}{dt}\left[\sum_{i=3}\mu_{ih}(t)\right] = k'_+\mu_{3l}(t) - k'_-\mu_{3h}(t) \qquad [3.14]$$

and for the mass concentration

$$\frac{dm}{dt} = \frac{d}{dt}\left[\sum_{i=3} i\mu_{ih}(t)\right] = \left[\sum_{i=3}(k_+\mu(t) - k_-)\mu_{ih}(t)\right] + 3\left[k'_+\mu_{3l}(t) - k'_-\mu_{3h}(t)\right] + k_-\mu_{3h}(t) \qquad [3.15]$$

The increasing rate of the total mass m_h participating in helical polymers or the decreasing rate of $\mu + m_l$ of monomers and linear polymers can be calculated as

$$\begin{aligned}-[d(\mu + m_l)/dt] = dm_h/dt = [k_+\mu(t) - k_-]\int dt\,[k'_+\mu_{3l}(t) - k'_-\mu_{3h}(t)] \\ +3[k'_+\mu_{3l}(t) - k'_-\mu_{3h}(t)] \\ +k_-\mu_{3h}(t)\end{aligned} \qquad [3.16]$$

Assuming that μ_{3l} and μ_{3h} are proportional to μ^3 (the polymerization–depolymerization reaction is more rapid than helix formation), we have for $m_l \ll \mu$

$$-\mathrm{d}\mu/\mathrm{d}t = (k_+\mu - k_-)\int c\mu^3 \mathrm{d}t \qquad [3.17]$$

where c is a constant.

Equation (3.17) shows how free monomer concentration decreases in time during the aggregation process.

When $k_+\mu \gg k_-$, the differential equation can be solved

$$\ln\frac{\left[1+(1-x^3)^{1/2}\right]\left[1-(1-x_0{}^3)^{1/2}\right]}{\left[1-(1-x^3)^{1/2}\right]\left[1+(1-x_0{}^3)^{1/2}\right]} = 3\alpha t \qquad [3.18]$$

where $x = x_0^3(\mu/\mu_0)^3$, $\alpha = \gamma^{1/2}x_0^{-3/2}$, $\gamma = (2/3)k_+c\mu_0^3$, $x_0^{-3} = 1 + k_+^3 h_0^2/\gamma$, and μ_0 is the initial concentration of monomers. If the initial concentration of helical nuclei h_0 is negligible, we have

$$\ln\frac{\left[1+(1-\mu^3/\mu_0^3)^{1/2}\right]}{\left[1-(1-\mu^3/\mu_0^3)^{1/2}\right]} \approx \ln\left[4\mu_0^3/\mu^3 - 1\right] = 3\alpha t \qquad [3.19]$$

As indicated by Eq. (3.19), the concentration of free monomer μ decreases exponentially as a power law of μ_0.

5. TIME EVOLUTION OF FIBRILS

A master equation can be used to describe the time evolution of the concentration $\mu(t,j)$ of aggregates of length j[17]

$$\begin{aligned}\frac{\partial\mu(t,j)}{\partial t} &= 2m(t)k_+\mu(t,j-1) - 2m(t)k_+\mu(t,j)\\ &+k_-(j-1)\mu(t,j) + 2k_-\sum_{i=j+1}\mu(t,i) + k_n m(t)^{n_c}\delta_{j,n_c}\end{aligned} \qquad [3.20]$$

where $m(t)$ is the concentration of monomers. The first term in Eq. (3.20) accounts for the increase in the number of filaments of length j due to the addition of monomers of either end of filament of length $j-1$ (Fig. 3.1C). The term $2m(t)k_+\mu(t,j)$ describes the decrease in the number of filaments of length j growing further to length $j+1$, while $k_-(j-1)\mu(t,j)$ reflects the possibility of a filament of length j breaking at any of its $j-1$ internal links. The term $2k_-\sum_{i=j+1}\mu(t,i)$ accounts for the fact that there are two links in any filament of length $i>j$ where breakage leads to a filament of length j, while

$k_n m(t)^{n_c} \delta_{j,n_c}$ represents the spontaneous formation of growth nuclei of size n_c (Fig. 3.1C). Hence, we have for the total number concentration

$$\frac{dP(t)}{dt} = k_-[M(t) - (2n_c - 1)P(t)] + k_n m(t)^{n_c} \quad [3.21]$$

and for the total mass concentration

$$\frac{dM(t)}{dt} = 2[m(t)k_+ - n_c(n_c - 1)k_-/2]P(t) + n_c k_n m(t)^{n_c} \quad [3.22]$$

Using fixed-point analysis, the system can be integrated as

$$P(t) = \frac{m_{\text{tot}}}{2n_c - 1} - \frac{\dfrac{m_{\text{tot}} k_- \exp[-(-2n_c - 1)k_- t]}{\kappa}}{Ei(-C_+ e^{\kappa t}) + \exp[(-2n_c - 1)k_- t]B_2} \quad [3.23]$$

and

$$M(t) = m_{\text{tot}}\left(1 - \exp\left[-C_+ \exp(\kappa t) + C_- \exp(-\kappa t) + k_n m_{\text{tot}}^{n_c - 1} k_-^{-1}\right]\right) \quad [3.24]$$

where $\kappa = \sqrt{2m_{\text{tot}} k_+ k_-}$ is the rate of multiplication of filament population, $C_\pm = k_+ P(0)/\kappa \pm M(0)/(2m_{\text{tot}}) \pm \left(k_n m_{\text{tot}}^{n_c - 1}\right)/(2k_-)$ and $m_{\text{tot}} = M(t) + m(t)$. Considering the steepest slope of the kinetic trace, we have $v_{\max} = \kappa/\log(1/C_+)$.

The variable $\kappa = \sqrt{2m_{\text{tot}} k_+ k_-}$ defines the lag phase, which exists only if the growth rate $m_{\text{tot}}\kappa/e$ is maximal at $t_{\max} = \kappa^{-1} \log(1/C_+)$. More generally, the parameter κ, which corresponds to the rate of multiplication of the population of fragments, emerges as the most important quantity describing the overall properties of systems that self-assemble by processes that involve elongation and fragmentation. In a regime where secondary nucleation through fragmentation of filaments is an effective source of filaments than primary nucleation $\left(k_-/\left(k_n m_{\text{tot}}^{n_c - 1}\right) \gg 1\right)$, observables such as the lag time and maximal growth rate depend primarily on just the single parameter κ[17] (Fig. 3.2A).

Very recently, a systematic investigation of bulk experimental measurements has highlighted the relevance of secondary pathways, other than fragmentation, in driving the overall aggregation reaction.[18] The analysis reveals a crucial role to existing aggregates, which would be able to accelerate the production of further aggregates, resulting in positive feedback type of mechanism. The approach, based on the standard model of filamentous growth first presented by Oosawa in the 1960s[16] and extended by Eaton

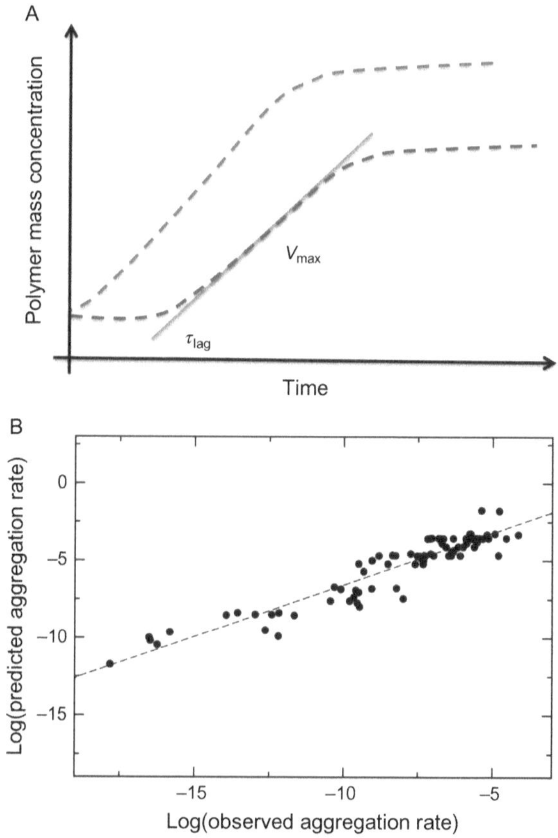

Figure 3.2 Aggregation kinetics. (A) Models for linear and helical polymerization (Eq. 3.18) can reproduce v_{max} in the exponential phase (green curve); equations based on nucleation, elongation, and fragmentation (Eq. 3.21) are used to describe lag phases (red curve). (B) The aggregation rates v_{max} can be accurately predicted using phenomenological formulas (Eq. 3.34). (For interpretation of the references to color in this figure legend, the reader is referred to the online version of this chapter.)

in the 1980s,[19] provides a new framework which allows derivation of closed form analytical solutions for the lower principal moments of the fibril length distribution, as well as a range of accompanying scale laws.

6. THE AGGREGATION RATE

Beaven *et al.* have shown, already in 1969,[20] that under a critical concentration, the process of aggregation of glucagon is slow and that polymerization occurs more readily at high concentrations. More recently, Ruschak and Miranker have reported for the islet amyloid peptide that the rate of fibril elongation increases with monomer concentration, being the slope for the fitting very close to 1.[21] These findings are in good agreement with the fact that the rate of multiplication of filament population (Section 5) is proportional to the square of protein (monomer and fibril) in solution. In general, concentration, temperature, ionic strength, and pH are essential factors influencing the process of aggregation and can be regarded as *extrinsic contributions* to distinguish them from the *intrinsic contributions* that are inherent properties of the polypeptide chain, dependent on the amino acid position. With good approximation, the aggregation rate can be assumed to increase with temperature and concentration because the probability of collision and elongation of polypeptide chains increases with temperature and concentration. Although aggregation rate and temperature are not expected to correlate above physiological values,[22] the use of linear dependences is preferable for the small extent of experimental accessible values. In agreement with quasi-elastic light-scattering experiments of fibrillogenesis of the amyloid-β protein, the aggregation rate could be assumed to be proportional to the concentration c for Ref. 23 and to be independent of the concentration above the critical value $c = c^*$.[24]

7. INTRINSIC DETERMINANTS OF PROTEIN AGGREGATION

A significant correlation was reported between the changes in the aggregation rates resulting from single amino acid mutations and their effect on physicochemical properties such as hydrophobicity, charge, and the propensity to adopt α-helical or β-sheet secondary structures.[25] The different factors were included in an equation that predicts changes in aggregation rates relative to the wild-type protein under denaturing conditions[25]

$$\log(v_{\mathrm{wt}}/v_{\mathrm{mut}}) = \alpha_{\mathrm{hydr}}\Delta I^{\mathrm{hydr}} + \alpha_{\mathrm{ss}}\Delta I^{\mathrm{ss}} + \alpha_{\mathrm{ch}}\Delta I^{\mathrm{ch}} \qquad [3.25]$$

In this equation, $\log(v_{\mathrm{wt}}/v_{\mathrm{mut}})$ represents the logarithm of the ratio between v_{wt} and v_{mut}, the aggregation rates of wild-type and mutant sequence, respectively, and ΔI^{hydr}, ΔI^{ss}, and ΔI^{ch} represent the change in hydrophobicity, I^{hydr}, secondary structure propensity, I^{ss}, and electrostatic charge, I^{ch}, upon mutation. The parameters α were obtained by fitting the individual terms of Eq. (3.25) to match predicted and experimental changes in aggregation rates upon mutation.[25] Equation (3.25) was shown to reproduce to a remarkable extent ($r = 0.8$) the changes in the aggregation rates observed experimentally for single amino acid substitutions for a series of peptides and proteins, including many associated with disease. It was also shown that other methods can be derived without fitting coefficients, by defining the values of the parameters from general considerations, for instance, according to aromatic contributions, secondary structure propensities, and solvent-accessible areas.[26] The following equation predicts the effect of a mutation on aggregation rate without the use of fitting parameters[26]

$$\nu_{\text{mut}}/\nu_{\text{wt}} = \phi_h \phi_\beta \phi_a \phi_c \quad [3.26]$$

The factor ϕ_h captures most of the apolar and polar interactions. An amino acid is called *p* if its side chain carries a charge or dipole; otherwise, it is called *a*. For mutations that involve the same type of amino acid (a → a) or(p → p)

$$\phi_h = \begin{cases} \text{ASA}^{\text{a}}_{\text{mut}}/\text{ASA}^{\text{a}}_{\text{wt}} & \text{a} \rightarrow \text{a} \\ \text{ASA}^{\text{p}}_{\text{wt}}/\text{ASA}^{\text{p}}_{\text{mut}} & \text{p} \rightarrow \text{p} \end{cases} \quad [3.27]$$

where ASA^{a} and ASA^{p} are the apolar and polar water-accessible surface areas of the amino acid chains.[26]

For mutations that involve different types of amino acids (a → p or p → a)

$$\phi_h = \begin{cases} 1/D_{\text{mut}} & \text{a} \rightarrow \text{p} \\ D_{\text{wt}} & \text{p} \rightarrow \text{a} \end{cases} \quad [3.28]$$

where D is the magnitude of the dipole of the amino acid side chains.

The factor ϕ_β is related to the ratio of β-propensity

$$\phi_\beta = \frac{\beta_{\text{mut}}}{\beta_{\text{wt}}} \quad [3.29]$$

Functions ϕ_a and ϕ_c approximate the effect of aromatic residues A and total charge C

$$\phi_a \phi_c = \exp[\Delta A - \Delta|C|/2] \quad [3.30]$$

The very high accuracy obtained with these simple mathematical formulas ($r > 0.85$) motivated the development of a series of sequence-based methods.[27,28]

8. PREDICTION OF AGGREGATION RATES

Considering that physicochemical properties of amino acids are important factors for aggregation,[25,26] we and others investigated whether such properties can be used to predict not only the changes in aggregation rates of peptides and proteins upon amino acid substitutions but also the overall aggregation rates starting from the knowledge of their amino acid sequences.[11,29]

In standard *in vitro* experiments, such extrinsic factors include the physicochemical parameters that define the environment of the polypeptides, such as pH, temperature, ionic strength, and protein and denaturant concentrations. Additionally, in order to study the relationship between aggregation and disease, it is important to consider also factors relevant to *in vivo* experiments, including the interactions with cellular components such as molecular chaperones, proteases that generate or process the amyloidogenic precursors, and the effectiveness of quality control mechanisms, as the ubiquitin-proteasome system. All these factors are absent from Eq. (3.25), which therefore is of limited use for the direct prediction of experimentally measured aggregation rates because the intrinsic (i.e., sequence-dependent) aggregation rates will be strongly modulated by extrinsic (i.e., sequence-independent) factors *in vivo*.

The aggregation propensity π_{il} of an l-residue segment starting at position i in a protein sequence can be evaluated as[29]

$$\pi_{il} = \phi_{il}\Phi_{il} \qquad [3.31]$$

The factor Φ_{il} contains exponential functions and is position dependent

$$\Phi_{il} = \exp[A_{il} + B_{il} + C_{il}] \qquad [3.32]$$

where A_{il}, B_{il}, and C_{il} are functions of the aromaticity, β-propensity, and charge. The factor ϕ_{il} depends almost exclusively on the amino acid composition

$$\phi_{il} = \left[\prod_{j=i}^{i+l-1}\left(\frac{S_j^{\mathrm{a}}}{S^{\mathrm{a}}}\theta^{\uparrow\uparrow} + \frac{S_j^{\mathrm{p}}}{S^{\mathrm{p}}}\theta^{\uparrow\downarrow}\right)\frac{S^{\mathrm{t}}}{S_j^{\mathrm{t}}}\frac{\sigma}{\sigma_j}\right] \qquad [3.33]$$

where S_j^{a}, S_j^{p}, S_j^{t}, and σ_j—weighted by their average over the 20 standard amino acids—are the side chain apolar, polar, total water-accessible surface area, and solubility, respectively. The functions $\theta^{\uparrow\uparrow}$ and $\theta^{\uparrow\downarrow}$ include positional effects and reflect the parallel or antiparallel tendency to aggregate if the majority of residues is apolar or polar, respectively.

Considering the high correlation between measured and predicted changes in aggregation rate upon single point mutations,[29] it is possible to utilize π_{il} to predict the absolute rate (Fig. 3.2B)

$$v_{il} = \alpha(c, T)\pi_{il} \tag{3.34}$$

where $\alpha(c,T)$ is introduced to take into account concentration and temperature dependence. Linear relationships between aggregation rates and concentration as well as temperature are assumed in the physiological range.[29]

The aggregation process of peptide and proteins depends strongly on the specific regions of their amino acid sequences whose aggregation propensities are particularly high. The definition of the intrinsic aggregation rate π_{il} enables the aggregation propensity profiles to be calculated in order to identify the aggregation-prone regions.[10,11]

The aggregation propensity profile can be reformulated by introducing the position-dependent score p_i^{agg}. For a given residue i, the p_i^{agg} score is calculated as

$$p_i^{\mathrm{agg}} = \alpha_{\mathrm{h}} p_{\mathrm{h}} + \alpha_{\mathrm{s}} p_{\mathrm{s}} + \alpha_{\mathrm{hyd}} p_{\mathrm{hyd}} \tag{3.35}$$

where p_{h} and p_{s} are the propensities for α-helix and β-sheet formation, respectively, and p_{hyd} is the hydrophobicity.[11] These propensities can be combined in a linear way with coefficients α determined as described below. The p_i^{agg} values are combined to provide a score, A_i^{p}, which describes the intrinsic propensity for aggregation as a function of the complete amino acid sequence.[10,11] At each position i along the sequence, we define the profile A_i^{p} as an average over a window of seven residues

$$A_i^{\mathrm{p}} = \frac{1}{7}\sum_{j=-3}^{3} p_{i+j}^{\mathrm{agg}} + \alpha_{\mathrm{pat}} I_i^{\mathrm{pat}} + \alpha_{\mathrm{gk}} I_i^{\mathrm{gk}} \tag{3.36}$$

where I_i^{pat} is the term that takes into account the presence of specific patterns of alternating hydrophobic and hydrophilic residues[30] and I_i^{gk} is the term that takes into account the gatekeeping effect of individual charges c_i[11]

$$I_i^{\mathrm{gk}} = \sum_{j=-10}^{10} c_{i+j} \tag{3.37}$$

The parameters α were fitted using a Monte Carlo optimization.[10,11] In order to compare the intrinsic propensity profiles, we normalize A_i^{p} by considering the average (μ_{A}) and the standard deviation (σ_{A}) of A_i^{p} at each position i for random sequences. The normalized intrinsic aggregation propensity profile is defined as

$$Z_i^{\mathrm{agg}} = \frac{A_i^{\mathrm{p}} - \mu}{\sigma} \tag{3.38}$$

where we calculated the average μ and the standard deviation σ over random sequences

$$\mu = \frac{1}{(N-8)N_S} \sum_{k=1}^{N_S} \sum_{i=4}^{N-4} A_i^p(S_k),$$
$$\sigma^2 = \frac{1}{(N-8)N_S} \sum_{k=1}^{N_S} \sum_{i=4}^{N-4} \left(A_i^p(S_k) - \mu\right)^2 \qquad [3.39]$$

In these formulas, we considered N_S random sequences of length N, and we verified that μ and σ are essentially constant for values of N ranging from 50 to 1000. Random sequences were generated by using the amino acid frequencies of the Uniprot database.

9. PREDICTION OF AGGREGATION-PRONE REGIONS IN NATIVE STATES OF PROTEINS

When a protein is folded, the propensity to form amyloid structures is often inversely related to the stability of its native state.[31] This finding suggests that regions with a high intrinsic propensity for aggregation may be buried inside stable and often highly cooperative structural elements, and therefore unable in such states to form the specific intermolecular interactions that lead to aggregation, although, following mutations that destabilize the native structure, they might acquire this ability.[9] A region of a polypeptide sequence should meet two fundamental conditions in order to promote aggregation: (i) it should have a high intrinsic aggregation propensity and (ii) it should be sufficiently unstructured or unstable to have the opportunity to form intermolecular interactions upon becoming exposed to the solvent through structural fluctuations.[32]

In order to be able to take into consideration the tendency of a given region of a protein sequence to adopt a folded conformation, we introduced the CamP method, which provides a position-dependent score, denoted as ln P_i, predicting the local structural stability at that position.[32] This method enables the high accuracy prediction from the knowledge of amino acid sequence of the regions that are buried in the native state of a protein and of the protection factors for native hydrogen exchange.[32] By combining the predictions of the intrinsic aggregation propensity profiles with those for folding into stable structures, it is possible to account for the influence of the structural context on the aggregation propensities. A new aggregation propensity profile $\tilde{Z}_i^{agg}$ can be defined by modulating the intrinsic aggregation propensity profile Z_i^{agg} with the local stability score[32] $\ln P_i$

$$\tilde{Z}_i^{agg} = Z_i^{agg}\left(1 - \frac{\ln P_i}{\ln P_{max}}\right) \quad [3.40]$$

where $\ln P_{max}$ is the maximal value that this parameter can reach. These modulations on the Z_i^{agg} profile are made only when $Z_i^{agg} > 0$ since we consider only the effects on the regions of high intrinsic aggregation propensity, which are those that effectively drive the aggregation process.

From the $\tilde{Z}_i^{agg}$ score, it is possible to define an overall aggregation propensity $\tilde{Z}^{agg}$ score by summing over all the amino acids of a sequence that have aggregation propensities higher than those of random sequences[11]

$$\tilde{Z}^{agg} = \frac{\sum_{i=1}^{N} \tilde{Z}_i^{agg} \vartheta\left(\tilde{Z}_i^{agg}\right)}{\sum_{i=1}^{N} \vartheta\left(\tilde{Z}_i^{agg}\right)} \quad [3.41]$$

10. LIFE ON THE EDGE—THE ROLE OF PROTEIN CONCENTRATION IN PROMOTING AGGREGATION

Investigating the physicochemical determinants of protein aggregation, we found a remarkable anticorrelation between the expression levels of human genes *in vivo* and the aggregation rates of proteins measured *in vitro*[33] (Fig. 3.3A). A simple principle can be recognized behind our finding that an evolutionary pressure acts to decrease the risk of aggregation in a way that proteins with higher aggregation propensity are generally less expressed. We observe that a "functional amyloid" like *Pmel17*[34] does not follow the trend as no evolutionary pressure is acting on a protein whose aggregation is beneficial for the organism. Thus, aggregation propensities are precisely tuned by evolutionary selection to levels that enable them to be functional at the concentrations required for optimal performance (Fig. 3.3A).

It is possible to speculate on the mathematical representation of the relationship between expression levels and aggregation rates, by defining the quantity

$$p = mv \quad [3.42]$$

In Eq. (3.42), m represents the expression level and v the aggregation rate. As m and v have practically the same range of variability (Fig. 3.3A), we can assume that $p \approx cst$. In analogy with classical mechanics, p can be regarded as

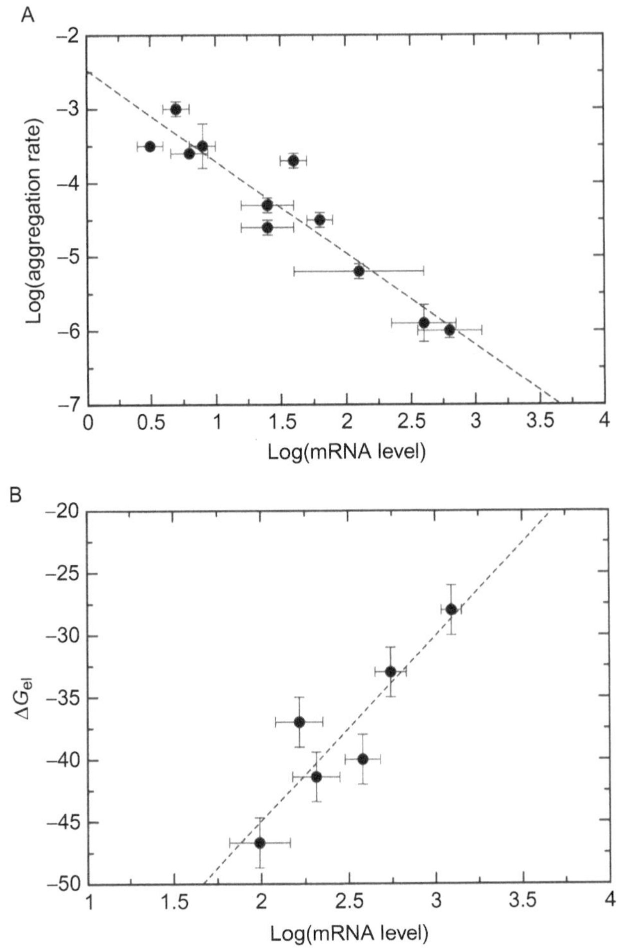

Figure 3.3 Life on the edge. (A) Experimental aggregation rates and mRNA expression levels are strongly anticorrelated. (B) Free energies in the metastable state correlate with mRNA expression levels.

the momentum and $f = \mathrm{d}p/\mathrm{d}t$ represents the associated force. Since $\mathrm{d}p/\mathrm{d}t \approx 0$, we can consider the cell in homeostasis. In the presence of perturbations, $\mathrm{d}p = v\mathrm{d}m + m\mathrm{d}v \neq 0$ and additional forces are required to modulate expression levels and aggregation rates. Indeed, the term $v\mathrm{d}m$ must be linked to regulatory networks[35,36] and $m\mathrm{d}v$ should be associated to molecular chaperones that intervene in order to prevent the formation of nonnative insoluble intermediates when folding into the native state is challenged.[13]

When the concentration M_S of a protein exceeds its critical value (see Eq. 3.6–3.7)[37]

$$M_S^{max} = \exp[\Delta G_{el}/KT] \qquad [3.43]$$

the native state is not thermodynamically stable and a protein can in principle lower its overall free energy through amyloid formation, in the same manner in which other types of molecules that exceed their solubility limit have a tendency to form insoluble amorphous or crystalline structures.[37] Do living systems operate under conditions of metastability under normal circumstances? In order to answer this question, we analyzed values from the literature for various critical concentrations.[37] We observe that mRNA expression levels strongly correlate with these critical concentrations (Fig. 3.3B). Our results have profound implications on our understanding of the thermodynamics and kinetics of protein molecules and point to the fact that evolution favors solubility over aggregation.[37]

It should be noted that we assumed a correlation between protein and RNA abundances in our analyses. The correlation between protein concentrations and mRNA expression levels is very well known for bacteria and fungi.[29] However, higher eukaryotes often require substantial posttranscriptional modifications to yield the final amount of protein. To reduce the effect of these modifications in our analysis, we used median scaling and quantile normalization of gene expression levels in different tissues. Accordingly, expression levels were averaged over all the tissues in which a gene was found expressed.[33] This procedure reduces the effect that regulatory processes have on protein expression, because tissue-specific cofactors are averaged out together with environmental conditions.

11. CONCLUSIONS

In this chapter, we described a series of methods for predicting the aggregation of proteins based on their physicochemical properties. The methodology presented is based on the idea that sequences determine protein behavior *in vitro*, in the cases of the folding, misfolding, and aggregation processes, as well as *in vivo*, in the cases of cellular toxicity, solubility, and interactions with chaperones that arise upon protein misfolding.

Our results reveal stringent conditions on the activities of proteins that are dictated by fundamental physicochemical properties. Based on these findings, it is possible to build a theoretical framework to predict which factors contribute most to the aggregation and toxicity of globular proteins, natively unfolded polypeptide chains, and systems that contain both folded and unfolded domains.

A wide number of diseases have been associated to protein misfolding and aggregation. Besides the actual aggregation process, several events that take place both upstream (i.e., mutations, oxidative stress, etc.) and downstream (e.g., promiscuous interactions, chaperones activation) can modify the onset and the severity of such debilitating pathologies, increasing consistently their degree of complexity.[36,38] Quantitative tools are required in order to address such complexity and identify relevance of each factor involved. A theoretical framework like the one proposed in this chapter works in this direction by allowing to describe quantitatively the contribution of the different amino acids to the aggregation process and ultimately to the onset of disease.

Most importantly, the possibility provided by the different theoretical approaches is of significant value in developing rational approaches to avoid aggregation in the biotechnology industry, as well as to understand which are the crucial factors to target in order to prevent this process from happening *in vivo*.

REFERENCES

1. Tandford C, Buckley 3rd CE, De PK, Lively EP. Effect of ethylene glycol on the conformation of gama-globulin and beta-lactoglobulin. *J Biol Chem* 1962;**237**:1168–71.
2. Tanford C. Protein denaturation. C. Theoretical models for the mechanism of denaturation. *Adv Protein Chem* 1970;**24**:1–95.
3. Serpell LC, Sunde M, Benson MD, Tennent GA, Pepys MB, Fraser PE. The protofilament substructure of amyloid fibrils. *J Mol Biol* 2000;**300**:1033–9.
4. Serpell LC. Alzheimer's amyloid fibrils: structure and assembly. *Biochim Biophys Acta* 2000;**1502**:16–30.
5. Fändrich M, Meinhardt J, Grigorieff N. Structural polymorphism of Alzheimer Abeta and other amyloid fibrils. *Prion* 2009;**3**:89–93.
6. Kopito RR. Aggresomes, inclusion bodies and protein aggregation. *Trends Cell Biol* 2000;**10**:524–30.
7. Schlieker C, Bukau B, Mogk A. Prevention and reversion of protein aggregation by molecular chaperones in the E. coli cytosol: implications for their applicability in biotechnology. *J Biotechnol* 2002;**96**:13–21.
8. Kelly JW. The alternative conformations of amyloidogenic proteins and their multi-step assembly pathways. *Curr Opin Struct Biol* 1998;**8**:101–6.
9. Dobson CM. Protein misfolding, evolution and disease. *Trends Biochem Sci* 1999;**24**: 329–32.
10. DuBay KF, Pawar AP, Chiti F, Zurdo J, Dobson CM, Vendruscolo M. Prediction of the absolute aggregation rates of amyloidogenic polypeptide chains. *J Mol Biol* 2004;**341**:1317–26.
11. Tartaglia GG, Pawar AP, Campioni S, Dobson CM, Chiti F, Vendruscolo M. Prediction of aggregation-prone regions in structured proteins. *J Mol Biol* 2008;**380**:425–36.
12. Bukau B, Weissman J, Horwich A. Molecular chaperones and protein quality control. *Cell* 2006;**125**:443–51.
13. Hartl FU, Hayer-Hartl M. Molecular chaperones in the cytosol: from nascent chain to folded protein. *Science* 2002;**295**:1852–8.

14. Schubert U, Antón LC, Gibbs J, Norbury CC, Yewdell JW, Bennink JR. Rapid degradation of a large fraction of newly synthesized proteins by proteasomes. *Nature* 2000;**404**:770–4.
15. Oosawa F, Asakura S. *Thermodynamics of the polymerization of protein*. Waltham, MA, USA: Academic Press Inc.; 1975.
16. Oosawa F, Kasai M. A theory of linear and helical aggregations of macromolecules. *J Mol Biol* 1962;**4**:10–21.
17. Knowles TPJ, Waudby CA, Devlin GL, Cohen SIA, Aguzzi A, Vendruscolo M, et al. An analytical solution to the kinetics of breakable filament assembly. *Science* 2009;**326**: 1533–7.
18. Cohen SIA, Vendruscolo M, Dobson CM, Knowles TPJ. From macroscopic measurements to microscopic mechanisms of protein aggregation. *J Mol Biol* 2012;**421**:160–71.
19. Ferrone FA, Hofrichter J, Sunshine HR, Eaton WA. Kinetic studies on photolysis-induced gelation of sickle cell hemoglobin suggest a new mechanism. *Biophys J* 1980; **32**:361–80.
20. Beaven GH, Gratzer WB, Davies HG. Formation and structure of gels and fibrils from glucagon. *Eur J Biochem* 1969;**11**:37–42.
21. Ruschak AM, Miranker AD. Fiber-dependent amyloid formation as catalysis of an existing reaction pathway. *Proc Natl Acad Sci USA* 2007;**104**:12341–6.
22. Massi F, Straub JE. Energy landscape theory for Alzheimer's amyloid beta-peptide fibril elongation. *Proteins* 2001;**42**:217–29.
23. Kusumoto Y, Lomakin A, Teplow DB, Benedek GB. Temperature dependence of amyloid beta-protein fibrillization. *Proc Natl Acad Sci USA* 1998;**95**:12277–82.
24. Lomakin A, Chung DS, Benedek GB, Kirschner DA, Teplow DB. On the nucleation and growth of amyloid beta-protein fibrils: detection of nuclei and quantitation of rate constants. *Proc Natl Acad Sci USA* 1996;**93**:1125–9.
25. Chiti F, Stefani M, Taddei N, Ramponi G, Dobson CM. Rationalization of the effects of mutations on peptide and protein aggregation rates. *Nature* 2003;**424**:805–8.
26. Tartaglia GG, Cavalli A, Pellarin R, Caflisch A. The role of aromaticity, exposed surface, and dipole moment in determining protein aggregation rates. *Protein Sci* 2004;**13**: 1939–41.
27. Sánchez de Groot N, Pallarés I, Avilés FX, Vendrell J, Ventura S. Prediction of "hot spots" of aggregation in disease-linked polypeptides. *BMC Struct Biol* 2005;**5**:18.
28. Fernandez-Escamilla A-M, Rousseau F, Schymkowitz J, Serrano L. Prediction of sequence-dependent and mutational effects on the aggregation of peptides and proteins. *Nat Biotechnol* 2004;**22**:1302–6.
29. Tartaglia GG, Cavalli A, Pellarin R, Caflisch A. Prediction of aggregation rate and aggregation-prone segments in polypeptide sequences. *Protein Sci* 2005;**14**: 2723–34.
30. Xiong H, Buckwalter B, Shieh H, Hecht M. Periodicity of polar and nonpolar amino acids is the major determinant of secondary structure in self-assembling oligomeric peptides. *Proc Natl Acad Sci* 1995;**92**:6349–53.
31. Chiti F, Taddei N, Bucciantini M, White P, Ramponi G, Dobson CM. Mutational analysis of the propensity for amyloid formation by a globular protein. *EMBO J* 2000;**19**:1441–9.
32. Tartaglia GG, Cavalli A, Vendruscolo M. Prediction of local structural stabilities of proteins from their amino acid sequences. *Structure* 2007;**15**:139–43.
33. Tartaglia GG, Pechmann S, Dobson CM, Vendruscolo M. Life on the edge: a link between gene expression levels and aggregation rates of human proteins. *Trends Biochem Sci* 2007;**32**:204–6.
34. Fowler DM, Koulov AV, Alory-Jost C, Marks MS, Balch WE, Kelly JW. Functional amyloid formation within mammalian tissue. *PLoS Biol* 2006;**4**:e6.

35. Luscombe NM, Babu MM, Yu H, Snyder M, Teichmann SA, Gerstein M. Genomic analysis of regulatory network dynamics reveals large topological changes. *Nature* 2004;**431**:308–12.
36. Cirillo D, Agostini F, Klus P, Marchese D, Rodriguez S, Bolognesi B, et al. Neurodegenerative diseases: quantitative predictions of protein-RNA interactions. *RNA* 2013;**19**:129–40.
37. Baldwin AJ, Knowles TPJ, Tartaglia GG, Fitzpatrick AW, Devlin GL, Shammas SL, et al. Metastability of native proteins and the phenomenon of amyloid formation. *J Am Chem Soc* 2011;**133**:14160–3.
38. Johnson R, Noble W, Tartaglia GG, Buckley NJ. Neurodegeneration as an RNA disorder. *Prog Neurobiol* 2012;**99**:293–315.

CHAPTER FOUR

Structural Aspects of Amyloid Formation

Xavier Salvatella[*,†]
[*]ICREA, Barcelona, Spain
[†]Joint BSC-IRB Research Programme in Computational Biology, IRB Barcelona, Barcelona, Spain

Contents

Abstract

Amyloid fibrils are highly organized and generally insoluble protein aggregates rich in β secondary structure that can be formed by a wide range of sequences. They have been the object of intense scrutiny because their formation has been associated with a number of neurodegenerative disorders such as Alzheimer's, Parkinson's, Huntington's, and Creutzfeldt–Jakob's diseases. As a consequence of these efforts, much is now known about the properties of proteins that render them prone to form amyloid fibrils, about the mechanism of fibrillation, about the molecular structures of the fibrils, and about the forces that stabilize them. The relationship between the structural properties of the monomeric protein and those of the corresponding aggregate has been, in particular, intensively studied. In this chapter, we will provide an account of current knowledge on this intriguing relationship and provide the reader with key references about this topic.

1. INTRODUCTION

Amyloid fibrils are highly organized and generally insoluble protein aggregates that can be formed by a wide range of polypeptide sequences.[1] They have been the object of intense study because their formation by

Progress in Molecular Biology and Translational Science, Volume 117
ISSN 1877-1173
http://dx.doi.org/10.1016/B978-0-12-386931-9.00004-0

specific proteins has been associated with a number of neurodegenerative disorders such as Alzheimer's disease (AD), Parkinson's disease (PD), Huntington's disease, and Creutzfeldt–Jakob's disease (CJD).[2]

The fibrils are rich in β secondary structure, with the β strands perpendicular to the axis of the fibril, and are stabilized mainly by the intermolecular hydrogen bonds that connect the various strands (Fig. 4.1).[3–5] The structural properties of amyloid fibrils are described in some detail in Section 2.

The determination of the structure of amyloid fibrils at high resolution is a very challenging endeavor because the fibrils are not amenable to the

Figure 4.1 Schematic representation of the general structural properties of amyloid fibrils which illustrates the perpendicular orientation of the β-strands relative to the fibrillar axis.[3] (For color version of this figure, the reader is referred to the online version of this chapter.)

conventional techniques of structural biology such as X-ray crystallography and solution nuclear magnetic resonance (NMR) spectroscopy. In spite of this, it has been possible to obtain very detailed structures of the fibrils formed by a small number of proteins, and their study has led to a very detailed understanding of the various possible ways in which the polypeptide backbone can be structured in these intriguing materials.[6–8]

In parallel to these developments, there has been a very intense effort at clarifying whether amyloid formation is causative of the various diseases to which it has been associated or is, instead, a mere consequence of the accumulation of protein that occurs in these conditions. Although the issue is far from settled, it is very likely that the causes and consequences of amyloid formation will vary from disease to disease.[2]

In neurodegenerative diseases such as AD and PD, it seems quite well established that aggregation leads to neurotoxicity not to the accumulation of amyloid fibrils formed by the peptide Aβ but, rather, to the formation of soluble oligomers of this peptide either by oligomerization of the monomeric peptide or by fragmentation of the fibrils themselves.[9]

Instead, in other diseases such as lysozyme systemic amyloidosis, a hereditary disease caused by mutations in the lysozyme gene that render the protein unstable, or hemodialysis-related amyloidosis, which is caused by the accumulation of β2-microglobulin in the blood of patients of renal failure treated by hemodialysis, it seems to be caused by the fibrils.[10]

It is now well established that the mechanism by which amyloid fibrils form involves primary and secondary nucleation steps and accounting for these two phenomena leads to predicted aggregation kinetics that closely match the experimental ones (Fig. 4.2A).[13] Primary nucleation is caused by the association of various monomers to form a nucleus capable of recruiting additional monomers in a process known as elongation.[14] Secondary nucleation increases the number of species capable of recruiting monomers either by the fragmentation of the fibrils or by branching, a process in which monomers are recruited laterally by a fibril (Fig. 4.2B).[13,15]

The nucleated nature of the mechanism of amyloid formation explains why the rate of aggregation can be very significantly increased by the presence of preformed fibrils in the aggregating solution. In this scenario, the preformed fibrils are immediately capable of elongating by recruitment of additional monomers, that is, they seed the formation of fibrils and effectively bypass the primary nucleation step.[14]

That preformed fibrils can accelerate fibril formation has important consequences in diseases where such species act as infectious agents that cause

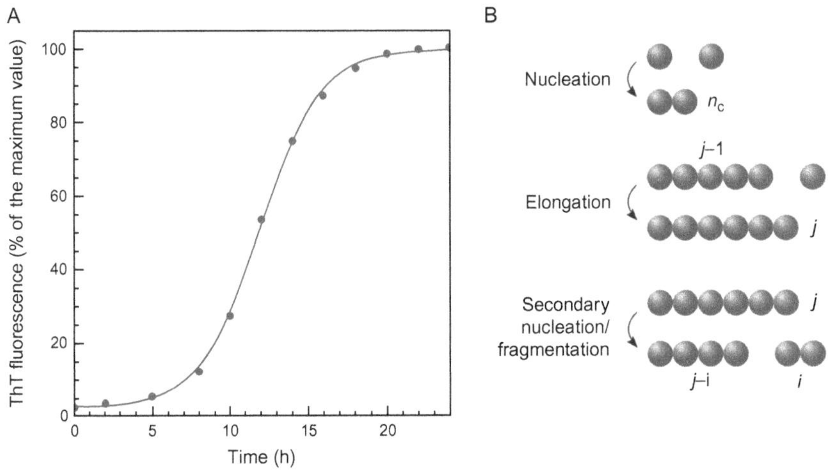

Figure 4.2 (A) Time course of amyloid formation by human lysozyme at pH 7.5 and 60 °C monitored by measuring the fluorescence emission of the dye thioflavin T at 485 nm, with excitation at 440 nm. This dye binds specifically to proteins adopting the amyloid conformation and in doing so changes its fluorescence properties.[11] (B) Schematic representation of the general mechanism of amyloid formation with well-defined primary and secondary nucleation and elongation steps.[12] (For color version of this figure, the reader is referred to the online version of this chapter.)

and/or accelerate the aggregation of the protein in the form of amyloid fibril in a host. Proteins whose aggregation *in vivo* is found to be transmissible are called prions, from *proteinaceous* and *infectious*.[16]

CJD, for example, is related to the formation of amyloid fibrils by the human prion protein (PrP), which in its native state is a globular protein rich in α secondary structure (PrP^C).[17] Although the disease is generally sporadic, it can be caused by contamination with very small amounts of prion protein[18] in the amyloid conformation (PrP^{Sc}) during surgical procedures in patients that would normally not suffer from the disease.[19]

The neurological disorder kuru, described only in the Fore tribe in Papua New Guinea, was due to the practice of cannibalism by individuals of this tribe during funeral ceremonies, in which they consumed body parts of deceased tribe members. Similarly, variant CJD is a variant of CJD that could be caused by the consumption of beef suffering from bovine spongiform encephalopathy by humans, as has been shown to be the case in macaques.[20]

Past and current interest in amyloid fibrils is mainly due to the involvement of these aggregates in a number of important diseases, but such structures can also have physiological functions that are not connected with any specific disorder.[21] In certain yeasts, for example, the aggregation of specific proteins, the yeast prions, in well-defined amyloid morphologies and their transmission within colonies is used as an epigenetic mechanism where information is not encoded in the DNA sequence but, rather, in the morphology of the amyloid fibril.[22]

There is recent evidence that transmissibility, that is, the ability of prions to be transported from cell to cell, from tissue to tissue, and even from organism to organism and to template amyloid formation at a remote site may be a general property of amyloid fibrils. It has been suggested that such mechanisms can play an important role in the spread of neurodegenerative diseases such as AD[23] and PD[24] in the brain of patients, and there is great interest in understanding the mechanisms used by amyloid fibrils to migrate within tissues.[25]

Although it had been initially suggested that only certain sequences linked to specific diseases and rich in β secondary structure possessed the ability to form amyloid fibrils, it was soon realized that this was in fact a property of most, if not all, polypeptide sequences.[1] Experiments showed that proteins in their native state have α structure[26] as well as polyamino acids such as polyGlu and polyLys[27] could also form amyloid fibrils with β secondary structure, indicating that the ability of forming amyloid fibrils is not encoded in specific polypeptide sequences and is instead a property of the peptide backbone.

Although all peptide sequences are capable of forming amyloid fibrils, not all sequences have the same propensity to do so. Studies on the effect of point mutations on the fibrillation rate of certain proteins sequences have revealed that a few physicochemical parameters of the sequence such as charge, secondary structure propensity, and hydrophobicity can explain much of the variance of aggregation rates.[28] Extensions of these initial studies using aggregation kinetics data available in the literature lead to the development of algorithms that can predict the relative aggregation rates of different sequences and the aggregation propensity of different residues within a given sequence.[29–31]

The formation of amyloid fibrils by a polypeptide involves profound structural changes (Fig. 4.3) and there has been great interest in determining to what extent the structural properties of the monomeric protein influence the mechanism and therefore the rate of amyloid formation.[32] In cases where the protein-forming amyloids is globular it has been shown that it is, in most cases, necessary to destabilize the native structure either by mutation or truncation and/or by changes in solution conditions for the protein to form amyloid fibrils.[33,34] In cases where the protein is instead intrinsically disordered, amyloid formation can occur directly from the natively disordered state.[35]

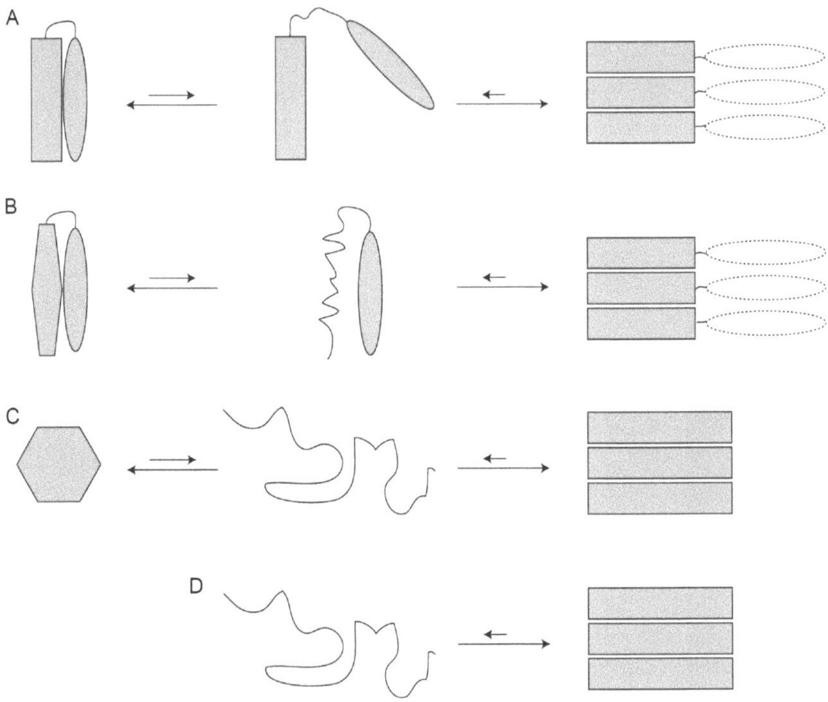

Figure 4.3 Proteins need to expose unsatisfied hydrogen bonds and/or hydrophobic residues in order to establish the intermolecular noncovalent interactions that initiate amyloid formation. In globular proteins this can occur by breaking well-defined intramolecular interactions (A) as well as by transient local (B) or global (C) unfolding. In intrinsically disordered proteins it can in principle occur spontaneously (D) as these systems are devoid of cooperatively formed secondary and tertiary structure. The structure of the region not undergoing conformational change during amyloid is dashed to represent that although it does not adopt the cross-β structure its structural and dynamical properties are likely affected by amyloid formation.

In both cases, spectroscopic studies have indicated that the precursor states are often rich in transient secondary and tertiary structure.[32] These investigations are obviously important for understanding the process of amyloid formation, but especially relevant from a drug discovery perspective. A scenario where the partially folded states need to form for amyloid formation to occur, where the partially folded states represent on-pathway intermediates, suggests that preventing their formation or association by small molecules could be a potential therapeutic approach for disorders where amyloid formation causes the disease. A scenario where these states are instead off-pathway suggests that stabilizing them may be, at least in principle, a potential therapeutic approach that would deplete the solution of the species that cause the formation of toxic species.

2. STRUCTURAL PROPERTIES OF AMYLOID FIBRILS

2.1. Actual structures

Much as the structure of globular proteins—that can be described in a hierarchical fashion in terms of primary, secondary, tertiary, and quaternary structure—the structure of amyloid fibrils can be described in terms of the structure adopted by the monomer in the fibrillar context, the relative structure of different monomers in a protofilament, and the relative arrangement of protofilaments in a fibril as shown in Fig. 4.4.[38,39]

High-resolution structures of amyloid fibrils[6–8] determined by using techniques described below reveal that these form extended β-sheets stabilized essentially by intermolecular hydrogen bonds involving mainly backbone atoms running parallel to the axis of the fibril in a structure defined as cross-β shown in Fig. 4.4B.[3,37] In addition, they have shown that one face of these extended β-sheets is exposed to water, whereas the other one is tightly

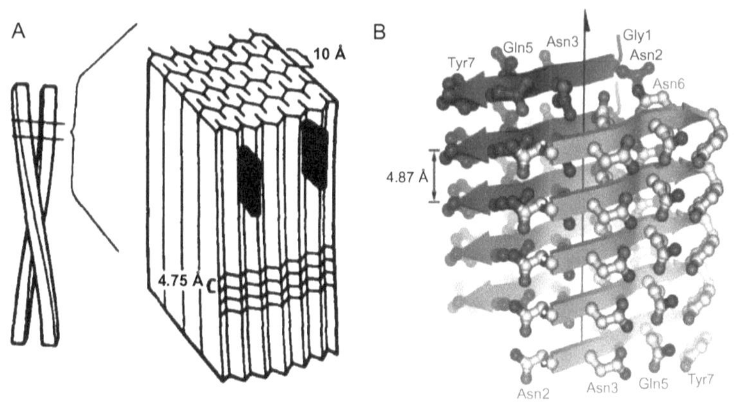

Figure 4.4 (A) Schematic hierarchical structure of an amyloid fibril composed of two protofilaments of five extended β-sheets with an indication of the two most important dimensions that can be determined experimentally using X-ray fiber diffraction (Fig. 4.7) and of the possible binding sites of dyes that specifically bind to amyloid fibrils such as Congo red and Thioflavin T.[37] (B) High-resolution structure of the cross-β spine that stabilizes the microcrystal formed by the peptide GNNQQNY, that is a fragment of the yeast prion Sup35 with a high propensity to form amyloid fibrils. In the crystal, the strands are arranged in parallel and in register.[6] (For color version of this figure, the reader is referred to the online version of this chapter.)

packed against another such β-sheet with a dry interface stabilized by van der Waals and hydrophobic interactions between interdigitated side chains.[6,7]

In fibrils formed by short peptides, the interacting sheets are formed by different molecules,[6] so that pairs of sheets form one protofilament, as is the case for the GNNQQNY peptide shown in Fig. 4.4B. In fibrils formed by long proteins, the various interacting β-sheets of a protofilament can be formed by different regions of sequence of a given protein as is the case for the yeast prion HET-s[8] and as has been suggested to be the case for the protein α-synuclein.[8,40]

This knowledge gathered from inspection of the high-resolution structures has been confirmed by alternative biophysical techniques that measure the resistance of amyloid fibrils to depolymerization by mechanical forces, by high temperature, or by the use of chaotropic agents that weaken the noncovalent interactions that stabilize the fibrils.[11,41–44] It is important to mention that the term protofibril, which has been used in the literature, does not refer to a structural element of amyloid fibrils but, rather, to metastable and distinct nonfibrillar structures populated during amyloid formation.

It is often the case that not all regions of sequence of a protein that forms amyloid fibrils become involved in the formation of cross-β structure. In these cases, it is convenient to differentiate between residues in core regions, defined as those that adopt the cross-β structure,[3] and residues in noncore regions, which can adopt other types of structures or remain disordered and do not contribute stabilizing interactions.[11,45,46] This is an important distinction as there is mounting evidence that the presence of noncore regions can underlie the toxicity of certain amyloid fibrils by perhaps interacting with the cell membrane or specific components of the cellular machinery.[11,47]

Contrary to what is the case for conventional protein structures, which are significantly robust to modest changes in sequence and solution conditions, amyloid fibrils are notoriously polymorphic.[11,48–52] A given sequence will therefore give rise to different amyloid morphologies depending on the exact nature of the conditions in which the monomeric protein is incubated such as ionic strength, pH, temperature, and concentration[11,48] as well as the type and degree of stirring of the aggregating solution.[50] The consequences of this property for biology and biomedicine are widespread as it implies that the structural properties of the fibrils formed by a particular protein can depend on the milieu where amyloid formation occurs and is therefore likely to be influenced by a multitude of fluctuating variables.

Amyloid polymorphism can be due to differences in molecular structure[11,53] and to variations in the packing of extended β-sheets and/or protofilaments in the amyloid fibrils.[49,54] That one given sequence can give rise to different molecular structures and different amyloid morphologies is likely to be a reflection of the rather generic interactions that stabilize the cross-β structure of amyloid fibrils, which involve backbone atoms and are therefore common to all amino acids except proline. As has been noted[6], the interactions between extended β-sheets are even less specific, thus multiplying the number of possible amyloid morphologies that a given polypeptide sequence can adopt.

That many amyloid morphologies are possible for a given molecular structure has been particularly well illustrated by the work of the Eisenberg group,[54] in which it was shown that eight different relative orientations of two extended β-sheets formed by short amyloidogenic peptides are possible. These differ in whether the β-sheet is parallel or antiparallel, in whether the two monomers facing one another in different sheets are in the parallel or antiparallel relative orientation and finally, in whether the interacting surfaces are face to face. Of these $2^3 = 8$ possibilities, five have been found to occur by the Eisenberg group in microcrystals formed by various sequences corresponding to fragments of amyloidogenic proteins.

As protofilaments and, ultimately, amyloid fibrils are formed by long β-sheets where β-strands are perpendicular to the fibrillar axis, fibrils display a twist that is clearly noticeable in microscopy images[55,56] and that is a characteristic of each amyloid morphology. In a particularly interesting study of amyloid polymorphism, it was shown how the fibrils formed *in vitro* by Aβ(1–40), a 40-residue peptide related to AD, can have various degrees of twist, reflecting that Aβ(1–40) can form different types of fibrils, of similar stability, under a particular set of conditions.[49]

The polymorphism of amyloid fibrils is one of the mechanistic bases for the phenomenon of prion strains that underlies the ability of prions to act as infectious agents and to transfer epigenetic traits.[41,42,57]

2.2. Methods for the structural analysis of amyloid fibrils

The structural analysis of amyloid fibrils is challenging and necessarily multidisciplinary because no technique provides information about all structural properties of interest. The most fundamental structural property, that is the molecular structure of the monomer embedded in the fibril, is particularly challenging to characterize because the tools commonly used to determine the structures of globular proteins at high resolution, that are X-ray crystallography and solution NMR spectroscopy, are not suited, at least in a conventional sense, to determine the structures of the fibrils.

As already mentioned, it has been possible to obtain microcrystals of amyloidogenic peptides under conditions where amyloid fibrils form and their study using X-ray diffraction has led to high-resolution descriptions of the intermolecular hydrogen-bonding interactions that stabilize the microcrystals.[6] As these are obtained under conditions that are similar to those that lead to amyloid formation, and have properties that match those determined for fibrils using low resolution techniques, it is reasonable to assume that the cross-β spines of these structures are equivalent. Although this powerful approach has been key to understand the structural features of the cross-β motif, it cannot be used to characterize the atomic structures of the fibrils formed by full-length proteins because these present long-range order in only one dimension, corresponding to the fibrillar axis, and cannot therefore be crystallized.

Conventional solution NMR as applied to determine the structure and the dynamics of soluble globular proteins cannot be used to determine the molecular structures of amyloid fibrils. This is because fibrils formed *in vitro* have dimensions that render the NMR signals too broad to be detected due to the relationship that exists between the lifetime of the NMR excited state and the rate of rotational diffusion of the structure under study. This is the case even for fibrils that have been fragmented by ultrasonication with the aim of decreasing their size to render them amenable to solution NMR methods.[58,59]

It has thus been challenging to use solution NMR to investigate the structures of core regions, but this technique has found significant use in the characterization of noncore regions due to their flexible nature. This increases the lifetime of the excited state to the extent that it allows detection of the NMR signals as is the case for flexible regions of intact ribosomes.[60,61] Assignment of the resonances of the noncore regions immediately identifies a method to identify them, to monitor their interactions with other molecules[46], and to characterize the size distribution of the fibrils contributing to the NMR signal by an analysis of their diffusion properties using pulse field gradient techniques.[62,63] In occasions where the signals of the noncore regions cannot be observed using solution NMR, it has proved possible to observe them by using high-resolution magic angle spinning NMR probes.[64]

Solid state NMR is an extremely powerful technique for characterizing the molecular structure of amyloid fibrils and other insoluble protein species.[65] In this technique, various strategies are used to alleviate the substantial problems that are associated with the measurement of NMR spectra of solids due to the influence of anisotropic interactions such as dipole–dipole interactions and the chemical shift anisotropy.

The main strategy is the very fast rotation of solid samples around an axis that is tilted 54.7°, the *magic* angle, relative to the static magnetic field of the spectrometer. When this rotation is sufficiently fast, the spectra obtained using solid state NMR can be similar in resolution to those obtained using solution NMR, allowing the characterization of the structure and dynamics of insoluble protein aggregates.

In favorable cases, it is possible to obtain NMR spectra of amyloid fibrils of impressive resolution and to measure NMR parameters that provide structural information, such as chemical shifts, at atomic resolution.[66] These can then be used to determine molecular structures by using computational approaches, equivalent to those used to determine structures for globular proteins, that identify conformations that fit the NMR parameters and are reasonable according to molecular mechanics force fields.

The first structural studies of amyloid fibrils that used solid state NMR focused on determining the molecular structure of the proteins and therefore provided little information about the relative placement of monomers in the β-sheet. They provided no information on the relative position of the sheets in the protofilament or on the position of protofilaments in the fibril (Fig. 4.5).[66–68] Since then, improvements in methodology have allowed the measurement of increasing numbers of intermolecular distances, which have allowed deriving atomic resolution structures for a small number of amyloid fibrils such as various amyloid morphologies formed by the peptide Aβ[7,50] and the yeast prion HET-s[8] (Fig. 4.6).

In addition to X-ray crystallography and NMR spectroscopy, a range of other biophysical techniques can be used to obtain very valuable structural information about amyloid fibrils, while requiring much less sample and experiment time. Circular dichroism (CD) is particularly powerful for the qualitative characterization of the secondary structure content of amyloid fibrils when the fibrils are soluble, which is the case for a large number of amyloid morphologies.

The CD spectrum of amyloids presents a weak band that is typical of the β secondary structure at ca. 215 nm[69] as shown in Fig. 4.7A for the amyloid fibrils formed by human lysozyme.[11] Care must be taken due to an effect called differential absorption flattening, which can cause a decrease in the mean residue ellipticity and a red shift of the CD bands that is due to the presence of suspended solids in the CD sample.[70,71] For this reason, the deconvolution of the CD spectrum of amyloid fibrils to obtain the fraction of residues in each of the various possible secondary structures is rarely carried out. Instead, as this technique is quite sensitive and gives good signal

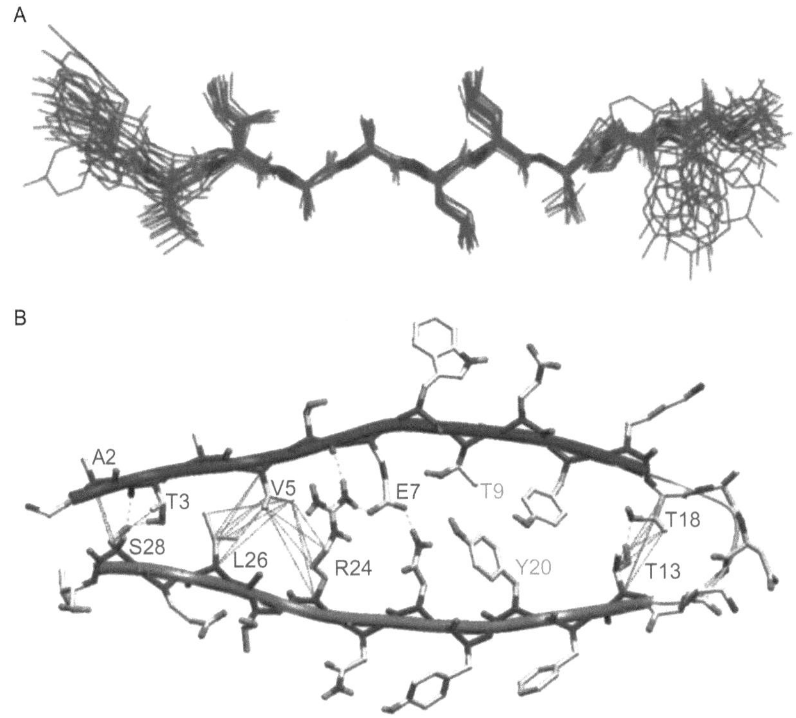

Figure 4.5 Molecular structures determined from solid state NMR data, (A) the amyloidogenic peptide TTR(105–115)[67] by the Griffin group. (B) A WW domain[68] by the Oschkinat and Fersht groups. (For color version of this figure, the reader is referred to the online version of this chapter.)

to noise ratios at the low concentrations used to generate amyloid fibrils *in vitro*, it is generally used to monitor the conformational changes associated with amyloid formation *in vitro* in a time-resolved fashion.

Attenuated total reflectance used in conjunction with Fourier transform infra-red spectroscopy (ATR–FTIR) is particularly powerful for the analysis of the secondary structure of amyloid fibrils.[72] The amide I band between 1600 and 1700 cm^{-1}, which is due to the vibration of the CO bond coupled to the bending of the NH bond and the stretching of the CN bond, is very sensitive to the structure and the hydrogen bonding status of the peptide plane and can be used to analyze the secondary structure of proteins.[73] The amide I bands corresponding to various types of secondary structures have been measured and reported and it is therefore possible to determine their contribution to the experimental ATR–FTIR spectrum by fitting synthetic spectra to the experimental one (Fig. 4.7B).[74–76] A key feature of this

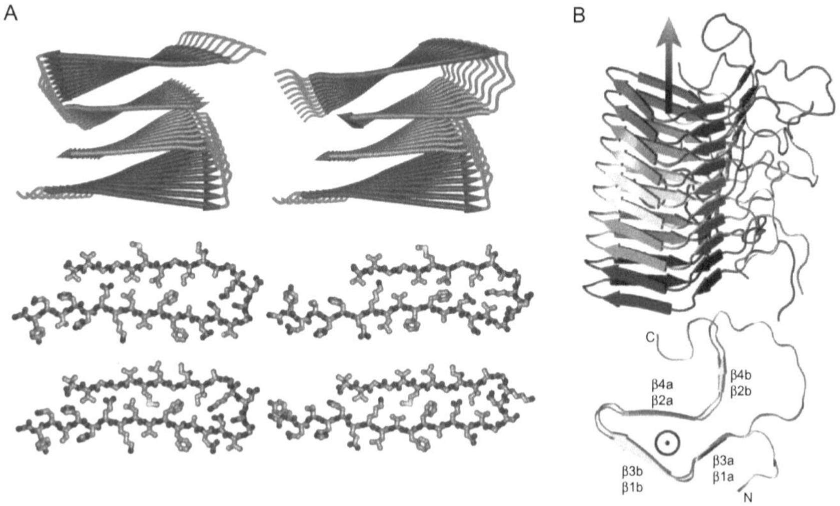

Figure 4.6 Structures of amyloid fibrils formed by full-length proteins determined to date using solid state NMR data reporting on the molecular structure of the protein in the fibrillar context as well as on the quaternary structure that stabilizes the fibrils. (A) Structures corresponding to two different amyloid morphologies formed by the Aβ peptide solved by the Tycko group,[7] which differ in the relative orientation of two extended β-sheets. (B) Structure of the amyloid fibrils formed by HET-s solved by the Meier group[8] in which the protein adopts the topology of a solenoid where both intra and intermolecular hydrogen bonds contribute to the stability of the fibril; five different molecules of Het-s are represented in this figure. (For color version of this figure, the reader is referred to the online version of this chapter.)

technique that renders it particularly powerful for the study of amyloids, is that it can allow differentiating parallel from antiparallel β-sheets.[73,77,78] Although the parallel topology is much more common than the antiparallel one, the latter has been shown to occur and has been linked to the cytotoxic effect of certain amyloid fibrils.[78]

X-ray fiber diffraction is a very powerful tool for the analysis of the structural properties of amyloid fibrils because it provides information that is very challenging to obtain by other methods.[3,37] When the fibrils are aligned, their diffraction pattern when they are exposed to X-rays provides information about the distance between β-strands in the β-sheet (which is typically 4.8 Å) and about the distance between extended β-sheets in the protofilament of the fibril (which is typically ca. 10 Å). Information about the former is obtained from the meridional reflection, whereas about the latter is obtained from the equatorial reflection (Fig. 4.7C). The equatorial reflection is quite characteristic of the various morphologies that a given

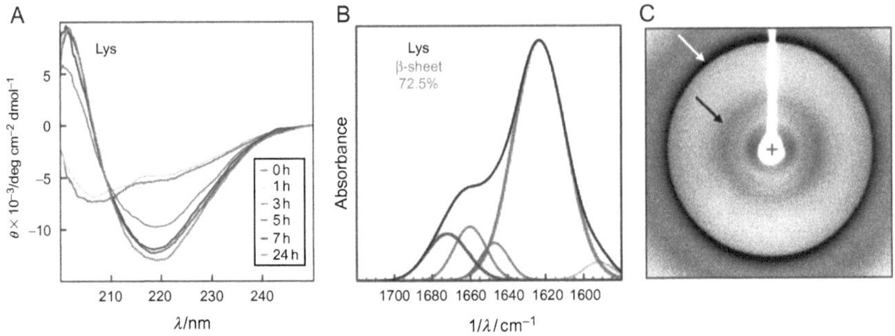

Figure 4.7 Examples of the use of biophysical tools to investigate the structural properties of the amyloid fibrils formed by lysozyme;[11,36] (A) circular dichroism spectra recorded during aggregation at pH 2 and 60°; (B) ATR–FTIR with an indication of the contribution of residues in different conformations to the FTIR signal where orange corresponds to β-strands, green to helices, and disordered residues and gray to side chains; (C) X-ray fiber diffraction where the black arrow points to the equatorial reflection that reports on the intersheet distance (typically 10 Å, Fig. 4.4) and the white arrow points to the meridional reflection that reports on the interstrand distance (typically 4.75 Å, Fig. 4.4). (See Color Insert.)

protein can form and has been related to the volume of the side chains involved in the dry interface.[27]

An essential step in the characterization of the structural properties of amyloid fibrils is the identification of the core regions, that is, those regions of sequence that form the cross-β structure and contribute to the stabilization of the fibrils. This is important because the size of the core is related to the stability of the fibrils, which has been related to their cytotoxic effects.[79] As already mentioned, the noncore regions can readily be identified by assigning the resonances of fibrils that are detectable by using solution NMR, but this requires that the fibrils are soluble, that the noncore regions are sufficiently flexible for their efficient detection, and that it is possible to assign these resonances by, for example, comparison of the spectrum of the aggregated protein with that of the soluble protein.[46,64]

In cases when these conditions are not fulfilled, it is possible to reveal the identity of the noncore regions indirectly by using limited proteolysis[80] and hydrogen–deuterium exchange. The former technique takes advantage of the fact that proteases cannot in general cleave peptide bonds connecting residues involved in the cross-β structure due to steric hindrance.[81] Incubating amyloid fibrils with a protease will only lead to proteolysis if noncore regions are present and the isolation and identification of the peptides thus produced, typically by high-pressure liquid chromatography coupled to mass spectrometry, can lead to their correct identification.[82–86]

Hydrogen–deuterium exchange relies instead on the observation that hydrogen atoms involved in hydrogen bonds exchange with deuterium atoms, when in D_2O, at a much slower rate than hydrogen atoms that are solvent exposed.[87,88] Since hydrogen bonds are a feature of the cross-β structure adopted by residues in the core of amyloid fibrils, it is possible to differentiate core and noncore regions by incubating the fibrils in D_2O and determining the degree of exchange of the various residues either by solution NMR or by a combination of proteolytic digestion and MS.[45]

In this experiment, the hydrogens of the backbone amides (NH groups) of residues in the noncore regions are exchanged by deuterium (to yield ND groups) by incubation of the fibrils in D_2O for a well-defined amount of time. At the end of the experiment, the exchange is quenched by snap freezing the solution in liquid nitrogen followed by lyophilization and dissolution of the resulting material in 5% buffered aqueous DMSO. When the protein is enriched in ^{15}N, it is possible to determine the degree of exchange of the different residues by measuring the intensity of the resonances corresponding to the various NH groups using solution NMR. In the spectrum, residues in noncore regions yield weak NMR signals because the corresponding NH groups are exchanged to ND, whereas residues in core regions show the same intensity as samples that have not been exposed to D_2O. It is possible to obtain similar information, although at lower resolution, by digesting the disaggregated protein with a protease and characterizing the degree of exchange of the resulting peptides using MS; in this case, the degree of exchange is determined by the increase in mass resulting from the replacement of H by D.

3. STRUCTURAL PROPERTIES OF AMYLOID PRECURSORS

3.1. Amyloid precursors

As mentioned in the introduction, the formation of amyloid fibrils follows a mechanism with well-defined—primary and secondary—nucleation and elongation steps. Primary nucleation and elongation are bimolecular reactions that are driven by the formation of the noncovalent interactions stabilizing amyloid fibrils that were described in Section 2. As it is widely accepted that globular proteins have evolved to maximize their solubility by minimizing the exposure of hydrophobic side chains and unsatisfied hydrogen bond donors and acceptors[89], it is reasonable to expect that proteins with a well-defined and persistent structure will have to undergo conformational changes that expose such functionalities before amyloid formation can occur.[4,90,91]

It is, in principle, possible to suggest three types of conformational changes that can lead to amyloid formation by globular proteins (Fig. 4.4).[4] The simplest case would involve a well-defined conformational transition such as the transient separation of two folded domains, tightly associated in the native structure, to populate an alternative conformation; the second one would involve the local unfolding of an element of secondary structure or domain to populate a partially unfolded state; the third one would involve the cooperative unfolding of the native structure to populate the fully unfolded state.

Any process that increases the free energy of the native cooperatively folded state will decrease the free energy difference between the native and the nonnative states mentioned above and has the potential to facilitate aggregation by increasing the population of species with unsatisfied chemical functionalities. Processes of this type include drastic changes in solution conditions such as pH changes, temperature jumps, the addition of cosolvents or solutes at high concentration, and the introduction of mutations that remove stabilizing interactions from the network of contacts that stabilize the native state.

It is important to mention that in proteins that are devoid of persistent secondary and tertiary structure, such as intrinsically disordered proteins (IDPs),[92] all residues of the protein have a similar degree of exposure to the solvent and other molecules of solute, and it is therefore in principle not necessary that a conformational change occurs prior to primary nucleation and/or elongation. It has nevertheless been suggested that certain IDPs such as α-synuclein must break a number of long-range *protective* transient tertiary interactions before amyloid formation can be initiated[93] similarly to what is the case for globular proteins.

An accurate structural characterization of the mechanisms by which proteins aggregate into amyloid fibrils requires a structural characterization of the three types of nonnative states mentioned above as well as of the disordered native state of IDPs. Similarly to what is the case for amyloid fibrils, this is a very challenging task because these states are often only transiently populated under the equilibrium conditions that lead to aggregation.[91] In addition, they typically present substantial disorder[90] and are therefore not amenable to the conventional tools of structural biology.[94]

3.2. Methods for the structural characterization of flexible proteins

Solution NMR is a spectroscopy that is particularly well suited to the study of precursors of amyloid fibrils because it has the ability of providing residue-specific NMR parameters that report on the structural and dynamical properties of flexible proteins.[95,96] It can provide information about the amplitude and, in some cases, timescale of conformational changes such as those schematically presented in Fig. 4.4A. It has in general been the technique of choice for the characterization of precursor states.[70,93,97–101]

The structural interpretation of NMR parameters in systems presenting disorder is challenging because the values of the parameters are time and ensemble averages and do not correspond to specific conformations.[94] In this case, using the type of approaches that are normally used to derive average structures can lead to unphysical conformations that may not be sampled in solution and therefore may not be representative of the structural properties of the precursor state that are of interest to study the mechanism of amyloid formation.

In this scenario, it is necessary to combine the structural information contained in the NMR parameters with motional models[102,103] or molecular simulations[98,104–107] to derive conformational ensembles that collectively agree with the NMR data. The use of these computational tools is indispensable because the NMR parameters do not contain sufficient information to identify all conformations contributing to the average measured experimentally, that is, the number of degrees of freedom of conformational ensembles is in most cases too large and the NMR parameters too sparse for the ensembles to be determined in a conventional sense.[108]

Many approaches have been used in the recent literature to exploit the structural information contained in NMR parameters in terms of conformational ensembles. The most direct approach involves the back calculation of the NMR parameters, using appropriate equations, from ensembles generated *a priori* by either using motional models or molecular simulations, typically molecular dynamics (MD).[102,109] When there is agreement between the back-calculated NMR parameters and their experimental counterparts, the conformational ensemble can be considered a reliable representation of the structural heterogeneity of the amyloid precursor.

In cases where the agreement is not quantitative, it is possible to use the NMR data to bias the motional model or the molecular simulations so that the resulting conformational ensemble is compatible with experiment. Several methods are available for this and they range from selection methods that identity sub-ensembles, taken from the ensemble obtained *a priori,* that match experiment,[99,110] to ensemble molecular simulations using Monte Carlo techniques or MD restrained by using empiric quadratic potentials that penalize configurations with NMR parameters that deviate from experiment.[111]

The type of NMR parameters that can be used for these studies include: (i) NOEs, that depend on the distribution of interresidue distances[112], (ii) order parameters, S^2 that depend on the distribution of orientations of backbone NH bond vectors in the molecular frame[113], (iii) scalar and residual dipolar couplings that report on the distribution of backbone dihedral angles,[104,106,107] and (iv) paramagnetic relaxation enhancements (PREs) that depend on the distribution of the distance separating a paramagnetic center attached to a Cys side chain and the NH groups of the protein.[98] NOEs and order parameters are particularly appropriate for globular proteins, whereas PREs are more suitable for disordered proteins due to their long range. Scalar and dipolar couplings can be used for both globular and disordered proteins provided that the changes in the alignment of the protein as its structure fluctuates are taken into account explicitly in the calculation of the dipolar couplings.[111]

It is also possible to analyze the structural properties of precursor states with techniques that do not aim at producing conformational ensembles but instead provide a qualitative description of which residues of the protein experience a conformational change. One such technique that is widely used is hydrogen/deuterium exchange monitored by NMR, which has already been introduced in Section 2 as it can also be used to analyze the structural properties of amyloid fibrils.[87,88]

In its simplest implementation the intensity of the NMR signals of the backbone NH of residues of the aggregating protein are monitored in a time resolved fashion when the protein is in a D_2O-based buffer. If aggregation does not occur during the experiment, the intensity of the NMR lines decreases exponentially with a time constant, called protection factor, that depends on the extent to which the NH group is protected from exchange by native hydrogen bonds and other noncovalent interactions. When this experiment is used to analyze the properties of amyloid precursors it immediately identifies the residues undergoing local conformational changes.

3.3. Structural studies of amyloid precursors

In spite of the difficulty of this type of studies, in which proteins have to be kept in solution for the duration of the experiments, a small number of amyloid precursors have been studied by using the techniques described above. The structural properties of both globular and IDPs have been analyzed and the information obtained has provided important data about the mechanism of amyloid formation. To illustrate the range of mechanisms that are possible, we will in this section discuss three systems, corresponding to the three possible types of precursors presented in Fig. 4.3.

The acylphosphatase from Sulfolobus solfataricus[114] (Sso AcP) is a model protein that has been extensively studied by the Chiti group.[115–119] It is not related to any disease but its aggregation *in vitro* provides a good example of a protein that requires a relatively minor conformational change for amyloid formation to occur. This protein aggregates when it is incubated in ca. 20% trifluoroethanol (TFE),[115] where it populates nonnative conformations that are enzymatically active and have a structure, as assessed by CD and NMR spectroscopies, that is very similar to that of the native state.[114,120] Structural as well as kinetic evidence shows that the unfolding of Ssp AcP is not required for aggregation[115] and the conformational change that allows establishing the first intermolecular interactions has been recently characterized by using a multidisciplinary approach that involved protein engineering techniques[118,121] as well as solution NMR.[122]

The former technique had established that aggregation in ca. 20% TFE is caused by interactions between the N-terminal disordered tail of a protein molecule and a β-strand (β4) at the edge of the β-sheet of a second protein molecule. Solution NMR was used to study the conformational changes occurring in a variant of the protein lacking the N-terminal disordered tail when it was incubated under amyloidogenic conditions. It was found that addition of TFE caused chemical shift changes in residues in the vicinity of strand β4, indicative of a subtle conformational change (Fig. 4.8A).[122]

In addition, hydrogen/deuterium exchange experiments monitored by NMR indicated an increase in the dynamics of the protein that was especially noticeable in the region corresponding to the β-strand known to be involved in the early stages of aggregation. Overall, the experiments showed that Sso AcP experiences a significant increase in its dynamics in ca. 20% TFE and a subtle conformational change in a β-strand of the protein and makes it possible for it to establish intermolecular interactions with the N-terminal disordered of a second protein molecule.[122]

Human lysozyme forms amyloid fibrils in the spleen, liver, and kidneys of patients of hereditary lysozyme nonneuropathic systemic amyloidosis.[125,126] This disease is caused by mutations in the lysozyme gene that destabilize the native structure of this protein and cause it to aggregate into amyloid fibrils. Lysozyme is a protein composed of two domains, a main domain rich in α secondary structure and a secondary one rich in β secondary structure (Fig. 4.8B). The folding and misfolding properties of lysozyme have been studied extensively by the Dobson group.[127]

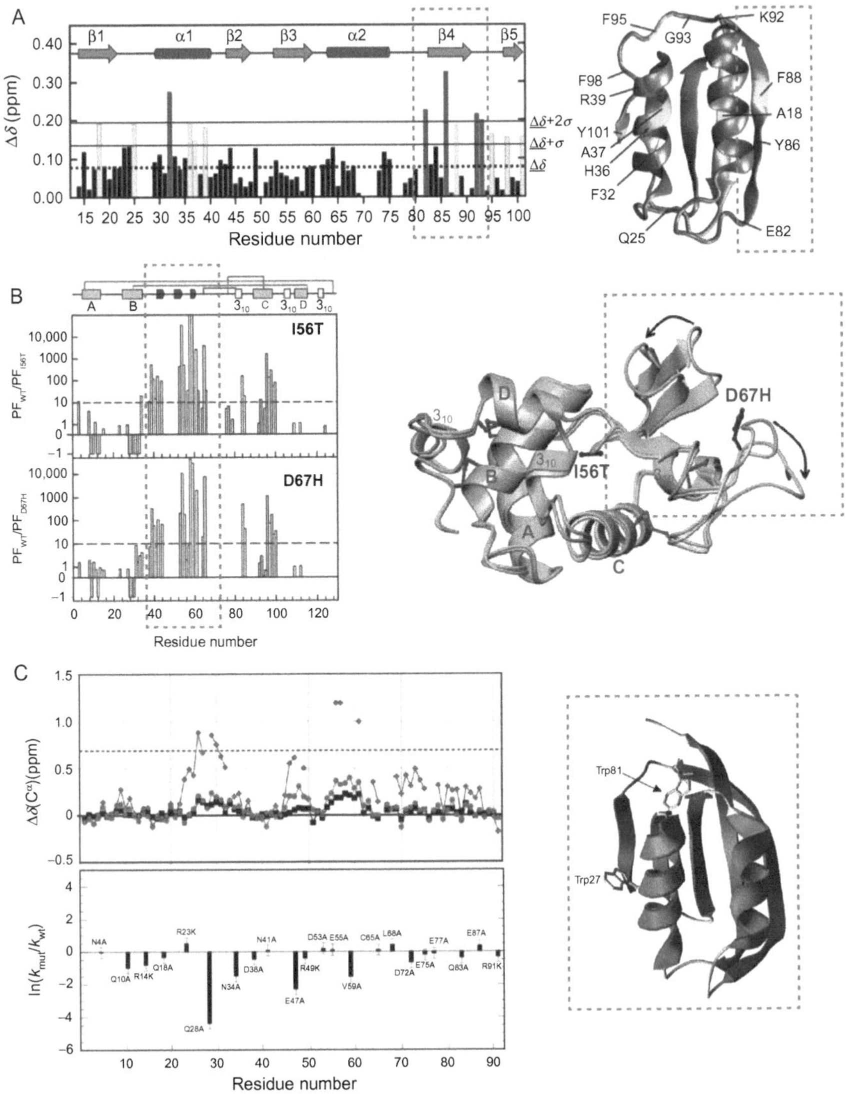

Figure 4.8 Illustration of three types of structural change that can cause amyloid formation by a globular protein where the region of sequence and structure that undergoes the structural change is highlighted, for each case, in blue. (A) Structure of Sso AcP, with an indication of the residues that experience significant chemical shift changes when the protein is incubated under aggregating conditions, and plot of the chemical shift change as a function of residue number. In the plot, the various residues are labeled according to the size of the change of chemical shift. The analysis reveals that the conformational change responsible for rendering the protein aggregation-prone occurs in the vicinity of strand β4; further experiments showed that local unfolding does not occur in this site.[122] (B) Structure of human lysozyme and changes

Most of the mutations found in patients of the disease are found in the β domain or in the interface between the two domains. Experiments carried out using hydrogen/deuterium exchange monitored by MS have shown that they decrease the cooperativity of the unfolding transition of lysozyme and therefore destabilize the β domain. Hydrogen/deuterium exchange experiments monitored by NMR revealed that this destabilization of the β domain causes its unfolding under the mild denaturing (pH 7.4, $T=60\ ^{\circ}C$) conditions that lead to amyloid formation (Fig. 4.8B).[97] A multidisciplinary analysis of the cross-β core of the amyloid fibrils formed under such conditions showed that it corresponds to the region of structure found to be disordered in the precursor, indicating that in this case the structural properties of the monomeric protein determine the molecular structure and the morphology of the fibrils.[11]

HypF-N (Fig. 4.8C) is a model protein that has been used by the Chiti group to investigate the relevance of transient structure for the mechanism of amyloid formation as well as to establish structure–activity relationships in protein oligomers.[70,124,128–133] This protein aggregates at pH 2 from its *molten globular* acid-denatured state, which is rich in transient secondary and tertiary structure as characterized by NMR. With the aim of identifying whether this transient structure played an important role in the mechanism of aggregation the kinetics of amyloid formation were measured for a series of mutants. Some of the mutants were in regions involved in transient structure whereas some other ones were in regions that were disordered and exposed in the precursor state.[70]

The result of this study was that the effect of the mutations on the aggregation rate could be predicted simply by considering the effect of the mutations on the physico-chemical properties of the proteins regardless of whether the particular amino acid was involved in the transient structure characterized by CD and NMR (Fig. 4.8C). An important implication of

in the protection factors of the various NH groups of human lysozyme induced by the mutations found in patients of lysozyme nonneuropathic systemic amyloidosis. The results illustrate that the mutations decrease the stability and local unfolding of the β domain of this protein under the conditions that lead to amyloid formation.[123] (C) Plot of the secondary Cα chemical shifts of HypF-N determined using solution NMR, which report on the helical propensity, at 0 M (green), 2 M (red), and 6 M (blue) urea—at 0 M urea there is significant helical propensity in various regions of sequence—and plot of the relative impact of various mutations on the rate of amyloid formation. A comparison affords that the transient secondary structure does not greatly affect aggregation rate.[124] (See Color Insert.)

this work is that the transient secondary and tertiary structure often observed in amyloidogenic states of proteins does not necessarily play a role in the mechanism of aggregation that is, does not define an on-pathway intermediate and is likely to be simply a consequence of the propensity of the sequence to establish interresidue interactions.[70]

Human lysozyme is a particularly interesting system because the mechanism by which it forms fibrils has been shown to depend on the conditions of incubation. As we have seen under mild denaturing conditions, where the protein is partially unfolded, it forms aggregates where only fraction of the backbone adopts the cross-β structure. However, under strongly denaturing conditions (pH 2, $T=25$ °C), where the protein is essentially fully denatured, most of the amino acids of this protein become structured in the cross-β conformation to yield fibrils that are remarkably stable to depolymerization due to the substantial size of their core.[11] These observations indicate that the presence of cooperatively folded structure in the precursor state can act as an important constraint and lead to fibrils that are partially structured in the cross-β conformation.

4. CONCLUSIONS

The availability of tools for characterizing both amyloid fibrils and their monomeric precursors is beginning to provide important mechanistic information about this biologically and biomedically important process. In many cases, the monomeric precursors possess transient secondary and tertiary interactions under the conditions that lead to amyloid formation and in some of these such transient interactions have been shown to be similar to those occurring in the amyloid fibrils. Detailed investigations of some of such systems have however shown that the impact of the transient interactions in the kinetics of amyloid formation is likely to be minimal, that is, does not affect the mechanism. In systems where the structure is cooperatively folded, however, the impact on the structure of the fibrils and, presumably, on the mechanism of amyloid formation can be very important. In spite of these recent advances, the number of proteins for which both the structural properties of the precursor state and those of the amyloid fibrils have been analyzed at high resolution is small. It is expected that the widespread use of the methods discussed in this chapter will provide a more conclusive account of the relationship between the structure of the precursor state and that of the fibrils resulting from its aggregation.

REFERENCES

1. Dobson CM. Protein misfolding, evolution and disease. *Trends Biochem Sci* 1999;**24**:329–32.
2. Chiti F, Dobson CM. Protein misfolding, functional amyloid, and human disease. *Annu Rev Biochem* 2006;**75**:333–66.
3. Sunde M, Serpell LC, Bartlam M, Fraser PE, Pepys MB, Blake CC. Common core structure of amyloid fibrils by synchrotron X-ray diffraction. *J Mol Biol* 1997;**273**:729–39.
4. Nelson R, Eisenberg D. Recent atomic models of amyloid fibril structure. *Curr Opin Struct Biol* 2006;**16**:260–5.
5. Greenwald J, Riek R. Biology of amyloid: structure, function, and regulation. *Structure* 2010;**18**:1244–60.
6. Nelson R, Sawaya MR, Balbirnie M, Madsen AØ, Riekel C, Grothe R, et al. Structure of the cross-beta spine of amyloid-like fibrils. *Nature* 2005;**435**:773–8.
7. Petkova AT, Yau W-M, Tycko R. Experimental constraints on quaternary structure in Alzheimer's beta-amyloid fibrils. *Biochemistry* 2006;**45**:498–512.
8. Wasmer C, Lange A, Van Melckebeke H, Siemer AB, Riek R, Meier BH. Amyloid fibrils of the HET-s(218–289) prion form a beta solenoid with a triangular hydrophobic core. *Science* 2008;**319**:1523–6.
9. Lansbury PT, Lashuel HA. A century-old debate on protein aggregation and neurodegeneration enters the clinic. *Nature* 2006;**443**:774–9.
10. Pepys MB. Amyloidosis. *Annu Rev Med* 2006;**57**:223–41.
11. Mossuto MF, Dhulesia A, Devlin G, Frare E, Kumita JR, de Laureto PP, et al. The non-core regions of human lysozyme amyloid fibrils influence cytotoxicity. *J Mol Biol* 2010;**402**:783–96.
12. Cohen SIA, Vendruscolo M, Dobson CM, Knowles TPJ. From macroscopic measurements to microscopic mechanisms of protein aggregation. *J Mol Biol* 2012;**421**:160–71.
13. Knowles TPJ, Waudby CA, Devlin GL, Cohen SIA, Aguzzi A, Vendruscolo M, et al. An analytical solution to the kinetics of breakable filament assembly. *Science* 2009;**326**:1533–7.
14. Jarrett JT, Lansbury PT. Seeding "one-dimensional crystallization" of amyloid: a pathogenic mechanism in Alzheimer's disease and scrapie? *Cell* 1993;**73**:1055–8.
15. Ruschak AM, Miranker AD. Fiber-dependent amyloid formation as catalysis of an existing reaction pathway. *Proc Natl Acad Sci U S A* 2007;**104**:12341–6.
16. Prusiner SB. Prions. *Proc Natl Acad Sci U S A* 1998;**95**:13363–83.
17. Pan KM, Baldwin M, Nguyen J, Gasset M, Serban A, Groth D, et al. Conversion of alpha-helices into beta-sheets features in the formation of the scrapie prion proteins. *Proc Natl Acad Sci U S A* 1993;**90**:10962–6.
18. Prusiner SB. Novel proteinaceous infectious particles cause scrapie. *Science* 1982;**216**: 136–44.
19. Lane KL, Brown P, Howell DN, Crain BJ, Hulette CM, Burger PC, et al. Creutzfeldt-Jakob disease in a pregnant woman with an implanted dura mater graft. *Neurosurgery* 1994;**34**:737–9.
20. Lasmézas CI, Deslys JP, Demaimay R, Adjou KT, Lamoury F, Dormont D, et al. BSE transmission to macaques. *Nature* 1996;**381**:743–4.
21. Fowler DM, Koulov AV, Balch WE, Kelly JW. Functional amyloid—from bacteria to humans. *Trends Biochem Sci* 2007;**32**:217–24.
22. Halfmann R, Jarosz DF, Jones SK, Chang A, Lancaster AK, Lindquist S. Prions are a common mechanism for phenotypic inheritance in wild yeasts. *Nature* 2012;**482**:363–8.
23. Stöhr J, Watts JC, Mensinger ZL, Oehler A, Grillo SK, DeArmond SJ, et al. Purified and synthetic Alzheimer's amyloid beta (Aβ) prions. *Proc Natl Acad Sci USA* 2012;**109**:11025–30.

24. Volpicelli-Daley LA, Luk KC, Patel TP, Tanik SA, Riddle DM, Stieber A, et al. Exogenous α-synuclein fibrils induce Lewy body pathology leading to synaptic dysfunction and neuron death. *Neuron* 2011;**72**:57–71.
25. Frost B, Diamond MI. Prion-like mechanisms in neurodegenerative diseases. *Nat Rev Neurosci* 2009;**11**:1–5.
26. Fändrich M, Fletcher MA, Dobson CM. Amyloid fibrils from muscle myoglobin. *Nature* 2001;**410**:165–6.
27. Fändrich M, Dobson CM. The behaviour of polyamino acids reveals an inverse side chain effect in amyloid structure formation. *EMBO J* 2002;**21**:5682–90.
28. Chiti F, Stefani M, Taddei N, Ramponi G, Dobson CM. Rationalization of the effects of mutations on peptide and protein aggregation rates. *Nature* 2003;**424**:805–8.
29. Pawar AP, Dubay KF, Zurdo J, Chiti F, Vendruscolo M, Dobson CM. Prediction of "aggregation-prone" and "aggregation-susceptible" regions in proteins associated with neurodegenerative diseases. *J Mol Biol* 2005;**350**:379–92.
30. Fernández-Escamilla A-M, Rousseau F, Schymkowitz J, Serrano L. Prediction of sequence-dependent and mutational effects on the aggregation of peptides and proteins. *Nat Biotechnol* 2004;**22**:1302–6.
31. Tartaglia GG, Vendruscolo M. The Zyggregator method for predicting protein aggregation propensities. *Chem Soc Rev* 2008;**37**:1395–401.
32. Kelly JW. Alternative conformations of amyloidogenic proteins govern their behavior. *Curr Opin Struct Biol* 1996;**6**:11–7.
33. Guijarro JI, Sunde M, Jones JA, Campbell ID, Dobson CM. Amyloid fibril formation by an SH3 domain. *Proc Natl Acad Sci U S A* 1998;**95**:4224–8.
34. Chiti F, Webster P, Taddei N, Clark A, Stefani M, Ramponi G, et al. Designing conditions for in vitro formation of amyloid protofilaments and fibrils. *Proc Natl Acad Sci U S A* 1999;**96**:3590–4.
35. Dobson CM. Protein folding and misfolding. *Nature* 2003;**426**:884–90.
36. Mossuto MF, Bolognesi B, Guixer B, Dhulesia A, Agostini F, Kumita JR, et al. Disulfide bonds reduce the toxicity of the amyloid fibrils formed by an extracellular protein. *Angew Chem Int Ed* 2011;**50**:7048–51.
37. Sunde M, Blake CC. The structure of amyloid fibrils by electron microscopy and X-ray diffraction. *Adv Protein Chem* 1997;**50**:123–59.
38. Tycko R. Characterization of amyloid structures at the molecular level by solid state nuclear magnetic resonance spectroscopy. *Methods Enzymol* 2006;**413**:103–22.
39. Eisenberg D, Nelson R, Sawaya MR, Balbirnie M, Sambashivan S, Ivanova MI, et al. The structural biology of protein aggregation diseases: fundamental questions and some answers. *Acc Chem Res* 2006;**39**:568–75.
40. Vilar M, Chou H-T, Lührs T, Maji SK, Riek-Loher D, Verel R, et al. The fold of alpha-synuclein fibrils. *Proc Natl Acad Sci U S A* 2008;**105**:8637–42.
41. Tanaka M, Chien P, Naber N, Cooke R, Weissman JS. Conformational variations in an infectious protein determine prion strain differences. *Nature* 2004;**428**:323–8.
42. Tanaka M, Collins SR, Toyama BH, Weissman JS. The physical basis of how prion conformations determine strain phenotypes. *Nature* 2006;**442**:585–9.
43. Toyama BH, Kelly MJS, Gross JD, Weissman JS. The structural basis of yeast prion strain variants. *Nature* 2007;**449**:233–7.
44. Knowles TP, Fitzpatrick AW, Meehan S, Mott HR, Vendruscolo M, Dobson CM, et al. Role of intermolecular forces in defining material properties of protein nanofibrils. *Science* 2007;**318**:1900–3.
45. Hoshino M, Katou H, Hagihara Y, Hasegawa K, Naiki H, Goto Y. Mapping the core of the beta(2)-microglobulin amyloid fibril by H/D exchange. *Nat Struct Mol Biol* 2002;**9**:332–6.

46. Baldwin AJ, Bader R, Christodoulou J, MacPhee CE, Dobson CM, Barker PD. Cytochrome display on amyloid fibrils. *J Am Chem Soc* 2006;**128**:2162–3.
47. Liu C, Zhao M, Jiang L, Cheng P-N, Park J, Sawaya MR, et al. Out-of-register β-sheets suggest a pathway to toxic amyloid aggregates. *Proc Natl Acad Sci U S A* 2012;**109**:20913–8.
48. Zurdo J, Guijarro JI, Jiménez JL, Saibil HR. Dependence on solution conditions of aggregation and amyloid formation by an SH3 domain. *J Mol Biol* 2001;**311**:325–40.
49. Meinhardt J, Sachse C, Hortschansky P, Grigorieff N, Fändrich M. Abeta(1–40) fibril polymorphism implies diverse interaction patterns in amyloid fibrils. *J Mol Biol* 2009;**386**:869–77.
50. Petkova AT, Leapman RD, Guo Z, Yau W-M, Mattson MP, Tycko R. Self-propagating, molecular-level polymorphism in Alzheimer's beta-amyloid fibrils. *Science* 2005;**307**:262–5.
51. Wasmer C, Soragni A, Sabaté R, Lange A, Riek R, Meier BH. Infectious and non-infectious amyloids of the HET-s(218–289) prion have different NMR spectra. *Angew Chem Int Ed* 2008;**47**:5839–41.
52. Verel R, Tomka IT, Bertozzi C, Cadalbert R, Kammerer RA, Steinmetz MO, et al. Polymorphism in an amyloid-like fibril-forming model peptide. *Angew Chem Int Ed* 2008;**47**:5842–5.
53. Paravastu AK, Leapman RD, Yau W-M, Tycko R. Molecular structural basis for polymorphism in Alzheimer's beta-amyloid fibrils. *Proc Natl Acad Sci U S A* 2008; **105**:18349–54.
54. Sawaya MR, Sambashivan S, Nelson R, Ivanova MI, Sievers SA, Apostol MI, et al. Atomic structures of amyloid cross-beta spines reveal varied steric zippers. *Nature* 2007;**447**:453–7.
55. Jiménez JL, Guijarro JI, Orlova E, Zurdo J, Dobson CM, Sunde M, et al. Cryo-electron microscopy structure of an SH3 amyloid fibril and model of the molecular packing. *EMBO J* 1999;**18**:815–21.
56. Jiménez JL, Nettleton EJ, Bouchard M, Robinson CV, Dobson CM, Saibil HR. The protofilament structure of insulin amyloid fibrils. *Proc Natl Acad Sci U S A* 2002; **99**:9196–201.
57. Tanaka M, Chien P, Yonekura K, Weissman JS. Mechanism of cross-species prion transmission: an infectious conformation compatible with two highly divergent yeast prion proteins. *Cell* 2005;**121**:49–62.
58. Chatani E, Lee Y, Yagi H, Yoshimura Y, Naiki H, Goto Y. Ultrasonication-dependent production and breakdown lead to minimum-sized amyloid fibrils. *Proc Natl Acad Sci U S A* 2009;**106**:11119–24.
59. Yoshimura Y, Sakurai K, Lee Y-H, Ikegami T, Chatani E, Naiki H, et al. Direct observation of minimum-sized amyloid fibrils using solution NMR spectroscopy. *Prot Sci* 2010;**19**:2347–55.
60. Christodoulou J, Larsson G, Fucini P, Connell SR, Pertinhez TA, Hanson CL, et al. Heteronuclear NMR investigations of dynamic regions of intact Escherichia coli ribosomes. *Proc Natl Acad Sci U S A* 2004;**101**:10949–54.
61. Hsu S-TD, Fucini P, Cabrita LD, Launay H, Dobson CM, Christodoulou J. Structure and dynamics of a ribosome-bound nascent chain by NMR spectroscopy. *Proc Natl Acad Sci U S A* 2007;**104**:16516–21.
62. Baldwin AJ, Christodoulou J, Barker PD, Dobson CM, Lippens G. Contribution of rotational diffusion to pulsed field gradient diffusion measurements. *J Chem Phys* 2007;**127**:114505.
63. Baldwin AJ, Anthony-Cahill SJ, Knowles TPJ, Lippens G, Christodoulou J, Barker PD, et al. Measurement of amyloid fibril length distributions by inclusion of rotational motion in solution NMR diffusion measurements. *Angew Chem Int Ed* 2008;**47**:3385–7.

64. Skora L, Becker S, Zweckstetter M. Characterization of amyloid fibrils of human beta-2-microglobulin by high-resolution magic-angle spinning NMR. *Chembiochem* 2010;**11**:1829–32.
65. Tycko R. Insights into the amyloid folding problem from solid-state NMR. *Biochemistry* 2003;**42**:3151–9.
66. Jaroniec CP, MacPhee CE, Astrof NS, Dobson CM, Griffin RG. Molecular conformation of a peptide fragment of transthyretin in an amyloid fibril. *Proc Natl Acad Sci USA* 2002;**99**:16748–53.
67. Jaroniec CP, MacPhee CE, Bajaj VS, McMahon MT, Dobson CM, Griffin RG. High-resolution molecular structure of a peptide in an amyloid fibril determined by magic angle spinning NMR spectroscopy. *Proc Natl Acad Sci U S A* 2004;**101**:711–6.
68. Ferguson N, Becker J, Tidow H, Tremmel S, Sharpe TD, Krause G, et al. General structural motifs of amyloid protofilaments. *Proc Natl Acad Sci U S A* 2006;**103**:16248–53.
69. Johnson WC. Protein secondary structure and circular dichroism: a practical guide. *Proteins* 1990;**7**:205–14.
70. Calloni G, Lendel C, Campioni S, Giannini S, Gliozzi A, Relini A, et al. Structure and dynamics of a partially folded protein are decoupled from its mechanism of aggregation. *J Am Chem Soc* 2008;**130**:13040–50.
71. Castiglioni E, Abbate S, Longhi G, Gangemi R. Wavelength shifts in solid-state circular dichroism spectra: a possible explanation. *Chirality* 2007;**19**:491–6.
72. Seshadri S, Khurana R, Fink AL. Fourier transform infrared spectroscopy in analysis of protein deposits. *Methods Enzymol* 1999;**309**:559–76.
73. Miyazawa T, Blout ER. The infrared spectra of polypeptides in various conformations: amide I andamide II bands. *J Am Chem Soc* 1961;**83**:712–9.
74. Krimm S, Bandekar J. Vibrational spectroscopy and conformation of peptides, polypeptides, and proteins. *Adv Protein Chem* 1986;**38**:181–364.
75. Surewicz WK, Mantsch HH, Chapman D. Determination of protein secondary structure by Fourier transform infrared spectroscopy: a critical assessment. *Biochemistry* 1993;**32**:389–94.
76. Oberg KA, Ruysschaert J-M, Goormaghtigh E. Rationally selected basis proteins: a new approach to selecting proteins for spectroscopic secondary structure analysis. *Prot Sci* 2003;**12**:2015–31.
77. Cerf E, Sarroukh R, Tamamizu-Kato S, Breydo L, Derclaye S, Dufrêne YF, et al. Antiparallel beta-sheet: a signature structure of the oligomeric amyloid beta-peptide. *Biochem J* 2009;**421**:415–23.
78. Berthelot K, Ta HP, Géan J, Lecomte S, Cullin C. In vivo and in vitro analyses of toxic mutants of HET-s: FTIR antiparallel signature correlates with amyloid toxicity. *J Mol Biol* 2011;**412**:137–52.
79. Milanesi L, Sheynis T, Xue W-F, Orlova EV, Hellewell AL, Jelinek R, et al. Direct three-dimensional visualization of membrane disruption by amyloid fibrils. *Proc Natl Acad Sci* 2012;**109**:20455–60.
80. Fontana A, Polverino de Laureto P, De Filippis V, Scaramella E, Zambonin M. Limited proteolysis in the study of protein conformation. In: Sterchi EE, Stöcker W, editors. *Proteolytic enzymes: tools and targets*. Heidelberg: Springer Verlag; 1999. p. 257–84.
81. Kheterpal I, Zhou S, Cook KD, Wetzel R. Abeta amyloid fibrils possess a core structure highly resistant to hydrogen exchange. *Proc Natl Acad Sci U S A* 2000; **97**:13597–601.
82. Kheterpal I, Williams A, Murphy C, Bledsoe B, Wetzel R. Structural features of the Abeta amyloid fibril elucidated by limited proteolysis. *Biochemistry* 2001;**40**:11757–67.
83. Miake H, Mizusawa H, Iwatsubo T, Hasegawa M. Biochemical characterization of the core structure of alpha-synuclein filaments. *J Biol Chem* 2002;**277**:19213–9.

84. Baxa U, Taylor KL, Wall JS, Simon MN, Cheng N, Wickner RB, et al. Architecture of Ure2p prion filaments: the N-terminal domains form a central core fiber. *J Biol Chem* 2003;**278**:43717–27.
85. Polverino de Laureto P, Taddei N, Frare E, Capanni C, Costantini S, Zurdo J, et al. Protein aggregation and amyloid fibril formation by an SH3 domain probed by limited proteolysis. *J Mol Biol* 2003;**334**:129–41.
86. Frare E, Mossuto MF, Polverino de Laureto P, Dumoulin M, Dobson CM, Fontana A. Identification of the core structure of lysozyme amyloid fibrils by proteolysis. *J Mol Biol* 2006;**361**:551–61.
87. Englander SW, Mayne L. Protein folding studied using hydrogen-exchange labeling and two-dimensional NMR. *Annu Rev Biophys Biomol Struct* 1992;**21**:243–65.
88. Englander SW. Protein folding intermediates and pathways studied by hydrogen exchange. *Annu Rev Biophys Biomol Struct* 2000;**29**:213–38.
89. Monsellier E, Chiti F. Prevention of amyloid-like aggregation as a driving force of protein evolution. *EMBO Rep* 2007;**8**:737–42.
90. Uversky VN, Fink AL. Conformational constraints for amyloid fibrillation: the importance of being unfolded. *BBA-Proteins Proteom* 2004;**1698**:131–53.
91. Chiti F, Dobson CM. Amyloid formation by globular proteins under native conditions. *Nat Chem Biol* 2009;**5**:15–22.
92. Dunker AK, Silman I, Uversky VN, Sussman JL. Function and structure of inherently disordered proteins. *Curr Opin Struct Biol* 2008;**18**:756–64.
93. Bertoncini CW, Jung Y-S, Fernandez CO, Hoyer W, Griesinger C, Jovin TM, et al. Release of long-range tertiary interactions potentiates aggregation of natively unstructured alpha-synuclein. *Proc Natl Acad Sci U S A* 2005;**102**:1430–5.
94. Vendruscolo M. Determination of conformationally heterogeneous states of proteins. *Curr Opin Struct Biol* 2007;**17**:15–20.
95. Palmer AG. NMR characterization of the dynamics of biomacromolecules. *Chem Rev* 2004;**104**:3623–40.
96. Mittermaier A, Kay LE. New tools provide new insights in NMR studies of protein dynamics. *Science* 2006;**312**:224–8.
97. Canet D, Last AM, Tito P, Sunde M, Spencer A, Archer DB, et al. Local cooperativity in the unfolding of an amyloidogenic variant of human lysozyme. *Nat Struct Mol Biol* 2002;**9**:308–15.
98. Dedmon MM, Lindorff-Larsen K, Christodoulou J, Vendruscolo M, Dobson CM. Mapping long-range interactions in alpha-synuclein using spin-label NMR and ensemble molecular dynamics simulations. *J Am Chem Soc* 2005;**127**:476–7.
99. Bernadó P, Bertoncini CW, Griesinger C, Zweckstetter M, Blackledge M. Defining long-range order and local disorder in native alpha-synuclein using residual dipolar couplings. *J Am Chem Soc* 2005;**127**:17968–9.
100. Mukrasch MD, Markwick PRL, Biernat J, Bergen Mv, Bernadó P, Griesinger C, et al. Highly populated turn conformations in natively unfolded tau protein identified from residual dipolar couplings and molecular simulation. *J Am Chem Soc* 2007;**129**: 5235–43.
101. Mukrasch MD, Bibow S, Korukottu J, Jeganathan S, Biernat J, Griesinger C, et al. Structural polymorphism of 441-residue tau at single residue resolution. *PLoS Biol* 2009;**7**:e34.
102. Bernadó P, Blanchard L, Timmins P, Marion D, Ruigrok RWH, Blackledge M. A structural model for unfolded proteins from residual dipolar couplings and small-angle x-ray scattering. *Proc Natl Acad Sci U S A* 2005;**102**:17002–7.
103. Bouvignies G, Bernadó P, Meier S, Cho K-i, Grzesiek S, Bru¨schweiler R, et al. Identification of slow correlated motions in proteins using residual dipolar and hydrogen-bond scalar couplings. *Proc Natl Acad Sci U S A* 2005;**102**:13885–90.

104. Clore GM, Schwieters CD. Amplitudes of protein backbone dynamics and correlated motions in a small alpha/beta protein: correspondence of dipolar coupling and heteronuclear relaxation measurements. *Biochemistry* 2004;**43**:10678–91.
105. Richter B, Gsponer J, Várnai P, Salvatella X, Vendruscolo M. The MUMO (minimal under-restraining minimal over-restraining) method for the determination of native state ensembles of proteins. *J Biomol NMR* 2007;**37**:117–35.
106. Lange OF, Lakomek N-A, Farès C, Schröder GF, Walter KFA, Becker S, et al. Recognition dynamics up to microseconds revealed from an RDC-derived ubiquitin ensemble in solution. *Science* 2008;**320**:1471–5.
107. Fenwick RB, Esteban-Martín S, Richter B, Lee D, Walter KFA, Milovanovic D, et al. Weak long-range correlated motions in a surface patch of ubiquitin involved in molecular recognition. *J Am Chem Soc* 2011;**133**:10336–9.
108. Esteban-Martín S, Bryn Fenwick R, Salvatella X. Synergistic use of NMR and MD simulations to study the structural heterogeneity of proteins. *WIREs Comput Mol Sci* 2012;**2**:466–78.
109. Smith LJ, Bolin KA, Schwalbe H, MacArthur MW, Thornton JM, Dobson CM. Analysis of main chain torsion angles in proteins: prediction of NMR coupling constants for native and random coil conformations. *J Mol Biol* 1996;**255**:494–506.
110. Nodet G, Salmon L, Ozenne V, Meier S, Jensen MR, Blackledge M. Quantitative description of backbone conformational sampling of unfolded proteins at amino acid resolution from NMR residual dipolar couplings. *J Am Chem Soc* 2009;**131**:17908–18.
111. Esteban-Martín S, Fenwick RB, Salvatella X. Refinement of ensembles describing unstructured proteins using NMR residual dipolar couplings. *J Am Chem Soc* 2010;**132**:4626–32.
112. Bonvin AMJJ, Brünger AT. Conformational variability of solution nuclear magnetic resonance structures. *J Mol Biol* 1995;**250**:80–93.
113. Best RB, Vendruscolo M. Determination of protein structures consistent with NMR order parameters. *J Am Chem Soc* 2004;**126**:8090–1.
114. Corazza A, Rosano C, Pagano K, Alverdi V, Esposito G, Capanni C, et al. Structure, conformational stability, and enzymatic properties of acylphosphatase from the hyperthermophile Sulfolobus solfataricus. *Proteins* 2006;**62**:64–79.
115. Plakoutsi G, Taddei N, Stefani M, Chiti F. Aggregation of the Acylphosphatase from Sulfolobus solfataricus: the folded and partially unfolded states can both be precursors for amyloid formation. *J Biol Chem* 2004;**279**:14111–9.
116. Plakoutsi G, Bemporad F, Calamai M, Taddei N, Dobson CM, Chiti F. Evidence for a mechanism of amyloid formation involving molecular reorganisation within native-like precursor aggregates. *J Mol Biol* 2005;**351**:910–22.
117. Bemporad F, Vannocci T, Varela L, Azuaga AI, Chiti F. A model for the aggregation of the acylphosphatase from Sulfolobus solfataricus in its native-like state. *BBA-Proteins Proteom* 2008;**1784**:1986–96.
118. Soldi G, Bemporad F, Chiti F. The degree of structural protection at the edge beta-strands determines the pathway of amyloid formation in globular proteins. *J Am Chem Soc* 2008;**130**:4295–302.
119. Pagano K, Bemporad F, Fogolari F, Esposito G, Viglino P, Chiti F, et al. Structural and dynamics characteristics of acylphosphatase from Sulfolobus solfataricus in the monomeric state and in the initial native-like aggregates. *J Biol Chem* 2010;**285**:14689–700.
120. Plakoutsi G, Bemporad F, Monti M, Pagnozzi D, Pucci P, Chiti F. Exploring the mechanism of formation of native-like and precursor amyloid oligomers for the native acylphosphatase from Sulfolobus solfataricus. *Structure* 2006;**14**:993–1001.
121. Bemporad F, Chiti F. "Native-like aggregation" of the acylphosphatase from Sulfolobus solfataricus and its biological implications. *FEBS Lett* 2009;**583**:2630–8.

122. Bemporad F, De Simone A, Chiti F, Dobson CM. Characterizing intermolecular interactions that initiate native-like protein aggregation. *Biophys J* 2012;**102**:2595–604.
123. Dumoulin M, Canet D, Last AM, Pardon E, Archer DB, Muyldermans S, et al. Reduced global cooperativity is a common feature underlying the amyloidogenicity of pathogenic lysozyme mutations. *J Mol Biol* 2005;**346**:773–88.
124. Campioni S, Mossuto MF, Torrassa S, Calloni G, Polverino de Laureto P, Relini A, et al. Conformational properties of the aggregation precursor state of HypF-N. *J Mol Biol* 2008;**379**:554–67.
125. Pepys MB, Hawkins PN, Booth D, Vigushin DM, Tennent GA, Soutar AK, et al. Human lysozyme gene mutations cause hereditary systemic amyloidosis. *Nature* 1993;**362**:553–7.
126. Booth D, Sunde M, Bellotti V, Robinson CV, Hutchinson WL, Fraser PE, et al. Instability, unfolding and aggregation of human lysozyme variants underlying amyloid fibrillogenesis. *Nature* 1997;**385**:787–93.
127. Dumoulin M, Kumita JR, Dobson CM. Normal and aberrant biological self-assembly: insights from studies of human lysozyme and its amyloidogenic variants. *Acc Chem Res* 2006;**39**:603–10.
128. Chiti F, Bucciantini M, Capanni C, Taddei N, Dobson CM, Stefani M. Solution conditions can promote formation of either amyloid protofilaments or mature fibrils from the HypF N-terminal domain. *Prot Sci* 2001;**10**:2541–7.
129. Bucciantini M, Giannoni E, Chiti F, Baroni F, Formigli L, Zurdo J, et al. Inherent toxicity of aggregates implies a common mechanism for protein misfolding diseases. *Nature* 2002;**416**:507–11.
130. Bucciantini M. Prefibrillar amyloid protein aggregates share common features of cytotoxicity. *J Biol Chem* 2004;**279**:31374–82.
131. Relini A, Torrassa S, Rolandi R, Gliozzi A, Rosano C, Canale C, et al. Monitoring the process of HypF fibrillization and liposome permeabilization by protofibrils. *J Mol Biol* 2004;**338**:943–57.
132. Calloni G, Zoffoli S, Stefani M, Dobson CM, Chiti F. Investigating the effects of mutations on protein aggregation in the cell. *J Biol Chem* 2005;**280**:10607–13.
133. Marcon G, Plakoutsi G, Canale C, Relini A, Taddei N, Dobson CM, et al. Amyloid formation from HypF-N under conditions in which the protein is initially in its native state. *J Mol Biol* 2005;**347**:323–35.

SECTION II

Oligomerization in Seven-Transmembrane Receptors

CHAPTER FIVE

Quaternary Structure Predictions and Structural Communication Features of GPCR Dimers

Francesca Fanelli, Michele Seeber, Angelo Felline, Daniele Casciari, Francesco Raimondi

Department of Life Sciences, Dulbecco Telethon Institute, University of Modena and Reggio Emilia, Modena, Italy

Contents

Abstract

In spite of the ever-increasing evidence that G protein-coupled receptors (GPCRs) form dimers/oligomers, the biological role(s) and structural architecture of homologous and heterologous receptor aggregation are, however, far from being clarified.

Progress in Molecular Biology and Translational Science, Volume 117
ISSN 1877-1173
http://dx.doi.org/10.1016/B978-0-12-386931-9.00005-2

This chapter reviews the insights gained so far, at multiscale levels of resolution, on GPCR dimerization/oligomerization from *in vitro* experiments, structure predictions, and structure determinations. Focus is put on the achievement by the FiPD-based approach, which proved effective in predicting the supramolecular organization of membrane proteins including GPCRs. The combination of FiPD-based quaternary structure predictions with molecular simulations and analyses can be a valuable tool to infer the effects of dimerization on the structural communication features of a receptor dimer/oligomer bound to functionally different ligands.

Ultimately, the integration between atomistic and mesoscopic simulations is expected to be a promising tool to unveil functioning mechanisms that involve intricate protein networks.

1. INTRODUCTION

1.1. G protein-coupled receptors

G protein-coupled receptors (GPCRs) constitute the largest superfamily of signal transduction membrane proteins, which mediate responses of a variety of bioactive molecules, including biogenic amines, amino acids, peptides, lipids, nucleotides, and proteins (reviewed in Refs. 1–5). As a result, GPCRs play a crucial role in many essential physiological processes as diverse as neurotransmission, cellular metabolism, secretion, cell growth, immune defense, and differentiation (reviewed in Refs. 1–5). They have also recently emerged as involved in tumor growth and metastasis as well as in the pathology of Alzheimer's disease.[6–8]

These receptors share an up-down bundle of seven transmembrane (TM) helices connected by three intracellular (I1, I2, and I3) and three extracellular loops (E1, E2, and E3), an extracellular N-terminus and an intracellular C-terminus.

High-resolution information on GPCRs from X-ray crystallography currently concerns (a) rhodopsin in its dark (inactive), constitutively active, and photoactivated states;[9–13] (b) β_2- and β_1-adrenergic receptors (ARs) bound to agonists, partial agonists, inverse agonists, antagonists, a nanobody, and heterotrimeric Gs;[14–22] (c) A_{2A} adenosine receptor ($A_{2A}R$) bound to an antagonist, a number of agonists, and an allosteric inverse agonist antibody;[23–27] (d) squid rhodopsin bound to 11-*cis* retinal;[28] (e) CXCR4 chemokine receptor bound to small molecule and cyclic peptide antagonists;[29] (f) D_3 dopamine receptor (D_3R) in complex with a D2/D3 selective antagonist;[30] (g) the H_1 histamine receptor (H_1R) in complex with an antagonist;[31] (h) the m2- and m3-muscarinic receptors bound to antagonists;[32,33]

(i) the δ-, κ-, and μ-opioid receptors (ORs) bound to antagonists;[34–36] (j) the nociceptin/orphanin FQ receptor in complex with a peptide mimetic antagonist;[37] and (k) a lipid GPCR in complex with a sphingolipid antagonist.[38]

GPCR function is based on structural communication as they receive and transmit extracellular signals inside the cell. Upon activation by extracellular signals, they activate the α-subunit in heterotrimeric guanine nucleotide binding proteins (G proteins) by catalyzing the exchange of bound GDP for GTP, that is, they act as guanine nucleotide exchange factors (GEFs) specialized for the Gα family of the Ras GTPase superfamily.[39] Thus, GPCRs are allosteric proteins that transform extracellular signals into promotion of nucleotide exchange in intracellular G proteins.

1.2. GPCR dimerization/oligomerization

The classical idea that GPCRs function as monomeric entities has been unsettled by the emerging concept of GPCR dimerization (reviewed in Refs. 40–56). GPCRs have classically been assumed to exist and function as monomeric entities, and the paradigms of ligand binding and signal transduction were based on this hypothesis. What suddenly stirred general interest on oligomerization was evidence showing that direct receptor–receptor interactions can rescue functional activity in complementation experiments. As shown by Maggio and coworkers, pairs of GPCR chimeras or truncated fragments, which were inactive when individually expressed, regained binding and signaling activity if coexpressed in the same membrane.[57,58] On this line, using mouse lutropin (LH) receptor (LHR) as a model GPCR, it was demonstrated that transgenic mice coexpressing binding-deficient and signaling-deficient forms of LHR can reestablish normal LH actions through intermolecular functional complementation of the mutant receptors in the absence of functional wild-type receptors.[59] This, *per se*, did not establish which role oligomerization plays in GPCR function, but it definitely demonstrated that receptor–receptor interaction could affect the agonist binding pocket and the G protein interacting interface of GPCRs, as importantly, perhaps, as the intramolecular interactions that directly drive receptor activation. The study on mouse LHR, indeed, provides *in vivo* evidence for the physiological relevance of intermolecular cooperation in GPCR signaling.[59]

Following early findings by Maggio and coworkers, studies on GPCR dimerization have been appearing at a steadily increasing pace.[40–49,51–56,60–74] Although their existence is now largely accepted, their functional importance remains more enigmatic and in some cases even controversial (reviewed in Refs. 40–49,75,76). These conclusions can also be inferred from the review article by Terrillon and Bouvier, concerning the state of the art in our understanding of the role of dimerization of GPCRs in the five different steps of their half-life cycle, that is, ontogeny, ligand-promoted regulation, pharmacological diversity, signal transduction, and internalization.[47]

1.3. Role of GPCR dimerization/oligomerization in ontogeny

As for ontogeny, GPCRs of family C were the most generous so far in yielding clues on the possible role of oligomerization. $GABABR_1$ and $GABABR_2$ receptors represent an emblematic case (reviewed in Ref. 77). Each of the two genes makes an incompetent GPCR protein, as the first is not properly glycosylated or inserted in the membrane and the second has no signaling activity. However, when both are expressed in the same cells, which physiologically occurs in brain neurons,[78,79] they make a functional GABA-responsive heterodimer.[78–81] The structural requirement in this case was identified in the C-terminus, where there are recognized sequences prone to make coiled-coil interactions,[82] that is in contrast with another member of family C receptors, the glutamate receptor type 1, where dimerization occurs in the N-terminus.[83] This large "flytrap" domain has structural similarity to the glutamate-binding domain of glutamate receptor channels, where they, in fact, form stable oligomeric forms.[1] Although the examples above seem "special" cases, evidence that GPCRs dimerize in order to act as molecular chaperones, that is, to catalyze their own folding and transport to the cell surface, was also found for family A GPCRs, such as the V_2 vasopressin receptors[84,85] and the chemokine receptor CCR5.[86] See also Ref. 50 for a review. Homodimerization/oligomerization of the 5-HT_{2C} receptor in the endoplasmic reticulum (ER) and Golgi of intact living cells was also demonstrated, suggesting that dimer/oligomer formation is a naturally occurring step in 5-HT_{2C} serotonin receptor maturation and processing.[87]

1.4. Effects of ligand binding on GPCR dimerization/oligomerization

A role for dimerization in GPCR ontogeny does not exclude the possibility that, once the receptor has reached the cell surface, its oligomeric state can be dynamically regulated by ligand. The question whether dimerization influences ligand-induced activation/regulation of GPCRs still remains to be answered. In fact, some studies suggest that ligand binding can regulate the dimer by either promoting or inhibiting its formation. In this respect, it was inferred that the β_2AR exists in dynamic equilibrium between monomers and higher order oligomers in living cells, with the average size of the oligomer being a tetramer and with inverse agonists promoting higher order oligomerization.[88] Furthermore, fluorescence spectroscopy approaches allowed to detect pirenzepine-induced dimerization of the M_1-muscarinic receptor,[89] as well as ligand regulation of the quaternary organization of cell surface M_3-muscarinic receptor.[90] Moreover, drug-modulated hetero-oligomerization in living neuronal cells was demonstrated for the A_{2A}R and the D_2 dopamine receptor (D_2R).[91] In contrast, other studies concluded that homodimerization and heterodimerization are constitutive processes that are not modulated by ligand binding (reviewed in Ref. 47). In this respect, atomic force microscopy (AFM) measurements showed that rhodopsin and opsin form constitutive dimers in dark-adaptive retinal membrane.[48,92] Along this line, experiments based on fluorescence resonance energy transfer (FRET) and bioluminescence energy transfer (BRET) reveal that many GPCRs exist as oligomers or at least as closely packed clusters in the membranes of living cells.[40–47,60,69,93–95]

1.5. Role of GPCR dimerization/oligomerization in the GEF activity

As for the question whether dimerization serves to the GEF function, studies contributed very little insight so far. Elegant experiments consisting of a combination of mass spectrometry after chemical cross-linking and neutron scattering in solution showed that the complex formed between the purified, activated leukotriene B_4 receptor BLT1 and $Gi_{\alpha2\beta1\gamma2}$ corresponds to a pentameric assembly of one dimeric receptor and one heterotrimeric G protein.[96] A successive study on a stabilized and purified heterodimer of BLT1 indicated that although ligand binding to one protomer in the heterodimer is associated with cross-conformational changes, a trans-activation mechanism where the ligand-free subunit would trigger GDP/GTP exchange is unfeasible.[97] A recent study demonstrated the importance of G protein in the maintenance and regulation of human somatostatin receptor dimers.[98] Furthermore, a functional complementation study enabling control on the components of a signaling unit indicated that for the D_2R, the minimal signaling unit is two receptors and one G protein.[99] For the 5-HT_4 serotonin receptor, it was shown that activation of one protomer in a dimer is sufficient to stimulate G proteins.[100] However, coupling efficiency was two times higher when both protomers were activated. Expression of combinations of the 5-HT_4

receptor, in which both protomers were able to bind to agonists but only one could couple to G proteins, suggested that upon agonist occupancy, protomers did not independently couple to G proteins but rather that only one G protein is activated. Asymmetric activation of GPCR dimers would imply that the minimal functional unit is two receptors and one heterotrimeric G protein.[99,101] In contrast to the evidences summarized above, for a number of GPCRs, *in vitro* experiments showed that one receptor molecule is sufficient to activate a G protein and bind arrestin[102] (reviewed in Ref. 103). Indeed, rhodopsin,[104,105] the β_2AR,[106] and the neurokinin-1 receptor[107] activate their cognate G proteins in the monomeric form. Moreover, supramolecular organization of rhodopsin,[104] neurotensin 1 (NTS1)[108] receptor, and leukotriene BLT2[109] receptor reduces G protein coupling. Along this line, the crystal structure of agonist-bound β_2AR and nucleotide-free heterotrimeric Gs shows a 1:1 receptor:G protein stoichiometry.[22]

Dimerization/oligomerization may be a strategy to diversify and extend the signaling properties that are intrinsic in each individual receptor gene (reviewed also in Refs. 110–112). ORs have been extensively studied in this regard.[113–115] It was shown that coexpressed δ- and κ-ORs can generate a different binding pattern and synergistic effects on MAP-kinase.[116] Similarly, mixes of μ- and δ-ORs gain special signaling and binding properties if jointly expressed.[117] A recent study demonstrated differential response to morphine of the oligomeric state of μ-OR in the presence of δ-OR.[118] It was found that the μ subtype exists primarily as a dimer that oligomerizes with δ into tetramers and morphine promotes the dissociation of these tetramers. Thus, it is possible that the complex combination of OR subtypes generated by studies in the precloning era and never matched by the genes that were actually found is explained by the fact that such receptors differ in signaling properties when they exist in homomeric or heteromeric form.[113,114] Similar results have been presented for many other receptor types and suggest a general trend.[40,43,47] Supramolecular arrays containing mixtures of different receptors may constitute specialized signaling patches of the plasma membrane, and their differential distribution in cell regions may have fundamental roles in fine-tuning the complex signaling networks of the central nervous system. This phenomenon, if true, poses a daunting challenge but also new opportunities in the design of new drugs. Indeed, GPCR heterodimers may have an important impact on drug discovery.[61,62,72,119–126] Emerging studies indeed provide evidence for tissue-specific and disease-specific receptor heteromerization (reviewed in Ref. 124). This suggests that heteromers represent novel drug targets for the identification of selective compounds with potentially fewer side effects. Furthermore, GPCR dimers may provide the necessary targets to increase the breadth and depth of receptors available for therapeutic interventions (reviewed in Ref. 120). Differential pharmacology, function, and regulation of GPCR heterodimers and heterooligomers suggest means to selectively target GPCRs in different tissues. They also hint that the mechanism of function of several pharmacological agents might be different *in vivo* than anticipated from simple ligand-screening programs that rely on heterologous expression of a single GPCR (reviewed in Ref. 119). It was demonstrated that the heterodimerization between the 5-HT_{2A} serotonin receptors and the metabotropic glutamate receptors may be involved in the altered cortical processes of schizophrenia, providing a promising new target for the treatment of psychosis.[126] Physiological relevance was demonstrated also for the heterodimerization of the Gq-coupled Angiotensin 1 receptor (AT_1R) and the Gi-coupled type 1 cannabinoid receptor.[125] Thus, allosterism across GPCR homo- or heteromers, whether dimers or higher order oligomers, can be considered as an additional topographical landscape that may open perspectives toward novel therapeutic approaches.[127] However, the question whether there are allosteric interactions across the receptor–receptor interface of a heterodimer that modulate the binding properties of the heterodimer components and thereby change their pharmacology still awaits an answer.[128] Collectively, allosteric communication across dimers and the related asymmetric/symmetric activation may have implications in multiple signaling pathways.[99,101,129,130] In this respect, a study on the BLT1 receptor highlighted asymmetric conformational changes in the receptor dimer only in the presence of the G protein, suggesting that the interaction of the G protein with the receptor dimer prevents a symmetric functioning of the dimer.[131] On the same line, *in vitro* evidence on the D_2R showed that a functional unit made of two receptors and one G protein is maximally activated by agonist binding to a single protomer, suggestive of an asymmetrically activated dimer.[99]

1.6. Effects of GPCR dimerization/oligomerization in G protein-independent signaling, protein scaffolding, and trafficking

One signaling pathway, in which dimerization/oligomerization may be crucial player, is G protein-independent signaling of GPCRs to mitogen-activated protein kinases. Arrestins, a family of proteins that were previously thought to be exclusively devoted to disconnect receptors from G proteins,[132] are now known to act also as recruiting adaptors that divert GPCR signaling toward intermediate kinases such as ERK1, ERK2, and others.[133,134] Recent AFM experiments on rhodopsin indicate that the photoreceptor forms dimers in the plasma membrane, with cytosolic protrusions that are located 3.7 nm apart.[92] This distance matches that of two evident grooves that are present in the crystal structure of arrestin.[92,135] This would suggest that GPCRs must be in dimeric form to bind arrestin. *In vitro* experiments, based on complementation of individually nonfunctional GPCRs, support this hypothesis, providing evidence that binding of β-arrestin-1 to m_3-muscarinic receptors requires paired stimulation of two receptor components within the same receptor dimer.[49,136,137] The consequences of arrestin-mediated signaling *in vivo* are not entirely understood yet, but studies of opioid effects in mice bearing deletions of arrestin genes suggest that arrestin signaling may be important for the control of tolerance and dependence.[138,139] However, as already stated above, a number of studies indicate that receptor dimerization is not needed to arrestin binding (reviewed in Ref. 103).[102]

It has been also proposed that GPCR oligomerization may provide the structural basis for organizing distinct cell compartments along the plasma membrane where different extracellular signals may be perceived and discriminated.[140] In the case of rhodopsin, it has been also postulated that oligomerization would serve to exclude rhodopsin molecules from phototransduction, a mechanism of adaptation.[141] Along this line, the molecular organization of the ternary complex involving melatonin 1 receptor (MT_1), Gi protein, and RGS20 was monitored in its basal and activated state by BRET between probes inserted at multiple sites of the complex.[142] The study allowed to infer a model wherein one Gi and one RGS20 protein bind to separate protomers of MT_1 dimers in a pre-associated complex that rearranges upon agonist activation. The study highlighted the advantage of GPCRs organized as dimers in which each protomer fulfils its specific task by binding to different GPCR-interacting proteins. Thus, regulated protein–protein interactions are key features of many aspects of GPCR function and there is now increasing evidence for GPCRs acting as part of multicomponent units comprising a variety of signaling and scaffolding molecules, organized in supramolecular signaling assemblies (signalosomes or transducisomes).[4,143,144]

Finally, it was suggested that heterodimerization could affect agonist-promoted GPCR endocytosis, a well-characterized process classically involved in signal attenuation. For many documented heterodimers, stimulation of only one of the protomers was sufficient to promote cointernalization of the two receptors (reviewed in Ref. 47). By contrast, receptors that do not undergo efficient agonist-promoted endocytosis were found to act as dominant negatives for endocytosis-prone receptors after heterodimerization (reviewed in Ref. 47). Although of significant potential interest, the physiological consequences of these observations on the regulation of GPCR desensitization/resensitization cycles remain to be determined. Collectively, dimerization may serve to affect receptor mobility at the cell surface and in intracellular trafficking.[54]

The extremely numerous experimental evidences for GPCR dimerization/oligomerization, however, provided very little insight into the architecture of the supramolecular receptor assemblies (reviewed in Refs. 40–56). This chapter reviews the insights gained so far into the architecture of GPCR dimers from X-ray crystallography, *in vitro* experiments as well as sequence-based and structure-based prediction methods (see Refs. 145,146). Focus will be put on quaternary structure predictions by a computational protocol relying on the combination of the ZDOCK protein–protein docking software with the FiPD docking analysis software.[147]

2. INSIGHTS FROM *IN VITRO* EXPERIMENTS INTO THE GPCR REGIONS INVOLVED IN RECEPTOR–RECEPTOR INTERACTION

The first high-resolution insights into the likely architecture of GPCR dimers came from AFM measurements, which led to the building of a semi-empirical structural model of a higher order rhodopsin oligomer.[135] According to this model, two monomers of rhodopsin interact with each other through E2, I2, H4, and H5 of both monomers. For family A GPCRs, recurrent evidence for the involvement of H6 in the inter-monomer interface came from early experiments with synthetic peptides holding the sequence of that helix, which appeared to inhibit homodimerization of either β_2AR or D_2R.[148,149] For the D_2R, a peptide from the H7 sequence proved to exert an inhibitory effect as well.[149] The involvement of H5 and H6 from D_2R in the heterodimerization with the A_{2A}R was indicated by BRET experiments using a $D_{2[1-4,7]}$R chimera, containing H5, H6, I3, and E3 from the D_1R sequence, a receptor that does not dimerize with A_{2A}R.[150] In fact, in contrast to the wild-type D_2R, the $D_{2[1-4,7]}$R chimera was not able to compete for the specific BRET between A_{2A}R and D_2R.[150] Experiments on purified leukotriene B_4 receptor provided evidence for the central role of H6 in stabilizing the receptor homodimer.[96] Also, H4 has been suggested to mediate intermonomer contacts on the basis of the results of cysteine cross-linking experiments on the D_2R.[151] Successive experiments highlighted potential rearrangements of the D_2R dimer architecture depending on the receptor functional state, the H4–H5 and H4–H4 contact dimers being, respectively, associated with inactive and active states.[152] In a successive study, the same authors provided evidence that D_2R forms higher order oligomers in living cells and that H1 and H8 form a second symmetric interface in addition to the previously identified H4 interface.[153] Truncation mutants by successive deletion of TM helices from amino and/or carboxyl terminations of the D_2R suggested that, in addition to H4, other receptor portions mediate homodimerization.[154] Cysteine cross-linking experiments on the δ-OR highlighted H4 and/or H5 as well at the homodimer interface.[155]

The involvement of H1 and H8 in GPCR oligomerization was recently proposed for the β_2AR as well.[88] *In vitro* site-directed mutagenesis driven by computational modeling also highlighted H1 and H8 as central in the interface of thromboxane A2 (TP) receptor dimers.[156]

In vitro disulphide-trapping experiments on the 5-HT_{2c} serotonin receptor indicated the existence of dimer architectures insensitive and sensitive to the functional states of the receptor.[157] In this respect, state-insensitive architectures were suggested to involve H1–H1 contacts, whereas state-sensitive architectures were suggested to involve H4–H5 contacts.[157]

In vitro experiments aimed at inducing disulphide-mediated dimerization of the α-factor pheromone receptor, Ste2p, highlighted H7–H7 dimers that decrease in the presence of the α-factor ligand, and H1–H1 dimers not affected by ligand binding.[158] A similar investigation on the same receptor highlighted the involvement of H1 and H4 in receptor oligomerization.[159]

Interactions between H1 and H4 were suggested to mediate the homodimerization of the CCR5 receptor, on the basis of the experimental observation that the combination of two point mutations, that is, I52(1.54)V and V150(4.47)A, impaired receptor function and prevented FRET, differently from the case of the wild type.[160] Furthermore, FRET experiments using either various $\alpha_{1b}AR$ fragments or α_{1b}/β_2 chimeras suggested that in the whole $\alpha_{1b}AR$, H1 and H7, which are adjacent in the helix bundle, act in concert in favoring receptor homooligomerization, with H1 being the prominent component of the interface.[161] Other FRET-based experiments on the same GPCR suggested the formation of H1–H1 and H4–H4 contacts in an oligomeric organization of the receptor.[162] Curiously, *in vitro* experiments on the M_3-muscarinic receptor indicated that C140(3.25) and C220 in E2, involved in a disulphide bond,[33] can also participate in the formation of intermolecular disulfide bonds in a dimer.[163] This was inferred from the observation that C140(3.25)A and C220A mutations prevented the formation of disulfide-linked receptor aggregates.[163]

X-ray determinations provide some information on the likely architecture of selected receptor dimers. Rhodopsin crystals show dimeric architectures characterized by H1–H1 or H8–H8 contacts all compatible with the AFM images.[164] A similar architecture was also found for the constitutively active opsin apoprotein.[11] With respect to the β_2AR, crystal packing in the presence of cholesterol shows a significant involvement of the hydrophobic molecule in the intermonomer interface.[15] In this framework, protein–protein contacts are minimal and include a pair of salt bridges between K60$^{(1.59)}$ and E338 from H8.[15] The crystal structure of CXCR4 in complex with the IT1t antagonist shows a dimer interface essentially contributed by amino acids from the extracellular halves of H5 and H6, with emphasis on the former (Fig. 5.1).[29] Additional contacts involving the cytosolic ends of H3, H4, H5, and I2 also contribute to the interface.[29] This dimer architecture characterizes the highest resolved complex with IT1t, encoded as 3ODU, as well as the lower resolution complexes encoded as 3OE8 and 3OE9.[29] The 3OE8 structure is a trimeric assembly characterized by an additional interface involving H5 and H6 from one protomer and H1 from the other protomer. The structure of μ-OR shows receptor molecules intimately associated into pairs along the crystallographic twofold axis through two different interfaces.[36] The first interface is a more limited parallel association mediated by H1, H2, and H8, whereas the second and more prominent interface involves H5 and H6.[36] Finally, in the crystallographic dimer of the κ-OR bound to the JDTic antagonist, the extracellular half of the interface relies on H1–H1 and H1–H2 contacts from both monomers, whereas the cytosolic end is made by H8–H8 contacts (Fig. 5.1).[35]

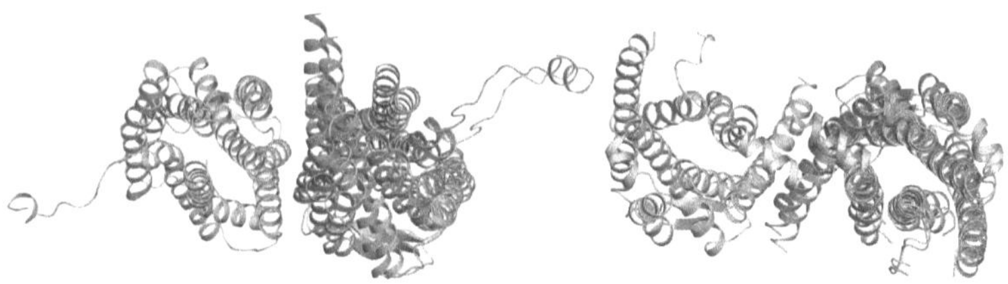

Figure 5.1 Superimposed structures of crystallographic (green) and predicted dimers of the CXCR4 (left) and κ-OR (right). The Cα-Root Mean Square Deviation (Cα-RMSDs) are 0.98 and 0.93 Å, for CXCR4 and κ-OR, respectively. (For interpretation of the references to color in this figure legend, the reader is referred to the online version of this chapter.)

Although the involvement of the intracellular regions cannot be excluded,[165] consensus emerges on H1, H4, and H8 being involved in GPCR dimerization/oligomerization (reviewed also in Ref. 166) (Table 5.1). However, it should be considered also the possibility that different dimeric architectures may exist in a dynamic equilibrium, as recently inferred from a study on m3-muscarinic receptor dimerization/oligomerization.[179] Furthermore, different receptor portions are expected to be simultaneously involved in the formation of high-order oligomers.

Table 5.1 GPCR portions involved in dimer/oligomer interfaces

Receptor	**Interface**[a]	**Source**[b]
In vitro		
Rhodopsin	I2, H4, H5	135
β_2AR	H6	148
D_2R	H6	149
Leukotriene B4 receptor	H6	96
D_2R–A_{2A}R	H5, I3, H6	150
D_2R	H4, H5	151,152,154
D_2R	H1, H8	153
δ-OR	H4, H5	155
β_2AR	H1, H8	88
TP	H1, H8	156
5-HT_{2C}	H1, H4, H5	157
Ste2p	H1, H4, H7	159
CCR5	H1, H4	160
α_{1b}AR	H1, H4, H7	161
m3 muscarinic	E2	163
X-ray crystallography		
Rhodopsin	H1, H8	164
opsin	H1, H8	11
β_2AR	H8	15
CXCR4	I2, H4, H5, H6[c]	29

Table 5.1 GPCR portions involved in dimer/oligomer interfaces—cont'd

Receptor	Interface	Source
μ-OR	H1, H2, H5, H6, H8	36
κ-OR	H1, H2, H8	35
Sequence-based predictions		
δ–μ ORs	$(H4, H5, H6)_{\delta}$–$H1_{\mu}$	167
δ-ORs	H4, H5	168
κ-OR	H5	168
μ-OR	H1	168
CCR5	H1, H4	160
Amine GPCRs	H4, H5, H6	169
Rhodopsin	H4, H5, I2	170
D_2R	H4	170
β_2AR	H6	170
m3 muscarinic	H3, H2, H4, I3	170
Structure-based predictions		
LHR	H4	171
NTS1	H1, H2, H4	172
TP	H1, H8	156
$A_{2A}R$	H1, H2, H4	173
D_2R–$A_{2A}R$	$(H5–H6)_{D2R}$–$(H4, Ctail)_{A2AR}$ or I2, H4, H5 from both receptors	150
D_2R–$A_{2A}R$	$H1_{D2R}$–$H4, H5_{A2AR}$ or I2, H4, H5 from both receptors	174
δ–μ ORs	$(H1, H5)_{\delta}$–$(H1, H7)_{\mu}$ or $(H4, H5)_{\delta}$–$(H6, H7)_{\mu}$	175
$5\text{-}HT_{1A}$–mGluR2	I2, H4, H5 from both receptors	176
A_3R	H1, H2, H4, H5	177
$5\text{-}HT_4$	H2, H3, H4	178

[a]Receptor portions, which contribute the most to the interface.
[b]Bibliographic source.
[c]Contacts between H5 and H6 from one protomer and H1 from the other protomer are additionally present in the lower resolution trimer encoded as 3OE8.

3. COMPUTATIONAL MODELING OF GPCR DIMERIZATION/OLIGOMERIZATION: SEQUENCE-BASED METHODS

Sequence-based and docking-based approaches investigated dimerization in different GPCRs, highlighting the involvement of different helices in the intermonomer interfaces (reviewed here below and in Refs. 180–184). Gouldson and coworkers have been pioneers in studying GPCR dimerization by using computational modeling and bioinformatics tools (reviewed also in Refs. 181,183,185,186).[187,188] They initially proposed a mechanism of receptor activation involving domain swapping, essentially supported by the results by Maggio and coworkers on chimeric M_3-muscarinic and α_2-ARs.[57] The authors employed sequence analyses like the correlated mutation (CM)[189] and evolutionary trace (ET)[190,191] methods[110,188] to infer the receptor portions involved in dimerization/oligomerization. The CM and ET methods allowed them to, respectively, examine the occurrence of both CMs and class-conserved residues. The CM method has been shown to provide information about interdomain contacts.[189] The correlation is interpreted as a result of the tendency of positions in proteins to mutate in a coordinated manner if the interface has to be preserved for structural or functional reasons. Thus, sequence changes occurring during evolution at the interface of dimerization of a given monomer A would be compensated by changes in the interacting monomer B, to preserve the interaction interface. The ET method is another approach to determining functional sites for a protein, given its 3D structure and a multiple sequence alignment.[190,191] It bears some similarities to CM analysis, as the ET residues may also be correlated, but it has the advantage that conserved residues are also included in the analysis.[190,191] The basic assumptions of the ET method are as follows: (a) that within a multiple sequence alignment the protein family retains its fold, on the basis of the idea that proteins that have evolved from a common ancestor will show similar backbone structure;[192] (b) that the location of the functional sites is conserved; (c) that these sites have distinctly lower mutation rates because of the evolutionary pressures of residues defining the functional sites; and (d) that this lower mutation rate is punctuated by mutation events that cause divergence.[190] For the ARs, CM analysis on 50 aligned sequences showed that the CMs do accumulate at the H5–H6 interface.[188] The remaining external correlated residues on H1, H2, and H7 were thought to be involved in the formation of a H1–H7 dimer intermediate or in the formation of higher order oligomers.[188] The ET method was applied to over 700 aligned

GPCR sequences.[110] The method predicted the occurrence of functionally important clusters of residues on the external faces of H5 and H6 for each family or subfamily of receptors; similar clusters were observed on H2 and H3. The probability that these clusters are not random was determined using Monte Carlo (MC) techniques. The cluster on H5 and H6 is consistent with both H5–H6 contact- and H5–H6 domain-swapped dimer formation. The observation of functionally important clusters of residues on H2 and H3 raised some possible interpretations, including heterodimerization and oligomerization.[110] In a successive report, being aware of some experimental evidence against domain swapping as a general mechanism of receptor dimerization,[193,194] and conscious that their computational approaches were unable to distinguish between contact- and domain-swapped dimers, the authors proposed that domain-swapped and contact dimers are essentially equivalent in their ability to signal, and this could underlie any failure to observe domain swapping. The role of domain swapping in GPCR dimerization still remains an open question, as it is apparently supported also by recent *in vitro* evidences,[195] although it remains inconsistent with oligomerization. The same team, more recently, combined the ET method with the entropy method to improve predictions.[196] They, indeed, recognized that the CM analysis, while being mathematically well defined, had the limit to predict relatively few amino acids, inconsistent with the expected extensions of protein–protein interfaces.[196] On the other hand, the ET method, although able to predict protein–protein contact areas, was, however, subjective. While earlier work tended to favor H6 as the most likely dimerization interface, the novel analyses favored H4.[196]

The CM analysis carried out by other researchers predicted that CCR5 homodimerization would involve H1–H4 contacts.[197]

A modification of the CM approach, the so-called subtractive correlated mutation method, was used to predict homo- and heterodimer interfaces in the opioid subfamily of GPCRs.[167,168] Application of the method to δ- and μ-ORs showed that most of the correlated residues of the δ-OR are located on the outer (lipid facing) surface of H4, H5, and H6, whereas, in the μ-OR, H1 is the helix that is likely to be involved in the heterodimerization with δ-OR. Since H4 and H6 cannot participate simultaneously in the intermonomer interface, these results were thought consistent with higher order oligomers. The same study applied to homodimerization of the δ-, κ-, and μ-ORs suggested that (a) H4 and/or H5 would participate in the formation of the δ homodimers, (b) H5 would participate in the formation of κ homodimers, and (c) H1 would participate in homodimerization of the μ subtype.[168]

The combination of CMA and automatic methods, such as the level entropy and the sequence space automatization methods, aimed at detecting amino acid positions that could have some functional significance for the whole family and, at the same time, are specific for each subfamily (i.e., tree-determinant positions),[198] was used to predict the interface in CCR5 homodimers, finding contacts between H1 and H4.[160]

A hidden-site class model of evolution, which employs different substitution matrixes to represent substitutions in different parts of the protein, was used to predict possible dimerization interfaces in aminergic GPCRs.[169] The approach predicted the involvement of H5 and H6 in most aminergic subfamilies and H4 and H5 in the muscarinic and opsin subfamilies.[169]

Predictions of oligomerization interfaces in a number of family A GPCRs belonging to the opsin, dopamine, adrenergic, and muscarinic acetylcholine subfamilies were also done by a multistep method made of two convergent sets of *in silico* experiments, one sequence-based and the other structure-based.[199] The sequence-based set of experiments consisted of multiple sequence alignments and search for conservation patterns. The structure-based set was achieved by reducing the 3D coordinates of the rhodopsin structure to a bidimensional plane and individuating the residues on the molecular surface of the rhodopsin monomer, in particular, the lipid-facing ones. The integration of the two sets of experiments led to 2D maps of conserved lipid-exposed residue clusters, which were interpreted as potentially involved in the dimerization interface. The predicted interfaces differed among subtypes. The following domains were predicted as implicated in homodimerization of selected receptor subtypes: (a) H4, H5, and I2, for rhodopsin; (b) H4, for the D_2R; (c) H6, for the β_2AR; and (d) amino acids from the cytosolic extension of H3 as well as from I2, H4, and I3, for the m_3-muscarinic receptor.[199] The prediction method by Nemoto and Toh is now available through the G protein-coupled receptor interface partners (GRIP) server (http://grip.cbrc.jp/GRIP/index.html).[170]

Collectively, the different sequence-based methods found consensus in predicting H4, H5, and H6 as the portions most likely involved in GPCR oligomerization (reviewed also in Refs. 145,200) (Table 5.1). Despite this general consensus, it is clear that the mean weakness of such methods resides in the fact that explicit interatomic interactions are completely neglected, thus lowering the resolution level of predictions.[196]

4. COMPUTATIONAL MODELING OF GPCR DIMERIZATION/OLIGOMERIZATION: STRUCTURE-BASED PREDICTIONS

So far, only very few predictions of the architecture of GPCRs dimers/oligomers relied on docking sampling and a significant part of such predictions were obtained by a computational protocol developed in our laboratory.[145,150,156,171–174,201] Such a protocol was based on the integration of protein-protein docking by the ZDOCK software[202] and an analysis software, FiPD, which helps minimizing false positives (visit "Software" at http://www.csbl.unimore.it).[147] The following two paragraphs describe the approach, hereafter defined as FiPD-based approach/method, and the latest achievements thereof.

The FiPD-based approach was employed also by another group to predict likely heterodimers of the δ- and μ-ORs.[175] The most likely heterodimer interface was predicted to be formed by H4, H5 of δ and H1, H7 of μ, with emphasis on the H4–H1 contacts. Another likely interface resulted to be contributed by H4, H5 of δ and H6, H7 of μ, with emphasis on the H4–H6 contacts.[175]

Rosetta docking was used as well for quaternary structure predictions of GPCRs.[176] In this respect, selection of the most reliable heterodimer between a β_2AR-based model of the 5-HT_{2A} serotonin receptor and a rhodopsin-based model of the class C metabotropic glutamate receptor 2 (mGluR2) was done upon visual inspection and structural similarity (in terms of RMSD) to the semiempirical model of rhodopsin (1N3M).[135]

The fit-centroids-normal method was employed to predict more likely homodimers of a rhodopsin-based model of the A_3 adenosine receptor.[177] In this respect, an initial dimeric model was achieved by superimposing each protomer with the semiempirical oligomeric model of rhodopsin. This initial dimer served to generate various TM contact dimers, (H1–H2, H2–H3, H2–H4, H3–H4, H4–H5, H5–H6, H6–H7, and H7–H1) by means of the fit-centroids-normal method. Following MD simulations, the most energetically favorable contact models were identified, which were the H4–H5 and H1–H2 dimers, the former resulting as the most favored according to the extension of the contact surface.

A number of GPCR dimers were predicted by manual docking.[203,204] In more detail, a tetramer of the V_2R was manually built based upon information from *in vitro* experiments and then subjected to MD simulations in explicit membrane/water. The study was aimed at highlighting residues responsible for the oligomer stability.[203] Manual docking was employed as well to build a putative homodimer of the δ-OR, characterized by an H4–H4 interface.[204]

The computational models of GPCR dimerization are being used also to assist the design of bivalent ligands.[174,205] In this respect, Russo and coworkers provided direct evidence for the functional interaction of both pharmacophores of bivalent ligands with the 5-HT_4 receptor.[178,205] To estimate the optimal length of the spacer, the authors docked the bivalent ligands into a computational model of 5-HT4 receptor homodimer, built by means of the GRAMM software.[178] The predicted interface in the homodimer involved H2, H3, and H4.

4.1. The FiPD-based approach to predict the quaternary structures of membrane proteins

We have proposed a computational strategy to predict the quaternary structure of TM α-helical proteins.[147] The approach consists of rigid-body docking using a version of the ZDOCK program devoid of desolvation as a component of the docking score.[202] It does not employ symmetry constraints either for improving sampling or in the filtering step. Furthermore, there are no size limitations for the systems under study, which are not limited to the TM regions but include the loop regions as well. The only requirement with this approach is the structural model of the protomer and the knowledge of a set of Cα-atoms, which lie at the two lipid/water interfaces, defining two parallel planes. In the target monomer, these two planes must be parallel to the *xy* plane and, hence, perpendicular to the *z*-axis to make the membrane topology filter work properly. This is due to the fact that ZDOCK expresses its docking solutions in terms of an *x*, *y*,*z*-translation and a $R_zR_xR_z$-rotation of the probe. If both target and probe are properly oriented, that is, with the interface Cα-atoms lying in the *xy* plane, the translation along the *z*-axis can be considered as an offset out of the membrane, and the R_x component of the rotation as a deviation from the original orientation in the membrane. Wrong membrane topology of the docked proteins generates errors in the filtering stage. If these planes are parallel to the *xy* plane and, hence, perpendicular to the *z*-axis, the orientation of the monomer is considered good and no reorientation is needed. In contrast, if such planes are not parallel to the *xy* plane, the monomer needs a reorientation. The latter can be done through the FiPD software (visit "Software" at http://www.csbl.unimore.it). Membrane topology predictors from a single predictor such as PRODIV-TMHMM_0.91 or consensus prediction from more than one source are worth using to individuate the interface Cα-atoms lying in the *xy* plane. For the membrane topology filter to work properly, the two docked structural models must have the appropriate orientation with respect to the putative membrane.

The structural models subjected to docking simulations can be either crystal structures or minimized averages from MD simulations carried out on selected models from comparative modeling.

In the case of homodimerization, two identical copies of the structural receptor model are docked together, that is, one monomer is used as a fixed protein (target) and the other as a mobile protein (probe). For predicting heterodimers, the structural model of receptor A is taken as a target, whereas the structural model of receptor B is taken as a probe and/or vice versa. A rotational sampling interval of 6° is recommended, (i.e., dense sampling), and the best 4000 solutions are retained and ranked according to the ZDOCK score. These solutions are then filtered by using the FiPD software,[147] the "membrane topology" filter, which discards all the solutions that violate the membrane topology requirements. The membrane topology filter indeed discards all the solutions characterized by a deviation angle from the original *z*-axis, that is, tilt angle, and a displacement of the geometrical center along the *z*-axis, that is, *z*-offset, above defined threshold values. In the case of GPCRs, tilt angle and *z*-offset thresholds of 0.4 rad and 6.0 Å, respectively, are worth using. Selection of quite permissive filtering cutoffs is necessary to avoid removal of reliable solutions as a consequence of possible small input deviations from the proper membrane topology. Following such filtering, discarded solutions generally constitute more than 94% of the solutions selected according to the docking score.

The filtered solutions from each run are merged with the target protein, leading to an equivalent number of dimers that are clustered using a Cα-RMSD threshold generally equal to 3.0 Å for each pair of superimposed dimers. All the amino acid residues in the dimer are included in Cα-RMSD calculations. Cluster analysis is based on a quality threshold-like (QT-like) clustering algorithm[206] implemented in both the FiPD and Wordom software.[147,207] The QT-like algorithm first calculates the Cα-RMSD for each superimposed pair of dimers/oligomers, and then it computes the number of neighbors for each dimer/oligomer according to the chosen Cα-RMSD threshold. The dimer/oligomer with the highest number of neighbors is considered as the center of the first cluster. All the neighbors of this configuration are removed from the ensemble of configurations to be counted only once. The center of the second cluster is then determined in the same way as for the first cluster, and this procedure is repeated until each structure is assigned to a cluster. Since the filtering cutoffs of the membrane topology parameters are intentionally quite permissive, inspection of the cluster centers (i.e., the solutions with the highest number of neighbors in each cluster) often serves as a final filter to discard remaining false positives, leading to a reduction of the reliable solutions to about 1% of the total 4000 solutions. The best scored docking solution/s from the most populated and reliable cluster/s is/are finally chosen.

Predictions of higher order oligomers can be done either by selecting solutions from reliable clusters (i.e., which can coexist in a multimeric complex) or by running a new docking by using the predicted dimer or, successively, the predicted higher order oligomer as a target and the protomer as a probe. Filtering and cluster analysis then follow the same procedure as that employed for predicting dimers.

The strength of the approach stands in its independence from the size of the system, symmetry information, and extension of the water-soluble domains.[147] Benchmarks of the approach were carried out on (a) the tetrameric potassium channel (Kch, 384 amino acids), (b) the pentameric MscL (540 amino acids) and eptameric MscS (1771 amino acids) mechanosensitive channels, and (c) the trimeric bacteriorhodopsin (698 amino acids). In all test cases the method led to native-like quaternary structures, that is, with C_α-RMSDs lower than 2.5 Å from the native oligomer.[147] The prediction protocol was also successfully probed on TM β-barrels like the OMPLA protein.[208]

The availability of the crystal structures of the CXCR4 (PDB code: 3ODU)[29] and the κ-OR (PDB code: 4DJH)[35] dimers let us validate the approach on GPCRs as well (Fanelli *et al.*, manuscript in preparation). In both cases, the protomers from each dimeric complex were docked either against themselves (i.e., A vs. A and B vs. B) or between each other (i.e., A vs. B).

For both receptors, the best filtered solution according to the docking score fell in one of the three most populated clusters, which were almost equipopulated. Such solution turned out to be a native-like one (Fig. 5.1), thus strengthening the accuracy of the prediction method.

4.2. Application of the FiPD-based approach to selected GPCRs

The FiPD-based protein–protein docking protocol has been employed for quaternary structure predictions of a number of GPCRs, including members of the amine,[174] hormone,[171] peptide,[172] prostanoid,[156] and purine,[173,174] subfamilies.

In this framework, quaternary structure predictions on a computational model of the LHR emphasized the role of H4, H5, and H6, with prominence to H4, in mediating intermonomer interactions.[171]

In a study on a computational model of the NTS1 receptor, the FiPD-based approach was combined with a protocol developed for estimating mutational effects on the binding free energy of Glycophorin A (GpA) homodimers[209] to help selection of the most reliable dimer architecture.[172] This was made possible by the availability of *in vitro*-determined intermonomer binding affinities for the NTS1, which represented a unique case within the rhodopsin family.[108] Binding affinity estimations were done following an adjustment of the original correlative model between ZD-score, which is essentially a size/shape descriptor, and free energy of dimerization of 21 GpA mutants (i.e., $\Delta G° = 6.6(\pm 1.3) - 33.2(\pm 3.1)ZD_w$, $N = 22$, $\sigma = 0.55$, $R = 0.92$, $p < 0.0001$, where ZD_w is the ZDOCK score normalized by the interface GpA amino acids). This adjustment consisted in dividing the ZD-score by the number of native interface residues and served to make the correlative model almost independent of the extension of the interface and, therefore, transferable to membrane proteins that, like GPCRs, hold different sizes compared to GpA. The FiPD-based approach finally predicted two possible dimer architecture: (a) DIM1 characterized by contacts between H1 and both H2 and H4 (and vice versa) as well as between E1 from both dimers; and (b) DIM2, characterized by H4–H4 and E2–E2 contacts. The equation reported above was employed to estimate the dimerization free energy of the two predicted dimers of NTS1, that is, DIM1 and DIM2. Prior to predictions, the ZDOCK scores concerning DIM1 and DIM2 (i.e., 17.48 and 15.96, respectively) were divided by the respective number of interface amino acids (i.e., 33 and 32, respectively). The estimated $\Delta G°$ values for DIM1 and DIM2 were −11.0 and −10.0 kcal/mol, respectively, the former being in slightly better agreement with *in vitro* determinations, which found a monomer–monomer K_d comprised between 2 and 20 nM (i.e., -10.5 kcal/mol $< \Delta G° < -12$ kcal/mol (personal communication of Dr. Reinhard Grisshammer)).[108] In the desirable future cases, in which *in vitro* intermonomer binding affinities will be available for other GPCRs, the correlative model will work as an additional criterion for helping selection of the most likely dimers.

Another application of the FiPD-based protocol included the design of mutants of the TP receptor α and β isoforms unable to talk between each other.[156] The study indicated that likely TP dimers are characterized by contacts between hydrophobic residues in H1 from both protomers, and, to a lower extent, by H1–H2,E1, and H8–H8 contacts.[156]

Docking-based predictions of D_2R–$A_{2A}R$ heterodimers were carried out by means of two different docking programs. The study predicted two alternative interfaces, one involving H5, I3, and H6 from D_2R and H3 and H4 from the $A_{2A}R$ and the other contributed by H3, I2, H4, and H5 from both receptors.[150] In that study, the structures of both receptors had been achieved by comparative modeling, by using the crystal structure of rhodopsin as a template. In a more recent study, D_2R–$A_{2A}R$ heterodimers were predicted by means the FiPD-based approach.[174] The D_2R was modeled by using the crystal structure of the β_2AR as a template, whereas the crystallographic structure encoded as 3EML was used for the $A_{2A}R$ protomer. The FiPD-based approach predicted two different heterodimers, one characterized by H1, H2, and H4 contacts from both receptors and the other by contacts between H1 from D_2R and both H4 and H5 from $A_{2A}R$. The H3–H5 dimer from earlier predictions and the two novel predicted architectures were used to probe the length of the shortest heterobivalent ligand in a list.[174] The docking study predicted that even when the shortest bivalent ligand adopts a partially extended conformation, it can be enough to allow the two pharmacophoric parts to bind to the respective docking sites in the D_2R–$A_{2A}R$ heterodimer.

The crystal structure of the $A_{2A}R$ bound to the ZM241385 (ZMA) antagonist was used to predict likely homodimers of the $A_{2A}R$.[173] The docking poses selected as the most reliable, that is, DIM1, DIM2, and DIM3 were symmetrical (Fig. 5.2). DIM1 is essentially characterized by contacts between (a) the extracellular ends of H1 and H3, (b) the cytosolic end of H1, and (c) the extracellular half of H2. DIM2 is essentially characterized by contacts between (a) H1 and H4, (b) the extracellular ends of H2 and H3, and (c) the extracellular half of H2. Finally, DIM3 is essentially characterized by contacts between (a) the cytosolic half of H6 from monomer A and the cytosolic halves of both H6 and H7 from monomer B; (b) the extracellular halves of H6 and H7; and (c) H8 and both IL3 and the cytosolic extension of H6. Based upon the regions that contribute the most to the three different interfaces, DIM1, DIM2, and DIM3 were respectively, indicated as H1–H1/H2–H2, H1–H4/H2–H2, and H6–H6/H6–H7 dimers.

Incidentally, the three predicted dimers were subjected to MD simulations and then to protein structure network (PSN) analyses in order to infer the effects of dimerization on the structural communication features of the protomer (see Section 5).[173]

In summary, quaternary structure predictions by the FiPD-based approach on computational models of the LHR, NTS1, TP, $A_{2A}R$, and

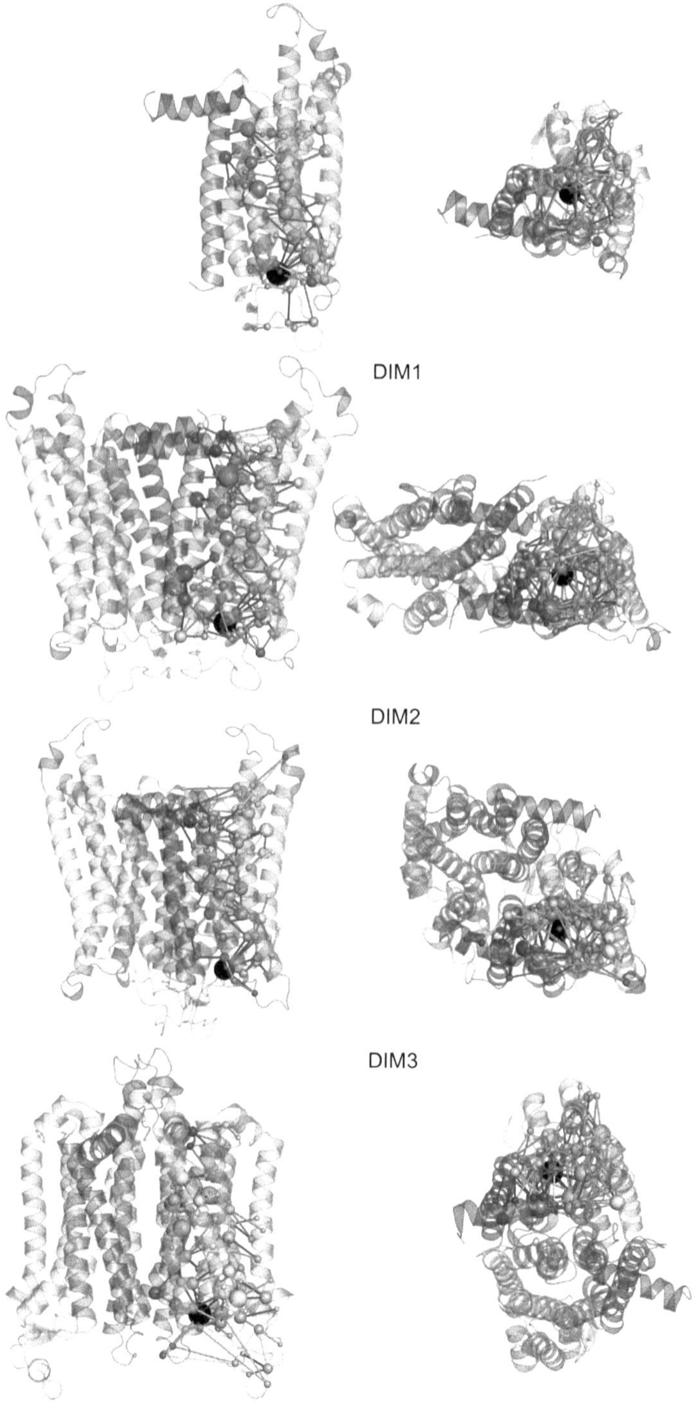

Figure 5.2 Continued

D_2R emphasize the role of H1 and H4 in mediating dimerization. This is consistent with ever-increasing evidence from structure determinations and *in vitro* experiments (see above and in Ref. 145).

5. GRAPH THEORY-BASED INVESTIGATION OF THE STRUCTURAL COMMUNICATION IN GPCR DIMERS

GPCRs are allosteric proteins whose functioning fundamentals are the communication between the two poles of the helix bundle, that is, the extracellular side receives and transfers extracellular signals to the intracellular side deputed to recognize and activate the G protein transducer. Functionally different ligands like agonists, inverse agonists, and antagonists as well as different homo/heterooligomeric states, that is, from dimer to higher order oligomer, are likely to exert differential impacts on such communication.

The representation of GPCR structures as networks of interacting amino acids can be a meaningful way to decipher the impact of ligand and dimerization/oligomerization on the molecular communication, which is intrinsic to the protein fold and likely serves to protein stability and function. The crystallographic structure of the $A_{2A}R$ simulated in its empty and antagonist-bound monomeric and dimeric forms was investigated in terms of information processing networks, in which privileged residues are deputed to receive and propagate information. This was achieved by means of the graph theory-based PSN analysis implemented in the Wordom software.[207] In detail, it composes of the PSN and PSN-PATH modules, which compute

Figure 5.2 Average minimized structures concerning the $A_{2A}R$ in a monomeric ZMA-bound form as well as in three different dimeric forms. Pictures refer to results published elsewhere.[173] Both a side view parallel to the membrane surface (left) and a top view from the intracellular side in a direction perpendicular to the membrane surface (right) are shown. The ZMA ligand is shown in black sticks. The images show the effects of dimerization on the protein structure graph (PSG) characterizing the receptor protomer bound to the ZMA antagonist. In this respect, the spheres centered on the Cα-atoms concern node pairs in the PSG, which are linked in more than 50% of frames in a 10 ns trajectory. Nodes are colored according to their location (i.e., according to the different receptor regions). In this respect, helices 1, 2, 3, 4, 5, 6, 7, and 8 are, respectively, blue, orange, green, pink, yellow, cyan, violet, and red. I1 and E1 are lime, I2 and E2 are gray, and I3 and E3 are magenta. The diameter of the sphere is proportional to the number of links made by the considered node, with the lowest value corresponding to one link. Link color refers to the frequency of the link. In this respect, cyan, green, yellow, orange, and red correspond, respectively, to the following frequency ranges: $50 \leq F \leq 60\%$, $60 \leq F \leq 70\%$, $70 \leq F \leq 80\%$, $80 \leq F \leq 90\%$, and $F \geq 90\%$. (See Color Insert.)

network components (e.g., nodes, hubs, links, shortest communication pathways, etc.) on MD trajectories. With this approach, the dynamics of the system is taken into account in terms of occurrence of network components and correlated motions.[173,210,211]

The dynamic network of interactions characterizing the $A_{2A}R$–ZMA complex in its monomeric state (MONO) was compared with that of the same complex in three different dimeric forms (DIM1, DIM2, and DIM3, see the paragraph above), as well as with that of the apo-monomer ($MONO_{APO}$).

Taken as a whole, the results of this investigation emphasized the role of highly conserved amino acids in H1, H2, H6, and H7 in maintaining the structure network of the $A_{2A}R$ through their persistent hub (i.e., nodes involved in at least four links) behavior, in line with a study on LHR and rhodopsin.[210,212] Consistently, such amino acids participate in the cytosolic halves of the most frequent paths, suggesting a fundamental role in the information flow between the two poles of the helix bundle. The most frequent communication paths do not include the E/DRY motif in the cytosolic extension of H3, an important receptor recognition portion for the G protein.[12,213] Lack of communication between antagonist and G protein binding sites is likely a feature of the functional state of the receptor, that is, inactive, and does not depend on whether the simulated ligand–receptor complex is in its monomeric or dimeric forms. The inactive-state features of the simulated complexes are strengthened by the observation that they all share interactions between R3.50 and both D3.49 and E6.30, similar to the crystal structure of dark rhodopsin.[9] Based on such consensus from different MD simulations, the lack of the R3.50–E6.30 interaction in the crystal structure is likely due to an artifact, like, for example, the presence of T4-lysozyme, whereas the interhelical salt bridge in the DRY is expected to be a feature of the inactive state of the $A_{2A}R$, similar to rhodopsin and other homologous GPCRs.[9,214] It is worth noting that here the amino acids have been labeled according to a position-linked nomenclature that allows comparisons within the rhodopsin family GPCRs.[215]

Collectively, the composition of the structure networks and of the most frequent paths highlights distinctive structural/functional roles for the DRY motif and the highly conserved amino acids in the cytosolic ends of H1, H2, and H7.

Moreover, the H1–H1/H2–H2 and H1–H4/H2–H2 symmetric dimers retain the typology of the most frequent communication paths seen in the complexed form of the monomer, but increasing significantly the number of hub-involving links and paths compared to the MONO form. A remarkable result is that $A_{2A}R$ dimerization affects the communication networks intrinsic to the receptor structure in a way dependent on the dimer architecture (Fig. 5.2). In this respect, the H1–H4/H2–H2 architecture shows the most diffuse communication among all the ZMA-complexed forms. In contrast, the H6–H6/H6–H7 dimer is suggested to impair ligand-mediated communication, since it is characterized by a significant reduction in the total number of paths compared to the MONO form and by the complete loss of ZMA-involving high frequency paths (i.e., found in $\geq$30% of the MD trajectory frames). The impairing effect of DIM3 architecture is expected to be even more marked for the agonist-bound forms, where it would hinder H6 movements required for receptor activation.[11]

In conclusion, the study suggests that certain dimeric architectures retain the most recurrent communication paths with respect to the monomeric antagonist-bound form but enhancing path numbers and frequencies, whereas some others impair ligand-mediated communication. Ligand binding affects the network as well. Overall, the communication network that pertains to the functional dynamics of a GPCR is expected to be influenced by ligand functionality, oligomeric order, and architecture of the supramolecular assembly.

6. CONCLUSIONS

In spite of the ever-increasing evidence that GPCRs form dimers/oligomers, the biological role(s) of homologous and heterologous receptor aggregation is/are, however, far from being clarified. Likewise, knowledge about the most likely architectures of GPCR dimers is still ill-defined. System's dependent variability in dimer architecture emerges from both *in vitro* experiments and sequence- or structure-based predictions. Collectively, consensus emerges on H1 and H4 being involved in GPCR dimerization/oligomerization.

The FiPD-based approach proved effective in predicting the quaternary structure of membrane proteins including GPCRs. MD simulations followed by PSN analysis on the FiPD-predicted GPCR dimers can be a valuable tool to infer the effects of dimerization on the structural communication features of a receptor when bound to functionally different ligands. Ultimately, these computational experiments may help unveiling the structural determinants and functional meaning of GPCR aggregation in supramolecular working units.

While protein–protein docking simulations are essential to gain insight, at the atomic level, into the architecture of the intermonomer interfaces, a mesoscopic MC simulation approach, such as that proposed by Woolf and Linderman, might create a more direct link with *in vitro* experiments on GPCR oligomerization and localization in selected membrane compartments.[216,217] By incorporating information from *in vitro* experiments, the computational approach by Woolf and Linderman has begun to address the problem of how the kinetics of protein–protein binding affects the overall organization of the membrane and how dimerization affects the global protein organization.[216,217] Preliminary results by this approach suggest that changes in dimerization interactions affect cellular physiology. The possibility of predicting the ligand's ability to induce homo- or heterodimerization is another intriguing potential of this approach in drug discovery. More recently, using a combination of stochastic (MC) and deterministic modeling, Fallahi-Sichani and Linderman proposed a novel mechanism for lipid raft partitioning of GPCRs based on reversible dimerization of receptors and demonstrated that such localization can affect GPCR signaling.[218] Briefly, the kinetic MC model served to determine the effect of a ligand-induced change in the dimerization status of receptors on localization within low diffusivity microdomains (lipid rafts) at the cell surface, and to estimate the timescale and level of receptor clustering and declustering. On the other hand, the deterministic ordinary differential equation model based on the collision coupling model served to study the effect of receptor localization within lipid rafts on downstream signaling events. Linking the two approaches allowed the authors to study and analyze G protein activation while incorporating the effects of receptor organization. Based on the results of simulations, the authors proposed a mechanism by which dimerization-inducing or dimerization-inhibiting characteristics of ligands can influence GPCR signaling by controlling receptor organization on the cell membrane.[218]

The integration between atomistic and mesoscopic simulations is expected to be a promising tool to unravel functioning mechanisms that involve intricate protein networks.

ACKNOWLEDGMENTS

This study was supported by an Airc-Italy grant [IG10740] and a Telethon-Italy grant [GGP11210/S00068TELC].

Graphics were drawn by means of Pymol (http://www.pymol.org/).

REFERENCES

1. Bockaert J, Pin JP. Molecular tinkering of G protein-coupled receptors: an evolutionary success. *EMBO J* 1999;**18**:1723–9.
2. Gether U. Uncovering molecular mechanisms involved in activation of G protein-coupled receptors. *Endocr Rev* 2000;**21**:90–113.
3. Lefkowitz RJ. The superfamily of heptahelical receptors. *Nat Cell Biol* 2000;**2**:E133–E136.
4. Pierce KL, Premont RT, Lefkowitz RJ. Seven-transmembrane receptors. *Nat Rev Mol Cell Biol* 2002;**3**:639–50.
5. Kristiansen K. Molecular mechanisms of ligand binding, signaling, and regulation within the superfamily of G-protein-coupled receptors: molecular modeling and mutagenesis approaches to receptor structure and function. *Pharmacol Ther* 2004;**103**:21–80.
6. Dorsam RT, Gutkind JS. G-protein-coupled receptors and cancer. *Nat Rev Cancer* 2007;**7**:79–94.
7. Lappano R, Maggiolini M. G protein-coupled receptors: novel targets for drug discovery in cancer. *Nat Rev Drug Discov* 2011;**10**:47–60.
8. Thathiah A, De Strooper B. The role of G protein-coupled receptors in the pathology of Alzheimer's disease. *Nat Rev Neurosci* 2011;**12**:73–87.
9. Palczewski K. G protein-coupled receptor rhodopsin. *Annu Rev Biochem* 2006; **75**:743–67.
10. Choe HW, Kim YJ, Park JH, Morizumi T, Pai EF, Krauss N, et al. Crystal structure of metarhodopsin II. *Nature* 2011;**471**:651–5.
11. Park JH, Scheerer P, Hofmann KP, Choe HW, Ernst OP. Crystal structure of the ligand-free G-protein-coupled receptor opsin. *Nature* 2008;**454**:183–7.
12. Scheerer P, Park JH, Hildebrand PW, Kim YJ, Krauss N, Choe HW, et al. Crystal structure of opsin in its G-protein-interacting conformation. *Nature* 2008;**455**:497–502.
13. Standfuss J, Edwards PC, D'Antona A, Fransen M, Xie G, Oprian DD, et al. The structural basis of agonist-induced activation in constitutively active rhodopsin. *Nature* 2011;**471**:656–60.
14. Rasmussen SG, Choi HJ, Rosenbaum DM, Kobilka TS, Thian FS, Edwards PC, et al. Crystal structure of the human beta(2) adrenergic G-protein-coupled receptor. *Nature* 2007;**450**:383–7.
15. Cherezov V, Rosenbaum DM, Hanson MA, Rasmussen SG, Thian FS, Kobilka TS, et al. High-resolution crystal structure of an engineered human beta2-adrenergic G protein-coupled receptor. *Science* 2007;**318**:1258–65.
16. Warne T, Serrano-Vega MJ, Baker JG, Moukhametzianov R, Edwards PC, Henderson R, et al. Structure of a beta1-adrenergic G-protein-coupled receptor. *Nature* 2008;**454**:486–91.
17. Hanson MA, Cherezov V, Griffith MT, Roth CB, Jaakola VP, Chien EY, et al. A specific cholesterol binding site is established by the 2.8 A structure of the human beta2-adrenergic receptor. *Structure* 2008;**16**:897–905.
18. Wacker D, Fenalti G, Brown MA, Katritch V, Abagyan R, Cherezov V, et al. Conserved binding mode of human beta(2) adrenergic receptor inverse agonists and antagonist revealed by X-ray crystallography. *J Am Chem Soc* 2010;**132**:11443–5.
19. Rasmussen SG, Choi HJ, Fung JJ, Pardon E, Casarosa P, Chae PS, et al. Structure of a nanobody-stabilized active state of the beta(2) adrenoceptor. *Nature* 2011;**469**:175–80.
20. Rosenbaum DM, Zhang C, Lyons JA, Holl R, Aragao D, Arlow DH, et al. Structure and function of an irreversible agonist-beta(2) adrenoceptor complex. *Nature* 2011;**469**:236–40.
21. Warne T, Moukhametzianov R, Baker JG, Nehme R, Edwards PC, Leslie AG, et al. The structural basis for agonist and partial agonist action on a beta(1)-adrenergic receptor. *Nature* 2011;**469**:241–4.

22. Rasmussen SG, Devree BT, Zou Y, Kruse AC, Chung KY, Kobilka TS, et al. Crystal structure of the beta(2) adrenergic receptor-Gs protein complex. *Nature* 2011;**469**: 175–81.
23. Jaakola VP, Griffith MT, Hanson MA, Cherezov V, Chien EY, Lane JR, et al. The 2.6 angstrom crystal structure of a human A2A adenosine receptor bound to an antagonist. *Science* 2008;**322**:1211–7.
24. Lebon G, Warne T, Edwards PC, Bennett K, Langmead CJ, Leslie AG, et al. Agonist-bound adenosine A2A receptor structures reveal common features of GPCR activation. *Nature* 2011;**474**:521–5.
25. Xu F, Wu H, Katritch V, Han GW, Jacobson KA, Gao ZG, et al. Structure of an agonist-bound human A2A adenosine receptor. *Science* 2011;**332**:322–7.
26. Dore AS, Robertson N, Errey JC, Ng I, Hollenstein K, Tehan B, et al. Structure of the adenosine A(2A) receptor in complex with ZM241385 and the xanthines XAC and caffeine. *Structure* 2011;**19**:1283–93.
27. Hino T, Arakawa T, Iwanari H, Yurugi-Kobayashi T, Ikeda-Suno C, Nakada-Nakura Y, et al. G-protein-coupled receptor inactivation by an allosteric inverse-agonist antibody. *Nature* 2012;**482**:237–40.
28. Murakami M, Kouyama T. Crystal structure of squid rhodopsin. *Nature* 2008;**453**: 363–7.
29. Wu B, Chien EY, Mol CD, Fenalti G, Liu W, Katritch V, et al. Structures of the CXCR4 chemokine GPCR with small-molecule and cyclic peptide antagonists. *Science* 2010;**330**:1066–71.
30. Chien EY, Liu W, Zhao Q, Katritch V, Han GW, Hanson MA, et al. Structure of the human dopamine D3 receptor in complex with a D2/D3 selective antagonist. *Science* 2010;**330**:1091–5.
31. Shimamura T, Shiroishi M, Weyand S, Tsujimoto H, Winter G, Katritch V, et al. Structure of the human histamine H1 receptor complex with doxepin. *Nature* 2011;**475**:65–70.
32. Haga K, Kruse AC, Asada H, Yurugi-Kobayashi T, Shiroishi M, Zhang C, et al. Structure of the human M2 muscarinic acetylcholine receptor bound to an antagonist. *Nature* 2012;**482**:547–51.
33. Kruse AC, Hu J, Pan AC, Arlow DH, Rosenbaum DM, Rosemond E, et al. Structure and dynamics of the M3 muscarinic acetylcholine receptor. *Nature* 2012;**482**:552–6.
34. Granier S, Manglik A, Kruse AC, Kobilka TS, Thian FS, Weis WI, et al. Structure of the delta-opioid receptor bound to naltrindole. *Nature* 2012;**485**:400–4.
35. Wu H, Wacker D, Mileni M, Katritch V, Han GW, Vardy E, et al. Structure of the human kappa-opioid receptor in complex with JDTic. *Nature* 2012;**485**:327–32.
36. Manglik A, Kruse AC, Kobilka TS, Thian FS, Mathiesen JM, Sunahara RK, et al. Crystal structure of the micro-opioid receptor bound to a morphinan antagonist. *Nature* 2012;**485**:321–6.
37. Thompson AA, Liu W, Chun E, Katritch V, Wu H, Vardy E, et al. Structure of the nociceptin/orphanin FQ receptor in complex with a peptide mimetic. *Nature* 2012;**485**:395–9.
38. Hanson MA, Roth CB, Jo E, Griffith MT, Scott FL, Reinhart G, et al. Crystal structure of a lipid G protein-coupled receptor. *Science* 2012;**335**:851–5.
39. Oldham WM, Hamm HE. Structural basis of function in heterotrimeric G proteins. *Q Rev Biophys* 2006;**39**:117–66.
40. Bouvier M. Oligomerization of G-protein-coupled transmitter receptors. *Nat Rev Neurosci* 2001;**2**:274–86.
41. Milligan G. Oligomerisation of G-protein-coupled receptors. *J Cell Sci* 2001;**114**: 1265–71.

42. Rios CD, Jordan BA, Gomes I, Devi LA. G-protein-coupled receptor dimerization: modulation of receptor function. *Pharmacol Ther* 2001;**92**:71–87.
43. George SR, O'Dowd BF, Lee SP. G-protein-coupled receptor oligomerization and its potential for drug discovery. *Nat Rev Drug Discov* 2002;**1**:808–20.
44. Agnati LF, Ferre S, Lluis C, Franco R, Fuxe K. Molecular mechanisms and therapeutical implications of intramembrane receptor/receptor interactions among heptahelical receptors with examples from the striatopallidal GABA neurons. *Pharmacol Rev* 2003;**55**:509–50.
45. Franco R, Canals M, Marcellino D, Ferre S, Agnati L, Mallol J, et al. Regulation of heptaspanning-membrane-receptor function by dimerization and clustering. *Trends Biochem Sci* 2003;**28**:238–43.
46. Kroeger KM, Pfleger KD, Eidne KA. G-protein coupled receptor oligomerization in neuroendocrine pathways. *Front Neuroendocrinol* 2003;**24**:254–78.
47. Terrillon S, Bouvier M. Roles of G-protein-coupled receptor dimerization. *EMBO Rep* 2004;**5**:30–4.
48. Park PS, Filipek S, Wells JW, Palczewski K. Oligomerization of G protein-coupled receptors: past, present, and future. *Biochemistry* 2004;**43**:15643–56.
49. Maggio R, Novi F, Scarselli M, Corsini GU. The impact of G-protein-coupled receptor hetero-oligomerization on function and pharmacology. *FEBS J* 2005;**272**: 2939–46.
50. Bulenger S, Marullo S, Bouvier M. Emerging role of homo- and heterodimerization in G-protein-coupled receptor biosynthesis and maturation. *Trends Pharmacol Sci* 2005;**26**: 131–7.
51. Milligan G. G protein-coupled receptor hetero-dimerization: contribution to pharmacology and function. *Br J Pharmacol* 2009;**158**:5–14.
52. Albizu L, Cottet M, Kralikova M, Stoev S, Seyer R, Brabet I, et al. Time-resolved FRET between GPCR ligands reveals oligomers in native tissues. *Nat Chem Biol* 2010;**6**:587–94.
53. Ferre S, Franco R. Oligomerization of G-protein-coupled receptors: a reality. *Curr Opin Pharmacol* 2010;**10**:1–5.
54. Lohse MJ. Dimerization in GPCR mobility and signaling. *Curr Opin Pharmacol* 2010;**10**:53–8.
55. Milligan G. The role of dimerisation in the cellular trafficking of G-protein-coupled receptors. *Curr Opin Pharmacol* 2010;**10**:23–9.
56. Palczewski K. Oligomeric forms of G protein-coupled receptors (GPCRs). *Trends Biochem Sci* 2010;**35**:595–600.
57. Maggio R, Vogel Z, Wess J. Coexpression studies with mutant muscarinic/adrenergic receptors provide evidence for intermolecular "cross-talk" between G-protein-linked receptors. *Proc Natl Acad Sci USA* 1993;**90**:3103–7.
58. Maggio R, Barbier P, Fornai F, Corsini GU. Functional role of the third cytoplasmic loop in muscarinic receptor dimerization. *J Biol Chem* 1996;**271**:31055–60.
59. Rivero-Muller A, Chou YY, Ji I, Lajic S, Hanyaloglu AC, Jonas K, et al. Rescue of defective G protein-coupled receptor function in vivo by intermolecular cooperation. *Proc Natl Acad Sci USA* 2010;**107**:2319–24.
60. Persani L, Calebiro D, Bonomi M. Technology Insight: modern methods to monitor protein-protein interactions reveal functional TSH receptor oligomerization. *Nat Clin Pract Endocrinol Metab* 2007;**3**:180–90.
61. Dalrymple MB, Pfleger KD, Eidne KA. G protein-coupled receptor dimers: functional consequences, disease states and drug targets. *Pharmacol Ther* 2008;**118**:359–71.
62. Wang J, Norcross M. Dimerization of chemokine receptors in living cells: key to receptor function and novel targets for therapy. *Drug Discov Today* 2008;**13**: 625–32.

63. Rives ML, Vol C, Fukazawa Y, Tinel N, Trinquet E, Ayoub MA, et al. Crosstalk between GABAB and mGlu1a receptors reveals new insight into GPCR signal integration. *EMBO J* 2009;**28**:2195–208.
64. Ferre S, Baler R, Bouvier M, Caron MG, Devi LA, Durroux T, et al. Building a new conceptual framework for receptor heteromers. *Nat Chem Biol* 2009;**5**:131–4.
65. Franco R. G-protein-coupled receptor heteromers or how neurons can display differently flavoured patterns in response to the same neurotransmitter. *Br J Pharmacol* 2009; **158**:23–31.
66. Maggio R, Aloisi G, Silvano E, Rossi M, Millan MJ. Heterodimerization of dopamine receptors: new insights into functional and therapeutic significance. *Parkinsonism Relat Disord* 2009;**15**(Suppl. 4):S2–S7.
67. Dorsch S, Klotz KN, Engelhardt S, Lohse MJ, Bunemann M. Analysis of receptor oligomerization by FRAP microscopy. *Nat Methods* 2009;**6**:225–30.
68. Harding PJ, Attrill H, Boehringer J, Ross S, Wadhams GH, Smith E, et al. Constitutive dimerization of the G-protein coupled receptor, neurotensin receptor 1, reconstituted into phospholipid bilayers. *Biophys J* 2009;**96**:964–73.
69. Ambrosio M, Lohse MJ. Microscopy: GPCR dimers moving closer. *Nat Chem Biol* 2010;**6**:570–1.
70. Cottet M, Albizu L, Perkovska S, Jean-Alphonse F, Rahmeh R, Orcel H, et al. Past, present and future of vasopressin and oxytocin receptor oligomers, prototypical GPCR models to study dimerization processes. *Curr Opin Pharmacol* 2010;**10**:59–66.
71. van Rijn RM, Whistler JL, Waldhoer M. Opioid-receptor-heteromer-specific trafficking and pharmacology. *Curr Opin Pharmacol* 2010;**10**:73–9.
72. Marshall FH, Foord SM. Heterodimerization of the GABAB receptor-implications for GPCR signaling and drug discovery. *Adv Pharmacol* 2010;**58**:63–91.
73. Zhang M, Feng X, Guan R, Hebert TE, Segaloff DL. A cell surface inactive mutant of the human lutropin receptor (hLHR) attenuates signaling of wild-type or constitutively active receptors via heterodimerization. *Cell Signal* 2009;**21**:1663–71.
74. Hwang JR, Baek MW, Sim J, Choi HS, Han JM, Kim YL, et al. Intermolecular crosstalk between NTR1 and NTR2 neurotensin receptor promotes intracellular sequestration and functional inhibition of NTR1 receptors. *Biochem Biophys Res Commun* 2010; **391**:1007–13.
75. Chabre M, Deterre P, Antonny B. The apparent cooperativity of some GPCRs does not necessarily imply dimerization. *Trends Pharmacol Sci* 2009;**30**:182–7.
76. Szidonya L, Cserzo M, Hunyady L. Dimerization and oligomerization of G-protein-coupled receptors: debated structures with established and emerging functions. *J Endocrinol* 2008;**196**:435–53.
77. Pin JP, Comps-Agrar L, Maurel D, Monnier C, Rives ML, Trinquet E, et al. G-protein-coupled receptor oligomers: two or more for what? Lessons from mGlu and GABAB receptors. *J Physiol* 2009;**587**:5337–44.
78. White JH, Wise A, Main MJ, Green A, Fraser NJ, Disney GH, et al. Heterodimerization is required for the formation of a functional GABA(B) receptor. *Nature* 1998;**396**:679–82.
79. Kaupmann K, Malitschek B, Schuler V, Heid J, Froestl W, Beck P, et al. GABA(B)-receptor subtypes assemble into functional heteromeric complexes. *Nature* 1998;**396**: 683–7.
80. Marshall FH, Jones KA, Kaupmann K, Bettler B. GABAB receptors—the first 7TM heterodimers. *Trends Pharmacol Sci* 1999;**20**:396–9.
81. Mohler H, Fritschy JM. GABAB receptors make it to the top—as dimers. *Trends Pharmacol Sci* 1999;**20**:87–9.
82. Margeta-Mitrovic M, Jan YN, Jan LY. A trafficking checkpoint controls GABA(B) receptor heterodimerization. *Neuron* 2000;**27**:97–106.

83. Kunishima N, Shimada Y, Tsuji Y, Sato T, Yamamoto M, Kumasaka T, et al. Structural basis of glutamate recognition by a dimeric metabotropic glutamate receptor. *Nature* 2000;**407**:971–7.
84. Zhu X, Wess J. Truncated V2 vasopressin receptors as negative regulators of wild-type V2 receptor function. *Biochemistry* 1998;**37**:15773–84.
85. Morello JP, Salahpour A, Laperriere A, Bernier V, Arthus MF, Lonergan M, et al. Pharmacological chaperones rescue cell-surface expression and function of misfolded V2 vasopressin receptor mutants. *J Clin Invest* 2000;**105**:887–95.
86. Benkirane M, Jin DY, Chun RF, Koup RA, Jeang KT. Mechanism of transdominant inhibition of CCR5-mediated HIV-1 infection by ccr5delta32. *J Biol Chem* 1997; **272**:30603–6.
87. Herrick-Davis K, Weaver BA, Grinde E, Mazurkiewicz JE. Serotonin 5-HT2C receptor homodimer biogenesis in the endoplasmic reticulum: real-time visualization with confocal fluorescence resonance energy transfer. *J Biol Chem* 2006;**281**:27109–16.
88. Fung JJ, Deupi X, Pardo L, Yao XJ, Velez-Ruiz GA, Devree BT, et al. Ligand-regulated oligomerization of beta(2)-adrenoceptors in a model lipid bilayer. *EMBO J* 2009;**28**:3315–28.
89. Ilien B, Glasser N, Clamme JP, Didier P, Piemont E, Chinnappan R, et al. Pirenzepine promotes the dimerization of muscarinic M1 receptors through a three-step binding process. *J Biol Chem* 2009;**284**:19533–43.
90. Alvarez-Curto E, Ward RJ, Pediani JD, Milligan G. Ligand regulation of the quaternary organization of cell surface M3 muscarinic acetylcholine receptors analyzed by fluorescence resonance energy transfer (FRET) imaging and homogeneous time-resolved FRET. *J Biol Chem* 2010;**285**:23318–30.
91. Vidi PA, Chemel BR, Hu CD, Watts VJ. Ligand-dependent oligomerization of dopamine D(2) and adenosine A(2A) receptors in living neuronal cells. *Mol Pharmacol* 2008;**74**:544–51.
92. Fotiadis D, Liang Y, Filipek S, Saperstein DA, Engel A, Palczewski K. Atomic-force microscopy: rhodopsin dimers in native disc membranes. *Nature* 2003;**421**:127–8.
93. Maurel D, Comps-Agrar L, Brock C, Rives ML, Bourrier E, Ayoub MA, et al. Cell-surface protein-protein interaction analysis with time-resolved FRET and snap-tag technologies: application to GPCR oligomerization. *Nat Methods* 2008;**5**: 561–7.
94. Ciruela F, Vilardaga JP, Fernandez-Duenas V. Lighting up multiprotein complexes: lessons from GPCR oligomerization. *Trends Biotechnol* 2010;**28**:407–15.
95. Pisterzi LF, Jansma DB, Georgiou J, Woodside MJ, Chou JT, Angers S, et al. Oligomeric size of the m2 muscarinic receptor in live cells as determined by quantitative fluorescence resonance energy transfer. *J Biol Chem* 2010;**285**:16723–38.
96. Baneres JL, Parello J. Structure-based analysis of GPCR function: evidence for a novel pentameric assembly between the dimeric leukotriene B4 receptor BLT1 and the G-protein. *J Mol Biol* 2003;**329**:815–29.
97. Damian M, Mary S, Martin A, Pin JP, Baneres JL. G protein activation by the leukotriene B4 receptor dimer. Evidence for an absence of trans-activation. *J Biol Chem* 2008;**283**:21084–92.
98. Grant M, Kumar U. The role of G-proteins in the dimerisation of human somatostatin receptor types 2 and 5. *Regul Pept* 2006;**159**:3–8.
99. Han Y, Moreira IS, Urizar E, Weinstein H, Javitch JA. Allosteric communication between protomers of dopamine class A GPCR dimers modulates activation. *Nat Chem Biol* 2009;**5**:688–95.
100. Pellissier LP, Barthet G, Gaven F, Cassier E, Trinquet E, Pin JP, et al. G protein activation by serotonin type 4 receptor dimers: evidence that turning on two protomers is more efficient. *J Biol Chem* 2011;**286**:9985–97.

101. Rovira X, Pin JP, Giraldo J. The asymmetric/symmetric activation of GPCR dimers as a possible mechanistic rationale for multiple signalling pathways. *Trends Pharmacol Sci* 2010;**31**:15–21.
102. Bayburt TH, Vishnivetskiy SA, McLean MA, Morizumi T, Huang CC, Tesmer JJ, et al. Monomeric rhodopsin is sufficient for normal rhodopsin kinase (GRK1) phosphorylation and arrestin-1 binding. *J Biol Chem* 2011;**286**:1420–8.
103. Gurevich VV, Gurevich EV. How and why do GPCRs dimerize? *Trends Pharmacol Sci* 2008;**29**:234–40.
104. Bayburt TH, Leitz AJ, Xie G, Oprian DD, Sligar SG. Transducin activation by nanoscale lipid bilayers containing one and two rhodopsins. *J Biol Chem* 2007; **282**:14875–81.
105. Ernst OP, Gramse V, Kolbe M, Hofmann KP, Heck M. Monomeric G protein-coupled receptor rhodopsin in solution activates its G protein transducin at the diffusion limit. *Proc Natl Acad Sci USA* 2007;**104**:10859–64.
106. Whorton MR, Bokoch MP, Rasmussen SG, Huang B, Zare RN, Kobilka B, et al. A monomeric G protein-coupled receptor isolated in a high-density lipoprotein particle efficiently activates its G protein. *Proc Natl Acad Sci USA* 2007;**104**:7682–7.
107. Meyer BH, Segura JM, Martinez KL, Hovius R, George N, Johnsson K, et al. FRET imaging reveals that functional neurokinin-1 receptors are monomeric and reside in membrane microdomains of live cells. *Proc Natl Acad Sci USA* 2006;**103**:2138–43.
108. White JF, Grodnitzky J, Louis JM, Trinh LB, Shiloach J, Gutierrez J, et al. Dimerization of the class A G protein-coupled neurotensin receptor NTS1 alters G protein interaction. *Proc Natl Acad Sci USA* 2007;**104**:12199–204.
109. Arcemisbehere L, Sen T, Boudier L, Balestre MN, Gaibelet G, Detouillon E, et al. Leukotriene BLT2 receptor monomers activate the G(i2) GTP-binding protein more efficiently than dimers. *J Biol Chem* 2010;**285**:6337–47.
110. Dean MK, Higgs C, Smith RE, Bywater RP, Snell CR, Scott PD, et al. Dimerization of G-protein-coupled receptors. *J Med Chem* 2001;**44**:4595–614.
111. Park PS, Palczewski K. Diversifying the repertoire of G protein-coupled receptors through oligomerization. *Proc Natl Acad Sci USA* 2005;**102**:8793–4.
112. Sakai T, Aoyama M, Kusakabe T, Tsuda M, Satake H. Functional diversity of signaling pathways through G protein-coupled receptor heterodimerization with a species-specific orphan receptor subtype. *Mol Biol Evol* 2010;**27**:1097–106.
113. Jordan BA, Cvejic S, Devi LA. Opioids and their complicated receptor complexes. *Neuropsychopharmacology* 2000;**23**:S5–S18.
114. Levac BA, O'Dowd BF, George SR. Oligomerization of opioid receptors: generation of novel signaling units. *Curr Opin Pharmacol* 2002;**2**:76–81.
115. Waldhoer M, Fong J, Jones RM, Lunzer MM, Sharma SK, Kostenis E, et al. A heterodimer-selective agonist shows in vivo relevance of G protein-coupled receptor dimers. *Proc Natl Acad Sci USA* 2005;**102**:9050–5.
116. Jordan BA, Devi LA. G-protein-coupled receptor heterodimerization modulates receptor function. *Nature* 1999;**399**:697–700.
117. George SR, Fan T, Xie Z, Tse R, Tam V, Varghese G, et al. Oligomerization of mu- and delta-opioid receptors. Generation of novel functional properties. *J Biol Chem* 2000;**275**:26128–35.
118. Golebiewska U, Johnston JM, Devi L, Filizola M, Scarlata S. Differential response to morphine of the oligomeric state of mu-opioid in the presence of delta-opioid receptors. *Biochemistry* 2011;**50**:2829–37.
119. Milligan G. G-protein-coupled receptor heterodimers: pharmacology, function and relevance to drug discovery. *Drug Discov Today* 2006;**11**:541–9.
120. Panetta R, Greenwood MT. Physiological relevance of GPCR oligomerization and its impact on drug discovery. *Drug Discov Today* 2008;**13**:1059–66.

121. Casado V, Cortes A, Mallol J, Perez-Capote K, Ferre S, Lluis C, et al. GPCR homomers and heteromers: a better choice as targets for drug development than GPCR monomers? *Pharmacol Ther* 2009;**124**:248–57.
122. Albizu L, Moreno JL, Gonzalez-Maeso J, Sealfon SC. Heteromerization of G protein-coupled receptors: relevance to neurological disorders and neurotherapeutics. *CNS Neurol Disord Drug Targets* 2010;**9**:636–50.
123. Maggio R, Millan MJ. Dopamine D2-D3 receptor heteromers: pharmacological properties and therapeutic significance. *Curr Opin Pharmacol* 2010;**10**:100–7.
124. Rozenfeld R, Devi LA. Receptor heteromerization and drug discovery. *Trends Pharmacol Sci* 2010;**31**:124–30.
125. Rozenfeld R, Gupta A, Gagnidze K, Lim MP, Gomes I, Lee-Ramos D, et al. AT1R-CBR heteromerization reveals a new mechanism for the pathogenic properties of angiotensin II. *EMBO J* 2011;**30**:2350–63.
126. Gonzalez-Maeso J, Ang RL, Yuen T, Chan P, Weisstaub NV, Lopez-Gimenez JF, et al. Identification of a serotonin/glutamate receptor complex implicated in psychosis. *Nature* 2008;**452**:93–7.
127. Smith NJ, Milligan G. Allostery at G protein-coupled receptor homo- and heteromers: uncharted pharmacological landscapes. *Pharmacol Rev* 2010;**62**:701–25.
128. Birdsall NJ. Class A GPCR heterodimers: evidence from binding studies. *Trends Pharmacol Sci* 2010;**31**:499–508.
129. Springael JY, Urizar E, Costagliola S, Vassart G, Parmentier M. Allosteric properties of G protein-coupled receptor oligomers. *Pharmacol Ther* 2007;**115**:410–8.
130. Maurice P, Kamal M, Jockers R. Asymmetry of GPCR oligomers supports their functional relevance. *Trends Pharmacol Sci* 2011;**32**:514–20.
131. Damian M, Martin A, Mesnier D, Pin JP, Baneres JL. Asymmetric conformational changes in a GPCR dimer controlled by G-proteins. *EMBO J* 2006;**25**:5693–702.
132. Krupnick JG, Benovic JL. The role of receptor kinases and arrestins in G protein-coupled receptor regulation. *Annu Rev Pharmacol Toxicol* 1998;**38**:289–319.
133. Luttrell LM, Roudabush FL, Choy EW, Miller WE, Field ME, Pierce KL, et al. Activation and targeting of extracellular signal-regulated kinases by beta-arrestin scaffolds. *Proc Natl Acad Sci USA* 2001;**98**:2449–54.
134. Hall RA, Lefkowitz RJ. Regulation of G protein-coupled receptor signaling by scaffold proteins. *Circ Res* 2002;**91**:672–80.
135. Liang Y, Fotiadis D, Filipek S, Saperstein DA, Palczewski K, Engel A. Organization of the G protein-coupled receptors rhodopsin and opsin in native membranes. *J Biol Chem* 2003;**278**:21655–62.
136. Novi F, Scarselli M, Corsini GU, Maggio R. The paired activation of the two components of the muscarinic M3 receptor dimer is required for induction of ERK1/2 phosphorylation. *J Biol Chem* 2004;**279**:7476–86.
137. Novi F, Stanasila L, Giorgi F, Corsini GU, Cotecchia S, Maggio R. Paired activation of two components within muscarinic M3 receptor dimers is required for recruitment of beta-arrestin-1 to the plasma membrane. *J Biol Chem* 2005;**280**:19768–76.
138. Bohn LM, Lefkowitz RJ, Gainetdinov RR, Peppel K, Caron MG, Lin FT. Enhanced morphine analgesia in mice lacking beta-arrestin 2. *Science* 1999;**286**:2495–8.
139. Bohn LM, Gainetdinov RR, Lin FT, Lefkowitz RJ, Caron MG. Mu-opioid receptor desensitization by beta-arrestin-2 determines morphine tolerance but not dependence. *Nature* 2000;**408**:720–3.
140. Maggio R, Innamorati G, Parenti M. G protein-coupled receptor oligomerization provides the framework for signal discrimination. *J Neurochem* 2007;**103**:1741–52.
141. Govardovskii VI, Korenyak DA, Shukolyukov SA, Zueva LV. Lateral diffusion of rhodopsin in photoreceptor membrane: a reappraisal. *Mol Vis* 2009;**15**:1717–29.

142. Maurice P, Daulat AM, Turecek R, Ivankova-Susankova K, Zamponi F, Kamal M, et al. Molecular organization and dynamics of the melatonin MT receptor/RGS20/G(i) protein complex reveal asymmetry of receptor dimers for RGS and G(i) coupling. *EMBO J* 2010;**29**:3646–59.
143. Brady AE, Limbird LE. G protein-coupled receptor interacting proteins: emerging roles in localization and signal transduction. *Cell Signal* 2002;**14**:297–309.
144. Joiner ML, Lise MF, Yuen EY, Kam AY, Zhang M, Hall DD, et al. Assembly of a beta2-adrenergic receptor—GluR1 signalling complex for localized cAMP signalling. *EMBO J* 2010;**29**:482–95.
145. Fanelli F, De Benedetti PG. Update 1 of: computational modeling approaches to structure-function analysis of G protein-coupled receptors. *Chem Rev* 2011;**111**: PR438–PR535.
146. Selent J, Kaczor AA. Oligomerization of G protein-coupled receptors: computational methods. *Curr Med Chem* 2011;**18**:4588–605.
147. Casciari D, Seeber M, Fanelli F. Quaternary structure predictions of transmembrane proteins starting from the monomer: a docking-based approach. *BMC Bioinformatics* 2006;**7**:340.
148. Hebert TE, Moffett S, Morello JP, Loisel TP, Bichet DG, Barret C, et al. A peptide derived from a beta2-adrenergic receptor transmembrane domain inhibits both receptor dimerization and activation. *J Biol Chem* 1996;**271**:16384–92.
149. Ng GY, O'Dowd BF, Lee SP, Chung HT, Brann MR, Seeman P, et al. Dopamine D2 receptor dimers and receptor-blocking peptides. *Biochem Biophys Res Commun* 1996;**227**:200–4.
150. Canals M, Marcellino D, Fanelli F, Ciruela F, De Benedetti P, Goldberg SR, et al. Adenosine A2A-dopamine D2 receptor-receptor heteromerization: qualitative and quantitative assessment by fluorescence and bioluminescence energy transfer. *J Biol Chem* 2003;**278**:46741–9.
151. Guo W, Shi L, Javitch JA. The fourth transmembrane segment forms the interface of the dopamine D2 receptor homodimer. *J Biol Chem* 2003;**278**:4385–8.
152. Guo W, Filizola M, Weinstein H, Javitch JA. Crosstalk in G protein-coupled receptors: changes at the transmembrane homodimer interface determine activation. *Proc Natl Acad Sci USA* 2005;**102**:17495–500.
153. Guo W, Urizar E, Kralikova M, Mobarec JC, Shi L, Filizola M, et al. Dopamine D2 receptors form higher order oligomers at physiological expression levels. *EMBO J* 2008;**27**:2293–304.
154. Lee SP, O'Dowd BF, Rajaram RD, Nguyen T, George SR. D2 dopamine receptor homodimerization is mediated by multiple sites of interaction, including an intermolecular interaction involving transmembrane domain 4. *Biochemistry* 2003;**42**:11023–31.
155. Johnston JM, Aburi M, Provasi D, Bortolato A, Urizar E, Lambert NA, et al. Making structural sense of dimerization interfaces of delta opioid receptor homodimers. *Biochemistry* 2011;**50**:1682–90.
156. Fanelli F, Mauri M, Capra V, Raimondi F, Guzzi F, Ambrosio M, et al. Light on the structure of thromboxane A2 receptor heterodimers. *Cell Mol Life Sci* 2011;**68**:3109–20.
157. Mancia F, Assur Z, Herman AG, Siegel R, Hendrickson WA. Ligand sensitivity in dimeric associations of the serotonin 5HT2c receptor. *EMBO Rep* 2008;**9**:363–9.
158. Kim H, Lee BK, Naider F, Becker JM. Identification of specific transmembrane residues and ligand-induced interface changes involved in homo-dimer formation of a yeast G protein-coupled receptor. *Biochemistry* 2009;**48**:10976–87.
159. Wang HX, Konopka JB. Identification of amino acids at two dimer interface regions of the alpha-factor receptor (Ste2). *Biochemistry* 2009;**48**:7132–9.

160. Hernanz-Falcon P, Rodriguez-Frade JM, Serrano A, Juan D, del Sol A, Soriano SF, et al. Identification of amino acid residues crucial for chemokine receptor dimerization. *Nat Immunol* 2004;**5**:216–23.
161. Stanasila L, Perez JB, Vogel H, Cotecchia S. Oligomerization of the alpha 1a- and alpha 1b-adrenergic receptor subtypes. Potential implications in receptor internalization. *J Biol Chem* 2003;**278**:40239–51.
162. Milligan G, Pediani JD, Canals M, Lopez-Gimenez JF. Oligomeric structure of the alpha1b-adrenoceptor: comparisons with rhodopsin. *Vision Res* 2006;**46**:4434–41.
163. Zeng F, Wess J. Molecular aspects of muscarinic receptor dimerization. *Neuropsychopharmacology* 2000;**23**:S19–S31.
164. Lodowski DT, Salom D, Le Trong I, Teller DC, Ballesteros JA, Palczewski K, et al. Crystal packing analysis of Rhodopsin crystals. *J Struct Biol* 2007;**158**:455–62.
165. Navarro G, Ferre S, Cordomi A, Moreno E, Mallol J, Casado V, et al. Interactions between intracellular domains as key determinants of the quaternary structure and function of receptor heteromers. *J Biol Chem* 2010;**285**:27346–59.
166. Parker MS, Sah R, Park EA, Sweatman T, Balasubramaniam A, Sallee FR, et al. Oligomerization of the heptahelical G protein coupling receptors: a case for association using transmembrane helices. *Mini Rev Med Chem* 2009;**9**:329–39.
167. Filizola M, Olmea O, Weinstein H. Prediction of heterodimerization interfaces of G-protein coupled receptors with a new subtractive correlated mutation method. *Protein Eng* 2002;**15**:881–5.
168. Filizola M, Weinstein H. Structural models for dimerization of G-protein coupled receptors: the opioid receptor homodimers. *Biopolymers* 2002;**66**:317–25.
169. Soyer OS, Dimmic MW, Neubig RR, Goldstein RA. Dimerization in aminergic G-protein-coupled receptors: application of a hidden-site class model of evolution. *Biochemistry* 2003;**42**:14522–31.
170. Nemoto W, Fukui K, Toh H. GRIP: a server for predicting interfaces for GPCR oligomerization. *J Recept Signal Transduct Res* 2009;**29**:312–7.
171. Fanelli F. Dimerization of the lutropin receptor: insights from computational modeling. *Mol Cell Endocrinol* 2007;**260–262**:59–64.
172. Casciari D, Dell'Orco D, Fanelli F. Homodimerization of neurotensin 1 receptor involves helices 1, 2, and 4: insights from quaternary structure predictions and dimerization free energy estimations. *J Chem Inf Model* 2008;**48**:1669–78.
173. Fanelli F, Felline A. Dimerization and ligand binding affect the structure network of A(2A) adenosine receptor. *Biochim Biophys Acta* 2011;**1808**:1256–66.
174. Soriano A, Ventura R, Molero A, Hoen R, Casado V, Cortes A, et al. Adenosine A2A receptor-antagonist/dopamine D2 receptor-agonist bivalent ligands as pharmacological tools to detect A2A-D2 receptor heteromers. *J Med Chem* 2009;**52**:5590–602.
175. Liu X, Kai M, Jin L, Wang R. Computational study of the heterodimerization between mu and delta receptors. *J Comput Aided Mol Des* 2009;**23**:321–32.
176. Bruno A, Guadix AE, Costantino G. Molecular dynamics simulation of the heterodimeric mGluR2/5HT(2A) complex. An atomistic resolution study of a potential new target in psychiatric conditions. *J Chem Inf Model* 2009;**49**:1602–16.
177. Kim SK, Jacobson KA. Computational prediction of homodimerization of the A3 adenosine receptor. *J Mol Graph Model* 2006;**25**:549–61.
178. Soulier JL, Russo O, Giner M, Rivail L, Berthouze M, Ongeri S, et al. Design and synthesis of specific probes for human 5-HT4 receptor dimerization studies. *J Med Chem* 2005;**48**:6220–8.
179. McMillin SM, Heusel M, Liu T, Costanzi S, Wess J. Structural basis of m3 muscarinic receptor dimer/oligomer formation. *J Biol Chem* 2011;**286**:28584–98.
180. Nemoto W, Toh H. Membrane interactive alpha-helices in GPCRs as a novel drug target. *Curr Protein Pept Sci* 2006;**7**:561–75.

181. Vohra S, Chintapalli SV, Illingworth CJ, Reeves PJ, Mullineaux PM, Clark HS, et al. Computational studies of Family A and Family B GPCRs. *Biochem Soc Trans* 2007;**35**:749–54.
182. Filizola M. Increasingly accurate dynamic molecular models of G-protein coupled receptor oligomers: Panacea or Pandora's box for novel drug discovery? *Life Sci* 2010;**86**:590–7.
183. Simpson LM, Taddese B, Wall ID, Reynolds CA. Bioinformatics and molecular modelling approaches to GPCR oligomerization. *Curr Opin Pharmacol* 2010;**10**:30–7.
184. Guidolin D, Ciruela F, Genedani S, Guescini M, Tortorella C, Albertin G, et al. Bioinformatics and mathematical modelling in the study of receptor-receptor interactions and receptor oligomerization Focus on adenosine receptors. *Biochim Biophys Acta* 2011;**1808**:1267–83.
185. Gouldson PR, Higgs C, Smith RE, Dean MK, Gkoutos GV, Reynolds CA. Dimerization and domain swapping in G-protein-coupled receptors: a computational study. *Neuropsychopharmacology* 2000;**23**:S60–S77.
186. Higgs C, Reynolds CA. Modelling G-protein coupled receptors. In: Eriksson LA, editor. *Theoretical and computational chemistry*, vol. 9. Amsterdam: Elsevier; 2001. p. 341–77.
187. Gouldson PR, Bywater RP, Reynolds CA. Correlated mutations amongst the external residues of G-protein coupled receptors. *Biochem Soc Trans* 1997;**25**:529S.
188. Gouldson PR, Snell CR, Bywater RP, Higgs C, Reynolds CA. Domain swapping in G-protein coupled receptor dimers. *Protein Eng* 1998;**11**:1181–93.
189. Pazos F, Helmer-Citterich M, Ausiello G, Valencia A. Correlated mutations contain information about protein-protein interaction. *J Mol Biol* 1997;**271**:511–23.
190. Lichtarge O, Bourne HR, Cohen FE. An evolutionary trace method defines binding surfaces common to protein families. *J Mol Biol* 1996;**257**:342–58.
191. Madabushi S, Gross AK, Philippi A, Meng EC, Wensel TG, Lichtarge O. Evolutionary trace of G protein-coupled receptors reveals clusters of residues that determine global and class-specific functions. *J Biol Chem* 2004;**279**:8126–32.
192. Chothia C, Lesk AM. The relation between the divergence of sequence and structure in proteins. *EMBO J* 1986;**5**:823–6.
193. Hadac EM, Ji Z, Pinon DI, Henne RM, Lybrand TP, Miller LJ. A peptide agonist acts by occupation of a monomeric G protein-coupled receptor: dual sites of covalent attachment to domains near TM1 and TM7 of the same molecule make biologically significant domain-swapped dimerization unlikely. *J Med Chem* 1999;**42**:2105–11.
194. Schulz A, Grosse R, Schultz G, Gudermann T, Schoneberg T. Structural implication for receptor oligomerization from functional reconstitution studies of mutant V2 vasopressin receptors. *J Biol Chem* 2000;**275**:2381–9.
195. Bakker RA, Dees G, Carrillo JJ, Booth RG, Lopez-Gimenez JF, Milligan G, et al. Domain swapping in the human histamine H1 receptor. *J Pharmacol Exp Ther* 2004;**311**:131–8.
196. Thummer RP, Campbell MP, Dean MK, Frusher MJ, Scott PD, Reynolds CA. Entropy and oligomerization in GPCRs. *J Mol Neurosci* 2005;**26**:113–22.
197. de Juan D, Mellado M, Rodriguez-Frade JM, Hernanz-Falcon P, Serrano A, del Sol A, et al. A framework for computational and experimental methods: identifying dimerization residues in CCR chemokine receptors. *Bioinformatics* 2005;**21**(Suppl. 2):ii13–ii18.
198. del Sol Mesa A, Pazos F, Valencia A. Automatic methods for predicting functionally important residues. *J Mol Biol* 2003;**326**:1289–302.
199. Nemoto W, Toh H. Prediction of interfaces for oligomerizations of G-protein coupled receptors. *Proteins* 2005;**58**:644–60.
200. Filizola M, Weinstein H. The study of G-protein coupled receptor oligomerization with computational modeling and bioinformatics. *FEBS J* 2005;**272**:2926–38.

201. Fanelli F, Menziani MC, Scheer A, Cotecchia S, De Benedetti P. Theoretical study on receptor/G protein recognition: new insights into the mechanism of the α1b-adrenergic receptor activation. *Int J Quantum Chem* 1999;**73**:71–83.
202. Chen R, Li L, Weng Z. ZDOCK: an initial-stage protein-docking algorithm. *Proteins* 2003;**52**:80–7.
203. Witt M, Slusarz MJ, Ciarkowski J. Molecular modeling of Vasopressin V2 Receptor tetramer in hydrated lipid membrane. *QSAR Comb Sci* 2008;**27**:684–93.
204. Provasi D, Johnston JM, Filizola M. Lesson from free-energy simulations of delta-opioid receptor homodimers involving the fourth transmembrane helix. *Biochemistry* 2010;**49**:6771–6.
205. Russo O, Berthouze M, Giner M, Soulier JL, Rivail L, Sicsic S, et al. Synthesis of specific bivalent probes that functionally interact with 5-HT(4) receptor dimers. *J Med Chem* 2007;**50**:4482–92.
206. Heyer LJ, Kruglyak S, Yooseph S. Exploring expression data: identification and analysis of coexpressed genes. *Genome Res* 1999;**9**:1106–15.
207. Seeber M, Felline A, Raimondi F, Muff S, Friedman R, Rao F, et al. Wordom: a user-friendly program for the analysis of molecular structures, trajectories, and free energy surfaces. *J Comput Chem* 2011;**32**:1183–94.
208. Dell'Orco D, Casciari D, Fanelli F. Quaternary structure predictions and estimation of mutational effects on the free energy of dimerization of the OMPLA protein. *J Struct Biol* 2008;**163**:155–62.
209. Dell'Orco D, De Benedetti PG, Fanelli F. In silico screening of mutational effects on transmembrane helix dimerization: insights from rigid-body docking and molecular dynamics simulations. *J Phys Chem B* 2007;**111**:9114–24.
210. Angelova K, Felline A, Lee M, Patel M, Puett D, Fanelli F. Conserved amino acids participate in the structure networks deputed to intramolecular communication in the lutropin receptor. *Cell Mol Life Sci* 2011;**68**:1227–39.
211. Raimondi F, Felline A, Portella G, Orozco M, Fanelli F. Light on the structural communication in Ras GTPases. *J Biomol Struct Dyn* 2013;**31**:142–57.
212. Fanelli F, Seeber M. Structural insights into retinitis pigmentosa from unfolding simulations of rhodopsin mutants. *FASEB J* 2010;**24**:3196–209.
213. Fanelli F, Dell'Orco D. Rhodopsin activation follows precoupling with transducin: inferences from computational analysis. *Biochemistry* 2005;**44**:14695–700.
214. Fanelli F, De Benedetti PG. Computational modeling approaches to structure-function analysis of G protein-coupled receptors. *Chem Rev* 2005;**105**:3297–351.
215. Ballesteros JA, Weinstein H. Integrated methods for the construction of three-dimensional models and computational probing of structure-function relations in G protein-coupled receptors. *Methods Neurosci.* 1995;**25**:366–428.
216. Woolf PJ, Linderman JJ. An algebra of dimerization and its implications for G-protein coupled receptor signaling. *J Theor Biol* 2004;**229**:157–68.
217. Woolf PJ, Linderman JJ. Self organization of membrane proteins via dimerization. *Biophys Chem* 2003;**104**:217–27.
218. Fallahi-Sichani M, Linderman JJ. Lipid raft-mediated regulation of G-protein coupled receptor signaling by ligands which influence receptor dimerization: a computational study. *PLoS One* 2009;**4**:e6604.

CHAPTER SIX

Challenges in the Development of Heteromer-GPCR-Based Drugs

Rafael Franco[*,†], **Eva Martínez-Pinilla**[*], **Ana Ricobaraza**[*], **Peter J. McCormick**[†,‡,§]

[*]Applied Medical Research Center (CIMA), University of Navarra, Pamplona, Spain
[†]Department of Biochemistry and Molecular Biology, University of Barcelona, Barcelona, Spain
[‡]Centro de Investigación Biomédica en Red de Enfermedades Neurodegenerativas (CIBERNED), Madrid, Spain
[§]Institute of Biomedicine of the University of Barcelona (IBUB), Barcelona, Spain

Contents

Abstract

G-protein-coupled receptors are targets of a variety of diseases. Drug screening has been classically performed assuming the occurrence of monomeric receptors. As more and more receptor heteromers are identified, the challenge is now to develop screening assays to select heteromer-specific drugs. These drugs may, for instance, be able to interact preferentially with prerather than with postsynaptic receptors. Heteromer-based drug discovery opens new perspectives in both Academic pursuits and for the Pharmaceutical industry.

1. INTRODUCTION

The G-protein-receptor (GPCR) superfamily is subdivided in six classes (A to F). Rhodopsin-like class A includes the highest number of targets for currently marketed drugs designed to combat a wide variety of diseases.

Progress in Molecular Biology and Translational Science, Volume 117
ISSN 1877-1173
http://dx.doi.org/10.1016/B978-0-12-386931-9.00006-4

The reluctance to consider that class A GPCRs may form dimers has given way to the concept that almost every member of the family may form homo- and/or heterodimers. The description of the ever-growing number of identified dimers and of the evidence for dimerization is out of the scope of the present report that will focus only in class A members. It is worth mentioning that support for dimerization of class A members has been boosted by the recent resolution of the structure of the μ-opioid receptor that crystallizes as a twofold symmetrical dimer through a four-helix bundle motif formed by transmembrane segments 5 and 6.[1]

Challenges in this young field are quite precise. One of them is to understand how homo- or heterodimers are formed and disrupted and how regulation of these processes occurs. Another challenge is to identify new heteromer pairs and to demonstrate the exact function of each of the already identified heteromers. The function of "monomeric" receptors has often relied on the use of "knock-out" (KO) animals which are of little use in the heteromer field. Animals defective in a given GPCR are in fact defective in all the heteromers containing the GPCR, and therefore, one has to be cautious about the interpretation of data taken from KO animals. They have been, however, instrumental, as controls, that is, to show that a given heterodimer (detected by any of the available means) disappears in the animal defective for one of the receptor partners in the heteromer.

An additional challenge is knowing the molecular architecture of heteromers. What is the exact size of the complexes? And what is the stoichiometry between protomers and G-proteins? On the one hand, the minimum size for a homomer is considered to be two (two equal protomers forming a homodimer), but for now, it is not well established whether a heteromer is consisting of two different protomers or is, for instance, composed of two homodimers to give a complex of four protomers. On the other hand, it is likely that the assumed 1 GPCR:1 G-protein stoichiometry, that is, one G protein per protomer/monomer, will soon be revisited. Clustering and internalization of receptors have been studied for "monomeric" receptors expressed in heterologous cells and in cell lines with endogenous expression.[2–5] The occurrence of heteromers should drive experiments to address how heteromerization affects receptor trafficking and, conversely, how receptor activation by agonists may alter the size (number of protomers) of the heteromeric complex. The participation of other proteins interacting with GPCRs[6–9] in the trafficking of heteromers is something to take into account as it may impact on heteromer formation/disruption regulation.

Solving these challenges will likely require the development of *ad hoc* techniques, as the existing ones cannot answer satisfactorily the many questions surrounding GPCR heteromerization. In fact, a remarkable recent development, the so-called proximity ligation assay (PLA), has allowed the easy detection of heteromers in natural sources. Such detection is crucial to know the exact role of heteromers in health and disease and to know whether heteromers may be selected as therapeutic targets. Until the appearance of PLA, which can be performed on tissue slices,[10,11] heteromer detection in natural sources was dependent on the identification of the heteromer fingerprint, something that requires relatively complex biochemical assays and careful analysis of data (see Ref. 12 for details).

Finally, another challenge that becomes an opportunity is the development of heteromer-selective drugs. Despite around a 30% of drugs on the market targeting GPCRs, it is very well known that few new candidates reach the market. Oligomerization of GPCRs may provide a new way that, hopefully, will lead to greater success in providing new therapeutic tools based on heteromer selectivity. As these drugs would preferentially target pathological cells expressing heteromers, the obvious advantages would be more precise targeting and less side effects. Heteromer-based drug discovery is the challenge that will be covered in this chapter.

2. GPCR HETEROMERS AS THERAPEUTIC TARGETS

One of the challenges in the development of heteromer-based drugs is to both identify heteromers as targets and identify heteromers in disease. Whereas identification of heteromers in natural sources is growing and there are several potential heteromers that may be targets for a variety of diseases, there are few reports showing whether a given heteromer is still present in diseased conditions. Therefore, before considering a heteromer as target for disease, its occurrence in the target cells of the patient should be addressed. Alternatively, a heteromer may appear in the course of the disease thus becoming a potential target that in healthy tissues would not be detectable. A first example was the higher expression in preeclampsia of the AT_1 receptor for the vasopressor angiotensin II and the B_2 receptor for the vasodepressor bradykinin.[13] Also in a rat model of preeclampsia, it has been shown that the AT_1 receptor and the α_{1D} adrenoceptor heteromerize and that whereas it is barely detectable in the aorta of healthy pregnant rats, its expression is markedly increased in preeclamptic rats.[14] Therefore, drugs acting on heteromer-containing angiotensin AT_1 receptors could eventually be of interest for preventing this disease that, depending on the geographic area, may occur in ~16% of pregnancies.[15,16] To our knowledge, no reports have appeared on the alteration of heteromers when a disease appears. We have recent data (Franco *et al.*, submitted) indicating that in both a rat and a monkey model of Parkinson's disease, heteromers consisting of dopamine D_2 and adenosine A_{2A} receptors are still present. These heteromers are present in striatal cells from healthy animals and also from parkinsonian animals. This means that A_{2A}/D_2 receptor heteromers may be a target for Parkinson's disease. Interestingly, the most used substitutive therapy, which consists of the administration of L-DOPA that is *in vivo* converted to dopamine, is targeting A_{2A}/D_2 receptor heteromers in GABAergic striatal neurons. Moreover, A_{2A} antagonists, which are in clinical studies (phases II–III) for Parkinson's disease, would also target heteromers. In fact, many drugs acting on GPCRs may be actually targeting heteromers.

For the above mentioned reasons, to select a heteromer as a therapeutic target would require the use of human samples taken from biobanks or biopsies. In these samples, the heteromers may be detected by their fingerprint, which consist of an interprotomer cross talk resulting in a shift in the competition curves obtained in radioligand-binding assays (see Ref. 17 for details). Alternatively heteromers may be identified by the recently developed PLA. Although it requires the availability of specific antibodies for each of the interacting receptors in the heteromer, this type of assay allows heteromer detection in tissue slices and for this reason becomes the more straightforward approach (see Ref. 18 for review). Once the heteromer is selected as a target, the second and more difficult task consists of the development of heteromer-selective drugs. For that purpose, drug screens must use biological material expressing the heteromer and must differentiate heteromer-selective drugs from protomer-selective drugs.

3. BIASED SIGNALING

The concept of bias in signaling was first discussed nearly 20 years ago under the umbrella of "agonist trafficking," where different agonists could lead to different G proteins trafficked to the receptor.[19–21] Despite its rather simplistic description, the basic concept has shown striking resilience, and today is a major field in GPCR biology. When the "agonist trafficking" concept was first discussed, it was generally thought that there was only one receptor involved in ligand binding. However, recent structural data demonstrating that GPCRs have different ligand-modulated signaling states, and that higher order oligomers are important for activity, suggest that GPCRs are not single on/off switches.[1,22,23] Instead, they serve as processors interpreting different inputs and providing a given output for every input based on the state of the receptor. How does this occur? As a protein, the receptor samples various conformations and energy states. This principle has been demonstrated in a variety of studies[24–30] over the past decade. Depending on the conditions, the energy landscape may change such that a given conformation which in some conditions may be the lowest energy state is now no longer the most stable state. On the face of it, this concept may seem an impediment to rational drug design. However, this flexibility in the energy landscape can be taken advantage of by designing and employing ligands which can stabilize certain conformations of the receptor. Two elegant experimental examples of this is the work done by Mary *et al.*[31] on the ghrelin receptor and by Zocher *et al.* with the β_2-adrenergic receptor.[32] In the context of heteromers, it has been proposed that one of the functions of receptor oligomerization may be to stabilize certain receptor folds.[33]

Currently, there are two types of biased signaling that have been demonstrated, the more common is that of arrestin-based signaling. This is where a given ligand–receptor complex, signals through β-arrestin molecules and not through activation of G proteins. This has been demonstrated for a variety of receptors. As the concept of arrestin-based signaling is still relatively new, the evidence for its biological relevance is small but growing.[34] A recent report showing that current antipsychotic drugs show differences in G-protein and arrestin-based signaling and that those that do signal via arrestin are indeed more potent is a major stride toward understanding the role arrestin-based signaling may play at the physiological level.[35] This well designed study strongly suggests that arrestin signaling may indeed play an important part of how the inputs into GPCRs are propagated. There are multiple examples of molecules that are arrestin-biased ligands, summarized by Whalen *et al.*[36] In terms of heteromers, there has long been the hypothesis that one of the functions of GPCR heteromers is to influence ligand binding of the partner receptor in such a way that could also alter arrestin recruitment. This theory has been recently supported by evidence that the CXCR4–CXCR7 association leads to arrestin-biased signaling.[37] Challenges to demonstrating clear biased signaling versus simple cross talk are highlighted by the works of Cervantes *et al.* and Saulière *et al.*[38,39] and discussed by Prezeau *et al.*[40] suggesting that one must delineate cross talk at the signaling level from true effects due to protein–protein interactions of heteromers.

In addition to ligand-driven arrestin signaling, there are also reports of receptors that signal only via arrestin signaling and not via G proteins. This appears to be the case for the chemokine receptor CXCR7.[41] In addition, CXCR7 is able to form heterodimers with a sister receptor CXCR4[42] and can lead to preferential recruitment of arrestin.[43]

Conversely, there are also receptors and ligands that exhibit G-protein-biased signaling. Typically, this is where no arrestin recruitment or arrestin–associated signaling (e.g., MAPK) is observed, or when knocking down arrestins seems to have no effect on receptor function. Examples of this include Innamorati *et al.* and a recent study by Busnelli *et al.* that screened for functionally selective compounds that promoted G-protein signaling for the oxytocin receptor.[44,45]

In the case of heteromers, it must be considered that they are made up of at least two and possibly four functional targets.[33] The affinities are most likely not equal in a heteromer pair. In addition, it is becoming more clear that there is most likely one or at maximum two G proteins able to bind the heteromer, and this G-protein could be unique to either partner receptor.[46] In some ways, heteromers may be easier to target than single receptors. How? If the hypothesis that single receptors sample different energy states is true then heteromers most likely are more restricted and could in theory sample less states thus providing more stable structures to bind but higher resolution screens must be designed to address this as well as molecular dynamics studies complemented with oligomeric X-ray structures. Incorporating techniques like fragment-based studies or mutations that stabilize certain states could be advantageous to identifying such heteromer-targeting compounds.[47,48]

4. ALLOSTERIC MODULATORS

A somewhat related topic to biased signaling born from some of the same early discussions of "agonist trafficking" is the action and identification of allosteric modulators. A subset of GPCR-activating ligands, termed allosteric agonists and positive allosteric modulators, has also been shown to activate specific downstream signaling pathways. This section does not discuss the specific techniques or approaches to identifying allosteric modulators as these topics have been thoroughly reviewed previously.[49,50] Instead, we focus on pointing out recent advances that may help clarify how to identify allosteric modulators. The classical model of allosterism is the Monod–Wyman–Changeux, also referred to as a "two-state model,"[51] which says in simplest terms that it should be possible to discover allosteric drugs that preferentially favor either active or inactive receptor states and can selectively modulate the properties of cobound ligands in a manner that correlates with the efficacy (positive, negative, or neutral) of such ligands. As a receptor can adopt more than two states, it should be then possible to identify compounds that bias signaling toward a given pathway in a ligand-specific manner.[52]

GPCRs transduce sensory signals of external origin such as photons, odors or pheromones, and endogenous signals, including neurotransmitters, (neuro)peptides, proteases, glycoprotein hormones, and ions, to the cytoplasmic side of the cell membrane.[53] In the ligand-free basal state, GPCRs exist in an equilibrium of conformations, each stabilized by a network of intramolecular interactions.[54] Agonists binding at the native (or orthosteric) site stabilize conformational changes in cytoplasmic domains that increase receptor signaling. Comparison of the crystal structure of inactive rhodopsin with that of the ligand-free opsin, which contains several distinctive features of the presumed active state, shows key structural changes that occur in the cytoplasmic G-protein-coupling domains. Conversely, inverse agonists decrease the basal, agonist-independent level of signaling by stabilizing different conformational changes.

The development of agents selective for specific GPCR subtypes has been hampered by the considerable structural similarity of their drug-binding pockets. In recent years, there have been remarkable advances in the discovery of novel ligands for GPCRs that act at allosteric sites that are topographically distinct from the orthosteric site, to regulate receptor function, efforts that have been guided by the landslide of recent X-ray structures of GPCRs.[22,55] Ligand binding to an allosteric-binding site might enhance (positive allosteric modulators) or decrease (negative allosteric modulators) the response of orthosteric agonists.[56,57] These compounds provide high selectivity, novel modes of efficacy and may lead to novel therapeutic agents for the treatment of multiple human disorders.

The main identification of allosteric modulators comes via ligand-binding assays. The use of these assays is a challenge. In terms of allosterism and heteromers, the challenge comes with the fact that classical models for fitting the data are insufficient. Why? It is generally accepted and the previous section on biased signaling supports, that heteromer formation can have a major impact on stabilizing receptor conformations. Thus, heteromer formation can directly impact ligand binding. This certitude demands that new models be developed for interpreting ligand-binding assays to identify allosteric modulators. Recently, attempts have been made to incorporate heteromer formation into new models for ligand binding.[58–62] Further, refinement of these models may be required, but they clearly demonstrate that classical models of ligand binding are less relevant when studying receptor heteromers.

One of the best characterized and most successful receptor families from a drug discovery point of view for the identification of allosteric modulators that also forms heteromers is the mGluR family of receptors.[63,64] Other success stories include the muscarinic receptors, adenosine receptors, and the chemokine receptor, CXCR4, reviewed in Ref. 57 and 65.

5. DUAL VERSUS BIVALENT DRUGS

Dual ligands are those that may interact with two different receptors but not necessarily forming heteromers. Dual drugs have been developed as potential therapeutic drugs in those cases for which activation and/or blockade of two different receptors may be of therapeutic benefit. To our knowledge, the most recent study concerns the design and characterization of optimized adenosine A_{2A}/A_1 receptor antagonists for the treatment of Parkinson's disease[66] for which dual ligands targeting both adenosine and dopamine receptors have also been reported.[67] The consensus in the GPCR heteromer research field is that, by definition, a heteromer has to be a functionally different entity, that is, that the function of the heteromer is not just the result of summing up the function of two different receptors.[12] One interesting example is the A_1–A_{2A} receptor[68] that is a sensor of adenosine concentration and acts in a completely opposite way when the concentration of the neuromodulator is low versus when it is high. Another relevant example is the D_1–D_2 receptor heterodimer that couples to G_q, whereas D_1 couples to G_s and D_2 couples to $G_{i.}$[69] This heteromer engages calcium-mediated signaling, which is less abundant than cAMP-mediated signaling in dopamine receptor containing neurons in the CNS. Taken together, this information indicates that dual drugs, as they are conceived, cannot be considered heteromer-selective drugs. There may be some exceptions as SKF-83959, which was developed as dopamine D_1 receptor agonist but is a dual drug able to engage the G_q-mediated signaling via the D_1–D_2 receptor heteromer (see below).

The group of Portoghese were pioneers in the design of bivalent molecules that were tested in biological systems expressing homo- and/or heterodimers of opioid receptors. In fact, as early as 1982, the first bivalent ligands were synthesized and tested as opioid receptor agonists and antagonists (see Ref. 70). Bivalent ligands consist of two moieties consisting on agonist/antagonists linked by a spacer of different length. As indicated below, ligands with a precise spacer length interact preferentially with receptor dimers. Bivalent ligands result in relatively big molecules that would not readily cross the blood–brain barrier thus making difficult their use to combat neurological and/or neuropsychiatric diseases. However, the spacer may be easily cleaved off *in vivo* to release the two moieties. After cleavage, these bivalent ligands would no longer interact selectively with dimers thus giving the impression that if they were to reach the market bivalent ligands would act more as dual drugs than as dimer-selective molecules.

Although the earliest studies on GPCR bivalent ligands did not necessarily aim to target dimeric complexes (see Ref. 71 for details), bivalent ligands have been instrumental to demonstrate the occurrence of dimers in natural sources. This difficult task has been eased by the development of the PLA (see earlier), but until then, it relied on the so-called heteromer fingerprint. Provided one has a bivalent ligand with a suitable spacer length, the detection of the fingerprint for given homo- or heterodimer becomes quite simple. In fact, the affinity of the bivalent ligand for a receptor in a dimer is higher than that of each of the moieties that are present in the bivalent molecule. Therefore, competition experiments in binding assays using a selective radioligand allows detection of dimers if a bivalent ligand has more affinity (lower K_D or IC_{50}) than the "monovalent" counterparts. The higher affinity is due to a higher local concentration when one of the moieties lies in one of the protomers forming the dimer. This indirect way to detect dimers has been used to identify in natural sources (brain striatum) the occurrence of adenosine A_{2A} and dopamine D_2 heteromers[72] and, in transfected cells, of dopamine D_2[73] and serotonin 5-HT_4[74] receptor homodimers. Interestingly, in 1996 Halazy *et al.*[75] used bivalent ligands constituted by two serotonin moieties to show that the affinity of the compounds (those of suitable spacer size) was higher than that of serotonin. These results indicate that unwantedly the ligands were detecting homodimers of serotonin receptors in cells transfected with the 1A, 1B, or 1D subtype. The 10-fold higher affinity of bivalents with linkers shorter than 40 atoms also suggests the occurrence of homodimers of human melacortin 4 receptors expressed in a heterologous system.[76] Finally, it is worth noting that bivalent ligands gave insights into dimer structure. On the one hand, the differential binding capacity of bivalents of different linker size gave an estimation of the distance between the two orthosteric-binding sites in a dimer. On the other hand, combination of binding data using bivalents with different linker size with modeling of receptor dimers served to provide insight into the structural organization of the protomers in the dimer. For opioid receptors, this approach was used as early as in 2001 to show that dimers with a transmembrane 4/5 interface were more likely than those with a 5/6 interface.[70] By far, opioid receptor bivalent ligands have been the subject of extensive development giving a myriad of compounds interacting with opioid receptor homo- or heterodimers. The most recent report is intended to explore the putative heterodimerization of μ-opioid and chemokine CCR5 receptors.[77]

6. SCREENING

Screens for homodimers are at present of little interest due to the fact that many receptors are forming dimers, and therefore, the actual screening procedures are likely targeting homodimers expressed in the transfected cells used for HTS. Even chemokine-mediated cell activation that was thought to be due to the binding of a monomeric chemokine to its monomeric receptor is more complex than initially predicted; several studies suggest that chemokines can dimerize and that their receptors are found as dimers and/or higher order oligomers at the cell surface.[78–80] Interestingly, synthetic bivalent ligands with rigid linkers can predict the dimer form of CXCR4 and be applied to molecular probes in cancerous cells.[81] In any case, the most challenging aspect is the design of heteromer-selective drugs, and therefore, this section is devoted to possible approaches to screen for heteromer-selective small drugs.

The discovery of the current heteromer-selective drugs has not resulted from a common and universally validated screening approach. Yet a bunch of very promising drugs have been discovered as interacting preferentially with heteromers. It should be noted that "heteromer" drugs may have more affinity for a given receptor in the conformation encountered in the heterodimer than in the conformation existing in the monomeric or homodimeric receptor.[69,82] At present, many of the reported heteromer-selective drugs are likely interacting with orthosteric centers. An exhaustive review of the literature is out of the scope of this chapter, and therefore, just a few examples will be given to show the potential of these types of drugs. SNC-80 was the first nonpeptide drug developed in 1994 that was regarded as a highly selective agonist for the δ-opioid receptor. Recently, however, it has been shown that SNC-80 selectively activates the μ/δ-opioid receptor heteromer strongly indicating that its antinoceptive potential is mediated by the heteromer.[82] Similarly, a benzodiazepine developed as agonist for dopamine D_1- and D_1-like receptors, SKF-83959, is a kind of mix dual (see earlier) and D_1/D_2-heteromer-selective compound. As expected, the compound binds in the heteromer to D_1, but it also binds to the G_q-coupled D_2 receptor in the heteromer. In fact, SKF-83959 was identified as a high-affinity partial agonist for a pertussis toxin-resistant D_2 receptor within the complex. Thus, SKF-83959, which does not bind to Gi-coupled receptors, allowed to demonstrate that heteromeric D_1–D_2 receptor signaling complex in brain is coupled to $G_{q/11}$ and requires agonist binding to both receptors for G-protein activation and intracellular calcium release.[69,83]

6.1. Direct binding-based approach

In the future, it is advisable that more robust and general methods for the screen of these compounds would be developed in both Academy and Pharmaceutical/Biotech industry. The use of bivalent ligands to be screened against the newly developed drugs has been suggested. For instance, Kühhorn *et al.*,[73] based on the potential involvement of homomerization of dopamine D_2 receptors in the pathophysiology of schizophrenia, have suggested that bivalent probes for the D_2 receptors may serve as promising tools for the discovery of atypical antipsychotics. However, the previous need to develop suitable bivalent ligands with convenient spacer arm lengths makes this possibility little attractive when drug discovery starts from scratch and no bivalent ligand is available for the heteromer of interest. The most straightforward approach for drug screening would be similar to that currently used for drugs targeting a single receptor. The development of heteromer-selective ligands would require, however, the use of three different biological materials. If using heterologous cells expressing the receptors, three different cell lines would be required, that is, one expressing one receptor, one expressing the other, and a third expressing both receptors at appropriate proportions for optimal heteromer formation. This would mean to, at least, triplicate the time for HTS comparing with the current screen procedures using just one receptor expressed in one cell line (Fig. 6.1A). It is a matter of cost/benefit and time will tell whether heteromer-selective drugs reaching the pharmaceutical market would drive an important need of the search for this kind of molecules. New techniques to minimize the time spent on screening would also be highly valuable in the overall context.

The basic binding-based approach to select heteromer-selective drugs is well defined by Orru *et al.*[84] who used it to profile adenosine A_{2A} receptor antagonists targeting pre- or postsynaptic receptors (Fig. 6.1B). As in the striatum, A_{2A}/D_2

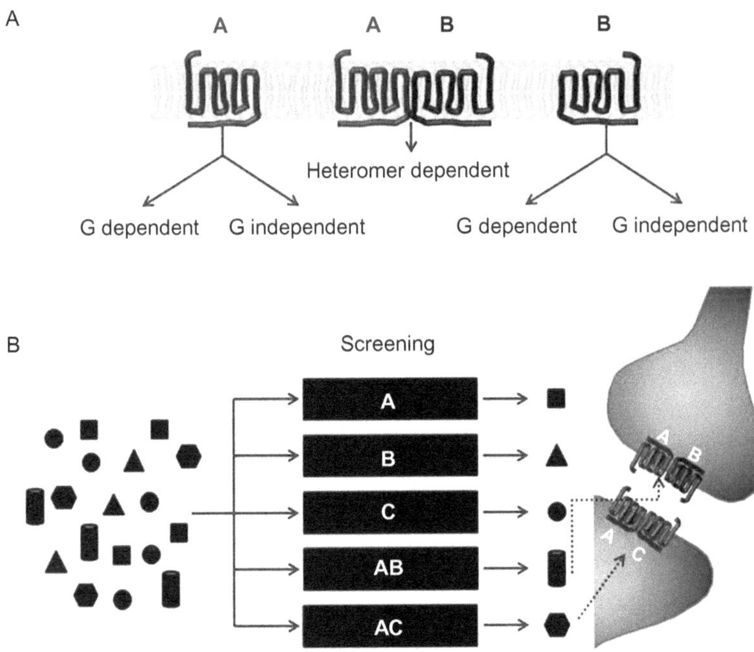

Figure 6.1 Heteromer-based screening. (Panel A) The heteromer has specific signaling pathways different from those of the individual receptors. (Panel B) Screening of compounds may be based on systems expressing individual receptors (A, B, or C) but also receptor heteromers (AB or AC). For instance, it is likely that heteromer-based screening will be able to find compounds more selective for presynaptic (AB) than for postsynaptic (AC) receptor heteromers or *vice versa*. (For color version of this figure, the reader is referred to the online version of this chapter.)

heteromers are located postsynaptically whereas A_1/A_{2A} heteromers are located presynaptically, A_{2A}/D_2-heteromer-selective heteromers would be preferentially useful in the therapy of Parkinson's disease whereas A_1/A_{2A}-heteromer-selective drugs would be potentially useful to combat Huntington's disease. Six structurally diverse antagonists selective for A_{2A} were tested in membranes isolated from four different CHO cell lines stably expressing similar amounts of receptors. Two were expressing either A_1 or A_{2A} receptors, and two were expressing A_1/A_{2A} or A_{2A}/D_2 heterodimers. Competition studies using a selective A_1 or A_{2A} radioligand led to the discovery of differences in the binding affinity depending on the heteromeric context of the A_{2A} receptor. It should be noted that it may happen that screening using just a single concentration of the drug does not give rise to significant differences, meaning that confirmation of heteromer selectivity requires a full competition curve using

8–12 different concentrations of the ligand being tested. Apart from significantly different equilibrium constant values, this approach led to the unexpected finding for a reported neutral antagonist, SCH-442416 that in A_{2A}/D_2-receptor-expressing cells but not in other cells competition curves were biphasic for SCH-442416 that is reported to be a neutral antagonist and, therefore, it should lead to a monophasic competition curve. This finding is important as heteromer-selective ligands may be identified by comparing the shape of the different competition curves obtained using the different cells. Only in A_{2A}/D_2-receptor-containing membranes SCH-442416 behaves as an atypical A_{2A} receptor antagonist proved by a very high negative cooperativity on binding to the receptors and a marked loss of affinity. In contrast, the behavior of the compound in cells expressing A_{2A} or A_1/A_{2A} was similar. These findings confirm that the orthosteric site of a receptor in heteromers may markedly change and shift affinities from the nanomolar to the micromolar range (or vice versa). The results also may explain why A_{2A} receptor antagonists may be potential antiparkinsonian drugs whereas others, such as SCH-442416, may be detrimental for Parkinson's but promising for dyskinetic disorders, obsessive–compulsive disorders or drug addiction.[84]

One important aspect to be considered is the correct handling of binding data to be able to detect the changes in the binding parameters due to heteromer-selective drugs. On the one hand, it is advisable to fit data using an approach able to distinguish between noncomplex (monophasic) and complex (biphasic) behavior in drug binding. As above mentioned, it would not be correct to fit data of SCH-442416 binding assuming that it always will give rise to monophasic competition curves. The assumption of complex binding behavior must be always taken into account when competition experiments are performed in cells expressing GPCR heteromers. On the other hand, data may be analyzed using an *ad hoc* model, the dimer-based model described elsewhere,[85] instead of fitting data to the model assuming the coexistence of two (high and low) affinity states. Fitting data to the dimer model can be easily performed as detailed described by Casadó *et al.*[58,59] Although the description of the model is out of the scope of the present chapter, one of the many advantages of its use must be pointed out as it allows a powerful way to compare the tested drugs. In fact, apart from the equilibrium constant(s) of the binding, the model allows calculation of a parameter (denoted as B_{50}) that is the concentration of drug that reduces the radioligand binding a 50%. This parameter is very robust as it is equally useful in both noncomplex- (monophasic) and complex (biphasic)-binding behaviors. The equivalent IC_{50} value is robust in noncomplex cases but becomes of controversial value for selecting lead compounds when the screened drugs exhibit biphasic competition curves or a mix of monophasic and biphasic competition curves (see Ref. 58).

In summary, direct screening must be based on binding approaches, although it is advisable to try quicker and easier approaches (see below) to make a preselection of compounds. It should be noted that the future will quite likely help in shortening the time of performance of binding assays. Interestingly, part of the complexity of the approach comes from the use of radioligands and the need of a filtration step, which is critical for accurate calculation of the kinetic/thermodynamic parameters of the binding. In fact, a homogeneous assay has been recently developed[86] for GPCR ligand-binding assay that based on time-resolved fluorescence resonance energy transfer (TR-FRET). The method is easy to implement for GPCRs with peptide or protein ligands as the TR-FRET acceptor must be covalently linked to the ligand. This nonradioactive-labeled compound would be similar to the radioligand in the radioactive-based classical assays. Small nonpeptide/nonprotein ligands selective for a variety of GPCRs and having a covalently linked TR-FRET acceptor are becoming commercially available. In this approach, the receptor must be modified to be able to (after cell expression) covalently link the suitable pair of TR-FRET. Provided the labeled ligand is available (commercially or synthesized in house), the technique is readily available to perform binding to a single GPCR. The method is robust and gives parameters totally comparable to those obtained using radioligand binding. If a TR-FRET-based assay is further developed to study binding to heteromers, it would likely become the method of choice to do in a homogenous and quick manner screening of heteromer-selective molecules.

6.2. Indirect signaling-based approach

Although the direct approach based on radioligand binding is the most suitable, it is time consuming and expensive. Therefore, it would be advisable to perform previous screenings to select a few compounds to be tested in the binding assays. This would allow one to know the actual affinity for the heteromer and to calculate the heteromer selectivity that is calculated by the ratio of affinity toward the heteromer versus each affinity toward the two protomers. A first such screening would be quicker and less expensive and has necessarily to be based on a differential signaling property of the heteromer. The prototypical example is the one provided by the dopamine D_1/D_2 receptor heteromer that unlike D_1 or D_2 receptors couples to G_q and to calcium-mediated signaling. In cases where heteromers are still coupled to G_s or G_i and even to both Gs and Gs, signaling-based screening becomes less advisable. The alternative is to look for more distal signaling molecules that may be specifically/differentially produced via the heteromer.

A screening method used for "monomeric" GPCR may also be used for screening on heteromers. The β-arrestin-based screening has become quite popular and is based in the detection by bioluminescence resonance energy transfer (BRET) of the time course of the interaction of a GPCR with β-arrestin. A so-called enhanced BRET methodology (see Ref. 87 for review) enables live cell compound screening and provides a new level of sensitivity for monitoring specific transient, weak or hardly detectable protein–protein complexes, including agonist-independent GPCR/β-arrestin interactions. This type of assay has been modified to be able to detect β-arrestin binding to receptor heteromers (see Ref. 88), and therefore, it may be potentially used for identification of heteromer-selective drugs. It must be noted that for this purpose the assays must use the same number of cells as in the case in binding-based assays, that is, expressing the individual receptors and the receptor heteromer. Also, for BRET-based assays, the receptors must be conveniently fused to donor and acceptor BRET pairs and the fusion proteins tested for correct functionality, that is, similar behavior than the receptor not fused to any other protein.

7. CONCLUSIONS

Despite the fact that heteromers are not a new concept in the field of GPCRs, their exploitation for the purposes of drug targeting have to date not been realized. Here, we have highlighted some of the outstanding challenges that need to be resolved to bring us further toward that realization. These challenges include a clear perspective on the stoichiometry and stability of heteromers as well as how these two may change in the presence of ligands. In addition, we do not understand how are heteromers formed and what controls their formation. Finally, and perhaps most importantly, there is the challenge to identify what biological functions are impacted by heteromers. To meet these challenges, it will be necessary to apply and refine new models for calculating binding constants. In addition, new tools and techniques that allow for the detection of heteromers in tissue are paramount. Finally, to target heteromers, the development of new classes of compounds that include allosteric modulators that act via the partner receptor or the design of multivalent ligands will increase the chances of successful therapeutic drugs.

ACKNOWLEDGMENTS

Support from the Spanish Ministry of Economía y Competitividad (SAF2012 39875-C02-01, SAF2010-18472) and Fundación de Investigación Médica Applicada (FIMA). P. J. M. is a Ramón y Cajal fellow.

REFERENCES

1. Manglik A, Kruse AC, Kobilka TS, Thian FS, Mathiesen JM, Sunahara RK, et al. Crystal structure of the μ-opioid receptor bound to a morphinan antagonist. *Nature* 2012;**485**:321–6.
2. Claing A, Laporte SA, Caron MG, Lefkowitz RJ. Endocytosis of G protein-coupled receptors: roles of G protein-coupled receptor kinases and beta-arrestin proteins. *Prog Neurobiol* 2002;**66**:61–79 [Review].
3. Franco R, Casadó V, Ciruela F, Mallol J, Lluis C, Canela EI. The cluster-arranged cooperative model: a model that accounts for the kinetics of binding to A1 adenosine receptors. *Biochemistry* 1996;**35**:3007–15.
4. Franco R, Canals M, Marcellino D, Ferré S, Agnati L, Mallol J, et al. Regulation of heptaspanning-membrane-receptor function by dimerization and clustering. *Trends Biochem Sci* 2003;**28**:238–43.
5. Escriche M, Burgueno J, Ciruela F, Canela EI, Mallol J, Enrich C, et al. Ligand-induced caveolae-mediated internalization of A(1) adenosine receptors: morphological evidence of endosomal sorting and receptor recycling. *Exp Cell Res* 2003;**285**:72–90.
6. Saura CA, Mallol J, Canela EI, Lluis C, Franco R. Adenosine deaminase and A1 adenosine receptors internalize together following agonist-induced receptor desensitization. *J Biol Chem* 1998;**273**:17610–7.
7. Sarrió S, Casadó V, Escriche M, Ciruela F, Mallol J, Canela EI, et al. The heat shock cognate protein hsc73 assembles with A(1) adenosine receptors to form functional modules in the cell membrane. *Mol Cell Biol* 2000;**20**:5164–74.
8. Burgueño J, Blake DJ, Benson MA, Tinsley CL, Esapa CT, Canela EI, et al. The adenosine A(2A) receptor interacts with the actin-binding protein alpha-actinin. *J Biol Chem* 2003;**278**:37545–52.
9. Saura C, Ciruela F, Casado V, Canela EI, Mallol J, Lluis C, et al. Adenosine deaminase interacts with A(1) adenosine receptors in pig brain cortical membranes. *J Neurochem* 1996;**66**:1675–82.
10. Söderberg O, Gullberg M, Jarvius M, Ridderstråle K, Leuchowius KJ, Jarvius J, et al. Direct observation of individual endogenous protein complexes in situ by proximity ligation. *Nat Methods* 2006;**3**:995–1000.
11. Gullberg M, Andersson AC. Visualization and quantification of protein-protein interactions in cells and tissues. *Nat Methods* 2010;**7**:5–6.
12. Ferré S, Baler R, Bouvier M, Caron MG, Devi LA, Durroux T, et al. Building a new conceptual framework for receptor heteromers. *Nat Chem Biol* 2009;**5**:131–4.
13. AbdAlla S, Lother H, el Massiery A, Quitterer U. Increased AT(1) receptor heterodimers in preeclampsia mediate enhanced angiotensin II responsiveness. *Nat Med* 2001;**7**:1003–9.
14. González-Hernández ML, Godínez-Hernández D, Bobadilla-Lugo RA, López-Sánchez P. Angiotensin-II type 1 receptor (AT1R) and alpha-1D adrenoceptor form a heterodimer during pregnancy-induced hypertension. *Auton Autacoid Pharmacol* 2010;**30**:167–72.

15. Kaaja R, Kinnunen T, Luoto R. Regional differences in the prevalence of pre-eclampsia in relation to the risk factors for coronary artery disease in women in Finland. *Eur Heart J* 2005;**26**:44–50.
16. Lydakis C, Beevers M, Beevers DG, Lip GY. The prevalence of pre-eclampsia and obstetric outcome in pregnancies of normotensive and hypertensive women attending a hospital specialist clinic. *Int J Clin Pract* 2001;**55**:361–7.
17. Franco R, Casadó V, Cortés A, Mallol J, Ciruela F, Ferré S, et al. G-protein-coupled receptor heteromers: function and ligand pharmacology. *Br J Pharmacol* 2008;**153** (Suppl. 1):S90–S98 [Review].
18. Trifilieff P, Rives ML, Urizar E, Piskorowski RA, Vishwasrao HD, Castrillon J, et al. Detection of antigen interactions ex vivo by proximity ligation assay: endogenous dopamine D2-adenosine A2A receptor complexes in the striatum. *Biotechniques* 2011;**51**:111–8.
19. González S, Moreno-Delgado D, Moreno E, Pérez-Capote K, Franco R, Mallol J, et al. Circadian-related heteromerization of adrenergic and dopamine D4 receptors modulates melatonin synthesis and release in the pineal gland. *PLoS Biol* 2012;**10**: e1001347.
20. Callen L, Moreno E, Barroso-Chinea P, Moreno-Delgado D, Cortes A, Mallol J, et al. Cannabinoid receptors CB1 and CB2 form functional heteromers in the brain. *J Biol Chem* 2012;**287**:20851–65.
21. Kenakin T. Agonist-receptor efficacy. II. Agonist trafficking of receptor signals. *Trends Pharmacol Sci* 1995;**16**:232–8.
22. Granier S, Kobilka B. A new era of GPCR structural and chemical biology. *Nat Chem Biol* 2012;**8**:670.
23. Nygaard R, Zou Y, Dror RO, Mildorf TJ, Arlow DH, Manglik A, et al. The dynamic process of β(2)-adrenergic receptor activation. *Cell* 2013;**153**:532–42.
24. West GM, Chien EYT, Katritch V, Gatchalian J, Chalmers MJ, Stevens RC, et al. Ligand-dependent perturbation of the conformational ensemble for the GPCR β2 adrenergic receptor revealed by HDX. *Structure* 2011;**19**:1424–32.
25. Kahsai AW, Xiao K, Rajagopal S, Ahn S, Shukla AK, Sun J, et al. Multiple ligand-specific conformations of the β2-adrenergic receptor. *Nat Chem Biol* 2011;**7**:692–700.
26. Ye S, Zaitseva E, Caltabiano G, Schertler GFX, Sakmar TP, Deupi X, et al. Tracking G-protein-coupled receptor activation using genetically encoded infrared probes. *Nature* 2010;**464**:1386–9.
27. Yao XJ, Ruiz GV, Whorton MR, Rasmussen SGF, DeVree BT, Deupi X, et al. The effect of ligand efficacy on the formation and stability of a GPCR-G protein complex. *Proc Natl Acad Sci USA* 2009;**106**:9501–6.
28. Yao X, Parnot C, Deupi X, Ratnala VRP, Swaminath G, Farrens D, et al. Coupling ligand structure to specific conformational switches in the β2-adrenoceptor. *Nat Chem Biol* 2006;**2**:417–22.
29. Swaminath G, Xiang Y, Lee TW, Steenhuis J, Parnot C, Kobilka BK. Sequential binding of agonists to the β2 adrenoceptor. Kinetic evidence for intermediate conformational states. *J Biol Chem* 2004;**279**:686–91.
30. MacKinnon AC, Waters C, Jodrell D, Haslett C, Sethi T. Bombesin and substance P analogues differentially regulate G-protein coupling to the bombesin receptor. Direct evidence for biased agonism. *J Biol Chem* 2001;**276**:28083–91.
31. Mary S, Damian M, Louet M, Floquet N, Fehrentz J-A, Marie J, et al. Ligands and signaling proteins govern the conformational landscape explored by a G protein-coupled receptor. *Proc Natl Acad Sci USA* 2012;**109**:8304–9.
32. Zocher M, Fung JJ, Kobilka BK, Müller DJ. Ligand-specific interactions modulate kinetic, energetic, and mechanical properties of the human β2 adrenergic receptor. *Structure* 2012;**20**:1391–402.

33. Maurice P, Kamal M, Jockers R. Asymmetry of GPCR oligomers supports their functional relevance. *Trends Pharmacol Sci* 2011;**32**:514–20.
34. Luttrell LM, Gesty-Palmer D. Beyond desensitization: physiological relevance of arrestin-dependent signaling. *Pharmacol Rev* 2010;**62**:305–30.
35. Allen JA, Yost JM, Setola V, Chen X, Sassano MF, Chen M, et al. Discovery of β-arrestin–biased dopamine D2 ligands for probing signal transduction pathways essential for antipsychotic efficacy. *Proc Natl Acad Sci* 2011;**108**:18488–93.
36. Whalen EJ, Rajagopal S, Lefkowitz RJ. Therapeutic potential of β-arrestin- and G protein-biased agonists. *Trends Mol Med* 2011;**17**:126–39.
37. Décaillot FM, Kazmi MA, Lin Y, Ray-Saha S, Sakmar TP, Sachdev P. CXCR7/CXCR4 heterodimer constitutively recruits β-arrestin to enhance cell migration. *J Biol Chem* 2011;**286**:32188–97.
38. Cervantes D, Crosby C, Xiang Y. Arrestin orchestrates crosstalk between G protein-coupled receptors to modulate the spatiotemporal activation of ERK MAPK. *Circ Res* 2009;**106**:1.
39. Saulière A, Bellot M, Paris H, Denis C, Finana F, Hansen JT, et al. Deciphering biased-agonism complexity reveals a new active AT1 receptor entity. *Nat Chem Biol* 2012;**8**:622–30.
40. Prezeau L, Rives M-L, Comps-Agrar L, Maurel D, Kniazeff J, Pin J-P. Functional crosstalk between GPCRs: with or without oligomerization. *Curr Opin Pharmacol* 2010;**10**:6–13.
41. Rajagopal S, Kim J, Ahn S, Craig S, Lam CM, Gerard NP, et al. β-arrestin- but not G protein-mediated signaling by the "decoy" receptor CXCR7. *Proc Natl Acad Sci* 2010;**107**:628–32.
42. Levoye A, Balabanian K, Baleux F, Bachelerie F, Lagane B. CXCR7 heterodimerizes with CXCR4 and regulates CXCL12-mediated G protein signalling. *Blood* 2009;**113**:6085–93.
43. Decaillot FM, Kazmi MA, Lin Y, Ray-Saha S, Sakmar TP, Sachdev P. CXCR7/CXCR4 heterodimer constitutively recruits {beta}-arrestin to enhance cell migration. *J Biol Chem* 2011;**286**:32188–97.
44. Innamorati G, Giannone F, Guzzi F, Rovati GE, Accomazzo MR, Chini B, et al. Heterotrimeric G proteins demonstrate differential sensitivity to beta-arrestin dependent desensitization. *Cell Signal* 2009;**21**:1135–42.
45. Busnelli M, Saulière A, Manning M, Bouvier M, Galés C, Chini B. Functional selective oxytocin-derived agonists discriminate between individual G protein family subtypes. *J Biol Chem* 2012;**287**:3617–29.
46. Han Y, Moreira IS, Urizar E, Weinstein H, Javitch JA. Allosteric communication between protomers of dopamine class A GPCR dimers modulates activation. *Nat Chem Biol* 2009;**5**:688–95.
47. Mason JS, Bortolato A, Congreve M, Marshall FH. New insights from structural biology into the druggability of G protein-coupled receptors. *Trends Pharmacol Sci* 2012;**33**:249–60.
48. de Kloe GE, Bailey D, Leurs R, de Esch IJP. Transforming fragments into candidates: small becomes big in medicinal chemistry. *Drug Discov Today* 2009;**14**:630–46.
49. Smith NJ, Milligan G. Allostery at G protein-coupled receptor homo- and heteromers: uncharted pharmacological landscapes. *Pharmacol Rev* 2010;**62**:701–25.
50. De los Frailes M, Diez E. Screening technologies for G protein-coupled receptors: from HTS to uHTS. *Methods Mol Biol* 2009;**552**:15–37.
51. Leff P. The two-state model of receptor activation. *Trends Pharmacol Sci* 1995;**16**:89–97.
52. Canals M, Sexton PM, Christopoulos A. Allostery in GPCRs: "MWC" revisited. *Trends Biochem Sci* 2011;**36**:663–72.

53. Kristiansen K. Molecular mechanisms of ligand binding, signaling, and regulation within the superfamily of G-protein-coupled receptors: molecular modeling and mutagenesis approaches to receptor structure and function. *Pharmacol Ther* 2004;**103**:21–80.
54. Kobilka B, Schertler GFX. New G-protein-coupled receptor crystal structures: insights and limitations. *Trends Pharmacol Sci* 2008;**29**:79–83.
55. Digby GJ, Conn PJ, Lindsley CW. Orthosteric- and allosteric-induced ligand-directed trafficking at GPCRs. *Curr Opin Drug Discov Devel* 2010;**13**:587–94.
56. May LT, Leach K, Sexton PM, Christopoulos A. Allosteric modulation of G protein-coupled receptors. *Annu Rev Pharmacol Toxicol* 2007;**47**:1–51.
57. Langmead CJ, Christopoulos A. Allosteric agonists of 7TM receptors: expanding the pharmacological toolbox. *Trends Pharmacol Sci* 2006;**27**:475–81.
58. Casadó V, Cortés A, Mallol J, Pérez-Capote K, Ferré S, Lluis C, et al. GPCR homomers and heteromers: a better choice as targets for drug development than GPCR monomers? *Pharmacol Ther* 2009;**124**:248–57.
59. Casadó V, Cortés A, Ciruela F, Mallol J, Ferré S, Lluis C, et al. Old and new ways to calculate the affinity of agonists and antagonists interacting with G-protein-coupled monomeric and dimeric receptors: the receptor-dimer cooperativity index. *Pharmacol Ther* 2007;**116**:343–54.
60. Franco R, Casadó V, Mallol J, Ferrada C, Ferré S, Fuxe K, et al. The two-state dimer receptor model: a general model for receptor dimers. *Mol Pharmacol* 2006; **69**:1905–12.
61. Franco R, Casadó V, Cortés A, Pérez-Capote K, Mallol J, Canela E, et al. Novel pharmacological targets based on receptor heteromers. *Brain Res Rev* 2008;**58**:475–82.
62. Gomes I, Ijzerman AP, Ye K, Maillet EL, Devi LA. G protein-coupled receptor heteromerization: a role in allosteric modulation of ligand binding. *Mol Pharmacol* 2011;**79**:1044–52.
63. González-Maeso J, Ang RL, Yuen T, Chan P, Weisstaub NV, López-Giménez JF, et al. Identification of a serotonin/glutamate receptor complex implicated in psychosis. *Nature* 2008;**452**:93–7.
64. Lindsley CW, Hopkins CR. Metabotropic glutamate receptor 4 (mGlu4)-positive allosteric modulators for the treatment of Parkinson's disease: historical perspective and review of the patent literature. *Expert Opin Ther Pat* 2012;**22**:461–81.
65. Müller CE, Schiedel AC, Baqi Y. Allosteric modulators of rhodopsin-like G protein-coupled receptors: opportunities in drug development. *Pharmacol Ther* 2012;**135**:292–315.
66. Shook BC, Rassnick S, Wallace N, Crooke J, Ault M, Chakravarty D, et al. Design and characterization of optimized adenosine A_{2A}/A_1 receptor antagonists for the treatment of Parkinson's disease. *J Med Chem* 2012;**55**:1402–17.
67. Vendrell M, Angulo E, Casadó V, Lluis C, Franco R, Albericio F, et al. Novel ergopeptides as dual ligands for adenosine and dopamine receptors. *J Med Chem* 2007;**50**:3062–9.
68. Ciruela F, Casadó V, Rodrigues RJ, Luján R, Burgueño J, Canals M, et al. Presynaptic control of striatal glutamatergic neurotransmission by adenosine A1-A2A receptor heteromers. *J Neurosci* 2006;**26**:2080–7.
69. Rashid AJ, So CH, Kong MM, Furtak T, El-Ghundi M, Cheng R, et al. D1-D2 dopamine receptor heterooligomers with unique pharmacology are coupled to rapid activation of Gq/11 in the striatum. *Proc Natl Acad Sci USA* 2007;**104**:654–9.
70. Portoghese PS. From models to molecules: opioid receptor dimers, bivalent ligands, and selective opioid receptor probes. *J Med Chem* 2001;**44**:2259–69.
71. Valant C, Robert LJ, Sexton PM, Christopoulos A. The best of both worlds? Bitopic orthosteric/allosteric ligands of g protein-coupled receptors. *Annu Rev Pharmacol Toxicol* 2012;**52**:153–78.

72. Soriano A, Ventura R, Molero A, Hoen R, Casadó V, Cortés A, et al. Adenosine A2A receptor-antagonist/dopamine D2 receptor-agonist bivalent ligands as pharmacological tools to detect A2A-D2 receptor heteromers. *J Med Chem* 2009;**52**:5590–602.
73. Kühhorn J, Hübner H, Gmeiner P. Bivalent dopamine D2 receptor ligands: synthesis and binding properties. *J Med Chem* 2011;**54**:4896–903.
74. Russo O, Berthouze M, Giner M, Soulier JL, Rivail L, Sicsic S, et al. Synthesis of specific bivalent probes that functionally interact with 5-HT(4) receptor dimers. *J Med Chem* 2007;**50**:4482–92.
75. Halazy S, Perez M, Fourrier C, Pallard I, Pauwels PJ, Palmier C, et al. Serotonin dimers: application of the bivalent ligand approach to the design of new potent and selective 5-HT(1B/1D) agonists. *J Med Chem* 1996;**39**:4920–7.
76. Handl HL, Sankaranarayanan R, Josan JS, Vagner J, Mash EA, Gillies RJ, et al. Synthesis and evaluation of bivalent NDP-alpha-MSH(7) peptide ligands for binding to the human melanocortin receptor 4 (hMC4R). *Bioconjug Chem* 2007;**18**:1101–9.
77. Yuan Y, Arnatt CK, Li G, Haney KM, Ding D, Jacob JC, et al. Design and synthesis of a bivalent ligand to explore the putative heterodimerization of the mu opioid receptor and the chemokine receptor CCR5. *Org Biomol Chem* 2012;**10**:2633–46.
78. Thelen M, Muñoz LM, Rodríguez-Frade JM, Mellado M. Chemokine receptor oligomerization: functional considerations. *Curr Opin Pharmacol* 2009;**10**:38–43.
79. Muñoz LM, Lucas P, Holgado BL, Barroso R, Vega B, Rodríguez-Frade JM, et al. Receptor oligomerization: a pivotal mechanism for regulating chemokine function. *Pharmacol Ther* 2011;**131**:351–8.
80. Springael JY, Urizar E, Parmentier M. Dimerization of chemokine receptors and its functional consequences. *Cytokine Growth Factor Rev* 2005;**16**:611–23.
81. Tanaka T, Nomura W, Narumi T, Masuda A, Tamamura H. Bivalent ligands of CXCR4 with rigid linkers for elucidation of the dimerization state in cells. *J Am Chem Soc* 2010;**132**:15899–901.
82. Metcalf MD, Yekkirala AS, Powers MD, Kitto KF, Fairbanks CA, Wilcox GL, et al. The δ opioid receptor agonist SNC80 selectively activates heteromeric μ-δ opioid receptors. *ACS Chem Neurosci* 2012;**3**:505–9.
83. Hasbi A, Fan T, Alijaniaram M, Nguyen T, Perreault ML, O'Dowd BF, et al. Calcium signaling cascade links dopamine D1-D2 receptor heteromer to striatal BDNF production and neuronal growth. *Proc Natl Acad Sci USA* 2009;**106**:21377–82.
84. Orru M, Bakesova´ J, Brugarolas M, Quiroz C, Beaumont V, Goldberg SR, et al. Striatal pre- and postsynaptic profile of adenosine A2A receptor antagonists. *PLoS One* 2011;**6**: e16088.
85. Franco R, Casadó V, Mallol J, Ferré S, Fuxe K, Cortés A, et al. Dimer-based model for heptaspanning membrane receptors. *Trends Biochem Sci* 2005;**30**:360–6.
86. Hu LA, Zhou T, Hamman BD, Liu Q. A homogeneous G protein-coupled receptor ligand binding assay based on time-resolved fluorescence resonance energy transfer. *Assay Drug Dev Technol* 2008;**6**:543–50.
87. Kocan M, Dalrymple MB, Seeber RM, Feldman BJ, Pfleger KD. Enhanced BRET technology for the monitoring of agonist-induced and agonist-independent interactions between GPCRs and β-arrestins. *Front Endocrinol* 2010;**1**:12.
88. See HB, Seeber RM, Kocan M, Eidne KA, Pfleger KD. Application of G protein-coupled receptor-heteromer identification technology to monitor β-arrestin recruitment to G protein-coupled receptor heteromers. *Assay Drug Dev Technol* 2011;**9**:21–30.

CHAPTER SEVEN

Di/Oligomerization of GPCRs—Mechanisms and Functional Significance

Adolfo Rivero-Müller*, Kim Carol Jonas†, Aylin Carla Hanyaloglu†, Ilpo Huhtaniemi*, †

*Department of Physiology, Institute for Biomedicine, University of Turku, Turku, Finland

†Department of Surgery and Cancer, Institute of Reproductive and Developmental Biology, Hammersmith Campus, Imperial College London, London, United Kingdom

Contents

Abstract

Initially considered as strict and rigid monomers, GPCRs have shown an ever-increasing flexibility in the number of intermolecular interactions they can participate in. Few rules can be applied to all members of this receptor superfamily, as the functional roles of resulting dimers or oligomers vary widely from biosynthesis, functional complementation, enhancement and control of signaling, positive and negative cooperativity, signal desensitization, membrane trafficking, and pharmacological diversity, among others. Receptor dimerization/oligomerization is a challenging phenomenon to disentangle even *in vitro* and thus *in vivo* evidence for its functional significance requires employment of a variety of cutting-edge techniques. Nevertheless, the evidence for GPCR

Progress in Molecular Biology and Translational Science, Volume 117
ISSN 1877-1173
http://dx.doi.org/10.1016/B978-0-12-386931-9.00007-6

di/oligomerization at the physiological level is mounting and has begun to shed light on an intricate network of interactions that challenge the previous knowledge on how this family of receptors function. Here, we summarize the evidence for GPCR di/oligomerization in living organisms with a particular focus of their involvement in health and disease.

1. INTRODUCTION

For more than 30 years, G protein-coupled receptors (GPCRs) have been observed to form aggregates when visualized by Western blotting and immunoprecipitation assays. However, the existence of functionally meaningful di/oligomeric GPCR structures has remained controversial until recently. Although for receptor families, such as tyrosine kinase, TNF-α, cytokine, and intracellular steroid receptors, di/oligomerization is the accepted paradigm for agonist-mediated activation,[1,2] GPCRs were until the 1990s considered as monomeric, single units, activated by a single ligand, that associate with heterotrimeric G proteins to evoke its intracellular signaling response(s).

One of the first suggestions that GPCR di/oligomers exist was postulated by Nigel Birdsall in 1982, who suggested that: "...a drug A to receptor A allosterically affect receptor B in such a way as to change the binding of drug B to receptor B?"[3] This was concluded from evidence gathered by Watanabe and co-workers in 1978 when they discovered that ligand activation of one receptor could alter the affinity of another receptor for its ligand. It was shown that the β2-adrenergic receptor (β2AR) lost affinity for its agonist, isoproterenol (ISO), in the presence of GTP. Surprisingly, methacholine, a nonselective muscarinic receptor agonist, could restore the affinity of β2AR for ISO in the presence of GTP, an effect that methacholine exerted without direct contact with the β2AR but through its own receptor, the muscarinic receptor (M_3).[4]

With the progress of biochemical techniques in the early 1990s, the view that GPCRs could form dimers/oligomers began to change when several groups produced the first evidence that GPCR dimers influenced the functionality of the receptors involved. The β2AR, dopamine D_2 receptor, and metabotropic glutamate (mGlu) receptor were all found as high molecular weight complexes corresponding to predicted dimer/oligomer molecular assembly.[5–7] This prompted many to reconsider if GPCRs might form higher-order structural complexes that could be necessary for their activation.

The search for di/oligomers, usually by co-immunoprecipitation (see Box 7.1), resulted in the discovery not only of homodimers but also of heterodimers. Heterodimers of closely related GPCRs, such as GABA receptors B_1 and B_2,[8–10] and the δ-opioid and κ-opioid receptors[11] were first described, followed by the detection of heterodimers of distantly related GPCRs, such as the angiotensin AT_1 and bradykinin B_2 receptors,[12] and aggregates of homo- and heterodimers such as the $GABA_B$ and mGlu (see below).

BOX 7.1 Methods

Co-immunoprecipitation

An antibody-based technique that analyzes the physical interactions between two receptors. This technique uses specific antibodies against one receptor protomer and probing for the other receptor counterpart. Cells are disrupted and the antibody-bound receptor is bound to immunoglobulin affinity beads by centrifugation or magnetic attraction. The unbound proteins are washed away and the receptor–antibody complex is then resolved using a denaturing polyacrylamide gel (PAGE). Proteins are then transferred from the gel to a nitrocellulose membrane. An antibody against the second dimerizing receptor, native or differentially tagged, is then used to visualize the latter receptor and to detect if this receptor is associated with the former.

FRET

In order to assess receptor–receptor interactions on the membrane of living or fixed cells, this microscopic technique is based on the principle that a chromophore can transfer energy to another chromophore if they are close enough to each other. One chromophore (donor, e.g., cyan fluorescent protein, CFP) in its excitation state transfers energy when the second chromophore (acceptor, e.g., yellow fluorescent protein, YFP) is in its proximity; the distance between the protomers must be 10–100 Å and this can be calculated by the energy transfer efficiency. Consequently, the donor's fluorescent intensity reduces while the acceptor fluorescence increases in intensity. The excitation spectrum of the acceptor must overlap the emission spectrum of the donor.

BRET

BRET is based on a naturally occurring phenomenon observed in sea creatures such as the jellyfish *Aequorea victoria* and sea pansy *Renilla reniformis*. BRET occurs between a luminescent donor, *Renilla* luciferase (RLuc) or one of its variants, and a fluorescent acceptor which is a variant of green fluorescent protein (GFP), such as YFP. When RLuc and YFP are in close proximity (<100 Å), the energy generated by the enzymatic degradation of RLuc's substrate (coelenterazine or a derivative) is transferred to YFP, emitting its characteristic fluorescence wavelength. The advantage that BRET has over other resonance energy transfer techniques like FRET is that it does not require a conventional light source to initiate energy transfer; thus, it has an improved signal-to-noise ratio and is ideal for light-sensitive proteins.

However, dimerization/oligomerization is still not considered the dogma for all GPCRs, particularly those of Class A, where contradictory interpretations of their existence, functional significance, and the modes of their dimerization, such as *kiss-and-run* interactions and transactivation, have fuelled the debate.[13–16] Certainly not all GPCRs form dimers for their basic functions, since monomers of both the β2AR and rhodopsin receptors, when reconstituted into high-density lipoprotein particles, could couple to G proteins.[a,17,18] Nevertheless, as we shall see, evidence at physiological level fully supports their existence.[b]

This chapter discusses the evidence for GPCR dimerization, the methods used to study these interactions in living cells, and the physiological impact dimerization brings. As we shall see, GPCR dimerization plays many roles in physiological and pathophysiological processes.

2. DIMERS IN LIVING CELLS

With the advent of molecular biology techniques, researchers began to clone and mutate GPCRs in order to determine the structural and functional contributions of different receptor domains. Such techniques utilized naturally occurring and laboratory-generated mutations that have shed light on many of the interactions between GPCR protomers of homo- and heterodimers.

Early mutational studies of GPCRs suggested that the interacting interface is through the 7-transmembrane domain (7TMD)[19] which upon ligand binding undergoes a conformational change, switching from an "off state" to an "active state".[20] However, because this conformation–activation appears to be true for all known GPCRs, independent of dimeric or monomeric states, it does not appear to be a dimerization-specific phenomenon.

Attempting to detect and visualize GPCR dimers, researchers have opted for biophysical such as bioluminescence resonance energy transfer or fluorescence resonance energy transfer (FRET) (see Box 7.1; Fig. 7.1A) and have corroborated the existence of dimers in living cells (Refs. 21 and 22, respectively). Although these results corroborated the existence of GPCR dimers, they also provided puzzling questions as GPCR dimers were

[a] However, the preparation mainly containing rhodopsin dimers/oligomers appeared to be more functionally active.

[b] For practicality, we will from now on collectively call the interactions between GPCRs as *dimerization* and, if relevant, differentiate between homo- and heterodimers or dimers and oligomers.

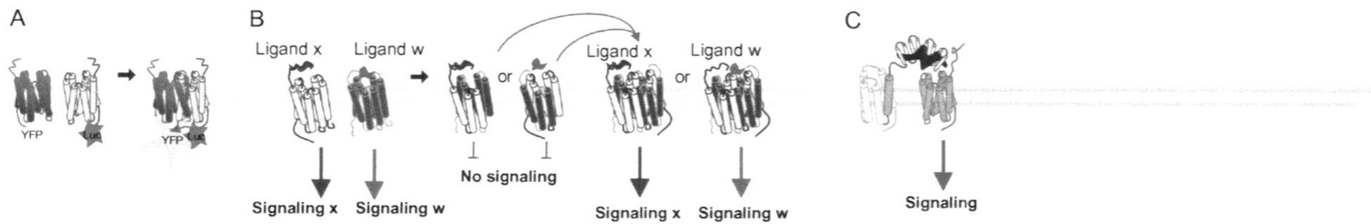

Figure 7.1 Models to study GPCR dimerization. Physical and functional interactions between receptors are analyzed by different methods, such as FRET and BRET (A), mutant receptor chimeras (B), or, in the case of receptors with large ECDs, by using the extracellular domain of a receptor fused with a membrane protein (could also be a different receptor) to transactivate its homologue TMD in a ligand-binding-deficient protomer (C). (For color version of this figure, the reader is referred to the online version of this chapter.)

found to (1) be present prior to ligand activation, (2) dissociate on ligand activation, or (3) remain unchanged in the presence or absence of ligand. Some of the interactions were so stable that they could still be detected by denaturing Western blots, while others were transient or very short lived (*kiss-and-run*). Nevertheless, GPCR dimerization appears more common than previously thought, adding an extra layer of complexity to their function.

3. OPERATIONAL DIMERS

The best evidence that dimers have functional significance has been demonstrated with Class C GPCRs. The $GABA_B$ receptors form *obligatory heterodimers*, where one of the protomers, $GABA_{B2}$, is unresponsive to ligand, acting "solely" as a chaperone to ensure correct trafficking and translocation of the $GABA_{B1}$ protomer to the cell surface, where it binds ligand and generates the intracellular signaling cascades[8] (Fig. 7.2A). It was also discovered that this interaction is more complex than just one receptor carrying the functional unit to the cell membrane, as the $GABA_{B1/B2}$ heterodimer is also indispensable for receptor G protein binding and signal activation.[23]

Thus, it was considered that stable dimer formation, through the TMDs, was required for GPCR functionality. Yet, as we discuss below, this concept was too simplistic as there are a multitude of ways in which GPCRs can interact and become activated. In an attempt to systematize these interactions, we can artificially separate dimers into several groups depending on their characteristics:

- through TMD or intracellular domains (*core dimerization*)
- *through ECDs* (extracellular domains)
- *negative and positive cooperativity* (allosteric regulation)
- through ECD to TMD (*intermolecular cooperation*).

3.1. Core dimerization

Since naturally occurring dimers form stable interactions, researchers have focused on determining how and where GPCR protein–protein interactions take place. Interactions between the TMDs were proven in a landmark experiment by Maggio *et al.*[19] who created chimeric receptors of the first five TMs of the α_2-AR or M_3 with the last two TMs of the M_3 or α_2-AR, respectively. They discovered that the chimeric receptors were only functional when coexpressed with their respective receptor counterpart,

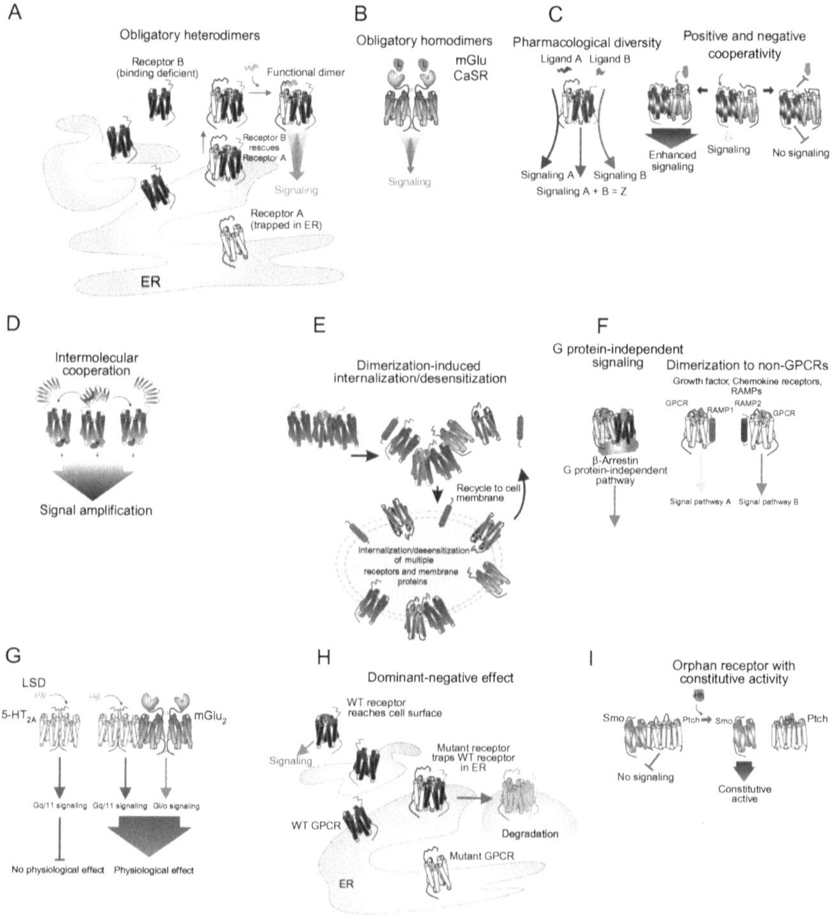

Figure 7.2 Roles of dimerization at physiological level. Dimerization of a transport-deficient receptor by a ligand-deficient receptor which results in an *obligatory heterodimer* (A). Similarly, some receptors form stable dimers through the ECDs, where the functional form is the homodimer (B). Dimers might change the intracellular responses (*pharmacological diversity*) or affect the affinity of the partnering protomer for its ligand: *positive* or *negative cooperativity* (C). A single activated receptor can transactivate (*intermolecular cooperation*) several neighboring receptors amplifying the signaling (D). Desensitization and internalization of GPCRs, and other membrane components, after activation of a receptor are induced by dimerization (E). Some signaling cascades are only induced when receptors form dimers with other GPCRs or other membrane proteins, for example, RAMPs (F). The hallucinogenic effect of LSD, which binds to the 5-HT_{2A} receptor, occurs only when dimerization with the $mGlu_2$ receptor (G). A mutant receptor unable to transit to the cell membrane can also dimerize and trap its WT counterpart, dominant-negative effect (H). A constitutively active orphan GPCR (Smo) is regulated by a dimerizing partner (Ptch); upon ligand binding (Hh) to Ptch, Smo is released and so its constitutive activity (I). L, ligand; ER, endoplasmic reticulum. *Panel G: Adapted from Ref. 94.* (For color version of this figure, the reader is referred to the online version of this chapter.)

partially rescuing ligand binding and signaling. This showed direct evidence of receptor–receptor interactions and functional rescue (Fig. 7.1B).

Further studies of receptor heterodimerization have been conducted to explore receptor transactivation. Using chimeras of mutant histamine H1 and α_{1b}-AR, with one receptor moiety coupling to a nonfunctional G protein (blocking any further G protein coupling at this receptor) and the other inactive mutant receptor moiety coupled to the WT G protein but unable to activate it, they observed transactivation between the G protein bound to the C-terminal of the inactive receptor and the intracellular domains of the other receptor protomer,[24] which could only be demonstrated by functional dimerization of these receptors.

It is possible that protomer distances in a dimer are crucial for activation of signal transduction, for example, G protein coupling, and by forcing monomers into close proximity so they rearrange into an active state. The luteinizing hormone/chorionic gonadotrophin receptor (LHCGR) has been shown to cluster in lipid rafts which, when exposed to ligand, increase in molecular weight.[25] Disruption of such lipid rafts using methyl-β-cyclodextrin decreased the signaling of these receptors, while receptor aggregation induced by a cross-linking antibody triggered receptor signaling in the absence of ligand.[26] Likewise, forced aggregation of several other receptors is sufficient for receptor activation,[27,28] supporting the functional significance of GPCR dimerization.

3.2. Dimerization through extracellular domains

While some members of Class C receptors such as the sweet and umami taste receptor heterodimers appear to function through similar mechanisms as the GABA receptors,[29] others such as the mGlu and calcium-sensing receptors (CaSR) have been shown to form *forced homodimers*, where signaling can only be achieved by a complex of two receptors[30] (Fig. 7.2B). Interestingly, these receptors with large ligand-binding ECDs achieve dimerization through interactions of the ECD of both receptors (Fig. 7.2B). The ECD domains of the $mGlu_1$,[31,32] $mGlu_4$,[33] and CaSRs[34] form homodimers in solution. This is also supported by structural X-ray crystallography data, determined both in the presence and absence of ligand.[35]

However, transient *kiss-and-run* interactions between the ECD of a receptor, when occupied by ligand, and the TMD of a neighboring receptor can result in activation of the latter receptor or activation of several neighboring receptor(s), resulting in signal amplification (Fig. 7.2D). However, such short-lived interactions are more difficult to detect and understand from a functional point of view (see Section 3.4 for further discussion).

3.3. Positive and negative cooperativity (allosteric regulation)

The occupancy of one protomer of a dimer can influence the affinity of a ligand for the other protomer of the dimer, a phenomenon known as *cooperativity*. *Positive cooperativity* is the result of activation of both protomers by specific agonists, where the first ligand–receptor coupling enhances the affinity of the second receptor to ligand, resulting in an "*active–active*" conformational change and full signal generation with a left shift of the dose–response curve. An elegant example of this is the M_3 homodimer. This receptor complex requires ligand occupation and activation of both protomers of the dimer for β-arrestin 1 recruitment.[36] This phenomenon has also been demonstrated for the δ-opioid receptor (δ-OR) and μ-opioid receptor (μ-OR) heterodimer, where agonist occupation of both ligand-binding sites can efficiently activate MAP kinase signaling, not otherwise observed.[11] *Negative cooperativity* is observed when one protomer is engaged by its ligand, blocking or decreasing the affinity of the second receptor for its ligand or intracellular signaling partners, resulting in a right shift of the dose–response curve. This has been observed among receptors with short ECDs, supporting the idea that physical interaction between individual TMDs occurs. Yet, as we shall discuss, GPCRs with large ECDs also can display negative cooperativity. Hence, dimerization can have both a positive and negative regulatory role in intracellular signaling responses (Fig. 7.2C).

Unexpectedly, Urizar *et al.* found that some Class A receptors with large ECDs: the glycoprotein hormone receptors [LHCGR, follicle-stimulating hormone receptor (FSHR), and thyroid-stimulating hormone receptor (TSHR)] can form dimers that act through negative cooperativity.[37] The same seems to apply to the mGlu receptors that form strict homodimers *in vitro* and *in vivo*. Yet, mGlu homodimers are also found within large clusters of $GABA_B$ heterodimers, and these large complexes function as negative regulators of $GABA_B$ G protein coupling.[38] These studies support the notion that an ECD–ligand complex can function as a protomer while the TMD translates ECD activation to intracellular responses.

3.4. Intermolecular cooperation

As previously discussed, in GPCRs with large ECDs responsible for ligand binding, such as the glycoprotein receptors and members of Class C receptors with globular ECDs ("Venus flytrap," VFT), the ECD could feasibly interact with the TMD of the same receptor and also with those of neighboring receptors. This raises the question: do the ECD and TMD function as *separate entities*? The first evidence for this type of interaction was obtained by mutation and protomer truncation studies where the ECD was either mutated or cleaved in order to determine if it could activate its neighboring counterpart. The laboratory of Tae Ji created many mutants and chimeras of the gonadotrophin receptors—FSHR and LHCGR—where the ECD of one receptor fused to a cell membrane anchor (lipoprotein or single TMD) could transactivate the TMD of a receptor lacking ligand-binding capacity[39–42] (Fig. 7.1C). This has also been confirmed in other studies using chimeras of glycoprotein receptors, LHCGR/FSHR, LHCGR/TSHR,[37,43] by crystallography data,[44] and FRET analysis of the LHCGR.[45]

Thus, complementary receptor protomers can rescue, at least partially, receptor functional deficiencies, and in the case of the glycoprotein hormone receptors, the ECD and the TMD can function as separate entities. Following this idea, Karges *et al.* elegantly designed an LHCGR mutant where the ECD was cleaved by a protease before the receptor reached the cell membrane. This cleaved ECD, only attached to the TMD by disulfide bonds, was able to bind hormone and activate the Gs coupling of the remaining 7TM fraction to levels equivalent to that of the WT LHCGR.[46] The finding that in insects, some pheromone receptors, Class C GPCRs, have a soluble ECD strongly supports this concept.[31] In this case, the ECD, or *soluble domain* to be more exact, changes conformation upon binding to a pheromone; the new conformation can activate the TMD at the cell membrane of target cells.

Additionally, Vassart *et al.* suggested that the ECD could also function as an *inverse agonist*, blocking further conformational changes in the TMD. It was proposed that ligand binding changes the conformation of the ECD releasing the activational "break." Furthermore, the new ECD conformation fully activates TMD-driven signaling, as appears to be the case with the TSHR.[47] This could also well be true for heterodimers where the ECD of one receptor, in the presence or absence of ligand, blocks the associated receptor protomer from becoming activated. However, so far there is little evidence supporting this.

4. PHARMACOLOGICAL DIVERSITY OF DIMERIZATION

Dimerization can impact on key processes that regulate receptor functionality such as desensitization (Fig. 7.2E), which acts to prevent *excitotoxicity*, a pathological process resulting in cell damage or death caused by excessive stimulation. Physiological examples of this have been demonstrated for numerous receptors. Upon activation of a single rhodopsin receptor, all the other cell surface rhodopsin receptor molecules become phosphorylated and thus temporarily desensitized.[48] The CCR5 and other GPCR subtypes can be conjointly phosphorylated and internalized when one of the subtypes is activated.[49]

The desensitization pattern of vasopressin $V_{1A}R$ and V_2R receptors differs depending on which protomer of the dimer is activated. Although endocytosis of both receptors is β-arrestin dependent, $V_{1A}R$ dissociates quickly from β-arrestin and is recycled back to the cell membrane while V_2R remains bound to β-arrestin and accumulates in endosomal compartments and subsequently degraded. When coexpressed, the two vasopressin receptors form heterodimers that are internalized together. The endocytotic fate of the heterodimers is determined by the receptor protomer that is activated. If $V_{1A}R$ is selectively activated, the heteromer is rapidly internalized and recycled back to the cell membrane. However, if a nonselective agonist is used to activate both protomers, then both receptors accumulate in endosomes with β-arrestin and are targeted for degradation. This was further supported by using a constitutively internalized V_2R mutant which, in the absence of agonist, could also internalize the $V_{1A}R$.[50]

Another example where heterodimerization can alter cell surface receptor levels concerns dopamine D_1 and D_2 heteromers. Different levels of cell surface expression of dopamine receptors D_1 and D_2 are observed when they are coexpressed; D_1 expression decreases in the presence of D_2, which in turn increases its membrane expression. These receptors form heterodimers in which both receptor types become internalized upon selective activation of one of the types. Receptor D_2, however, does not internalize if expressed alone.[51]

GPCRs can form dimers with other membrane receptors or proteins, and this supports the notion that dimerization works to promote pharmacological diversity. The elegant work of McLatchie *et al.* in *Xenopus* oocytes demonstrated that the calcitonin-receptor-like receptor (CRLR) is functionally dependent on the receptor activity-modifying proteins (RAMPs). Originally considered two different GPCRs: when CRLR associates with RAMP1, it functions as the calcitonin gene-related peptide receptor; yet, when associated with RAMP2, it responds specifically to adrenomedullin[52] (Fig. 7.2F).

5. *IN VIVO* EVIDENCE OF DIMERIZATION

The first *in vivo* evidence of functional GPCR dimers was obtained with the Class C receptors. The $GABA_B$ receptors were found as functional heterodimers *in vivo* using FRET detection of native $GABA_B$ receptors.[9,53]

The physiological and mechanistic function of $GABA_B$ heterodimers has been meticulously explored by Schwenk *et al.* finding that a subfamily of the KCTD (potassium channel tetramerization domain-containing) proteins increase the agonist potency and changed the signaling responses, promoting receptor desensitization of the $GABA_B$ receptor heterodimers.[54] Moreover, dimerization-interfering proteins, such as 14-3-3 zeta protein, are involved in pathological situations. Using animal models for chronic pain, Laffray *et al.* demonstrated that when 14-3-3 zeta is overexpressed under chronic pain conditions *in vivo*, it interacts with $GABA_{B1}$ subunit impairing the receptor's signaling responses and thus the pain-relieving effects of $GABA_B$ dimers.[55]

Utilizing a range of direct and indirect approaches, it has been demonstrated that Class A GPCR heterodimers are physiologically significant. A pharmacological agonist for the μ-OR which also functions as an antagonist of the δ-OR, a receptor involved in the tolerance and dependence of μ-OR activation (e.g., by morphine), blocks the side effects of morphine without reducing analgesia.[56] Moreover, heterodimers of the cannabinoid type 1 receptor, which couples to Gαi, and angiotensin type 1 receptor (AT_1R), which couples to Gαq, enhanced the mitogenic signaling and profibrogenic activity of AT_1R in ethanol-treated hepatic stellate cells. This is the main cell type responsible for hepatic fibrosis leading to liver cirrhosis.[57]

To elucidate if our model GPCR, the LHCGR, could form functional dimers *in vivo*, we collaborated with the group of Tae Ji in the creation of complementary mutant LHCGRs, with one mutant receptor unable to bind hormone but retaining G protein coupling ability and the other binding proficient but signaling deficient. We found that the point mutation C22A was sufficient to abolish ligand binding. The complementary signaling-deficient mutant required more drastic deletions. It is known that most activating and several inactivating mutations are located in the 6th and 7th TM helices of the TMD,[58] and thus we decided to delete both TM6 and 7. We included a prolactin secretion signal to this receptor (TM67Δ) to promote its targeting to the cell membrane despite its structural deficiency. It was confirmed that both mutant receptors could traffic to the cell surface, and functional testing demonstrated that the binding-deficient LHCGR (C22A) was unable to bind hCG while the signaling-deficient LHCGR (TM67Δ) was able to bind hCG but not able to signal. Interestingly, coexpression of both mutants in the human embryonic kidney cell line (HEK293) resulted in partial rescue of cAMP generation. Encouraged by these *in vitro* results, we created the same mutations in bacterial artificial chromosomes (BACs) containing the entire mouse LHCGR gene, so the receptors would have the correct spatiotemporal expression in the physiological context. The mutations were performed by *recombineering*[59] in *E. coli*, and the mutant BACs were purified and injected into mouse embryos.

Transgene positive mice were then crossed with the LHCGR knockout (KO) mouse line, previously created in our laboratory.[60] The male mice carrying no endogenous WT LHCGR gene (LuRKO) present with arrested postnatal sexual maturation, cryptorchid testes, small and poorly developed accessory sex organs, Leydig cell hypoplasia, spermatogenic arrest at the round spermatid stage, and very low testosterone production. Each mutant alone in the LuRKO background had the exact same phenotype as the LuRKO mice (Fig. 7.3A–C). We also confirmed that the mutant receptors could be detected on the cell surface and that the TM67Δ mutant could bind hormone. Strikingly, when both mutants were coexpressed in the LuRKO background, the phenotype was reversed in these male mice resulting in the restoration of full spermatogenesis and testis size despite lower testosterone levels in comparison to WT animals[61] (Fig. 7.3D–E).

Indeed, a receptor can potentiate the intracellular responses of the partnering receptor in a dimer. This is the case for the dopamine receptor-3 (D_3), which forms heterodimers with dopamine receptor-1 (D_1). Stimulation of both receptors by pharmacological agonists potentiates D_1 receptor-mediated behavioral effect (locomotion activation), which was absent in D_3-KO mice.[62]

Animal models provide an excellent tool for studying the role of dimerization in a physiological context. The hallucinogenic effects of 1-(2,5-dimethoxy-4-iodophenyl)-2-aminopropane (DOI) or lysergic acid diethylamide (LSD) are mainly exerted by activation of the serotonin 5-HT_{2A} receptor (5-HT_{2A}). Yet this receptor, itself a homodimer, forms oligomeric complexes with $mGlu_2$ dimers. Much surprise was generated by the finding that both receptors are required for DOI or LSD to enact their hallucinogenic activity as neither the 5-HT_{2A}-KO nor the $mGlu_2$-KO mice responded by the head-twitch behavior characteristic of hallucinogenic compounds[63,64] (Fig. 7.2G).

The importance of dimerization at the physiological level becomes evident by the existence of diseases clearly caused by GPCR dimerization.

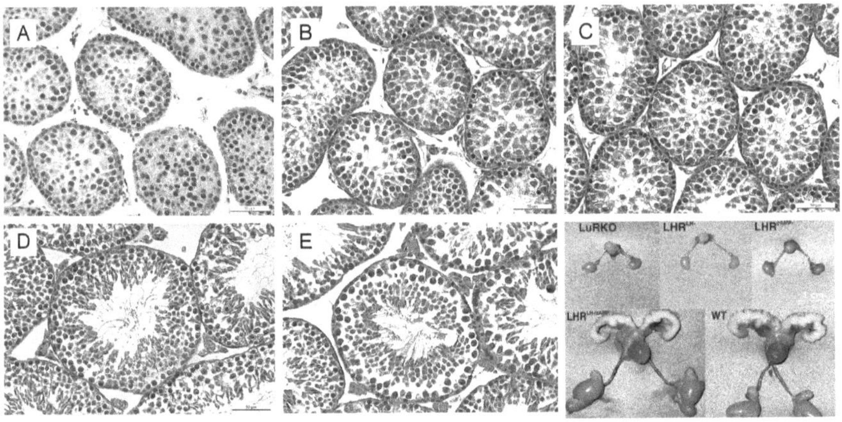

Figure 7.3 Rescue of spermatogenesis and sexual development by intermolecular cooperation of mutant LHCGRs. Similar to the system represented in Fig. 7.1C, a mutated LHCGR unable to generate signaling responses (LHR^{cAMP-}) can bind its ligand (LH) and activate an LHCGR lacking ligand-binding capacity (LHR^{LH-}). The intermolecular cooperation between these receptors partially rescue receptor function in the absence of the WT receptor (LHCGR knockout, LuRKO, A), and thus the phenotype of male mice expressing both mutants ($LHR^{LH-/cAMP-}$), while each mutant alone (B and C, respectively) is totally inactive. Full spermatogenesis is rescued by intermolecular cooperation in $LHR^{LH-/cAMP-}$ (D) as compared with WT (E) mice. Macroscopic pictures of gonads and accessory sex organs of LHCGR knockout (LuRKO), LHR^{LH-}, LHR^{cAMP-}, $LHR^{LH-/cAMP-}$, and WT showing the extent of rescue by intermolecular cooperation. *Modified from Ref. 61.* (For color version of this figure, the reader is referred to the online version of this chapter.)

6. DIMERIZATION IN HEALTH AND DISEASE

Probably, the first example of a disease-causing dimerization was documented with Class A receptors. Patients suffering from preeclampsia, a form of gestational hypertension connected to hypersensitivity to the vasoconstrictor angiotensin II, have the type 1 angiotensin II receptor (AT_1) and increased expression of the bradykinin B_2 receptor (B_2) in platelets and omental vessels. While AT_1 is the sole receptor for angiotensin II, heterodimerization with B_2 receptor observed in these patients results in an increase in intracellular signaling and resistance to the reduction of AT_1 receptor responsiveness to oxidative stress. This reduction ultimately results in vasoconstriction, which raises blood pressure and oxidative stress, and in turn blocks the action of AT_1 receptor homodimers but not the activation of the pathological heterodimers.[65]

Inhibition of GPCR signaling through inhibition of a protomer within a heterodimeric partner can occur in pathophysiological conditions. During cardiac failure, the blockade of the AT_1 would be beneficial due to vasodilatation of heart vessels. AT_1 can heterodimerize with both βAR subtypes, and the pharmacological inhibition of βARs by β-blockers results in inhibition of AT_1 function increasing the survival rate in animal models by regulating the two receptors involved in the pathophysiology of ventricular dysfunction.[66] Thus, the functional inhibition of a protomer through antagonists directed to its dimeric partner could provide an opportunity for therapeutic intervention.

It has been documented that some mutant receptors can act by "hijacking" the WT counterpart in heterozygote carriers, a process called the *dominant-negative effect*[67–71] (Fig. 7.2H). This provides evidence that dimerization often takes place during receptor biosynthesis and that these interactions are crucial for receptor function. Importantly, it implicates receptor dimerization in human disease.

6.1. Targeting dimers

Rescuing the expression of a GPCR either from endoplasmic reticulum (ER) retention by its mutated counterpart or from misfolding by utilizing decoys or small molecules, including allosteric modulators, presents a tantalizing target for drug discovery.[72–74] An example of this has been recently demonstrated for the melanocortin 4 receptor (MC_4R), a receptor when mutated is often associated with obesity. Most of the MC_4R mutants are intracellularly retained and targeted for degradation due to misfolding. As many of these mutations are in the TM domains, distinct from its ligand-binding region, trafficking of the receptors to the cell surface could restore at least partial activity of the receptors. The studies of Ward *et al.*[75] and Rene *et al.*[76] have provided evidence that this is possible. They screened compound libraries for small pharmaceuticals (pharmacological chaperones) that could rescue MC_4R expression to the cell membrane and found that with some mutants this was possible. So far, the rescue has been determined by measuring selected signaling pathways in cells; thus, *in vivo* experiments are essential to determine the extent of the pharmacological rescue.

Splice variants of GPCRs, resulting in only a fragment of the full receptor, can also affect the function of the WT receptor or even form heterodimers with unique pharmacology. This was the case for the μ-OR KO mice where deletion of exon 1 resulted in two differentially responsive groups, those who did not respond to morphine (full gene KO) and those where a truncated form of the μ-OR, missing exon 1, coded TM1, was expressed by an upstream promoter.[77] A synthetic opiate that acts on this truncated receptor was a potent analgesic yet without the side effects of other opiates (respiratory depression, addiction, and constipation). Similarly, a 5TM truncated ghrelin receptor, and its ligand the ghrelin receptor polypeptide (GhSR1b), not only affects the constitutive activity of the full receptor (GhSR) by dimeriation[78] but also generates novel pharmacology by dimerizing with the neurotensin receptor 1 (NTS_1), a heterodimer able to respond to neuromedin U.[79]

Adenosine A_{2A} receptor agonists have been used in animal models and patients suffering from Parkinson's disease with encouraging results.[80–82] The mechanism of action takes place at the A_{2A}/dopamine receptor-2 (D_2) heterodimers, where A_{2A} counteracts D_2-postsynaptic effects during dopamine depletion, reducing motor symptoms and slowing neurodegeneration.

As described earlier, some of the clearest evidence for dimerization at the physiological level comes from interactions between two distinct receptors generating differential responses to the agonist of one protomer. Since there are hundreds of receptors, and up to one-third of nonodorant GPCRs with unknown ligand and thus unknown function, we focus on their pharmacological importance in Section 6.2.

6.2. Deorphanization of receptors

Orphan GPCRs, receptors for which no ligand has yet been identified, might nevertheless function as signaling modulators by interacting with other GPCRs. The $GABA_{B2}$ receptor could be considered as an orphan receptor since it has no ligand-binding properties yet functions as a dimerizing partner. Also, the taste receptors $T1R_1$ and $T1R_2$ were discovered by Charles Zuker's group as orphan receptors, but $T1R_1$ and $T1R_2$ form obligatory heterodimers with the $T1R_3$ receptor that determines each heterodimer's function, $T1R_2$-$T1R_3$ for sweet taste and $T1R_1$-$T1R_3$ for umami taste.[29,83–85]

In *Drosophila*, the orphan receptor 83b (DOR_{83b})—a conserved and ubiquitously expressed odorant receptor with no apparent affinity for odorants—is essential for the transport, maintenance, and functional response of odorant receptors in olfactory neurons.[86,87] In mammals, the orphan receptor GPR_{50}, a melatonin-like receptor, constitutively dimerizes with the circadian neurohormone melatonin receptors MT_1 and MT_2. Interestingly, GPR_{50}

only affects MT_1 function, which instead of modulating signaling or activity becomes inhibited by this interaction.[88] There are also constitutively active orphan receptors, several of viral origin, whose activity is regulated by dimerization. Smoothened (Smo), the receptor-mediating Hedgehog's (Hh) effects during embryo development, is constitutively active, and this activity is inhibited by the 12TM receptor Patched (Ptch)—the actual receptor for Hh. Upon binding of Hh to Ptch, Smo is released gaining its constitutive functions[89] (Fig. 7.2I). This reminds us that there are potentially many other forms of receptor interactions and more *deorphanizations* are yet to be discovered, as there remain hundreds of unliganded receptors.

7. CONCLUSION

At the physiological level, GPCRs dimerize for several reasons: as a form of quality control mechanism upon their biosynthesis, to diversify receptor conformations and state of activity, to control strength and duration of cell signaling, and to diversify their functions and responses (Fig. 7.2).

In principle, however, there are many possibilities of what dimerization controls or modulates. As *Homo sapiens* have approximately the same number of genes as rodents, rodents having many more GPCRs than humans (~1300 vs. ~800) (olfactory receptors overrepresented in the former),[90] and as humans are far more complex than rodents, we can conclude that the combinations of GPCRs, and other protomers, play an important role in the function and diversification of signaling generated by these vital receptors. It is evolution's fine-tuning that has achieved an astonishing number of receptor combinations, a complexity far beyond our original conception, which will provide exciting scientific conundrums and possible discoveries for future generations of biologists and life scientists to come.

REFERENCES

1. Schlessinger J. Cell signaling by receptor tyrosine kinases. *Cell* 2000;**103**:211–25.
2. Weiss A, Schlessinger J. Switching signals on or off by receptor dimerization. *Cell* 1998;**94**:277–80.
3. Birdsall NJM. Can different receptors interact directly with each other? *Trends Neurosci* 1982;**5**:137–8.
4. Watanabe AM, McConnaughey MM, Strawbridge RA, Fleming JW, Jones LR, Besch Jr HR. Muscarinic cholinergic receptor modulation of beta-adrenergic receptor affinity for catecholamines. *J Biol Chem* 1978;**253**:4833–6.
5. Hebert TE, Moffett S, Morello JP, Loisel TP, Bichet DG, Barret C, et al. A peptide derived from a beta2-adrenergic receptor transmembrane domain inhibits both receptor dimerization and activation. *J Biol Chem* 1996;**271**:16384–92.

6. Ng GY, O'Dowd BF, Lee SP, Chung HT, Brann MR, Seeman P, et al. Dopamine D2 receptor dimers and receptor-blocking peptides. *Biochem Biophys Res Commun* 1996;**227**:200–4.
7. Romano C, Yang WL, O'Malley KL. Metabotropic glutamate receptor 5 is a disulfide-linked dimer. *J Biol Chem* 1996;**271**:28612–6.
8. White JH, Wise A, Main MJ, Green A, Fraser NJ, Disney GH, et al. Heterodimerization is required for the formation of a functional GABA(B) receptor. *Nature* 1998;**396**:679–82.
9. Jones KA, Borowsky B, Tamm JA, Craig DA, Durkin MM, Dai M, et al. GABA(B) receptors function as a heteromeric assembly of the subunits GABA(B)R1 and GABA (B)R2. *Nature* 1998;**396**:674–9.
10. Kaupmann K, Malitschek B, Schuler V, Heid J, Froestl W, Beck P, et al. GABA(B)-receptor subtypes assemble into functional heteromeric complexes. *Nature* 1998;**396**:683–7.
11. Jordan BA, Devi LA. G-protein-coupled receptor heterodimerization modulates receptor function. *Nature* 1999;**399**:697–700.
12. AbdAlla S, Lother H, Quitterer U. AT1-receptor heterodimers show enhanced G-protein activation and altered receptor sequestration. *Nature* 2000;**407**:94–8.
13. James JR, Oliveira MI, Carmo AM, Iaboni A, Davis SJ. A rigorous experimental framework for detecting protein oligomerization using bioluminescence resonance energy transfer. *Nat Methods* 2006;**3**:1001–6.
14. Bouvier M, Heveker N, Jockers R, Marullo S, Milligan G. BRET analysis of GPCR oligomerization: newer does not mean better. *Nat Methods* 2007;**4**:3–4, author reply 4.
15. Salahpour A, Masri B. Experimental challenge to a 'rigorous' BRET analysis of GPCR oligomerization. *Nat Methods* 2007;**4**:599–600, author reply 601.
16. Gurevich VV, Gurevich EV. GPCR monomers and oligomers: it takes all kinds. *Trends Neurosci* 2008;**31**:74–81.
17. Whorton MR, Bokoch MP, Rasmussen SG, Huang B, Zare RN, Kobilka B, et al. A monomeric G protein-coupled receptor isolated in a high-density lipoprotein particle efficiently activates its G protein. *Proc Natl Acad Sci USA* 2007;**104**:7682–7.
18. Bayburt TH, Vishnivetskiy SA, McLean MA, Morizumi T, Huang CC, Tesmer JJ, et al. Monomeric rhodopsin is sufficient for normal rhodopsin kinase (GRK1) phosphorylation and arrestin-1 binding. *J Biol Chem* 2010;**286**:1420–8.
19. Maggio R, Vogel Z, Wess J. Coexpression studies with mutant muscarinic/adrenergic receptors provide evidence for intermolecular "cross-talk" between G-protein-linked receptors. *Proc Natl Acad Sci USA* 1993;**90**:3103–7.
20. Parnot C, Kobilka B. Toward understanding GPCR dimers. *Nat Struct Mol Biol* 2004;**11**:691–2.
21. Angers S, Salahpour A, Joly E, Hilairet S, Chelsky D, Dennis M, et al. Detection of beta 2-adrenergic receptor dimerization in living cells using bioluminescence resonance energy transfer (BRET). *Proc Natl Acad Sci USA* 2000;**97**:3684–9.
22. Overton MC, Blumer KJ. G-protein-coupled receptors function as oligomers in vivo. *Curr Biol* 2000;**10**:341–4.
23. Robbins MJ, Calver AR, Filippov AK, Hirst WD, Russell RB, Wood MD, et al. GABA (B2) is essential for g-protein coupling of the GABA(B) receptor heterodimer. *J Neurosci* 2001;**21**:8043–52.
24. Carrillo JJ, Pediani J, Milligan G. Dimers of class A G protein-coupled receptors function via agonist-mediated trans-activation of associated G proteins. *J Biol Chem* 2003;**278**:42578–87.
25. Horvat RD, Barisas BG, Roess DA. Luteinizing hormone receptors are self-associated in slowly diffusing complexes during receptor desensitization. *Mol Endocrinol* 2001;**15**:534–42.

26. Smith SM, Lei Y, Liu J, Cahill ME, Hagen GM, Barisas BG, et al. Luteinizing hormone receptors translocate to plasma membrane microdomains after binding of human chorionic gonadotropin. *Endocrinology* 2006;**147**:1789–95.
27. Conn PM, Rogers DC, Stewart JM, Niedel J, Sheffield T. Conversion of a gonadotropin-releasing hormone antagonist to an agonist. *Nature* 1982;**296**:653–5.
28. Hebert TE, Loisel TP, Adam L, Ethier N, Onge SS, Bouvier M. Functional rescue of a constitutively desensitized beta2AR through receptor dimerization. *Biochem J* 1998;**330** (Pt 1):287–93.
29. Xu H, Staszewski L, Tang H, Adler E, Zoller M, Li X. Different functional roles of T1R subunits in the heteromeric taste receptors. *Proc Natl Acad Sci USA* 2004;**101**:14258–63.
30. Pin JP, Neubig R, Bouvier M, Devi L, Filizola M, Javitch JA, et al. International Union of Basic and Clinical Pharmacology. LXVII. Recommendations for the recognition and nomenclature of G protein-coupled receptor heteromultimers. *Pharmacol Rev* 2007;**59**:5–13.
31. Okamoto T, Sekiyama N, Otsu M, Shimada Y, Sato A, Nakanishi S, et al. Expression and purification of the extracellular ligand binding region of metabotropic glutamate receptor subtype 1. *J Biol Chem* 1998;**273**:13089–96.
32. Tsuji Y, Shimada Y, Takeshita T, Kajimura N, Nomura S, Sekiyama N, et al. Cryptic dimer interface and domain organization of the extracellular region of metabotropic glutamate receptor subtype 1. *J Biol Chem* 2000;**275**:28144–51.
33. Han G, Hampson DR. Ligand binding to the amino-terminal domain of the mGluR4 subtype of metabotropic glutamate receptor. *J Biol Chem* 1999;**274**:10008–13.
34. Goldsmith PK, Fan GF, Ray K, Shiloach J, McPhie P, Rogers KV, et al. Expression, purification, and biochemical characterization of the amino-terminal extracellular domain of the human calcium receptor. *J Biol Chem* 1999;**274**:11303–9.
35. Kunishima N, Shimada Y, Tsuji Y, Sato T, Yamamoto M, Kumasaka T, et al. Structural basis of glutamate recognition by a dimeric metabotropic glutamate receptor. *Nature* 2000;**407**:971–7.
36. Novi F, Stanasila L, Giorgi F, Corsini GU, Cotecchia S, Maggio R. Paired activation of two components within muscarinic M3 receptor dimers is required for recruitment of beta-arrestin-1 to the plasma membrane. *J Biol Chem* 2005;**280**:19768–76.
37. Urizar E, Montanelli L, Loy T, Bonomi M, Swillens S, Gales C, et al. Glycoprotein hormone receptors: link between receptor homodimerization and negative cooperativity. *EMBO J* 2005;**24**:1954–64.
38. Pin JP, Comps-Agrar L, Maurel D, Monnier C, Rives ML, Trinquet E, et al. G-protein-coupled receptor oligomers: two or more for what? Lessons from mGlu and GABAB receptors. *J Physiol* 2009;**587**:5337–44.
39. Ji I, Lee C, Jeoung M, Koo Y, Sievert GA, Ji TH. Trans-activation of mutant follicle-stimulating hormone receptors selectively generates only one of two hormone signals. *Mol Endocrinol* 2004;**18**:968–78.
40. Ji I, Lee C, Song Y, Conn PM, Ji TH. Cis- and trans-activation of hormone receptors: the LH receptor. *Mol Endocrinol* 2002;**16**:1299–308.
41. Lee C, Ji I, Ryu K, Song Y, Conn PM, Ji TH. Two defective heterozygous luteinizing hormone receptors can rescue hormone action. *J Biol Chem* 2002;**277**:15795–800.
42. Lee C, Ji IJ, Ji TH. Use of defined-function mutants to access receptor-receptor interactions. *Methods* 2002;**27**:318–23.
43. Osuga Y, Hayashi M, Kudo M, Conti M, Kobilka B, Hsueh AJ. Co-expression of defective luteinizing hormone receptor fragments partially reconstitutes ligand-induced signal generation. *J Biol Chem* 1997;**272**:25006–12.
44. Fan QR, Hendrickson WA. Structure of human follicle-stimulating hormone in complex with its receptor. *Nature* 2005;**433**:269–77.

45. Roess DA, Horvat RD, Munnelly H, Barisas BG. Luteinizing hormone receptors are self-associated in the plasma membrane. *Endocrinology* 2000;**141**:4518–23.
46. Karges B, Gidenne S, Aumas C, Haddad F, Kelly PA, Milgrom E, et al. Zero-length cross-linking reveals that tight interactions between the extracellular and transmembrane domains of the luteinizing hormone receptor persist during receptor activation. *Mol Endocrinol* 2005;**19**:2086–98.
47. Vassart G, Pardo L, Costagliola S. A molecular dissection of the glycoprotein hormone receptors. *Trends Biochem Sci* 2004;**29**:119–26.
48. Binder BM, Biernbaum MS, Bownds MD. Light activation of one rhodopsin molecule causes the phosphorylation of hundreds of others. A reaction observed in electropermeabilized frog rod outer segments exposed to dim illumination. *J Biol Chem* 1990;**265**:15333–40.
49. Huttenrauch F, Pollok-Kopp B, Oppermann M. G protein-coupled receptor kinases promote phosphorylation and beta-arrestin-mediated internalization of CCR5 homo- and hetero-oligomers. *J Biol Chem* 2005;**280**:37503–15.
50. Terrillon S, Barberis C, Bouvier M. Heterodimerization of V1a and V2 vasopressin receptors determines the interaction with beta-arrestin and their trafficking patterns. *Proc Natl Acad Sci USA* 2004;**101**:1548–53.
51. So CH, Varghese G, Curley KJ, Kong MM, Alijaniaram M, Ji X, et al. D1 and D2 dopamine receptors form heterooligomers and cointernalize after selective activation of either receptor. *Mol Pharmacol* 2005;**68**:568–78.
52. McLatchie LM, Fraser NJ, Main MJ, Wise A, Brown J, Thompson N, et al. RAMPs regulate the transport and ligand specificity of the calcitonin-receptor-like receptor. *Nature* 1998;**393**:333–9.
53. Comps-Agrar L, Kniazeff J, Norskov-Lauritsen L, Maurel D, Gassmann M, Gregor N, et al. The oligomeric state sets GABA(B) receptor signalling efficacy. *EMBO J* 2011;**30**: 2336–49.
54. Schwenk J, Metz M, Zolles G, Turecek R, Fritzius T, Bildl W, et al. Native GABA(B) receptors are heteromultimers with a family of auxiliary subunits. *Nature* 2010;**465**:231–5.
55. Laffray S, Bouali-Benazzouz R, Papon MA, Favereaux A, Jiang Y, Holm T, et al. Impairment of GABAB receptor dimer by endogenous 14-3-3zeta in chronic pain conditions. *EMBO J* 2012;**31**:3239–51.
56. Daniels DJ, Lenard NR, Etienne CL, Law PY, Roerig SC, Portoghese PS. Opioid-induced tolerance and dependence in mice is modulated by the distance between pharmacophores in a bivalent ligand series. *Proc Natl Acad Sci USA* 2005;**102**:19208–13.
57. Rozenfeld R, Gupta A, Gagnidze K, Lim MP, Gomes I, Lee-Ramos D, et al. AT1R-CB(1)R heteromerization reveals a new mechanism for the pathogenic properties of angiotensin II. *EMBO J* 2011;**30**:2350–63.
58. Huhtaniemi IT, Themmen AP. Mutations in human gonadotropin and gonadotropin-receptor genes. *Endocrine* 2005;**26**:207–17.
59. Zhang Y, Buchholz F, Muyrers JP, Stewart AF. A new logic for DNA engineering using recombination in Escherichia coli. *Nat Genet* 1998;**20**:123–8.
60. Zhang FP, Poutanen M, Wilbertz J, Huhtaniemi I. Normal prenatal but arrested postnatal sexual development of luteinizing hormone receptor knockout (LuRKO) mice. *Mol Endocrinol* 2001;**15**:172–83.
61. Rivero-Müller A, Chou Y-Y, Ji I, Lajic S, Hanyaloglu AC, Jonas K, et al. Rescue of defective G protein-coupled receptor function in vivo by intermolecular co-operation. *Proc Natl Acad Sci USA* 2010;**107**:2319–24.
62. Marcellino D, Ferre S, Casado V, Cortes A, Le Foll B, Mazzola C, et al. Identification of dopamine D1-D3 receptor heteromers. Indications for a role of synergistic D1-D3 receptor interactions in the striatum. *J Biol Chem* 2008;**283**:26016–25.

63. Gonzalez-Maeso J, Weisstaub NV, Zhou M, Chan P, Ivic L, Ang R, et al. Hallucinogens recruit specific cortical 5-HT(2A) receptor-mediated signaling pathways to affect behavior. *Neuron* 2007;**53**:439–52.
64. Moreno JL, Holloway T, Albizu L, Sealfon SC, Gonzalez-Maeso J. Metabotropic glutamate mGlu2 receptor is necessary for the pharmacological and behavioral effects induced by hallucinogenic 5-HT2A receptor agonists. *Neurosci Lett* 2011;**493**:76–9.
65. AbdAlla S, Lother H, El Massiery A, Quitterer U. Increased AT(1) receptor heterodimers in preeclampsia mediate enhanced angiotensin II responsiveness. *Nat Med* 2001;**7**:1003–9.
66. Barki-Harrington L, Luttrell LM, Rockman HA. Dual inhibition of beta-adrenergic and angiotensin II receptors by a single antagonist: a functional role for receptor-receptor interaction in vivo. *Circulation* 2003;**108**:1611–8.
67. Brothers SP, Cornea A, Janovick JA, Conn PM. Human loss-of-function gonadotropin-releasing hormone receptor mutants retain wild-type receptors in the endoplasmic reticulum: molecular basis of the dominant-negative effect. *Mol Endocrinol* 2004;**18**:1787–97.
68. Calebiro D, de Filippis T, Lucchi S, Covino C, Panigone S, Beck-Peccoz P, et al. Intracellular entrapment of wild-type TSH receptor by oligomerization with mutants linked to dominant TSH resistance. *Hum Mol Genet* 2005;**14**:2991–3002.
69. McElvaine AT, Mayo KE. A dominant-negative human growth hormone-releasing hormone (GHRH) receptor splice variant inhibits GHRH binding. *Endocrinology* 2006;**147**:1884–94.
70. Leanos-Miranda A, Ulloa-Aguirre A, Janovick JA, Conn PM. In vitro coexpression and pharmacological rescue of mutant gonadotropin-releasing hormone receptors causing hypogonadotropic hypogonadism in humans expressing compound heterozygous alleles. *J Clin Endocrinol Metab* 2005;**90**:3001–8.
71. Janovick JA, Knollman PE, Brothers SP, Ayala-Yanez R, Aziz AS, Conn PM. Regulation of G protein-coupled receptor trafficking by inefficient plasma membrane expression: molecular basis of an evolved strategy. *J Biol Chem* 2006;**281**:8417–25.
72. Conn PM, Ulloa-Aguirre A. Trafficking of G-protein-coupled receptors to the plasma membrane: insights for pharmacoperone drugs. *Trends Endocrinol Metab* 2010;**21**:190–7.
73. Ulloa-Aguirre A, Janovick JA, Brothers SP, Conn PM. Pharmacologic rescue of conformationally-defective proteins: implications for the treatment of human disease. *Traffic* 2004;**5**:821–37.
74. Newton CL, Whay AM, McArdle CA, Zhang M, van Koppen CJ, van de Lagemaat R, et al. Rescue of expression and signaling of human luteinizing hormone G protein-coupled receptor mutants with an allosterically binding small-molecule agonist. *Proc Natl Acad Sci USA* 2011;**108**:7172–6.
75. Ward NA, Hirst S, Williams J, Findlay JB. Pharmacological chaperones increase the cell-surface expression of intracellularly retained mutants of the melanocortin 4 receptor with unique rescuing efficacy profiles. *Biochem Soc Trans* 2012;**40**:717–20.
76. Rene P, Le Gouill C, Pogozheva ID, Lee G, Mosberg HI, Farooqi IS, et al. Pharmacological chaperones restore function to MC4R mutants responsible for severe early-onset obesity. *J Pharmacol Exp Ther* 2010;**335**:520–32.
77. Majumdar S, Grinnell S, Le Rouzic V, Burgman M, Polikar L, Ansonoff M, et al. Truncated G protein-coupled mu opioid receptor MOR-1 splice variants are targets for highly potent opioid analgesics lacking side effects. *Proc Natl Acad Sci USA* 2011;**108**:19778–83.
78. Leung PK, Chow KB, Lau PN, Chu KM, Chan CB, Cheng CH, et al. The truncated ghrelin receptor polypeptide (GHS-R1b) acts as a dominant-negative mutant of the ghrelin receptor. *Cell Signal* 2007;**19**:1011–22.

79. Takahashi K, Furukawa C, Takano A, Ishikawa N, Kato T, Hayama S, et al. The neuromedin U-growth hormone secretagogue receptor 1b/neurotensin receptor 1 oncogenic signaling pathway as a therapeutic target for lung cancer. *Cancer Res* 2006;**66**:9408–19.
80. Kanda T, Jackson MJ, Smith LA, Pearce RK, Nakamura J, Kase H, et al. Adenosine A2A antagonist: a novel antiparkinsonian agent that does not provoke dyskinesia in parkinsonian monkeys. *Ann Neurol* 1998;**43**:507–13.
81. Xu K, Bastia E, Schwarzschild M. Therapeutic potential of adenosine A(2A) receptor antagonists in Parkinson's disease. *Pharmacol Ther* 2005;**105**:267–310.
82. Schwarzschild MA, Agnati L, Fuxe K, Chen JF, Morelli M. Targeting adenosine A2A receptors in Parkinson's disease. *Trends Neurosci* 2006;**29**:647–54.
83. Hoon MA, Adler E, Lindemeier J, Battey JF, Ryba NJ, Zuker CS. Putative mammalian taste receptors: a class of taste-specific GPCRs with distinct topographic selectivity. *Cell* 1999;**96**:541–51.
84. Nelson G, Hoon MA, Chandrashekar J, Zhang Y, Ryba NJ, Zuker CS. Mammalian sweet taste receptors. *Cell* 2001;**106**:381–90.
85. Adler E, Hoon MA, Mueller KL, Chandrashekar J, Ryba NJ, Zuker CS. A novel family of mammalian taste receptors. *Cell* 2000;**100**:693–702.
86. Neuhaus EM, Gisselmann G, Zhang W, Dooley R, Stortkuhl K, Hatt H. Odorant receptor heterodimerization in the olfactory system of Drosophila melanogaster. *Nat Neurosci* 2005;**8**:15–7.
87. Benton R, Sachse S, Michnick SW, Vosshall LB. Atypical membrane topology and heteromeric function of Drosophila odorant receptors in vivo. *PLoS Biol* 2006;**4**:e20.
88. Levoye A, Dam J, Ayoub MA, Guillaume JL, Couturier C, Delagrange P. The orphan GPR50 receptor specifically inhibits MT1 melatonin receptor function through heterodimerization. *EMBO J* 2006;**25**:3012–23.
89. Riobo NA, Lu K, Emerson Jr CP. Hedgehog signal transduction: signal integration and cross talk in development and cancer. *Cell Cycle* 2006;**5**:1612–5.
90. Gloriam DE, Fredriksson R, Schioth HB. The G protein-coupled receptor subset of the rat genome. *BMC Genomics* 2007;**8**:338.

FURTHER READING

91. Gurevich VV, Gurevich EV. How and why do GPCRs dimerize? *Trends Pharmacol Sci* 2008;**29**:234–40.
92. Angers S, Salahpour A, Bouvier M. Dimerization: an emerging concept for G protein-coupled receptor ontogeny and function. *Annu Rev Pharmacol Toxicol* 2002;**42**:409–35.
93. Terrillon S, Bouvier M. Roles of G-protein-coupled receptor dimerization. *EMBO Rep* 2004;**5**:30–4.
94. Gonzalez-Maeso J. GPCR oligomers in pharmacology and signaling. *Mol Brain* 2011;**4**:20.
95. Haack KK, Tougas MR, Jones KT, El-Dahr SS, Radhakrishna H, McCarty NA. A novel bioassay for detecting GPCR heterodimerization: transactivation of beta 2 adrenergic receptor by bradykinin receptor. *J Biomol Screen* 2010;**15**:251–60.
96. Janovick JA, Ulloa-Aguirre A, Conn PM. Evolved regulation of gonadotropin-releasing hormone receptor cell surface expression. *Endocrine* 2003;**22**:317–27.
97. Ulloa-Aguirre A, Janovick JA, Miranda AL, Conn PM. G-protein-coupled receptor trafficking: understanding the chemical basis of health and disease. *ACS Chem Biol* 2006;**1**:631–8.
98. Ritter SL, Hall RA. Fine-tuning of GPCR activity by receptor-interacting proteins. *Nat Rev Mol Cell Biol* 2009;**10**:819–30.

99. Milligan G. G protein-coupled receptor dimerization: function and ligand pharmacology. *Mol Pharmacol* 2004;**66**:1–7.
100. Bai M. Dimerization of G-protein-coupled receptors: roles in signal transduction. *Cell Signal* 2004;**16**:175–86.
101. Conn PM, Ulloa-Aguirre A, Ito J, Janovick JA. G protein-coupled receptor trafficking in health and disease: lessons learned to prepare for therapeutic mutant rescue in vivo. *Pharmacol Rev* 2007;**59**:225–50.
102. Leanos-Miranda A, Ulloa-Aguirre A, Ji TH, Janovick JA, Conn PM. Dominant-negative action of disease-causing gonadotropin-releasing hormone receptor (GnRHR) mutants: a trait that potentially coevolved with decreased plasma membrane expression of GnRHR in humans. *J Clin Endocrinol Metab* 2003;**88**:3360–7.
103. Wise H. The roles played by highly truncated splice variants of G protein-coupled receptors. *J Mol Signal* 2012;**7**:13.

CHAPTER EIGHT

G Protein-Coupled Receptor Heterocomplexes in Neuropsychiatric Disorders

José L. Moreno[*], **Terrell Holloway**[*], **Javier González-Maeso**[*,†,‡]
[*]Department of Psychiatry, Mount Sinai School of Medicine, One Gustave L. Levy Place, New York, USA
[†]Department of Neurology, Mount Sinai School of Medicine, One Gustave L. Levy Place, New York, USA
[‡]Friedman Brain Institute, Mount Sinai School of Medicine, One Gustave L. Levy Place, New York, USA

Contents

Abstract

G protein-coupled receptors (or GPCRs) represent the largest family of membrane proteins in the human genome and are the target of approximately half of all therapeutic drugs. GPCRs contain a conserved structure of seven transmembrane domains. Their amino terminus is located extracellularly, whereas the carboxy terminus extends into the cytoplasm. Accumulating evidence suggests that GPCRs exist and function as monomeric entities. Nevertheless, more recent findings indicate that GPCRs can also form dimers or even higher order oligomers. The differential pharmacological and signaling properties of GPCR heteromeric complexes hint that their physiological effects may be different as compared to those obtained in tissue cultures that express a particular GPCR. In this chapter, we review current data on the role of GPCR heteromerization in receptor signaling, as well as its potential implication in neuropsychiatric disorders such as schizophrenia, depression, and Parkinson's disease.

Progress in Molecular Biology and Translational Science, Volume 117
ISSN 1877-1173
http://dx.doi.org/10.1016/B978-0-12-386931-9.00008-8

1. INTRODUCTION

Among the different proteins found in cell membranes responsible for transmitting information from the extracellular environment into the cytoplasm, there are those ones belonging to the superfamily of G protein-coupled receptors (GPCRs).[1–4] It has been described in the human genome about 800 members of GPCRs,[5,6] and they are the target of nearly half of the drugs currently in use for the treatment of different human diseases.[3,7] GPCRs are single polypeptides embedded in the membrane with the amino terminus remaining in the extracellular side and the carboxy terminus in the cytoplasm.[8–11] Until recently, GPCRs have been assumed to function as monomers that, upon agonist binding, initiate the activation of heterotrimeric G proteins and consequently modulate the function of intracellular signaling cascades downstream. This is supported by findings that demonstrate functional G protein coupling of a single purified family A GPCR reconstituted into a phospholipid bilayer.[12,13] During the past two decades, however, there has been much evidence to suggest that GPCRs exist as dimers and higher order oligomers.[14–19] Of particular interest is the demonstration of heteromerization between two GPCRs belonging to different families. These protein complexes have been shown to affect diverse aspects of receptor function.[20–24] This chapter reviews some of the latest advances in our understanding of structure and function of GPCR heteromers, and their role in neuropsychiatric disorders such as schizophrenia, depression, and Parkinson's disease.

2. STRUCTURE OF GPCR HETEROMERS

The three major families of GPCRs include family A (rhodopsin-like), family B (glucagon-related receptors), and family C (metabotropic glutamate (mGlu)-related receptors).[1,5,6] Much evidence indicates that family C GPCRs, including $GABA_B$[25–31] and mGlu receptors[30,32–36], exist and function as constitutive dimers. The $GABA_B$ receptor has been shown to be a heterodimer of $GABA_B$-R1 and $GABA_B$-R2, each of which is unable to form a functional $GABA_B$ receptor. The domains responsible for $GABA_B$ heterodimeric receptor formation are located in the C-termini of both $GABA_B$-R1 and $GABA_B$-R2 receptors. Recent findings also suggest that mGlu receptors are expressed at the plasma membrane as strict dimers, and not higher order oligomers, that are covalently linked through a disulfide bond. In marked contrast, however, family A GPCRs are assembled as homomers and heteromers through structural mechanisms that involve noncovalent interactions between amino acid residues located in transmembrane (TM) regions. As an example, it has been suggested using biochemical and biophysical approaches such as coimmunoprecipitation of epitope-tagged receptors and fluorescence resonance energy transfer (FRET) that α_{1b}-adrenergic receptors are assembled into oligomeric structures with symmetric contact points involving residues located in TM domains 1 and 4[14,37–39] (see also Refs.40 and 41 for reviews about FRET and other techniques based on energy transfer). Expression of α_{1b}-adrenergic receptors as higher order oligomers was suggested based on the use of three-color FRET (3-FRET). Similar findings were obtained with the dopamine D2 receptor. Thus, the combination of bioluminescence/fluorescence complementation and energy transfer provided evidence that dopamine D2 receptors are expressed as higher order oligomers in the plasma membrane.[15,38,39] The use of cysteine cross-linking experiments also pointed toward the location of TM1 and TM4 at the symmetrical interfaces of the dopamine D2 receptor oligomers (i.e., TM1–TM1 interface and TM4–TM4 interface).[15,38,39] The critical role of TM4 in the formation of GPCR complexes is further supported by recent findings with the serotonin 5-HT_{2A} and the metabotropic glutamate 2 (mGlu2) receptors.[20,23,24] It was demonstrated that 5-HT_{2A} and mGlu2, but not mGlu3, form a GPCR heteromer in tissue culture. The differences between mGlu2 and mGlu3 receptors to be assembled as a GPCR heteromer with the 5-HT_{2A} receptor provided the rational for the design of mGlu2/mGlu3 chimeras that showed differences in their capacity to be expressed in close proximity with coexpressed 5-HT_{2A} receptors. The authors suggested that three residues located at the intracellular end of TM4 of mGlu2 are necessary to form a GPCR heteromer with the 5-HT_{2A} in HEK293 cells.[24] Together, these findings suggest that TM4 and TM1 contribute to the formation of GPCR homo- and heterocomplexes.

Another important question related to structure and function of GPCRs concerns the stability of dimeric/oligomeric states.[42] The use of fluorescence recovery after photobleaching (FRAP) has demonstrated transient interactions on a timescale of seconds between the components of the β1-adrenergic receptor homomer. On the contrary, the β2-adrenergic receptor has been shown to be expressed as a stable oligomer in HEK293 cells.[43] A dynamic equilibrium between monomers and homodimers/oligomers has also been recently suggested using FRAP in HEK293 cells expressing dopamine D2 receptors.[44] Similarly, the use of total internal reflection fluorescence microscopy in living cells point toward a dynamic nature of muscarinic M1 receptor dimer formation. Thus, muscarinic M1 receptors are expressed as monomers or dimers, but not oligomers. It was also shown that muscarinic M1 receptors do not form constitutive dimers, with ~30% of the receptor molecules in close proximity at any given time.[17]

3. ROLE OF GPCR HETEROCOMPLEXES IN NEUROPSYCHIATRIC DISORDERS

Although is has been clearly established in reconstituted systems that single GPCR molecules are able to couple with and activate heterotrimeric G proteins, recent evidence also demonstrates close molecular proximity between different GPCR molecules that affects functional outcomes such as ligand binding profiles, patterns of G protein coupling, and subcellular location (Fig. 8.1). This has opened in recent years a new field of research that adds complexity to the mechanisms underlying the physiological and behavioral responses induced by signaling pathways downstream GPCRs (Table 8.1). As certain receptor subtypes, such as dopamine, serotonin, glutamate, and adenosine receptors, have been shown to form GPCR heteromers *in vitro* in tissue culture and *in vivo* in animal models, a better understanding of their structure, neuroanatomical location, and physiological function may represent a new target for the design of drugs for the treatment of neuropsychiatric disorders such as schizophrenia, depression, suicidal behavior, and drug abuse.[45–48] Here, we will review recent findings related to GPCR heteromers and their role in central nervous system (CNS) function.

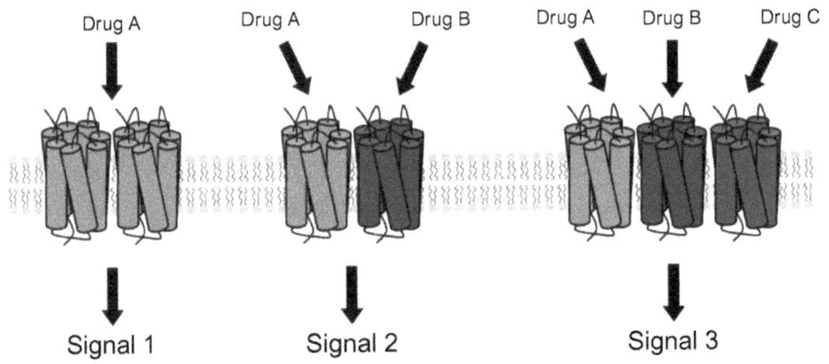

Figure 8.1 Schematic model of functional responses induced by GPCR homomers as compared to GPCR heteromers. Combinations of drugs modulate different patterns of signaling responses, and a cross talk between the components of the GPCR heteromeric complex leads to unique signaling properties. (For color version of this figure, the reader is referred to the online version of this chapter.)

Table 8.1 Examples of GPCR heterocomplexes with potential roles in neuropsychiatric disorders

GPCR heterocomplex	Experimental approaches	Neuropsychiatric relevance
A1–A2A	Cell lines, striatum (IP, BRET, FRET)	Drug tolerance
A1–D1	Cell lines, striatum (IP, BRET, BP)	Involved in PD and cocaine addiction
A1–mGlu1	Cell lines (IP)	Involved in schizophrenia
A1–$5HT_{2A}$	Cortex (BP)	Involved in schizophrenia
A2A–D2	Cell lines, striatum (IP, BRET, FRET, Phar, BP)	Reversal of PD deficits, putative antipsychotic, and anticocaine properties
A2A–mGlu5	Cell lines, striatum (IP)	Possible PD, schizophrenia, and cocaine treatment
A2A–D2–mGlu5	Cell lines, striatum (IP, SRET)	Possible PD, schizophrenia, and cocaine treatment
A2A–CB1	Cell lines, striatum (IP, BRET, BP)	
A2A–CB1–D2	Cell lines (BP, SRET)	Involved in schizophrenia
α_{2A}–MOR	Cell lines, striatum (IP, BRET, FRET)	Involved in addiction
CB1–D2	Cell lines, striatum (IP, FRET, Phar, BP)	Possible PD, addiction, and schizophrenia treatment
CRH_1–V_{1b}	Cell lines (IP, BRET, FRET)	Involved in mood disorder, depression
D1–D2	Cell lines, striatum (IP, FRET)	Possible PD, schizophrenia, and treatment
D1–D3	Cell lines (BRET, FRET, Phar, BP)	Cocaine addiction
D2–D3	Cell lines, striatum (IP, BP)	Involved in PD and schizophrenia
D2–H_3	Cell lines, striatum (BRET, Phar, BP)	Involved in PD
D4–MOR	Striatum (Phar)	Acquisition during drug addiction and risks for drug use initiation
mGlu2–$5HT_{2A}$	Cell lines, cortex (IP, BRET, FRET, Phar, BP)	Possible schizophrenia treatment

IP, coimmunoprecipitation; BRET, bioluminescence resonance energy transfer; FRET, fluorescence resonance energy transfer; SRET, sequential resonance energy transfer; Phar, pharmacological allosteric effects in radioligand binding assays; PD, Parkinson's disease; BP, behavioral paradigms.

4. ADENOSINE AND DOPAMINE RECEPTORS

4.1. Adenosine A1 and dopamine D1 receptors

Expression of adenosine A1 receptor and dopamine D1 receptor as a GPCR heteromer has been demonstrated using approaches such as coimmunoprecipitation, bioluminescence resonance energy transfer (BRET), and FRET.[49–53] It has also been shown that adenosine A1 receptor and dopamine D1 receptor are colocalized in soma and dendritic regions of cortical neurons. Selective adenosine A1 receptor antagonists reverse the hyperlocomotor behavioral effects induced by dopamine D1 receptor agonists, and opposite effects were shown after administration of adenosine A1 receptor agonists. These findings suggest adenosine A1 receptor as a potential target to modulate dopamine D1 receptor-dependent signaling. Importantly, suboptimal doses of adenosine A1 receptor antagonists potentiate the therapeutic-like effects of dopamine D1 antagonists in rodent models of Parkinson's disease.

4.2. Adenosine A2A and dopamine D2 receptors

Kjell Fuxe at the Karolinska Institute and collaborators provided the first evidence of a ligand binding interaction between adenosine A2A and dopamine D2 receptors in rat striatum,[54,55] and they suggested intramembrane receptor/receptor interaction as a potential mechanism involved in this pharmacological cross talk. Based on these findings, they demonstrated the existence of adenosine A2A–dopamine D2 receptor heteromers using a variety of approaches such as coimmunoprecipitation, BRET, and FRET *in vitro* and in tissue culture.[56–60] An electrostatic interaction between the intracellular loop 3 of dopamine D2 receptor and the C-terminal tail of adenosine A2A receptor has been proposed as necessary for this protein–protein interaction.[60] Behavioral and microdialysis assays in whole animal models suggest the functional role of the adenosine A2A–dopamine D2 receptor heteromer in whole animal models through a mechanism that involves their coexpression in striatopallidal GABAergic neurons and nucleus accumbens.[61–72] Thus, adenosine A2A receptor antagonists reduce dopamine D2 receptor-dependent signaling and pharmacological properties,[54,73–75] and enhance therapeutic effects in rodent models of Parkinson's disease.[65,76–81] The adenosine A2A receptor antagonist istradefylline (KW-6002) has been used in animal models of Parkinson's disease.[77] Importantly, clinical studies with istradefylline showed a symptomatic improvement in Parkinson's disease patients.[82,83] Bivalent ligands that bind both dopamine D2 and adenosine A2A receptors have also been proposed as a new approach to treat Parkinson's disease.[84]

The potential role of the adenosine A2–dopamine D2 receptor heteromer as a target of antipsychotic drugs has been suggested using the hyperlocomotor activity induced by amphetamine and the noncompetitive NMDA receptor antagonist phencyclidine as rodent model of schizophrenia and psychosis.[85,86] The adenosine A2A receptor agonist CGS21680 reduces the hyperlocomotor activity induced by amphetamine or phencyclidine.[87] The authors provide evidence suggesting that this antipsychotic-like behavioral effect may be mediated by inhibition of dopamine D2 receptor-dependent signaling through the adenosine A2A–dopamine D2 receptor heteromer in striatopallidal GABAergic neurons.[88]

4.3. Adenosine A2A, dopamine D2, and mGlu5 receptors

In addition to adenosine and dopamine transmission, glutamate transmission plays an important role in the function of striatal GABAergic efferent neurons originating in the nucleus accumbens. In 1984, L-glutamate was found to reduce the affinity of dopamine D2 receptor agonist binding sites in rat striatal membrane preparations, providing the first evidence of the existence of glutamate receptor–dopamine D2 receptor heteromeric complexes.[89] It was also shown that adenosine A2A receptor agonists and group I mGlu receptor agonists synergistically reduce the affinity of dopamine D2 receptor agonists binding in striatal membrane preparations.[90] In line with these results, adenosine A2A receptor and mGlu5 receptor colocalize in primary cultures of rat striatal neurons[91] and in striatal glutamate nerve terminals.[92,93] Similarly, adenosine A2A receptors and dopamine D2 receptors colocalize with mGlu5 receptors in rat striatum.[90] Additional data showed that adenosine A2A receptor and mGlu5 receptor agonists act synergistically to increase extracellular levels of GABA in the nucleus accumbens, which also potentiate the inhibitory effects of the dopamine D2 receptor.[94] In Parkinson's disease, glutamate transmission is overactive mostly due to the reduced inhibitory effect of the dopamine D2 receptor,[95] and antiparkinsonian drugs regulate functional responses that require adenosine A2A and mGlu5 receptors.[96,97]

Moreover, recent data suggest that adenosine A2A antagonists and mGlu5 antagonists induce antiparkinsonian-like effects in animal models acting through the adenosine A2A–dopamine D2–mGlu5 receptor oligomer.[98] This is further supported by findings showing that the antiparkinsonian-like effects of mGlu5 receptor antagonists need coexpression of adenosine A2A and dopamine D2 receptors.[96] Since the intramembrane adenosine A2A–dopamine D2 receptor functional interaction is positively regulated by mGlu5 receptor agonists,[90,93,99,100] together, these results suggest that adenosine A2A receptor antagonists and mGlu5 receptor antagonists induce antiparkinsonian-like effects through a mechanism that requires expression of adenosine A2A, mGlu5, and dopamine D2 as a GPCR oligomeric functional unit.[98] Based on these findings, the authors proposed the combination of adenosine A2A and mGlu5 receptor antagonists to enhance their antiparkinsonian-like effects in rodent models of motor deficits, and suggested that this nondopaminergic therapy may avoid the adverse effects of dopaminergic drugs such as dyskinesia and cognitive dysfunction.[97,101,102]

4.4. Adenosine A2A, dopamine D2, and cannabinoid CB1 receptors

The cannabinoid system represents an important inhibitory neuroregulator acting in the CNS.[103–107] Cannabinoid CB1 receptors are coexpressed with the dopamine D2 receptor ventral striatopallidal GABAergic neurons, and with the adenosine A2A receptor in corticostriatal glutamatergic terminals.[108–114] Activation of cannabinoid CB1 receptors by WIN55,212-2 leads to motor suppression in rodent models.[108] Importantly, coimmunoprecipitation and BRET experiments showed that CB1 and adenosine A2A receptors interact together in close molecular proximity in living cells and in rat striatum.[115] It was also suggested that this GPCR heteromer is functional since some of the CB1 receptor-dependent locomotor effects were affected in adenosine A2A receptor knockout mice.[116] Biochemical and cellular signaling assays in SH-SY5Y neuroglioblastoma cells together with behavioral tests in mice indicated that some of the effects induced by activation of the cannabinoid CB1 receptor in striatum depend upon adenosine A2A receptor-dependent signaling.[108,115] Several studies have reported antagonistic interactions between cannabinoid CB1 and dopamine D2 receptors.[117–119] Together with coimmunoprecipitation and FRET experiments,[46] the reduction of the affinity of dopamine D2 receptor agonists in the presence of cannabinoid CB1 receptor agonists point toward expression of these two receptors as a GPCR heteromer.[45] It has also been suggested that the antagonistic interactions between cannabinoid CB1 and dopamine D2 receptors as a GPCR heteromer may also involve the adenosine A2A receptor.[120–122] This is supported by the demonstration that adenosine A2A receptors directly interact with both dopamine D2[56] and cannabinoid CB1[115] receptors. The importance of this functional cross talk has been further suggested by recent findings using a method that combined biomolecular fluorescence complementation (BiFC) and BRET techniques and showed expression of dopamine D2, cannabinoid CB1, and adenosine A2A receptors as a higher order GPCR oligomeric complex in living cells.[123–125]

In postmortem human brain of schizophrenic subjects, radioligand binding assays showed alterations in expression of adenosine A2A,[126] dopamine D2,[127] and cannabinoid CB1[128] receptors. In rodent models, chronic antipsychotic treatment with the atypical antipsychotic drug clozapine induces downregulation of cannabinoid CB1 receptor expression in nucleus accumbens,[129,130] which has been suggested to be a consequence of a compensatory mechanism that reduces the endocannabinoid-mediated suppression of GABA release.[130] Together, these findings suggest that the adenosine A2A–cannabinoid CB1–dopamine D2 receptor heteromer may represent a potential target for new antipsychotic compounds.

4.5. Serotonin and glutamate receptors

Serotonin (5-HT) is one of the most ancient signaling molecules in evolution, and is involved in biological processes such as learning and memory, mood, food intake, sleep, reproduction, circadian rhythm, thermoregulation, pain perception, and social behavior.[131,132] Serotonin receptors also play an important role in cardiovascular, gastrointestinal, and endocrine function.[132] Glutamate is the primary excitatory neurotransmitter in the CNS.[133] Pyramidal neurons represent approximately 80% of the neurons of the cortex, and glutamate serves as the neurotransmitter of the cortical pyramidal cells.[134,135] Glutamate receptors are classified either as ion channel receptors (ionotropic) or metabotropic GPCRs.[136–138] mGlu receptors have an important function in synaptic modulation throughout the CNS.[135]

Recent findings demonstrate that 5-HT_{2A} and mGlu2 receptors colocalize at a subcellular level in mouse cortical pyramidal neurons.[20,23,24] It has also been shown that these two receptors form a GPCR heteromer in HEK293 cells with consequences on pharmacology, signaling, and behavioral effects of drugs that bind to either 5-HT_{2A} or mGlu2 receptors.[20,24] Radioligand binding assays showed that drugs that activate the mGlu2 receptor increase the affinity of hallucinogenic drugs for the 5-HT_{2A} receptor, and that drugs that activate the 5-HT_{2A} receptor decrease the affinity of agonists for the mGlu2 receptor.[20] Using ion channels in *Xenopus* oocytes as markers of $G_{q/11}$-dependent and $G_{i/o}$-dependent G protein signaling, recent findings demonstrate that both serotonergic and glutamatergic ligands balance the pattern of G protein signaling downstream of the 5-HT_{2A}–mGlu2 receptor heteromer in a way that predicts their antipsychotic or propsychotic potential, which may provide the basis for the rationale design of new antipsychotic drugs that affect the function of this serotonin–glutamate receptor heteromer.[23]

The behavioral effects of 5-HT_{2A} and mGlu2 as a GPCR heteromer have been recently demonstrated using hallucinogenic 5-HT_{2A} receptor agonists as a rodent model of psychosis. Hallucinogenic 5-HT_{2A} agonists, such as lysergic acid diethylamide (LSD) and mescaline, induce head-twitch behavior in mice, and this behavior is absent in 5-HT_{2A} receptor knockout mice.[139,140] It has been shown that the head-twitch behavioral response is significantly reduced in mGlu2 knockout mice.[141] More importantly, viral-mediated overexpression of wild-type mGlu2 receptor in frontal cortex rescues the effects of LSD-like drugs in mGlu2 knockout mice,[24,141] and this does not occur by overexpression of mGlu2ΔTM4N, a construct that is unable to form a GPCR heteromer with the 5-HT_{2A} receptor.[24] These observations suggest that it is 5-HT_{2A}–mGlu2 receptor complex, and not the 5-HT_{2A} alone, which is the molecular target responsible for psychoactive-like behavioral effects of LSD-like hallucinogenic drugs in mouse models of pychosis.[23,24,141]

Expression of the components of the 5-HT_{2A}–mGlu2 receptor complex has been shown to be dysregulated (i.e., increased 5-HT_{2A} receptor and decreased mGlu2 receptor) in postmortem human brain of antipsychotic-free schizophrenic subjects.[20] More recent findings provide evidence that the allosteric binding cross talk between 5-HT_{2A} receptor and mGlu2 receptor as a GPCR heterocomplex is upregulated in postmortem schizophrenia brain.[24,142] It is then possible that 5-HT_{2A}–mGlu2 receptor complex-dependent signaling effects may integrate serotonin and glutamate signaling, and therefore contribute to the abnormalities of thought and behavior in schizophrenia patients.

5. CONCLUSION

Understanding the structure and function of GPCR heteromers is essential for the discovery of mechanisms that define the signaling cross talk between the components of these protein complexes. Given that a large number of GPCRs are expressed in individual cells at any given time, further work is needed to better unravel the molecular rules that govern GPCR heteromeric formation, as well as the stability and lifetime of such structural units. Here, we have focused our attention on the potential role of GPCR heteromers in the treatment of neuropsychiatric disorders such as schizophrenia and Parkinson's disease. Despite these advances, further research efforts will focus on a better understanding of the structure, pharmacology, and behavioral function of GPCR heteromers *in vitro* and in animal models, with the final goal of developing new therapeutic drugs that specifically affect their function.

ACKNOWLEDGMENT

We thank the National Institutes of Health for grant support of our GPCR research.

REFERENCES

1. Pierce KL, Premont RT, Lefkowitz RJ. Seven-transmembrane receptors. *Nat Rev Mol Cell Biol* 2002;**3**:639–50.
2. Kristiansen K. Molecular mechanisms of ligand binding, signaling, and regulation within the superfamily of G-protein-coupled receptors: molecular modeling and mutagenesis approaches to receptor structure and function. *Pharmacol Ther* 2004;**103**:21–80.
3. Lefkowitz RJ. Historical review: a brief history and personal retrospective of seven-transmembrane receptors. *Trends Pharmacol Sci* 2004;**25**:413–22.
4. Perez DM. The evolutionarily triumphant G-protein-coupled receptor. *Mol Pharmacol* 2003;**63**:1202–5.
5. Fredriksson R, Lagerstrom MC, Lundin LG, Schioth HB. The G-protein-coupled receptors in the human genome form five main families. Phylogenetic analysis, paralogon groups, and fingerprints. *Mol Pharmacol* 2003;**63**:1256–72.
6. Foord SM, Bonner TI, Neubig RR, Rosser EM, Pin JP, Davenport AP, et al. International Union of Pharmacology. XLVI. G protein-coupled receptor list. *Pharmacol Rev* 2005;**57**:279–88.
7. Kim Ka, von Zastrow M. Old drugs learn new tricks: insights from mammalian trace amine receptors. *Mol Pharmacol* 2001;**60**:1165–7.
8. Rosenbaum DM, Rasmussen SG, Kobilka BK. The structure and function of G-protein-coupled receptors. *Nature* 2009;**459**:356–63.
9. Cherezov V, Rosenbaum DM, Hanson MA, Rasmussen SG, Thian FS, Kobilka TS, et al. High-resolution crystal structure of an engineered human beta2-adrenergic G protein-coupled receptor. *Science* 2007;**318**:1258–65.
10. Rasmussen SG, Choi HJ, Rosenbaum DM, Kobilka TS, Thian FS, Edwards PC, et al. Crystal structure of the human beta2 adrenergic G-protein-coupled receptor. *Nature* 2007;**450**:383–7.
11. Rosenbaum DM, Zhang C, Lyons JA, Holl R, Aragao D, Arlow DH, et al. Structure and function of an irreversible agonist-beta(2) adrenoceptor complex. *Nature* 2011;**469**:236–40.
12. Whorton MR, Bokoch MP, Rasmussen SG, Huang B, Zare RN, Kobilka B, et al. A monomeric G protein-coupled receptor isolated in a high-density lipoprotein particle efficiently activates its G protein. *Proc Natl Acad Sci USA* 2007;**104**:7682–7.

13. Whorton MR, Jastrzebska B, Park PS, Fotiadis D, Engel A, Palczewski K, et al. Efficient coupling of transducin to monomeric rhodopsin in a phospholipid bilayer. *J Biol Chem* 2008;**283**:4387–94.
14. Lopez-Gimenez JF, Canals M, Pediani JD, Milligan G. The alpha1b-adrenoceptor exists as a higher-order oligomer: effective oligomerization is required for receptor maturation, surface delivery, and function. *Mol Pharmacol* 2007;**71**:1015–29.
15. Guo W, Urizar E, Kralikova M, Mobarec JC, Shi L, Filizola M, et al. Dopamine D2 receptors form higher order oligomers at physiological expression levels. *EMBO J* 2008;**27**:2293–304.
16. Fung JJ, Deupi X, Pardo L, Yao XJ, Velez-Ruiz GA, Devree BT, et al. Ligand-regulated oligomerization of beta(2)-adrenoceptors in a model lipid bilayer. *EMBO J* 2009;**28**:3315–28.
17. Hern JA, Baig AH, Mashanov GI, Birdsall B, Corrie JE, Lazareno S, et al. Formation and dissociation of M1 muscarinic receptor dimers seen by total internal reflection fluorescence imaging of single molecules. *Proc Natl Acad Sci USA* 2010;**107**:2693–8.
18. Milligan G. G protein-coupled receptor hetero-dimerization: contribution to pharmacology and function. *Br J Pharmacol* 2009;**158**:5–14.
19. Prezeau L, Rives ML, Comps-Agrar L, Maurel D, Kniazeff J, Pin JP. Functional crosstalk between GPCRs: with or without oligomerization. *Curr Opin Pharmacol* 2010;**10**:6–13.
20. Gonzalez-Maeso J, Ang RL, Yuen T, Chan P, Weisstaub NV, Lopez-Gimenez JF, et al. Identification of a serotonin/glutamate receptor complex implicated in psychosis. *Nature* 2008;**452**:93–7.
21. Vilardaga JP, Nikolaev VO, Lorenz K, Ferrandon S, Zhuang Z, Lohse MJ. Conformational cross-talk between alpha2A-adrenergic and mu-opioid receptors controls cell signaling. *Nat Chem Biol* 2008;**4**:126–31.
22. Urizar E, Yano H, Kolster R, Gales C, Lambert N, Javitch JA. CODA-RET reveals functional selectivity as a result of GPCR heteromerization. *Nat Chem Biol* 2011;**7**:624–30.
23. Fribourg M, Moreno JL, Holloway T, Provasi D, Baki L, Mahajan R, et al. Decoding the signaling of a GPCR heteromeric complex reveals a unifying mechanism of action of antipsychotic drugs. *Cell* 2011;**147**:1011–23.
24. Moreno JL, Muguruza C, Umali A, Mortillo S, Holloway T, Pilar-Cuellar F, et al. Identification of three residues essential for 5-HT2A-mGlu2 receptor heteromerization and its psychoactive behavioral function. *J Biol Chem* 2012;**287**:44301–19.
25. Comps-Agrar L, Kniazeff J, Norskov-Lauritsen L, Maurel D, Gassmann M, Gregor N, et al. The oligomeric state sets GABA(B) receptor signalling efficacy. *EMBO J* 2011;**30**:2336–49.
26. Kaupmann K, Huggel K, Heid J, Flor PJ, Bischoff S, Mickel SJ, et al. Expression cloning of GABA(B) receptors uncovers similarity to metabotropic glutamate receptors. *Nature* 1997;**386**:239–46.
27. Kaupmann K, Malitschek B, Schuler V, Heid J, Froestl W, Beck P, et al. GABA(B)-receptor subtypes assemble into functional heteromeric complexes. *Nature* 1998;**396**:683–7.
28. White JH, Wise A, Main MJ, Green A, Fraser NJ, Disney GH, et al. Heterodimerization is required for the formation of a functional GABA(B) receptor. *Nature* 1998;**396**:679–82.
29. Jones KA, Borowsky B, Tamm JA, Craig DA, Durkin MM, Dai M, et al. GABA(B) receptors function as a heteromeric assembly of the subunits GABA(B)R1 and GABA(B)R2. *Nature* 1998;**396**:674–9.
30. Margeta-Mitrovic M, Jan YN, Jan LY. A trafficking checkpoint controls GABA(B) receptor heterodimerization. *Neuron* 2000;**27**:97–106.

31. Galvez T, Duthey B, Kniazeff J, Blahos J, Rovelli G, Bettler B, et al. Allosteric interactions between GB1 and GB2 subunits are required for optimal GABA(B) receptor function. *EMBO J* 2001;**20**:2152–9.
32. Tsuji Y, Shimada Y, Takeshita T, Kajimura N, Nomura S, Sekiyama N, et al. Cryptic dimer interface and domain organization of the extracellular region of metabotropic glutamate receptor subtype 1. *J Biol Chem* 2000;**275**:28144–51.
33. El Moustaine D, Granier S, Doumazane E, Scholler P, Rahmeh R, Bron P, et al. Distinct roles of metabotropic glutamate receptor dimerization in agonist activation and G-protein coupling. *Proc Natl Acad Sci USA* 2012;**109**:16342–7.
34. Rives ML, Vol C, Fukazawa Y, Tinel N, Trinquet E, Ayoub MA, et al. Crosstalk between GABAB and mGlu1a receptors reveals new insight into GPCR signal integration. *EMBO J* 2009;**28**:2195–208.
35. Maurel D, Comps-Agrar L, Brock C, Rives ML, Bourrier E, Ayoub MA, et al. Cell-surface protein-protein interaction analysis with time-resolved FRET and snap-tag technologies: application to GPCR oligomerization. *Nat Methods* 2008;**5**:561–7.
36. Kniazeff J, Bessis AS, Maurel D, Ansanay H, Prezeau L, Pin JP. Closed state of both binding domains of homodimeric mGlu receptors is required for full activity. *Nat Struct Mol Biol* 2004;**11**:706–13.
37. Carrillo JJ, Lopez-Gimenez JF, Milligan G. Multiple interactions between transmembrane helices generate the oligomeric alpha1b-adrenoceptor. *Mol Pharmacol* 2004;**66**:1123–37.
38. Guo W, Shi L, Filizola M, Weinstein H, Javitch JA. Crosstalk in G protein-coupled receptors: changes at the transmembrane homodimer interface determine activation. *Proc Natl Acad Sci USA* 2005;**102**:17495–500.
39. Guo W, Shi L, Javitch JA. The fourth transmembrane segment forms the interface of the dopamine D2 receptor homodimer. *J Biol Chem* 2003;**278**:4385–8.
40. James JR, Oliveira MI, Carmo AM, Iaboni A, Davis SJ. A rigorous experimental framework for detecting protein oligomerization using bioluminescence resonance energy transfer. *Nat Methods* 2006;**3**:1001–6.
41. Johnsson K. Visualizing biochemical activities in living cells. *Nat Chem Biol* 2009;**5**:63–5.
42. Vilardaga JP, Agnati LF, Fuxe K, Ciruela F. G-protein-coupled receptor heteromer dynamics. *J Cell Sci* 2010;**123**:4215–20.
43. Dorsch S, Klotz KN, Engelhardt S, Lohse MJ, Bunemann M. Analysis of receptor oligomerization by FRAP microscopy. *Nat Methods* 2009;**6**:225–30.
44. Fonseca JM, Lambert NA. Instability of a class a G protein-coupled receptor oligomer interface. *Mol Pharmacol* 2009;**75**:1296–9.
45. Albizu L, Moreno JL, Gonzalez-Maeso J, Sealfon SC. Heteromerization of G protein-coupled receptors: relevance to neurological disorders and neurotherapeutics. *CNS Neurol Disord Drug Targets* 2010;**9**:636–50.
46. Fuxe K, Marcellino D, Rivera A, Diaz-Cabiale Z, Filip M, Gago B, et al. Receptor-receptor interactions within receptor mosaics. Impact on neuropsychopharmacology. *Brain Res Rev* 2008;**58**:415–52.
47. Fuxe K, Borroto-Escuela DO, Marcellino D, Romero-Fernandez W, Frankowska M, Guidolin D, et al. GPCR heteromers and their allosteric receptor-receptor interactions. *Curr Med Chem* 2012;**19**:356–63.
48. Moreno JL, Sealfon SC, Gonzalez-Maeso J. Group II metabotropic glutamate receptors and schizophrenia. *Cell Mol Life Sci* 2009;**66**:3777–85.
49. Gines S, Hillion J, Torvinen M, Le Crom S, Casado V, Canela EI, et al. Dopamine D1 and adenosine A1 receptors form functionally interacting heteromeric complexes. *Proc Natl Acad Sci USA* 2000;**97**:8606–11.
50. Ferre S, Ciruela F, Quiroz C, Lujan R, Popoli P, Cunha RA, et al. Adenosine receptor heteromers and their integrative role in striatal function. *ScientificWorldJournal* 2007;**7**:74–85.

51. Franco R, Casado V, Cortes A, Mallol J, Ciruela F, Ferre S, et al. G-protein-coupled receptor heteromers: function and ligand pharmacology. *Br J Pharmacol* 2008;**153** (Suppl. 1):S90–S98.
52. Salamone JD, Farrar AM, Font L, Patel V, Schlar DE, Nunes EJ, et al. Differential actions of adenosine A1 and A2A antagonists on the effort-related effects of dopamine D2 antagonism. *Behav Brain Res* 2009;**201**:216–22.
53. Schindler M, Harris CA, Hayes B, Papotti M, Humphrey PP. Immunohistochemical localization of adenosine A1 receptors in human brain regions. *Neurosci Lett* 2001;**297**:211–5.
54. Ferre S, von Euler G, Johansson B, Fredholm BB, Fuxe K. Stimulation of high-affinity adenosine A2 receptors decreases the affinity of dopamine D2 receptors in rat striatal membranes. *Proc Natl Acad Sci USA* 1991;**88**:7238–41.
55. Ferre S, Popoli P, Gimenez-Llort L, Finnman UB, Martinez E, Scotti de Carolis A, et al. Postsynaptic antagonistic interaction between adenosine A1 and dopamine D1 receptors. *Neuroreport* 1994;**6**:73–6.
56. Canals M, Marcellino D, Fanelli F, Ciruela F, de Benedetti P, Goldberg SR, et al. Adenosine A2A-dopamine D2 receptor-receptor heteromerization: qualitative and quantitative assessment by fluorescence and bioluminescence energy transfer. *J Biol Chem* 2003;**278**:46741–9.
57. Torvinen M, Kozell LB, Neve KA, Agnati LF, Fuxe K. Biochemical identification of the dopamine D2 receptor domains interacting with the adenosine A2A receptor. *J Mol Neurosci* 2004;**24**:173–80.
58. Navarro G, Aymerich MS, Marcellino D, Cortes A, Casado V, Mallol J, et al. Interactions between calmodulin, adenosine A2A, and dopamine D2 receptors. *J Biol Chem* 2009;**284**:28058–68.
59. Hillion J, Canals M, Torvinen M, Casado V, Scott R, Terasmaa A, et al. Coaggregation, cointernalization, and codesensitization of adenosine A2A receptors and dopamine D2 receptors. *J Biol Chem* 2002;**277**:18091–7.
60. Ciruela F, Burgueno J, Casado V, Canals M, Marcellino D, Goldberg SR, et al. Combining mass spectrometry and pull-down techniques for the study of receptor heteromerization. Direct epitope-epitope electrostatic interactions between adenosine A2A and dopamine D2 receptors. *Anal Chem* 2004;**76**:5354–63.
61. Gluck MR, Santana LA, Granson H, Yahr MD. Novel dopamine releasing response of an anti-convulsant agent with possible anti-Parkinson's activity. *J Neural Transm* 2004;**111**:713–24.
62. Fuxe K, Ferre S, Genedani S, Franco R, Agnati LF. Adenosine receptor-dopamine receptor interactions in the basal ganglia and their relevance for brain function. *Physiol Behav* 2007;**92**:210–7.
63. Rosin DL, Hettinger BD, Lee A, Linden J. Anatomy of adenosine A2A receptors in brain: morphological substrates for integration of striatal function. *Neurology* 2003;**61**:S12–S18.
64. Schiffmann SN, Fisone G, Moresco R, Cunha RA, Ferre S. Adenosine A2A receptors and basal ganglia physiology. *Prog Neurobiol* 2007;**83**:277–92.
65. Shen HY, Coelho JE, Ohtsuka N, Canas PM, Day YJ, Huang QY, et al. A critical role of the adenosine A2A receptor in extrastriatal neurons in modulating psychomotor activity as revealed by opposite phenotypes of striatum and forebrain A2A receptor knock-outs. *J Neurosci* 2008;**28**:2970–5.
66. Varty GB, Hodgson RA, Pond AJ, Grzelak ME, Parker EM, Hunter JC. The effects of adenosine A2A receptor antagonists on haloperidol-induced movement disorders in primates. *Psychopharmacology (Berl)* 2008;**200**:393–401.
67. El Yacoubi M, Ledent C, Parmentier M, Costentin J, Vaugeois JM. Adenosine A2A receptor knockout mice are partially protected against drug-induced catalepsy. *Neuroreport* 2001;**12**:983–6.

68. Li W, Dai S, An J, Xiong R, Li P, Chen X, et al. Genetic inactivation of adenosine A2A receptors attenuates acute traumatic brain injury in the mouse cortical impact model. *Exp Neurol* 2009;**215**:69–76.
69. Matsuya T, Takuma K, Sato K, Asai M, Murakami Y, Miyoshi S, et al. Synergistic effects of adenosine A2A antagonist and L-DOPA on rotational behaviors in 6-hydroxydopamine-induced hemi-Parkinsonian mouse model. *J Pharmacol Sci* 2007;**103**:329–32.
70. Mori A, Shindou T. Modulation of GABAergic transmission in the striatopallidal system by adenosine A2A receptors: a potential mechanism for the antiparkinsonian effects of A2A antagonists. *Neurology* 2003;**61**:S44–S48.
71. Chen JF, Huang Z, Ma J, Zhu J, Moratalla R, Standaert D, et al. A(2A) adenosine receptor deficiency attenuates brain injury induced by transient focal ischemia in mice. *J Neurosci* 1999;**19**:9192–200.
72. Perier C, Marin C, Bonastre M, Tolosa E, Hirsch EC. AMPA receptor antagonist LY293558 reverses preproenkephalin mRNA overexpression in the striatum of 6-OHDA-lesioned-rats treated with L-dopa. *Eur J Neurosci* 2002;**16**:2236–40.
73. Dasgupta S, Ferre S, Kull B, Hedlund PB, Finnman UB, Ahlberg S, et al. Adenosine A2A receptors modulate the binding characteristics of dopamine D2 receptors in stably cotransfected fibroblast cells. *Eur J Pharmacol* 1996;**316**:325–31.
74. Salim H, Ferre S, Dalal A, Peterfreund RA, Fuxe K, Vincent JD, et al. Activation of adenosine A1 and A2A receptors modulates dopamine D2 receptor-induced responses in stably transfected human neuroblastoma cells. *J Neurochem* 2000;**74**:432–9.
75. Kull B, Ferre S, Arslan G, Svenningsson P, Fuxe K, Owman C, et al. Reciprocal interactions between adenosine A2A and dopamine D2 receptors in Chinese hamster ovary cells co-transfected with the two receptors. *Biochem Pharmacol* 1999;**58**:1035–45.
76. Kim DS, Palmiter RD. Adenosine receptor blockade reverses hypophagia and enhances locomotor activity of dopamine-deficient mice. *Proc Natl Acad Sci USA* 2003;**100**:1346–51.
77. Chase TN, Bibbiani F, Bara-Jimenez W, Dimitrova T, Oh-Lee JD. Translating A2A antagonist KW6002 from animal models to parkinsonian patients. *Neurology* 2003;**61**: S107–S111.
78. Fuxe K, Marcellino D, Genedani S, Agnati L. Adenosine A(2A) receptors, dopamine D(2) receptors and their interactions in Parkinson's disease. *Mov Disord* 2007;**22**:1990–2017.
79. Jenner P. Istradefylline, a novel adenosine A2A receptor antagonist, for the treatment of Parkinson's disease. *Expert Opin Investig Drugs* 2005;**14**:729–38.
80. Schwarzschild MA, Agnati L, Fuxe K, Chen JF, Morelli M. Targeting adenosine A2A receptors in Parkinson's disease. *Trends Neurosci* 2006;**29**:647–54.
81. Shiozaki S, Ichikawa S, Nakamura J, Kitamura S, Yamada K, Kuwana Y. Actions of adenosine A2A receptor antagonist KW-6002 on drug-induced catalepsy and hypokinesia caused by reserpine or MPTP. *Psychopharmacology (Berl)* 1999;**147**:90–5.
82. Bara-Jimenez W, Sherzai A, Dimitrova T, Favit A, Bibbiani F, Gillespie M, et al. Adenosine A(2A) receptor antagonist treatment of Parkinson's disease. *Neurology* 2003;**61**:293–6.
83. Hauser RA, Hubble JP, Truong DD. Randomized trial of the adenosine A(2A) receptor antagonist istradefylline in advanced PD. *Neurology* 2003;**61**:297–303.
84. Soriano A, Ventura R, Molero A, Hoen R, Casado V, Cortes A, et al. Adenosine A2A receptor-antagonist/dopamine D2 receptor-agonist bivalent ligands as pharmacological tools to detect A2A-D2 receptor heteromers. *J Med Chem* 2009;**52**:5590–602.
85. Rimondini R, Ferre S, Ogren SO, Fuxe K. Adenosine A2A agonists: a potential new type of atypical antipsychotic. *Neuropsychopharmacology* 1997;**17**:82–91.
86. Malec D, Poleszak E. Involvement of adenosine receptors in dizocilpine-induced motor activity in mice. *Pharmacol Rep* 2006;**58**:101–6.

87. Sills TL, Azampanah A, Fletcher PJ. The adenosine A2A agonist CGS 21680 reverses the reduction in prepulse inhibition of the acoustic startle response induced by phencyclidine, but not by apomorphine and amphetamine. *Psychopharmacology (Berl)* 2001;**156**:187–93.
88. Ferre S. Adenosine-dopamine interactions in the ventral striatum. Implications for the treatment of schizophrenia. *Psychopharmacology (Berl)* 1997;**133**:107–20.
89. Fuxe K, Celani MF, Martire M, Zini I, Zoli M, Agnati LF. L-Glutamate reduces the affinity of [3H]N-propylnorapomorphine binding sites in striatal membranes. *Eur J Pharmacol* 1984;**100**:127–30.
90. Ferre S, Popoli P, Rimondini R, Reggio R, Kehr J, Fuxe K. Adenosine A2A and group I metabotropic glutamate receptors synergistically modulate the binding characteristics of dopamine D2 receptors in the rat striatum. *Neuropharmacology* 1999;**38**:129–40.
91. Fuxe K, Agnati LF, Jacobsen K, Hillion J, Canals M, Torvinen M, et al. Receptor heteromerization in adenosine A2A receptor signaling: relevance for striatal function and Parkinson's disease. *Neurology* 2003;**61**:S19–S23.
92. Rodrigues RJ, Alfaro TM, Rebola N, Oliveira CR, Cunha RA. Co-localization and functional interaction between adenosine A(2A) and metabotropic group 5 receptors in glutamatergic nerve terminals of the rat striatum. *J Neurochem* 2005;**92**:433–41.
93. Ferre S, Karcz-Kubicha M, Hope BT, Popoli P, Burgueno J, Gutierrez MA, et al. Synergistic interaction between adenosine A2A and glutamate mGlu5 receptors: implications for striatal neuronal function. *Proc Natl Acad Sci USA* 2002;**99**:11940–5.
94. Diaz-Cabiale Z, Vivo M, Del Arco A, O'Connor WT, Harte MK, Muller CE, et al. Metabotropic glutamate mGlu5 receptor-mediated modulation of the ventral striopallidal GABA pathway in rats. Interactions with adenosine A(2A) and dopamine D(2) receptors. *Neurosci Lett* 2002;**324**:154–8.
95. Lindsley CW, Hopkins CR. Metabotropic glutamate receptor 4 (mGlu4)-positive allosteric modulators for the treatment of Parkinson's disease: historical perspective and review of the patent literature. *Expert Opin Ther Pat* 2012;**22**:461–81.
96. Kachroo A, Orlando LR, Grandy DK, Chen JF, Young AB, Schwarzschild MA. Interactions between metabotropic glutamate 5 and adenosine A2A receptors in normal and parkinsonian mice. *J Neurosci* 2005;**25**:10414–9.
97. Coccurello R, Breysse N, Amalric M. Simultaneous blockade of adenosine A2A and metabotropic glutamate mGlu5 receptors increase their efficacy in reversing Parkinsonian deficits in rats. *Neuropsychopharmacology* 2004;**29**:1451–61.
98. Cabello N, Gandia J, Bertarelli DC, Watanabe M, Lluis C, Franco R, et al. Metabotropic glutamate type 5, dopamine D2 and adenosine A2a receptors form higher-order oligomers in living cells. *J Neurochem* 2009;**109**:1497–507.
99. Popoli P, Pezzola A, Torvinen M, Reggio R, Pintor A, Scarchilli L, et al. The selective mGlu(5) receptor agonist CHPG inhibits quinpirole-induced turning in 6-hydroxydopamine-lesioned rats and modulates the binding characteristics of dopamine D(2) receptors in the rat striatum: interactions with adenosine A(2a) receptors. *Neuropsychopharmacology* 2001;**25**:505–13.
100. Nishi A, Liu F, Matsuyama S, Hamada M, Higashi H, Nairn AC, et al. Metabotropic mGlu5 receptors regulate adenosine A2A receptor signaling. *Proc Natl Acad Sci USA* 2003;**100**:1322–7.
101. Marino MJ, Valenti O, Conn PJ. Glutamate receptors and Parkinson's disease: opportunities for intervention. *Drugs Aging* 2003;**20**:377–97.
102. Xu J, He J, Castleberry AM, Balasubramanian S, Lau AG, Hall RA. Heterodimerization of alpha 2A- and beta 1-adrenergic receptors. *J Biol Chem* 2003;**278**:10770–7.
103. Ashton JC, Appleton I, Darlington CL, Smith PF. Immunohistochemical localization of cerebrovascular cannabinoid CB1 receptor protein. *J Cardiovasc Pharmacol* 2004;**44**:517–9.

104. Benito C, Kim WK, Chavarria I, Hillard CJ, Mackie K, Tolon RM, et al. A glial endogenous cannabinoid system is upregulated in the brains of macaques with simian immunodeficiency virus-induced encephalitis. *J Neurosci* 2005;**25**:2530–6.
105. Hohmann AG, Herkenham M. Localization of central cannabinoid CB1 receptor messenger RNA in neuronal subpopulations of rat dorsal root ganglia: a double-label in situ hybridization study. *Neuroscience* 1999;**90**:923–31.
106. Smith TH, Sim-Selley LJ, Selley DE. Cannabinoid CB1 receptor-interacting proteins: novel targets for central nervous system drug discovery? *Br J Pharmacol* 2010;**160**:454–66.
107. Aguado T, Romero E, Monory K, Palazuelos J, Sendtner M, Marsicano G, et al. The CB1 cannabinoid receptor mediates excitotoxicity-induced neural progenitor proliferation and neurogenesis. *J Biol Chem* 2007;**282**:23892–8.
108. Tebano MT, Martire A, Chiodi V, Pepponi R, Ferrante A, Domenici MR, et al. Adenosine A2A receptors enable the synaptic effects of cannabinoid CB1 receptors in the rodent striatum. *J Neurochem* 2009;**110**:1921–30.
109. Pickel VM, Chan J, Kearn CS, Mackie K. Targeting dopamine D2 and cannabinoid-1 (CB1) receptors in rat nucleus accumbens. *J Comp Neurol* 2006;**495**:299–313.
110. Yin HH, Lovinger DM. Frequency-specific and D2 receptor-mediated inhibition of glutamate release by retrograde endocannabinoid signaling. *Proc Natl Acad Sci USA* 2006;**103**:8251–6.
111. Julian MD, Martin AB, Cuellar B, Rodriguez De Fonseca F, Navarro M, Moratalla R, et al. Neuroanatomical relationship between type 1 cannabinoid receptors and dopaminergic systems in the rat basal ganglia. *Neuroscience* 2003;**119**:309–18.
112. Fusco FR, Martorana A, Giampa C, De March Z, Farini D, D'Angelo V, et al. Immunolocalization of CB1 receptor in rat striatal neurons: a confocal microscopy study. *Synapse* 2004;**53**:159–67.
113. Kofalvi A, Rodrigues RJ, Ledent C, Mackie K, Vizi ES, Cunha RA, et al. Involvement of cannabinoid receptors in the regulation of neurotransmitter release in the rodent striatum: a combined immunochemical and pharmacological analysis. *J Neurosci* 2005;**25**:2874–84.
114. Matyas F, Yanovsky Y, Mackie K, Kelsch W, Misgeld U, Freund TF. Subcellular localization of type 1 cannabinoid receptors in the rat basal ganglia. *Neuroscience* 2006;**137**:337–61.
115. Carriba P, Ortiz O, Patkar K, Justinova Z, Stroik J, Themann A, et al. Striatal adenosine A2A and cannabinoid CB1 receptors form functional heteromeric complexes that mediate the motor effects of cannabinoids. *Neuropsychopharmacology* 2007;**32**:2249–59.
116. Soria G, Castane A, Berrendero F, Ledent C, Parmentier M, Maldonado R, et al. Adenosine A2A receptors are involved in physical dependence and place conditioning induced by THC. *Eur J Neurosci* 2004;**20**:2203–13.
117. Meschler JP, Howlett AC. Signal transduction interactions between CB1 cannabinoid and dopamine receptors in the rat and monkey striatum. *Neuropharmacology* 2001;**40**:918–26.
118. Jarrahian A, Watts VJ, Barker EL. D2 dopamine receptors modulate Galpha-subunit coupling of the CB1 cannabinoid receptor. *J Pharmacol Exp Ther* 2004;**308**:880–6.
119. Martin AB, Fernandez-Espejo E, Ferrer B, Gorriti MA, Bilbao A, Navarro M, et al. Expression and function of CB1 receptor in the rat striatum: localization and effects on D1 and D2 dopamine receptor-mediated motor behaviors. *Neuropsychopharmacology* 2008;**33**:1667–79.
120. Kearn CS, Blake-Palmer K, Daniel E, Mackie K, Glass M. Concurrent stimulation of cannabinoid CB1 and dopamine D2 receptors enhances heterodimer formation: a mechanism for receptor cross-talk? *Mol Pharmacol* 2005;**67**:1697–704.

121. Marcellino D, Carriba P, Filip M, Borgkvist A, Frankowska M, Bellido I, et al. Antagonistic cannabinoid CB1/dopamine D2 receptor interactions in striatal CB1/D2 heteromers. A combined neurochemical and behavioral analysis. *Neuropharmacology* 2008;**54**:815–23.
122. Ferre S, Goldberg SR, Lluis C, Franco R. Looking for the role of cannabinoid receptor heteromers in striatal function. *Neuropharmacology* 2009;**56**(Suppl. 1):226–34.
123. Navarro G, Carriba P, Gandia J, Ciruela F, Casado V, Cortes A, et al. Detection of heteromers formed by cannabinoid CB1, dopamine D2, and adenosine A2A G-protein-coupled receptors by combining bimolecular fluorescence complementation and bioluminescence energy transfer. *ScientificWorldJournal* 2008;**8**:1088–97.
124. Carriba P, Navarro G, Ciruela F, Ferre S, Casado V, Agnati L, et al. Detection of heteromerization of more than two proteins by sequential BRET-FRET. *Nat Methods* 2008;**5**:727–33.
125. Kerppola TK. Bimolecular fluorescence complementation: visualization of molecular interactions in living cells. *Methods Cell Biol* 2008;**85**:431–70.
126. Deckert J, Brenner M, Durany N, Zochling R, Paulus W, Ransmayr G, et al. Up-regulation of striatal adenosine A(2A) receptors in schizophrenia. *Neuroreport* 2003;**14**:313–6.
127. Guillin O, Abi-Dargham A, Laruelle M. Neurobiology of dopamine in schizophrenia. *Int Rev Neurobiol* 2007;**78**:1–39.
128. Dean B, Sundram S, Bradbury R, Scarr E, Copolov D. Studies on [3H]CP-55940 binding in the human central nervous system: regional specific changes in density of cannabinoid-1 receptors associated with schizophrenia and cannabis use. *Neuroscience* 2001;**103**:9–15.
129. Sundram S, Copolov D, Dean B. Clozapine decreases [3H] CP 55940 binding to the cannabinoid 1 receptor in the rat nucleus accumbens. *Naunyn Schmiedebergs Arch Pharmacol* 2005;**371**:428–33.
130. Uriguen L, Garcia-Fuster MJ, Callado LF, Morentin B, La Harpe R, Casado V, et al. Immunodensity and mRNA expression of A2A adenosine, D2 dopamine, and CB1 cannabinoid receptors in postmortem frontal cortex of subjects with schizophrenia: effect of antipsychotic treatment. *Psychopharmacology (Berl)* 2009;**206**:313–24.
131. Kriegebaum C, Gutknecht L, Schmitt A, Lesch KP, Reif A. Serotonin now: part 1. Neurobiology and developmental genetics. *Fortschr Neurol Psychiatr* 2010;**78**:319–31.
132. Nichols DE, Nichols CD. Serotonin receptors. *Chem Rev* 2008;**108**:1614–41.
133. Hollmann M, Heinemann S. Cloned glutamate receptors. *Annu Rev Neurosci* 1994;**17**:31–108.
134. Conn PJ, Pin JP. Pharmacology and functions of metabotropic glutamate receptors. *Annu Rev Pharmacol Toxicol* 1997;**37**:205–37.
135. Benarroch EE. Metabotropic glutamate receptors: synaptic modulators and therapeutic targets for neurologic disease. *Neurology* 2008;**70**:964–8.
136. Schoepp DD, Conn PJ. Metabotropic glutamate receptors. *Pharmacol Biochem Behav* 2002;**74**:255–6.
137. Jingami H, Nakanishi S, Morikawa K. Structure of the metabotropic glutamate receptor. *Curr Opin Neurobiol* 2003;**13**:271–8.
138. Ferraguti F, Shigemoto R. Metabotropic glutamate receptors. *Cell Tissue Res* 2006;**326**: 483–504.
139. Gonzalez-Maeso J, Yuen T, Ebersole BJ, Wurmbach E, Lira A, Zhou M, et al. Transcriptome fingerprints distinguish hallucinogenic and nonhallucinogenic 5-hydroxytryptamine 2A receptor agonist effects in mouse somatosensory cortex. *J Neurosci* 2003;**23**:8836–43.

140. Gonzalez-Maeso J, Weisstaub NV, Zhou M, Chan P, Ivic L, Ang R, et al. Hallucinogens recruit specific cortical 5-HT(2A) receptor-mediated signaling pathways to affect behavior. *Neuron* 2007;**53**:439–52.
141. Moreno JL, Holloway T, Albizu L, Sealfon SC, Gonzalez-Maeso J. Metabotropic glutamate mGlu2 receptor is necessary for the pharmacological and behavioral effects induced by hallucinogenic 5-HT2A receptor agonists. *Neurosci Lett* 2011;**493**:76–9.
142. Muguruza C, Moreno JL, Umali A, Callado LF, Meana JJ, Gonzalez-Maeso J. Dysregulated 5-HT2A receptor binding in postmortem frontal cortex of schizophrenic subjects. *Eur Neuropsychopharmacol* 2012; http://dx.doi.org/10.1016/j.euroneuro.2012.10.006 [Epub ahead of print].

CHAPTER NINE

Disease-Specific Heteromerization of G-Protein-Coupled Receptors That Target Drugs of Abuse

Ivone Gomes[*], **Wakako Fujita**[*], **Moraje V. Chandrakala**[*], **Lakshmi A. Devi**[*,†]

[*]Department of Pharmacology and Systems Therapeutics, Mount Sinai School of Medicine, New York, USA
[†]Department of Neuroscience and Friedman Brain Institute, Mount Sinai School of Medicine, New York, USA

Contents

Abstract

Drugs of abuse such as morphine or marijuana exert their effects through the activation of G-protein-coupled receptors (GPCRs), the opioid and cannabinoid receptors, respectively. Moreover, interactions between either of these receptors have been shown to be involved in the rewarding effects of drugs of abuse. Recent advances in the field, using a variety of approaches, have demonstrated that many GPCRs, including opioid, cannabinoid, and dopamine receptors, can form associations between different receptor subtypes or with other GPCRs to form heteromeric complexes. The formation of these complexes, in turn, leads to the modulation of the properties of individual protomers. The development of tools that can selectively disrupt GPCR heteromers as well as monoclonal antibodies that can selectively block signaling by specific heteromer pairs has indicated that heteromers involving opioid, cannabinoid, or dopamine receptors may play a role in various disease states. In this review, we describe evidence for opioid,

Progress in Molecular Biology and Translational Science, Volume 117
ISSN 1877-1173
http://dx.doi.org/10.1016/B978-0-12-386931-9.00009-X

cannabinoid, and dopamine receptor heteromerization and the potential role of GPCR heteromers in pathophysiological conditions.

1. INTRODUCTION

Morphine, an alkaloid extracted from the poppy *Papaver somniferum*, and Δ^9-tetrahydrocannabinol (also known as THC or more commonly as marijuana), the main psychoactive extract from *Cannabis sativum*, are commonly abused drugs due to their mood-altering effects. Morphine is the most commonly used drug in the treatment of chronic pain, while THC is used in the alleviation of neuropathic pain and for the reduction of eye pressure in glaucoma patients.[1–5] However, long-term use of these drugs leads to a number of undesirable side effects, including the development of tolerance and addiction. In the latter context, several lines of evidence demonstrate the involvement of the dopaminergic system in the rewarding effects of drugs of abuse (reviewed in Ref. 6). Both morphine and THC exert their physiological effects via activation of $G_{\alpha i/o}$-coupled receptors, the opioid and cannabinoid receptors, respectively.[1,3–5] In an effort to develop drugs that target these receptors and exert the same physiological effects as morphine or THC but without the undesirable side effects associated with chronic drug use, several groups have investigated whether protein–protein interactions, including receptor heteromerization, could modulate the function of individual receptors. Here, we describe heteromerization as associations between different subtypes of a receptor or between two different receptors. Although heteromers involving opioid, cannabinoid, or dopamine receptors have been described, not much is known about their physiological role. In this context, studies using peptides that disrupt specific heteromers, or antibodies that selectively block heteromer signaling, suggest that GPCR heteromers may play a role in a number of pathologies.[7–11] While additional GPCR heteromers associated with other drugs of abuse have been reported, for the sake of brevity and conciseness, in this review we have focused only on heteromers that contain either opioid or cannabinoid receptors as one of the partners. We have also included dopamine receptors since they are the primary effectors of the action of stimulants such as cocaine and amphetamines. These are described below along with the putative role of these heteromers in the development of tolerance and other disease states.

1.1. Opioid receptors

Opioid receptors are members of the rhodopsin family of GPCRs and their signal transduction involves activation of $G_{\alpha i/o}$ proteins, downstream inhibition of adenylyl cyclase activity, the opening of inwardly rectifying K^+ channels, and inhibition of Ca^{2+} channels.[1] Molecular cloning studies have identified three types of opioid receptors μ-, δ-, and κ-opioid receptors (reviewed in Ref. 2).

Opioid receptors can be activated by endogenous peptides derived from the processing of the large precursor proteins, proopiomelanocortin, proenkephalin, and prodynorphin, by prohormone convertases via the classic peptide-processing pathway[12] to generate mainly β-endorphin, Met- and Leu-enkephalin, and dynorphins.[13] It is generally thought that enkephalins exert their biological effects primarily by activating δ opioid receptors (DOR), β-endorphin by activating μ opioid receptors (MOR), and dynorphins by activating κ opioid receptors (KOR).[13,14] In addition, opioid receptors can also be activated by drugs of abuse such as morphine and heroin.[15] Agonist-mediated activation of opioid receptors leads to the modulation of many biological functions including pain perception, motivation, locomotion, hormone secretion, and reward.[1,2] Studies using genetic knockdown have further elucidated the role of each opioid receptor type. These studies have implicated MOR in the analgesic effects of morphine and in the rewarding effects resulting from other nonopioid drugs of abuse including cannabinoids,[16–20] DOR in the regulation of emotional responses including anxiety and depression,[21] and KOR in the dysphoric effects of opioids and cannabinoids as well as in stress-induced relapse.[22,23] Together, these studies reveal a broad role for opioid receptors in normal physiology and disease.

1.2. Cannabinoid receptors

Like opioid receptors, cannabinoid receptors are also members of the rhodopsin family of GPCRs and signal via the $G_{\alpha i/o}$-signaling pathway, to activate similar downstream signaling cascades.[3–5] The two main subtypes of cannabinoid receptors are CB_1R and CB_2R.[3] Recently, the orphan receptor, GPR55, has been proposed as a putative cannabinoid receptor subtype (reviewed in Ref. 24).

CB_1R and CB_2R are activated by endogenous ligands derived from membrane lipids termed "endocannabinoids" which together with the receptors constitute the "endocannabinoid system" (for review, see Refs. 25–27). The two major lipid-based endocannabinoids are anandamide and 2-arachidonoylglycerol (2-AG)[3] that are generated by phospholipases[28–32] and degraded by fatty acid amide hydrolase,[33,34] cyclo-oxygenases, lipooxygenases,[35,36] and monoacylglycerol lipase.[37] Recently, endogenous peptidic agonists of cannabinoid receptors have been identified.[38–42] These peptides, RVD-hemopressin and VD-hemopressin, are derived from the α-chain of hemoglobin[38] and appear to activate a signal transduction pathway distinct from that of classic CB_1R ligands.[38] However, very little information is available about the enzymatic steps involved in either the generation or degradation of RVD-hemopressin.

Studies show that the endocannabinoid system plays an important role in many physiological processes including pain perception, learning and memory, appetite control, motor coordination, lipogenesis, and modulation of immune responses (for reviews, see Refs. 25–27,43,44). Studies with knockout mice indicate that the CB_1R subtype plays a role in the central effects of cannabinoids, including mood disorders, learning and memory, motor coordination, feeding, control of pain, and stress responses, and in the reinforcing and rewarding effects of drugs of abuse (reviewed in Refs. 45,46). Conversely, the CB_2R subtype plays a role in the peripheral effects of cannabinoids, including the development and function of the immune system, embryonic development, bone homeostasis, and pain perception, and plays an antifibrogenic role in liver diseases (reviewed in Ref. 47). These studies imply a broad role for CB_1R in central nervous system physiology and CB_2R in immune system function.

1.3. Dopamine receptors

Like opioid and cannabinoid receptors, dopamine receptors are members of the family A of GPCRs. Based on their ability to modulate cAMP levels, these receptors are classified as D1- or D2-like receptors (reviewed in Ref. 48). D1-like receptors couple to $G_{\alpha s/olf}$ proteins and are thought to be exclusively present at the presynapse (reviewed in Ref. 48). Activation of D1-like receptors leads to increases in cAMP levels, activation of cAMP-dependent protein kinases, increase in the phosphorylation levels of DARPP-32 (a molecule involved in the integration of cellular signaling responses to neurotransmitters) and ERK1/2 (the latter particularly in medium spiny neurons, MSNs), increases in L-type Ca^{2+} channel and NMDA receptor activity, inhibition of N- or P/Q-type Ca^{2+} channel and K^+ channel activity, enhancement of AMPA currents and modulation of GABA currents (reviewed in Refs. 48–50). D2-like receptors couple to

$G_{\alpha i/o}$ proteins and are present both at the pre- and postsynapse (reviewed in Ref. 48). Activation of D2-like receptors decreases cAMP levels and the activity of protein kinase A, decreases the phosphorylation levels of DARPP-32, inhibits ERK1/2 phosphorylation in striatum, decreases the activity of L-, N-, and P/Q-type Ca^{2+} channels, and activates K^{+} channels (reviewed in Refs. 48,49). There is some evidence suggesting that, in addition to $G_{\alpha s/olf}$ and $G_{\alpha i/o}$-mediated signaling, dopamine receptors can also signal via $G_{\alpha q}$- and G-protein-independent mechanisms (reviewed in Refs. 48–50). However, the mechanisms underlying $G_{\alpha q}$-mediated signaling are not completely understood (reviewed in Refs. 48–50). Studies show that dopamine receptors play a role in the modulation of locomotor activity, olfaction, retinal processes, cardiovascular, renal, and immune functions, memory and learning, feeding, sleep regulation, and in reward behaviors, particularly those related to drugs of abuse (reviewed in Ref. 48). Five subtypes of dopamine receptor have been identified. While D1 and D5 dopamine receptors belong to the D1-like receptor type, D2, D3, and D4 receptors belong to the D2-like receptor type (reviewed in Refs. 48–50).

As the name suggests, the endogenous agonist of dopamine receptors is dopamine. Dopamine can be synthesized from phenylalanine and tyrosine by tyrosine hydroxylase (reviewed in Ref. 50). Following synthesis in the cytosol, dopamine is taken into storage vesicles by vesicular monoamine transporters (reviewed in Ref. 50). Dopamine is released at the synapse by Ca^{2+}-mediated fusion of dopamine-containing vesicles with neuronal membranes (reviewed in Ref. 50). The synaptic activity of released dopamine is terminated by its reuptake into presynaptic nerve terminals by dopamine transporters (reviewed in Ref. 50). In the mammalian brain, four major dopaminergic pathways have been identified. These include the nigrostriatal, mesolimbic, mesocortical, and tuberoinfundibular systems (reviewed in Refs. 48–50). Interestingly, it has been shown that MSNs in the striatum and nucleus accumbens can be phenotypically distinguished depending on the expression of D1 or D2 dopamine receptors. D1 dopamine receptors (D1R) are expressed by MSNs that project into the medial globus pallidus and the substantia nigra pars reticulata, D2 dopamine receptors (D2R) are expressed by MSNs that project into the lateral globus pallidus (reviewed in Refs. 48–50), and a small percentage of MSNs coexpress D1R and D2R (reviewed in Refs. 48–50).

In order to elucidate the physiological roles of different dopamine receptor subtypes, knockout mice have been generated. Data with these mice have been the subject of excellent reviews (see Refs. 50–53) and are not described in detail here. Briefly, studies with D1R knockout mice suggest that the latter exhibit alterations in the motivational or attentive aspects of behavior.[52] Interestingly, D1R knockout mice exhibit a decrease in the locomotor effects of cocaine.[54] Studies with D2R knockout mice suggest that they may represent an animal model for Parkinson's disease as they recapitulate the extrapyramidal symptoms of the disease.[50] Interestingly, while the release of dopamine elicited by drugs of abuse such as morphine and cocaine is higher in D2R knockout mice than in wild-type animals (reviewed in Ref. 55), the rewarding effects of these drugs are attenuated in the knockout mice (reviewed in Ref. 55). D3R knockout mice exhibit increased activities in novel and exploratory environments, in rearing behavior, and deficits in spatial working memory compared to wild-type mice (reviewed in Ref. 50). These mice also exhibited a decrease in methamphetamine-mediated locomotion[56] and behavioral sensitization to ethanol but not to amphetamines.[56,57] Moreover, the D3R knockout mice exhibit attenuation in the development of tolerance and precipitated withdrawal to chronic morphine administration.[58] Studies with D4R knockout mice show that this receptor subtype may be involved in the modulation of novelty-related exploration and avoidance behavior to unconditioned stimuli (reviewed in Ref. 50). With regard to drugs of abuse, these mice are more responsive to the locomotor effects of ethanol, cocaine, or amphetamines than wild-type mice (reviewed in Ref. 50). In addition, D4R alleles associated with novelty seeking behavior have been correlated with drug abuse, attention deficit hyperactivity disorder, and gambling (reviewed in Ref. 50). Studies show that D5R knockout animals exhibit hypertension suggesting that this dopamine receptor subtype may be involved in the regulation of blood pressure (reviewed in Ref. 50). Taken together, these studies reveal a role for dopamine receptors in normal physiology and in drug abuse.

2. RECEPTOR HETEROMERIZATION

In the past decade, the idea that GPCRs associate to form heteromeric complexes with novel pharmacology has emerged (for reviews, see Refs. 59–65). Initial evidence for GPCR heteromerization came from studies showing that activation of a GPCR modulated the function of another GPCR.[66–70] The possibility that GPCRs could form heteromers was supported by elegant complementation studies. These included studies using chimeras of the α_{2c}-adrenergic and m3 muscarinic receptors (α_{2c}/m3 and m3/α_{2c}) that, when expressed individually, could not bind to either radiolabeled adrenergic or muscarinic ligands but when coexpressed could.[71] Similarly, coexpression of two different mutants of the angiotensin AT1 receptor that individually could not bind to angiotensin led to the restoration of angiotensin binding upon coexpression of these mutant receptors.[72]

The first direct evidence for GPCR heteromerization came from studies with the metabotropic $GABA_B$ receptors. The expression of functional $GABA_B$ receptors at the cell surface requires the coexpression of both the $GABA_{B1}$ and $GABA_{B2}$ protein subunits.[73–76] Another example for the requirement for heteromerization for receptor function is provided by taste receptors where associations between $T1R_1$ and $T1R_2$ form a receptor that recognizes sweet tasting molecules, while those between $T1R_1$ and $T1R_3$ recognize "umami" tasting molecules.[77–79]

The availability of epitope-tagged GPCRs for immunoisolation of interacting complexes to examine whether two GPCRs are in close proximity for direct interaction has shown that a large number of GPCRs can form heteromers (for reviews, see Refs. 59–61). Scientists in the field have applied the following criteria in order to ascertain that two GPCRs form heteromers: (i) colocalization of the two interacting receptors in the same cellular and subcellular compartment; (ii) immunoisolation of interacting receptors, using receptor-selective antibodies, from wild-type animal tissue that expresses both receptors (but not from animals lacking one of the receptors); (iii) proximity-based assays showing that the two receptors are in close proximity to interact only in cells/tissues expressing both receptors and not in cells/tissues lacking individual receptors; and (iv) unique pharmacological properties as determined using highly selective reagents targeting the heteromer.

2.1. Methods to study GPCR heteromerization

2.1.1 Techniques to study GPCR heteromerization in vitro

i. *Coimmunoprecipitation assays*: In this technique, interacting complexes in tissues are detected using antibodies selective for individual protomers. Alternatively, cells are cotransfected with the GPCRs of interest, each of which is differentially epitope tagged (e.g., with a Flag or myc epitope) and antibodies to the epitope tags used for the assay. Tissues or cells are then solubilized with a buffer-containing detergent, and antibodies to an individual receptor protomer or to the epitope tag on one of the receptors (e.g., the Flag epitope) are used to immunoprecipitate that receptor.[80,81] The immunoprecipitates are subjected to Western blot analysis and probed for the presence of the second receptor using antibodies selective for that receptor or to the epitope tag (e.g., the myc epitope) present on the receptor.[80,81] The detection of a signal on Western blots indicates that the two receptors form interacting complexes. Given the hydrophobic nature of GPCRs, appropriate controls are used in order to rule out the artifactual detection of interacting complexes. These include carrying out immunoprecipitation studies with tissue from animals lacking one receptor or with cells that express only one receptor, mixing cells that individually express each receptor prior to solubilization and immunoprecipitation, using different detergents for solubilization and/or using cross-linking reagents to cross-link cell surface proteins prior to solubilization and immunoprecipitation.[80,81] This strategy was first used to demonstrate opioid receptor heteromers in tissues[7] or in cells that coexpress both receptors.[82]

ii. *Bioluminescence resonance energy transfer assay*: In order for two receptors to form heteromers, they have to be in close proximity to interact in live cells. One of the earliest proximity-based assays used to examine GPCR heteromerization was bioluminescence resonance energy transfer (BRET) (reviewed in Ref. 83). In this assay, the C-terminus of one of the receptors is fused to *Renilla reniformis* luciferase (Rluc), while the C-terminus of the other receptor is fused to a genetic variant of green fluorescent protein (GFP) such as yellow fluorescent protein (YFP) (reviewed in Ref. 83). Cells are transfected with both receptors and then treated with the luciferase substrate, coelenterazine h. The action of luciferase on its substrate leads to the emission of light at 480 nm which can excite YFP which then emits light at 530 nm only if the two receptors are less than 100 Å apart (reviewed in Ref. 83). An advantage of this technique is that it does not require conventional light illumination, and this eliminates artifactual results due to autofluorescence (reviewed in Ref. 83). BRET assay was used to demonstrate proximity between opioid receptor types as well as between opioid and other receptors (Table 9.1). Several generations of BRET have been developed. $BRET^1$ uses YFP or enhanced GFP in combination with Rluc and coelenterazine h and has the advantage that the BRET signal can be detected for up to 1 h (reviewed in Ref. 83). $BRET^2$ uses a GFP variant that is excited at ~400 nm and DeepBlueC as the Rluc substrate. This allows significant separation between the excitation and emission spectra of GFP. The $BRET^2$ signal decays

Table 9.1 Opioid receptor heteromers

Heteromer pair	Detection techniques	Changes in heteromer properties			References
		Binding	Signaling	Trafficking	
DOR–α_{2A}AR	Co-IP, BRET	Not reported	Synergy in inhibition of neuropeptide release; ↑ DOR-mediated neurite outgrowth	Not reported	84,85
DOR–β_2AR	Co-IP, BRET	Not changed	Not changed	β_2AR internalized by DOR agonists	86,87
DOR–CB_1R	Co-IP, BRET	Not reported	↓ CB_1R-mediated signaling	Change in CB_1R localization	88,89
DOR–CXCR4	Co-IP, FRET	Not changed	Nonfunctional	Not changed	90
DOR–D1R	Immuno-EM	Not reported	↓ DOR signaling by D1R agonists	DOR redistribution by cocaine withdrawal	91,92
DOR–KOR	Co-IP, BRET	↓ Affinity for selective DOR or KOR ligands; binding cooperativity with DOR + KOR ligands	↑ With a combination of DOR + KOR agonists	Altered DOR trafficking	82,86
DOR–MOR	Co-IP, BRET	↓ Affinity for receptor-selective agonists; allosteric modulation of MOR binding by DOR ligands and vice versa	Switch from $G_{\alpha i}$ to $G_{\alpha z}$; heteromer coupling to β-arrestin; potentiation of MOR binding by DOR ligands and vice versa	MOR and DOR endocytosed independently of each other; some DOR agonists induce heteromer internalization	7,93–103
DOR–SNSR	BRET	Not reported	Preferential $G_{\alpha q}$ signaling	Not reported	104
KOR–β_2AR	Co-IP	Not changed	↓ β_2AR signaling	Altered	105

Continued

Table 9.1 Opioid receptor heteromers—cont'd

Heteromer pair	Detection techniques	Changes in heteromer properties			References
		Binding	Signaling	Trafficking	
KOR–APJ	Co-IP, BRET	Not reported	↑ By agonists to either receptor	Not reported	106
MOR–α_{2A}AR	Co-IP, BRET, FRET	Not reported	α_{2A}R ↑ MOR signaling; α_{2A}R ligands ↓ MOR signaling	Agonist to one receptor does not cause heteromer internalization	107–109
MOR–CB_1R	Co-IP, BRET, FRET	Not reported	CB_1R ↓ MOR signaling; MOR ↓ CB_1R signaling	Not reported	89,110
MOR–CCR5	Co-IP	Not reported	MOR agonist ↓ CCR5 signaling; CCR5 agonist ↓ MOR signaling	Not reported	111,112
MOR–GRPR	Co-IP	Not reported	MOR-mediated Ca^{2+} signaling only in cells expressing the heteromer	MOR-mediated GRPR internalization	113
MOR–KOR	Co-IP, BRET	↓ Affinity for MOR-selective ligands	Not reported	Not reported	82,114,115
MOR–NK-1	Co-IP, BRET	Not changed	Not changed	MOR internalization by NK-1 agonist and vice versa	116
MOR–OFQ/N	Co-IP	↑ Affinity for MOR-selective ligands	↓ Potency of MOR ligands	Not reported	117,118
MOR–SSTR2	Co-IP	Not changed	Not changed	↑ Heteromer internalization by sSSTR2 agonist	119

rapidly; however, the use of Rluc variants, Rluc2 or Rluc8, ensures a high BRET signal that can last for hours depending on the donor and acceptor combinations chosen (reviewed in Ref. 83). $BRET^3$ makes use of RLuc8 and a variant of red fluorescent protein, mOrange, as donor and acceptor pairs. $BRET^3$ exhibits the most red-shifted emission wavelength (564 nm), and consequently yields a several-fold improvement in light output compared with other BRET assays (reviewed in Ref. 83).

iii. *Fluorescence resonance energy transfer assay*: Another technique commonly used to determine close proximity between two GPCRs is fluorescence resonance energy transfer (FRET). Here, cells are cotransfected with pairs of receptors of interest; one of the receptors is fused to a fluorescent donor and the other to a fluorescent acceptor (reviewed in Ref. 120). The most commonly used fluorescent donor is cyan fluorescent protein (CFP), while the fluorescent acceptor is YFP (reviewed in Ref. 120). Light excitation at 400 nm causes CFP to emit light at 480 nm which can then excite YFP, if it is in sufficiently close proximity, causing it to emit light at 580 nm (reviewed in Ref. 120). Given that the FRET efficiency is inversely proportional to the sixth power of the distance between the two fluorophores, a FRET signal is observed only if the donor–acceptor pair is less than 100 Å apart (reviewed in Ref. 120). An advantage of FRET is that it is amenable to confocal microscopy. Problems with FRET leading to poor signal-to-noise ratio include (i) cross talk, wherein light used to excite the donor also excites the acceptor fluorophore; (ii) bleed through, due to partial overlap between the spectra of the donor and acceptor fluorescent proteins; and (iii) photobleaching (reviewed in Ref. 120). Several modifications of the FRET technique have been developed to improve its signal-to-noise ratio including photobleaching FRET (pbFRET), photoquenching FRET, fluorescence lifetime imaging microscopy (FLIM), and time-resolved FRET. Several excellent reviews describe these techniques in detail.[120–124]

Several techniques to label GPCRs for FRET studies have been developed including (i) labeling one receptor with a FlAsH tag (a 15-amino acid polypeptide tag with a tetracysteine core (CCXXCC) that covalently binds to a fluorogenic bisarsenical fluorescein ligand whose fluorescence increases upon binding to the polypeptide tag) and fusing the other receptor with CFP, (ii) using mTurquoise instead of CFP for FLIM experiments, (iii) using Eu^{3+}- and Tb^{3+} cryptate in combination with a fluorescent protein, and (iv) using SNAP- and CLIP tags (reviewed in Ref. 124). Moreover, FRET sensors have been developed to examine ligand binding to a GPCR, receptor activation, conformational changes in receptor dimers, receptor trafficking, and receptor interaction with signaling molecules (reviewed in Ref. 124).

iv. *Protein complementation*: In recent years, protein-fragment complementation assays have been used to examine GPCR heteromerization in living cells. In this approach, cells are transfected with one GPCR fused to a fragment of luciferase and another GPCR fused to a complementary fragment (reviewed in Ref. 121). Reconstitution of active luciferase enzyme would occur only if the two receptors were in close proximity. This approach is known as bimolecular luminescence complementation (BiLC) (reviewed in Ref. 121). A combination of protein complementation with the BRET assay[125] has proven to be amenable to the development of sensors to study not only receptor heteromerization but also conformational changes following receptor activation, receptor interactions with signaling molecules, and how the latter are modulated by ligand treatment (reviewed in Ref. 83). Moreover, recent advances in BRET technology have enabled these assays to be carried out in live animals.[126]

Another protein complementation approach uses a receptor fused with a fragment of YFP (nonfluorescent) and another receptor fused with a complementary YFP fragment (nonfluorescent). YFP fluorescence is reconstituted only when the two receptors are in close proximity. This approach is known as bimolecular fluorescence complementation (BiFC) (reviewed in Ref. 121). The advantage of BiFC over BiLC is that it can provide information about the subcellular localization of receptor interactions (reviewed in Ref. 121). Moreover, a combination of BiFC or BiLC with either BRET or FRET has been used to demonstrate the presence of GPCR trimers or tetramers.[127–130]

v. *Atomic force microscopy and crystallography*: Biophysical approaches such as atomic force microscopy (AFM) and X-ray crystallography have been useful in demonstrating the existence of GPCR dimers. AFM uses a scanning tunneling microscope to measure the movement of a cantilever beam over the sample surface.[131] Using this approach, studies showed that rhodopsin, the prototypical family A GPCR, was arranged in dimeric arrays.[132] X-Ray crystallography, a method used for determining the arrangement of atoms within a molecule, can provide information about the structural organization of a GPCR. Until very recently, the high-resolution crystal structures for only a few GPCRs were available. These included the structures for rhodopsin and extracellular domains of metabotropic glutamate receptors, which indicated that these receptors could exist as dimers or higher-order structures.[133,134] The obstacles in obtaining high-resolution structures for most GPCRs included low expression levels in endogenous tissues, difficulties in expressing the receptor at high enough levels in heterologous cells, structural instability following detergent solubilization, as well as multiple receptor conformational states. Recent approaches aimed at stabilizing a GPCR into a single conformational state, through the use of either monoclonal antibodies that stabilize the conformation of intracellular loop 3 or by replacing a portion of this loop with T4-lysozyme, together with crystallization in the presence of a receptor antagonist, have permitted the determination of high-resolution structures for a number of GPCRs including those for opioid receptors.[135–142] In this context, it is to be noted that the recent MOR and chemokine receptor, CXCR4, crystal structures reveal a dimeric receptor arrangement consistent with the idea that these receptors could exist as higher-order structures *in vivo*.[137,139]

2.1.2 Techniques to study GPCR heteromerization in vivo

i. *Ligand-aided FRET*: The presence of GPCRs in endogenous tissue can be detected by FRET assays using fluorescently labeled ligands to each receptor protomer. Since both agonists and antagonists can be used in such studies, they provide additional information on the conformational state of the receptor.[143] This approach makes use of fluorophores compatible with TR-FRET to label the ligands in order to obtain a high signal-to-noise ratio.[143] Thus, ligands are labeled with donor–acceptor pairs (typical examples of donor–acceptor pairs optimal for TR-FRET include Eu-PBBP, Lumi4-Tb, or Alex-64-d1), and the assay is then carried out in tissue patches.[143] Such an approach was used to detect the presence of oxytocin receptor dimers/oligomers in mammalian gland.[143]

ii. *Antibody-aided FRET*: The presence of GPCR heteromers in live cells can be detected by a combination of selective antibodies and FRET. For example, in live cells expressing epitope-tagged receptors, secondary antibodies labeled with either fluorescein or rhodamine were used in pbFRET experiments to monitor the formation of SSTR1–SSTR5 and D2R–SSTR5 heteromers.[144,145] In another study, antibodies to epitope tags on the receptor were labeled with either Eu^{3+} or allophycocyanin protein and used in TR-FRET experiments to demonstrate the presence of DOR oligomers at the cell surface.[146] An extension of the latter approach could be used to examine GPCR heteromerization in endogenous tissue by labeling monoclonal antibodies that are highly selective to individual receptors in the heteromer with complementary TR-FRET donor–acceptor pairs. Such an approach was used to detect the presence of $GABA_B$ receptor tetramers on brain membranes.[147]

iii. *Proximity-based ligation assays (PLAs)*: Recently, PLAs have been used to detect the presence of GPCR heteromers in endogenous tissue.[148,149] This is an antibody-based approach where tissue sections of interest are first incubated with anti-GPCR selective antibodies.[148,149] The antibody to each protomer in the heteromeric complex under investigation is raised in a different species (rabbit or mouse). Sections are then incubated with species-specific secondary antibodies coupled to complementary oligonucleotide sequences (PLA probes).[148,149] When the antibodies to the two receptors are in close proximity (<17 nm), the PLA probes can be ligated using ligase and a ligation solution comprising of two oligonucleotides that will hybridize with the two PLA probes to form a closed circle.[148,149] The ligated closed circle is then amplified by polymerase in a rolling-circle amplification reaction in the presence of fluorescently labeled oligonucleotides.[148,149] Hybridization of the fluorescently labeled nucleotides appears as a fluorescent spot that can then be visualized by fluorescence microscopy.[148,149] This assay has been used to detect the presence of CB_1R–CB_2R receptor heteromers in the pineal body and nucleus accumbens and of dopamine D2–adenosine A_{2A} receptor heteromers in the striatum.[148,149]

iv. *Heteromer-selective antibodies*: Monoclonal antibodies that selectively recognize a GPCR heteromer are useful tools that can be used to not only determine the tissue distribution of the heteromer but also its possible physiological role. A subtractive immunization strategy has been described for the development of such antibodies.[7,9,10,88] In this approach, mice are first made tolerant to nonspecific cell membrane epitopes by treatment with cyclophosphamide.[7,9,10,88] Mice are then immunized with membranes from cells expressing the heteromer of interest.[7,9,10,88] Antibodies selective for MOR–DOR, DOR–CB_1R, or DOR–AT1R heteromers have been generated using this approach.[7,9,10,88] Studies using these antibodies show that the antibodies are able to selectively block signaling by specific heteromers and that MOR–DOR heteromers are increased in brain regions involved in pain processing under conditions leading to the development of tolerance to morphine, while DOR–CB_1R heteromer levels are increased in specific brain regions during neuropathic pain, and DOR–AT1R heteromer levels are increased during ethanol-induced liver fibrosis.[7,9,10,88]

2.2. Opioid receptor heteromerization

Early radiolabeled binding studies, as well as studies measuring the antinociceptive effects of selective opioid receptor ligands, suggested the presence of multiple opioid receptor subtypes.[150,151] However, to date, the cDNAs for only MOR, DOR, and KOR have been identified.[2,20] This suggested the possibility that the multiple opioid receptor subtypes detected by binding and antinociceptive studies could arise from the formation of heteromeric complexes between opioid receptor subtypes (MOR, DOR, KOR) or between the latter and other GPCRs. The availability of cDNAs for epitope-tagged receptors together with coimmunoprecipitation and/or proximity-based assays has enabled studies showing that opioid receptors can form heteromers with a number of GPCRs (see Table 9.1). These are described below.

DOR–KOR heteromers: The first opioid receptor heteromer to be detected and characterized involved DOR and KOR.[82] Coimmunoprecipitation studies, using differentially epitope-tagged receptors, showed that DOR and KOR could form interacting complexes.[82] BRET studies showed that both receptors were in close enough proximity (<100 Å) in live cells to directly interact.[86] Analysis of the binding, signaling, and trafficking properties of DOR–KOR heteromers revealed that the latter exhibited properties distinct from that of the individual receptors. For example, the binding affinities of selective DOR and KOR ligands were reduced in the DOR–KOR heteromer; however, a combination of DOR and KOR ligands exhibited cooperativity in binding that in the case of agonists was translated into a potentiation in signaling responses.[82] This binding cooperativity was also observed with a combination of DOR and KOR antagonists.[82] Bivalent ligands consisting of DOR and KOR antagonists (naltrindole and 5′-guanidinonaltrindole (5′-GNTI), respectively) linked by spacers of variable length were used to target the DOR–KOR heteromer.[152] One of these bivalent ligands, KDN-21, exhibited a higher affinity and selectivity for DOR–KOR heteromers compared to DOR and KOR homomers.[153] Pharmacological and binding data indicate that KDN-21 binds to DOR–KOR heteromers of the DOR1–KOR2 phenotype.[152,153] Interestingly, a study identified 6′-guanidinonaltrindole (6′-GNTI) as a DOR–KOR heteromer-selective agonist that induced antinociception when administered to the spinal cord but not to the brain.[154] However, a recent study has suggested that 6′-GNTI is a KOR-biased ligand[155]; thus, it remains to be seen if 6′-GNTI represents a DOR–KOR heteromer-biased ligand. Studies that suggest a role for DOR–KOR heteromers in pain pathways are described in Section 3.

DOR–MOR heteromers: This is the most extensively studied opioid receptor heteromer due to its role in morphine-mediated antinociception and in the development of tolerance and dependence to opiates. A comprehensive body of evidence suggested the possibility of DOR–MOR heteromerization (reviewed in Ref. 156). Noteworthy are radiolabeled studies indicating that (i) there is an allosteric modulation of radiolabeled binding to DOR by MOR ligands[157,158]; (ii) intracerebroventricular potentiation of morphine-mediated antinociception by nonantinociceptive doses of DOR agonists which can be blocked by the DOR antagonists ICI174,864 and naltrindole-5′-isothiocyanate[159]; (iii) potentiation of acute morphine-mediated antinociception by DOR antagonists and inhibition of development of tolerance or reduction of already developed tolerance to morphine[160]; (iv) prevention of the development of tolerance and dependence in mice by pretreatment with DOR antagonists[161]; and (v) lack of development of tolerance to morphine in DOR knockout mice.[162]

Evidence that MOR and DOR form heteromeric complexes *in vitro* came from coimmunoprecipitation and BRET studies using differentially epitope-tagged receptors.[93–95] In addition, coimmunoprecipitation studies using receptor-selective antibodies revealed the presence of DOR–MOR heteromers in spinal cord membranes from wild-type but not DOR knockout mice.[94] More recently, antibodies that selectively recognize DOR–MOR heteromers were used to detect the presence of these heteromers in brain regions of wild-type but not MOR or DOR knockout mice.[7]

DOR–MOR heteromers exhibit distinct pharmacological properties from that of individual receptors.[93–95] Low nonsignaling doses of DOR or MOR ligands potentiate the binding and signaling of MOR or DOR receptors, respectively.[93,94,96] Moreover, the potency of highly selective synthetic MOR or DOR ligands is decreased, while the affinity for endomorphin-1 and DOR-selective agonists is increased in DOR–MOR heteromers as compared to MOR or DOR.[95,97]

Studies show that DOR–MOR heteromerization leads to changes in the signaling pathways activated by the heteromers compared to individual receptor homomers. In this context, a study showed that the DOR–MOR heteromer associates with pertussis-insensitive $G_{\alpha z}$ instead of the pertussis-sensitive $G_{\alpha i}$ and this interaction occurs in the endoplasmic reticulum.[98,99] Another study reported that activation of a protomer in the DOR–MOR heteromer leads to a switch in signaling from $G_{\alpha i}$- (as it occurs in receptor homomers) to β-arrestin-mediated signal transduction leading to changes in the spatiotemporal dynamics of ERK1/2 phosphorylation and ultimately differential activation of transcription factors.[100] It has also been reported that while the MOR-selective agonist, DAMGO, stimulates Ca^{2+} signaling in cells expressing DOR–MOR heteromers by activating phospholipase C and releasing Ca^{2+} from intracellular stores, it inhibits Ca^{2+} signaling in GH3 cells expressing only MOR by inhibiting adenylyl cyclase activity and voltage-gated Ca^{2+} channels and by activating inwardly rectifying K^{137+}. Taken together, these studies indicate that heteromerization increases the repertoire of signaling for opioid receptors.

A few studies have examined how DOR–MOR heteromerization modulates the trafficking of individual receptors. One study reported that treatment with receptor-selective agonists (DAMGO or deltorphin II but not DPDPE or DSLET) led to internalization of the DOR–MOR heteromer.[99] However, another study examining receptor turnover in cells that expressed increasing levels of DOR with constant levels of MOR reported that the two receptors endocytosed independently of each other in the context of the heteromer.[102] Finally, a recent study reported that methadone could promote the internalization of the DOR–MOR heteromer targeting it to the degradation machinery.[103] Together, these data suggest that the internalization of the DOR–MOR heteromer could be a ligand type-dependent event. Recent studies suggest a role for these heteromers in pathways leading to development of tolerance to morphine; this is discussed in Section 3.

DOR–CB_1R heteromers: Early studies examining interactions between DOR and CB_1R showed that (i) long-term exposure to a DOR agonist attenuated CB_1R-mediated stimulation of G-protein activity in N18TG2 neuroblastoma cells that coexpress DOR and CB_1R[163]; (ii) the DOR antagonist, naltrindole, blocked Δ^9-THC-mediated anxiolytic responses[105]; (iii) CB_1R levels as well as agonist-mediated G-protein activity are increased in the substantia nigra of mice lacking DOR as compared to wild-type animals[164]; and (iv) DOR activity is increased in the caudate putamen of CB_1R knockout mice compared to wild-type mice.[165]

DOR–CB_1R heteromers are detected in HEK-293 and in Neuro 2A cells coexpressing both receptors by coimmunoprecipitation and BRET assays.[88,89] DOR and CB_1R are colocalized in cortical neurons as examined by immunofluorescence studies with receptor-selective antibodies.[88] Interestingly, heteromerization leads to a differential recruitment of trafficking (adaptor) proteins; CB_1R associates with the adaptor protein AP-3 in the absence of DOR and with AP-2 in the presence of DOR.[88] This, in turn, leads to differential localization of CB_1R from intracellular vesicles, in the absence of DOR, to the cell surface in the presence of DOR.[88] In addition, lentiviral-mediated knockdown of DOR in F11 cells that endogenously coexpress CB_1R and DOR leads to a decrease in cell surface localization of CB_1R.[88] Heteromerization also alters CB_1R signaling; the potency of the CB_1R agonist, Hu-210, is decreased in cells expressing DOR–CB_1R in assays that measure G-protein activity or ERK1/2 phosphorylation, and in these cells, knockdown of DOR leads to an increase in the potency of Hu-210.[88] Moreover, DOR–CB_1R heteromerization leads to a PLC-mediated recruitment of β-arrestin 3 to the heteromeric complex, and to the activation of novel signaling pathways that promote cell survival.[88] Recent studies suggest a role for these heteromers in pain pathways; this is discussed in Section 3.

DOR–CXCR4 heteromers: Given that both DOR and chemokine CXCR4 receptors are expressed in immune cells, coimmunoprecipitation and FRET assays were used to determine whether both receptors could form heteromers.[90] These studies showed that DOR and CXCR4 form interacting complexes.[90] Moreover, the studies showed that simultaneous activation of DOR and CXCR4 in cells coexpressing both receptors did not affect cell viability, chemokine binding, or receptor internalization but led to the formation of nonfunctional complexes.[90] This suggests that DOR–CXCR4 heteromerization provides a means to silence CXCR4 function and could be important in the regulation of immune function.

DOR–sensory neuron-specific receptor heteromers: Sensory neuron-specific receptors (SNSRs) are mainly expressed in dorsal root ganglia in the spinal cord, where they mediate pronociceptive responses.[166] Given that DOR and SNSR-4 share localization in dorsal root ganglia, can be activated by the bovine adrenal medulla peptide 22 (BAM22), and their functions are antagonistic to each other, led to studies examining whether both receptors could form heteromeric complexes. BRET assays in HEK-293 cells coexpressing differential epitope-tagged receptors indicated that both receptors were in sufficiently close proximity to directly interact in live cells.[104]

In cells expressing DOR–SNSR-4 heteromers, activation of SNSR-4 alone led to $G_{\alpha q}$-mediated signaling, while activation of DOR alone resulted in $G_{\alpha i}$-mediated signaling. However, coactivation of both receptors by receptor-specific agonists led to a preferential signaling via $G_{\alpha q}$.[104] Interestingly, the endogenous ligands for SNSR are derived from the proteolytic processing of the proenkephalin precursor that also generates the antinociceptive endogenous DOR ligands, Leu- and Met-enkephalin.[167] Moreover, one of these peptides, BAM22, has been shown to bind to both DOR and SNSR-4.[167] Activation of DOR–SNSR-4 heteromers with BAM22 leads to signaling via $G_{\alpha q}$. This is the first demonstration of an endogenous peptide as a bivalent agonist of a GPCR heteromer.

DOR–β_2 adrenergic receptor heteromers: Interactions between DOR, a $G_{\alpha i}$-coupled receptor, and the β_2 adrenergic receptor (β_2AR), a $G_{\alpha s}$-coupled receptor, were suggested by studies showing that in the isolated isovolumic rat heart the DOR agonist, Leu-enkephalin, significantly inhibited β_2AR-mediated positive ionotropic effects and stimulation of cAMP levels, effects that were blocked by the opioid antagonist, naloxone, and by pertussis toxin treatment.[168] Moreover, in isolated rat ventricular myocytes, nanomolar concentrations of Leu-enkephalin could block β_2AR-mediated increases in sarcolemmal L-type Ca^{2+} currents, cytosolic Ca^{2+} transients, and contraction.[169] Coimmunoprecipitation and BRET assays using differentially epitope-tagged receptors demonstrated the formation of DOR–β_2AR heteromers in HEK-293 cells.[86,87] While the binding and signaling properties of individual protomers in the DOR–β_2AR heteromer were not changed by heteromerization, individual receptor trafficking was, in that DOR agonists could internalize β_2AR only in cells coexpressing both receptors.[87]

DOR–α_{2A} adrenergic receptor heteromers: A number of studies have suggested interactions between DOR and α_{2A} adrenergic receptors (α_{2A}AR). These include studies showing that (i) a selective imidazole/α_2AR agonist synergizes with the DOR agonist, deltorphin II, to block substance P elicited nociception[170]; (ii) the synergy between agonists for the two receptors was observed in α_{2C}AR and in MOR knockout mice but not in α_{2A}AR knockout mice suggesting an involvement of α_{2A}AR and of DOR in these interactions[171–173]; and (iii) cotreatment of synaptosomes isolated from the spinal cord or of spinal slices with α_{2A}AR and DOR agonists led to potentiation of inhibition of K^+-stimulated release of neuropeptides.[84,174]

Studies using differentially epitope-tagged receptors expressed in heterologous cells together with coimmunoprecipitation and BRET assays show that α_{2A}AR and DOR are in close proximity (<100 Å apart) to form interacting complexes.[85] Not much is known about how DOR–α_{2A}AR heteromerization modulates the pharmacological, signaling, and trafficking properties of this heteromer pair. One study suggested the involvement of protein kinase C in heteromer-mediated synergy in inhibition of calcitonin gene-related peptide release from spinal cord slices,[84] while another indicated that the presence of α_{2A}AR is sufficient to potentiate DOR-mediated neurite outgrowth in Neuro 2A cells.[85] The latter study indicated that just the presence of α_{2A}AR is sufficient to modulate DOR activity; whether the converse also occurs and how ligand occupancy of both protomers in the DOR–α_{2A}AR heteromer modulates individual receptor function need to be further evaluated.

DOR–D1R heteromers: Interactions between DOR and D1R were suggested by studies showing (i) that alterations in dopaminergic neurotransmission modulate the endogenous levels of opioid peptides[175–177]; (ii) attenuation of DOR-mediated signaling by the D1R agonist, SKF 82958[178]; and (iii) enhancement of climbing behavior in mice by a combination of DOR and D1R agonists.[179]

Immunoelectron microscopy detected the presence of DOR and D1R immunoreactivity in a subset of dendrites in the dorsolateral striatum and nucleus accumbens of rats, indicating that both receptors were present in the same neuron.[91,92] Moreover, withdrawal from cocaine, a drug that blocks dopamine transporter activity and thereby alters dopamine neurotransmission by increasing and prolonging signaling by postsynaptic dopamine receptors caused a redistribution of DOR in the nucleus accumbens.[91] Although these colocalization studies indicate that DOR–D1R could form heteromers, this needs to be further evaluated by studies using epitope-tagged receptors, coimmunoprecipitation, and proximity-based assays.

MOR–KOR heteromers: Very few studies have examined heteromerization between MOR and KOR. An early study using coimmunoprecipitation of differentially epitope-tagged receptors did not detect the presence of interacting MOR–KOR complexes.[82] This does not rule out the formation of MOR–KOR heteromers if the latter are not stable under the detergent conditions used in the coimmunoprecipitation studies. Another study used BRET to show that MOR and KOR are in close proximity to interact in live cells.[114] A recent study reported the formation of MOR–KOR heteromers in spinal cord membranes of female, but not male, rats and showed that it is dependent on the stage of the estrous cycle.[115] Interestingly, this study identified dynorphin 1–17 as an MOR–KOR heteromer-selective ligand.[115]

MOR–α_{2A} adrenergic heteromers: Studies describing synergistic interactions between the antinociceptive effects of MOR and α_{2A}AR agonists[180,181] and with α_{2A}AR knockout mice that exhibit a reduction in the analgesic potency of morphine[173] suggested the possibility of MOR–α_{2A}AR heteromerization.

Differential epitope-tagged receptors together with coimmunoprecipitation, BRET, or FRET assays demonstrated the formation of MOR–α_{2A}AR heteromers.[107,108] Studies examining heteromer-mediated signaling showed that while the presence of α_{2A}AR is sufficient to enhance MOR signaling, coactivation of both receptors in the heteromeric complex leads to a significant attenuation of signaling responses.[107] Proximity-based assays (FRET, BRET) showed that, in cells coexpressing MOR–α_{2A}AR heteromers, treatment with morphine induces conformational changes in norepinephrine occupied α_{2A}AR leading to a decrease in $G_{\alpha i}$-mediated signaling.[108]

Finally, a study examining the trafficking of MOR–α_{2A}AR heteromers showed that treatment with the agonist to one receptor protomer did not lead to cointernalization of the heteromeric partner receptor.[109] However, given that, in addition to heteromers, the cells express individual receptor homomers, gives rise to the possibility that the agonists are in fact causing internalization of the homomer rather than the targeted receptor in the heteromer. Thus further studies are required to evaluate the trafficking properties of MOR–α_{2A}AR heteromers.

MOR–CB_1R heteromers: Several studies have described functional interactions between MOR and CB_1R (reviewed in Ref. 43). These include studies showing that (i) Δ^9-THC enhances the potency of morphine in animal models[4]; (ii) conditioned place preference to Δ^9-THC is not observed in MOR knockout mice[17]; and (iii) CB_1R knockout mice exhibit a reduction in the reinforcing properties of morphine as well as in the severity of withdrawal symptoms to this opioid.[182]

Coimmunoprecipitation, BRET, and FRET assays using differentially epitope-tagged receptors detected the presence of MOR–CB_1R heteromers in heterologous expressing systems.[89,110] While the presence of unliganded CB_1R does not affect MOR-mediated signaling, ligand binding to CB_1R decreases MOR-mediated signaling.[89] Conversely, ligand binding to MOR decreases CB_1R-mediated signaling.[89] These interactions had physiological relevance as observed by inhibition of neurite outgrowth in Neuro2A cells that coexpress MOR and CB_1R following cotreatment with agonists to both receptors.[89] Electrophysiological studies with oocytes coexpressing MOR and CB_1R and where one of the receptors is fused to a chimeric G_{qi5} showed that agonists to either receptor elicited Ca^{2+}-activated Cl^{-} currents thus demonstrating the formation of a functional receptor heteromer.[110] However, this study did not examine the effect of coadministration of agonists to both receptors on Ca^{2+}-activated Cl^{-} currents.

MOR–chemokine receptor 5 heteromers: Studies have described the presence of MOR in immune cells and that receptor activation by morphine increases the expression of chemokine receptor 5 (CCR5) suggesting interactions between these two receptors.[183,184] Coimmunoprecipitation studies using differentially epitope-tagged receptors showed that MOR–CCR5 interacting complexes could be detected only in cells coexpressing both receptors.[111,112] Examination of the properties of the MOR–CCR5 heteromer showed that treatment with either the MOR agonist, DAMGO, or the CCR5 agonist, RANTES, induced chemotaxis in cells coexpressing both receptors.[111] However, pretreatment with DAMGO inhibited RANTES-induced chemotaxis, increased CCR5 phosphorylation, and decreased G-protein coupling to CCR5.[111] The converse was observed when cells were pretreated with RANTES[111] suggesting cross-desensitization of MOR and CCR5 within the heteromeric complex.

MOR–gastrin-releasing peptide receptor heteromers: Studies showing that (i) MOR agonist administration induces itching (also known as pruritus)[185–187]; (ii) scratching responses to itch-inducing stimuli including morphine administration are abrogated in mice lacking gastrin-releasing peptide receptor (GRPR)[113,188,189]; and (iii) colocalization of MOR and GRPR in the dorsal horn of the spinal cord[113] led to the examination of interactions between the MOR isoform, MOR1D (comprising exons 1–3 and 8–9 of the *Oprm* gene) and GRPR.[113]

Direct interactions between MOR1D and GRPR were suggested by coimmunoprecipitation studies carried out in heterologous cells expressing epitope-tagged receptors or with spinal cord membranes that endogenously express both receptors.[113] Examination of the signaling properties of the MOR1D–GRPR heteromer showed that morphine-induced intracellular Ca^{2+} release only in cells coexpressing both receptors.[113] Interestingly, treatment with morphine enhanced GRPR internalization, while treatment with the GRPR agonist had no effect on MOR internalization in cells coexpressing both receptors.[113]

MOR–orphanin FQ/nociceptin receptor heteromers: Interactions between the orphanin FQ/nociceptin receptor (OFQ/N) and MOR have also been described. These include studies showing that an OFQ/N receptor antagonist potentiated the analgesic effects of morphine both in naïve mice and in mice that were tolerant to this opioid.[190,191] Coimmunoprecipitation studies using differentially epitope-tagged receptors showed that MOR and OFQ/N receptors could form interacting complexes in cells coexpressing both receptors.[117,118] MOR–OFQ/N heteromers exhibited an enhanced affinity for MOR-selective ligands.[117] Interestingly, MOR–OFQ/N receptor heteromerization attenuated the potency of MOR-mediated inhibition of adenylyl cyclase activity and ERK1/2 phosphorylation.[118] Moreover, treatment with the OFQ/N receptor agonist, nociceptin, led to desensitization of both MOR- and OFQ/N receptor-mediated signaling, while treatment with the MOR agonist, DAMGO, desensitized only MOR-mediated signaling.[118] These studies suggest that MOR–OFQ/N receptor heteromerization may play a role in modulating MOR receptor function.

MOR–NK-1 receptor heteromers: Substance P exerts its effects through the activation of neurokinin-1 (NK-1) receptors. Both MOR and NK-1 receptors have been detected in dorsal horn neurons of the spinal cord as well as in brain regions implicated in mood disorders, stress responses as well as those that mediate the rewarding effects of drugs of abuse.[192,193] In addition, treatment with morphine leads to an increase in the expression of NK-1 receptors in cortical neurons leading to a greater substance P-mediated Ca^{2+} mobilization than in untreated neurons.[194] Moreover, activation of NK-1 receptors modulates neuronal morphine-mediated endocytosis and MOR desensitization.[195] These observations together with those showing that the rewarding effects of morphine are not present in NK-1 receptor knockout mice led to studies examining whether these two receptors formed heteromeric complexes.[196] Coimmunoprecipitation and BRET assays using differentially epitope-tagged receptors demonstrate the presence of MOR–NK-1 receptor heteromers in HEK-293 cells coexpressing both receptors.[116] MOR–NK-1 receptor heteromerization did not lead to changes in the binding and signaling properties of individual protomers.[116] However, treatment with the MOR-selective agonist, DAMGO, led to phosphorylation and cointernalization of NK-1 receptors, while treatment with substance P induced the phosphorylation and cointernalization of MOR.[116] Moreover, in cells coexpressing MOR and NK-1 receptors but not in cells expressing only MOR, cotreatment with substance P and DAMGO-induced cointernalization of both receptors along with β-arrestin into the same endosomal compartment leading to a delay in the recycling and resensitization kinetics of MOR.[116]

MOR–somatostatin 2A receptor heteromers: Another receptor pair investigated for the possible formation of heteromeric complexes involved

MOR and somatostatin 2A receptors (sst2a) based on the fact that the latter exhibited 38% homology with the amino acid sequence of MOR.[119] Coimmunoprecipitation studies using differentially epitope-tagged receptors demonstrated the presence of interacting MOR–sst2a receptor complexes in HEK-293 cells coexpressing both receptors.[119] MOR–sst2a receptor heteromerization did not affect protomer-mediated ligand binding and signaling; however, it modulated agonist-mediated receptor trafficking.[119] In this context, in cells coexpressing MOR and sst2a, the sst2a receptor agonist promoted the internalization of both receptors, while the MOR agonist promoted only the internalization of MOR.[119] Although it is possible that MOR–sst2a heteromers do not undergo internalization following treatment with MOR agonists (the observed internalization being that of MOR homomers), it is to be noted that MOR agonists desensitize sst2a receptor signaling and sst2a receptor agonists desensitize MOR-mediated signaling.[119]

KOR–β_2AR heteromers: Both KOR and β_2AR are present in the heart and treatment with the β_2AR agonist, isoproterenol, has been shown to increase radiolabeled U69,593 binding to KOR present in rat cardiac sarcolemma.[197,198] This suggested interactions between KOR and β_2AR. Coimmunoprecipitation studies using differentially epitope-tagged receptors expressed in HEK-293 cells indicated that KOR and β_2AR formed interacting complexes.[87] While KOR–β_2AR heteromerization did not affect the pharmacology of each protomer in the complex, it did modulate trafficking and signaling as examined by observing the phosphorylation of ERK1/2 in cells coexpressing both receptors.[87] For example, treatment with a β_2AR agonist reduced β_2AR internalization in cells coexpressing KOR–β_2AR, but could induce internalization in cells expressing only β_2AR.[87] Moreover, β_2AR-mediated increases in ERK1/2 phosphorylation are significantly attenuated in cells coexpressing KOR–β_2AR heteromers. Together, these results indicate an important physiological role for KOR–β_2AR heteromerization in modulation of β_2AR function.

KOR–apelin receptor heteromers: The apelin receptor (APJ) binds to the apelin peptide and plays a role in the modulation of several functions including cardiovascular functions, neuroprotection, pain perception, and regulation of food and water intake.[199,200] Given that both KOR and APJ exhibit a similar distribution and function in the cardiovascular and nervous system, studies examined the possible heteromerization between these two receptors.[106] Coimmunoprecipitation, colocalization, and BRET assays show that KOR and APJ can form interacting complexes in HEK-293 cells.[106]

KOR–APJ heteromerization leads to potentiation of ERK1/2 phosphorylation following treatment with either the APJ or KOR agonist compared to what is observed in cells expressing only homomers for either receptor.[106] Moreover, KOR–APJ heteromerization leads to a decrease in the activity of protein kinase A, an increase in the activity of protein kinase C, and an increase in cell proliferation.[106]

2.3. Cannabinoid receptor heteromerization

Compared to opioid receptors, fewer studies have examined heteromerization of cannabinoid receptors. While heteromerization between opioid receptor subtypes and CB_1R has been described in Section 2.1, in this section we describe cannabinoid receptor heteromers that do not involve opioid receptors (see Table 9.2).

CB_1R–adenosine 2A receptor heteromers: Several studies suggested interactions between CB_1R and adenosine 2A receptors ($A_{2A}R$). These include studies showing that (i) activation of $A_{2A}R$ blocks, while inhibition of $A_{2A}R$ promotes, CB_1R-mediated effects on evoked glutamate release[210]; (ii) the rewarding and aversive effects of Δ^9-THC were significantly reduced in $A_{2A}R$ knockout mice[211]; (iii) $A_{2A}R$ antagonist treatment blocks the synergy between MOR and CB_1R for CRE-mediated gene transcription[212]; (iv) locomotor behavior induced by administration of $A_{2A}R$ antagonists is attenuated by CB_1R antagonists or in CB_1R knockout mice[213]; and (v) low doses of $A_{2A}R$ antagonists attenuated while higher doses potentiated the self-administration of Δ^9-THC and anandamide.[214]

The formation of interacting CB_1R–$A_{2A}R$ complexes in rat striatum was detected by coimmunoprecipitation studies using receptor-selective antibodies.[201] BRET using differentially epitope receptors demonstrated that CB_1R and $A_{2A}R$ are in close proximity to interact in HEK-293 cells.[201] Studies carried out in SH-SY5Y neuroblastoma cells, that endogenously express CB_1R and $A_{2A}R$, show that CB_1R activation inhibits forskolin-mediated increases in cAMP levels, and that the latter is blocked by an $A_{2A}R$ antagonist.[201] Moreover, $A_{2A}R$-mediated increases in cAMP levels are blocked by a CB_1R antagonist.[201]

CB_1R–AT1 angiotensin receptor heteromers: Early studies demonstrating an upregulation of CB_1R in liver cells that have resident AT1 angiotensin receptors (AT1R) prompted the examination of interactions between these two receptors.[215–218] The heteromerization of CB_1R and AT1R was demonstrated by coimmunoprecipitation, colocalization, and BRET assays using

Table 9.2 Cannabinoid receptor heteromers[a]

Heteromer pair	Detection techniques	Changes in heteromer properties			References
		Binding	Signaling	Trafficking	
CB_1R–$A_{2A}R$	Co-IP, BRET	Not reported	CB_1R antagonist ↓ $A_{2A}R$ signaling; $A_{2A}R$ antagonist ↓ CB_1R signaling	Not reported	201
CB_1R–β_2AR	BRET	Not reported	Alteration in signaling of individual protomers	CB_1R internalization by β_2AR agonist	202
CB_1R–AT1R	Co-IP, BRET	Not reported	AT1R signals via $G_{\alpha i}$ instead of $G_{\alpha q}$; CB_1R agonists ↑ AT1R signaling; CB_1R antagonists ↓ AT1R signaling	Not reported	10
CB_1R–CB_2R	BRET, proximity ligation assay	Not reported	Antagonistic interactions	Not reported	148
CB_1R–D2R	Co-IP	Not reported	CB_1R signals via $G_{\alpha s}$ in the presence of D2R agonist	Not reported	203–206
CB_1R–OX1R	Co-IP, TR-FRET	Not reported	CB_1R ↑ OX1R signaling	OX1R agonist ↑ heteromer internalization	207,208
CB_1R–D2R–$A_{2A}R$	BRET + FRET; bimolecular fluorescence complementation + BRET	Not reported	Not reported	Not reported	129,209

[a]This table does not include opioid–cannabinoid receptor heteromers. These are described in Table 9.1.

differentially epitope-tagged receptors.[10] Examination of the signaling properties of the CB_1R–AT1R heteromer shows that, in cells coexpressing both receptors, the AT1R agonist induces a rapid, robust dose-dependent increase in ERK1/2 phosphorylation that is drastically reduced either by RNAi-mediated knockdown of CB_1R or by inhibiting the activity of diacylglycerol lipase, the enzyme responsible for the synthesis of the endocannabinoid 2-AG.[10] In addition, CB_1R agonists potentiate while antagonists block AT1R-mediated increases in ERK1/2 phosphorylation.[10] Interestingly, an antibody that selectively recognizes the CB_1R–AT1R heteromer blocks AT1R-mediated ERK signaling in cells coexpressing both receptors.[10] Moreover, in cells expressing the CB_1R–AT1R heteromer, CB_1R-mediated ERK1/2 phosphorylation is only observed following activation of AT1R by a very low dose of agonist that by itself does not induce ERK1/2 phosphorylation.[10]

Examination of receptor G-protein coupling shows that, in cells expressing the CB_1R–AT1R heteromer, AT1R-mediated ERK1/2 phosphorylation occurs via $G_{\alpha i}$ instead of $G_{\alpha q}$. Moreover, AT1R-mediated intracellular Ca^{2+} mobilization occurs via $G_{\alpha q}$ but requires the presence of CB_1R since it is significantly attenuated by RNAi-mediated knockdown of CB_1R.[10] Interestingly, activation of CB_1R did not increase intracellular Ca^{2+} mobilization, although the CB_1R agonist could potentiate AT1R-mediated effects.[10] Taken together, these studies show that CB_1R–AT1R heteromerization provides a mechanism for modulation of AT1R function by CB_1R that could be of clinical relevance in pathologies involving these receptors (see Section 3).

CB_1R–β_2AR heteromers: Based on coexpression of CB_1R and β_2AR in several tissues and on inhibition of noradrenergic neurotransmission by CB_1R a study examined whether both receptors formed interacting complexes.[202] BRET assays using differentially epitope-tagged receptors showed that both receptors were in close proximity to form complexes in live cells.[202] The BRET signal was increased in the presence of a CB_1R antagonist suggesting that the latter either caused a change in receptor conformation (bringing in close proximity the acceptor and donor pairs) or caused an increase in the surface levels of this heteromer.[202] Heteromerization between CB_1R and β_2AR was shown to lead to alterations in the signaling properties of individual receptors in that it led to an increase in CB_1R-mediated ERK1/2 phosphorylation and a decrease in CREB phosphorylation compared to cells expressing only CB_1R.[202] Moreover, β_2AR-mediated ERK1/2 phosphorylation was attenuated by an inverse

agonist but not by a neutral agonist to CB_1R.[202] CB_1R–β_2AR heteromerization modulated the trafficking properties of individual receptors in that treatment with a β_2AR receptor agonist induced CB_1R internalization in cells expressing both receptors.[202]

CB_1R–CB_2R heteromers: A recent report has described heteromerization between CB_1R and CB_2R *in vivo*.[148] The authors used BRET assays with differentially epitope-tagged receptors in HEK-293 cells to show that CB_1R and CB_2R are in close proximity in live cells.[148] They also carried out a proximity ligation assay in SH-SY5Y neuroblastoma cells (endogenously express CB_1R) transfected with epitope-tagged CB_2R, in pinealocytes obtained from rat pineal glands, and in rat nucleus accumbens slices to show that both receptors are in close proximity (<17 nm) to form interacting complexes.[148] The authors examined the signaling properties of CB_1R–CB_2R heteromers in SH-SY5Y cells transfected with CB_2R and reported that a CB_2R agonist induces Akt/PKB phosphorylation only in cells that coexpressed both receptors, and that this is significantly attenuated by cotreatment with the CB_1R agonist.[148] Interestingly, this translated into a physiological response given that CB_2R-mediated neurite outgrowth in cells expressing CB_1R–CB_2R heteromers was blocked by the CB_1R agonist.[148] These results suggest that the interactions between CB_1R and CB_2R are antagonistic in nature.

CB_1R–dopamine D2 receptor heteromers: A number of studies have reported interactions between CB_1R and D2R. These include studies showing that (i) the D2R agonist, quinpirole, mediated motor behavior is stimulated by the CB_1R antagonist, SR141716A[203]; (ii) G-protein coupling to CB_1R is modulated by D2R in cells coexpressing both receptors[204,205]; and (iii) in primary striatal neuronal cultures treatment either with the CB_1R agonist, Hu-210, or with the D2R agonist, quinpirole, leads to inhibition of forskolin-stimulated cAMP levels,[204] whereas coadministration of Hu-210 and quinpirole resulted in increases in cAMP levels mediated via CB_1R stimulation of $G_{\alpha s}$.[204] Interestingly, while the presence of D2R is required for CB_1R coupling to $G_{\alpha s}$, persistent activation of D2R switches the coupling from $G_{\alpha s}$ to $G_{\alpha i}$.[205] The formation of CB_1R–D2R heteromeric complexes was demonstrated by coimmunoprecipitation studies using differentially epitope-tagged receptors that show that cotreatment with subsaturating concentrations of agonists to both receptors causes an increase in the levels of CB_1R–D2R heteromers.[206]

CB_1R–orexin-1 receptor heteromers: Coimmunoprecipitation studies along with homogenous TR-FRET assays using SNAP- and CLIP-tagged

substrates were used to show that CB_1R and orexin-1 receptors (OX1R) form multimeric complexes.[207] Examination of the trafficking properties of this heteromer showed that the OX1R agonist was more efficient at promoting heteromer internalization compared to OX1R internalization.[207] In heterologous cells coexpressing CB_1R and OX1R, a $G_{q/11}$-coupled receptor, treatment with the OX1R agonist led to a potent and long-lasting increase in ERK1/2 phosphorylation compared to cells expressing only OX1R, and that this increase was blocked by a CB_1R antagonist or by pertussis toxin pretreatment.[208] Moreover, coadministration of CB_1R and OX1R agonists did not result in additive effects suggesting that just the presence of CB_1R was sufficient to modulate OX1R signaling.[208] Given the involvement of CB_1R and OX1R in feeding, targeting this heteromer pair could lead to the identification of novel therapeutics for the treatment of obesity.

CB_1R–D2R–$A_{2A}R$ heteromers: Above, we described studies showing heteromerization between CB_1R and D2R and between CB_1R and $A_{2A}R$. In addition, D2R–$A_{2A}R$ heteromerization has been reported.[219] This led to the hypothesis that the three receptors could associate to form an oligomeric complex. This was tested with the sequential use of BRET and FRET assays.[209] BiFC assay was combined with BRET to show that the three receptors were in close proximity to interact.[129] Although this finding is exciting, further studies are needed to determine whether these heterotrimers are present endogenously, and if so, what is their tissue distribution and physiological role.

2.4. Dopamine receptor heteromerization

Several studies have examined heteromerization between dopamine receptor subtypes and between the latter and other GPCRs. Since heteromerization between dopamine receptors and opioid receptor subtypes or CB_1R have been described in Sections 2.2 and 2.3, we describe dopamine receptor heteromers that do not involve opioid or cannabinoid receptors in this section (see Table 9.3).

D1R–D2R heteromers: Interactions between D1R and D2R were suggested by the colocalization of D1R and D2R in MSNs in the striatum (reviewed in Refs. 48–50). Further evidence for direct interaction was suggested by the findings that (i) there is a requirement of D1R activation for D2R-mediated locomotor activity[241]; (ii) enhancement of the acute effects of cocaine requires activation of D1R and of D2R[242]; and

Table 9.3 Dopamine receptor heteromers[a]

Heteromer pair	Detection techniques	Changes in heteromer properties: Binding	Signaling	Trafficking	References
D1R–D2R	Co-IP, FRET, protein complementation assay	Not changed	$G_{\alpha q}$-mediated signaling by the heteromer	Not reported	125,220–224
D1R–D3R	Co-IP, BRET	↑ Dopamine binding	↑ Dopamine signaling	Heteromer internalized only by agonists to both receptors	225
D1R–A_{1A}R	Co-IP	Not reported	↓ D1R signaling by combination of agonists to both receptors	Not reported	226
D2R–D3R	Co-IP, functional complementation	Not reported	↑ Potency with preferential protomer agonists	Not reported	227,228
D2R–A_{2A}R	Co-IP, BRET, FRET	Not reported	D2R ↓ efficacy and potency of A_{2A}R agonists	Heteromer internalization by agonist to either protomer	219,229–232
D2R–GSHR1a	TR-FRET	Not reported	D2R signals via $G_{\alpha q}$ and not $G_{\alpha i}$	Not reported	233
D2R–NTS1	IP + fluorescence spectroscopy	↓ Affinity for D2R agonist in the presence of NTS1 agonist	↑ Potency of D2R in the presence of NTS1 agonist	NTS1 agonist ↑ D2R internalization	234,235

D2R–OTR	BRET, proximity ligation assays	↑ D2R antagonist binding by low doses of OTR agonist	↑ Efficacy and potency of D2R signaling by OTR agonist	Not reported	236
D2R–SSTR2	Co-IP, pbFRET	Modulation of binding properties of individual protomers	SSTR2 agonist ↑ D2R signaling	D2R agonist ↑ SSTR2 internalization	237
D2R–SSTR5	Functional complementation, pbFRET	↑ Affinity for each protomer agonist	Agonist to one protomer ↑ signaling by agonist to second protomer	Not reported	144
D2R–5-HT_{2A}R	Co-IP	D2R agonist ↑ affinity of 5-HT_{2A}R agonist	D2R agonist ↑ potency of hallucinogenic 5-HT_{2A}R agonist	Not reported	238
D2R–mGluR5–A_{2A}R	Co-IP, BRET + bimolecular fluorescence complementation	Not reported	Not reported	Not reported	127
D4R–α_{1B}AR	Co-IP, BRET	D4R agonist ↓ binding displacement by α_{1B}AR agonist	D4R agonist ↓ α_{1B}AR signaling	Not reported	239
D4R–β_1AR	Co-IP, BRET	D4R agonist ↓ binding displacement by β_1AR agonist	D4R agonist ↓ β_1AR signaling	Not reported	239
D4R–β_2AR	Co-IP, BRET	Not reported	Not reported	Not reported	240

[a]This table does not include heteromers involving opioid or cannabinoid receptors. These are described in Tables 9.1 and 9.2.

(iii) potentiation in the release of arachidonic acid is observed upon coactivation of D1R and D2R in cells expressing both receptors.[243]

Evidence for direct interactions between D1R and D2R came from coimmunoprecipitation and confocal FRET studies carried out with rat striatal membranes, primary rat striatal neuronal cultures or with cells coexpressing epitope-tagged receptors.[220–224] In addition, protein complementation studies where coexpression of D1R fused to a fragment of luciferase and of D2R fused to a complementary luciferase fragment resulted in the reconstitution of active luciferase supported D1R–D2R heteromerization.[125] Signaling assays showed that while activation of D1R led to $G_{\alpha s/olf}$-mediated signaling and of D2R to $G_{\alpha i}$-mediated signaling, coactivation of both receptors led to increases in intracellular Ca^{2+} levels.[220,221] Since the increase in intracellular Ca^{2+} levels is not observed in cells that express only D1R or D2R this suggested that the D1R–D2R heteromer can couple to $G_{\alpha q}$ to induce Ca^{2+}-mediated signaling.[220,221] Interestingly, occupancy of D1R in cells expressing D1R–D2R heteromers results in rapid desensitization of heteromer-mediated Ca^{2+} signaling that is regulated by both the catalytic and RGS domains of GRK2.[223]

The D1R agonist, SKF83959, has been proposed as a D1R–D2R heteromer-selective ligand based on the finding that (i) it did not increase adenylyl cyclase activity in cells coexpressing D1R–D2R or expressing only D1R; (ii) it increased intracellular Ca^{2+} release in cells coexpressing D1R–D2R, and this effect was potentiated by coadministration of quinpirole; and (iii) it increased [^{35}S]GTPγS binding to $G_{\alpha q}$ and to $G_{\alpha i}$ in cells coexpressing D1R–D2R.[221] Finally, D1R–D2R-mediated signaling by SKF83959 in primary striatal neurons via $G_{\alpha q}$ was found to lead to activation of both cytosolic and nuclear Ca^{2+}/calmodulin-dependent kinase IIα resulting in increased expression of BDNF and ultimately rapid neuronal maturation and differentiation.[222] Moreover, treatment with SKF83959 increased grooming behavior in rats and decreased the phosphorylation of the GluR1 subunit of AMPA receptors via Ca^{2+}/calmodulin-dependent kinase IIα in the nucleus accumbens.[224]

D1R–D3R heteromers: Interactions between D1R and D3R were suggested by studies showing that (i) D1R agonists could induce the expression of mRNA and increase D3R levels, and that this could be blocked by a D1R antagonist[244,245]; (ii) the coactivation of D1R and D3R caused a potentiation in the expression of substance P mRNA[246,247]; (iii) D3R knockout mice exhibit a decrease in D1R-mediated expression of the immediate early gene, c-fos[248]; and (iv) cross talk between D1R and

D3R may play a role in rearing habituation and reward behavior to low doses of cocaine.[249]

Evidence for direct interactions came from coimmunoprecipitation studies showing that D1R and D3R form interacting complexes in cells coexpressing both receptors and in rat striatal membranes.[225] BRET assays showed that the two receptors were in close proximity in living cells.[225] Treatment with dopamine increased the BRET signal, and this effect could be blocked by expression of a dominant negative mutant of dynamin leading to the suggestion that the increased signal could be due to internalized D1R–D3R heteromers.[225] Binding studies with a radiolabeled D1R antagonist showed that dopamine binds with higher affinity to D1R in the presence of D3R.[225] This is translated into more potent dopamine-mediated stimulation of adenylyl cyclase activity in cells coexpressing D1R and D3R than in cells expressing only D1R.[225] Examination of the trafficking properties of D1R–D3R heteromer showed that treatment with either a D1R or a D3R agonist did not induce heteromer internalization; however, cotreatment with agonists to both receptors induced the internalization of the D1R–D3R heteromer.[225]

D2R–D3R heteromers: Functional complementation studies suggested that D2R and D3R could form heterodimers.[227] For example, expression of D2R with a mutant D3R receptor (that cannot signal) restores the ability of D3R agonists to inhibit adenylyl cyclase activity.[227] Coimmunoprecipitation assays with differentially epitope-tagged receptors showed that D2R and D3R could form interacting complexes.[227] Signaling assays show that preferential D3R versus D2R agonists (pramipexole, ropinirole, or S32504) were more potent at inhibiting the activity of a chimeric adenylyl cyclase (does not respond to D3R activation) in cells coexpressing D2R and D3R compared to cells expressing only D2R (reviewed in Ref. 228). In contrast, preferential D2R agonists such as pergolide showed a marginal increase in potency in cells coexpressing the D2R–D3R heteromer while no changes were observed for the full D2R/D3R agonist, quinpirole (reviewed in Ref. 228). Interestingly, partial D2R/D3R agonists with antipsychotic activity did not inhibit adenylyl cyclase activity in cells coexpressing both receptors and also could block quinpirole-mediated inhibition of cAMP levels (reviewed in Ref. 228). This suggests that in cells expressing D2R–D3R heteromers partial D2R/D3R agonists can function as antagonists (reviewed in Ref. 228).

D1R–adenosine $A_{1A}R$ heteromers: Interactions between D1R and $A_{1A}R$ receptors were suggested by studies showing that (i) $A_{1A}R$ activation leads to a decrease in D1R-mediated locomotion and oral dyskinesia in animals,[250] while $A_{1A}R$ antagonists enhance D1R-mediated locomotion[251]; (ii) radiolabeled D1R antagonist affinity is decreased by $A_{1A}R$ agonist[252]; (iii) $A_{1A}R$ agonists modulate D1R-mediated electroencephalographic arousal[11]; and (iv) in cells coexpressing $A_{1A}R$ and D1R, $A_{1A}R$ antagonists cause an increase in both basal- and dopamine-mediated increases in cAMP levels while an $A_{1A}R$ agonist blocks dopamine-mediated effects.[253]

Coimmunoprecipitation studies using receptor-selective antibodies detected the presence of D1R–$A_{1A}R$ interacting complexes in cells coexpressing both receptors.[226] In fibroblast cells that coexpress D1R and $A_{1A}R$, pretreatment with agonists to both receptors decreases D1R-mediated increases in cAMP levels, while pretreatment with either receptor agonist does not. Additional studies using proximity-based assays to confirm the formation of D1R–$A_{1A}R$ heteromers *in vivo* as well as studies to evaluate whether the D1R–$A_{1A}R$ heteromer exhibits novel properties compared to individual receptor homomers need to be carried out.

D2R–adenosine $A_{2A}R$ heteromers: Interactions between D2R and $A_{2A}R$ are suggested by studies showing that (i) D2R and $A_{2A}R$ are colocalized in GABAergic striatopallidal neurons[254]; (ii) selective $A_{2A}R$ agonists decrease D2R binding and signaling[229,255,256]; (iii) Blockade or knockdown of D2R impairs locomotive behavior, and this effect can be attenuated by $A_{2A}R$ antagonists[257,258]; and (iv) antagonists of D2R produce catalepsy and this is blocked by $A_{2A}R$ antagonists.[259] The formation of interacting D2R–$A_{2A}R$ complexes was demonstrated by coimmunoprecipitation studies and by proximity-based assays such as BRET and FRET.[219,230–232] Investigation of the pharmacological properties of the D2R–$A_{2A}R$ in cells coexpressing both receptors showed that D2R agonists caused a decrease in the affinity for radiolabeled $A_{2A}R$ agonist binding.[229] Moreover, signaling assays showed that D2R agonists decreased the efficacy and potency of $A_{2A}R$ agonists in cells coexpressing both receptors.[229] In addition, studies showed that treatment with either $A_{2A}R$ or the D2R agonists causes internalization of the D2R–$A_{2A}R$ heteromer.[219,260]

Bivalent ligands comprising of a D2R agonist joined by spacers of different lengths (26–118 atoms) to an $A_{2A}R$ antagonist have been used to show that in striatal membranes coexpressing both receptors, these ligands cause a greater displacement of radiolabeled ligand binding to either D2R or $A_{2A}R$ than the corresponding monovalent ligands.[261]

D2R–ghrelin receptor heteromers: A recent study examined D2R and ghrelin receptor (GHSR1a) interactions because of their robust colocalization to a subset of hypothalamic neurons.[233] Time-resolved FRET carried out in heterologous cells and with hypothalamic membrane preparations demonstrated that the two receptors are in close proximity to form heteromers.[233] In cells coexpressing D2R and GHSR1a, activation of D2R led to increases in intracellular Ca^{2+} release and not to $G_{\alpha i}$-mediated signaling and this could be blocked by a D2R or a GHSR1a antagonist.[233] Moreover, cotreatment with agonists to both receptors led to additive effects on intracellular Ca^{2+} release.[233] In addition, cross-desensitization of Ca^{2+} release is observed in cells coexpressing both receptors in that a 30-min pretreatment with agonist to one receptor attenuated intracellular Ca^{2+} release mediated via the other receptor.[233] Studies with wild-type and ghrelin receptor knockout mice showed that administration of a D2R agonist caused a reduction in food intake in wild-type but not in knockout mice.[233] This reduction in food intake can be blocked by a GHSR1a antagonist.[233] These results indicate that D2R–GHSR1a heteromerization could play an important role in the modulation of food intake.

D2R–neurotensin receptor heteromers: Interactions between dopamine and neurotensin receptors were demonstrated by studies showing that (i) type 1 neurotensin receptors (NTS1) are expressed in substantia nigra and ventral tegmental area dopaminergic neurons[262]; (ii) NTS1 has excitatory effects on dopaminergic neurons present in the midbrain[262]; (iii) neurotensin causes a modest decrease in the affinity of D2R for receptor agonists[262]; (iv) D2R autoreceptor function is decreased by NTS1 activation via a protein kinase C- and Ca^{2+}-dependent mechanism[262]; and (v) activation of D2R attenuates NTS1-mediated Ca^{2+} mobilization.[262]

A combination of immunoprecipitation of the long isoform of D2R ($D2R_L$) with fluorescent spectroscopy for detection of the presence of C-terminally epitope tagged (with enhanced CFP) NTS1 in the immunoprecipitate indicated that the two receptors formed interacting complexes only in cells coexpressing both receptors.[234] Radiolabeled binding experiments with membrane preparations from cells coexpressing $D2R_L$ and NTS1 showed that the NTS1 agonist decreased the affinity of radiolabeled agonist binding to $D2R_L$.[234] However, dopamine ligands (agonist or antagonist) had no effect on radiolabeled agonist binding to NTS1.[234] Signaling assays showed that the potency of $D2R_L$-mediated signaling is increased in the presence of neurotensin.[234] Internalization assays showed that in cells coexpressing $D2R_L$ and NTS1 treatment with NTS1 agonist increased $D2R_L$ internalization via a PKC-dependent mechanism.[235]

D2R–oxytocin receptor heteromers: Studies showing (i) the involvement of striatal oxytocin receptors (OTR) in pair bond formation in monogamous prairie female voles and in regulating social and emotional behaviors in animals (rodents, sheep, humans)[263–267]; (ii) the involvement of ascending dopaminergic neurons in social attachment behavior[1,2]; (iii) that a D2R antagonist can block partner preference[268,269]; and (iv) administration of a D2R agonist into the nucleus accumbens can induce partner preference and this is blocked by an OTR antagonist[268,269] suggest interactions between D2R and OTR.

BRET assays in cells coexpressing both receptors showed that D2R and OTR are in close proximity in live cells.[236] *In situ* proximity ligation assays detected the presence of D2R–OTR heteromers in the neuropil of nucleus accumbens and in the dorsal striatum of rats.[236] Binding assays with nucleus accumbens membrane preparations showed that a low, but not high, concentration of the OTR agonist increase the maximal number of binding sites of a radiolabeled D2R antagonist and this is blocked by an OTR antagonist.[236] Signaling assays with nucleus accumbens membrane preparations showed that the OTR agonist increases the efficacy and potency of D2R-mediated G-protein activity.[236] These studies indicate that D2R–OTR heteromers may play a role in modulating social and emotional behavior.

D2R–somatostatin receptor heteromers: Interactions between dopamine and somatostatin receptors were suggested by studies showing that (i) a decrease of somatostatin levels attenuated dopamine receptor-mediated stereotypy as well as locomotor responses to amphetamine[270]; (ii) dopamine enhances somatostatin receptor-mediated inhibition of adenylyl cyclase activity[271,272]; and that (iii) a dopamine receptor agonist attenuates the chronic effects of a long-acting somatostatin analogue.[273] This led investigators to examine heteromerization between D2R and somatostatin receptor subtypes.

Direct interactions between D2R and somatostatin receptor 2 subtype (SSTR2) was examined by coimmunoprecipitation and pbFRET assays that demonstrated that these two receptors formed complexes in heterologous cells coexpressing both receptors as well as in primary striatal neuronal cultures.[237] Binding studies showed that in cells coexpressing D2R and SSTR2 a combination of agonists to both receptors displaced radiolabeled antagonist binding to D2R more efficiently than the D2R agonist alone.[237] Interestingly, in these cells, the SSTR2 agonist could dose-dependently displace radiolabeled antagonist binding to D2R.[237] Also, the total number of binding sites for the radiolabeled ligand binding to SSTR2 was increased in the presence of dopamine.[237] Examination of signaling properties of the D2R–SSTR2 heteromer showed that D2R-mediated inhibition of cAMP levels was lower for some agonists (dopamine) and not others (quinpirole) than that observed in cells that expressed only D2R.[237] Moreover, the presence of the SSTR2 agonist caused an increase in the potency of quinpirole and in both the potency and efficacy of dopamine in inhibiting cAMP levels.[237] Examination of the trafficking properties of the D2R–SSTR2 heteromer showed that treatment with D2R agonist enhanced agonist-mediated internalization of SSTR2.[237]

Heteromerization between D2R and SSTR5 was suggested by the functional rescue of a C-tail deletion mutant of human SSTR5.[144] Only when coexpressed with D2R does agonist activation of the mutant SSTR5 lead to inhibition in forskolin-mediated increases in cAMP levels.[144] pbFRET supported the formation of D2R–SSTR5 heteromers.[144] Interestingly, treatment with agonists to either receptor or cotreatment with agonists to both receptors led to an increase in FRET efficiency, while a decrease in the latter was observed with D2R antagonist.[144] This would suggest that either agonist treatment leads to an increase in the level of D2R–SSTR5 heteromers or, alternatively, it changes receptor conformation leading to a decrease in the distance between the fluorescent probes.[144] The pharmacological properties of the D2R–SSTR5 heteromer were found to be distinct from those of individual receptor homomers in that the heteromers exhibited a greater affinity for the binding of each protomer agonist.[144] Moreover, occupancy of one of the receptors in the D2R–SSTR5 heteromer led to an increase in the binding affinity and signaling by the agonist to the second receptor.[144]

D2R–serotonin receptor heteromers: Studies showing that typical antipsychotic drugs used to treat schizophrenia are D2R antagonists and that some symptoms of schizophrenia are similar to those observed with hallucinogenic serotonin receptor agonists led to the examination of heteromerization between D2R and serotonin 5-HT_{2A} receptors (5-HT_{2A}R).[274,275]

Coimmunoprecipitation studies show that D2R and 5-HT_{2A}R form interacting complexes.[238] Examination of the pharmacological properties of the D2R–5-HT_{2A}R heteromer in cells or in striatal membranes coexpressing both receptors show that a D2R agonist increased the binding affinity of a 5-HT_{2A}R agonist, and this is blocked by a D2R antagonist.[238] The presence of D2R had no effect on serotonin-mediated inositol phosphate production, but it increased the potency of a hallucinogenic 5-HT_{2A}R agonist.[238] Interestingly, D2R activation had no effect on serotonin-mediated inositol phosphate production, but decreased the potency of the hallucinogenic 5-HT_{2A}R agonist.[238] Behavioral studies with 5-HT_{2A}R knockout mice show that the antipsychotic effects of the D2R antagonist, haloperidol, require the presence of 5-HT_{2A}R.[238]

D4R–α adrenergic receptor heteromers: Heteromer formation between D4R and α adrenergic receptors (αAR) was examined based on studies showing that the D4R gene is the most prominently expressed dopamine receptor gene in the rat pineal gland, and that its mRNA expression is circadian and controlled by adrenergic signaling.[276]

Coimmunoprecipitation and BRET studies with differentially epitope-tagged receptors show that D4R and α_{1B}AR form interacting complexes.[239] Pharmacological characterization of D4R–adrenergic receptor heteromers shows that the D4R agonist decreases the ability of an α_{1B}AR agonist to displace radiolabeled antagonist binding to α_{1B}AR.[239] Signaling assays in cells expressing the D4R–α_{1B}AR heteromer or in pineal gland preparations show that pretreatment or cotreatment with a selective D4R agonist leads to an inhibition of α_{1B}AR-mediated phosphorylation of ERK1/2 and of Akt/PKB.[239] Interestingly, a D4R receptor antagonist could also block α_{1B}AR-mediated signaling and the converse also occurred in that adrenergic antagonists could block D4R-mediated signaling.[239] Given that D4R expression is circadian, and that α_{1B}AR plays a role in modulating melatonin synthesis and release from the pineal gland, the formation of D4R–α_{1B}AR heteromers provides a means for regulating melatonin production and activity.[239]

D4R–β_1 adrenergic receptor heteromers: Studies examining D4R–αAR heteromerization also examined whether D4R and β_1 adrenergic receptors (β_1AR) formed heteromers. Coimmunoprecipitation and BRET studies with differentially epitope-tagged receptors show that D4R and β_1AR form interacting complexes.[239] Pharmacological characterization of D4R–β_1AR heteromers shows that the D4R agonist decreases the ability of a β_1AR agonist to displace radiolabeled antagonist binding to β_1AR.[239] Signaling assays in cells expressing either the D4R–β_1AR heteromer or in pineal gland preparations show that pretreatment or cotreatment with a selective D4R agonist leads to an inhibition of β_1AR-mediated phosphorylation of ERK1/2 and of Akt/PKB.[239] Interestingly, a D4R receptor antagonist could also block β_1AR-mediated signaling and the converse also occurred in that adrenergic antagonists could block D4R-mediated signaling.[239] These studies suggest that like the D4R–α_{1B}AR heteromer the D4R–β_1AR heteromer could also play a role in regulating melatonin production and activity.[239]

D4R–β_2 adrenergic receptor heteromers: Based on studies showing (i) colocalization of D2R-like receptors with β-AR in rat primary cortical neurons and in prefrontal cortex[277] and that (ii) D2R-like receptor activation inhibited β-AR-mediated increases in cAMP levels,[277] heteromerization between D4R and β_2-AR was examined.

Coimmunoprecipitation, BRET, and binding assays with β_2-AR immunoprecipitates, obtained from detergent solubilized mouse brain extracts using alprenolol coupled to Sepharose beads, show that D4R and β_2-AR form interacting complexes and are in close proximity in cells or tissues coexpressing both receptors.[240] However, further studies are needed to evaluate whether the D4R–β_2-AR heteromer exhibits novel properties compared to individual receptor homomers.

D2R–metabotropic glutamate receptor 5–adenosine$_{2A}$ receptor heteromers: The possibility of formation of heteromers involving D2R, metabotropic glutamate receptor 5 (mGluR5), and adenosine$_{2A}$ receptors (A_{2A}R) was suggested by studies showing (i) similar localization of these receptors in striatopallidal neurons[278–280]; (ii) that a mGluR5 agonist inhibited D2R-mediated contralateral turning behavior, and this was potentiated by an A_{2A}R agonist and attenuated by an A_{2A}R antagonist[281]; and (iii) that a mGluR5 agonist reduced the high affinity state for the D2R agonist, and this was potentiated by an A_{2A}R agonist.[281]

Coimmunoprecipitation studies carried out with cells coexpressing D2R, A_{2A}R, and mGluR5, and also with rat striatal membranes, show that these receptors form interacting complexes.[127] BRET assays with cells expressing a constant ratio of A_{2A}R tagged with luciferase and D2R tagged with YFP show that increasing amounts of mGluR5 can disrupt the BRET signal.[127] Similarly, increasing amounts of A_{2A}R could disrupt the BRET signal between D2R tagged with luciferase and mGluR5 tagged with YFP.[127] This suggests that the three receptors are in close proximity to interact.[127] The formation of D2R–A_{2A}R–mGluR5 was confirmed by a combination of BRET and BiFC assays.[127] For this, cells were transfected with A_{2A}R tagged with luciferase, mGluR5 fused with an N-terminal fragment of YFP and D2R fused with a C-terminal fragment of YFP.[127] A BRET signal was observed only when all three receptors were expressed in a cell suggesting that these receptors are in close proximity to form heterotrimers.[127] Further studies are needed in order to elucidate whether the D2R–A_{2A}R–mGluR5 heterotrimer exhibits novel properties compared to individual receptor homomers or heterodimers.

3. HETEROMERS IN DISEASE

Although it has become increasingly evident that GPCRs can form heterodimers,[61,63,64,282] very little information is available about the physiological role of GPCR heteromers. Pharmacological and coimmunoprecipitation studies using receptor-selective antibodies suggest that some receptor heteromers may play a role in the disease state. However, care has to be taken in the interpretation of such data given that such studies do not rule out a contribution of receptor homomers to the disease state. The recent development of antibodies that selectively recognize receptor heteromers, but not receptor homomers, and of TAT-based peptides that can disrupt heteromer formation, has provided us with tools to study the role of GPCR heteromers in the disease state.[7–10] In the following sections, we describe a few cases where a role for GPCR heteromers involving opioid, cannabinoid, and dopamine receptors in pathology has been implicated.

Analgesia and addiction: Antibodies that selectively recognize the MOR–DOR heteromer detect increased heteromer levels in brain regions involved in pain processing following chronic morphine administration under a paradigm that leads to the development of tolerance.[7] Taken with the fact that MOR–DOR heteromers signal via the β-arrestin-mediated signaling cascade[100] and that β-arrestin or DOR knockout mice do not develop tolerance to morphine[162,283] would suggest that MOR–DOR heteromers play a role in the development of tolerance to this opioid. This is supported by a study showing that the administration of a MOR transmembrane 1 TAT peptide fusion protein, which selectively disrupts the MOR–DOR heteromer, resulted in an enhanced analgesic response to morphine with a concomitant decrease in tolerance to the drug.[8]

Studies have also examined whether the DOR–MOR heteromer could be a target for the development of therapeutics that are as efficacious as morphine but lack the associated adverse side effects. In this context, a study showed that low doses of the DOR antagonists, TIPPψ or naltrindole, could potentiate intrathecal morphine-mediated analgesia.[94,160] Moreover, DOR antagonist administration to animals exhibiting tolerance to morphine led to a restoration of the analgesic response and potency of morphine.[160] Based on similar observations, bivalent ligands targeting the DOR–MOR heteromer were developed.[284] These bivalent ligands comprised a MOR agonist pharmacophore, oxymorphone, joined to a DOR antagonist pharmacophore, naltrindole, by spacers of variable lengths (16–21 atoms).[284] The bivalent compounds exhibited antinociceptive activity.[284] Moreover, compounds with spacer lengths between 19 and 21 atoms did not exhibit antinociceptive tolerance and physical dependence and were less rewarding than morphine.[284,285]

An involvement of DOR–KOR heteromers in thermal allodynia was suggested by studies demonstrating the presence of these heteromers in peripheral sensory neurons.[286] Moreover, the antiallodynic effects of the

KOR agonist, 6′-GNTI, to prostaglandin E2-induced thermal allodynia were blocked by either DOR or KOR antagonists.[286] This, together with the finding that the antiallodynic effects of the DOR agonist, DPDPE, were potentiated by a DOR–KOR heteromer-selective antibody supports a role for the DOR–KOR heteromer in modulating thermal allodynia.[286]

A recent study has implicated the CB_1R–DOR heteromer in neuropathic pain.[9] This study found that levels of CB_1R and DOR are increased in distinct brain regions following peripheral nerve lesion.[9] An antibody selective for the CB_1R–DOR heteromer detected increased heteromer levels in the cortex, hypothalamus, and striatum of lesioned animals.[9] In addition, CB_1R–DOR heteromerization led to a decrease in DOR activity in the cortex of lesioned animals[9] and this decrease could be attenuated by low, nonsignaling doses of CB_1R ligands.[9] Thus, the CB_1R–DOR heteromer is a novel target for the treatment of neuropathic pain.

Liver fibrosis: Antibodies that selectively recognize the CB_1R–AT1R heteromer detect a significant upregulation of this heteromer in hepatic stellate cells of rats chronically treated with ethanol.[10] This leads to increased AT1R-mediated signaling that can be blocked by a CB_1R antagonist or by inhibiting the synthesis of the endocannabinoid, 2-AG.[10] In addition, the AT1R-mediated increase in profibrogenic markers in hepatic stellate cells of rats chronically treated with ethanol was completely blocked by a CB_1R antagonist.[10] These results suggest that the profibrogenic activity of AT1R in ethanol-induced liver fibrosis requires CB_1R activity. Thus, the CB_1R–AT1R heteromer represents a novel therapeutic target for the treatment of this disease.

Psychiatric disorders: An involvement of dopamine receptor subtype heteromers in major depression has been suggested by studies showing an increase in D1R–D2R interacting complexes in the striatum of postmortem individuals that suffered from major depression.[287] Interestingly, bilateral infusion into the rat prefrontal cortex of a TAT peptide that disrupts D1R–D2R heteromers results in a decrease in the levels of D1R–D2R heteromers and leads to antidepressant-like effects in the forced swim test.[287] This suggests an involvement of D1–D2 receptor heteromers in the pathology associated with major depression.

Involvement of D1R–D2R heteromers in schizophrenia was suggested by studies showing that clozapine, a drug used to treat this condition, binds to both D1R and D2R but that the high affinity binding to D1R decreases in the presence of D2R.[288,289] Moreover, FRET assays showed that clozapine appears to modulate the formation of D1R–D2R heteromers in that high concentrations of the drug increase the FRET signal between D1R and D2R while low concentrations attenuate this signal.[288,289] Further support for the involvement D1R–D2R heteromers in schizophrenia came from studies showing that acute treatment with a heteromer-selective agonist increased the striatal expression of brain-derived neurotrophic factor, a molecule that has been linked to schizophrenia (for review, see Ref. 290).

An involvement of D2R–5-HT_{2A}R heteromers in schizophrenia is suggested by studies showing that coexpression of D2R in cells increases the signaling potency of a hallucinogenic 5-HT_{2A}R agonist, but activation of D2R decreases it.[238] Moreover, the antipsychotic effects of the D2R antagonist, haloperidol, are decreased in 5-HT_{2A}R knockout mice[238] suggesting that the D2R–5-HT_{2A}R heteromer is a target of this drug.

An involvement of CB_1R–D2R–A_{2A}R heteromers in schizophrenia has been suggested based on studies showing upregulation of CB_1R, D2R, and A_{2A}R in the postmortem brains of schizophrenics.[291–293] Moreover, in animal models of schizophrenia, treatment with antipsychotics decreases CB_1R levels in the nucleus accumbens.[294] These reports together with studies describing CB_1R–D2R and CB_1R–A_{2A}R interactions (described in Section 2) and studies showing that these three receptors can form heteromers in live cells suggest that the CB_1R–D2R–A_{2A}R heteromer could be a novel therapeutic target in the treatment of schizophrenia.

Parkinson's disease: The involvement of D2R–A_{2A}R heteromers in Parkinson's disease is suggested by studies showing that the antagonistic D2R–A_{2A}R interactions were increased in the striatal membranes from rat models of Parkinson's disease (reviewed in Ref. 295). In addition, studies indicate that A_{2A}R antagonists have anti-Parkinsonian effects (reviewed in Ref. 295). For example, A_{2A}R antagonist administration could attenuate dopamine-mediated neuronal degeneration in animal models of Parkinson's disease (reviewed in Ref. 295). Given that D2R–A_{2A}R heteromers are constitutively formed, it is possible that A_{2A}R antagonists targeting the heteromer could increase D2R-mediated signaling leading, in turn, to an attenuation of Parkinson's disease symptoms (reviewed in Ref. 295).

The heterotrimer involving D2R–mGluR5–A_{2A}R could also be a novel therapeutic target in the treatment of Parkinson's disease (reviewed in Ref. 295). This is supported by studies showing that mGluR5 antagonist-mediated induction of motor activity requires the presence of both D2R and A_{2A}R and that a combination of mGluR5 and A_{2A}R antagonists could potentiate locomotor behavior in reserpinized mice (reviewed in Ref. 295).

Another heterotrimer that could be a novel target for the treatment of Parkinson's disease involves CB_1R–D2R–$A_{2A}R$ (reviewed in Ref. 295). Low doses of CB_1R and $A_{2A}R$ antagonists could synergize to potentiate D2R-mediated anti-Parkinsonian effects during the early stages of Parkinson's disease when a substantial release of dopamine still occurs (reviewed in Ref. 295).

4. CONCLUSIONS

Recent evidence indicates that some GPCR heteromers including those associated with drugs of abuse are upregulated under pathophysiological conditions. However, the physiological role of these heteromers in pathology is not clearly understood due to the lack of tools to distinguish between heteromer and homomer-mediated effects. For this, heteromer-specific ligands, agonists and/or antagonists, are needed to elucidate the role of GPCR heteromers in normal and pathophysiological conditions. Although heteromer-specific ligands for a few GPCR heteromers have been reported they are mostly receptor agonists and need to be extensively evaluated for their heteromer selectivity in both *in vitro* and *in vivo* assays. Moreover, the availability of heteromer-selective antagonists would help in the evaluation of heteromer levels in the normal and diseased states as well as in studies examining heteromer pharmacology. The recent development of antibodies that selectively recognize a GPCR heteromer in endogenous tissue, and of TAT peptides that can specifically disrupt the heteromer of interest, could help to elucidate the role of GPCR heteromers in pathological conditions.[7–10] This is clearly shown in the case of MOR–DOR heteromers where heteromer-selective antibodies demonstrate that these heteromers are increased in brain areas involved in pain perception following chronic treatment with morphine under a paradigm that leads to the development of tolerance.[7] In addition, studies with TAT peptides that selectively disrupt the MOR–DOR heteromer suggest a role for the heteromer to keep morphine-mediated signaling via MOR in the desensitized state.[8]

Studies demonstrating that GPCR heteromers are upregulated during a disease state suggest that these heteromers could be novel targets for the development of therapeutics to treat these diseases. If heteromers are disease-specific and/or exhibit unique tissue specificity, they would serve as ideal drugs for the development of therapeutics with potentially lesser side effects than the ones that are currently available.

ACKNOWLEDGMENTS

We thank Jonathan Wardman and Steven D. Stockton, Jr. for critical reading and helpful comments. L. A. D. is supported by NIH grants DA008863 and DA019521.

REFERENCES

1. Kieffer BL. Recent advances in molecular recognition and signal transduction of active peptides: receptors for opioid peptides. *Cell Mol Neurobiol* 1995;**15**:615–35.
2. Kieffer BL. Opioids: first lessons from knockout mice. *Trends Pharmacol Sci* 1999;**20**: 19–26.
3. Howlett AC, Barth F, Bonner TI, Cabral G, Casellas P, Devane WA, et al. International Union of Pharmacology. XXVII. Classification of cannabinoid receptors. *Pharmacol Rev* 2002;**54**:161–202.
4. Cichewicz DL. Synergistic interactions between cannabinoid and opioid analgesics. *Life Sci* 2004;**74**:1317–24.
5. Vigano D, Rubino T, Parolaro D. Molecular and cellular basis of cannabinoid and opioid interactions. *Pharmacol Biochem Behav* 2005;**81**:360–8.
6. Sulzer D. How addictive drugs disrupt presynaptic dopamine neurotransmission. *Neuron* 2011;**69**:628–49.
7. Gupta A, Mulder J, Gomes I, Rozenfeld R, Bushlin I, Ong E, et al. Increased abundance of opioid receptor heteromers after chronic morphine administration. *Sci Signal* 2010;**3**:ra54.
8. He SQ, Zhang ZN, Guan JS, Liu HR, Zhao B, Wang HB, et al. Facilitation of mu-opioid receptor activity by preventing delta-opioid receptor-mediated codegradation. *Neuron* 2011;**69**:120–31.
9. Bushlin I, Gupta A, Stockton Jr SD, Miller LK, Devi LA. Dimerization with cannabinoid receptors allosterically modulates delta opioid receptor activity during neuropathic pain. *PLoS One* 2012;**7**:e49789.
10. Rozenfeld R, Gupta A, Gagnidze K, Lim MP, Gomes I, Lee-Ramos D, et al. AT1R-CBR heteromerization reveals a new mechanism for the pathogenic properties of angiotensin II. *EMBO J* 2011;**30**:2350–63.
11. Popoli P, Ferre S, Pezzola A, Reggio R, deCarolis AS, Fuxe K. Stimulation of adenosine A(1) receptors prevents the EEG arousal due to dopamine D-1 receptor activation in rabbits. *Eur J Pharmacol* 1996;**305**:123–6.
12. Berezniuk I, Fricker LD. Endogenous opioids. In: Pasternak GW, editor. *The opiate receptors*. vol. 23, New York: Humana Press; 2011. p. 93–120.
13. Akil H, Watson SJ, Young E, Lewis ME, Khachaturian H, Walker JM. Endogenous opioids: biology and function. *Annu Rev Neurosci* 1984;**7**:223–55.
14. Suh HH, Tseng LF. Delta but not mu-opioid receptors in the spinal cord are involved in antinociception induced by beta-endorphin given intracerebroventricularly in mice. *J Pharmacol Exp Ther* 1990;**253**:981–6.
15. Sim LJ, Selley DE, Dworkin SI, Childers SR. Effects of chronic morphine administration on mu opioid receptor-stimulated [35S]GTPgammaS autoradiography in rat brain. *J Neurosci* 1996;**16**:2684–92.
16. Matthes HW, Maldonado R, Simonin F, Valverde O, Slowe S, Kitchen I, et al. Loss of morphine-induced analgesia, reward effect and withdrawal symptoms in mice lacking the mu-opioid-receptor gene. *Nature* 1996;**383**:819–23.
17. Ghozland S, Matthes HW, Simonin F, Filliol D, Kieffer BL, Maldonado R. Motivational effects of cannabinoids are mediated by mu-opioid and kappa-opioid receptors. *J Neurosci* 2002;**22**:1146–54.

18. Roberts AJ, McDonald JS, Heyser CJ, Kieffer BL, Matthes HW, Koob GF, et al. mu-Opioid receptor knockout mice do not self-administer alcohol. *J Pharmacol Exp Ther* 2000;**293**:1002–8.
19. Moles A, Kieffer BL, D'Amato FR. Deficit in attachment behavior in mice lacking the mu-opioid receptor gene. *Science* 2004;**304**:1983–6.
20. Kieffer BL, Evans CJ. Opioid receptors: from binding sites to visible molecules in vivo. *Neuropharmacology* 2009;**56**(Suppl. 1):205–12.
21. Filliol D, Ghozland S, Chluba J, Martin M, Matthes HW, Simonin F, et al. Mice deficient for delta- and mu-opioid receptors exhibit opposing alterations of emotional responses. *Nat Genet* 2000;**25**:195–200.
22. Simonin F, Valverde O, Smadja C, Slowe S, Kitchen I, Dierich A, et al. Disruption of the kappa-opioid receptor gene in mice enhances sensitivity to chemical visceral pain, impairs pharmacological actions of the selective kappa-agonist U-50,488H and attenuates morphine withdrawal. *EMBO J* 1998;**17**:886–97.
23. Land BB, Bruchas MR, Lemos JC, Xu M, Melief EJ, Chavkin C. The dysphoric component of stress is encoded by activation of the dynorphin kappa-opioid system. *J Neurosci* 2008;**28**:407–14.
24. Zhao P, Abood ME. GPR55 and GPR35 and their relationship to cannabinoid and lysophospholipid receptors. *Life Sci* 2012;**92**:453–7.
25. Di Marzo V, Petrosino S. Endocannabinoids and the regulation of their levels in health and disease. *Curr Opin Lipidol* 2007;**18**:129–40.
26. Battista N, Fezza F, Finazzi-Agro A, Maccarrone M. The endocannabinoid system in neurodegeneration. *Ital J Biochem* 2006;**55**:283–9.
27. Boyd ST. The endocannabinoid system. *Pharmacotherapy* 2006;**26**:218S–221S.
28. Okamoto Y, Morishita J, Tsuboi K, Tonai T, Ueda N. Molecular characterization of a phospholipase D generating anandamide and its congeners. *J Biol Chem* 2004;**279**:5298–305.
29. Simon GM, Cravatt BF. Endocannabinoid biosynthesis proceeding through glycerophospho-N-acyl ethanolamine and a role for alpha/beta-hydrolase 4 in this pathway. *J Biol Chem* 2006;**281**:26465–72.
30. Liu J, Wang L, Harvey-White J, Osei-Hyiaman D, Razdan R, Gong Q, et al. A biosynthetic pathway for anandamide. *Proc Natl Acad Sci USA* 2006;**103**:13345–50.
31. Prescott SM, Majerus PW. Characterization of 1,2-diacylglycerol hydrolysis in human platelets. Demonstration of an arachidonoyl-monoacylglycerol intermediate. *J Biol Chem* 1983;**258**:764–9.
32. Farooqui AA, Taylor WA, Horrocks LA. Separation of bovine brain mono- and diacylglycerol lipases by heparin sepharose affinity chromatography. *Biochem Biophys Res Commun* 1984;**122**:1241–6.
33. Cravatt BF, Demarest K, Patricelli MP, Bracey MH, Giang DK, Martin BR, et al. Supersensitivity to anandamide and enhanced endogenous cannabinoid signaling in mice lacking fatty acid amide hydrolase. *Proc Natl Acad Sci USA* 2001;**98**:9371–6.
34. Kathuria S, Gaetani S, Fegley D, Valino F, Duranti A, Tontini A, et al. Modulation of anxiety through blockade of anandamide hydrolysis. *Nat Med* 2003;**9**:76–81.
35. Hampson AJ, Hill WA, Zan-Phillips M, Makriyannis A, Leung E, Eglen RM, et al. Anandamide hydroxylation by brain lipoxygenase:metabolite structures and potencies at the cannabinoid receptor. *Biochim Biophys Acta* 1995;**1259**:173–9.
36. Yu M, Ives D, Ramesha CS. Synthesis of prostaglandin E2 ethanolamide from anandamide by cyclooxygenase-2. *J Biol Chem* 1997;**272**:21181–6.
37. Dinh TP, Carpenter D, Leslie FM, Freund TF, Katona I, Sensi SL, et al. Brain monoglyceride lipase participating in endocannabinoid inactivation. *Proc Natl Acad Sci USA* 2002;**99**:10819–24.

38. Gomes I, Grushko JS, Golebiewska U, Hoogendoorn S, Gupta A, Heimann AS, et al. Novel endogenous peptide agonists of cannabinoid receptors. *FASEB J* 2009;**23**: 3020–9.
39. Heimann AS, Gomes L, Dale CS, Pagano RL, Gupta A, de Souza LL, et al. Hemopressin is an inverse agonist of CB1 cannabinoid receptors. *Proc Natl Acad Sci USA* 2007;**104**:20588–93.
40. Gomes I, Dale CS, Casten K, Geigner MA, Gozzo FC, Ferro ES, et al. Hemoglobin-derived peptides as novel type of bioactive signaling molecules. *AAPS J* 2010;**12**: 658–69.
41. Gelman JS, Sironi J, Castro LM, Ferro ES, Fricker LD. Hemopressins and other hemoglobin-derived peptides in mouse brain: comparison between brain, blood, and heart peptidome and regulation in Cpefat/fat mice. *J Neurochem* 2010;**113**:871–80.
42. Gelman JS, Fricker LD. Hemopressin and other bioactive peptides from cytosolic proteins: are these non-classical neuropeptides? *AAPS J* 2010;**12**:279–89.
43. Bushlin I, Rozenfeld R, Devi LA. Cannabinoid-opioid interactions during neuropathic pain and analgesia. *Curr Opin Pharmacol* 2010;**10**:80–6.
44. Miller LK, Devi LA. The highs and lows of cannabinoid receptor expression in disease: mechanisms and their therapeutic implications. *Pharmacol Rev* 2011;**63**:461–70.
45. Valverde O, Torrens M. Cbi receptor-deficient mice as a model for depression. *Neuroscience* 2012;**204**:193–206.
46. Valverde O, Karsak M, Zimmer A. Analysis of the endocannabinoid system by using CB1 cannabinoid receptor knockout mice. *Handb Exp Pharmacol* 2005;**168**:117–45.
47. Buckley NE. The peripheral cannabinoid receptor knockout mice: an update. *Br J Pharmacol* 2008;**153**:309–18.
48. Beaulieu JM, Gainetdinov RR. The physiology, signaling, and pharmacology of dopamine receptors. *Pharmacol Rev* 2011;**63**:182–217.
49. Neve KA, Seamans JK, Trantham-Davidson H. Dopamine receptor signaling. *J Recept Signal Transduct Res* 2004;**24**:165–205.
50. Viggiano D, Ruocco LA, Sadile AG. Dopamine phenotype and behaviour in animal models: in relation to attention deficit hyperactivity disorder. *Neurosci Biobehav Rev* 2003;**27**:623–37.
51. Missale C, Nash SR, Robinson SW, Jaber M, Caron MG. Dopamine receptors: from structure to function. *Physiol Rev* 1998;**78**:189–225.
52. Drago J, Padungchaichot P, Accili D, Fuchs S. Dopamine receptors and dopamine transporter in brain function and addictive behaviors: insights from targeted mouse mutants. *Dev Neurosci* 1998;**20**:188–203.
53. Waddington JL, O'Tuathaigh C, O'Sullivan G, Tomiyama K, Koshikawa N, Croke DT. Phenotypic studies on dopamine receptor subtype and associated signal transduction mutants: insights and challenges from 10 years at the psychopharmacology-molecular biology interface. *Psychopharmacology (Berl)* 2005;**181**:611–38.
54. Xu M, Hu XT, Cooper DC, Moratalla R, Graybiel AM, White FJ, et al. Elimination of cocaine-induced hyperactivity and dopamine-mediated neurophysiological effects in dopamine D1 receptor mutant mice. *Cell* 1994;**79**:945–55.
55. De Mei C, Ramos M, Iitaka C, Borrelli E. Getting specialized: presynaptic and postsynaptic dopamine D2 receptors. *Curr Opin Pharmacol* 2009;**9**:53–8.
56. Zhu J, Chen YJ, Zhao N, Cao GF, Dang YH, Han W, et al. Distinct roles of dopamine D3 receptors in modulating methamphetamine-induced behavioral sensitization and ultrastructural plasticity in the shell of the nucleus accumbens. *J Neurosci Res* 2012;**90**:895–904.
57. Harrison SJ, Nobrega JN. Differential susceptibility to ethanol and amphetamine sensitization in dopamine D3 receptor-deficient mice. *Psychopharmacology (Berl)* 2009;**204**:49–59.

58. Li T, Hou Y, Cao W, Yan CX, Chen T, Li SB. Role of dopamine D3 receptors in basal nociception regulation and in morphine-induced tolerance and withdrawal. *Brain Res* 2012;**1433**:80–4.
59. Kaczor AA, Selent J. Oligomerization of G protein-coupled receptors: biochemical and biophysical methods. *Curr Med Chem* 2011;**18**:4606–34.
60. Tadagaki K, Jockers R, Kamal M. History and biological significance of GPCR heteromerization in the neuroendocrine system. *Neuroendocrinology* 2012;**95**:223–31.
61. Gonzalez-Maeso J. GPCR oligomers in pharmacology and signaling. *Mol Brain* 2011;**4**:20.
62. Gomes I, Jordan BA, Gupta A, Rios C, Trapaidze N, Devi LA. G protein coupled receptor dimerization: implications in modulating receptor function. *J Mol Med (Berl)* 2001;**79**:226–42.
63. Rios CD, Jordan BA, Gomes I, Devi LA. G-protein-coupled receptor dimerization: modulation of receptor function. *Pharmacol Ther* 2001;**92**:71–87.
64. Rozenfeld R, Devi LA. Exploring a role for heteromerization in GPCR signalling specificity. *Biochem J* 2011;**433**:11–8.
65. Rozenfeld R, Devi LA. Receptor heteromerization and drug discovery. *Trends Pharmacol Sci* 2010;**31**:124–30.
66. Rostene WH, Fischette CT, McEwen BS. Modulation by vasoactive intestinal peptide (VIP) of serotonin receptors in membranes from rat hippocampus. *J Neurosci* 1983;**3**:2414–9.
67. Fuxe K, Ferre S, Zoli M, Agnati LF. Integrated events in central dopamine transmission as analyzed at multiple levels. Evidence for intramembrane adenosine A2A/dopamine D2 and adenosine A1/dopamine D1 receptor interactions in the basal ganglia. *Brain Res Brain Res Rev* 1998;**26**:258–73.
68. Zoli M, Agnati LF, Hedlund PB, Li XM, Ferre S, Fuxe K. Receptor-receptor interactions as an integrative mechanism in nerve cells. *Mol Neurobiol* 1993;**7**:293–334.
69. Fuxe K, Agnati LF, Benfenati F, Cimmino M, Algeri S, Hokfelt T, et al. Modulation by cholecystokinins of 3H-spiroperidol binding in rat striatum: evidence for increased affinity and reduction in the number of binding sites. *Acta Physiol Scand* 1981;**113**:567–9.
70. Maggi A, U'Prichard DC, Enna SJ. beta-Adrenergic regulation of alpha 2-adrenergic receptors in the central nervous system. *Science* 1980;**207**:645–7.
71. Maggio R, Vogel Z, Wess J. Coexpression studies with mutant muscarinic/adrenergic receptors provide evidence for intermolecular "cross-talk" between G-protein-linked receptors. *Proc Natl Acad Sci USA* 1993;**90**:3103–7.
72. Monnot C, Bihoreau C, Conchon S, Curnow KM, Corvol P, Clauser E. Polar residues in the transmembrane domains of the type 1 angiotensin II receptor are required for binding and coupling. Reconstitution of the binding site by co-expression of two deficient mutants. *J Biol Chem* 1996;**271**:1507–13.
73. Jones KA, Borowsky B, Tamm JA, Craig DA, Durkin MM, Dai M, et al. GABA(B) receptors function as a heteromeric assembly of the subunits GABA(B)R1 and GABA(B)R2. *Nature* 1998;**396**:674–9.
74. Kaupmann K, Malitschek B, Schuler V, Heid J, Froestl W, Beck P, et al. GABA(B)-receptor subtypes assemble into functional heteromeric complexes. *Nature* 1998;**396**:683–7.
75. Kuner R, Kohr G, Grunewald S, Eisenhardt G, Bach A, Kornau HC. Role of heteromer formation in GABAB receptor function. *Science* 1999;**283**:74–7.
76. White JH, Wise A, Main MJ, Green A, Fraser NJ, Disney GH, et al. Heterodimerization is required for the formation of a functional GABA(B) receptor. *Nature* 1998;**396**:679–82.
77. Cui M, Jiang P, Maillet E, Max M, Margolskee RF, Osman R. The heterodimeric sweet taste receptor has multiple potential ligand binding sites. *Curr Pharm Des* 2006;**12**:4591–600.

78. Nelson G, Hoon MA, Chandrashekar J, Zhang Y, Ryba NJ, Zuker CS. Mammalian sweet taste receptors. *Cell* 2001;**106**:381–90.
79. Nelson G, Chandrashekar J, Hoon MA, Feng L, Zhao G, Ryba NJ, et al. An amino-acid taste receptor. *Nature* 2002;**416**:199–202.
80. Gomes I, Filipovska J, Jordan BA, Devi LA. Oligomerization of opioid receptors. *Methods* 2002;**27**:358–65.
81. Gomes I, Filipovska J, Devi LA. Opioid receptor oligomerization. Detection and functional characterization of interacting receptors. *Methods Mol Med* 2003;**84**:157–83.
82. Jordan BA, Devi LA. G-protein-coupled receptor heterodimerization modulates receptor function. *Nature* 1999;**399**:697–700.
83. Salahpour A, Espinoza S, Masri B, Lam V, Barak LS, Gainetdinov RR. BRET biosensors to study GPCR biology, pharmacology, and signal transduction. *Front Endocrinol (Lausanne)* 2012;**3**:105.
84. Overland AC, Kitto KF, Chabot-Dore AJ, Rothwell PE, Fairbanks CA, Stone LS, et al. Protein kinase C mediates the synergistic interaction between agonists acting at alpha(2)-adrenergic and delta-opioid receptors in spinal cord. *J Neurosci* 2009;**29**:13264–73.
85. Rios C, Gomes I, Devi LA. Interactions between delta opioid receptors and alpha-adrenoceptors. *Clin Exp Pharmacol Physiol* 2004;**31**:833–6.
86. Ramsay D, Kellett E, McVey M, Rees S, Milligan G. Homo- and hetero-oligomeric interactions between G-protein-coupled receptors in living cells monitored by two variants of bioluminescence resonance energy transfer (BRET): hetero-oligomers between receptor subtypes form more efficiently than between less closely related sequences. *Biochem J* 2002;**365**:429–40.
87. Jordan BA, Trapaidze N, Gomes I, Nivarthi R, Devi LA. Oligomerization of opioid receptors with beta 2-adrenergic receptors: a role in trafficking and mitogen-activated protein kinase activation. *Proc Natl Acad Sci USA* 2001;**98**:343–8.
88. Rozenfeld R, Bushlin I, Gomes I, Tzavaras N, Gupta A, Neves S, et al. Receptor heteromerization expands the repertoire of cannabinoid signaling in rodent neurons. *PLoS One* 2012;**7**:e29239.
89. Rios C, Gomes I, Devi LA. Mu opioid and CB1 cannabinoid receptor interactions: reciprocal inhibition of receptor signaling and neuritogenesis. *Br J Pharmacol* 2006;**148**:387–95.
90. Pello OM, Martinez-Munoz L, Parrillas V, Serrano A, Rodriguez-Frade JM, Toro MJ, et al. Ligand stabilization of CXCR4/delta-opioid receptor heterodimers reveals a mechanism for immune response regulation. *Eur J Immunol* 2008;**38**:537–49.
91. Ambrose-Lanci LM, Peiris NB, Unterwald EM, Van Bockstaele EJ. Cocaine withdrawal-induced trafficking of delta-opioid receptors in rat nucleus accumbens. *Brain Res* 2008;**1210**:92–102.
92. Ambrose LM, Gallagher SM, Unterwald EM, Van Bockstaele EJ. Dopamine-D1 and delta-opioid receptors co-exist in rat striatal neurons. *Neurosci Lett* 2006;**399**:191–6.
93. Gomes I, Jordan BA, Gupta A, Trapaidze N, Nagy V, Devi LA. Heterodimerization of mu and delta opioid receptors: a role in opiate synergy. *J Neurosci* 2000;**20**:RC110.
94. Gomes I, Gupta A, Filipovska J, Szeto HH, Pintar JE, Devi LA. A role for heterodimerization of mu and delta opiate receptors in enhancing morphine analgesia. *Proc Natl Acad Sci USA* 2004;**101**:5135–9.
95. George SR, Fan T, Xie Z, Tse R, Tam V, Varghese G, et al. Oligomerization of mu- and delta-opioid receptors. Generation of novel functional properties. *J Biol Chem* 2000;**275**:26128–35.
96. Gomes I, Ijzerman AP, Ye K, Maillet EL, Devi LA. G protein-coupled receptor heteromerization: a role in allosteric modulation of ligand binding. *Mol Pharmacol* 2011;**79**:1044–52.

97. Kabli N, Martin N, Fan T, Nguyen T, Hasbi A, Balboni G, et al. Agonists at the delta-opioid receptor modify the binding of micro-receptor agonists to the micro-delta receptor hetero-oligomer. *Br J Pharmacol* 2010;**161**:1122–36.
98. Fan T, Varghese G, Nguyen T, Tse R, O'Dowd BF, George SR. A role for the distal carboxyl tails in generating the novel pharmacology and G protein activation profile of mu and delta opioid receptor hetero-oligomers. *J Biol Chem* 2005;**280**:38478–88.
99. Hasbi A, Nguyen T, Fan T, Cheng R, Rashid A, Alijaniaram M, et al. Trafficking of preassembled opioid mu-delta heterooligomer-Gz signaling complexes to the plasma membrane: coregulation by agonists. *Biochemistry* 2007;**46**:12997–3009.
100. Rozenfeld R, Devi LA. Receptor heterodimerization leads to a switch in signaling: beta-arrestin2-mediated ERK activation by mu-delta opioid receptor heterodimers. *FASEB J* 2007;**21**:2455–65.
101. Charles AC, Mostovskaya N, Asas K, Evans CJ, Dankovich ML, Hales TG. Coexpression of delta-opioid receptors with micro receptors in GH3 cells changes the functional response to micro agonists from inhibitory to excitatory. *Mol Pharmacol* 2003;**63**:89–95.
102. Law PY, Erickson-Herbrandson LJ, Zha QQ, Solberg J, Chu J, Sarre A, et al. Heterodimerization of mu- and delta-opioid receptors occurs at the cell surface only and requires receptor-G protein interactions. *J Biol Chem* 2005;**280**:11152–64.
103. Milan-Lobo L, Whistler JL. Heteromerization of the mu- and delta-opioid receptors produces ligand-biased antagonism and alters mu-receptor trafficking. *J Pharmacol Exp Ther* 2011;**337**:868–75.
104. Breit A, Gagnidze K, Devi LA, Lagace M, Bouvier M. Simultaneous activation of the delta opioid receptor (deltaOR)/sensory neuron-specific receptor-4 (SNSR-4) hetero-oligomer by the mixed bivalent agonist bovine adrenal medulla peptide 22 activates SNSR-4 but inhibits deltaOR signaling. *Mol Pharmacol* 2006;**70**:686–96.
105. Berrendero F, Maldonado R. Involvement of the opioid system in the anxiolytic-like effects induced by Delta(9)-tetrahydrocannabinol. *Psychopharmacology (Berl)* 2002;**163**:111–7.
106. Li Y, Chen J, Bai B, Du H, Liu Y, Liu H. Heterodimerization of human apelin and kappa opioid receptors: roles in signal transduction. *Cell Signal* 2012;**24**:991–1001.
107. Jordan BA, Gomes I, Rios C, Filipovska J, Devi LA. Functional interactions between mu opioid and alpha 2A-adrenergic receptors. *Mol Pharmacol* 2003;**64**:1317–24.
108. Vilardaga JP, Nikolaev VO, Lorenz K, Ferrandon S, Zhuang Z, Lohse MJ. Conformational cross-talk between alpha2A-adrenergic and mu-opioid receptors controls cell signaling. *Nat Chem Biol* 2008;**4**:126–31.
109. Zhang YQ, Limbird LE. Hetero-oligomers of alpha2A-adrenergic and mu-opioid receptors do not lead to transactivation of G-proteins or altered endocytosis profiles. *Biochem Soc Trans* 2004;**32**:856–60.
110. Hojo M, Sudo Y, Ando Y, Minami K, Takada M, Matsubara T, et al. mu-Opioid receptor forms a functional heterodimer with cannabinoid CB1 receptor: electrophysiological and FRET assay analysis. *J Pharmacol Sci* 2008;**108**:308–19.
111. Chen CG, Li J, Bot G, Szabo I, Rogers TJ, Liu-Chen LY. Heterodimerization and cross-desensitization between the mu-opioid receptor and the chemokine CCR5 receptor. *Eur J Pharmacol* 2004;**483**:175–86.
112. Suzuki S, Chuang LF, Yau P, Doi RH, Chuang RY. Interactions of opioid and chemokine receptors: oligomerization of mu, kappa, and delta with CCR5 on immune cells. *Exp Cell Res* 2002;**280**:192–200.
113. Liu XY, Liu ZC, Sun YG, Ross M, Kim S, Tsai FF, et al. Unidirectional cross-activation of GRPR by MOR1D uncouples itch and analgesia induced by opioids. *Cell* 2011;**147**:447–58.

114. Wang D, Sun X, Bohn LM, Sadee W. Opioid receptor homo- and heterodimerization in living cells by quantitative bioluminescence resonance energy transfer. *Mol Pharmacol* 2005;**67**:2173–84.
115. Chakrabarti S, Liu NJ, Gintzler AR. Formation of mu-/kappa-opioid receptor heterodimer is sex-dependent and mediates female-specific opioid analgesia. *Proc Natl Acad Sci USA* 2010;**107**:20115–9.
116. Pfeiffer M, Kirscht S, Stumm R, Koch T, Wu DF, Laugsch M, et al. Heterodimerization of substance P and mu-opioid receptors regulates receptor trafficking and resensitization. *J Biol Chem* 2003;**278**:51630–7.
117. Pan YX, Bolan E, Pasternak GW. Dimerization of morphine and orphanin FQ/nociceptin receptors: generation of a novel opioid receptor subtype. *Biochem Biophys Res Commun* 2002;**297**:659–63.
118. Wang HL, Hsu CY, Huang PC, Kuo YL, Li AH, Yeh TH, et al. Heterodimerization of opioid receptor-like 1 and mu-opioid receptors impairs the potency of micro receptor agonist. *J Neurochem* 2005;**92**:1285–94.
119. Pfeiffer M, Koch T, Schroder H, Laugsch M, Hollt V, Schulz S. Heterodimerization of somatostatin and opioid receptors cross-modulates phosphorylation, internalization, and desensitization. *J Biol Chem* 2002;**277**:19762–72.
120. Gell DA, Grant RP, Mackay JP. The detection and quantitation of protein oligomerization. *Adv Exp Med Biol* 2012;**747**:19–41.
121. Ciruela F, Vilardaga JP, Fernandez-Duenas V. Lighting up multiprotein complexes: lessons from GPCR oligomerization. *Trends Biotechnol* 2010;**28**:407–15.
122. Day RN, Davidson MW. Fluorescent proteins for FRET microscopy: monitoring protein interactions in living cells. *Bioessays* 2012;**34**:341–50.
123. Padilla-Parra S, Tramier M. FRET microscopy in the living cell: different approaches, strengths and weaknesses. *Bioessays* 2012;**34**:369–76.
124. Lohse MJ, Nuber S, Hoffmann C. Fluorescence/bioluminescence resonance energy transfer techniques to study G-protein-coupled receptor activation and signaling. *Pharmacol Rev* 2012;**64**:299–336.
125. Urizar E, Yano H, Kolster R, Gales C, Lambert N, Javitch JA. CODA-RET reveals functional selectivity as a result of GPCR heteromerization. *Nat Chem Biol* 2011;**7**:624–30.
126. Dragulescu-Andrasi A, Chan CT, De A, Massoud TF, Gambhir SS. Bioluminescence resonance energy transfer (BRET) imaging of protein-protein interactions within deep tissues of living subjects. *Proc Natl Acad Sci USA* 2011;**108**:12060–5.
127. Cabello N, Gandia J, Bertarelli DC, Watanabe M, Lluis C, Franco R, et al. Metabotropic glutamate type 5, dopamine D2 and adenosine A2a receptors form higher-order oligomers in living cells. *J Neurochem* 2009;**109**:1497–507.
128. Gandia J, Galino J, Amaral OB, Soriano A, Lluis C, Franco R, et al. Detection of higher-order G protein-coupled receptor oligomers by a combined BRET-BiFC technique. *FEBS Lett* 2008;**582**:2979–84.
129. Navarro G, Carriba P, Gandia J, Ciruela F, Casado V, Cortes A, et al. Detection of heteromers formed by cannabinoid CB1, dopamine D2, and adenosine A2A G-protein-coupled receptors by combining bimolecular fluorescence complementation and bioluminescence energy transfer. *Scientific World J* 2008;**8**:1088–97.
130. Vidi PA, Chen JJ, Irudayaraj JMK, Watts VJ. Adenosine A(2A) receptors assemble into higher-order oligomers at the plasma membrane. *FEBS Lett* 2008;**582**:3985–90.
131. Binnig G, Quate CF, Gerber C. Atomic force microscope. *Phys Rev Lett* 1986;**56**:930–3.
132. Fotiadis D, Liang Y, Filipek S, Saperstein DA, Engel A, Palczewski K. The G protein-coupled receptor rhodopsin in the native membrane. *FEBS Lett* 2004;**564**:281–8.

133. Kunishima N, Shimada Y, Tsuji Y, Sato T, Yamamoto M, Kumasaka T, et al. Structural basis of glutamate recognition by a dimeric metabotropic glutamate receptor. *Nature* 2000;**407**:971–7.
134. Fotiadis D, Jastrzebska B, Philippsen A, Muller DJ, Palczewski K, Engel A. Structure of the rhodopsin dimer: a working model for G-protein-coupled receptors. *Curr Opin Struct Biol* 2006;**16**:252–9.
135. Cherezov V, Rosenbaum DM, Hanson MA, Rasmussen SG, Thian FS, Kobilka TS, et al. High-resolution crystal structure of an engineered human beta2-adrenergic G protein-coupled receptor. *Science* 2007;**318**:1258–65.
136. Hanson MA, Stevens RC. Discovery of new GPCR biology: one receptor structure at a time. *Structure* 2009;**17**:8–14.
137. Wu BL, Chien EYT, Mol CD, Fenalti G, Liu W, Katritch V, et al. Structures of the CXCR4 chemokine GPCR with small-molecule and cyclic peptide antagonists. *Science* 2010;**330**:1066–71.
138. Chien EYT, Liu W, Zhao QA, Katritch V, Han GW, Hanson MA, et al. Structure of the human dopamine D3 receptor in complex with a D2/D3 selective antagonist. *Science* 2010;**330**:1091–5.
139. Manglik A, Kruse AC, Kobilka TS, Thian FS, Mathiesen JM, Sunahara RK, et al. Crystal structure of the micro-opioid receptor bound to a morphinan antagonist. *Nature* 2012;**485**:321–6.
140. Wu H, Wacker D, Mileni M, Katritch V, Han GW, Vardy E, et al. Structure of the human kappa-opioid receptor in complex with JDTic. *Nature* 2012;**485**:327–32.
141. Granier S, Manglik A, Kruse AC, Kobilka TS, Thian FS, Weis WI, et al. Structure of the delta-opioid receptor bound to naltrindole. *Nature* 2012;**485**:400–4.
142. Thompson AA, Liu W, Chun E, Katritch V, Wu H, Vardy E, et al. Structure of the nociceptin/orphanin FQ receptor in complex with a peptide mimetic. *Nature* 2012;**485**:395–9.
143. Albizu L, Cottet M, Kralikova M, Stoev S, Seyer R, Brabet I, et al. Time-resolved FRET between GPCR ligands reveals oligomers in native tissues. *Nat Chem Biol* 2010;**6**:587–94.
144. Rocheville M, Lange DC, Kumar U, Patel SC, Patel RC, Patel YC. Receptors for dopamine and somatostatin: formation of hetero-oligomers with enhanced functional activity. *Science* 2000;**288**:154–7.
145. Rocheville M, Lange DC, Kumar U, Sasi R, Patel RC, Patel YC. Subtypes of the somatostatin receptor assemble as functional homo- and heterodimers. *J Biol Chem* 2000;**275**:7862–9.
146. McVey M, Ramsay D, Kellett E, Rees S, Wilson S, Pope AJ, et al. Monitoring receptor oligomerization using time-resolved fluorescence resonance energy transfer and bioluminescence resonance energy transfer—the human delta-opioid receptor displays constitutive oligomerization at the cell surface, which is not regulated by receptor occupancy. *J Biol Chem* 2001;**276**:14092–9.
147. Comps-Agrar L, Kniazeff J, Norskov-Lauritsen L, Maurel D, Gassmann M, Gregor N, et al. The oligomeric state sets GABA(B) receptor signalling efficacy. *EMBO J* 2011;**30**:2336–49.
148. Callen L, Moreno E, Barroso-Chinea P, Moreno-Delgado D, Cortes A, Mallol J, et al. Cannabinoid receptors CB1 and CB2 form functional heteromers in brain. *J Biol Chem* 2012;**287**:20851–65.
149. Trifilieff P, Rives ML, Urizar E, Piskorowski RA, Vishwasrao HD, Castrillon J, et al. Detection of antigen interactions ex vivo by proximity ligation assay: endogenous dopamine D2-adenosine A2A receptor complexes in the striatum. *Biotechniques* 2011;**51**:111–8.

150. Jiang Q, Takemori AE, Sultana M, Portoghese PS, Bowen WD, Mosberg HI, et al. Differential antagonism of opioid delta antinociception by [D-Ala2, Leu5, Cys6] enkephalin and naltrindole 5'-isothiocyanate: evidence for delta receptor subtypes. *J Pharmacol Exp Ther* 1991;**257**:1069–75.
151. Zukin RS, Eghbali M, Olive D, Unterwald EM, Tempel A. Characterization and visualization of rat and guinea pig brain kappa opioid receptors: evidence for kappa 1 and kappa 2 opioid receptors. *Proc Natl Acad Sci USA* 1988;**85**:4061–5.
152. Bhushan RG, Sharma SK, Xie Z, Daniels DJ, Portoghese PS. A bivalent ligand (KDN-21) reveals spinal delta and kappa opioid receptors are organized as heterodimers that give rise to delta(1) and kappa(2) phenotypes. Selective targeting of delta-kappa heterodimers. *J Med Chem* 2004;**47**:2969–72.
153. Xie Z, Bhushan RG, Daniels DJ, Portoghese PS. Interaction of bivalent ligand KDN21 with heterodimeric delta-kappa opioid receptors in human embryonic kidney 293 cells. *Mol Pharmacol* 2005;**68**:1079–86.
154. Waldhoer M, Fong J, Jones RM, Lunzer MM, Sharma SK, Kostenis E, et al. A heterodimer-selective agonist shows in vivo relevance of G protein-coupled receptor dimers. *Proc Natl Acad Sci USA* 2005;**102**:9050–5.
155. Rives ML, Rossillo M, Liu-Chen LY, Javitch JA. 6′-Guanidinonaltrindole (6′-GNTI) is a G protein-biased kappa-opioid receptor agonist that inhibits arrestin recruitment. *J Biol Chem* 2012;**287**:27050–4.
156. Costantino CM, Gomes I, Stockton SD, Lim MP, Devi LA. Opioid receptor heteromers in analgesia. *Expert Rev Mol Med* 2012;**14**:.
157. Rothman RB, Bowen WD, Herkenham M, Jacobson AE, Rice KC, Pert CB. A quantitative study of [3H]D-Ala2-D-Leu5-enkephalin binding to rat brain membranes. Evidence that oxymorphone is a noncompetitive inhibitor of the lower affinity delta-binding site. *Mol Pharmacol* 1985;**27**:399–409.
158. Rothman RB, Long JB, Bykov V, Jacobson AE, Rice KC, Holaday JW. Beta-FNA binds irreversibly to the opiate receptor complex: in vivo and in vitro evidence. *J Pharmacol Exp Ther* 1988;**247**:405–16.
159. Porreca F, Takemori AE, Sultana M, Portoghese PS, Bowen WD, Mosberg HI. Modulation of mu-mediated antinociception in the mouse involves opioid delta-2 receptors. *J Pharmacol Exp Ther* 1992;**263**:147–52.
160. Abul-Husn NS, Sutak M, Milne B, Jhamandas K. Augmentation of spinal morphine analgesia and inhibition of tolerance by low doses of mu- and delta-opioid receptor antagonists. *Br J Pharmacol* 2007;**151**:877–87.
161. Abdelhamid EE, Takemori AE. Characteristics of mu and delta opioid binding sites in striatal slices of morphine-tolerant and -dependent mice. *Eur J Pharmacol* 1991;**198**:157–63.
162. Zhu Y, King MA, Schuller AG, Nitsche JF, Reidl M, Elde RP, et al. Retention of supraspinal delta-like analgesia and loss of morphine tolerance in delta opioid receptor knockout mice. *Neuron* 1999;**24**:243–52.
163. Shapira M, Gafni M, Sarne Y. Independence of, and interactions between, cannabinoid and opioid signal transduction pathways in N18TG2 cells. *Brain Res* 1998;**806**:26–35.
164. Berrendero F, Mendizabal V, Murtra P, Kieffer BL, Maldonado R. Cannabinoid receptor and WIN 55 212-2-stimulated [35S]-GTPgammaS binding in the brain of mu-, delta- and kappa-opioid receptor knockout mice. *Eur J Neurosci* 2003;**18**:2197–202.
165. Uriguen L, Berrendero F, Ledent C, Maldonado R, Manzanares J. Kappa- and delta-opioid receptor functional activities are increased in the caudate putamen of cannabinoid CB1 receptor knockout mice. *Eur J Neurosci* 2005;**22**:2106–10.
166. Grazzini E, Puma C, Roy MO, Yu XH, O'Donnell D, Schmidt R, et al. Sensory neuron-specific receptor activation elicits central and peripheral nociceptive effects in rats. *Proc Natl Acad Sci USA* 2004;**101**:7175–80.

167. Lembo PM, Grazzini E, Groblewski T, O'Donnell D, Roy MO, Zhang J, et al. Proenkephalin A gene products activate a new family of sensory neuron-specific GPCRs. *Nat Neurosci* 2002;**5**:201–9.
168. Pepe S, Xiao RP, Hohl C, Altschuld R, Lakatta EG. 'Cross talk' between opioid peptide and adrenergic receptor signaling in isolated rat heart. *Circulation* 1997;**95**:2122–9.
169. Xiao RP, Pepe S, Spurgeon HA, Capogrossi MC, Lakatta EG. Opioid peptide receptor stimulation reverses beta-adrenergic effects in rat heart cells. *Am J Physiol Heart Circ Physiol* 1997;**272**:H797–H805.
170. Fairbanks CA, Posthumus IJ, Kitto KF, Stone LS, Wilcox GL. Moxonidine, a selective imidazoline/alpha(2) adrenergic receptor agonist, synergizes with morphine and deltorphin II to inhibit substance P-induced behavior in mice. *Pain* 2000;**84**:13–20.
171. Fairbanks CA, Stone LS, Kitto KF, Nguyen HO, Posthumus IJ, Wilcox GL. alpha(2C)-Adrenergic receptors mediate spinal analgesia and adrenergic-opioid synergy. *J Pharmacol Exp Ther* 2002;**300**:282–90.
172. Guo XH, Fairbanks CA, Stone LS, Loh HH. DPDPE-UK14,304 synergy is retained in mu opioid receptor knockout mice. *Pain* 2003;**104**:209–17.
173. Stone LS, MacMillan LB, Kitto KF, Limbird LE, Wilcox GL. The alpha2a adrenergic receptor subtype mediates spinal analgesia evoked by alpha2 agonists and is necessary for spinal adrenergic-opioid synergy. *J Neurosci* 1997;**17**:7157–65.
174. Riedl MS, Schnell SA, Overland AC, Chabot-Dore AJ, Taylor AM, Ribeiro-Da-Silva A, et al. Coexpression of alpha(2A)-adrenergic and delta-opioid receptors in substance P-containing terminals in rat dorsal horn (vol. 513, p. 385, 2009). *J Comp Neurol* 2009;**514**:674 674.
175. Daunais JB, McGinty JF. Acute and chronic cocaine administration differentially alters striatal opioid and nuclear transcription factor mRNAs. *Synapse* 1994;**18**:35–45.
176. George SR, Kertesz M. Met-enkephalin concentrations in striatum respond reciprocally to alterations in dopamine neurotransmission. *Peptides* 1987;**8**:487–92.
177. Hurd YL, Brown EE, Finlay JM, Fibiger HC, Gerfen CR. Cocaine self-administration differentially alters mRNA expression of striatal peptides. *Brain Res Mol Brain Res* 1992;**13**:165–70.
178. Unterwald EM, Cuntapay M. Dopamine-opioid interactions in the rat striatum: a modulatory role for dopamine D1 receptors in delta opioid receptor-mediated signal transduction. *Neuropharmacology* 2000;**39**:372–81.
179. Ito S, Mori T, Sawaguchi T. Differential effects of micro-opioid, delta-opioid and kappa-opioid receptor agonists on dopamine receptor agonist-induced climbing behavior in mice. *Behav Pharmacol* 2006;**17**:691–701.
180. Drasner K, Fields HL. Synergy between the antinociceptive effects of intrathecal clonidine and systemic morphine in the rat. *Pain* 1988;**32**:309–12.
181. Ossipov MH, Lopez Y, Bian D, Nichols ML, Porreca F. Synergistic antinociceptive interactions of morphine and clonidine in rats with nerve-ligation injury. *Anesthesiology* 1997;**86**:196–204.
182. Ledent C, Valverde O, Cossu G, Petitet F, Aubert JF, Beslot F, et al. Unresponsiveness to cannabinoids and reduced addictive effects of opiates in CB1 receptor knockout mice. *Science* 1999;**283**:401–4.
183. Chuang TK, Killam Jr KF, Chuang LF, Kung HF, Sheng WS, Chao CC, et al. Mu opioid receptor gene expression in immune cells. *Biochem Biophys Res Commun* 1995;**216**:922–30.
184. Miyagi T, Chuang LF, Doi RH, Carlos MP, Torres JV, Chuang RY. Morphine induces gene expression of CCR5 in human CEM x174 lymphocytes. *J Biol Chem* 2000;**275**:31305–10.
185. Ballantyne JC, Loach AB, Carr DB. Itching after epidural and spinal opiates. *Pain* 1988;**33**:149–60.

186. Chaney MA. Side effects of intrathecal and epidural opioids. *Can J Anaesth* 1995;**42**:891–903.
187. Hales P. Pruritus after epidural morphine. *Lancet* 1980;**2**:204.
188. Sun YG, Chen ZF. A gastrin-releasing peptide receptor mediates the itch sensation in the spinal cord. *Nature* 2007;**448**:700–3.
189. Sun YG, Zhao ZQ, Meng XL, Yin J, Liu XY, Chen ZF. Cellular basis of itch sensation. *Science* 2009;**325**:1531–4.
190. Rizzi A, Bigoni R, Marzola G, Guerrini R, Salvadori S, Regoli D, et al. The nociceptin/orphanin FQ receptor antagonist, [Nphe1]NC(1-13)NH2, potentiates morphine analgesia. *Neuroreport* 2000;**11**:2369–72.
191. King M, Chang A, Pasternak GW. Functional blockade of opioid analgesia by orphanin FQ/nociceptin. *Biochem Pharmacol* 1998;**55**:1537–40.
192. Aicher SA, Punnoose A, Goldberg A. mu-Opioid receptors often colocalize with the substance P receptor (NK1) in the trigeminal dorsal horn. *J Neurosci* 2000;**20**:4345–54.
193. Aicher SA, Sharma S, Cheng PY, Liu-Chen LY, Pickel VM. Dual ultrastructural localization of mu-opiate receptors and substance P in the dorsal horn. *Synapse* 2000;**36**:12–20.
194. Wan Q, Douglas SD, Wang X, Kolson DL, O'Donnell LA, Ho WZ. Morphine upregulates functional expression of neurokinin-1 receptor in neurons. *J Neurosci Res* 2006;**84**:1588–96.
195. Yu YJ, Arttamangkul S, Evans CJ, Williams JT, von Zastrow M. Neurokinin 1 receptors regulate morphine-induced endocytosis and desensitization of mu-opioid receptors in CNS neurons. *J Neurosci* 2009;**29**:222–33.
196. Murtra P, Sheasby AM, Hunt SP, De Felipe C. Rewarding effects of opiates are absent in mice lacking the receptor for substance P. *Nature* 2000;**405**:180–3.
197. Tai KK, Jin WQ, Chan TK, Wong TM. Characterization of [3H]U69593 binding sites in the rat heart by receptor binding assays. *J Mol Cell Cardiol* 1991;**23**:1297–302.
198. Ventura C, Bastagli L, Bernardi P, Caldarera CM, Guarnieri C. Opioid receptors in rat cardiac sarcolemma: effect of phenylephrine and isoproterenol. *Biochim Biophys Acta* 1989;**987**:69–74.
199. Lee DK, George SR, O'Dowd BF. Unravelling the roles of the apelin system: prospective therapeutic applications in heart failure and obesity. *Trends Pharmacol Sci* 2006;**27**:190–4.
200. Xu N, Wang H, Fan L, Chen Q. Supraspinal administration of apelin-13 induces antinociception via the opioid receptor in mice. *Peptides* 2009;**30**:1153–7.
201. Carriba P, Ortiz O, Patkar K, Justinova Z, Stroik J, Themann A, et al. Striatal adenosine A(2A) and Cannabinoid CB1 receptors form functional heteromeric complexes that mediate the motor effects of Cannabinoids. *Neuropsychopharmacology* 2007;**32**:2249–59.
202. Hudson BD, Hebert TE, Kelly MEM. Physical and functional interaction between CB1 cannabinoid receptors and beta(2)-adrenoceptors. *Br J Pharmacol* 2010;**160**:627–42.
203. Giuffrida A, Parsons LH, Kerr TM, Rodriguez de Fonseca F, Navarro M, Piomelli D. Dopamine activation of endogenous cannabinoid signaling in dorsal striatum. *Nat Neurosci* 1999;**2**:358–63.
204. Glass M, Felder CC. Concurrent stimulation of cannabinoid CB1 and dopamine D2 receptors augments cAMP accumulation in striatal neurons: evidence for a Gs linkage to the CB1 receptor. *J Neurosci* 1997;**17**:5327–33.
205. Jarrahian A, Watts VJ, Barker EL. D2 dopamine receptors modulate Galpha-subunit coupling of the CB1 cannabinoid receptor. *J Pharmacol Exp Ther* 2004;**308**:880–6.
206. Kearn CS, Blake-Palmer K, Daniel E, Mackie K, Glass M. Concurrent stimulation of cannabinoid CB1 and dopamine D2 receptors enhances heterodimer formation: a mechanism for receptor cross-talk? *Mol Pharmacol* 2005;**67**:1697–704.

207. Ward RJ, Pediani JD, Milligan G. Heteromultimerization of cannabinoid CB1 receptor and orexin OX1 receptor generates a unique complex in which both protomers are regulated by orexin A. *J Biol Chem* 2011;**286**:37414–28.
208. Hilairet S, Bouaboula M, Carriere D, Le Fur G, Casellas P. Hypersensitization of the orexin 1 receptor by the CB1 receptor—evidence for cross-talk blocked by the specific CB1 antagonist, SR141716. *J Biol Chem* 2003;**278**:23731–7.
209. Carriba P, Navarro G, Ciruela F, Ferre S, Casado V, Agnati L, et al. Detection of heteromerization of more than two proteins by sequential BRET-FRET. *Nat Methods* 2008;**5**:727–33.
210. Martire A, Tebano MT, Chiodi V, Ferreira SG, Cunha RA, Kofalvi A, et al. Presynaptic adenosine A2A receptors control cannabinoid CB1 receptor-mediated inhibition of striatal glutamatergic neurotransmission. *J Neurochem* 2011;**116**:273–80.
211. Soria G, Castane A, Berrendero F, Ledent C, Parmentier M, Maldonado R, et al. Adenosine A2A receptors are involved in physical dependence and place conditioning induced by THC. *Eur J Neurosci* 2004;**20**:2203–13.
212. Yao LN, McFarland K, Fan PP, Jiang Z, Ueda T, Diamond I. Adenosine A2a blockade prevents synergy between mu-opiate and cannabinoid CB1 receptors and eliminates heroin-seeking behavior in addicted rats. *Proc Natl Acad Sci USA* 2006;**103**:7877–82.
213. Lerner TN, Horne EA, Stella N, Kreitzer AC. Endocannabinoid signaling mediates psychomotor activation by adenosine A(2A) antagonists. *J Neurosci* 2010;**30**:2160–4.
214. Justinova Z, Ferre S, Redhi GH, Mascia P, Stroik J, Quarta D, et al. Reinforcing and neurochemical effects of cannabinoid CB1 receptor agonists, but not cocaine, are altered by an adenosine A2A receptor antagonist. *Addict Biol* 2011;**16**:405–15.
215. Lanthier N, Horsmans Y, Leclercq IA. The metabolic syndrome: how it may influence hepatic stellate cell activation and hepatic fibrosis. *Curr Opin Clin Nutr Metab Care* 2009;**12**:404–11.
216. Mallat A, Lotersztajn S. Endocannabinoids and liver disease. I. Endocannabinoids and their receptors in the liver. *Am J Physiol Gastrointest Liver Physiol* 2008;**294**:G9–G12.
217. Siegmund SV, Schwabe RF. Endocannabinoids and liver disease. II. Endocannabinoids in the pathogenesis and treatment of liver fibrosis. *Am J Physiol Gastrointest Liver Physiol* 2008;**294**:G357–G362.
218. Teixeira-Clerc F, Julien B, Grenard P, Tran Van Nhieu J, Deveaux V, Li L, et al. CB1 cannabinoid receptor antagonism: a new strategy for the treatment of liver fibrosis. *Nat Med* 2006;**12**:671–6.
219. Hillion J, Canals M, Torvinen M, Casado V, Scott R, Terasmaa A, et al. Coaggregation, cointernalization, and codesensitization of adenosine A2A receptors and dopamine D2 receptors. *J Biol Chem* 2002;**277**:18091–7.
220. Lee SP, So CH, Rashid AJ, Varghese G, Cheng R, Lanca AJ, et al. Dopamine D1 and D2 receptor co-activation generates a novel phospholipase C-mediated calcium signal. *J Biol Chem* 2004;**279**:35671–8.
221. Rashid AJ, So CH, Kong MM, Furtak T, El-Ghundi M, Cheng R, et al. D1-D2 dopamine receptor heterooligomers with unique pharmacology are coupled to rapid activation of Gq/11 in the striatum. *Proc Natl Acad Sci USA* 2007;**104**:654–9.
222. Hasbi A, Fan T, Alijaniaram M, Nguyen T, Perreault ML, O'Dowd BF, et al. Calcium signaling cascade links dopamine D1-D2 receptor heteromer to striatal BDNF production and neuronal growth. *Proc Natl Acad Sci USA* 2009;**106**:21377–82.
223. Verma V, Hasbi A, O'Dowd BF, George SR. Dopamine D1-D2 receptor heteromer-mediated calcium release is desensitized by D1 receptor occupancy with or without signal activation dual functional regulation by G protein-coupled receptor kinase 2. *J Biol Chem* 2010;**285**:35092–103.
224. Perreault ML, Hasbi A, Alijaniaram M, Fan T, Varghese G, Fletcher PJ, et al. The dopamine D1-D2 receptor heteromer localizes in dynorphin/enkephalin neurons increased

high affinity state following amphetamine and in schizophrenia. *J Biol Chem* 2010;**285**:36625–34.

225. Fiorentini C, Busi C, Gorruso E, Gotti C, Spano P, Missale C. Reciprocal regulation of dopamine D1 and D3 receptor function and trafficking by heterodimerization. *Mol Pharmacol* 2008;**74**:59–69.
226. Gines S, Hillion J, Torvinen M, Le Crom S, Casado V, Canela EI, et al. Dopamine D1 and adenosine A1 receptors form functionally interacting heteromeric complexes. *Proc Natl Acad Sci USA* 2000;**97**:8606–11.
227. Scarselli M, Novi F, Schallmach E, Lin R, Baragli A, Colzi A, et al. D2/D3 dopamine receptor heterodimers exhibit unique functional properties. *J Biol Chem* 2001;**276**: 30308–14.
228. Maggio R, Millan MJ. Dopamine D2-D3 receptor heteromers: pharmacological properties and therapeutic significance. *Curr Opin Pharmacol* 2010;**10**:100–7.
229. Trincavelli ML, Daniele S, Orlandini E, Navarro G, Casado V, Giacomelli C, et al. A new D-2 dopamine receptor agonist allosterically modulates A(2A) adenosine receptor signalling by interacting with the A(2A)/D-2 receptor heteromer. *Cell Signal* 2012;**24**:951–60.
230. Canals M, Marcellino D, Fanelli F, Ciruela F, de Benedetti P, Goldberg SR, et al. Adenosine A(2A)-dopamine D2 receptor-receptor heteromerization—qualitative and quantitative assessment by fluorescence and bioluminescence energy transfer. *J Biol Chem* 2003;**278**:46741–9.
231. Kamiya T, Saitoh O, Yoshioka K, Nakata H. Oligomerization of adenosine A(2A) and dopamine D-2 receptors in living cells. *Biochem Biophys Res Commun* 2003;**306**: 544–9.
232. Fuxe K, Canals M, Torvinen M, Marcellino D, Terasmaa A, Genedani S, et al. Intramembrane receptor-receptor interactions: a novel principle in molecular medicine. *J Neural Transm* 2007;**114**:49–75.
233. Kern A, Albarran-Zeckler R, Walsh HE, Smith RG. Apo-ghrelin receptor forms heteromers with DRD2 in hypothalamic neurons and is essential for anorexigenic effects of DRD2 agonism. *Neuron* 2012;**73**:317–32.
234. Koschatzky S, Tschammer N, Gmeiner P. Cross-receptor interactions between dopamine D(2L) and neurotensin NTS(1) receptors modulate binding affinities of dopaminergics. *ACS Chem Neurosci* 2011;**2**:308–16.
235. Thibault D, Albert PR, Pineyro G, Trudeau LE. Neurotensin triggers dopamine D2 receptor desensitization through a protein kinase C and beta-arrestin1-dependent mechanism. *J Biol Chem* 2011;**286**:9174–84.
236. Romero-Fernandez W, Borroto-Escuela DO, Agnati LF, Fuxe K. Evidence for the existence of dopamine d2-oxytocin receptor heteromers in the ventral and dorsal striatum with facilitatory receptor-receptor interactions. *Mol Psychiatry* 2012. http://dx.doi.org/10.1038/mp.2012.103.
237. Baragli A, Alturaihi H, Watt HL, Abdallah A, Kumar U. Heterooligomerization of human dopamine receptor 2 and somatostatin receptor 2 Co-immunoprecipitation and fluorescence resonance energy transfer analysis. *Cell Signal* 2007;**19**:2304–16.
238. Albizu L, Holloway T, Gonzalez-Maeso J, Sealfon SC. Functional crosstalk and heteromerization of serotonin 5-HT2A and dopamine D2 receptors. *Neuropharmacology* 2011;**61**:770–7.
239. Gonzalez S, Moreno-Delgado D, Moreno E, Perez-Capote K, Franco R, Mallol J, et al. Circadian-related heteromerization of adrenergic and dopamine D(4) receptors modulates melatonin synthesis and release in the pineal gland. *PLoS Biol* 2012;**10**:e1001347.
240. Rebois RV, Maki K, Meeks JA, Fishman PH, Hebert TE, Northup JK. D(2)-like dopamine and beta-adrenergic receptors form a signaling complex that integrates G(s)- and G(i)-mediated regulation of adenylyl cyclase. *Cell Signal* 2012;**24**:2051–60.

241. Dreher JK, Jackson DM. Role of D1 and D2 dopamine receptors in mediating locomotor activity elicited from the nucleus accumbens of rats. *Brain Res* 1989;**487**:267–77.
242. Kita K, Shiratani T, Takenouchi K, Fukuzako H, Takigawa M. Effects of D1 and D2 dopamine receptor antagonists on cocaine-induced self-stimulation and locomotor activity in rats. *Eur Neuropsychopharmacol* 1999;**9**:1–7.
243. Piomelli D, Pilon C, Giros B, Sokoloff P, Martres MP, Schwartz JC. Dopamine activation of the arachidonic acid cascade as a basis for D1/D2 receptor synergism. *Nature* 1991;**353**:164–7.
244. Levavi-Sivan B, Park BH, Fuchs S, Fishburn CS. Human D3 dopamine receptor in the medulloblastoma TE671 cell line: cross-talk between D1 and D3 receptors. *FEBS Lett* 1998;**439**:138–42.
245. Bordet R, Ridray S, Schwartz JC, Sokoloff P. Involvement of the direct striatonigral pathway in levodopa-induced sensitization in 6-hydroxydopamine-lesioned rats. *Eur J Neurosci* 2000;**12**:2117–23.
246. Ridray S, Griffon N, Mignon V, Souil E, Carboni S, Diaz J, et al. Coexpression of dopamine D1 and D3 receptors in islands of Calleja and shell of nucleus accumbens of the rat: opposite and synergistic functional interactions. *Eur J Neurosci* 1998;**10**:1676–86.
247. Schwartz JC, Diaz J, Bordet R, Griffon N, Perachon S, Pilon C, et al. Functional implications of multiple dopamine receptor subtypes: the D1/D3 receptor coexistence. *Brain Res Brain Res Rev* 1998;**26**:236–42.
248. Jung MY, Schmauss C. Decreased c-fos responses to dopamine D(1) receptor agonist stimulation in mice deficient for D(3) receptors. *J Biol Chem* 1999;**274**:29406–12.
249. Karasinska JM, George SR, Cheng R, O'Dowd BF. Deletion of dopamine D1 and D3 receptors differentially affects spontaneous behaviour and cocaine-induced locomotor activity, reward and CREB phosphorylation. *Eur J Neurosci* 2005;**22**:1741–50.
250. Ferre S, Popoli P, Gimenezllort L, Finnman UB, Martinez E, Decarolis AS, et al. Postsynaptic antagonistic interaction between adenosine a(1) and dopamine D-1 receptors. *Neuroreport* 1994;**6**:73–6.
251. Popoli P, Gimenez-Llort L, Pezzola A, Reggio R, Martinez E, Fuxe K, et al. Adenosine A1 receptor blockade selectively potentiates the motor effects induced by dopamine D1 receptor stimulation in rodents. *Neurosci Lett* 1996;**218**:209–13.
252. Ferre S, Popoli P, Tinner-Staines B, Fuxe K. Adenosine A1 receptor-dopamine D1 receptor interaction in the rat limbic system: modulation of dopamine D1 receptor antagonist binding sites. *Neurosci Lett* 1996;**208**:109–12.
253. Ferre S, Torvinen M, Antoniou K, Irenius E, Civelli O, Arenas E, et al. Adenosine A1 receptor-mediated modulation of dopamine D1 receptors in stably cotransfected fibroblast cells. *J Biol Chem* 1998;**273**:4718–24.
254. Fink JS, Weaver DR, Rivkees SA, Peterfreund RA, Pollack AE, Adler EM, et al. Molecular cloning of the rat A2 adenosine receptor: selective co-expression with D2 dopamine receptors in rat striatum. *Brain Res Mol Brain Res* 1992;**14**:186–95.
255. Ferre S, von Euler G, Johansson B, Fredholm BB, Fuxe K. Stimulation of high-affinity adenosine A2 receptors decreases the affinity of dopamine D2 receptors in rat striatal membranes. *Proc Natl Acad Sci USA* 1991;**88**:7238–41.
256. Ferre S, Fuxe K. Dopamine denervation leads to an increase in the intramembrane interaction between adenosine-A2 and dopamine-D2 receptors in the neostriatum. *Brain Res* 1992;**594**:124–30.
257. Aoyama S, Kase H, Borrelli E. Rescue of locomotor impairment in dopamine D2 receptor-deficient mice by an adenosine A2A receptor antagonist. *J Neurosci* 2000;**20**:5848–52.
258. Grondin R, Bedard PJ, Tahar AH, Gregoire L, Mori A, Kase H. Antiparkinsonian effect of a new selective adenosine A(2A) receptor antagonist in MPTP-treated monkeys. *Neurology* 1999;**52**:1673–7.

259. Hauber W, Neuscheler P, Nagel J, Muller CE. Catalepsy induced by a blockade of dopamine D-1 or D-2 receptors was reversed by a concomitant blockade of adenosine A(2A) receptors in the caudate-putamen of rats. *Eur J Neurosci* 2001;**14**:1287–93.
260. Torvinen M, Torri C, Tombesi A, Marcellino D, Watson S, Lluis C, et al. Trafficking of adenosine A2A and dopamine D2 receptors. *J Mol Neurosci* 2005;**25**:191–200.
261. Soriano A, Ventura R, Molero A, Hoen R, Casado V, Cortes A, et al. Adenosine A2A receptor-antagonist/dopamine D2 receptor-agonist bivalent ligands as pharmacological tools to detect A2A-D2 receptor heteromers. *J Med Chem* 2009;**52**:5590–602.
262. Jomphe C, Lemelin PL, Okano H, Kobayashi K, Trudeau LE. Bidirectional regulation of dopamine D2 and neurotensin NTS1 receptors in dopamine neurons. *Eur J Neurosci* 2006;**24**:2789–800.
263. Young LJ, Lim MM, Gingrich B, Insel TR. Cellular mechanisms of social attachment. *Horm Behav* 2001;**40**:133–8.
264. Young LJ, Wang Z. The neurobiology of pair bonding. *Nat Neurosci* 2004;**7**:1048–54.
265. Lee HJ, Macbeth AH, Pagani JH, Young 3rd. WS. Oxytocin: the great facilitator of life. *Prog Neurobiol* 2009;**88**:127–51.
266. Petersson M, Diaz-Cabiale Z, Angel Narvaez J, Fuxe K, Uvnas-Moberg K. Oxytocin increases the density of high affinity alpha(2)-adrenoceptors within the hypothalamus, the amygdala and the nucleus of the solitary tract in ovariectomized rats. *Brain Res* 2005;**1049**:234–9.
267. Neumann ID. Brain oxytocin: a key regulator of emotional and social behaviours in both females and males. *J Neuroendocrinol* 2008;**20**:858–65.
268. Aragona BJ, Liu Y, Curtis JT, Stephan FK, Wang Z. A critical role for nucleus accumbens dopamine in partner-preference formation in male prairie voles. *J Neurosci* 2003;**23**:3483–90.
269. Gingrich B, Liu Y, Cascio C, Wang Z, Insel TR. Dopamine D2 receptors in the nucleus accumbens are important for social attachment in female prairie voles (Microtus ochrogaster). *Behav Neurosci* 2000;**114**:173–83.
270. Martin-Iverson MT, Radke JM, Vincent SR. The effects of cysteamine on dopamine-mediated behaviors: evidence for dopamine-somatostatin interactions in the striatum. *Pharmacol Biochem Behav* 1986;**24**:1707–14.
271. Rodriguez-Sanchez MN, Puebla L, Lopez-Sanudo S, Rodriguez-Martin E, Martin-Espinosa A, Rodriguez-Pena MS, et al. Dopamine enhances somatostatin receptor-mediated inhibition of adenylate cyclase in rat striatum and hippocampus. *J Neurosci Res* 1997;**48**:238–48.
272. Izquierdo-Claros RM, Boyano-Adanez MC, Larsson C, Gustavsson L, Arilla E. Acute effects of D1- and D2-receptor agonist and antagonist drugs on somatostatin binding, inhibition of adenylyl cyclase activity and accumulation of inositol 1,4,5-trisphosphate in the rat striatum. *Brain Res Mol Brain Res* 1997;**47**:99–107.
273. Leblanc R, Gauthier S, Gauvin M, Quirion R, Palmour R, Masson H. Neurobehavioral effects of intrathecal somatostatinergic treatment in subhuman primates. *Neurology* 1988;**38**:1887–90.
274. Costa E. The role of serotonin in neurobiology. *Integr Neurobiol Schizophrenia* 1960;**2**:175–227.
275. Nichols DE. Hallucinogens. *Pharmacol Ther* 2004;**101**:131–81.
276. Kim JS, Bailey MJ, Weller JL, Sugden D, Rath MF, Moller M, et al. Thyroid hormone and adrenergic signaling interact to control pineal expression of the dopamine receptor D4 gene (Drd4). *Mol Cell Endocrinol* 2010;**314**:128–35.
277. Montezinho LP, Castro MM, Duarte CB, Penschuck S, Geraldes CF, Mork A. The interaction between dopamine D2-like and beta-adrenergic receptors in the prefrontal cortex is altered by mood-stabilizing agents. *J Neurochem* 2006;**96**:1336–48.

278. Ferre S, Agnati LF, Ciruela F, Lluis C, Woods AS, Fuxe K, et al. Neurotransmitter receptor heteromers and their integrative role in 'local modules': the striatal spine module. *Brain Res Rev* 2007;**55**:55–67.
279. Ferre S, Ciruela F, Woods AS, Lluis C, Franco R. Functional relevance of neurotransmitter receptor heteromers in the central nervous system. *Trends Neurosci* 2007;**30**:440–6.
280. Paquet M, Smith Y. Group I metabotropic glutamate receptors in the monkey striatum: subsynaptic association with glutamatergic and dopaminergic afferents. *J Neurosci* 2003;**23**:7659–69.
281. Popoli P, Pezzola A, Torvinen M, Reggio R, Pintor A, Scarchilli L, et al. The selective mGlu(5) receptor agonist CHPG inhibits quinpirole-induced turning in 6-hydroxydopamine-lesioned rats and modulates the binding characteristics of dopamine D(2) receptors in the rat striatum: interactions with adenosine A(2a) receptors. *Neuropsychopharmacology* 2001;**25**:505–13.
282. Satake H, Sakai T. Recent advances and perceptions in studies of heterodimerization between G protein-coupled receptors. *Protein Pept Lett* 2008;**15**:300–8.
283. Bohn LM, Lefkowitz RJ, Gainetdinov RR, Peppel K, Caron MG, Lin FT. Enhanced morphine analgesia in mice lacking beta-arrestin 2. *Science* 1999;**286**:2495–8.
284. Daniels DJ, Lenard NR, Etienne CL, Law PY, Roerig SC, Portoghese PS. Opioid-induced tolerance and dependence in mice is modulated by the distance between pharmacophores in a bivalent ligand series. *Proc Natl Acad Sci USA* 2005;**102**:19208–13.
285. Lenard NR, Daniels DJ, Portoghese PS, Roerig SC. Absence of conditioned place preference or reinstatement with bivalent ligands containing mu-opioid receptor agonist and delta-opioid receptor antagonist pharmacophores. *Eur J Pharmacol* 2007;**566**:75–82.
286. Berg KA, Rowan MP, Gupta A, Sanchez TA, Silva M, Gomes I, et al. Allosteric interactions between delta and kappa opioid receptors in peripheral sensory neurons. *Mol Pharmacol* 2012;**81**:264–72.
287. Pei L, Li S, Wang M, Diwan M, Anisman H, Fletcher PJ, et al. Uncoupling the dopamine D1-D2 receptor complex exerts antidepressant-like effects. *Nat Med* 2010;**16**:1393–5.
288. Dziedzicka-Wasylewska M, Faron-Gorecka A, Gorecki A, Kusemider M. Mechanism of action of clozapine in the context of dopamine D1–D2 receptor hetero-dimerization—a working hypothesis. *Pharmacol Rep* 2008;**60**:581–7.
289. Faron-Gorecka A, Gorecki A, Kusmider M, Wasylewski Z, Dziedzicka-Wasylewska M. The role of D1-D2 receptor hetero-dimerization in the mechanism of action of clozapine. *Eur Neuropsychopharmacol* 2008;**18**:682–91.
290. Perreault ML, O'Dowd BF, George SR. Dopamine receptor homooligomers and heterooligomers in schizophrenia. *CNS Neurosci Ther* 2011;**17**:52–7.
291. Guillin O, Abi-Dargham A, Laruelle M. Neurobiology of dopamine in schizophrenia. *Integr Neurobiol Schizophrenia* 2007;**78**:1–39.
292. Dean B, Sundram S, Bradbury R, Scarr E, Copolov D. Studies on [H-3]CP-55940 binding in the human central nervous system: regional specific changes in density of cannabinoid-1 receptors associated with schizophrenia and cannabis use. *Neuroscience* 2001;**103**:9–15.
293. Deckert J, Brenner M, Durany N, Zochling R, Paulus W, Ransmayr G, et al. Up-regulation of striatal adenosine A(2A) receptors in schizophrenia. *Neuroreport* 2003;**14**:313–6.
294. Sundram S, Copolov D, Dean B. Clozapine decreases [3H] CP 55940 binding to the cannabinoid 1 receptor in the rat nucleus accumbens. *Naunyn Schmiedebergs Arch Pharmacol* 2005;**371**:428–33.
295. Fuxe K, Marcellino D, Borroto-Escuela DO, Guescini M, Fernandez-Duenas V, Tanganelli S, et al. Adenosine-dopamine interactions in the pathophysiology and treatment of CNS disorders. *CNS Neurosci Ther* 2010;**16**:e18–e42.

SECTION III

Oligomerization in Ion Channels

CHAPTER TEN

Social Networking Among Voltage-Activated Potassium Channels

Lioubov I. Brueggemann, Saverio Gentile, Kenneth L. Byron
Department of Molecular Pharmacology and Therapeutics, Loyola University Chicago, Maywood, Illinois, USA

Contents

Abstract

Voltage-activated potassium channels (Kv channels) are ubiquitously expressed proteins that subserve a wide range of cellular functions. From their birth in the endoplasmic reticulum, Kv channels assemble from multiple subunits in complex ways that determine where they live in the cell, their biophysical characteristics, and their role in enabling different kinds of cells to respond to specific environmental signals to generate appropriate functional responses. This chapter describes the types of protein–protein interactions among pore-forming channel subunits and their auxiliary protein partners,

Progress in Molecular Biology and Translational Science, Volume 117
ISSN 1877-1173
http://dx.doi.org/10.1016/B978-0-12-386931-9.00010-6

as well as posttranslational protein modifications that occur in various cell types. This complex oligomerization of channel subunits establishes precise cell type-specific Kv channel localization and function, which in turn drives a diverse range of cellular signal transduction mechanisms uniquely suited to the physiological contexts in which they are found.

1. INTRODUCTION

1.1. The concept of social networking as it pertains to voltage-activated potassium channels in cells

As in human social networks, different biochemical components of cells interact in elaborate and complex ways to subserve all of the various aspects of cellular physiology. Proteins may be regarded as the primary players. They have diverse structures and functional roles, and as in human social networks, proteins rarely operate in isolation from other proteins. In this chapter, we examine a particular subset of proteins with a well-characterized function and consider how their "social" interactions with other proteins enable them to be healthy and productive members of the cellular environment or subvert their normal roles and lead them to a path of death and destruction. More specifically, we focus on voltage-activated potassium channels (Kv channels) and examine how protein–protein interactions modulate their expression, locations, and activities to produce exquisitely fine-tuned functional responses in various cell types.

1.2. Getting together to regulate K^+ flux

Potassium channels are proteinaceous structures which form membrane pores that can open or close and when open allow potassium ions (K^+) to pass through the pore. In most cases, these channels are inserted into the plasma membrane where they regulate the flux of K^+ between the cytosol and the extracellular space, thereby regulating membrane voltage and ionic balance to consequently affect a variety of cellular processes that depend on these factors. There are many families of potassium channels, but we focus on the voltage-activated (Kv) channel family, which is one of the largest and most diverse families of ion channels.[1] Kv channels are activated by membrane depolarization and in turn play important roles in cellular signal transduction. The human genome encodes 12 subfamilies (clans?) of Kv channel α subunits (numbered Kv1 through Kv12) that form channels with diverse functional and pharmacological characteristics and participate in a wide variety of physiological responses.[1] It is likely that every differentiated cell in the human body expresses some complement of Kv channels.

The pore-forming structure of Kv channels requires the coordinated assembly of four α subunits, each of which is an individual gene product.[2] This tetramerization may involve four identical α subunits (homotetramers) or assembly of different α subunits from the same or different clans of Kv channels (heterotetramers). In general, it is thought that members of a single clan of Kv channels tend to tetramerize within a limited social circle. For example, Kv1 clan members may hook up with other Kv1 α subunits (e.g., Kv1.1 with Kv1.4) but not with members of other clans, such as Kv2 or Kv4 α subunits.[2] There are several clans of modulatory Kv α subunits that are unable to form functional channels by themselves (Kv5, Kv6, Kv8, Kv9) but can produce functional channels by associating with Kv2 or Kv3 α subunits.[1,3,4] Many Kv channel α subunits also socialize with auxiliary protein subunits, which influence their trafficking and biophysical properties (Table 10.1). These include β channel subunits (e.g., KCNEs), K channel-interacting proteins (KChIPs), and calmodulin (CaM).[33–35] A compact oligomeric social circle is formed from these diverse subunits to form a protein complex that can perform its ion channel function appropriately in the context of the physiology of the cell in which it is found.

1.3. The early life of a Kv channel—Establishing an intimate social circle

Kv channels are born in the rough endoplasmic reticulum (ER). Their mRNAs are translated into a sequence of amino acids that are covalently linked by peptide bonds, with the nascent polypeptide emerging gradually into the lumen of the ER. As membrane-resident proteins, they insert into the ER membrane orienting as though the ER lumen were the extracellular space. For Kv1.3 channels, this process is thought to involve the initial insertion of the second of six membrane-spanning domains (S2), which then establishes the orientation of the remaining five membrane-spanning domains (S1, S3–S6), ultimately dictating critical aspects of the tertiary structure of a single α subunit.[36] The earliest social interactions involve the establishment of a quaternary structure, that is, the formation of a tetramer of α subunits and incorporation of auxiliary subunits, which also takes place within the ER membrane. This process may occur in concert with the peptide elongation and formation of secondary and tertiary structures of the individual subunits.[37] In particular, for Kv1–Kv4 subtypes, tetramerization involves a T1 domain located near the amino terminus of the polypeptide

Table 10.1 Some Auxiliary Subunits and Their Effects on Kv Channel Localization or Modulation

		Auxiliary subunit				
α Subunit	α Subunit	Soluble	Membrane	Scaffold	Effect of auxiliary subunit/scaffold	Reference
Kv1.1	Kv1.2	Kvβ2			Localization in axonal membrane	5,6
Kv1.1	Kv1.4				Localization in nonaxonal membrane	5
Kv1				PSD-95, SAP-97	Mediate channels clustering and stabilization at plasma membrane	5,7–9
Kv4		KChIP	DPP6/10		Increase surface expression and current densities, and speed up kinetics of inactivation and recovery from inactivation	10–12
Kv4.2	Kv4.3			SAP-97	Mediate CAMKII-dependent phosphorylation	13,14
Kv4.2				AKAP79	Mediate PKA-dependent phosphorylation and PP2B-dependent dephosphorylation	15
Kv7.1			KCNE1		Increase surface expression and current densities, slow down kinetic of activation, shift activation to more positive voltages, and change pharmacology	16–23
Kv7.1		CaM	KCNE1		Ca^{2+}-dependent increase in current density	24,25
Kv7.1			KCNE1	Yotiao/AKAP9	Mediate PKA-dependent phosphorylation	26
Kv7.2	Kv7.3				Increase surface expression and current densities	27–29
Kv7.2	Kv7.3	CaM			Folding, trafficking to the plasma membrane, and Ca^{2+}-dependent decrease in current density	30,31
Kv7.2	Kv7.3			AKAP79/150	PKC-dependent phosphorylation	32

chain,[38] which is formed early in the translation process and may therefore enable socialization to begin before the polypeptides are fully formed.[36] It is less clear how subunit assembly progresses for other Kv clans, such as Kv7, for which tetramerization involves an A-domain near the carboxy terminus (the last stretch of amino acids to be added to the polypeptide as it comes off the ribosome).[30,39] Most of what has been learned about these processes is based on expression systems such as human embryonic kidney (HEK) 293 cells and *Xenopus* oocytes, which may or may not reflect the mechanisms that govern ontogeny of native Kv channels in mammalian tissues.[37]

The assembly of a tetramer of α subunits is a crucial determinant of the ultimate fate of the protein complex, so it is important to consider the factors that influence this initial social networking process. If the tetrameric structure is achieved as the individual polypeptides are being translated from their mRNAs, we must envision a coordinated microdomain of ribosomes producing each of the α subunits in parallel and in close proximity to enable the correct assembly and folding as the nascent polypeptides are extended. For this to happen, the correct mRNAs must find their way to these ribosomal microarrays to provide the right templates for α subunit synthesis. Thus, according to this scenario, the subunit stoichiometry and the decision to form homomeric or heteromeric tetramers would be determined based on mRNA trafficking rather than on the availability of α subunit proteins.

If membrane insertion and tetrameric assembly occur after translation is completed, we could speculate that the newly translated, full-length α subunit polypeptides find their partners within the ER lumen like boys and girls at their first high school dance. In this case, their choice of partners may be based on the relative affinity for particular α subunits and on their abundance. If this is simply a stochastic process, we must imagine that heteromeric subunit assembly would still require the mRNAs from the different genes to be translated in reasonably close proximity within the rough ER to ensure a high probability that the different gene products would find each other and hook up. Thus, a microdomain for the translation of relevant mRNAs would still be an important aspect.

The relative abundance of mRNA transcripts for different α subunits on the cytosolic side of the ribosomal microdomain would likely be a common constraint in determining subunit stoichiometry, whether tetramers assemble during or after the translation process. The correct transcripts must be present at the ribosomes for coordinated translation to occur to provide the correct stoichiometry for subunit assembly. Relative mRNA abundance would likely also influence the incorporation of Kv β subunits into the channel complex. We are only beginning to understand the complex cellular mechanisms by which mRNA abundance is determined. Recent insights into the transcriptional regulation and switching mechanisms assigned to nongene DNA (based on results from the ENCODE Project) suggest that there are many processes involved. We might envision some coordination at the level of transcription (e.g., similar transcription rates for two genes to produce channels formed as a pair of heterodimers) or message processing (e.g., coordinated trafficking, stabilization of desirable mRNAs, or culling of unwanted mRNAs) to achieve appropriate levels of mRNAs for particular α subunits at ribosomal microdomains. Finally, a coordinated delivery of mRNAs to the ribosomes may be necessary in order to drive a specific stoichiometry of tetramer assembly within the ER. The details of these mechanisms have not been worked out for any proteins, let alone for Kv channel proteins.

1.4. What motivates the social networking?

Among the key questions that remain to be answered are what factors determine which α subunits come together to form a tetramer and how do auxiliary subunits influence these decisions. Although little is known about these earliest social interactions, chaperone proteins (which may include β subunits, KChIPs, potassium channel-associated protein (KChAP), heat-shock proteins, or other friends) and scaffolding proteins (like postsynaptic density 95 (PSD-95) and A-kinase anchoring proteins (AKAPs)) may help to assemble ion channels at specific plasma membrane locations and thereby influence how these proteins participate in cellular signal transduction or respond to pharmacological agents. Posttranslational modifications (e.g., glycosylation, phosphorylation, and ubiquitination) involve acquaintances encountered in specific cellular locations that leave their mark on the Kv channel protein, with important consequences for the function and fate of the channel complex. We consider some specific examples below.

2. Kv CHANNEL OLIGOMERIZATION: GATHERING, TRAVELING, AND SETTLING DOWN TOGETHER

2.1. Getting together in the right place

Kv channels are critically important for signal transduction in many cell types, but their role is best understood in excitable cells like neurons and myocytes. For neurons, their complex electrical behavior is shaped by Kv channels with distinct biophysical properties. This diversity of neuronal Kv channels is enhanced by oligomeric assembly of channel subunits.

Heterotetrameric Kv channels composed of Kv α subunits and associated auxiliary subunits[40] contribute to modulate cellular excitability at specific cellular sites.

The precise localization of ion channels to distinct subcellular compartments plays an important role in the control of neuronal excitability. Initially, this depends on which Kv α subunits are present in the tetramer. For example, neuronal Kv1.1 contributes to shaping the repolarization phase of axonal action potentials (APs).[41] Kv1.1 knockout mice exhibit spontaneous seizures throughout their adult life, associated with increased neuronal excitability and slowed AP repolarization.[42] Mutations in Kv1.1 that cause abnormalities in conductance or folding and trafficking of Kv1.1-containing channels are associated with human hereditary episodic ataxia/myokymia type I.[43] When expressed in heterologous systems, the neuronal Kv1.1 gives rise to rapidly activating, slowly inactivating, outward delayed rectifier K^+ currents.[44,45] However, despite the ability of Kv1.1 to form functional homotetramers in expression systems, Kv1.1 does not exist in its homotetrameric form in neurons.[46,47] This has been attributed to an ER retention motif in the outer pore region of Kv1.1 α subunit that serves as a binding side for an unidentified ER retention protein.[48–50] Considering that Kv1.1 is essential for normal axonal functions but does not exist in its homomeric form in neurons suggests that heteromerization of the Kv1.1 α subunits with other α subunits of the Kv1 family is a necessary event. Summarizing immunohistochemical studies, Vacher *et al.* reported that endogenous neuronal Kv1.1 in axons was colocalized with Kv1.2 and/or Kv1.4 α subunits.[40] Kv1.1/Kv1.2 heteromers localized in axonal compartments whereas Kv1.1/Kv1.4 heteromers do not, despite the ability of homomeric Kv1.4 channels to reside in axons.[5] These findings suggest that Kv1.2 but not Kv1.4 α subunits may serve as chaperones for Kv1.1, enabling the latter to exit the ER and form heterotetramers in the axonal membrane. However, Kv1.1/Kv1.4 heteromers can traffic to other plasma membrane compartments of the neuron, suggesting that Kv1.1 is dominant in this pairing in determining their subcellular location, that is, the association of Kv1.1 with Kv1.4 α subunits prevents the latter from getting to the axon in the form of heteromeric channels.[5]

Another example of obligatory heterotetramerization of Kv α subunits is the Kv7.2/Kv7.3 channel underlying neuronal M-currents. M-currents are slowly activating, noninactivating currents which are suppressed upon activation of muscarinic acetylcholine (ACh) receptors. This regulation by ACh serves an important and well-characterized modulatory role in neuronal excitability.[51] Mutations in either KCNQ2 or KCNQ3 genes (encoding Kv7.2 and Kv7.3, respectively) cause benign familial neonatal convulsions (BFNC).[52] In heterologous expression systems, currents through Kv7.3 homomeric channels are exceptionally small. This is attributed to an inhibitory effect of an alanine residue at position 315, which resides in the inner pore vestibule of the channel and suppresses the outward movement of K^+ through Kv7.3 homotetrameric channels.[27] In Kv7.2/Kv7.3 heteromers, this negative effect of A315 in Kv7.3 is neutralized by a threonine at the equivalent 276 position in Kv7.2.[27] Despite the relatively small current amplitude, the maximal open probability (P_o) of homomeric Kv7.3 channels is quite high (~0.9), indicating that 90% of the channels can be open simultaneously. In contrast, Kv7.2 homomers have a low maximal P_o of ~0.2.[53] It has been suggested that incorporation of Kv7.3 α subunits into Kv7.2-containing channels increases the maximal P_o to 0.3 and also increases the number of functional channels at the plasma membrane, thus increasing macroscopic M-current density up to 10-fold compared with M-currents measured from Kv7.2 homomers.[27–29]

2.2. Partnering up with β subunits

The interaction of Kv α subunits with Kv β subunits is also important in determining channel assembly, localization, and function. Cytosolic auxiliary β subunits play dual chaperone and modulatory roles in Kv channel physiology.[54,55] The chaperone function of β subunits includes trafficking and stabilization of the ion channel complex, which leads to increased surface expression and increased current densities (e.g., coexpression of Kv4.2 α subunits with KChIP3 β subunits increases current density by sixfold[10,11]). According to our current understanding, auxiliary β subunits associate with the T1 domain of the N-terminus of α subunits early in ontogenesis within the rough ER, with a most common stoichiometry of $T1_4{:}\beta_4$.[54,56] This process is thought to occur during translation, though it is not yet clear whether the translation of both α and β subunits are temporally synchronized or whether newly synthesizing α subunits get together with prebuilt β subunit tetramers. Kvβ2 tetramers form readily due to a positive cooperativity for Kvβ2 binding to itself.[54] The appearance of Kvβ2 tetramers prior to detection of Kv1.2 α subunits and the ability of Kvβ2 tetramers to be trafficked independently of α subunits support a potential scaffolding/chaperone role of the Kvβ2 tetramers for α subunit tetramerization and trafficking.[6,54,57] Axonal trafficking of Kv1.2-containing channels requires interaction of their N-terminal T1 tetramerization domains with cytosolic Kvβ2 auxiliary subunits.[58] Kvβ2, independently of the Kv1.2 α subunit, interacts with the KIF3/kinesin II microtubule-based motor and with the microtubule plus-end tracking protein EB1, which is associated with growing tips of microtubules. Both of those interactions are essential for the delivery of Kv1.2/Kvβ2-containing channels to the axonal membrane.[6]

In addition to *cytosolic* auxiliary β subunits, there are two families of *transmembrane* auxiliary β subunits which include dipeptidyl-aminopeptidase-like proteins (DPP6 and DPP10) and KCNE proteins (KCNE1–5, also known as MinK and MiRPs). DPP6 and DPP10 are transmembrane glycoproteins containing short cytoplasmic N-termini and long extracellular C-termini and lacking enzymatic activities.[59] DPP6/10 transmembrane orientation is opposite to KCNE β subunits, which have C-termini facing the cytosol.[60] The transmembrane partitioning of KCNEs and DPPs may enable them to interact with the membrane-spanning domains of Kv α subunits in a manner that is independent of the cytosolic Kv α subunit domains that associate with cytosolic β subunits. This raises the possibility that pore-forming Kv α subunits can interact with both cytosolic and transmembrane β subunits simultaneously.

One of the most interesting and well-studied examples of such interactions is the ternary protein complex mediating fast-transient subthreshold-activating somatodendritic A-type K^+ currents. The unique properties of the channels underlying these currents include fast activation and inactivation and fast recovery from inactivation. These properties influence membrane repolarization during AP firing and set the delay between membrane depolarization and spiking, thereby dictating the firing frequency in neurons.[61–64] Somatodendritic A-type currents are attributed to K^+ channels assembled from homo- or heteromultimeric Kv4 pore-forming α subunits. To produce mature subthreshold A-type channels, Kv4 subunits interact simultaneously and independently with cytosolic β subunits (KChIPs) and transmembrane β subunits DPP6/10 (Fig. 10.1).[11,65] This double interaction appears to be necessary for proper channel function. Each of the β subunits confers a different type of modulation on the Kv4 channels. Interaction with KChIPs is necessary for stabilization and trafficking of the channels from the ER to the plasma membrane.[10] But this interaction with KChIPs at the Kv4 N-terminal T1 domain slows down N-type inactivation of the channel, rendering the N-type inactivating "ball" more bulky and less mobile and thereby unmasking closed-state inactivation.[11] The complex of Kv4 with KChIPs recovers faster from inactivation than Kv4 alone, though not as fast as is required for proper functioning of somatodendritic A-type currents.

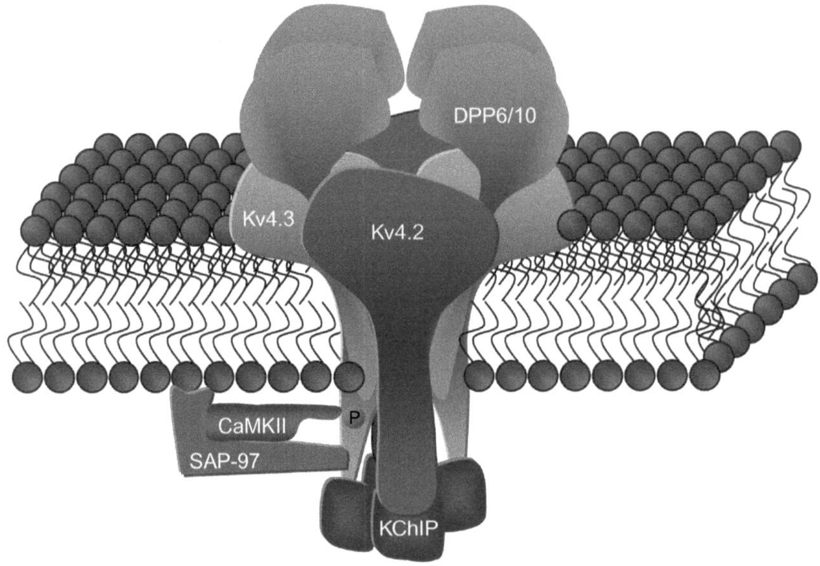

Figure 10.1 Schematic representation of Kv4 channel interaction with auxiliary subunits and scaffolding proteins. Kv4.2 and Kv4.3 α subunits, depicted in different shades of blue, form a transmembrane tetrameric protein complex. The cytosolic portions of Kv4 α subunits (N-terminal T1 domain not indicated on the schematic) interact with a tetrameric complex of soluble cytosolic KChIP subunits. Transmembrane DPP6/10 auxiliary subunits, depicted in orange with large extracellular C-terminal domains and short intracellular N-termini, interact with transmembrane domains of Kv4 α subunits. SAP-97 scaffolding protein interacts with the C-termini of Kv4 α subunits and positions CaMKII in close proximity to the channels to control phosphorylation of the Kv4 C-terminus. (For interpretation of the references to color in this figure legend, the reader is referred to the online version of this chapter.)

To speed up inactivation and recovery from inactivation of Kv4/KChIP A-type currents, transmembrane DPP6 or DPP10 β subunits incorporate into the complex. The transmembrane portions of DPP6 and DPP10 β subunits interact with S1–S2 portions of the Kv4 voltage-sensing domains, speeding up gating charge motion and shifting the charge–voltage (Q–V) curve to more negative voltages.[12] This negative shift of voltage dependence of inactivation in DPP6/10-containing Kv4 channels enables overall inactivation rates to increase.[11] DPP6-S (short splice variant) incorporation into the Kv4 channel complex also increases unitary conductance of the channel. This effect is attributed to two negative charges in the juxtamembrane cytoplasmic N-terminal region of DPP6-S.[66] But most importantly, DPP6/10 significantly accelerates the recovery from inactivation of A-type somatodendritic channels by mechanisms that are as yet not fully understood.[11] The electrophysiological characteristics conferred by the association of both transmembrane and cytosolic Kv β subunits enable appropriate fine-tuning of AP firing rates.

Transmembrane auxiliary β subunits of a second family identified earlier (KCNE1–5) are also important modulators of Kv channels. The most studied is the pairing between KCNE1 and Kv7.1. Kv7.1/KCNE1 forms the channel underlying cardiac I_{Ks}, the slow delayed rectifier current, which is important for termination of the cardiac AP.[16,17] When expressed alone, Kv7.1 α subunits form functional homotetrameric channels. However, coexpression with KCNE1 increases forward trafficking of the channels and also dramatically changes the biophysical and pharmacological characteristics of the current. KCNE1 incorporation induces a 10-fold decrease in activation rate, 20 mV positive shift of voltage dependence of activation, abolishes inactivation, increases single channel conductance four- to sevenfold, and changes the pharmacological properties of the channel.[16–23] Association of Kv7.1 and KCNE occurs in the ER early in the trafficking pathway but does not require cotranslation.[67] The predominant stoichiometry of the channel complex may be 4Kv7.1:2KCNE1 based on the binding of an irreversible charybdotoxin-based inhibitor to KCNE subunits in the channel complex.[68] However, varying stoichiometries have been observed, depending on the relative abundance of the transcripts in *Xenopus laevis* oocytes.[69] All three domains of KCNE: extracellular amino terminus, cytosolic carboxy terminus, and transmembrane domain are involved in modification of the biophysical properties of the Kv7.1/KCNE channel (Fig. 10.2), suggesting the existence of multiple Kv7.1–KCNE interaction sites (summarized by Sun *et al.*[70]).

Nakajo *et al.* found that both the voltage dependence and kinetics of channel gating depend on the relative densities of Kv7.1 and KCNE1, leading them to speculate that "the heart rhythm may be regulated by the relative expression of the auxiliary subunit and the resulting stoichiometry of the channel complex."[69] Expression of other KCNE family members, particularly KCNE4, KCNE5, and KCNE3, has also been detected in human heart samples, with different relative expression levels in normal and diseased hearts.[71] Heterologous expression of these genes profoundly altered I_{Ks} in an expression system, leading these authors to speculate that the relative abundance of different KCNE accessory subunits may be important for cardiac Kv channel function.[71]

2.3. CaM joins the gang

Calcium is a fundamental second messenger in all cells. The cytosolic concentration of free calcium ions ($[Ca^{2+}]_{cyt}$) often rises abruptly in response to stimuli such as activation of G_q-coupled receptors. The elevation of

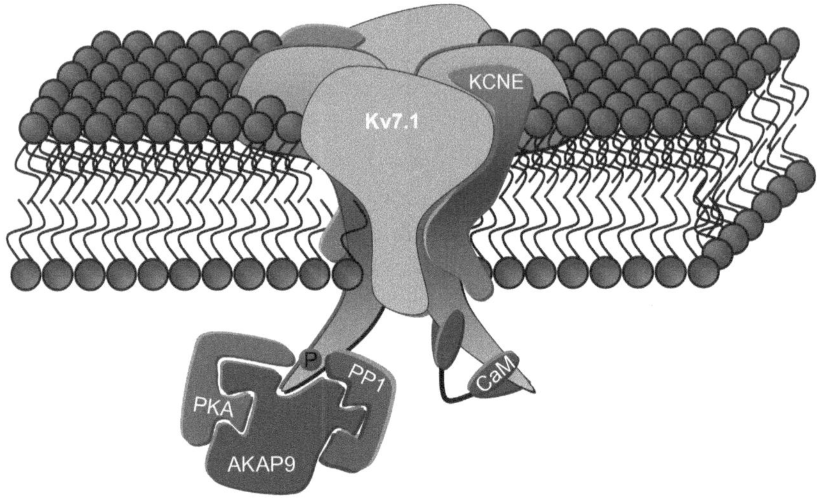

Figure 10.2 Schematic representation of Kv7.1 channel interaction with KCNE1 auxiliary subunits and scaffolding protein AKAP9. Kv7.1 subunits, depicted in blue, form a transmembrane tetrameric protein complex. Transmembrane KCNE1 auxiliary subunits, depicted in orange, interact with transmembrane, extracellular, and intracellular domains of Kv7.1 α subunits. Four EF hands of CaM, depicted in red, interact with C-terminal intracellular domains of Kv7.1 α subunits (shown only for one α subunit for clarity). The far distal portions of the C-terminal intracellular domain of Kv7.1 α subunit interact with a scaffolding protein AKAP9 (Yotiao). AKAP9 positions PKA and PP1 in close proximity to the channels to control phosphorylation of the Kv7.1 α subunit. (For interpretation of the references to color in this figure legend, the reader is referred to the online version of this chapter.)

$[Ca^{2+}]_{cyt}$ typically involves increased Ca^{2+} entry (through plasma membrane Ca^{2+} channels and nonselective cation channels) and release of intracellular Ca^{2+} (predominantly from the ER and mitochondria). Ca^{2+}-signaling responses often have an impact on the activities of K^+ channels in the plasma membrane. In particular, there are three types of Ca^{2+}-dependent potassium channels belonging to a family of K^+ channels different from Kv channels. These channels may be activated by an elevation of $[Ca^{2+}]_{cyt}$, either by directly sensing the change in $[Ca^{2+}]_{cyt}$ (Ca^{2+} binds directly to the channel protein) or indirectly via a Ca^{2+}-sensing auxiliary subunit. Voltage- and Ca^{2+}-dependent big conductance potassium channels (BK channels) possess two Ca^{2+}-binding sites on each of their tetrameric C-termini,[72] which enable them to detect an elevation of $[Ca^{2+}]_{cyt}$ and dramatically increase their activity. Voltage-independent intermediate conductance (IK) and small conductance (SK) potassium channels require CaM as a constitutively tethered auxiliary subunit for Ca^{2+}-dependent regulation.[73,74] CaM is a small (17 kDa) globular protein with two Ca^{2+}-binding lobes, the N- and C-lobes, which are connected by a flexible central linker. Each lobe has two EF-hand motifs that can bind one Ca^{2+} ion each. Thus, the binding of up to four Ca^{2+} ions to CaM alters its conformation and that signal is in turn transferred to the channels to regulate their activity.

CaM is a promiscuous Ca^{2+} sensor that can confer Ca^{2+} sensitivity to many cellular enzymes and ion channels,[75] including some members of the Kv channel family. The effect of Ca^{2+} binding in these cases is usually suppression of channel activity rather than enhancement of channel activity. There are only two Kv channels families that are appreciably sensitive to $[Ca^{2+}]_{cyt}$: Kv10 (*ether-á-go-go* or EAG) and Kv7. Using *Xenopus* oocytes as an expression system, Schönherr *et al.* demonstrated loss of Ca^{2+}-dependent inhibition of Kv10.1 channels in excised inside-out patches, but Ca^{2+} sensitivity was restored upon application of recombinant CaM, indicative of a transient association of CaM with the channel.[76] Using GST (glutathione *S*-transferase) fusion proteins containing different parts of the Kv10.1 C-terminus, Heinemann and colleagues also found that CaM binds to the middle part of the C-terminal domain as well as to an N-terminal domain, and both of these interactions are required for Ca^{2+}-induced suppression of Kv10 currents.[33,76]

Whereas Ca^{2+}-bound CaM associates transiently with Kv10 channels, Ca^{2+}-free CaM (ApoCaM) is found to be permanently tethered to Kv7 channels as an auxiliary subunit.[77,78] Association of CaM with the C-terminus of Kv7 channels likely occurs early in their biogenesis and appears to be Ca^{2+} independent.[78] Yeast two-hybrid analysis revealed that the association with CaM requires two proximal α helical domains (A and B) found in the C-terminus of all Kv7 channels, while distal α-helixes C and D are thought to form coiled-coil assemblies.[77–79] Helix A contains an IQ-like putative CaM-binding domain, whereas helix B contains two overlapping 1-5-10 consensus CaM-binding motifs. Point mutations in these CaM-binding motifs are able to disrupt the interactions.[77,78] All four EF hands of CaM are required for binding to the C-terminal AB region of Kv7 channels (Fig. 10.2).[77]

CaM also serves as a regulatory auxiliary subunit for Kv7 channels, but its role can be either inhibitory or stimulatory. Association of CaM with Kv7.2/7.3 channel α subunits mediates the observed inhibition of neuronal M-currents following a bradykinin-induced rise in $[Ca^{2+}]_{cyt}$.[31] In contrast, CaM has been implicated in the potentiation of Kv7.1-mediated I_{Ks} following an increase in $[Ca^{2+}]_{cyt}$ in guinea pig cardiomyocytes.[24] Moreover, mutations in the putative CaM-binding sites of Kv7.1 that disrupt C-terminal interactions with CaM are found to be associated with a reduction in I_{Ks} current densities and cardiac electrical abnormalities (long QT syndromes (LQTSs)).[25]

The role of CaM as an auxiliary subunit of Kv7 channels is not limited to establishing their Ca^{2+} sensitivities—CaM has also been found to be essential for proper folding and trafficking of the channels to the plasma membrane (as summarized by Haitin and Attali[30]). A mutation in Kv7.2 (R535G) that weakens CaM-binding results in ER retention of Kv7.2 channels, and this may account for its association with neonatal convulsions (BFNC).[80]

2.4. Kv channel macrocomplexes: Influencing and being influenced by the larger social network

Social interactions of Kv channels are not limited to auxiliary subunits but can also involve scaffolding proteins that gather multiple signaling proteins together in subcellular locations for specialized cellular responses. Scaffolding proteins, such as membrane-associated guanylate kinases (MAGUKs) and AKAPs, help to hold Kv channels in close proximity to regulatory proteins and restrain subcellular localization of the channel oligomeric complexes.

i. MAGUKs include PSD-95, synapse-associated protein 97 (SAP-97), and many other members. The prototypical member of the MAGUK family, PSD-95, which is found at high levels in neuronal postsynaptic densities, was initially identified as a binding partner of Kv1 α subunits using a yeast two-hybrid screen.[81] The PSD-95 Kv1 interaction involves the PDZ (*P*SD-95, *D*lg (disc large homolog), and *Z*O-1 (zona occludens)) domain of PSD-95 and a xS/TxV motif at the very end of the Kv1 α subunit C-terminus.[81,82] Later structural studies of MAGUKs, summarized by Zheng and colleagues,[83] determined that PSD-95 and other members of the *Drosophila* Dlg1-like group of MAGUKs contain three N-terminal PDZ domains followed by a Src-homology domain and a C-terminal guanylate kinase domain. The N-terminus may also contain a palmitoylation motif in some splice variants that renders them less mobile.[83] MAGUKs have been implicated in development, plasticity of synapses, and clustering of ionotropic 2-amino-3-(3-hydroxy-5-methyl-isoxazol-4-yl)propanoic acid and *N*-methyl-D-aspartate receptors.[83,84]

The role of PSD-95 in clustering of Kv 1 channels has mostly been studied using expression systems. The clustering was found to be dependent on cell surface expression of the channels but was independent of the subtype of Kv1 α subunit.[5,7] PSD-95 has been found to be a prominent component of Kv1 complexes purified from mammalian brain, but mutation of PSD-95 in mice did not change Kv1.1/Kv1.2 channel localization in juxtaparanodal regions of myelinated axons.[85] Thus, the physiological relevance of PSD-95/Kv1 α subunit interaction in the CNS remains to be elucidated.

Expression of MAGUKs is not limited to the CNS. In rat cerebral artery myocytes, endogenous PSD-95 and Kv1.2 α subunits interact based on coimmunoprecipitation and colocalization experiments.[8] Interestingly, downregulation of PSD-95 by antisense oligonucleotides induced a significant reduction in both Kv1.2 protein level and Kv1-mediated currents recorded in isolated myocytes. Myocytes in pressurized arteries treated with PSD-95 antisense oligonucleotides exhibited a 20 mV positive shift in resting membrane potential and the arteries had a significantly reduced basal diameter,[8] indicating that the association of PSD-95 with Kv1.2 is essential for Kv channel function and the maintenance of more negative membrane voltages that oppose myocyte contraction and cerebral artery constriction.

Another MAGUK protein, SAP-97, is abundantly expressed in cardiac myocytes where it interacts with Kv1.5, which underlies the cardiac ultrarapid (I_{Kur}) delayed rectifier K^+ current.[13,14] The general effect of SAP-97 appears to be to stabilize the channel proteins in clusters at the plasma membrane and thereby enhance the currents without altering their biophysical characteristics.[86] SAP-97 overexpression in neonatal rat cardiac myocytes resulted in the clustering of endogenous Kv1.5 subunits at myocyte–myocyte contacts and an increase in I_{Kur}.[9] The mobility of GFP-tagged Kv1.5 subunits expressed in either cardiac myocytes or Chinese hamster ovary (CHO) cells was also reduced by SAP-97 overexpression.[9]

SAP-97 also interacts with Kv4.2/4.3 channels, a major determinant of transient outward (I_{to}) K^+ current in cardiac myocytes. Based on a GST fusion protein-binding assay, the SAP-97 PDZ domain interacts with the binding motif Ser-Ala-Leu (SAL) at the end of the C-terminus of both Kv4.2 and Kv4.3 channels (Fig. 10.1). This interaction with SAP-97 has marked consequences for I_{to} in cardiac myocytes. The absence of SAP-97, or deletion of the SAP-97-binding motif SAL on the Kv4.3 α subunit, prevents regulation of I_{to} by CaM-dependent protein kinase II (CaMKII).[14] In an expression system, CaMKII-mediated phosphorylation of Ser at position 550 in the

C-terminal domain of Kv4.3 induces dramatic alterations in the biophysical characteristics of the Kv4.3 current. These include a twofold reduction of the inactivation rate, accompanied by a positive shift of both the activation and inactivation curves and acceleration of recovery from inactivation.[87] Inhibition of CaMKII activity produces opposite effects on the Kv4.3 current kinetics and gating and accelerates inactivation of endogenous I_{to} in cardiac myocytes.[14,87]

ii. AKAPs are also common associates of the Kv channels that serve to link the channels with other proteins and thereby help coordinate Kv channel-dependent signal transduction. AKAPs are structurally diverse but functionally similar proteins, defined on the basis of their ability to bind protein kinase A (PKA). AKAPs tether inactive PKA holoenzymes at defined locations within the cell where they are positioned to phosphorylate local substrates in response to kinase activation by cAMP. Among the most recently appreciated properties of different AKAPs is their ability to assemble several enzymes into signaling macrocomplexes.[88] From the cohort of Kv channels, only a few are known to interact with AKAPs. The short list includes the Kv4.2 α subunit of A-type channels and all five members of the Kv7 family. The consequences of those interactions are quite different, though in all cases phosphorylation of the channels is a factor.

In the case of Kv4.2, acute activation of PKA (as may occur upon the induction of long-term potentiation, a form of synaptic plasticity) leads to a decrease in Kv4.2 surface expression and a reduction in A-type K^+ current.[89] PKA-mediated phosphorylation of Kv4.2 at serine 552 (S552) on its C-terminal domain promoted channel internalization in dendritic spines of hippocampal neurons.[90] This highly localized event required interaction of Kv4.2 with AKAP79; disruption of AKAP79 interaction with Kv4.2 resulted in an increase in surface expression of Kv4.2 and increased A-type current density in cultured rat hippocampal neurons.[15] Based on coimmunoprecipitation of deletion fragments of AKAP79 and the Kv4.2 C-terminus, it was found that the interaction involves a MAGUK-binding domain of AKAP79 and the C-terminal region of Kv4.2 between amino acids 417 and 490.[15]

AKAP79 mediates not only phosphorylation of Kv4.2 at S552 by PKA but also its dephosphorylation by protein phosphatase 2B (calcineurin, CaN), providing a feedback mechanism for spatial and temporal regulation of A-type channels. Scott and colleagues first demonstrated that AKAP79 and CaN interact based on yeast two-hybrid analysis and that AKAP75 (the bovine homolog of AKAP79) exists in a ternary complex with CaN and the PKA regulatory subunit in bovine brain.[91] CaN and the regulatory subunit of PKA were also colocalized in rat hippocampal neurons, presumably via AKAP150 (the rat homolog of AKAP79).[91] In COS7 cells cotransfected with Kv4.2 and AKAP79, treatment with the CaN inhibitor, FK506, decreased surface expression of Kv4.2 by ~40%.[15] In electrophysiological recordings from cultured hippocampal neurons, peak A-type current density was also significantly decreased by FK506.[15]

The same AKAP proteins (AKAP79/150) also regulate phosphorylation of Kv7.2/7.3 channels underlying neuronal M-currents. These channels are localized mainly in the initial segment of axons.[40] In this case, upon activation of muscarinic acetylcholine receptors (mAChR) and consequent activation of protein kinase C (PKC), serine residues in helix B of the Kv7.2 C-terminus are phosphorylated, leading to suppression of M-currents.[32] AKAP79/150 anchored to the Kv7.2/7.3 channels is involved in this process. The N-terminal portion of AKAP150, residues 1–143, binds to the proximal part of the Kv7.2 C-terminus between amino acids 321–499.[32] At the same time, PKC binds to the N-terminal A-domain of AKAP150 between residues 31 and 51, thus overlapping with the Kv7.2–AKAP150-binding domain.[32,92] The AKAP150 anchoring complex may include mAChR based on coimmunoprecipitation of heterologously expressed AKAP150 and M1 muscarinic receptors[93] and fluorescence resonance energy transfer between AKAP150 and M1 receptors in CHO cells.[94] Thus, AKAP79/150 likely mediates spatial proximity of Kv7.2/7.3 channels with mAChR and PKC. A likely chain of events upon binding of ACh to the mAChR may include PLC activation, cleavage of phosphoinositol 4,5-bisphosphate (PIP_2) to produce diacylglycerol (DAG), activation of PKC bound to AKAP79/150 by DAG, and subsequent phosphorylation of Kv7.2 on Ser534/541.[32] That PKC-dependent phosphorylation reduces M-current amplitude and increases the sensitivity of Kv7.2/7.3 channels to muscarinic cholinergic regulation,[32] perhaps by decreasing the affinity of the channels for PIP_2.[95]

In contrast to neuronal members of the Kv7 family (Kv7.2–7.5), cardiac Kv7.1 channels underlying I_{Ks} interact with a different AKAP, yotiao (also known as AKAP9; Fig.10.2). Yotiao binds to the Kv7.1 C-terminus at a position without apparent homology to the Kv7.2–AKAP79/150-binding site. N- and C-terminal yotiao-binding domains interact with a Kv7.1 leucine zipper motif located in the coiled-coil helix D (amino acid residues 688–616) in human Kv7.1.[26] Yotiao positions PKA and protein phosphatase 1 (PP1) in close physical proximity to Kv7.1/KCNE1 channels, hence controlling phosphorylation of Kv7.1 at N-terminal Ser27 (Fig. 10.2).[26] This scaffolding role of yotiao is very important physiologically. Sympathetic autonomic regulation of heart rate involves activation of β-adrenergic receptors and cAMP-dependent activation of PKA. PKA-mediated phosphorylation of Kv7.1 leads to an increase in I_{Ks} amplitude and a reduction of AP duration, contributing to the well-known chronotropic effects of β-adrenergic receptor activation by catecholamines. One of the LQT mutations in the KCNQ1 gene, (G589D), prevents interaction of Kv7.1 with yotiao and renders the mutated Kv7.1 channels insensitive to PKA.[26]

2.5. Is it possible to change Kv channel partnerships in the plasma membrane?

How permanent are the Kv channel associations once they have been established within oligomeric complexes at the plasma membrane? Do the subunits that initially associate within the rough ER remain together through thick and thin until they are ultimately internalized and degraded as a group? Or can dissociation of subunits take place at the plasma membrane in response to physiological or pharmacological stimuli? An intriguing example of pharmacological dissociation of Kv1.1/Kvβ1 channel complexes by cortisone raises the possibility of posttranslational disruption of channel complexes.[96] As described earlier, association of Kv α and β subunits happens early in biogenesis and mature channels are incorporated into the plasma membrane as oligomers. In the case of Kv1.1 α subunits, coexpression of Kvβ1 subunits converts the slowly inactivating Kv1.1-mediated delayed rectifier K^+ currents into fast inactivating A-type currents,[55] indicating that incorporation of the Kvβ1 subunits into the Kv1.1 α subunit tetramer stably alters the biophysical characteristics of the channels. The modulation of current properties by Kvβ1 is attributed to introduction of N-type inactivation by the Kvβ1 N-terminal inactivating domain.[55] Pan and colleagues found that the tight interaction of Kv1.1/Kvβ1 can be disrupted by application of cortisone.[96] Cortisone, at micromolar concentrations, was able to convert A-type currents, indicative of Kv1.1/Kvβ1 oligomers, back to Kv1.1 homomeric-like sustained delayed rectifier currents. Cocrystallization of the Kvβ2 core (which is highly homologous to Kvβ1) with cortisone allowed identification of two binding sides:

one located close to the nicotinamide adenine dinucleotide phosphate cofactor-binding domain and therefore defined as the *enzymatic site* and another located in between the two neighboring Kv β subunits and defined as the *interface site*.[96] It turned out that binding of cortisone to the Kv β subunit *interface site* promoted dissociation of the T1 domain of the Kv1.1 α subunit and the Kv β subunit. This cortisone-induced dissociation was prevented by mutation in the *interface site* or by covalent linkage of the Kvβ1 C-terminus to the N-terminus of Kv1.1, which prevented Kvβ1 from being released from the Kv1.1 α subunit even when it falls off the T1 domain.[96] As the authors note, "the micromolar range EC_{50} value indicates that cortisone does not modulate channel activity under normal physiological conditions."[96] Nevertheless, the possibility of posttranslational disruption of Kv channel oligomeric complexes may have yet unidentified functional consequences.

3. THE DARK SIDE OF Kv CHANNEL SOCIAL NETWORKING

Since their discovery, Kv channels have been appreciated as important regulators of rapid cellular processes ranging from neurotransmission to muscle cell contraction. However, recent studies suggest a more sinister role for these proteins. The discovery that Kv channels may also control cell volume has led to an appreciation that the relative abundance of functional Kv channels and their interacting partners in particular subcellular locations can determine whether a cell lives or dies.

The sophisticated biochemical sequence leading to programmed cell death, also known as apoptosis, has been studied for decades. However, only recently, ionic homeostasis has been appreciated to play a fundamental role in this complicated cellular process. The cytosolic ionic environment in mammalian cells is predominated by K^+, Ca^{2+}, Na^+, and Cl^- ions. Among these, the cytosolic concentrations of K^+ and Na^+ are significantly higher than any other ions, with K^+ (140 mM) being about 14 times more concentrated than Na^+. K^+ flux across the plasma membrane not only plays a fundamental role in regulating membrane voltage, but changes in cytosolic K^+ concentration have a strong impact on cellular osmolarity, as water molecules passively follow K^+ movement across the surface membrane. It has been observed that, in the early stages of apoptosis, a massive efflux of K^+ occurs.[97] Consequently, morphological changes in the cell take place, including a reduction of cell volume (cell shrinkage). Activation of caspase 3, an essential step in a cascade of proapoptotic proteolytic events, involving several cysteine–aspartic acid proteases (caspases), has been reported to follow the reduction of intracellular K^+ concentration.[97] These same early apoptotic events were attenuated by inhibition of K^+ channel activity.[98–100] Pharmacological enhancement of Kv7 (KCNQ family) currents in cultured neurons triggers apoptosis.[101] Thus, in some circumstances, activation of plasma membrane Kv channels may be a harbinger of cell death.

Kv channels may also play an important role in programmed cell death in tumor cells. Several studies have reported that enhancing the activity of Kv channels accelerates apoptotic events in carcinoma cells induced by several agents such as ultraviolet light or superoxides.[102–104] Kv11 (EAG family) channels have been found to be expressed in several types of tumor cells in which they are proposed to play an important role in determining cell volume during H_2O_2-induced apoptosis.[104] Although the proapoptotic biochemical mechanisms triggering disruption of K^+ homeostasis still remain to be discovered, recent studies have revealed that modulation of Kv channels via ancillary subunits might play a role in determining programmed cell death.[105] As described earlier, Kv channel activity can be modulated by accessory proteins playing a variety of roles, including chaperoning and scaffolding. Overexpression of the KChAP has been found to dramatically enhance K^+ efflux in prostate cancer cells.[105] This cytosolic auxiliary β subunit binds the intracellular N-terminal domains of several Kv channels including Kv1.3–1.5, Kv2.1, Kv2.2, and Kv4.3.[1,106] In an expression system, KChAP enhanced surface membrane expression of Kv1.3, Kv2.1, Kv2.2, and Kv4.3 channels without affecting activation/inactivation kinetics or channel gating.[106–108] As a consequence of this KChAP chaperone function, one might speculate that cells with increased expression of KChAP may exhibit a higher efflux of K^+; efflux of water ensues, followed by cell shrinkage and finally apoptosis. There is considerable interest in potassium channels as novel drug targets to promote apoptosis of tumor cells for cancer therapy.[109]

Release of proapoptotic factors, for example, Bcl-2 family members, from mitochondria in response to stress-associated elements is one of the important mechanisms for programmed cell death.[110,111] It is well accepted that mitochondrial function is regulated by changes in electrical potential across the inner mitochondrial membrane. For example, depolarization of this membrane is associated with release of cytochrome *c*, an essential component of the electron transport chain, which, when released from the inner mitochondrial membrane, triggers a cascade of cytotoxic events, including caspase activation, that ultimately result in apoptosis.[112] Kv1.3 has been found to be expressed in mitochondria where its activity may modulate mitochondrial membrane potential.[113] Interestingly, it has been found that Kv1.3 channel activity can be inhibited by a Bcl-2 family member, Bax, upon induction of a variety of proapoptotic pathways.[114] It is proposed that the consequence of Bax-induced Kv1.3 inhibition is mitochondrial membrane depolarization and release of cytochrome *c*, leading to apoptosis.[114]

The ubiquitous presence of voltage-gated potassium channels in diverse cell types suggests that the regulation of their activity in response to proapoptotic stimuli might be considered as a generic mechanism in programming cell death. As in the case of KChAP association with Kv2.1, the oligomeric structure of the channels may play a fundamental role in determining whether these channels serve as friendly signal transducers for normal cellular physiology or whether their activity may be co-opted to deliver a lethal blow.

4. RESPONDING TO THE ENVIRONMENT: Kv CHANNEL OLIGOMERS AS SIGNAL TRANSDUCERS

Kv channels are increasingly recognized as important signal transduction intermediates in both excitable and nonexcitable cells. In most cases, their role in cellular signaling is initiated by posttranslational modifications of the channel proteins that alter their location or function. In their early development, posttranslational modification of the protein subunits can contribute to the subcellular localization of Kv channels. This process is incompletely elucidated but almost certainly involves a processive modulation of the channels as they travel from the ER to the Golgi apparatus before transport of the mature channels within post-Golgi carrier vesicles to the specific subcellular surface domain where their function is ultimately realized. Kv channels are assembled and core glycosylated in the ER and are further glycosylated in the Golgi apparatus.[37,56,57] N-linked glycosylation has been implicated in proper folding, surface trafficking, and modulation of Kv currents,[115–117] though in some cases glycosylation appears to be superfluous for these functions.[37] Other posttranslational modifications that likely play a role in subcellular localization of Kv channels include palmitoylation.

Reversible phosphorylation has also been found to have a strong impact on several types of Kv channels by regulating their localization, biophysical properties, and heteromeric oligomerization. Phosphorylation by PKA or PKC is a common mechanism for regulating Kv channel diversity and channel activity.[118] For example, PKA activation enhanced Kv7.2/Kv7.3 currents, an effect that was abrogated by mutation of an N-terminal consensus site for PKA (S52).[119] Modulation of Kv3.4 channel N-type inactivation by PKC was recently reported to regulate repolarization and duration of APs in dorsal root ganglion cells.[120] PKC also upregulated Kv1.3 channel activity in human lymphocytes, shifting the voltage dependence to enhance their activity at physiologically relevant membrane potentials.[121]

Biochemical signaling cascades involving reversible phosphorylation of Kv channels complexes can be severely altered in disease states. For example, in pyramidal neurons of a rat model of seizure, Kv2.1 channel phosphorylation is reduced, and this is associated with redistribution of the channels from their normal high density clusters on the somatodendritic membrane to a more uniform pattern on the soma and dendrites.[122] To investigate the mechanisms underlying this observation, Misonou and colleagues used cultured rat hippocampal neurons and determined that glutamate-induced (or KCl depolarization-dependent) calcium entry leads to activation of the calcium-activated Ser/Thr protein phosphatase, CaN, and the consequent dephosphorylation of Kv2.1.[122] This is followed by a redistribution of the channels in the surface membrane from a clustered to a more dispersed pattern; these effects are associated with a shift in the voltage dependence of activation of the Kv currents.[122] Activation of CaN appears to be both necessary and sufficient to induce all of these effects.

Signaling mechanisms promoting oligomerization can provide Kv channels an ability to respond to a large variety of extrinsic stimuli beyond changes in voltage. One of the most important mechanisms of intracellular signaling relies on the activity of protein kinases and phosphatases. Phosphorylation of channel subunits may affect the oligomerization of channel subunits. For example, the incorporation of Kv7.1/KCNE1 channels into a signaling complex with the AKAP yotiao is regulated by reversible phosphorylation mediated by PKA and dephosphorylation by PP1.[26] Formation of this macromolecular complex strongly impacts current activity by augmenting surface membrane density of the channel. For the A-type channel Kv4.2, phosphorylation of the α subunit is an important event that promotes association of the ancillary subunit KChIP3 without affecting gating or kinetics of the channel.[123]

While many phosphorylation studies have focused on Kv α subunits, phosphorylation of Kv β subunits can also have important functional consequences. For example, in *Xenopus* oocytes expressing Kv1.5 and Kvβ1.3, activation of PKA reduced channel inactivation, resulting in an increase in

Kv current. This effect of PKA appears to be mediated via phosphorylation of the β subunit.[124] In another example, a yeast two-hybrid approach revealed that the synaptonemal complex proteins required for meiotic chromosome synapsis, ZIP1 and ZIP2, act as a physical link between the Kvβ2 subunits and PKC zeta.[125] ZIP1 and ZIP2 interacted with both Kvβ1 and Kvβ2 but not with the N-terminal T1 domain of the Kv1a subunits.[125] Kvβ2 was a specific substrate for phosphorylation, which was increased much more in the presence of ZIP1 than with ZIP2.[125] The functional consequences of this phosphorylation were not determined.

Activation of the G protein-coupled M1 mAChR has been reported to inhibit Kv1.2 channel activity by PKC-dependent activation of the tyrosine kinase, Pyk2, leading to phosphorylation of the Kv1.2 α subunits.[126–128] This event was reversed by physical interaction of the channel with a tyrosine phosphatase, receptor tyrosine phosphatase α (RTPα). RTPα interacts with both the N- and C-termini of the channel,[126] consistent with other evidence for a close physical proximity of the N- and C-termini of Kv channels.[129] RTPα itself was also tyrosine phosphorylated and coimmunoprecipitated with Kv1.2 in a manner that was dependent on mAChR activation.[126] The identity of the specific tyrosine on Kv1.2 whose reversible phosphorylation is responsible for the suppression of channel activity is not known, though it was localized to the C-terminal cytosolic portion of Kv1.2.[127] The dynamic macromolecular assembly of signaling components appears to enable both the suppression of currents following mAChR activation as well as the rapid reversal of this effect, presumably to provide a temporally discrete signal suited to activate appropriate downstream effectors.

5. DIVERSITY WITHIN THE SOCIAL CIRCLE

The processes of alternative RNA splicing and RNA editing can contribute to increase protein diversity in Kv channel oligomers.[130] Following the example described earlier involving a ternary complex of Kv4.2, KChIP, and DPP6/10, it is important to recognize that alternative splicing of the transmembrane β subunit N-terminus can produce at least five DPP6 and four DPP10 isoforms, which differ only in their short intracellular domains, but modulate properties of Kv4 channels differently.[131–133] Thus, these multiple forms of transmembrane β subunits likely contribute to the diversity of somatodendritic A-type K^+ currents recorded from different neuronal populations.[65] A more dramatic example is found in *Drosophila*, where the *ether-à-go-go* gene (*eag*) transcripts can be differentially spliced to encode for an 80-kDa protein (Eag80) containing only the N-terminus and C-terminus of the Kv10 channel but lacking all the transmembrane domains that form the channel pore structure.[134] Activation of PKA or PKC appeared to be essential for the synthesis of eag80; however, as for other proteins, very little is known about biochemical signaling regulating alterative splicing of eag channels. The presence of the nonconductive Eag80 subunit which shares the C-terminal domain required for the formation of a tetramer of full-length Kv10 α subunits might be expected to alter the oligomeric structure or function of these Kv channels. Expression of Eag80 was associated with changes in cellular morphology, though the effects on Kv currents were not determined.[134] Splice variants have also been identified for mammalian *ether-à-go-go* channels (Kv10). bEAG1 and bEAG2 were initially identified based on a screening of a bovine genomic DNA library and subsequently detected in ganglion cells of bovine retina by *in situ* hybridization.[135] The bEAG2 polypeptide contains an additional insertion of 27 amino acids in the extracellular linker between transmembrane segments S3 and S4. Expression of either form in HEK 293 cells produced Kv currents with similar biophysical characteristics, except for slightly different activation kinetics and different sensitivities to Mg^{2+}.[135] It remains unclear what is the functional impact of having different splice variants of EAG in retinal ganglion cells, though it is generally accepted that alternative splicing and editing of mRNAs promote diversity in Kv channel oligomers that contribute to the specificity of cellular signal transduction.[130]

6. DRUGS IN THE NEIGHBORHOOD

Drugs can facilitate or interfere with social interactions among Kv channels and alter their ion conducting function in ways that depend on the social context in which they are presented. There are many examples of pharmacological responses that differ among Kv channels with different subunit stoichiometries. Among the best examples are the Kv7 (KCNQ) channels. The five Kv7 channel α subunit subtypes (Kv7.1–7.5) can each assemble as functional homotetramers, but there are also a number of potential heterotetrameric Kv7 channels assembled from different combinations of Kv7 α subunits. For example, among the most common neuronal Kv7 channels are heteromers of Kv7.2 and Kv7.3.[136] The antiepileptic drug retigabine, which exerts its antiepileptic effect by enhancing M-currents through Kv7.2/Kv7.3 channels in neurons, does not affect cardiac Kv7.1 channels.[137,138] The M-current-enhancing effect is attributed to a retigabine-binding motif found in Kv7.2–Kv7.5 channels near the channel gate, involving conserved amino acid residues (glycine (G)301 and tryptophan (W)236 in human Kv7.2), which enables retigabine to stabilize the open state of the channel.[139–141] The lack of W236 in Kv7.1 explains the lack of activity of retigabine at cardiac I_{Ks} channels, which are composed of Kv7.1 tetramers with up to four KCNE1 (MinK) Kv β subunits.[69] Even among the Kv7.2–Kv7.5 subtypes that are susceptible to enhancement by retigabine, the sensitivities vary depending on the subunit stoichiometry, with the following rank order of potency (based on determinations of the half-maximal effective concentration (EC_{50})): Kv7.3 > Kv7.2/Kv7.3 = Kv7.3/Kv7.5 > Kv7.2 > Kv7.4 = Kv7.5, covering the concentration range from 0.6 to 6.4 μM.[138]

More dramatic differences in pharmacology based on subunit stoichiometry have been observed in the responses of Kv7 channels to cyclooxygenase-2 (COX-2) inhibitors. Celecoxib, a widely used anti-inflammatory drug, was first reported to enhance the activity of Kv7.5 channels, an apparently COX-2-independent effect that might have relevance for its cardiovascular risk profile.[142] Subsequently, Du *et al.* reported that celecoxib also enhanced Kv7.2–7.4, Kv7.2/7.3, and Kv7.3/7.5 currents but inhibited Kv7.1 and Kv7.1/KCNE1.[143] Another clinically used COX-2 inhibitor, diclofenac, was first reported by Peretz *et al.* to activate Kv7.2/Kv7.3 channels by causing a hyperpolarizing shift of the voltage activation curve and by markedly slowing the deactivation kinetics.[144] Brueggemann *et al.* later found that diclofenac exerts diametrically opposite effects on channels composed entirely of either Kv7.4 or Kv7.5 subunits when exogenously expressed in A7r5 rat aortic smooth muscle cells.[145] Currents mediated by Kv7.4 homomers were robustly enhanced in the presence of 100 μM diclofenac, whereas Kv7.5 homomers were abruptly blocked by the same concentration of diclofenac; heteromeric Kv7.4/Kv7.5 channels exhibited an intermediate response to diclofenac.[145] These different pharmacological profiles were diagnostic in evaluating the effects of diclofenac on freshly dissociated rat mesenteric arterial myocytes, which express both Kv7.4 and Kv7.5 channels. The results suggested that the native channels in the arterial myocytes are composed primarily of Kv7.4/Kv7.5 heteromers rather than a mix of homomeric channels.[145]

There are other examples where differences in pharmacology have been used to determine subunit stoichiometry of native Kv channels. Delayed rectifier Kv currents in rabbit portal vein myocytes were inhibited by 4-AP and resistant to inhibition by charybdotoxin in a manner that was inconsistent with the effects of these drugs on either Kv1.2 or Kv1.5

homomeric channels but consistent with native channels composed of both subunits (i.e., Kv1.2/Kv1.5 heteromers).[146] Molecular biological and biochemical studies by the same laboratory provided further evidence that the native 4-AP-sensitive Kv channels exist as heteromultimeric Kv channel complexes composed of Kv1.2, Kv1.5, and Kvβ1.2 subunits.[147]

Pharmacological agents do not only discriminate among channels with different subunit stoichiometries, they may also alter Kv channel oligomerization and/or localization. One recently described example is based on an investigation of the effects of antipsychotic drugs (APDs) in mice. The conventional APD haloperidol and the atypical APDs clozapine and olanzapine were each administered to mice for a week, and in each case, they were found to upregulate the expression of Kv1.1, while simultaneously downregulating the expression of KChIP3 in the brains of the mice.[148] The findings raise the interesting possibility that these genomic effects on Kv channel subunit expression may underlie the common efficacy of these drugs as treatments for schizophrenia.

The Kv11.1 channel encoded by the human *ether-à-go-go*-related gene plays an important role in repolarization of the cardiac AP; its inhibition or dysfunction can lead to a condition called LQTS which is associated with potentially fatal ventricular arrhythmias. Kv11.1 is so commonly inhibited by drugs that pharmaceutical companies routinely screen new drug candidates for acute Kv11.1 blocking actions to weed out those that exhibit this potentially harmful effect. However, a reduction of Kv11.1 current does not always involve direct channel block that would be detected in this type of routine drug screening. Acute administration of pentamidine, a clinically used antiprotozoal agent, did not inhibit Kv11.1 currents in HEK-293 cells or CHO cells, but chronic administration of pentamidine was found to reduce the membrane expression of the channel protein (and hence reduced whole cell currents) at clinically relevant doses.[149,150] Thus, altered channel trafficking may be a contributing mechanism for the documented tendency of pentamidine to elicit LQTS in patients.[151,152]

7. CONCLUSIONS

Voltage-activated potassium channels revel in their collective coexistence. Their ability to engage in active social interactions, beyond satisfying their gregarious natures, serves a very important role in the dynamic concert of cell biology from birth to death.

REFERENCES

1. Gutman GA, Chandy KG, Grissmer S, Lazdunski M, McKinnon D, Pardo LA, et al. International Union of Pharmacology. LIII. Nomenclature and molecular relationships of voltage-gated potassium channels. *Pharmacol Rev* 2005;**57**:473–508.
2. Xu J, Yu W, Jan YN, Jan LY, Li M. Assembly of voltage-gated potassium channels. *J Biol Chem* 1995;**270**:24761–8.
3. Hugnot JP, Salinas M, Lesage F, Guillemare E, de Weille J, Heurteaux C, et al. Kv8.1, a new neuronal potassium channel subunit with specific inhibitory properties towards Shab and Shaw channels. *EMBO J* 1996;**15**:3322–31.
4. Kerschensteiner D, Monje F, Stocker M. Structural determinants of the regulation of the voltage-gated potassium channel Kv2.1 by the modulatory α-subunit Kv9.3. *J Biol Chem* 2003;**278**:18154–61.
5. Jenkins PM, McIntyre JC, Zhang L, Anantharam A, Vesely ED, Arendt KL, et al. Subunit-dependent axonal trafficking of distinct alpha heteromeric potassium channel complexes. *J Neurosci* 2011;**31**:13224–35.
6. Gu C, Zhou W, Puthenveedu MA, Xu M, Jan YN, Jan LY. The microtubule plus-end tracking protein EB1 is required for Kv1 voltage-gated K^+ channel axonal targeting. *Neuron* 2006;**52**:803–16.
7. Tiffany AM, Manganas LN, Kim E, Hsueh YP, Sheng M, Trimmer JS. PSD-95 and SAP97 exhibit distinct mechanisms for regulating K^+ channel surface expression and clustering. *J Cell Biol* 2000;**148**:147–58.
8. Joseph BK, Thakali KM, Pathan AR, Kang E, Rusch NJ, Rhee SW. Postsynaptic density-95 scaffolding of Shaker-type K^+ channels in smooth muscle cells regulates the diameter of cerebral arteries. *J Physiol* 2011;**589**:5143–52.
9. Abi-Char J, El-Haou S, Balse E, Neyroud N, Vranckx R, Coulombe A, et al. The anchoring protein SAP97 retains Kv1.5 channels in the plasma membrane of cardiac myocytes. *Am J Physiol Heart Circ Physiol* 2008;**294**:H1851–H1861.
10. An WF, Bowlby MR, Betty M, Cao J, Ling HP, Mendoza G, et al. Modulation of A-type potassium channels by a family of calcium sensors. *Nature* 2000;**403**:553–6.
11. Jerng HH, Kunjilwar K, Pfaffinger PJ. Multiprotein assembly of Kv4.2, KChIP3 and DPP10 produces ternary channel complexes with ISA-like properties. *J Physiol* 2005;**568**:767–88.
12. Dougherty K, Tu L, Deutsch C, Covarrubias M. The dipeptidyl-aminopeptidase-like protein 6 is an integral voltage sensor-interacting beta-subunit of neuronal $K_V4.2$ channels. *Channels (Austin)* 2009;**3**:122–8.
13. Godreau D, Vranckx R, Maguy A, Rucker-Martin C, Goyenvalle C, Abdelshafy S, et al. Expression, regulation and role of the MAGUK protein SAP-97 in human atrial myocardium. *Cardiovasc Res* 2002;**56**:433–42.
14. El-Haou S, Balse E, Neyroud N, Dilanian G, Gavillet B, Abriel H, et al. Kv4 potassium channels form a tripartite complex with the anchoring protein SAP97 and CaMKII in cardiac myocytes. *Circ Res* 2009;**104**:758–69.
15. Lin L, Sun W, Kung F, Dell'Acqua ML, Hoffman DA. AKAP79/150 impacts intrinsic excitability of hippocampal neurons through phospho-regulation of A-type K^+ channel trafficking. *J Neurosci* 2011;**31**:1323–32.
16. Barhanin J, Lesage F, Guillemare E, Fink M, Lazdunski M, Romey G. K_VLQT1 and lsK (minK) proteins associate to form the I_{Ks} cardiac potassium current. *Nature* 1996;**384**:78–80.
17. Sanguinetti MC, Curran ME, Zou A, Shen J, Spector PS, Atkinson DL, et al. Coassembly of K_VLQT1 and minK (IsK) proteins to form cardiac I_{Ks} potassium channel. *Nature* 1996;**384**:80–3.

18. Pusch M. Increase of the single-channel conductance of KvLQT1 potassium channels induced by the association with minK. *Pflugers Arch* 1998;**437**:172–4.
19. Sesti F, Goldstein SA. Single-channel characteristics of wild-type I_{Ks} channels and channels formed with two minK mutants that cause long QT syndrome. *J Gen Physiol* 1998;**112**:651–63.
20. Yang Y, Sigworth FJ. Single-channel properties of IKs potassium channels. *J Gen Physiol* 1998;**112**:665–78.
21. Wang HS, Brown BS, McKinnon D, Cohen IS. Molecular basis for differential sensitivity of KCNQ and I_{Ks} channels to the cognitive enhancer XE991. *Mol Pharmacol* 2000;**57**:1218–23.
22. Busch AE, Busch GL, Ford E, Suessbrich H, Lang HJ, Greger R, et al. The role of the IsK protein in the specific pharmacological properties of the IKs channel complex. *Br J Pharmacol* 1997;**122**:187–9.
23. Pusch M, Magrassi R, Wollnik B, Conti F. Activation and inactivation of homomeric KvLQT1 potassium channels. *Biophys J* 1998;**75**:785–92.
24. Nitta J, Furukawa T, Marumo F, Sawanobori T, Hiraoka M. Subcellular mechanism for Ca^{2+}-dependent enhancement of delayed rectifier K^+ current in isolated membrane patches of guinea pig ventricular myocytes. *Circ Res* 1994;**74**:96–104.
25. Ghosh S, Nunziato DA, Pitt GS. KCNQ1 assembly and function is blocked by long-QT syndrome mutations that disrupt interaction with calmodulin. *Circ Res* 2006;**98**:1048–54.
26. Marx SO, Kurokawa J, Reiken S, Motoike H, D'Armiento J, Marks AR, et al. Requirement of a macromolecular signaling complex for beta adrenergic receptor modulation of the KCNQ1-KCNE1 potassium channel. *Science* 2002;**295**:496–9.
27. Etxeberria A, Santana-Castro I, Regalado MP, Aivar P, Villarroel A. Three mechanisms underlie KCNQ2/3 heteromeric potassium M-channel potentiation. *J Neurosci* 2004;**24**:9146–52.
28. Li Y, Gamper N, Hilgemann DW, Shapiro MS. Regulation of Kv7 (KCNQ) K^+ channel open probability by phosphatidylinositol 4,5-bisphosphate. *J Neurosci* 2005;**25**: 9825–35.
29. Zaika O, Hernandez CC, Bal M, Tolstykh GP, Shapiro MS. Determinants within the turret and pore-loop domains of KCNQ3 K^+ channels governing functional activity. *Biophys J* 2008;**95**:5121–37.
30. Haitin Y, Attali B. The C-terminus of Kv7 channels: a multifunctional module. *J Physiol* 2008;**586**:1803–10.
31. Gamper N, Stockand JD, Shapiro MS. Subunit-specific modulation of KCNQ potassium channels by Src tyrosine kinase. *J Neurosci* 2003;**23**:84–95.
32. Hoshi N, Zhang JS, Omaki M, Takeuchi T, Yokoyama S, Wanaverbecq N, et al. AKAP150 signaling complex promotes suppression of the M-current by muscarinic agonists. *Nat Neurosci* 2003;**6**:564–71.
33. Ziechner U, Schönherr R, Born A-K, Gavrilova-Ruch O, Glaser RW, Malesevic M, et al. Inhibition of human ether à go-go potassium channels by Ca^{2+}/calmodulin binding to the cytosolic N- and C-termini. *FEBS J* 2006;**273**:1074–86.
34. Gamper N, Li Y, Shapiro MS. Structural requirements for differential sensitivity of KCNQ K^+ channels to modulation by Ca^{2+}/calmodulin. *Mol Biol Cell* 2005;**16**: 3538–51.
35. Pongs O, Schwarz JR. Ancillary subunits associated with voltage-dependent K^+ channels. *Physiol Rev* 2010;**90**:755–96.
36. Deutsch C. Potassium channel ontogeny. *Annu Rev Physiol* 2002;**64**:19–46.
37. Deal KK, Lovinger DM, Tamkun MM. The brain Kv1.1 potassium channel: in vitro and in vivo studies on subunit assembly and posttranslational processing. *J Neurosci* 1994;**14**:1666–76.

38. Robinson JM, Deutsch C. Coupled tertiary folding and oligomerization of the T1 domain of Kv channels. *Neuron* 2005;**45**:223–32.
39. Maljevic S, Lerche C, Seebohm G, Alekov AK, Busch AE, Lerche H. C-terminal interaction of KCNQ2 and KCNQ3 K^+ channels. *J Physiol* 2003;**548**:353–60.
40. Vacher H, Mohapatra DP, Trimmer JS. Localization and targeting of voltage-dependent ion channels in mammalian central neurons. *Physiol Rev* 2008;**88**:1407–47.
41. Kole MHP, Letzkus JJ, Stuart GJ. Axon initial segment Kv1 channels control axonal action potential waveform and synaptic efficacy. *Neuron* 2007;**55**:633–47.
42. Smart SL, Lopantsev V, Zhang CL, Robbins CA, Wang H, Chiu SY, et al. Deletion of the $K_V1.1$ potassium channel causes epilepsy in mice. *Neuron* 1998;**20**:809–19.
43. Browne DL, Gancher ST, Nutt JG, Brunt ER, Smith EA, Kramer P, et al. Episodic ataxia/myokymia syndrome is associated with point mutations in the human potassium channel gene, KCNA1. *Nat Genet* 1994;**8**:136–40.
44. Grissmer S, Nguyen AN, Aiyar J, Hanson DC, Mather RJ, Gutman GA, et al. Pharmacological characterization of five cloned voltage-gated K^+ channels, types Kv1.1, 1.2, 1.3, 1.5, and 3.1, stably expressed in mammalian cell lines. *Mol Pharmacol* 1994;**45**:1227–34.
45. Stuhmer W, Ruppersberg JP, Schroter KH, Sakmann B, Stocker M, Giese KP, et al. Molecular basis of functional diversity of voltage-gated potassium channels in mammalian brain. *EMBO J* 1989;**8**:3235–44.
46. Koch RO, Wanner SG, Koschak A, Hanner M, Schwarzer C, Kaczorowski GJ, et al. Complex subunit assembly of neuronal voltage-gated K+channels. Basis for high-affinity toxin interactions and pharmacology. *J Biol Chem* 1997;**272**:27577–81.
47. Shamotienko OG, Parcej DN, Dolly JO. Subunit combinations defined for K^+ channel Kv1 subtypes in synaptic membranes from bovine brain. *Biochem* 1997;**36**:8195–201.
48. Vacher H, Mohapatra DP, Misonou H, Trimmer JS. Regulation of Kv1 channel trafficking by the mamba snake neurotoxin dendrotoxin K. *FASEB J* 2007;**21**:906–14.
49. Manganas LN, Wang Q, Scannevin RH, Antonucci DE, Rhodes KJ, Trimmer JS. Identification of a trafficking determinant localized to the Kv1 potassium channel pore. *Proc Natl Acad Sci USA* 2001;**98**:14055–9.
50. Zhu J, Watanabe I, Gomez B, Thornhill WB. Determinants involved in Kv1 potassium channel folding in the endoplasmic reticulum, glycosylation in the Golgi, and cell surface expression. *J Biol Chem* 2001;**276**:39419–27.
51. Wang HS, Pan Z, Shi W, Brown BS, Wymore RS, Cohen IS, et al. KCNQ2 and KCNQ3 potassium channel subunits: molecular correlates of the M-channel. *Science* 1998;**282**:1890–3.
52. Charlier C, Singh NA, Ryan SG, Lewis TB, Reus BE, Leach RJ, et al. A pore mutation in a novel KQT-like potassium channel gene in an idiopathic epilepsy family. *Nat Genet* 1998;**18**:53–5.
53. Li Y, Gamper N, Shapiro MS. Single-channel analysis of KCNQ K^+ channels reveals the mechanism of augmentation by a cysteine-modifying reagent. *J Neurosci* 2004;**24**:5079–90.
54. Xu J, Yu W, Wright JM, Raab RW, Li M. Distinct functional stoichiometry of potassium channel beta subunits. *Proc Natl Acad Sci USA* 1998;**95**:1846–51.
55. Rettig J, Heinemann SH, Wunder F, Lorra C, Parcej DN, Dolly JO, et al. Inactivation properties of voltage-gated K^+ channels altered by presence of beta-subunit. *Nature* 1994;**369**:289–94.
56. Nagaya N, Papazian DM. Potassium channel alpha and beta subunits assemble in the endoplasmic reticulum. *J Biol Chem* 1997;**272**:3022–7.
57. Deutsch C. The birth of a channel. *Neuron* 2003;**40**:265–76.
58. Gu C, Jan YN, Jan LY. A conserved domain in axonal targeting of Kv1 (Shaker) voltage-gated potassium channels. *Science* 2003;**301**:646–9.

59. Wada K, Yokotani N, Hunter C, Doi K, Wenthold RJ, Shimasaki S. Differential expression of two distinct forms of mRNA encoding members of a dipeptidyl aminopeptidase family. *Proc Natl Acad Sci USA* 1992;**89**:197–201.
60. Takumi T, Ohkubo H, Nakanishi S. Cloning of a membrane protein that induces a slow voltage-gated potassium current. *Science* 1988;**242**:1042–5.
61. Llinas RR. The intrinsic electrophysiological properties of mammalian neurons: insights into central nervous system function. *Science* 1988;**242**:1654–64.
62. Baxter DA, Byrne JH. Ionic conductance mechanisms contributing to the electrophysiological properties of neurons. *Curr Opin Neurobiol* 1991;**1**:105–12.
63. Hille B. *Potassium channels and chloride channels*. Sunderland, MA: Sinauer Associates, Inc.; 2001.
64. Liss B, Franz O, Sewing S, Bruns R, Neuhoff H, Roeper J. Tuning pacemaker frequency of individual dopaminergic neurons by Kv4.3L and KChip3.1 transcription. *EMBO J* 2001;**20**:5715–24.
65. Maffie J, Rudy B. Weighing the evidence for a ternary protein complex mediating A-type K^+ currents in neurons. *J Physiol* 2008;**586**:5609–23.
66. Kaulin YA, De Santiago-Castillo JA, Rocha CA, Nadal MS, Rudy B, Covarrubias M. The dipeptidyl-peptidase-like protein DPP6 determines the unitary conductance of neuronal Kv4.2 channels. *J Neurosci* 2009;**29**:3242–51.
67. Vanoye CG, Welch RC, Tian C, Sanders CR, George Jr. AL. KCNQ1/KCNE1 assembly, co-translation not required. *Channels (Austin)* 2010;**4**:108–14.
68. Morin TJ, Kobertz WR. Counting membrane-embedded KCNE beta-subunits in functioning K^+ channel complexes. *Proc Natl Acad Sci USA* 2008;**105**:1478–82.
69. Nakajo K, Ulbrich MH, Kubo Y, Isacoff EY. Stoichiometry of the KCNQ1—KCNE1 ion channel complex. *Proc Natl Acad Sci USA* 2010;**107**:18862–7.
70. Sun X, Zaydman MA, Cui J. Regulation of voltage-activated K^+ channel gating by transmembrane beta subunits. *Front Pharmacol* 2012;**3**:1–10.
71. Lundquist AL, Manderfield LJ, Vanoye CG, Rogers CS, Donahue BS, Chang PA, et al. Expression of multiple KCNE genes in human heart may enable variable modulation of I_{Ks}. *J Mol Cell Cardiol* 2005;**38**:277–87.
72. Lee US, Cui J. BK channel activation: structural and functional insights. *Trends Neurosci* 2010;**33**:415–23.
73. Fanger CM, Ghanshani S, Logsdon NJ, Rauer H, Kalman K, Zhou J, et al. Calmodulin mediates calcium-dependent activation of the intermediate conductance K_{Ca} channel, IKCa1. *J Biol Chem* 1999;**274**:5746–54.
74. Xia XM, Fakler B, Rivard A, Wayman G, Johnson-Pais T, Keen JE, et al. Mechanism of calcium gating in small-conductance calcium-activated potassium channels. *Nature* 1998;**395**:503–7.
75. Saimi Y, Kung C. Calmodulin as an ion channel subunit. *Annu Rev Physiol* 2002;**64**:289–311.
76. Schönherr R, Lober K, Heinemann SH. Inhibition of human ether a go-go potassium channels by Ca^{2+}/calmodulin. *EMBO J* 2000;**19**:3263–71.
77. Yus-Najera E, Santana-Castro I, Villarroel A. The identification and characterization of a noncontinuous calmodulin-binding site in noninactivating voltage-dependent KCNQ potassium channels. *J Biol Chem* 2002;**277**:28545–53.
78. Wen H, Levitan IB. Calmodulin is an auxiliary subunit of KCNQ2/3 potassium channels. *J Neurosci* 2002;**22**:7991–8001.
79. Schwake M, Athanasiadu D, Beimgraben C, Blanz J, Beck C, Jentsch TJ, et al. Structural determinants of M-type KCNQ (Kv7) K^+ channel assembly. *J Neurosci* 2006;**26**:3757–66.
80. Etxeberria A, Aivar P, Rodriguez-Alfaro JA, Alaimo A, Villace P, Gomez-Posada JC, et al. Calmodulin regulates the trafficking of KCNQ2 potassium channels. *FASEB J* 2008;**22**:1135–43.

81. Kim E, Niethammer M, Rothschild A, Jan YN, Sheng M. Clustering of Shaker-type K^+ channels by interaction with a family of membrane-associated guanylate kinases. *Nature* 1995;**378**:85–8.
82. Doyle DA, Lee A, Lewis J, Kim E, Sheng M, MacKinnon R. Crystal structures of a complexed and peptide-free membrane protein-binding domain: molecular basis of peptide recognition by PDZ. *Cell* 1996;**85**:1067–76.
83. Zheng CY, Seabold GK, Horak M, Petralia RS. MAGUKs, synaptic development, and synaptic plasticity. *Neuroscientist* 2011;**17**:493–512.
84. Elias GM, Nicoll RA. Synaptic trafficking of glutamate receptors by MAGUK scaffolding proteins. *Trends Cell Biol* 2007;**17**:343–52.
85. Rasband MN, Park EW, Zhen D, Arbuckle MI, Poliak S, Peles E, et al. Clustering of neuronal potassium channels is independent of their interaction with PSD-95. *J Cell Biol* 2002;**159**:663–72.
86. Balse E, Steele DF, Abriel H, Coulombe A, Fedida D, Hatem SpN. Dynamic of ion channel expression at the plasma membrane of cardiomyocytes. *Physiol Rev* 2012;**92**:1317–58.
87. Sergeant GP, Ohya S, Reihill JA, Perrino BA, Amberg GC, Imaizumi Y, et al. Regulation of Kv4.3 currents by Ca^{2+}/calmodulin-dependent protein kinase II. *Am J Physiol Cell Physiol* 2005;**288**:C304–C313.
88. Michel JJ, Scott JD. AKAP mediated signal transduction. *Annu Rev Pharmacol Toxicol* 2002;**42**:235–57.
89. Kim J, Hoffman DA. Potassium channels: newly found players in synaptic plasticity. *Neuroscientist* 2008;**14**:276–86.
90. Hammond RS, Lin L, Sidorov MS, Wikenheiser AM, Hoffman DA. Protein kinase a mediates activity-dependent Kv4.2 channel trafficking. *J Neurosci* 2008;**28**:7513–9.
91. Coghlan VM, Perrino BA, Howard M, Langeberg LK, Hicks JB, Gallatin WM, et al. Association of protein kinase A and protein phosphatase 2B with a common anchoring protein. *Science* 1995;**267**:108–11.
92. Klauck TM, Faux MC, Labudda K, Langeberg LK, Jaken S, Scott JD. Coordination of three signaling enzymes by AKAP79, a mammalian scaffold protein. *Science* 1996;**271**: 1589–92.
93. Hoshi N, Langeberg LK, Scott JD. Distinct enzyme combinations in AKAP signalling complexes permit functional diversity. *Nat Cell Biol* 2005;**7**:1066–73.
94. Zhang J, Bal M, Bierbower S, Zaika O, Shapiro MS. AKAP79/150 signal complexes in G-protein modulation of neuronal ion channels. *J Neurosci* 2011;**31**:7199–211.
95. Hernandez CC, Falkenburger B, Shapiro MS. Affinity for phosphatidylinositol 4,5-bisphosphate determines muscarinic agonist sensitivity of Kv7 K^+ channels. *J Gen Physiol* 2009;**134**:437–48.
96. Pan Y, Weng J, Kabaleeswaran V, Li H, Cao Y, Bhosle RC, et al. Cortisone dissociates the Shaker family K^+ channels from their beta subunits. *Nat Chem Biol* 2008;**4**:708–14.
97. Bortner CD, Hughes FM, Cidlowski JA. A primary role for K^+ and Na^+ efflux in the activation of apoptosis. *J Biol Chem* 1997;**272**:32436–42.
98. Yu SP, Yeh CH, Sensi SL, Gwag BJ, Canzoniero LM, Farhangrazi ZS, et al. Mediation of neuronal apoptosis by enhancement of outward potassium current. *Science* 1997;**278**:114–7.
99. Wang X, Xiao AY, Ichinose T, Yu SP. Effects of tetraethylammonium analogs on apoptosis and membrane currents in cultured cortical neurons. *J Pharmacol Exp Ther* 2000;**295**:524–30.
100. Wei L, Yu SP, Gottron F, Snider BJ, Zipfel GJ, Choi DW. Potassium channel blockers attenuate hypoxia- and ischemia-induced neuronal death in vitro and in vivo. *Stroke* 2003;**34**:1281–6.
101. Zhou X, Wei J, Song M, Francis K, Yu SP. Novel role of KCNQ2/3 channels in regulating neuronal cell viability. *Cell Death Differ* 2011;**18**:493–505.

102. Wang L, Xu D, Dai W, Lu L. An ultraviolet-activated K^+ channel mediates apoptosis of myeloblastic leukemia cells. *J Biol Chem* 1999;**274**:3678–85.
103. Yu SP. Regulation and critical role of potassium homeostasis in apoptosis. *Prog Neurobiol* 2003;**70**:363–86.
104. Wang H, Zhang Y, Cao L, Han H, Wang J, Yang B, et al. HERG K^+ channel, a regulator of tumor cell apoptosis and proliferation. *Cancer Res* 2002;**62**:4843–8.
105. Wible BA, Wang L, Kuryshev YA, Basu A, Haldar S, Brown AM. Increased K^+ efflux and apoptosis induced by the potassium channel modulatory protein KChAP/PIAS3beta in prostate cancer cells. *J Biol Chem* 2002;**277**:17852–62.
106. Pourrier M, Schram G, Nattel S. Properties, expression and potential roles of cardiac K^+ channel accessory subunits: MinK, MiRPs, KChIP, and KChAP. *J Membr Biol* 2003;**194**:141–52.
107. Wible BA, Yang Q, Kuryshev YA, Accili EA, Brown AM. Cloning and expression of a novel K+channel regulatory protein, KChAP. *J Biol Chem* 1998;**273**: 11745–51.
108. Kuryshev YA, Wible BA, Gudz TI, Ramirez AN, Brown AM. KChAP/Kvbeta1.2 interactions and their effects on cardiac Kv channel expression. *Am J Physiol Cell Physiol* 2001;**281**:C290–C299.
109. Felipe A, Vicente R, Villalonga N, Roura-Ferrer M, Martinez-Marmol R, Sole L, et al. Potassium channels: new targets in cancer therapy. *Cancer Detect Prev* 2006;**30**:375–85.
110. Martinou JC, Youle RJ. Mitochondria in apoptosis: Bcl-2 family members and mitochondrial dynamics. *Dev Cell* 2011;**21**:92–101.
111. Burlacu A. Regulation of apoptosis by Bcl-2 family proteins. *J Cell Mol Med* 2003;**7**: 249–57.
112. Kroemer G, Reed JC. Mitochondrial control of cell death. *Nat Med* 2000;**6**:513–9.
113. Szabo I, Bock J, Jekle A, Soddemann M, Adams C, Lang F, et al. A novel potassium channel in lymphocyte mitochondria. *J Biol Chem* 2005;**280**:12790–8.
114. Szabo I, Bock J, Grassme H, Soddemann M, Wilker B, Lang F, et al. Mitochondrial potassium channel Kv1.3 mediates Bax-induced apoptosis in lymphocytes. *Proc Natl Acad Sci USA* 2008;**105**:14861–6.
115. Noma K, Kimura K, Minatohara K, Nakashima H, Nagao Y, Mizoguchi A, et al. Triple N-glycosylation in the long S5-P loop regulates the activation and trafficking of the Kv12.2 potassium channel. *J Biol Chem* 2009;**284**:33139–50.
116. Petrecca K, Atanasiu R, Akhavan A, Shrier A. N-linked glycosylation sites determine HERG channel surface membrane expression. *J Physiol* 1999;**515**:41–8.
117. Watanabe I, Wang HG, Sutachan JJ, Zhu J, Recio-Pinto E, Thornhill WB. Glycosylation affects rat Kv1.1 potassium channel gating by a combined surface potential and cooperative subunit interaction mechanism. *J Physiol* 2003;**550**:51–66.
118. Park K-S, Yang J-W, Seikel E, Trimmer JS. Potassium channel phosphorylation in excitable cells: providing dynamic functional variability to a diverse family of ion channels. *Physiology (Bethesda)* 2008;**23**:49–57.
119. Schroeder BC, Kubisch C, Stein V, Jentsch TJ. Moderate loss of function of cyclic-AMP-modulated KCNQ2/KCNQ3 K^+ channels causes epilepsy. *Nature* 1998;**396**: 687–90.
120. Ritter DM, Ho C, O'Leary ME, Covarrubias M. Modulation of Kv3.4 channel N-type inactivation by protein kinase C shapes the action potential in dorsal root ganglion neurons. *J Physiol* 2012;**590**:145–61.
121. Chung I, Schlichter LC. Native Kv1.3 channels are upregulated by protein kinase C. *J Membr Biol* 1997;**156**:73–85.
122. Misonou H, Mohapatra DP, Park EW, Leung V, Zhen D, Misonou K, et al. Regulation of ion channel localization and phosphorylation by neuronal activity. *Nat Neurosci* 2004;**7**:711–8.

123. Schrader LA, Anderson AE, Mayne A, Pfaffinger PJ, Sweatt JD. PKA modulation of Kv4.2-encoded A-type potassium channels requires formation of a supramolecular complex. *J Neurosci* 2002;**22**:10123–33.
124. Kwak Y-G, Hu N, Wei J, George AL, Grobaski TD, Tamkun MM, et al. Protein kinase A phosphorylation alters Kv1.3 subunit-mediated inactivation of the Kv1.5 potassium channel. *J Biol Chem* 1999;**274**:13928–32.
125. Gong J, Xu J, Bezanilla M, van Huizen R, Derin R, Li M. Differential stimulation of PKC phosphorylation of potassium channels by ZIP1 and ZIP2. *Science* 1999;**285**:1565–9.
126. Tsai W, Morielli AD, Cachero TG, Peralta EG. Receptor protein tyrosine phosphatase alpha participates in the m1 muscarinic acetylcholine receptor-dependent regulation of Kv1.2 channel activity. *EMBO J* 1999;**18**:109–18.
127. Felsch JS, Cachero TG, Peralta EG. Activation of protein tyrosine kinase PYK2 by the m1 muscarinic acetylcholine receptor. *Proc Natl Acad Sci USA* 1998;**95**:5051–6.
128. Huang XY, Morielli AD, Peralta EG. Tyrosine kinase-dependent suppression of a potassium channel by the G protein-coupled m1 muscarinic acetylcholine receptor. *Cell* 1993;**75**:1145–56.
129. Schulteis CT, Nagaya N, Papazian DM. Intersubunit interaction between amino- and carboxyl-terminal cysteine residues in tetrameric shaker K+channels. *Biochemistry* 1996;**35**:12133–40.
130. Jan LY, Jan YN. Voltage-gated potassium channels and the diversity of electrical signalling. *J Physiol* 2012;**590**:2591–9.
131. Nadal MS, Amarillo Y, Vega-Saenz de Miera E, Rudy B. Differential characterization of three alternative spliced isoforms of DPPX. *Brain Res* 2006;**1094**:1–12.
132. Takimoto K, Hayashi Y, Ren X, Yoshimura N. Species and tissue differences in the expression of DPPY splicing variants. *Biochem Biophys Res Commun* 2006;**348**:1094–100.
133. Jerng HH, Lauver AD, Pfaffinger PJ. DPP10 splice variants are localized in distinct neuronal populations and act to differentially regulate the inactivation properties of Kv4-based ion channels. *Mol Cell Neurosci* 2007;**35**:604–24.
134. Sun XX, Bostrom SL, Griffith LC. Alternative splicing of the eag potassium channel gene in Drosophila generates a novel signal transduction scaffolding protein. *Mol Cell Neurosci* 2009;**40**:338–43.
135. Frings S, Brull N, Dzeja C, Angele A, Hagen V, Kaupp UB, et al. Characterization of ether-à-go-go channels present in photoreceptors reveals similarity to IKx, a K^+ current in rod inner segments. *J Gen Physiol* 1998;**111**:583–99.
136. Delmas P, Brown DA. Pathways modulating neural KCNQ/M (Kv7) potassium channels. *Nat Rev Neurosci* 2005;**6**:850–62.
137. Wickenden AD, Yu W, Zou A, Jegla T, Wagoner PK. Retigabine, a novel anticonvulsant, enhances activation of KCNQ2/Q3 potassium channels. *Mol Pharmacol* 2000;**58**:591–600.
138. Gunthorpe MJ, Large CH, Sankar R. The mechanism of action of retigabine (ezogabine), a first-in-class K+channel opener for the treatment of epilepsy. *Epilepsia* 2012;**53**:412–24.
139. Wuttke TV, Seebohm G, Bail S, Maljevic S, Lerche H. The new anticonvulsant retigabine favors voltage-dependent opening of the Kv7.2 (KCNQ2) channel by binding to its activation gate. *Mol Pharmacol* 2005;**67**:1009–17.
140. Schenzer A, Friedrich T, Pusch M, Saftig P, Jentsch TJ, Grotzinger J, et al. Molecular determinants of KCNQ (K_v7) K^+ channel sensitivity to the anticonvulsant retigabine. *J Neurosci* 2005;**25**:5051–60.
141. Lange W, Geissendorfer J, Schenzer A, Grotzinger J, Seebohm G, Friedrich T, et al. Refinement of the binding site and mode of action of the anticonvulsant retigabine on KCNQ K^+ channels. *Mol Pharmacol* 2009;**75**:272–80.

142. Brueggemann LI, Mackie AR, Mani BK, Cribbs LL, Byron KL. Differential effects of selective cyclooxygenase-2 inhibitors on vascular smooth muscle ion channels may account for differences in cardiovascular risk profiles. *Mol Pharmacol* 2009;**76**:1053–61.
143. Du XN, Zhang X, Qi JL, An HL, Li JW, Wan YM, et al. Characteristics and molecular basis of celecoxib modulation on K_v7 potassium channels. *Br J Pharmacol* 2011;**164**:1722–37.
144. Peretz A, Degani N, Nachman R, Uziyel Y, Gibor G, Shabat D, et al. Meclofenamic acid and diclofenac, novel templates of KCNQ2/Q3 potassium channel openers, depress cortical neuron activity and exhibit anticonvulsant properties. *Mol Pharmacol* 2005;**67**:1053–66.
145. Brueggemann LI, Mackie AR, Martin JL, Cribbs LL, Byron KL. Diclofenac distinguishes among homomeric and heteromeric potassium channels composed of KCNQ4 and KCNQ5 subunits. *Mol Pharmacol* 2011;**79**:10–23.
146. Kerr PM, Clement-Chomienne O, Thorneloe KS, Chen TT, Ishii K, Sontag DP, et al. Heteromultimeric Kv1.2-Kv1.5 channels underlie 4-aminopyridine-sensitive delayed rectifier K^+ current of rabbit vascular myocytes. *Circ Res* 2001;**89**:1038–44.
147. Thorneloe KS, Chen TT, Kerr PM, Grier EF, Horowitz B, Cole WC, et al. Molecular composition of 4-aminopyridine-sensitive voltage-gated K^+ channels of vascular smooth muscle. *Circ Res* 2001;**89**:1030–7.
148. Duncan CE, Chetcuti AF, Schofield PR. Coregulation of genes in the mouse brain following treatment with clozapine, haloperidol, or olanzapine implicates altered potassium channel subunit expression in the mechanism of antipsychotic drug action. *Psychiatr Genet* 2008;**18**:226–39.
149. Kuryshev YA, Ficker E, Wang L, Hawryluk P, Dennis AT, Wible BA, et al. Pentamidine-induced long QT syndrome and block of hERG trafficking. *J Pharmacol Exp Ther* 2005; **312**:316–23.
150. Cordes JS, Sun Z, Lloyd DB, Bradley JA, Opsahl AC, Tengowski MW, et al. Pentamidine reduces hERG expression to prolong the QT interval. *Br J Pharmacol* 2005;**145**:15–23.
151. Wharton JM, Demopulos PA, Goldschlager N. Torsade de pointes during administration of pentamidine isethionate. *Am J Med* 1987;**83**:571–6.
152. Cortese LM, Gasser Jr. RA, Bjornson DC, Dacey MJ, Oster CN. Prolonged recurrence of pentamidine-induced torsades de pointes. *Ann Pharmacother* 1992;**26**:1365–9.

CHAPTER ELEVEN

Oligomerization of the Mitochondrial Protein VDAC1: From Structure to Function and Cancer Therapy

Varda Shoshan-Barmatz[*,†], **Dario Mizrachi**[*,†], **Nurit Keinan**[*,†]
[*]Department of Life Sciences, Ben-Gurion University of the Negev, Beer-Sheva, Israel
[†]The National Institute for Biotechnology in the Negev, Ben-Gurion University of the Negev, Beer-Sheva, Israel

Contents

Abstract

The voltage-dependent anion channel (VDAC1), lying in the mitochondrial outer membrane (OMM), mediates the transport of ions and metabolites, thus controlling the cross talk between mitochondria and the rest of the cell. VDAC1 has also been recognized as a key protein in mitochondria-mediated apoptosis, is the proposed target

Progress in Molecular Biology and Translational Science, Volume 117
ISSN 1877-1173
http://dx.doi.org/10.1016/B978-0-12-386931-9.00011-8

for the pro- and antiapoptotic Bcl-2-family of proteins and is involved in apoptotic protein release from the mitochondria. Questions, however, remain as to if and how VDAC1 mediates the transfer of apoptotic proteins across the OMM. Our recent studies suggest that upon apoptosis induction, VDAC1 oligomerizes to form a new large pore allowing the passage of a folded protein, like cytochrome *c*. This review provides insight into the central role of VDAC1 in mammalian cell life and death and emphasizes VDAC1 function in apoptosis, focusing on VDAC1 oligomerization as a key step in mitochondria-mediated apoptosis and key structural features of VDAC1 that mediate its apoptotic function.

ABBREVIATIONS

Cyto *c* cytochrome *c*
EGS ethylene glycol bis-(succinimidyl succinate)
IMM inner mitochondrial membrane
OMM outer mitochondrial membrane
PLB planar lipid bilayer
PTP permeability transition pore
RuR ruthenium red
VDAC voltage-dependent anion channel

1. OVERVIEW

Mitochondria, organelles long thought to serve solely as cellular energy generators, are now known to assume functions well beyond this critical bioenergetic role. Mitochondria play a crucial part in the modulation of the cellular redox status, osmotic regulation, pH control, calcium homeostasis, and signaling events. In addition, mitochondria mediate the responses of cells to multiple physiological and genetic stresses, interorganelle communication, and cell proliferation. Mitochondria also serve as a center for the regulation of programmed cell death or apoptosis.

To fulfill the various activities attributed to this organelle, mitochondria contain about 1000 different proteins with tissue-specific features. One of these proteins is the voltage-dependent anion channel, VDAC, located in the outer mitochondrial membrane (OMM), the barrier between the mitochondrial intermembrane space (IMS) and the cytoplasm. VDAC1 plays an important role as a controlled passage for adenine nucleotides, Ca^{2+}, and other metabolites into and out of mitochondria. As such, it has been recognized that VDAC plays a crucial role in regulating the metabolic and energetic functions of mitochondria (Fig. 11.1).

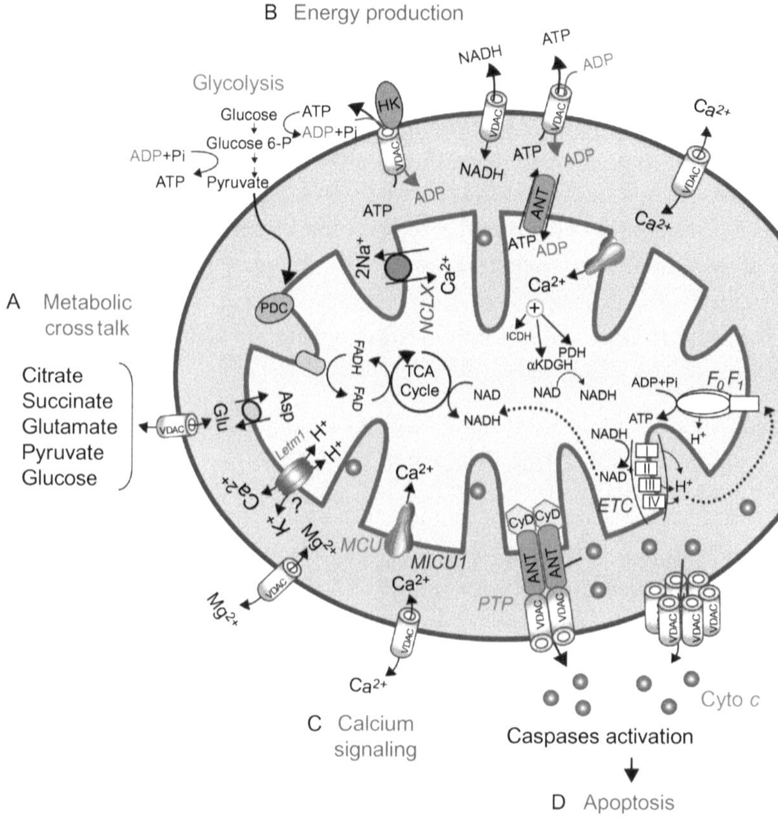

Figure 11.1 Schematic representation of VDAC1 as a multifunctional channel and convergence point for a variety of cell survival and cell death signals. (A) VDAC1 plays a central role controlling cross talk between mitochondria and the rest of the cell. VDAC1 also regulates cellular energy homeostasis (B) by transporting ATP/ADP and NADH between the intermembrane space and the cytosol and by binding HK. (C) VDAC1 is responsible for Ca^{2+} transport across the OMM. Also presented are the Ca^{2+} influx and efflux transport systems in the inner mitochondrial membranes. The uptake of Ca^{2+} into the matrix is mediated by a Ca^{2+}-selective transporter, the mitochondrial Ca^{2+} uniporter (MCU), regulated by a calcium-sensing accessory subunit (MICU1). Ca^{2+} efflux is mediated by NCLX, a Na^{+}/Ca^{2+} exchanger. High levels of matrix Ca^{2+} accumulation trigger the opening of the PTP, a fast Ca^{2+} release channel. Molecular fluxes are indicated by arrows. Ca^{2+}-mediated regulation of the tricarboxylic acid (TCA) cycle is also presented. The activation of pyruvate dehydrogenase (PDH), isocitrate dehydrogenase (ICDH), and α-ketoglutarate dehydrogenase (α KGDH) by intramitochondrial Ca^{2+}, leading to enhanced activity of the TCA cycle, is shown. The electron transport chain (ETC) and the ATP synthase (F_oF_1) are also presented. (D) In the apoptotic process, VDAC1 binds the apoptosis regulatory proteins, that is, Bcl-2 family members and HK, and is active in the release of Cyto *c* into the cytosol. The proposed PTP and VDAC1 oligomer as Cyto *c* release channels are presented. (For color version of this figure, the reader is referred to the online version of this chapter.)

It is also well accepted that mitochondria play an important role in the regulation of apoptosis via release of proapoptotic agents and/or disruption of cellular energy metabolism.[1–3] Given the importance of mitochondria as integrators and amplifiers of the death program, intense efforts have been directed at exploring the mechanisms of mitochondria-mediated apoptosis. Such efforts include studies on the release of proapoptotic effectors, the participation of Bcl-2 (B-cell leukemia/lymphoma 2) family members in apoptosis, as well as the involvement of reactive oxygen species (ROS) and Ca^{2+} in this event.[4–6] Still, one of the major open questions concerning apoptosis is related to how mitochondrial proapoptoic proteins are released and how this process is regulated.

2. MITOCHONDRIA AND APOPTOSIS

Apoptosis is a cellular process mediated by a large and diverse range of extracellular and/or intracellular signals that allow for the controlled and regulated death of cells. Apoptosis is a genetically regulated process important for tissue homeostasis and development in multicellular metazoans. Defects in the regulation of apoptosis are often associated with disease, such as neuronal degenerative diseases, tumorigenesis, autoimmune disorders, and viral infections.[7–9] Apoptosis can be induced in cells via a number of mechanisms that are either extracellular or intracellular in origin.

In general terms, apoptotic pathways can be subdivided into the extrinsic and the intrinsic pathways. In extrinsic pathways, apoptotic signals are initiated by ligand engagement of cell surface receptors, such as Fas and TNF receptors, as well as by signaling by hormones, growth factors, nitric oxide, or cytokines.[10,11] The intrinsic apoptotic pathway, on the other hand, requires the involvement of mitochondria. During transduction of an apoptotic signal into the cell, an alteration in mitochondrial membrane permeability occurs, causing the release of apoptogenic factors, such as cytochrome *c* (Cyto *c*), apoptosis-inducing factor (AIF), Smac/DIABLO, and endonuclease G from the IMS into the cytosol.[11–14] These proteins participate in complex processes resulting in the activation of proteases and nucleases, leading to protein and DNA degradation, and finally, cell death. Most notable among the released proteins is Cyto *c* that initiates apoptosis by binding to a central apoptotic regulator, Apaf-1, thereby promoting oligomerization of Apaf-1 and activation of caspase 9. Caspase 9 subsequently activates effector caspases, such as caspases 3, 6, and 7, encouraging execution of programmed cell death. How Cyto *c* and other apoptogenic factors are released from mitochondria is not clear, as will be discussed below.

A proapoptotic signal can be generated either inside the mitochondria, for example, by increased Ca^{2+} and/or ROS concentrations, or from outside by proapoptotic proteins, such as tBid and Bax. In the latter case, mitochondrial apoptosis induction is used to amplify the extramitochondrial proapoptotic message coming from death receptors, such as FAS, or increased p53 levels.

The pathways activated by extrinsic and intrinsic signals can overlap to some extent. Receptor binding by an extrinsic signal typically leads to recruitment of adapter proteins that promote caspase oligomerization and autoprocessing. The sensitivity of cells to any of the apoptosis stimuli can vary, depending on a number of factors, such as the expression level of pro- and antiapoptotic proteins (e.g., Bcl-2 proteins or inhibitors of apoptosis proteins), the severity of the stimulus, and the stage of the cell cycle.

3. BAX-, BAK-MEDIATED APOPTOSIS INVOLVES THEIR OLIGOMERIZATION

Oligomerization as a mechanism for the mediation of Cyto *c* release and apoptosis has been previously proposed for two key proteins in the mitochondrial pathway of apoptosis, namely, Bax and Bak (Fig. 11.2). In healthy cells, Bax resides in the cytosol in an inactive state. Upon apoptotic signaling, Bax translocates to the mitochondria. Insertion of Bax into the OMM is accompanied by conformational changes and oligomerization, resulting in mitochondrial OMM permeabilization and release of IMS proteins, such as Cyto *c*[15–17] (Fig. 11.2). It has been shown that TNF-α activation of caspase 8 resulted in the cleavage of Bid to form truncated tBid, which, in turn, interacts with Bax and Bak to generate Bax and Bak oligomers, forming complexes as large as 500 kDa.[18] At the OMM, tBid stimulates the conformational transition of Bax to its active monomeric membrane-integrated form through the insertion of the transmembrane segment and integration of helices 5 and 6 into the bilayer, with activated Bax undergoing autooligomerization to yield an OMM pore.[19,20] It is accepted that Bak follows a similar process,[21] where, following Bak activation, dimers are formed that then multimerize into larger oligomers.[22]

Another suggested mechanism for the activation of Bax/Bak by BH3-only molecules (BH3s) is based on the finding that Bax undergoes stepwise structural reorganization, leading to mitochondrial targeting and homooligomerization. Given this reorganization, it is thought that the α1 helix of Bax keeps the α9 helix engaged in the dimerization pocket,

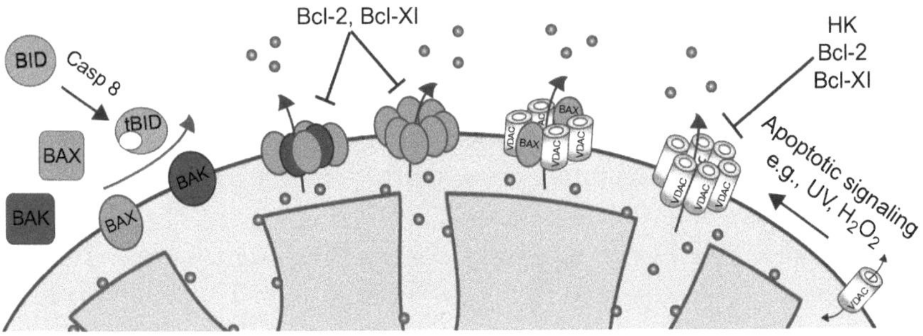

Figure 11.2 Bax- and Bak-mediated Cyto *c* release. The release of Cyto *c* from the mitochondrial intermembrane space is achieved by the activation of pore-forming proteins. Bax is located in the cytosol before translocation to the OMM. Bax and Bak are activated by the product of caspase 8-mediated cleavage of Bid. Upon activation, Bax and Bak begin the pore-formation process as homo- or hetero-oligomers. Bax and Bak translocation to the OMM and subsequent oligomerization are both marked by conformational activation. VDAC1 can be activated by apoptotic signaling to begin the pore-forming process. At least six VDAC1 monomers are required to form the new pore with a calculated diameter of 4 nm, considering an external monomer diameter of 4 nm. In the event that both Bax and VDAC1 are activated, a hetero-oligomer composed of Bax/VDAC1-formed pores appears. (For color version of this figure, the reader is referred to the online version of this chapter.)

rendering Bax as a monomer in the cytosol. The activator BH3s, tBid/Bim/PUMA, attack and expose the α1 helix of Bax, resulting in secondary disengagement of the α9 helix and mitochondrial insertion. Activator BH3s remain associated with the N-terminally exposed Bax through the BH1 domain to drive homooligomerization. Bak, as an integral mitochondrial membrane protein, can bypass the first activation step, explaining why its killing kinetics are faster than those of Bax.[23] These suggested pathways of Bax/Bak activation gained support with the finding that genetic deletion of *Bid*, *Bim*, and *Puma* prevented homooligomerization of Bax and Bak, and thus Cyto *c*-mediated activation of caspases, in response to diverse death signals.[24]

4. VDAC1 AS A GATEKEEPER OF MITOCHONDRIAL FUNCTION

4.1. VDAC isoforms, structure, and channel activity

Found at the OMM, VDAC1 mediates the complex interactions between mitochondria and other parts of the cell by transporting anions, cations, ATP, Ca^{2+}, and metabolites (Fig. 11.1). VDAC1 thus plays an important role in coordinating communication between mitochondrion and cytosol. An essential aspect of this management is the transient formation of complexes of VDAC1 with other proteins. Over the past few years, several hypotheses and potential mechanisms have been forwarded for and against a role for VDAC1 in OMM permeability and the subsequent release of apoptosis-promoting proteins (Fig. 11.1). In this section, the proposed structure, channel properties and function, and cellular expression of VDAC will be discussed.

4.1.1 VDAC isoforms and functions

Three different VDAC genes, encoding distinctly expressed isoforms, have been reported in humans, rat, and mice.[25–28] VDAC isoforms appear to serve specialized functions, since the more complex the organism, the more VDAC isoforms it possesses. For instance, *Neurospora crassa* encodes only one form of VDAC, *Saccharomyces cerevisiae* encodes two, while, as noted above, mice and humans each encode three isoforms.[28,29] The three genes that encode the distinct isoforms of mouse VDAC, *VDAC1–3*, have been characterized.[28] Each isoform is 65–70% identical to the others.[30] Phylogenetic analysis indicates that VDAC3 diverged from the primordial VDAC first, with VDAC1 and VDAC2 arising more recently.[31] Mammalian VDAC isoforms are present in a wide variety of tissues and are expressed at high levels in heart, kidney, brain, and skeletal muscle. VDAC2 and VDAC3 were localized to the outer dense fibers of the bovine sperm flagellum,[32] while the localization of VDAC2 to the acrosomal region of bovine spermatozoa was recently demonstrated.[33]

Both VDAC1 and VDAC2 are ubiquitously expressed, with VDAC3 showing a more restricted organ distribution.[28,34,35] In tissues that have been studied, VDAC1 represents the predominant form expressed.[36] However, the tissue-specific distribution and intracellular localization of the different VDAC isoforms are not well established.

The VDAC isoforms also appear to assume different physiological roles.[37,38] In mice, deletion of *VDAC1* and *VDAC2* reduces respiratory capacity,[39] the absence of VDAC3 causes male sterility, while a lack of both VDAC1 and VDAC3 causes growth retardation[40] and is associated with deficits in learning behavior and synaptic plasticity.[41] Embryonic fibroblasts deficient in VDAC2 but not cells lacking the more abundant VDAC1 exhibit enhanced Bak oligomerization and are more susceptible to apoptotic death induced by staurosporine (STS) and etoposide,[37] as supported by results suggesting VDAC2 to be a critical inhibitor of Bak-mediated apoptotic responses.[42]

In this review, we focus on VDAC1, as it is the abundant isoform in most cells and the most studied isoform.[43,44]

4.1.2 VDAC1 membrane topology and molecular structure

4.1.2.1 VDAC1 structure and the N-terminal domain function

In 2008, three independent technical approaches unraveled the three-dimensional structure of VDAC1 at atomic resolution.[45–47] The three structures were found to be almost identical, featuring 19 α-strands connected by flexible loops to form an β-barrel, along with a 25-residue-long N-terminal region[45–47] (see Fig. 11.3). In all three solved VDAC1 structures, a helical conformation is present within the N-terminal region, attached to the channel wall but not part of it.[45–47]

Several functional implications of this architecture were suggested. It was proposed that the N-terminal region lies inside the pore yet could move in the open space.[48] The mobility of the N-terminal region was further supported by studies showing that the N-terminal α-helix exhibits motion during voltage gating,[49–52] and that anti-VDAC1 antibodies raised against the N-terminal region of the protein interact with membranal VDAC1.[53–55]

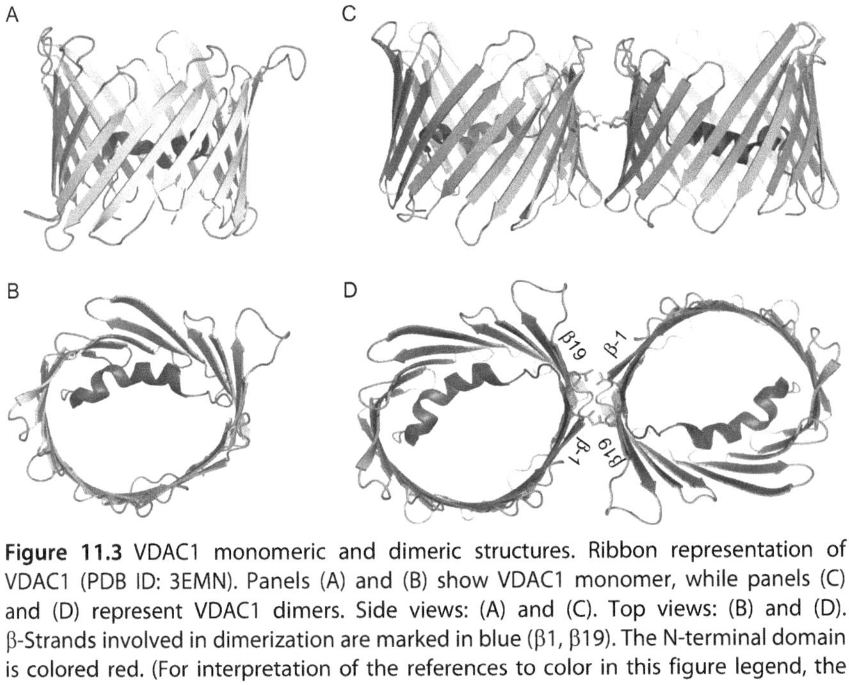

Figure 11.3 VDAC1 monomeric and dimeric structures. Ribbon representation of VDAC1 (PDB ID: 3EMN). Panels (A) and (B) show VDAC1 monomer, while panels (C) and (D) represent VDAC1 dimers. Side views: (A) and (C). Top views: (B) and (D). β-Strands involved in dimerization are marked in blue (β1, β19). The N-terminal domain is colored red. (For interpretation of the references to color in this figure legend, the reader is referred to the online version of this chapter.)

It is also shown that the N-terminal α-helix interacts with apoptosis-regulating proteins of the Bcl-2 family (i.e., Bax, Bcl-2, and Bcl-xL)[56–59] and hexokinase (HK),[56,60] and that movement of this VDAC1 segment out of the pore modulates the antiapoptotic activities of these binding partners. Moreover, cells expressing N-terminal domain-truncated VDAC1 are resistant to apoptosis.[56] These findings indicate that the N-terminal domain is required for apoptosis induction; interactions with HK-I, Bcl-xL, and Bcl-2; and protection against apoptosis.[56]

Recently,[49] we demonstrated that the VDAC1 N-terminal region lies inside the pore yet can be exposed outside the pore where it interacts with the N-terminal region of a second molecule to form a dimer. Moreover, specific mutations disrupting the α-helix structure or the flexible glycine-rich sequence of this domain affected N-terminal region translocation from the internal pore to the channel surface. The results suggest that N-terminal region mobility is involved in channel gating and interaction with anti-apoptotic proteins.[49]

4.1.2.2 VDAC1 oligomeric structure

Substantial evidence for the formation of higher-order VDAC1-containing complexes has been presented.[61–68] It has been demonstrated that VDAC1 can exist in different oligomeric states, namely, as monomers, dimers, trimers, tetramers, hexamers, and higher-order oligomers.

On the basis of electron microscopy-based analysis of the 2D crystallized OMM, it was noted that VDAC1 pores are organized into ordered arrays, each bearing six monomers with interchannel contacts.[69] Chemical cross-linking experiments of purified VDAC1 and membrane-embedded VDAC1 demonstrated the assembly of dimers, trimers, tetramers, and higher VDAC1 oligomers.[62] The supramolecular organization of VDAC has also been demonstrated using atomic force microscopy to study the native OMM.[64,65] In addition, the application of symmetry operators on the NMR-based structure of recombinant hVDAC1[45] indicated that it may form a parallel dimer, although analysis of the crystal packing of mouse VDAC1 revealed an anti-parallel dimer, which further assembles as hexamers.[70] Recently, it was suggested that the specific lipid composition of the OMM, that changes upon apoptogenic stimuli, significantly enhances VDAC1 oligomerization.[71]

The function of VDAC1 oligomers is not known. It was proposed that an organization of VDAC1 beyond the monomeric or dimeric forms may contribute to stabilizing the protein.[70] On the other hand, it was proposed that the oligomeric assembly of VDAC1 offers a platform for other proteins to oligomerize, such as HK.[62] HK-I assumes a tetrameric structure that is greatly enhanced when the enzyme is bound to mitochondria[72] or when it interacts with the mitochondria to inhibit permeability transition pore (PTP) opening.[55] Creatine kinase, when bound to VDAC1 at the IMS, forms high-order oligomers,[73,74] interacting with VDAC1 exclusively in the octameric state, with the dimeric state only showing weak affinity for VDAC1.[75]

VDAC1 oligomerization has been proposed to also play important physiological roles in the regulation of VDAC1 function, including apoptosis (see Section 6.1), and in mediating the binding of apoptosis-regulating proteins.[56,62,66]

4.1.3 Channel activity of VDAC1

The functional properties of VDAC1 have been examined following reconstitution into a planar lipid bilayer (PLB). The activity of the channel is reflected in the flow of ions (i.e., current) through a membrane that is otherwise a barrier to ion flow (Fig. 11.4). Bilayer-reconstituted VDAC1 allows voltage-dependent conductance and displays ion selectivity. The single-channel conductance of mitochondrial VDAC1 in 1 M KCl or NaCl ranges between 4 and 5 nS.[77–81] At low voltages (10 mV), the VDAC1 channel is stable in a long-lived open state (up to 2 h). At high positive or negative potentials (>40 mV), VDAC1 presents multiple substates with different

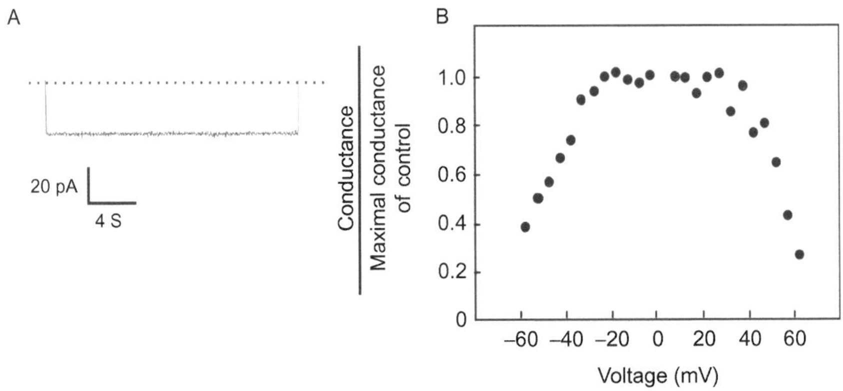

Figure 11.4 VDAC1 channel conductance. Rat liver purified VDAC1 was reconstituted into a PLB and recording of channel conductance was carried out as described previously.[76] (A) Currents through VDAC1 were obtained in response to voltage step from 0 to −10 mV. The broken line indicates the zero current level. (B) VDAC1 conductance was measured as a function of voltage, from 60 to −60 mV. The average steady-state conductance at a given voltage was normalized to the conductance at 10 mV.

ionic selectivities and permeabilities.[79,82] The bell-shaped VDAC1activity profile (Fig. 11.4B) led to characterization of the so-called closed state of the channel. In this low-conducting state, VDAC1 is more permeable to cations. In the fully open state, VDAC1 selectively conducts small ions (e.g., Cl^-, K^+, Na^+), yet also anions, such as glutamate[79] and ATP,[83] and large cations, such as acetylcholine, dopamine,[79] and Tris.[84]

The N-terminal α-helical segment of the channel has been proposed to act as the voltage sensor, gating the pore via its conformational changes or movements.[45,47–50,56]

5. VDAC1 AND APOPTOSIS: STRUCTURE–FUNCTION

5.1. Proposed mechanisms of VDAC1-mediated apoptosis

Several reviews have focused on the role of VDAC1 as an essential player in apoptosis, offering several hypotheses regarding the mechanism of mitochondria-mediated apoptosis implicating VDAC1.[38,85–93] However, none of the current models of mitochondrial membrane permeabilization can account for all of the experimental data. One model proposes that VDAC1 is a component of the PTP that is activated by proapoptotic stimuli.[12,90,91,94–96] Another model proposes that Bax interacts with VDAC1, resulting in Cyto *c* permeation through the OMM.[97,98] A third model proposes that closure of the VDAC1 channel prevents the efficient exchange of ATP and ADP between the cytosol and mitochondrial matrix, followed by swelling and membrane rupture.[99,100] A more recent model proposes VDAC1 oligomerization as mediating the release of proapoptotic proteins.[57,62,66,91] Indeed, the involvement of VDAC1 in the mechanism of Cyto *c* release from the mitochondrial IMS into the cytosol and the regulation of this process has encouraged research at various levels. While numerous studies have shown VDAC1 as a component of the apoptosis machinery,[38,85–93] other studies cast doubt on the contribution of VDAC1 to mitochondrial membrane permeability transition changes leading to cell death.[101,102] In addition, two views as to the relationship between the conducting state of the VDAC1 channel and cell death appear in the literature. While one view suggests that closure, rather than opening of VDAC1, leads to mitochondria outer membrane permeabilization and apoptosis,[99] another model supports a direct relationship between VDAC1-mediated channel opening and release of proapoptotic proteins.[38,85–93,97] These differing views may result from the involvement of two VDAC1-conducting channels, namely, an intrinsic pore (Section 4.1.3) and an intra-VDAC1 subunit channel (Section 6.1). In the following, findings related to VDAC1 involvement in apoptotic cell death are discussed.

6. VDAC1 AND Cyto *c* RELEASE

There is substantial evidence connecting VDAC1 to apoptosis and suggesting that VDAC1 is a critical player in the release of apoptogenic proteins from mitochondria in mammalian cells.[38,85–93] Cyto *c* release, cell death, and Bax–VDAC1 interaction are all inhibited by anti-VDAC1 antibodies.[103–105] The inhibition of Cyto *c* release and cell death by HK were both shown to specifically rely on interaction with native, but not mutated, VDAC1.[91,106] Similarly, ruthenium red (RuR) interacts with native, but not mutated, VDAC1 to prevent Cyto *c* release and cell death.[106–109] In addition, siRNA-mediated down-expression of VDAC1 prevented cell death induced by cisplatin and strongly reduces cisplatin-induced release of Cyto *c* and AIF and maturation of caspases-3,[110] while depletion of VDAC1 strongly suppresses the cisplatin-induced activation of Bax.[110] Finally, it is also been shown that reducing VDAC1 expression by siRNA attenuates endostatin-induced apoptosis.[111] A proposed mechanism for VDAC1 structure mediating Cyto *c* release is presented here.

6.1. VDAC1 oligomerization and the release of Cyto *c*

When considering models of VDAC1-mediated protein release, one should note the molecular sizes of the released proteins (12–100 kDa). The diameter of the VDAC1 pore at its high conductance state is about 2.6–3.0 nm, sufficient to move nucleotides and small molecules but insufficient to pass a folded protein like Cyto *c*. Yet, release of Cyto *c* via purified VDAC1 reconstituted into liposomes in which Cyto *c* was encapsulated has been demonstrated.[62,103] Therefore, the formation of a VDAC1-composed large channel serving as the Cyto *c* release channel has been proposed.[61,62,66]

Various studies using purified rat liver,[62] brain mitochondria,[55] or recombinant human VDAC1[57] have reported that both soluble purified and membrane-embedded VDAC1 can assemble into dimers, trimers, and tetramers in a dynamic process. In addition, a supramolecular organization of VDAC1 was also demonstrated.[64,65] VDAC1 oligomerization in VDAC1-reconstituted liposomes has been shown by applied FRET technology.[62] Here, purified VDAC1 was labeled with fluorescein isothiocyanate (FITC) or eosin-5-isothiovyanate (EITC) as donor and acceptor fluorophores, respectively, with enhancement in FRET signals reflecting association between FITC–VDAC1 and EITC–VDAC1 upon the formation of dimers or higher oligomeric combinations. Moreover, FRET intensity was decreased upon addition of an excess of unlabeled VDAC1 during the dissociation/reassociation process, suggesting VDAC1 to exist in a dynamic equilibrium between monomeric and higher-order oligomers. Thus, it was proposed that dynamic VDAC1 oligomerization mediates the formation of a large flexible pore between individual subunits of VDAC1, serving as a channel that allows Cyto *c* to cross the OMM.[56,61,62,91]

This suggestion has gained experimental support by the demonstration that apoptosis induction in cultured cells leads to VDAC1 oligomerization.[66] Apoptosis induction by various inducers was accompanied by an up to 20-fold increase in VDAC1 oligomerization, as revealed by chemical cross-linking, reflecting a shift in VDAC1 organization toward the oligomeric form[66] (see Fig. 11.5A and B). Apoptosis-mediated enhancement of VDAC1 oligomerization was obtained regardless of the cell type or apoptosis inducer used, including STS, curcumin, As_2O_3, etoposide, cisplatin, selenite, TNF-α, H_2O_2, or UV light, all affecting mitochondria yet acting via different mechanisms (Table 11.1). Moreover, correlation between the level of VDAC1 oligomerization and apoptosis, as a function of the concentration and time of exposure to the apoptosis stimuli, was observed. Conversely, the apoptosis inhibitor, 4,4′-diisothiocyanostilbene-2,2′-disulfonic acid (DIDS), prevented STS-induced VDAC1 oligomerization and apoptosis.[61,66,112]

A VDAC1 oligomeric state was also directly monitored in living cells using BRET2 (bioluminescence resonance energy transfer) technology, where an enhanced BRET signal (three- to five-fold) was observed upon apoptosis induction[66] (Fig. 11.5C and D). In BRET2 technology, VDAC1 proteins are tagged with either *Renilla* luciferase (RLuc) as the donor or a variant of GFP protein (GFP2) as acceptor. Energy transfer occurs when the donor and acceptor are spatially close (<10 nm), making the technique ideal for monitoring protein–protein interactions in biological systems. Enhancement of the BRET2 signal corresponds to the activation of VDAC1 oligomerization, while attenuation of apoptosis-enhanced BRET2 signal indicates inhibition of VDAC1 oligomerization. Indeed, BRET2 technology is applicable for use with the well-defined apoptosis inducers, STS, As_2O_3, and sodium selenite. All these apoptosis inducers enhanced the BRET2 signal (GFP2/Rluc ratio) in rVDAC1–Rluc- and rVDAC1–GFP2-expressing cells, relative to cells not exposed to an apoptosis inducer (Fig. 11.5C).

In addition, it was demonstrated that cells expressing a VDAC1 dimeric fusion protein comprising wild-type and RuR-insensitive E72Q-mutated

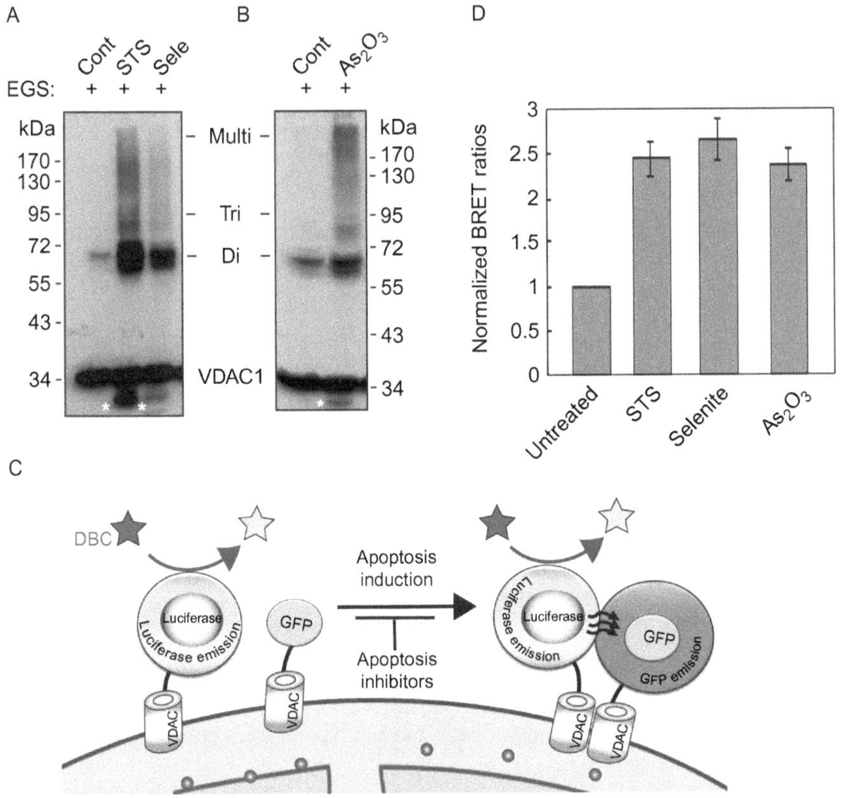

Figure 11.5 Apoptosis induction by STS, selenite, or As_2O_3 is associated with VDAC1 oligomerization. (A) and (B) Apoptosis induction triggers VDAC1 oligomerization. HeLa cells were exposed to the apoptosis-inducing reagent, STS (4 μM, 4 h), selenite (25 μM, 4 h), or As_2O_3 (30 μM, 18 h), washed twice with PBS and incubated (2.5 mg/ml) at 30 °C with 250 μM EGS for 15 min. The cells were then subjected to SDS-PAGE, followed by immunoblotting using anti-VDAC1 antibodies. VDAC1 monomers, dimers, trimers, tetramers, and multimers are indicated. The positions of molecular weight protein standards are provided. (C) Schematic representation of oligomerization detection by BRET2. Energy transfer between VDAC1–luciferase (Luc) (light-producing enzyme) as the donor and VDAC1–GFP2 (fluorophore) as the acceptor that occurs only when the donor and the acceptor are spatially close. The BRET signal is obtained when the two VDAC1 molecules interact physically. (D) The ratios of the BRET2 signals obtained in cells treated with STS (4 μM, 4 h), selenite (25 μM, 4 h), As_2O_3 (30 μM, 18 h), or with the relevant amount of DMSO are given. The shRNA-VDAC1 T-REx-293 cells were cotransfected with VDAC1–Rluc (0.2 μg) and rVDAC1–GFP2-encoding (0.8 μg) plasmids. As control, cells were transfected with the VDAC1–Rluc-encoding plasmid. Luciferase and GFP signals were measured following 48 h of transfection. BRET2 signals were calculated as described.[66] Luciferase activity was measured, following the addition of the substrate, DBC, as luminescence, while GFP fluorescence was measured at 510 nm. All readings were performed with an Infinite 200 ELISA reader. (For color version of this figure, the reader is referred to the online version of this chapter.)

Table 11.1 Summary of conditions and compounds inducing apoptosis and VDAC1 oligomerization

Agents/compounds inducing apoptosis and VDAC1 oligomerization	Conditions for induction and analysis	Refs.
UV irradiation	HEK-293 cells; UV exposure (15–60 s); EGS cross-linking and immunoblot (after 24 h)	66
TNF-α	HEK-293 cells;10–20 ng/ml TNF-α for 16–24 h; EGS cross-linking and immunoblot	66
Etoposide	HEK-293 cells; 2–5 μM etoposide for 16–24 h; EGS cross-linking and immunoblot	66
Staurosporin (STS)	HEK-293; HeLa, T47D, or A549 cells, 0.5–5 μM STS for 2.5–5 h or 0.1–0.2 μM STS for 16–24 h; EGS, DFDNB, or BMOE cross-linking and immunoblot or BRET	56,66,112–116
Cisplatin	HEK-293 cells; 40–60 μM cisplatin for 16–24 h; EGS cross-linking and immunoblot or BRET	66
Selenite	HEK-293 or HeLa cells; 8–10 μM selenite for 16–24 h; EGS cross-linking and immunoblot or BRET	66,112,115
H_2O_2	HEK-293 or HeLa cells; 0.5–1 mM H_2O_2 for 2–5 h; EGS cross-linking and immunoblot	66,112
Curcumin	HEK-293 cells; 40–60 μM curcumin for 16–24 h; EGS cross-linking and immunoblot	66
As_2O_3	HEK-293, HeLa, A549, or IM-9 cells; 2–30 μM As_2O_3 for 16–24 h; EGS or DSS cross-linking and immunoblot or BRET	66,105,115
VDAC1 overexpression	HEK-293 cells; 24–72 h transfection; EGS or BMOE cross-linking and immunoblot	56,66,116
Lipids, phosphatidyl glycerol	VDAC1 reconstituted into giant unilamellar vesicles, fluorescence cross-correlation spectroscopy	70,71

VDAC1 showed no protection by RuR against STS-induced apoptosis. This dominant-negative VDAC1 mutant reveals oligomeric VDAC1 to be the active unit in mitochondria-mediated apoptosis.[113]

Recently, using structural- and computational-based approaches, in combination with site-directed mutagenesis, cysteine replacement and chemical cross-linking, contact sites between VDAC1 molecules in dimers and higher oligomers were identified.[114] Dimeric VDAC1 with a contact site involving β-strands 1, 2, and 19 was identified. Moreover, the results suggest that VDAC1 exists as a dimer that undergoes conformational changes upon apoptosis induction to assemble into higher oligomeric states with contact sites also involving β-strands 8 and 16[114] (Fig. 11.6A).

The relationship between VDAC1 oligomerization and apoptosis is also reflected in studies attributing the apoptosis-inducing effect of As_2O_3 to an induction of VDAC1 homodimerization.[105] This effect was prevented by overexpression of the antiapoptotic protein, Bcl-xL.[105] Recently, hepatitis E virus ORF3 protein was found to upregulate VDAC1 expression, with ORF3-expressing cells showing higher levels of oligomeric VDAC1, as revealed by chemical cross-linking.[117] It was also demonstrated that VDAC1 oligomerization is encouraged in the presence of Cyto *c* encapsulated in VDAC1-reconstituted liposomes.[62]

6.2. VDAC1 N-terminal and the release of Cyto *c*

Previous studies suggest that the N-terminal region is not required for VDAC1 oligomerization but is important for Cyto *c* release and subsequent apoptosis.[56] The VDAC1 N-terminal region assumes a helical conformation (see Section 4.1.2.1). In the combined NMR/X-ray structure of human VDAC1, the helix comprises residues Tyr7 to Val17 and is folded horizontally inside the barrel wall (Fig. 11.6Ba).[45] Similar positioning is seen in the crystal structure of mouse VDAC1; however, here the helical region is formed by amino acids 6–20 and the hydrogen-bonding pattern is broken at Leu-10 and Gly-11, separating the helix into two segments.[47] On the other hand, residues 11–20 are difficult to observe in solution-state NMR, suggesting the dynamic behavior of this segment[45–47] (Fig. 11.6Bb). Increased mobility of the N-terminal region is provided by a glycine-rich domain,[49] connecting the N-terminal domain to β-strand 1. As proposed above, VDAC1 undergoes oligomerization upon apoptosis induction to form a large pore. The newly formed pore composed of VDAC1 β-barrels would form a hydrophobic channel that would not allow charged proteins, such as Cyto *c*, to pass. However, if the amphipathic N-terminal region of

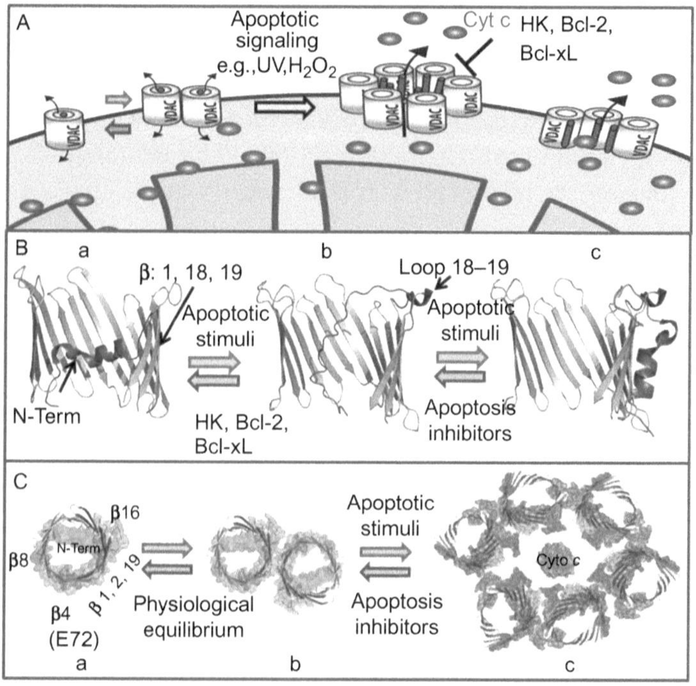

Figure 11.6 Proposed model of VDAC1 oligomerization. (A) A general overview of the VDAC1 oligomerization to form Cyto *c*-conducting channel as induced by apoptotic signaling. VDAC1 exists in equilibrium between the monomeric and dimeric states and is engaged in metabolite and ion exchange functions. Upon apoptosis induction, VDAC1 undergoes oligomerization concomitant with the translocation of the N-terminal domain (red rod) from the pore interior to the inside of a newly formed oligomeric VDAC1 pore (outside the channel). Because of its amphipathic characteristic, the N-terminal domain coats the hydrophobic oligomer pore with a hydrophilic surface, thus allowing Cyto *c* and other apoptogenic proteins to cross the OMM. HK and the anti-apoptotic proteins, Bcl-2 and Bcl-xL, inhibit this process. (B) VDAC1 N-terminal domain structure and location. (a) The crystal structure of VDAC1 (PDB ID: 3EMN) suggests a well-ordered N-terminal domain, interacting through its hydrophobic residues with the interior wall of the channel.[47] (b) The NMR-based structure of human VDAC1 (PDB ID: 2K4T)[46] reveals the unstructured conformation of the N-terminal that may exit the channel to interact with other proteins or to seek stability after apoptotic signaling induces changes in the channel. (c) Here, we propose that the N-terminal region confers stability by interacting with the closest unstable hydrophobic region of VDAC1 (β-strands 1, 2, 19). It should be noted that in the structure of human VDAC1, the loop between β-strands 18 and 19 is more helical, hinting at other changes in this area. Thus, the entire region of dimerization that is directly connected to the N-terminal region may be subjected to conformational activation to reprogram VDAC1 channel activity into a pore-forming unit that releases Cyto *c*. (C) Protein–protein interaction surfaces mapped on VDAC1. (a) Monomeric VDAC1 with the N-terminal domain (red) inside the pore. (b) Dimer formation involves β-sheets 1, 2, 19 (light blue). (c) Upon assembly of higher-order oligomers induced upon

(Continued)

each VDAC1 molecule is moved to line the external surface of the β-barrel, a hydrophilic pore would form (Fig. 11.6C). We propose that upon apoptosis induction, VDAC1 undergoes oligomerization, concomitant with the translocation of the N-terminal domain from the pore interior and into a newly formed oligomeric VDAC1 pore (Fig. 11.6C). This relocated N-terminal domain converts the hydrophobic pore to a pore that presents hydrophilic environment, thus allowing Cyto *c* and other apoptogenic proteins to cross the OMM.[49,56] The newly positioned N-terminal region is in close proximity to Glu72 and is also surrounded by other hydrophilic residues. This hydrophilic surface drives further oligomerization and the creation of a pore large enough to release Cyto *c* into the cytosol. Still, the exact molecular architecture of the Cyto *c*-conducting channel is yet to be determined.

6.3. Signaling mechanisms for induction of VDAC1 oligomerization

The precise mechanisms regulating Cyto *c* release remain unknown. Although substantial evidence support the formation of higher-order VDAC1-containing complexes upon apoptosis induction in cultured cells, the signaling mechanism responsible for induction of VDAC1 oligomerization is not yet known. The dynamic equilibrium that exists between monomers and oligomers can be influenced by intracellular Ca^{2+}, ROS, or ATP levels; by posttranslational modifications; and by associated proteins, such as HK-I or proteins from the Bcl-2 family.

Several studies have shown that Ca^{2+} is a mediator of cell death signaling and that anticancer drugs, such as thapsigargin, STS, selenite, cisplatin, and Ca^{2+} ionophores, induce apoptotic cell death and interfere with Ca^{2+} homeostasis.[118] In addition, mitochondrial Ca^{2+} is involved in apoptosis induction.[119] Ca^{2+} may mediate apoptosis via directly interacting with VDAC1 that mediates Ca^{2+} transport across the OMM and binds this cation.[76] However,

apoptosis induction, the N-terminal domain leaves the pore interior and interacts with the dimer surface (β-strands 1, 2, 19) adjacent to Glu72 and surrounded by other hydrophilic residues leading to dimer interface shifting to involve association of β-sheet 8 and 16 (purple and green, respectively). This newly formed hydrophilic surface drives further oligomerization with the amphipathic properties of the N-terminal domain leading to the formation of a pore large enough to allow release of Cyto *c* (light purple inside the pore) to the cytosolic space. β-Strand 4 (orange), that domain that harbors Glu72 and is associated with HK-, Bcl-xL- and Bcl-2-binding to VDAC1,[58,59,106] is in close proximity to the N-terminal domain. (For interpretation of the references to color in this figure legend, the reader is referred to the online version of this chapter.)

mitochondrial Ca^{2+} can affect VDAC1 oligomerization via a different mechanism. Most of the Cyto *c* is associated with the anionic cardiolipin (CL), such that Cyto *c* dissociation from CL is a prerequisite step for release of the protein. This could be induced by CL peroxidation,[120] by Ca^{2+} interaction with CL,[121] or by tBid.[122] It was also demonstrated that VDAC1 oligomerization is encouraged in the presence of Cyto *c*.[62] Thus, VDAC1 oligomerization can be regulated through its Ca^{2+} transport activity or via binding of Ca^{2+}.[76]

VDAC1 oligomerization can be induced by H_2O_2,[61,66] implying that the dynamic equilibrium between VDAC1 monomers and oligomers can be influenced by ROS. DIDS, an inhibitor of VDAC1 channel activity, was found to inhibit $O_2^{-\cdot}$-induced apoptosis[103,104,123,124] and VDAC1 oligomerization,[66] suggesting that $O_2^{-\cdot}$ induces Cyto *c* release via VDAC1-dependent permeabilization of the OMM.[103] Moreover, $O_2^{-\cdot}$ was found to evoke Cyto *c* release in VDAC1-reconstituted liposomes.[103] The translocator protein, TSPO, appears to be involved in the generation of ROS, leading to apoptosis induction.[125] In addition, TSPO–VDAC1 interactions are proposed to play a role in apoptosis,[126–128] where the close association of TSPO with VDAC1 is proposed to enhance the concentration of ROS generated by TSPO, leading to apoptosis induction.[129–131] The mechanism by which ROS lead to VDAC1 oligomerization is not clear. ROS may act by inducing upregulation of VDAC1 expression[132] or via modification of amino acid residues affecting VDAC1 conformation, leading to its oligomerization.

7. VDAC1 OVEREXPRESSION LEADS TO OLIGOMERIZATION AND INDUCTION OF CELL DEATH

7.1. Overexpression of VDAC from different sources resulted in apoptotic cell death

It has been demonstrated that apoptosis can be induced by overexpression of human, murine, yeast, and rice VDAC (osVDAC4) regardless of the cells used or VDAC source.[107,133,134] Overexpression of murine or rat VDAC1 or VDAC1–GFP or E72Q–mVDAC1 in U-937 cells resulted in cell death (70–85%) characterized by nuclear fragmentation.[107] Similar results were obtained upon overexpression of rice VDAC in the Jurkat T-cell line.[133] Such cell death was blocked by Bcl-2 or DIDS, also shown to inhibit VDAC1 channel activity.[123,135] Overexpression of mVDAC1 in Neuro-2a cells also induced apoptosis.[136] Moreover, overexpression of *Paralichthys olivaceus* VDAC in fish cells induced apoptosis. These findings suggest that VDAC is a conserved mitochondrial element of the death machinery.

7.2. VDAC1 expression in cancers is enhanced by proapoptotic drugs

Several cancer treatments were found to enhance expression of VDAC1. Upregulation of VDAC1 expression was noted in three different acute lymphoblastic leukemia (ALL) cell lines (697, Sup-B15, and RS4;11) following prednisolone treatment, an observation that can be explored for predicting eventual outcome in childhood ALL.[137] VDAC1 overexpression was observed in a cisplatin-sensitive cervix squamous cell carcinoma cell line (A431) when exposed to cisplatin, while in a cisplatin-resistant cell line (A431/Pt), such treatment resulted in downregulation of VDAC1.[138] Upregulation of VDAC1 has been reported upon UV irradiation of apoptosis-sensitive cells.[139] In A375 human malignant melanoma cells, arbutin (hydroquinone-*O*-β-D-glucopyranoside), a tyrosinase inhibitor and a potential antitumor agent, was found to induce apoptosis by inducing VDAC1 overexpression.[140,141] In addition, ROS were found to induce upregulation of VDAC1 that could be prevented by the ROS chelator, epigallocatechin.[132]

The relationship between VDAC1 expression levels and sensitivity to various treatments was presented in several studies. The PC3 and DU145 prostate cancer cell lines are relatively resistant to apoptosis as induced by G3139 and were found to express less VDAC1 than did G3139-sensitive LNCaP cells.[142] Similarly, reducing VDAC1 expression by siRNA efficiently prevented cisplatin-induced apoptosis and Bax activation in non-small cell lung cancer,[143] attenuated endostatin-induced apoptosis,[111] and inhibited selenite-induced PTP opening.[144] The anticancer activity of furanonaphthoquinones was increased upon VDAC1 overexpression and decreased upon silencing of VDAC1 expression by siRNA.[124,145] It has also been shown that CD45 expression is accompanied by elevated VDAC1 synthesis in myeloma cells sensitized to a diverse set of apoptotic stimuli acting via the mitochondrial pathways.[146]

Hepatitis E virus ORF3 protein was also found to upregulate VDAC1 expression levels.[117] It is known that radiation can also induce the upregulation of VDAC1 in LYas cells[139] and that increased expression of VDAC1 is correlated with uterine epithelial apoptosis after estrogen deprivation.[147] Recently, it was demonstrated that increased expression of VDAC1 sensitizes carcinoma cells to apoptosis induced by cisplatin, mechlorethamine, and its derivative, melphalan.[148] Finally, HEV-infected human hepatoma cells showed enhanced VDAC1 expression, leading to VDAC1 oligomerization and apoptosis.[117] Thus, cell death induction by reagents, such as arbutin, prednisolone, or cisplatin; by viral proteins; or by UV irradiation, as well as the correlation between drug effectiveness and VDAC1 expression, all suggest that the anticancer activities of these drugs and treatments are associated with VDAC1 levels. This idea is further supported by the finding that cisplatin-induced apoptosis is inhibited in cells silenced for VDAC1 expression.[144,149] In addition, endostatin-induced apoptosis was decreased upon silencing of VDAC1 expression and enhanced by overexpression of VDAC1.[111]

7.3. Proposed mechanism for cell death induction by VDAC1 overexpression

Taken together, the findings presented above suggest that the cellular expression level of VDAC1 is a crucial factor in the process of mitochondria-mediated apoptosis. We propose that the high levels of VDAC1 promote its oligomerization, leading to apoptosis.[61,66]

How can enhanced VDAC1 expression lead to cell death? It is possible that increased VDAC1 levels produce a significant increase in OMM leakage, in turn decreasing cell viability. Since cell death induced by VDAC1 overexpression is prevented by RuR,[107,108] Bcl-2, or DIDS,[133] all shown to interact with VDAC1, it seems that it is not a nonspecific increase in the OMM leakage but rather an increase in VDAC1 functionality that is responsible for apoptotic cell death. Moreover, cells overexpressing HK-I, shown to reduce VDAC1 channel conductance, are resistant to cell death induced by VDAC1 overexpression.[55,106,107] VDAC1 oligomerization was shown to be increased severalfold upon apoptosis induction.[66] Accordingly, an increase in VDAC1 oligomerization in cells overexpressing VDAC1 has been observed.[56] As presented in Section 6.1, dynamic VDAC1 oligomerization is involved in the release of Cyto *c* from mitochondria. As such, VDAC1 overexpression would encourage VDAC1 oligomerization and thus allow for release of proapoptotic proteins from the mitochondrial IMS and, subsequently, apoptosis.

8. VDAC1, BAX, AND BAK HETERO-OLIGOMERS MEDIATE Cyto *c* RELEASE

It has also been hypothesized that a protein-conducting channel may be formed within hetero-oligomers containing VDAC1 and the proapoptotic proteins, Bax or Bak.[91] Bax was demonstrated to increase the pore size of rat brain mitochondrial VDAC1 in the presence of tBid.[150] Similarly, in

PLBs, Bax and VDAC1, when combined, revealed single-channel conductance higher by factors of 4 and 10 over values obtained with VDAC1 and Bax alone, and that could be inhibited by Bcl-xL.[98] It has been shown that As_2O_3-induced Bax and VDAC1 oligomerization and the apoptosis-inducing effects of such rearrangements could be attributed to the induction of homodimerization of VDAC1 molecules.[151] In addition, injection of anti-VDAC1 antibodies into cells prevented Bax-induced Cyto *c* release, the interaction of Bax with VDAC1, and the triggering of cell death.[104] Independent evidence supporting a role for Bax–VDAC1 interaction in apoptosis was provided by the finding that HK-I and II can inhibit apoptosis by binding to VDAC1 and inhibiting VDAC1–Bax interaction.[152] It was further shown that Bax and VDAC1 can form a large pore that is permeable to Cyto *c*.[97,153] It was also proposed that oligomeric VDAC1, as the prime Cyto *c* release channel, is regulated by Bax.[154] In addition, Bax–VDAC1 complexes were identified in digitonin-solubilized cerebellar granule neurons.[155]

Bax–VDAC1 was found to form a protein complex in primary rat hepatocytes and in neonatal rat cerebellum upon acute ethanol exposure that induces oxidative stress and apoptosis.[156,157] Prevention of Bax–VDAC1 interactions by a microinjection of anti-VDAC1 antibodies effectively prevented hepatocyte apoptosis by ethanol.[156] In addition, cerebral ischemia in rat CNS resulted in elevated formation of Bax–VDAC1 heterodimers *in vivo* and in isolated brain mitochondria *in vitro*, with this triggering the release of Cyto *c* from isolated mitochondria.[158]

Interaction between VDAC2 and mitochondrial Bak in living cells was detected using a chemical cross-linking approach. This complex can be disrupted by tBid. VDAC2 thus inhibits BAK activation and mitochondrial apoptosis.[37] However, VDAC2 was reported as being required for tBID-induced mitochondrial apoptosis by recruiting newly synthesized BAK to the mitochondria.[159]

9. PROSPECTIVE

Mitochondrial apoptosis induction is known to effectively kill cells. Thus, one of the mechanisms by which chemotherapeutics destroy cancer cells is via the induction of apoptosis. Indeed, in recent years, it has become clear that a number of experimental chemotherapeutic agents directly affect mitochondria.[160]

The VDAC1 oligomeric assembly formed during apoptosis and possibly associated with the release of Cyto *c* and other apoptogenic factors, thus inducing cell death, are not yet defined. Future directions should characterize the VDAC1 oligomeric states and their dynamics, as these represent a target for novel therapies that could stabilize VDAC1 homooligomers, leading to apoptosis.

As presented above, it was demonstrated that some anticancer agents select VDAC1–3 isoforms as their targets. Furthermore, some of these drugs increase VDAC1 expression via an unknown mechanism. We propose that this increased level of VDAC1 shifts the equilibrium toward oligomeric VDAC1, thus leading to Cyto *c* release and subsequent apoptosis. Modulating VDAC1 oligomerization can lay down the framework for a novel approach to developing therapeutics for cancers and neurodegenerative diseases. The BRET2 technology developed by our group allows screening for compounds that may stimulate or inhibit apoptosis. Oligomerization inhibitors are expected to inhibit apoptosis induction and can thus be used to combat neurodegenerative diseases, while oligomerization enhancers that induce apoptosis can be used in cancer therapy (Fig. 11.7). Such VDAC1-targeted strategy is superior to approaches that target specific events in apoptosis regulation and should be successful in dealing with the variety of cell phenotypes that arise in a population of transformed cells. In other words, we contend that VDAC1 oligomerization is such a general common

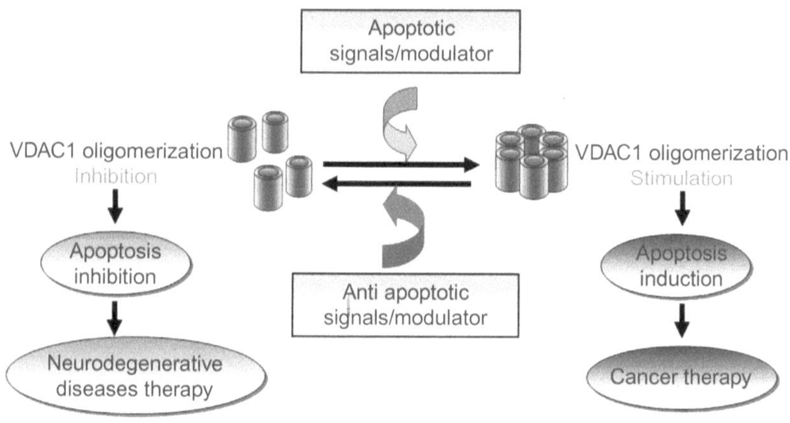

Figure 11.7 Modulation of VDAC1 oligomerization mediates apoptosis to offer novel cancer and neurodegenerative disease therapies. Targeting the VDAC1 oligomeric state provides a new strategy for selectively and directly manipulating apoptosis. This can serve as the basis for developing therapeutic applications for diseases in which apoptosis plays a significant role. Specifically, failure to undergo apoptosis in cancer and the increased apoptosis seen in neurodegenerative diseases would be countered. (For color version of this figure, the reader is referred to the online version of this chapter.)

denominator of mitochondria-mediated apoptosis that the chances of resistance to its disruption are very remote. As outlined above, the selectivity of this approach toward malignant cells is based on the overexpression of VDAC1 in cancer. Finally, interfering with VDAC1 oligomerization provides the means to overcome tumor chemoresistance and to lay down the foundations for more effective chemotherapy without or with fewer side effects.

ACKNOWLEDGMENT

This research was supported by a grant (649/09) from the Israel Science Foundation. Support from Phil and Sima Needleman is highly acknowledged.

REFERENCES

1. Gulbins E, Dreschers S, Bock J. Role of mitochondria in apoptosis. *Exp Physiol* 2003;**88**:85–90.
2. Kroemer G, Reed JC. Mitochondrial control of cell death. *Nat Med* 2000;**6**:513–9.
3. Desagher S, Martinou JC. Mitochondria as the central control point of apoptosis. *Trends Cell Biol* 2000;**10**:369–77.
4. Keeble JA, Gilmore AP. Apoptosis commitment—translating survival signals into decisions on mitochondria. *Cell Res* 2007;**17**:976–84.
5. Kroemer G, Galluzzi L, Brenner C. Mitochondrial membrane permeabilization in cell death. *Physiol Rev* 2007;**87**:99–163.
6. Le Bras M, Clement MV, Pervaiz S, Brenner C. Reactive oxygen species and the mitochondrial signaling pathway of cell death. *Histol Histopathol* 2005;**20**:205–19.
7. Hickman JA. Apoptosis and tumourigenesis. *Curr Opin Genet Dev* 2002;**12**:67–72.
8. Johnstone RW, Ruefli AA, Lowe SW. Apoptosis: a link between cancer genetics and chemotherapy. *Cell* 2002;**108**:153–64.
9. Yasuhara S, Asai A, Sahani ND, Martyn JA. Mitochondria, endoplasmic reticulum, and alternative pathways of cell death in critical illness. *Crit Care Med* 2007;**35**:S488–S495.
10. Chowdhury I, Tharakan B, Bhat GK. Current concepts in apoptosis: the physiological suicide program revisited. *Cell Mol Biol Lett* 2006;**11**:506–25.
11. Green DR, Evan GI. A matter of life and death. *Cancer Cell* 2002;**1**:19–30.
12. Crompton M. The mitochondrial permeability transition pore and its role in cell death. *Biochem J* 1999;**341**(Pt. 2):233–49.
13. Gottlieb RA. Role of mitochondria in apoptosis. *Crit Rev Eukaryot Gene Expr* 2000;**10**:231–9.
14. Wang GQ, Gastman BR, Wieckowski E, Goldstein LA, Gambotto A, Kim TH, et al. A role for mitochondrial Bak in apoptotic response to anticancer drugs. *J Biol Chem* 2001;**276**:34307–17.
15. Lovell JF, Billen LP, Bindner S, Shamas-Din A, Fradin C, Leber B, et al. Membrane binding by tBid initiates an ordered series of events culminating in membrane permeabilization by Bax. *Cell* 2008;**135**:1074–84.
16. Westphal D, Dewson G, Czabotar PE, Kluck RM. Molecular biology of Bax and Bak activation and action. *Biochim Biophys Acta* 2011;**1813**:521–31.
17. Shamas-Din A, Brahmbhatt H, Leber B, Andrews DW. BH3-only proteins: orchestrators of apoptosis. *Biochim Biophys Acta* 2011;**1813**:508–20.
18. Sundararajan R, Cuconati A, Nelson D, White E. Tumor necrosis factor-alpha induces Bax-Bak interaction and apoptosis, which is inhibited by adenovirus E1B 19K. *J Biol Chem* 2001;**276**:45120–7.

19. Shore GC. Apoptosis: it's BAK to VDAC. *EMBO Rep* 2009;**10**:1311–3.
20. Shore GC, Nguyen M. Bcl-2 proteins and apoptosis: choose your partner. *Cell* 2008;**135**:1004–6.
21. Ruffolo SC, Shore GC. BCL-2 selectively interacts with the BID-induced open conformer of BAK, inhibiting BAK auto-oligomerization. *J Biol Chem* 2003;**278**: 25039–45.
22. Dewson G, Kratina T, Czabotar P, Day CL, Adams JM, Kluck RM. Bak activation for apoptosis involves oligomerization of dimers via their alpha6 helices. *Mol Cell* 2009;**36**:696–703.
23. Kim H, Tu HC, Ren D, Takeuchi O, Jeffers JR, Zambetti GP, et al. Stepwise activation of BAX and BAK by tBID, BIM, and PUMA initiates mitochondrial apoptosis. *Mol Cell* 2009;**36**:487–99.
24. Ren D, Tu HC, Kim H, Wang GX, Bean GR, Takeuchi O, et al. BID, BIM, and PUMA are essential for activation of the BAX- and BAK-dependent cell death program. *Science* 2010;**330**:1390–3.
25. Hinsch KD, Asmarinah , Hinsch E, Konrad L. VDAC2 (porin-2) expression pattern and localization in the bovine testis. *Biochim Biophys Acta* 2001;**1518**:329–33.
26. Thomas L, Blachly-Dyson E, Colombini M, Forte M. Mapping of residues forming the voltage sensor of the voltage-dependent anion-selective channel. *Proc Natl Acad Sci USA* 1993;**90**:5446–9.
27. Messina A, Guarino F, Oliva M, van den Heuvel LP, Smeitink J, De Pinto V. Characterization of the human porin isoform 1 (HVDAC1) gene by amplification on the whole human genome: a tool for porin deficiency analysis. *Biochem Biophys Res Commun* 2000;**270**:787–92.
28. Sampson MJ, Lovell RS, Craigen WJ. The murine voltage-dependent anion channel gene family. Conserved structure and function. *J Biol Chem* 1997;**272**:18966–73.
29. Blachly-Dyson E, Zambronicz EB, Yu WH, Adams V, McCabe ER, Adelman J, et al. Cloning and functional expression in yeast of two human isoforms of the outer mitochondrial membrane channel, the voltage-dependent anion channel. *J Biol Chem* 1993;**268**:1835–41.
30. Xu X, Decker W, Sampson MJ, Craigen WJ, Colombini M. Mouse VDAC isoforms expressed in yeast: channel properties and their roles in mitochondrial outer membrane permeability. *J Membr Biol* 1999;**170**:89–102.
31. Sampson MJ, Lovell RS, Davison DB, Craigen WJ. A novel mouse mitochondrial voltage-dependent anion channel gene localizes to chromosome 8. *Genomics* 1996;**36**:192–6.
32. Hinsch KD, De Pinto V, Aires VA, Schneider X, Messina A, Hinsch E. Voltage-dependent anion-selective channels VDAC2 and VDAC3 are abundant proteins in bovine outer dense fibers, a cytoskeletal component of the sperm flagellum. *J Biol Chem* 2004;**279**:15281–8.
33. Triphan X, Menzel VA, Petrunkina AM, Cassara MC, Wemheuer W, Hinsch KD, et al. Localisation and function of voltage-dependent anion channels (VDAC) in bovine spermatozoa. *Pflugers Arch* 2008;**455**:677–86.
34. Shinohara Y, Ishida T, Hino M, Yamazaki N, Baba Y, Terada H. Characterization of porin isoforms expressed in tumor cells. *Eur J Biochem* 2000;**267**:6067–73.
35. Yu WH, Wolfgang W, Forte M. Subcellular localization of human voltage-dependent anion channel isoforms. *J Biol Chem* 1995;**270**:13998–4006.
36. Huizing M, Ruitenbeek W, van den Heuvel LP, Dolce V, Iacobazzi V, Smeitink JA, et al. Human mitochondrial transmembrane metabolite carriers: tissue distribution and its implication for mitochondrial disorders. *J Bioenerg Biomembr* 1998;**30**:277–84.
37. Cheng EH, Sheiko TV, Fisher JK, Craigen WJ, Korsmeyer SJ. VDAC2 inhibits BAK activation and mitochondrial apoptosis. *Science* 2003;**301**:513–7.

38. Shoshan-Barmatz V, De Pinto V, Zweckstetter M, Raviv Z, Keinan N, Arbel N. VDAC, a multi-functional mitochondrial protein regulating cell life and death. *Mol Aspects Med* 2010;**31**:227–85.
39. Wu S, Sampson MJ, Decker WK, Craigen WJ. Each mammalian mitochondrial outer membrane porin protein is dispensable: effects on cellular respiration. *Biochim Biophys Acta* 1999;**1452**:68–78.
40. Sampson MJ, Decker WK, Beaudet AL, Ruitenbeek W, Armstrong D, Hicks MJ, et al. Immotile sperm and infertility in mice lacking mitochondrial voltage-dependent anion channel type 3. *J Biol Chem* 2001;**276**:39206–12.
41. Weeber EJ, Levy M, Sampson MJ, Anflous K, Armstrong DL, Brown SE, et al. The role of mitochondrial porins and the permeability transition pore in learning and synaptic plasticity. *J Biol Chem* 2002;**277**:18891–7.
42. Chandra D, Choy G, Daniel PT, Tang DG. Bax-dependent regulation of Bak by voltage-dependent anion channel 2. *J Biol Chem* 2005;**280**:19051–61.
43. Yamamoto T, Yamada A, Watanabe M, Yoshimura Y, Yamazaki N, Yoshimura Y, et al. VDAC1, having a shorter N-terminus than VDAC2 but showing the same migration in an SDS-polyacrylamide gel, is the predominant form expressed in mitochondria of various tissues. *J Proteome Res* 2006;**5**:3336–44.
44. De Pinto V, Guarino F, Guarnera Andrea, Messina A, Reina S, Tomasello FM, et al. Characterization of human vdac isoforms: a peculiar function for vdac3? *Biochim Biophys Acta* 2010;**1797**:1268–75.
45. Bayrhuber M, Meins T, Habeck M, Becker S, Giller K, Villinger S, et al. Structure of the human voltage-dependent anion channel. *Proc Natl Acad Sci USA* 2008;**105**: 15370–5.
46. Hiller S, Garces RG, Malia TJ, Orekhov VY, Colombini M, Wagner G. Solution structure of the integral human membrane protein VDAC-1 in detergent micelles. *Science* 2008;**321**:1206–10.
47. Ujwal R, Cascio D, Colletier JP, Faham S, Zhang J, Toro L, et al. The crystal structure of mouse VDAC1 at 2.3 A resolution reveals mechanistic insights into metabolite gating. *Proc Natl Acad Sci USA* 2008;**105**:17742–7.
48. Hiller S, Wagner G. The role of solution NMR in the structure determinations of VDAC-1 and other membrane proteins. *Curr Opin Struct Biol* 2009;**19**:396–401.
49. Geula S, Ben-Hail D, Shoshan-Barmatz V. Structure-based analysis of VDAC1: N-terminus location, translocation, channel gating and association with anti-apoptotic proteins. *Biochem J* 2012;**444**:475–85.
50. Koppel DA, Kinnally KW, Masters P, Forte M, Blachly-Dyson E, Mannella CA. Bacterial expression and characterization of the mitochondrial outer membrane channel. Effects of n-terminal modifications. *J Biol Chem* 1998;**273**:13794–800.
51. Mannella CA. Minireview: on the structure and gating mechanism of the mitochondrial channel, VDAC. *J Bioenerg Biomembr* 1997;**29**:525–31.
52. Popp B, Court DA, Benz R, Neupert W, Lill R. The role of the N and C termini of recombinant Neurospora mitochondrial porin in channel formation and voltage-dependent gating. *J Biol Chem* 1996;**271**:13593–9.
53. Junankar PR, Dulhunty AF, Curtis SM, Pace SM, Thinnes FP. Porin-type 1 proteins in sarcoplasmic reticulum and plasmalemma of striated muscle fibres. *J Muscle Res Cell Motil* 1995;**16**:595–610.
54. Abu-Hamad S, Sivan S, Shoshan-Barmatz V. The expression level of the voltage-dependent anion channel controls life and death of the cell. *Proc Natl Acad Sci USA* 2006;**103**:5787–92.
55. Azoulay-Zohar H, Israelson A, Abu-Hamad S, Shoshan-Barmatz V. In self-defence: hexokinase promotes voltage-dependent anion channel closure and prevents mitochondria-mediated apoptotic cell death. *Biochem J* 2004;**377**:347–55.

56. Abu-Hamad S, Arbel N, Calo D, Arzoine L, Israelson A, Keinan N, et al. The VDAC1 N-terminus is essential both for apoptosis and the protective effect of anti-apoptotic proteins. *J Cell Sci* 2009;**122**:1906–16.
57. Shi Y, Chen J, Weng C, Chen R, Zheng Y, Chen Q, et al. Identification of the protein-protein contact site and interaction mode of human VDAC1 with Bcl-2 family proteins. *Biochem Biophys Res Commun* 2003;**305**:989–96.
58. Arbel N, Ben-Hail D, Shoshan-Barmatz V. Mediation of the anti-apoptotic activity of Bcl-xL upon interaction with VDAC1. *J Biol Chem* 2012;**287**:23152–61.
59. Arbel N, Shoshan-Barmatz V. Voltage-dependent anion channel 1-based peptides interact with Bcl-2 to prevent antiapoptotic activity. *J Biol Chem* 2010;**285**:6053–62.
60. Arzoine L, Zilberberg N, Ben-Romano R, Shoshan-Barmatz V. Voltage-dependent anion channel 1-based peptides interact with hexokinase to prevent its anti-apoptotic activity. *J Biol Chem* 2009;**284**:3946–55.
61. Shoshan-Barmatz V, Arbel N, Arzoine L. VDAC, the voltage-dependent anion channel: function, regulation & mitochondrial signaling in cell life and death. *Cell Sci* 2008;**4**:74–118.
62. Zalk R, Israelson A, Garty E, Azoulay-Zohar H, Shoshan-Barmatz V. Oligomeric states of the voltage-dependent anion channel and cytochrome c release from mitochondria. *Biochem J* 2005;**386**:73–83.
63. Zeth K, Meins T, Vonrhein C. Approaching the structure of human VDAC1, a key molecule in mitochondrial cross-talk. *J Bioenerg Biomembr* 2008;**40**:127–32.
64. Goncalves RP, Buzhynskyy N, Prima V, Sturgis JN, Scheuring S. Supramolecular assembly of VDAC in native mitochondrial outer membranes. *J Mol Biol* 2007;**369**:413–8.
65. Hoogenboom BW, Suda K, Engel A, Fotiadis D. The supramolecular assemblies of voltage-dependent anion channels in the native membrane. *J Mol Biol* 2007;**370**:246–55.
66. Keinan N, Tyomkin D, Shoshan-Barmatz V. Oligomerization of the mitochondrial protein voltage-dependent anion channel is coupled to the induction of apoptosis. *Mol Cell Biol* 2010;**30**:5698–709.
67. Malia TJ, Wagner G. NMR structural investigation of the mitochondrial outer membrane protein VDAC and its interaction with antiapoptotic Bcl-xL. *Biochemistry* 2007;**46**:514–25.
68. Shoshan-Barmatz V, Keinan N, Zaid H. Uncovering the role of VDAC in the regulation of cell life and death. *J Bioenerg Biomembr* 2008;**40**:183–91.
69. Guo XW, Smith PR, Cognon B, D'Arcangelis D, Dolginova E, Mannella CA. Molecular design of the voltage-dependent, anion-selective channel in the mitochondrial outer membrane. *J Struct Biol* 1995;**114**:41–59.
70. Ujwal R, Cascio D, Chaptal V, Ping P, Abramson J. Crystal packing analysis of murine VDAC1 crystals in a lipidic environment reveals novel insights on oligomerization and orientation. *Channels (Austin)* 2009;**3**:167–70.
71. Betaneli V, Petrov EP, Schwille P. The role of lipids in VDAC oligomerization. *Biophys J* 2012;**102**:523–31.
72. Xie G, Wilson JE. Tetrameric structure of mitochondrially bound rat brain hexokinase: a crosslinking study. *Arch Biochem Biophys* 1990;**276**:285–93.
73. Brdiczka D, Kaldis P, Wallimann T. In vitro complex formation between the octamer of mitochondrial creatine kinase and porin. *J Biol Chem* 1994;**269**:27640–4.
74. Stachowiak O, Schlattner U, Dolder M, Wallimann T. Oligomeric state and membrane binding behaviour of creatine kinase isoenzymes: implications for cellular function and mitochondrial structure. *Mol Cell Biochem* 1998;**184**:141–51.
75. Schlattner U, Dolder M, Wallimann T, Tokarska-Schlattner M. Mitochondrial creatine kinase and mitochondrial outer membrane porin show a direct interaction that is modulated by calcium. *J Biol Chem* 2001;**276**:48027–30.

76. Gincel D, Zaid H, Shoshan-Barmatz V. Calcium binding and translocation by the voltage-dependent anion channel: a possible regulatory mechanism in mitochondrial function. *Biochem J* 2001;**358**:147–55.
77. Benz R. Permeation of hydrophilic solutes through mitochondrial outer membranes: review on mitochondrial porins. *Biochim Biophys Acta* 1994;**1197**:167–96.
78. Colombini M. Structure and mode of action of a voltage dependent anion-selective channel (VDAC) located in the outer mitochondrial membrane. *Ann N Y Acad Sci* 1980;**341**:552–63.
79. Gincel D, Silberberg SD, Shoshan-Barmatz V. Modulation of the voltage-dependent anion channel (VDAC) by glutamate. *J Bioenerg Biomembr* 2000;**32**:571–83.
80. De Pinto V, Tommasino M, Benz R, Palmieri F. The 35 kDa DCCD-binding protein from pig heart mitochondria is the mitochondrial porin. *Biochim Biophys Acta* 1985;**813**:230–42.
81. Mannella CA, Guo XW, Cognon B. Diameter of the mitochondrial outer membrane channel: evidence from electron microscopy of frozen-hydrated membrane crystals. *FEBS Lett* 1989;**253**:231–4.
82. Hodge T, Colombini M. Regulation of metabolite flux through voltage-gating of VDAC channels. *J Membr Biol* 1997;**157**:271–9.
83. Rostovtseva T, Colombini M. VDAC channels mediate and gate the flow of ATP: implications for the regulation of mitochondrial function. *Biophys J* 1997;**72**:1954–62.
84. Benz R, Kottke M, Brdiczka D. The cationically selective state of the mitochondrial outer membrane pore: a study with intact mitochondria and reconstituted mitochondrial porin. *Biochim Biophys Acta* 1990;**1022**:311–8.
85. Tsujimoto Y, Shimizu S. The voltage-dependent anion channel: an essential player in apoptosis. *Biochimie* 2002;**84**:187–93.
86. De Pinto V, Messina A, Accardi R, Aiello R, Guarino F, Tomasello MF, et al. New functions of an old protein: the eukaryotic porin or voltage dependent anion selective channel (VDAC). *Ital J Biochem* 2003;**52**:17–24.
87. Vyssokikh MY, Brdiczka D. The function of complexes between the outer mitochondrial membrane pore (VDAC) and the adenine nucleotide translocase in regulation of energy metabolism and apoptosis. *Acta Biochim Pol* 2003;**50**:389–404.
88. Granville DJ, Gottlieb RA. The mitochondrial voltage-dependent anion channel (VDAC) as a therapeutic target for initiating cell death. *Curr Med Chem* 2003;**10**: 1527–33.
89. Lemasters JJ, Holmuhamedov E. Voltage-dependent anion channel (VDAC) as mitochondrial governator—thinking outside the box. *Biochim Biophys Acta* 2006;**1762**: 181–90.
90. Shoshan-Barmatz V, Gincel D. The voltage-dependent anion channel: characterization, modulation, and role in mitochondrial function in cell life and death. *Cell Biochem Biophys* 2003;**39**:279–92.
91. Shoshan-Barmatz V, Israelson A, Brdiczka D, Sheu SS. The voltage-dependent anion channel (VDAC): function in intracellular signalling, cell life and cell death. *Curr Pharm Des* 2006;**12**:2249–70.
92. Shoshan-Barmatz V, Ben-Hail D. VDAC, a multi-functional mitochondrial protein as a pharmacological target. *Mitochondrion* 2012;**12**:24–34.
93. Shoshan-Barmatz V, Golan M. Mitochondrial VDAC1: function in cell life and death and a target for cancer therapy. *Curr Med Chem* 2012;**19**:714–35.
94. Bernardi P. Mitochondrial transport of cations: channels, exchangers, and permeability transition. *Physiol Rev* 1999;**79**:1127–55.
95. Tsujimoto Y, Shimizu S. Role of the mitochondrial membrane permeability transition in cell death. *Apoptosis* 2007;**12**:835–40.

96. Lemasters JJ. Modulation of mitochondrial membrane permeability in pathogenesis, autophagy and control of metabolism. *J Gastroenterol Hepatol* 2007;**22**(Suppl. 1): S31–S37.
97. Shimizu S, Narita M, Tsujimoto Y. Bcl-2 family proteins regulate the release of apoptogenic cytochrome c by the mitochondrial channel VDAC. *Nature* 1999;**399**:483–7.
98. Shimizu S, Ide T, Yanagida T, Tsujimoto Y. Electrophysiological study of a novel large pore formed by Bax and the voltage-dependent anion channel that is permeable to cytochrome c. *J Biol Chem* 2000;**275**:12321–5.
99. Vander Heiden MG, Chandel NS, Li XX, Schumacker PT, Colombini M, Thompson CB. Outer mitochondrial membrane permeability can regulate coupled respiration and cell survival. *Proc Natl Acad Sci USA* 2000;**97**:4666–71.
100. Rostovtseva TK, Tan W, Colombini M. On the role of VDAC in apoptosis: fact and fiction. *J Bioenerg Biomembr* 2005;**37**:129–42.
101. Baines CP, Kaiser RA, Sheiko T, Craigen WJ, Molkentin JD. Voltage-dependent anion channels are dispensable for mitochondrial-dependent cell death. *Nat Cell Biol* 2007;**9**:550–5.
102. McCommis KS, Baines CP. The role of VDAC in cell death: friend or foe? *Biochim Biophys Acta* 2012;**1818**:1444–50.
103. Madesh M, Hajnoczky G. VDAC-dependent permeabilization of the outer mitochondrial membrane by superoxide induces rapid and massive cytochrome c release. *J Cell Biol* 2001;**155**:1003–15.
104. Shimizu S, Matsuoka Y, Shinohara Y, Yoneda Y, Tsujimoto Y. Essential role of voltage-dependent anion channel in various forms of apoptosis in mammalian cells. *J Cell Biol* 2001;**152**:237–50.
105. Zheng Y, Shi Y, Tian C, Jiang C, Jin H, Chen J, et al. Essential role of the voltage-dependent anion channel (VDAC) in mitochondrial permeability transition pore opening and cytochrome c release induced by arsenic trioxide. *Oncogene* 2004;**23**:1239–47.
106. Abu-Hamad S, Zaid H, Israelson A, Nahon E, Shoshan-Barmatz V. Hexokinase-I protection against apoptotic cell death is mediated via interaction with the voltage-dependent anion channel-1: mapping the site of binding. *J Biol Chem* 2008;**283**:13482–90.
107. Zaid H, Abu-Hamad S, Israelson A, Nathan I, Shoshan-Barmatz V. The voltage-dependent anion channel-1 modulates apoptotic cell death. *Cell Death Differ* 2005; **12**:751–60.
108. Israelson A, Zaid H, Abu-Hamad S, Nahon E, Shoshan-Barmatz V. Mapping the ruthenium red-binding site of the voltage-dependent anion channel-1. *Cell Calcium* 2008;**43**:196–204.
109. Israelson A, Abu-Hamad S, Zaid H, Nahon E, Shoshan-Barmatz V. Localization of the voltage-dependent anion channel-1 Ca2+-binding sites. *Cell Calcium* 2007;**41**:235–44.
110. Tajeddine N, Galluzzi L, Kepp O, Hangen E, Morselli E, Senovilla L, et al. Hierarchical involvement of Bak, VDAC1 and Bax in cisplatin-induced cell death. *Oncogene* 2008;**27**:4221–32.
111. Yuan S, Fu Y, Wang X, Shi H, Huang Y, Song X, et al. Voltage-dependent anion channel 1 is involved in endostatin-induced endothelial cell apoptosis. *FASEB J* 2008;**22**:2809–20.
112. Shoshan-Barmatz V, Keinan N, Abu-Hamad S, Tyomkin D, Aram L. Apoptosis is regulated by the VDAC1 N-terminal region and by VDAC oligomerization: release of cytochrome c AIF and Smac/Diablo. *Biochim Biophys Acta* 2010;**1797**:1281–91.
113. Mader A, Abu-Hamad S, Arbel N, Gutierrez-Aguilar M, Shoshan-Barmatz V. Dominant-negative VDAC1 mutants reveal oligomeric VDAC1 to be the active unit in mitochondria-mediated apoptosis. *Biochem J* 2010;**429**:147–55.

114. Geula S, Naveed H, Liang J, Shoshan-Barmatz V. Structure-based analysis of VDAC1 protein: defining oligomer contact sites. *J Biol Chem* 2012;**287**:2179–90.
115. Shoshan-Barmatz V, Mizrachi D. VDAC1: from structure to cancer therapy. *Front Oncol* 2012;**2**:164.
116. Aram L, Geula S, Arbel N, Shoshan-Barmatz V. VDAC1 cysteine residues: topology and function in channel activity and apoptosis. *Biochem J* 2010;**427**:445–54.
117. Moin SM, Panteva M, Jameel S. The hepatitis E virus Orf3 protein protects cells from mitochondrial depolarization and death. *J Biol Chem* 2007;**282**:21124–33.
118. Florea AM, Busselberg D. Anti-cancer drugs interfere with intracellular calcium signaling. *Neurotoxicology* 2009;**30**:803–10.
119. Rasola A, Bernardi P. Mitochondrial permeability transition in Ca(2+)-dependent apoptosis and necrosis. *Cell Calcium* 2011;**50**:222–33.
120. Shidoji Y, Hayashi K, Komura S, Ohishi N, Yagi K. Loss of molecular interaction between cytochrome c and cardiolipin due to lipid peroxidation. *Biochem Biophys Res Commun* 1999;**264**:343–7.
121. Huang Y, Liu L, Shi C, Huang J, Li G. Electrochemical analysis of the effect of Ca2+ on cardiolipin-cytochrome c interaction. *Biochim Biophys Acta* 2006;**1760**:1827–30.
122. Liu J, Weiss A, Durrant D, Chi NW, Lee RM. The cardiolipin-binding domain of Bid affects mitochondrial respiration and enhances cytochrome c release. *Apoptosis* 2004;**9**:533–41.
123. Shoshan-Barmatz V, Hadad N, Feng W, Shafir I, Orr I, Varsanyi M, et al. VDAC/porin is present in sarcoplasmic reticulum from skeletal muscle. *FEBS Lett* 1996;**386**:205–10.
124. Simamura E, Hirai K, Shimada H, Koyama J, Niwa Y, Shimizu S. Furanonaphthoquinones cause apoptosis of cancer cells by inducing the production of reactive oxygen species by the mitochondrial voltage-dependent anion channel. *Cancer Biol Ther* 2006;**5**:1523–9.
125. Veenman L, Shandalov Y, Gavish M. VDAC activation by the 18 kDa translocator protein (TSPO), implications for apoptosis. *J Bioenerg Biomembr* 2008;**40**:199–205.
126. Gavish M, Bachman I, Shoukrun R, Katz Y, Veenman L, Weisinger G, et al. Enigma of the peripheral benzodiazepine receptor. *Pharmacol Rev* 1999;**51**:629–50.
127. Levin E, Premkumar A, Veenman L, Kugler W, Leschiner S, Spanier I, et al. The peripheral-type benzodiazepine receptor and tumorigenicity: isoquinoline binding protein (IBP) antisense knockdown in the C6 glioma cell line. *Biochemistry* 2005;**44**:9924–35.
128. Veenman L, Levin E, Weisinger G, Leschiner S, Spanier I, Snyder SH, et al. Peripheral-type benzodiazepine receptor density and in vitro tumorigenicity of glioma cell lines. *Biochem Pharmacol* 2004;**68**:689–98.
129. Golani I, Weizman A, Leschiner S, Spanier I, Eckstein N, Limor R, et al. Hormonal regulation of peripheral benzodiazepine receptor binding properties is mediated by subunit interaction. *Biochemistry* 2001;**40**:10213–22.
130. Veenman L, Leschiner S, Spanier I, Weisinger G, Weizman A, Gavish M. PK 11195 attenuates kainic acid-induced seizures and alterations in peripheral-type benzodiazepine receptor (PBR) protein components in the rat brain. *J Neurochem* 2002;**80**:917–27.
131. Veenman L, Papadopoulos V, Gavish M. Channel-like functions of the 18-kDa translocator protein (TSPO): regulation of apoptosis and steroidogenesis as part of the host-defense response. *Curr Pharm Des* 2007;**13**:2385–405.
132. Jung JY, Han CR, Jeong YJ, Kim HJ, Lim HS, Lee KH, et al. Epigallocatechin gallate inhibits nitric oxide-induced apoptosis in rat PC12 cells. *Neurosci Lett* 2007;**411**:222–7.
133. Godbole A, Varghese J, Sarin A, Mathew MK. VDAC is a conserved element of death pathways in plant and animal systems. *Biochim Biophys Acta* 2003;**1642**:87–96.

134. Lu AJ, Dong CW, Du CS, Zhang QY. Characterization and expression analysis of Paralichthys olivaceus voltage-dependent anion channel (VDAC) gene in response to virus infection. *Fish Shellfish Immunol* 2007;**23**:601–13.
135. Shafir I, Feng W, Shoshan-Barmataz V. Voltage-dependent anion channel proteins in synaptosomes of the torpedo electric organ: immunolocalization, purification, and characterization. *J Bioenerg Biomembr* 1998;**30**:499–510.
136. Ghosh T, Pandey N, Maitra A, Brahmachari SK, Pillai B. A role for voltage-dependent anion channel Vdac1 in polyglutamine-mediated neuronal cell death. *PLoS One* 2007;**2**:e1170.
137. Jiang N, Kham SK, Koh GS, Suang Lim JY, Ariffin H, Chew FT, et al. Identification of prognostic protein biomarkers in childhood acute lymphoblastic leukemia (ALL). *J Proteomics* 2011;**74**:843–57.
138. Castagna A, Antonioli P, Astner H, Hamdan M, Righetti SC, Perego P, et al. A proteomic approach to cisplatin resistance in the cervix squamous cell carcinoma cell line A431. *Proteomics* 2004;**4**:3246–67.
139. Voehringer DW, Hirschberg DL, Xiao J, Lu Q, Roederer M, Lock CB, et al. Gene microarray identification of redox and mitochondrial elements that control resistance or sensitivity to apoptosis. *Proc Natl Acad Sci USA* 2000;**97**:2680–5.
140. Nawarak J, Huang-Liu R, Kao SH, Liao HH, Sinchaikul S, Chen ST, et al. Proteomics analysis of A375 human malignant melanoma cells in response to arbutin treatment. *Biochim Biophys Acta* 2009;**1794**:159–67.
141. Cheng SL, Liu RH, Sheu JN, Chen ST, Sinchaikul S, Tsay GJ. Toxicogenomics of A375 human malignant melanoma cells treated with arbutin. *J Biomed Sci* 2007;**14**: 87–105.
142. Lai JC, Tan W, Benimetskaya L, Miller P, Colombini M, Stein CA. A pharmacologic target of G3139 in melanoma cells may be the mitochondrial VDAC. *Proc Natl Acad Sci USA* 2006;**103**:7494–9.
143. Tajeddine N, Galluzzi L, Kepp O, Hangen E, Morselli E, Senovilla L, et al. Hierarchical involvement of Bak, VDAC1 and Bax in cisplatin-induced cell death. *Oncogene* 2008;**10**:4221–32.
144. Tomasello F, Messina A, Lartigue L, Schembri L, Medina C, Reina S, et al. Outer membrane VDAC1 controls permeability transition of the inner mitochondrial membrane in cellulo during stress-induced apoptosis. *Cell Res* 2009;**19**:1363–76.
145. Simamura E, Shimada H, Ishigaki Y, Hatta T, Higashi N, Hirai K. Bioreductive activation of quinone antitumor drugs by mitochondrial voltage-dependent anion channel 1. *Anat Sci Int* 2008;**83**:261–6.
146. Liu S, Ishikawa H, Tsuyama N, Li FJ, Abroun S, Otsuyama KI, et al. Increased susceptibility to apoptosis in CD45(+) myeloma cells accompanied by the increased expression of VDAC1. *Oncogene* 2006;**25**:419–29.
147. Takagi-Morishita Y, Yamada N, Sugihara A, Iwasaki T, Tsujimura T, Terada N. Mouse uterine epithelial apoptosis is associated with expression of mitochondrial voltage-dependent anion channels, release of cytochrome C from mitochondria, and the ratio of Bax to Bcl-2 or Bcl-X. *Biol Reprod* 2003;**68**:1178–84.
148. Sharaf el dein O, Gallerne C, Brenner C, Lemaire C. Increased expression of VDAC1 sensitizes carcinoma cells to apoptosis induced by DNA cross-linking agents. *Biochem Pharmacol* 2012;**83**:1172–82.
149. De Pinto V, Tomasello F, Messina A, Guarino F, Benz R, La Mendola D, et al. Determination of the conformation of the human VDAC1 N-terminal peptide, a protein moiety essential for the functional properties of the pore. *Chembiochem* 2007;**8**:744–56.
150. Banerjee J, Ghosh S. Bax increases the pore size of rat brain mitochondrial voltage-dependent anion channel in the presence of tBid. *Biochem Biophys Res Commun* 2004;**323**:310–4.

151. Yu J, Qian H, Li Y, Wang Y, Zhang X, Liang X, et al. Therapeutic effect of arsenic trioxide (As2O3) on cervical cancer in vitro and in vivo through apoptosis induction. *Cancer Biol Ther* 2007;**6**:580–6.
152. Pastorino JG, Shulga N, Hoek JB. Mitochondrial binding of hexokinase II inhibits Bax-induced cytochrome c release and apoptosis. *J Biol Chem* 2002;**277**:7610–8.
153. Shimizu S, Tsujimoto Y. Proapoptotic BH3-only Bcl-2 family members induce cytochrome c release, but not mitochondrial membrane potential loss, and do not directly modulate voltage-dependent anion channel activity. *Proc Natl Acad Sci USA* 2000;**97**: 577–82.
154. Debatin KM, Poncet D, Kroemer G. Chemotherapy: targeting the mitochondrial cell death pathway. *Oncogene* 2002;**21**:8786–803.
155. Huckabee DB, Jekabsons MB. Identification of Bax-voltage-dependent anion channel 1 complexes in digitonin-solubilized cerebellar granule neurons. *J Neurochem* 2011;**119**:1137–50.
156. Adachi M, Higuchi H, Miura S, Azuma T, Inokuchi S, Saito H, et al. Bax interacts with the voltage-dependent anion channel and mediates ethanol-induced apoptosis in rat hepatocytes. *Am J Physiol Gastrointest Liver Physiol* 2004;**287**:G695–G705.
157. Heaton MB, Siler-Marsiglio K, Paiva M, Kotler A, Rogozinski J, Kubovec S, et al. Ethanol influences on bax associations with mitochondrial membrane proteins in neonatal rat cerebellum. *Dev Neurobiol* 2013;**73**:127–41.
158. Cao G, Minami M, Pei W, Yan C, Chen D, O'Horo C, et al. Intracellular Bax translocation after transient cerebral ischemia: implications for a role of the mitochondrial apoptotic signaling pathway in ischemic neuronal death. *J Cereb Blood Flow Metab* 2001;**21**:321–33.
159. Roy SS, Ehrlich AM, Craigen WJ, Hajnoczky G. VDAC2 is required for truncated BID-induced mitochondrial apoptosis by recruiting BAK to the mitochondria. *EMBO Rep* 2009;**10**:1341–7.
160. Costantini P, Jacotot E, Decaudin D, Kroemer G. Mitochondrion as a novel target of anticancer chemotherapy. *J Natl Cancer Inst* 2000;**92**:1042–53.

CHAPTER TWELVE

Consequences of Dimerization of the Voltage-Gated Proton Channel

Susan M.E. Smith*, Thomas E. DeCoursey†
*Department of Pathology and Laboratory Medicine, Emory School of Medicine, Atlanta Georgia, USA
†Department of Molecular Biophysics and Physiology, Rush University, Chicago Illinois, USA

Contents

Abstract

The human voltage-gated proton channel, hH_V1, appears to exist mainly as a dimer. Teleologically, this is puzzling because each protomer retains the main properties that characterize this protein: proton conduction that is regulated by conformational (channel opening and closing) changes that occur in response to both voltage and pH. The H_V1 dimer is mainly linked by C-terminal coiled-coil interactions. Several types of mutations produce monomeric constructs that open approximately five times faster than the wild-type dimeric channel but with weaker voltage dependence. Intriguingly, the quintessential function of the H_V1 dimer, opening to allow H^+ conduction, occurs cooperatively. Both protomers undergo a conformational change, but both must undergo this

Progress in Molecular Biology and Translational Science, Volume 117
ISSN 1877-1173
http://dx.doi.org/10.1016/B978-0-12-386931-9.00012-X

transition before either can conduct. The teleological purpose of dimerization may be to steepen the voltage dependence of channel opening, at least in phagocytes. In other cells, the purpose is not understood. Finally, several single-celled species have H_V that are likely monomeric.

1. INTRODUCTION

Ion channels comprise a large family of membrane proteins that regulate the passage of ions, usually one particular species of ion, across cell or organelle membranes. Most ion channels are multimeric, but dimeric channels do occur, albeit infrequently. The voltage-gated proton channel (H_V1) gene (*HVCN1*) was identified in 2006,[1,2] and its dimeric nature was discovered shortly thereafter.[3–5] The main focus of this review is to survey the consequences of dimerization for the molecular and biological functions of these channels. On the molecular scale, proton channels open and close, conduct only protons when they are open, and characteristically are regulated closely by the pH gradient.[6] Voltage-gated proton channels in humans (hH_V1) have diverse functions in a variety of cells. They participate in pathogen killing by phagocytes,[7–11] histamine secretion by basophils,[12] surface pH regulation by airway epithelia,[13] capacitation and motility of sperm prior to fertilization,[14,15] B lymphocyte signaling,[16] and may exacerbate breast cancer metastasis[17] and brain damage in ischemic stroke.[18]

2. DIMERIZATION OF MEMBRANE PROTEINS

A number of ion channels function only when assembled into the multimer that forms the ion-conducting structure. For example, voltage-gated K^+ channels are tetramers that use two transmembrane (TM) helices (S5–S6) from each of the four subunits to produce a single, central K^+ selective pore (Fig. 12.1). In a few ion channels, including H_V1, individual monomeric units appear to be fully functional, yet the native channel assembles as a dimer or oligomer with properties distinct from the monomer. The high frequency of membrane protein homodimers (and higher order homo-oligomers) argues that there is a strong evolutionary advantage for this kind of quaternary structure. Commonly accepted "reasons" for dimerization of membrane (and soluble) proteins include increased stability, increased specificity of interaction with regulatory proteins, regulation or modulation of activity, and acquisition of cooperativity. H_V1 activity is modulated by at

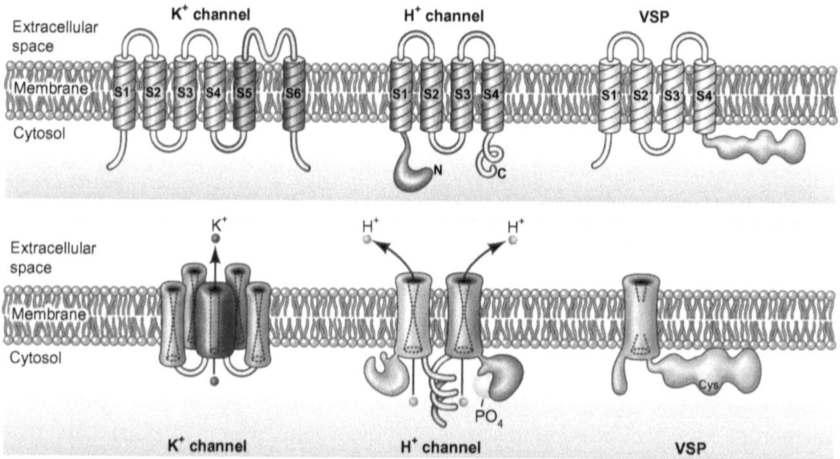

Figure 12.1 Architecture of three classes of VSD containing molecules. The top row shows the monomeric protein; the lower row shows the final assembled protein. The K^+ channel assembles as a tetramer,[19,20] with four separate VSD elements connecting to a single central pore (the pathway taken by K^+ as it permeates). The voltage-gated H^+ channel in many species, including mammals, assembles as a dimer,[3–5] although when constrained to exist in monomeric form, it retains its key properties of proton-specific conduction, voltage gating (opening upon depolarization of the membrane potential), and ΔpH dependence that strongly regulates the opening of the channel on the basis of pH. The VSP, a voltage sensitive phosphatase, is thought to exist and function as a monomer.[21] *Reprinted from Ref. 9, copyright 2010.* (See Color Insert.)

least one protein kinase, although no evidence exists one way or the other for differential specificity of the kinase for monomeric or dimeric H_V1. H_V1 dimers do act cooperatively, which has implications for physiological interplay with NADPH oxidase function in human neutrophils. Although not definitely established, dimer assembly does not appear to be involved in enhanced gating, the best characterized functional modulation of H_V1.

3. EVIDENCE THAT PROTON CHANNELS EXIST AS DIMERS

A variety of data supports the dimeric nature of the native H_V1 in humans, mice, and *Ciona intestinalis*.[3–5,22–26] The clearest and the most direct evidence that hH_V1 are dimers was provided by Tombola *et al.*,[5] who attached green fluorescent protein (GFP) to the channel molecule, and then observed photobleaching of individual channel molecules. As illustrated in Fig. 12.2, the fluorescence intensity of most channels decayed in two

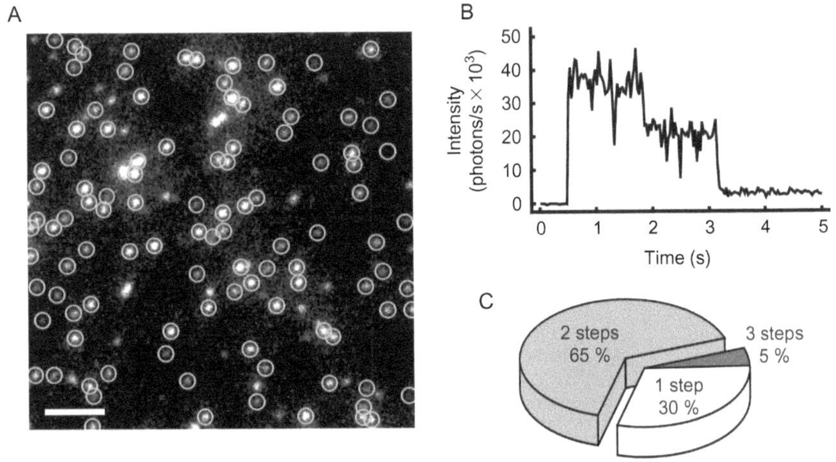

Figure 12.2 Evidence that the human proton channel, hH_V1, is a dimer. (A) shows GFP (green fluorescent protein)-tagged hH_V1 channels visualized under fluorescence microscopy. Circled spots were followed over time, as illustrated in (B), where the fluorescence intensity of one spot can be seen to decay in two distinct steps. The pie chart in (C) shows the frequency that tagged channels decayed with the indicated number of steps. *Reprinted from Ref. 5, copyright 2008, with permission from Elsevier.*

discrete steps. Additional evidence that hH_V1 is a dimer, and that each protomer has a separate conduction pathway, was provided by tandem dimers including an introduced Cys at a location that enables block of the channel by a methanethiosulfonate reagent. Wild-type (WT)–WT dimers were not blocked, WT–Cys dimers were half blocked, and Cys–Cys dimers were completely blocked.[5] Some impairment of dimerization was seen when the N-terminus was disrupted or replaced, but dimerization was completely prevented by C-terminus substitution.[5]

Koch *et al.*[3] demonstrated the multimeric nature of the mouse H_V1, mH_V1 (which they called mVSOP), by coimmunoprecipitation of the channel labeled with two different tags. They further used fluorescence resonance energy transfer (FRET) to determine the distance between tagged residues in an extracellular loop of *C. intestinalis* H_V1 (CiH_V1 or CiVSOP) protomers to be 42 Å, also consistent with their being multimers. They showed that the preferred multimeric configuration was a dimer by creating a linked heterodimer including a single introduced Cys residue. Lack of FRET in this construct showed that only two subunits are involved. Gonzalez *et al.*[22] demonstrated FRET in full-length CiH_V1, which disappeared in the N- and C-terminal truncated channel, showing that this construct was monomeric.

Another kind of evidence is provided by the cross-linker disuccinimidyl suberate (DSS), which produced a distinct band at the dimer position on western blots.[3,4,26] In one study, there were weak bands at positions corresponding to higher oligomers,[4] in the other not.[3] The higher order oligomer bands could be nonspecific artifacts, as suggested by Lee and coauthors, but at least some could also represent natural, weaker interactions. Overexpression, cross-linking, and Western blotting all suffer from artifacts, so results must be interpreted with caution; still, these two studies used different H_V1 proteins, cell lines, and methods, greatly increasing the probability that they represent normal interactions. Further increase in confidence is provided by the demonstration of the dimeric nature of native hH_V1 *in situ* in human neutrophils[26] using western blots. Faint bands at the dimer level, which were greatly enhanced by DSS treatment, were seen in lysates from human neutrophils, eosinophils, and monocytes.

3.1. Not all H_V1 are likely to be dimers

Direct evidence exists for the dimeric tendencies of H_V1 in humans, mice, and *C. intestinalis*. However, because the most important determinant of dimeric status is C-terminal coiled-coil interaction,[3–5,22–26] H_V1 lacking this property may exist as monomers. Of confirmed or predicted H_V1, those in the dinoflagellate *Karlodinium veneficum*[27] and in the diatoms *Phaeodactylum tricornutum*[28] and *Thalassiosira*, lack predicted coiled-coil domains and are presumed to be monomeric. H_V1 in coccolithophores, which are also single-celled eukaryotes, contain predicted C-terminal coiled-coil domains, which precludes any generalization regarding distinctions between oligomerization in multicellular and unicellular organisms. At present, there is no evidence that any accessory proteins are involved in or required for dimerization. The proton conduction that is the biological function of hH_V1 can be demonstrated when the purified protein is incorporated into membrane vesicles.[29] Further characterization of the physiological roles of H_V1 in various organisms will be needed to understand any distinction in the roles of dimeric versus monomeric channels.

4. COMPARISON OF THE PROPERTIES OF MONOMERIC AND DIMERIC CONSTRUCTS OF H_V1

Exploring the differences in behavior between monomeric and dimeric constructs of H_V1 is of intrinsic interest but may also provide clues to the evolutionary "purpose" of dimerization. A word of caution must be proffered before accepting the following comparison. Because (most) H_V1 normally exist as dimers, unnatural, and invariably extreme measures must be adopted to force it into monomeric status. For example, truncation of the N- and C-termini must be considered drastic measures. These manipulations themselves may have consequences other than simply producing monomeric channels.

4.1. What does "cooperative gating" mean for ion channels?

Cooperative gating of H_V1 has been discussed with the idea that the two protomers do not function independently. In order to understand the evidence that exists for cooperative gating in H_V1, we will first consider the idea of cooperativity as it relates to ion channels. Cooperative binding of substrates to proteins that have multiple-binding sites is a straightforward concept. A well-known example of (positive) cooperativity is that the first O_2 to bind the tetrameric hemoglobin molecule increases the O_2 affinity of the three remaining binding sites. A classic description of cooperativity in ion channels was provided by Hodgkin and Huxley.

Hodgkin and Huxley observed that both Na^+ and K^+ currents in squid axons activated (turned on) with a delay,[30] which they could explain by postulating multiple identical voltage-sensing elements, all of which must move before the conductance appears. A mathematical representation of this idea is to assume identical conformational changes in each element, each a simple first-order transition between a resting state (***R***) and an activated state (***A***) (Scheme 12.1):

$$\boldsymbol{R} \underset{\beta}{\overset{\alpha}{\rightleftharpoons}} \boldsymbol{A}$$

Scheme 12.1 Two state model.

where the forward and backward rate constants are α and β. On average, the movement of one element (protomer) after a step change of voltage follows an exponential time course, with time constant, $\tau_{act} = 1/(\alpha + \beta)$. Both rate constants are typically voltage-dependent, and the steady state probability of ***A***, the activated state, at a given voltage is $\alpha/(\alpha + \beta)$. If we assume that the channel can conduct only after both voltage-sensor domains (VSDs) in a dimeric channel reach state ***A***, then the opening process can be shown as in Scheme 12.2.

$$\boldsymbol{RR} \underset{\beta}{\overset{2\alpha}{\rightleftarrows}} \boldsymbol{RA} \underset{2\beta}{\overset{\alpha}{\rightleftarrows}} \boldsymbol{AA} = \text{OPEN}$$

Scheme 12.2 Hodgkin-Huxley model.

Scheme 12.2 describes the Hodgkin–Huxley (HH) model for a two component channel. In this scheme, the current will turn on with a sigmoid time course, and upon repolarization, will turn off with an exponential time course, because once either protomer moves to its ***R*** state, conductance ends.

Turning to H_V1, we have discussed evidence for dimers consisting of two identical protomers, each with a distinct conduction pathway and some means of responding to membrane potential changes. In the simplest case, movement of charged groups within the TM helices of the protein would produce a conformational change that directly results in conductance ("channel opening"). In this case, as soon as the first protomer "opened" there would be current. Nonindependence in this scenario could reflect interactions between protomers, such that opening of one facilitates (or inhibits) the opening of the other, exhibiting positive (or negative) cooperativity, in a sense similar to O_2 binding in hemoglobin.

A different class of nonindependence would arise if interactions within the H_V1 dimer resulted in conductance occurring only after both protomers "open." In the sense that conductance requires both members of the dimer to perform, this can be considered cooperative, but this phenomenon can also occur if each protomer moves independently of the other in response to voltage changes. The cooperativity in this example results not from interactions during the initial response to voltage but during a subsequent step during which the conductance appears.[31] This is the kind of cooperativity modeled by Hodgkin and Huxley.[32]

A slightly more complicated model was proposed by Gonzalez *et al.*[22] They postulated an "allosteric" mechanism, in which the $\boldsymbol{R} \rightarrow \boldsymbol{A}$ transition does not immediately result in conductance, even in a monomer, but instead an additional transition must occur: $\boldsymbol{R} \rightarrow \boldsymbol{A} \rightarrow \mathbf{O}$, where **O** is the open (conducting) state. Here, the $\boldsymbol{R} \rightarrow \boldsymbol{A}$ transition (horizontal steps) reflects the movement of the fourth TM helix (S4) in response to voltage into an activated configuration, and the $\boldsymbol{A} \rightarrow \mathbf{O}$ transition (vertical steps) reflects a conformational change that results in conductance (channel opening). For the H_V1 dimer, a simplified version of the allosteric model is shown in Scheme 12.3.

Here, the $\boldsymbol{C}_n$ states are closed (nonconducting), and the $\mathbf{O}_n$ states are open (conducting). The allosteric factor L (which also appears in the vertical

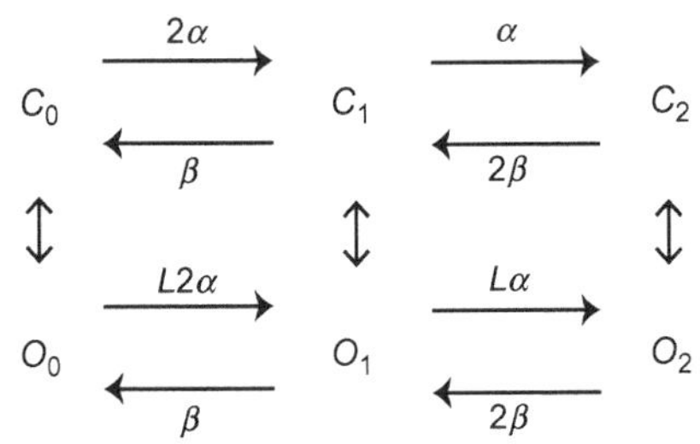

Scheme 12.3 Allosteric model.

transitions, but is not shown here) determines how much interaction exists between protomers. Aside from additional complexity, the allosteric model differs in one significant respect from the HH model. This point requires a preliminary discussion of gating charge.

4.2. What is gating charge and how is it measured?

A voltage-dependent ion channel opens in response to a change in membrane potential. This occurs because there are net charges within the membrane spanning regions of the protein that move in response to voltage changes. For example, the S4 region of most voltage-gated ion channels contains several positively charged Arg (or Lys) residues, and when the membrane is depolarized (by making the inside of the cell more positive), these positive charges are pushed outward. "Gating charge" refers to the number of elementary charges (e_0) in each channel molecule that must be transferred from one side of the membrane to the other during the opening process. If one charge travels across only half the membrane potential field, this would count as 0.5 e_0. One additional complication is that the membrane potential does not change uniformly, as was assumed by Goldman in his "constant field theory,"[33] but is focused at the narrowest part of the pore.[34–37] A standard method for estimating the effective gating charge is called the "limiting slope" method. The steady state conductance (which is assumed to reflect approximately the open probability of the channel, P_{open}) is plotted semilogarithmically against voltage. Almers showed that the slope at the most negative voltage range that can be detected is a good estimate of the effective gating charge for any gating model in which the channel traverses an arbitrary number of closed states to arrive at a single open state.[38] More detailed analysis of which kinds of models follow or deviate from this guideline is provided by Sigg and Bezanilla.[39] For Na^+[40] and *Shaker* K^+ channels, the total effective gating charge estimated by this method is 12–14 e_0.[41–44] To a rough approximation, this estimate corresponds well with the physical manifestation expected if four Arg residues from each of the four VSDs of the tetrameric channel, crossed most of the membrane electrical field. Returning to the models described earlier, the limiting slope method applies to the HH model (Scheme 12.2), but not strictly to the allosteric model (Scheme 12.3). The reason for this is that the allosteric model permits (a) channel opening before all the VSDs have moved and (b) transitions between open states. If these transitions involve movement of gating charge, then the limiting slope will underestimate the true gating charge.[39] An excellent example of this phenomenon is the BK channel, whose g–V relationship at moderately negative voltages first becomes steeper, but at larger negative voltages becomes much shallower.[45] An allosteric model described the gating of this channel well, and explained the anomalous limiting slope data, because channels can open before all the voltage sensors have moved and also because charge movement can occur during transitions between open states.[46] The error introduced by using the limiting slope approach depends on the degree of allosteric interaction; at one extreme, the allosteric model degenerates into a linear model (in which all channels follow the $C_0 \rightarrow C_1 \rightarrow C_2 \rightarrow O_2$ pathway in Scheme 12.3), in which case the limiting slope gives the correct value.

If we consider the K^+ channel in Fig. 12.1, the physical embodiment of the HH model (with movement of all four VSDs required before opening occurs) is clear. There are four identical VSDs, each of which must undergo a conformational change before the central pore is opened.[30,47,48] For H_V1, the physical interpretation is less obvious, because there are two separate pores.

4.3. Evidence that gating of the two protomers in H_V1 is "cooperative" (not independent)

Evidence already discussed shows that each protomer in H_V1 contains a separate conduction pathway. It has become clear, however, that the gating (opening and closing in response to voltage changes) of the two protomers does not occur independently. One line of evidence comes from coexpressing roughly equal amounts of WT hH_V1 and a mutant, E153C, that has the property that its proton conductance–voltage (g_H–V) relationship is shifted—50 mV compared with WT. Independent gating of the protomers should result in a component of negatively shifted channels. On the other hand, if both protomers must undergo a conformational change that precedes conduction of either, channels should open at the more restrictive (more positive) regime. When the experiment was performed, the resulting g_H–V relationship practically superimposed on that of the

WT channel, which has the more positive voltage requirement. Linked heterodimers shared the nearly WT g_H–V relationship seen in the coexpression studies. A puzzling feature of these results was the lack of a shoulder on the g_H–V relationship in the coexpression studies, which would be expected from the fraction of double mutant dimers.

Gonzalez *et al.*[22] also produced elegant evidence for cooperative gating (Fig. 12.3C). They labeled the outer end of the S4 domain with a fluorophore and used voltage-clamp fluorometry to observe S4 movement simultaneously with proton currents during voltage pulses. The fluorescence signal reflecting S4 movement (red) increased exponentially, whereas the proton current (black) increased with a sigmoid time course. In the traditional Hodgkin–Huxley[32] formalism of Scheme 12.2, the time course of the fluorescence signal raised to the second power (green) will predict the time course of the current if the two protomers gate independently and if both must undergo a conformational change before either conducts. It is a happy event when data cooperate so nicely with theoretical expectations!

Further evidence supporting this form of cooperative gating emerged when the Larsson group determined the effective gating charge of hH_V1. WT H_V1 channels have an effective gating charge of ~6 e_0 in rat[49,50] and human.[51] In CiH_V1, the WT channel had 5.9 e_0 but the monomeric construct had 2.7 e_0.[22] This result strongly supports the idea that both protomers must move before either can conduct. A recent study of the mouse Hv1, mH_V1, produced identical phenomenology: the dimeric channel had twice the effective gating charge as seen in monomeric constructs.[25] In this species, the gating charge in WT channels was 4 e_0, with 2 e_0 seen in monomeric constructs.

4.4. Activation kinetics differs between monomer and dimer

Koch *et al.*[3] produced monomeric mH_V1 (mouse) channels by truncating the C-terminus and also by truncating both C- and N-termini. The C-truncated channel (ΔC) opened 2.5 times faster, and the doubly truncated channel (ΔNΔC) opened five times faster. If we interpret the double truncation as more efficiently producing monomeric channels (i.e., that some dimers remain even with C-truncation), then the monomer opens five times faster. In a later study, C-truncation of mH_V1 accelerated channel activation, τ_{act}, by ~sixfold (Fujiwara *et al.*).[25]

Tombola *et al.*[5] produced monomeric chimerae by attaching the C-terminus alone or both the N- and C-termini from the CiVSP, a phosphatase from *C. intestinalis* thought to exist as a monomer,[21] onto hH_V1. A minor

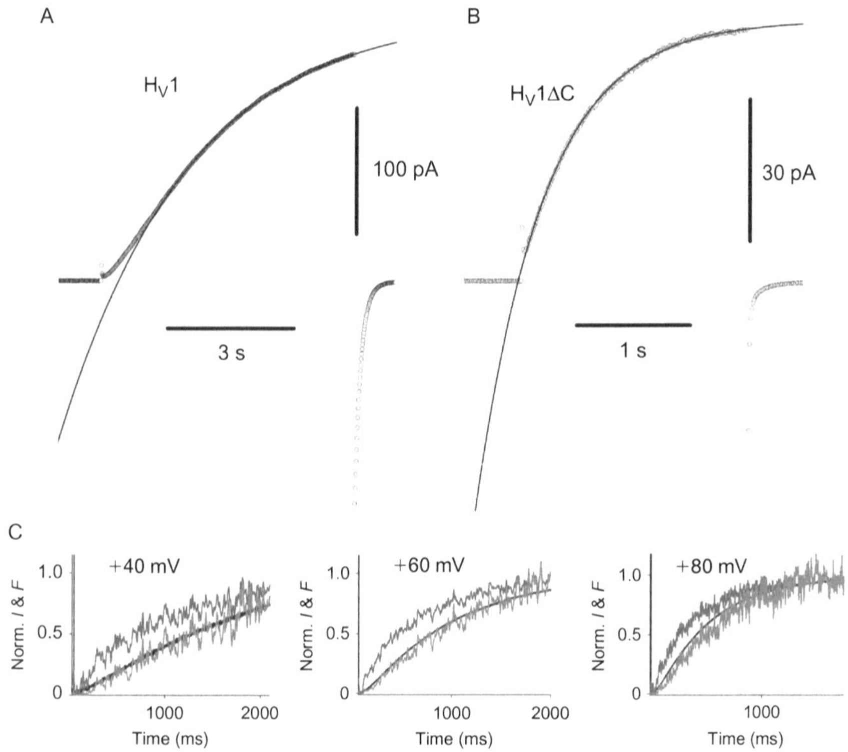

Figure 12.3 Gating kinetics of monomeric and dimeric hH_V1 and CiH_V1. The sigmoid activation of WT dimeric hH_V1 channels (A) is contrasted with the more rapid and exponential turn-on of C-truncated channels (B), presumed to be monomeric. The curves are single exponential fits. In (C), the kinetics of current activation of CiH_V1 at three voltages (black) is compared with the normalized fluorescence signal (red) emitted by a fluorophore tag attached to the top of the S4 domain. When the fluorescence signal is squared (see Scheme 12.2), it practically superimposes onto the current, consistent with a Hodgkin–Huxley-type gating mechanism in which both protomers must "activate" before either can conduct. *Figures A and B: Reprinted with permission from Ref. 24, copyright 2010. Figure C: Reprinted by permission from Ref. 22, copyright 2010.* (See Color Insert.)

effect of swapping the N-terminus alone was also observed. The chimera with both N- and C-termini from CiVSP grafted onto the hH_V1 TM domains activated substantially more rapidly, although no quantitative results were provided. The g_H–V relationship for this monomeric construct was 30% less steep and its midpoint voltage about 10 mV more positive.

Musset *et al.*[24] produced presumed monomeric hH_V1 by C-truncation ($hH_V1\Delta C$). Although independent confirmation that C-truncation produces mainly monomers was not obtained, the activation kinetics was well fitted by a single exponential function (Fig. 12.3B). Activation of the

C-truncated channel was 6.6 times faster than the WT dimer, measured in inside–out membrane patches at pH_o 7.5, pH_i 7.5. The difference was less pronounced in whole-cell data at pH_o 7.0, pH_i 6.5, with the $hH_V1\Delta C$ about three times faster than WT.[52] Consistent with the report of Tombola *et al.*,[5] the g_H–V relationship was somewhat less steep in the monomeric construct, with the midpoint shifted 10–15 mV more positive than WT. The $hH_V1\Delta C$ construct not only opened more rapidly but also the time course of turn-on of current was exponential (Fig. 12.3B), in contrast with the distinctly sigmoidal time course of WT currents (Fig. 12.3A).[24] This behavior is consistent with the HH-type model discussed earlier (Scheme 12.2) in which each protomer must activate before either can conduct. In summary, gating kinetics supports the idea of cooperative gating.

A mechanism for cooperative gating of mH_V1 was proposed recently by Fujiwara *et al.*[25] A number of mutations were carried out in the C-terminal domain. The short region between the C-terminal (intracellular) end of S4 and the start of the coiled-coil region was deleted, extended, or replaced with different amino acids. The resulting mutants fell into two groups: one group (including the monomeric ΔC) opened several times more rapidly and had an effective gating charge of 2 e_0 based on limiting slope measurements, half that of 4 e_0 in the other group (including WT channels). Cooperative gating is consistent with the observation that the slower opening phenotype had twice the gating charge. Several conclusions can be drawn. In general, coiled-coil dimer assembly inhibits activation of H_V1, in the sense that the dimer opens more slowly. Second, the mere fact that the channel was dimeric did not ensure cooperative gating, because a flexible linker (GGG) inserted between S4 and the C-terminus prevented cooperativity. Similarly, a dimer comprising two C-truncated protomers failed to exhibit cooperativity. A related conclusion is that interactions occurring within the TM regions evidently are not sufficient to ensure cooperativity, because these constructs were shown to be dimeric, yet exhibited no signs of cooperativity.

The Okamura group[25] proposed that the C-termini that are known to be linked in the dimer by coiled-coil interactions[4,23] directly modulate S4 movement that results in channel opening. The C-termini are considered to be part of a rigid rod extending through the S4 helix, so that movement of S4 in one protomer would be linked mechanically via the C-termini to the second S4 domain. It is not precisely clear what degree of independence is allowed in this model. Thus, if both entire C-terminus S4 domain assemblies were completely rigid, both protomers would be expected to move simultaneously. If the result of this concerted movement were direct activation of the conductance (opening of the channels), then the time course of current turn-on would be exponential, not sigmoid. Given that the movement of S4 observed directly by Gonzalez *et al.*[22] had an exponential time course, but the current has a sigmoid turn-on (Fig. 12.3C), one possibility is that a certain degree of flexibility in the structure allows a delay between the movement of the two S4 domains. Alternatively, there must be one or more additional steps that couple S4 movement to opening of the conduction pathway that would account for the delayed turn-on of the current. The Larsson group[22] could model their results either with the assumption (a) that the two protomers move independently and both must move before conduction occurs in either or (b) that there is a strong allosteric interaction that favors coordinated movement of both S4 domains over the movement of a single protomer. The latter model was also proposed by Tombola *et al.*[53]

Distinguishing these two classes of models may be difficult, especially when there is a sufficiently high degree of allosteric coupling, because the allosteric mechanism becomes indistinguishable from the HH model. Indirect evidence that seems to support the allosteric model comes from attempts to measure single channel current amplitude in H_V1. Unitary H^+ currents are $\sim 10^3$ smaller than those of most ion channels and consequently are very difficult to resolve directly. However, under favorable conditions (absence of other conductances, extremely high-resistance seals (up to 5 TΩ) which minimize noise,[54] maximized conductance induced by using low pH_i), Cherny *et al.*[55] were just able to resolve what appeared to be unitary currents, of $\sim$10 fA amplitude. An alternative method of measuring unitary currents involves analyzing current fluctuations resulting from stochastic opening and closing of channels.[56] Current variance was well resolved in excised membrane patches from human eosinophils and provided reproducible estimates of single channel conductance.[55] However, the values derived from current fluctuations were consistently about twofold smaller than those from direct observation. Concerted opening of both protomers would result in channel openings with a conductance double that of each monomer, while allosteric interaction would allow two separate steps. The current fluctuation results can be reconciled with the apparent double-sized unitary current events if allosteric interaction results in both protomers opening in rapid succession, so that the current appears to comprise a single step, rather than two steps. Because the unitary current amplitude was just on the borderline of detectability, caution should be exercised before putting too much weight on this kind of interpretation.

5. PROPOSALS FOR THE H_V1 DIMER INTERFACE

The first proposal[4] for the orientation of the hH_V1 dimer is shown in Fig. 12.4A. Cys residues were introduced at numerous locations, and cross-bridge formation was evaluated. Several cross bridges were detected at the extracellular end of S1 (red cylinders, extracellular is toward the top of Fig. 12.4) and in the S1–S2 linker; this was proposed to be the main interface region, in addition to the predicted coiled-coil interaction of the C-termini.[4] This dimer orientation is supported by FRET data obtained from *Ciona* Hv1[3] that indicated a distance of 42 Å between Cys^{243} on each protomer. Cys^{243} of CiHv1 is on the S2/S3 linker (connecting green and blue helices in Fig. 12.4), predicted to be far apart in a dimer connected at the S1 helices.

On the other hand, data consistent with more than one dimer interface also exist. In the Lee *et al.*[4] study, distinct crosslinks at hH_V1 position 194, also in the S2/S3 linker, were observed, indicating that this position attains close proximity a measurable fraction of the time, which would be unexpected if the only possible interface were through the S1 helices.

Additional evidence for another dimer interface comes from the use of Zn^{2+}, the most potent inhibitor of H_V channels,[57–60] which acts by preventing channel opening. Although sensitivity to Zn^{2+} has been examined almost religiously as a pathognomonic feature that must be confirmed in any newly discovered proton conductance in cells or species not

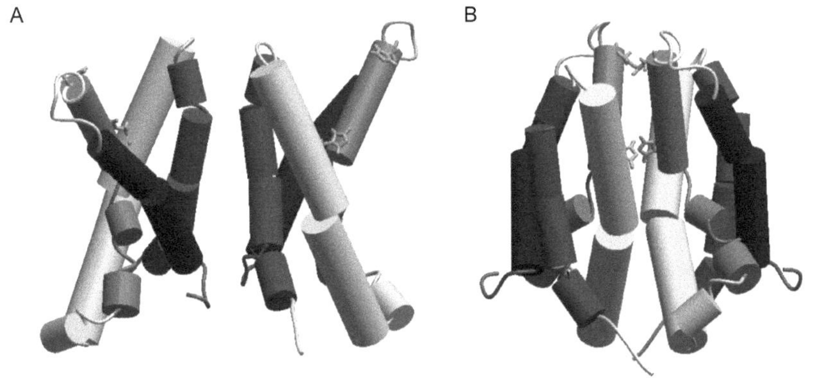

Figure 12.4 Two proposed dimer interfaces. Cross-linking studies indicate several points of attachment at the top (outer) end of S1 (red).[4] The binding of Zn^{2+} to various hH_V1 mutants with His^{140} or His^{193} (aqua) replaced by Ala suggested that high-affinity bidentate Zn^{2+} binding occurs at the dimer interface.[24] *Reprinted with permission from Ref. 24, copyright 2010.* (See Color Insert.)

previously studied, Zn^{2+} sensitivity is extremely sensitive to pH.[57] The sensitivity of the channel to externally applied Zn^{2+} becomes quite weak at low pH_o, strongly suggestive of competition between H^+ and Zn^{2+} for the binding site where Zn^{2+} exerts its inhibitory effects. By mathematically modeling the competition between H^+ and Zn^{2+} at different pH_o, Cherny and DeCoursey concluded that Zn^{2+} must be multiply coordinated, and that the pK_a of the coordinating groups was roughly 6–7, suggesting that two or more His residues formed the Zn^{2+}-binding site.[57]

When the hH_V1 gene was identified, and two externally accessible His residues (His^{140} and His^{193}) were shown to account for the inhibitory effects of Zn^{2+},[1] the obvious conclusion was that these two His residues most likely coordinated Zn^{2+} at its site of action. However, in a molecular model of the open-state monomer of hH_V1, His^{140} and His^{193} were too far apart to coordinate Zn^{2+} plausibly.[24] Protein:protein docking of the model structure indicated the possibility that His pairs from different protomers could approach close enough for Zn^{2+} coordination to occur in a dimer whose interface was between the S2/S3 helices.

Several types of experiments were carried out to test the hypothesis of interprotomer Zn^{2+} coordination. The first prediction of this hypothesis is that the affinity of Zn^{2+} for the dimeric channel should be greater than for the monomer, which would lack the possibility of bidentate coordination. Indeed, C-terminally truncated constructs of both human and mouse H_V1 (presumed to be monomeric—see earlier) exhibited significantly diminished response to Zn^{2+}, consistent with the prediction. This result conflicts with a report using a different H_V1 monomer construct.[53] Because this report was based on measurements under one set of conditions at one pH at a single voltage,[53] a chance combination of separate Zn^{2+} effects may have canceled each other out. Zn^{2+} has complex pharmacological effects, including profoundly slowing activation, shifting the g_H–V relationship positively, and probably also decreasing $g_{H,max}$.[57] It is also possible that the different monomeric constructs responded differently to Zn^{2+}.

Another set of experiments involved testing mutants and tandem constructs in which alanine replaced the histidines in several combinations. Figure 12.5 illustrates the constructs tested and their relative Zn^{2+} sensitivity, using the slowing of channel opening as the parameter of interest. A similar pattern was observed when the shift of the g_H–V relationship by Zn^{2+} was the parameter evaluated. As previously shown by Ramsey *et al.*,[1] mutation of His^{140} or His^{193} alone attenuated but did not eliminate the Zn^{2+} sensitivity, demonstrating that both residues contribute to WT Zn^{2+} sensitivity. The double

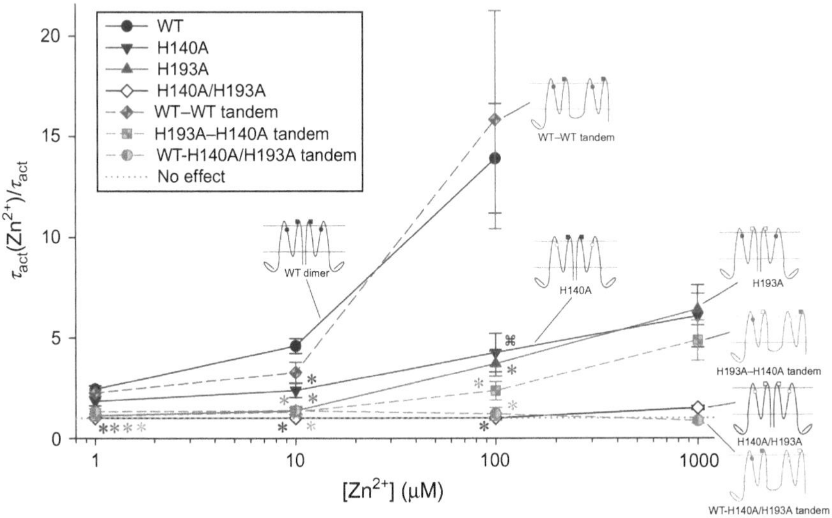

Figure 12.5 Evidence that Zn^{2+} binds with high affinity at the interface between protomers in the hH_V1 dimer. The slowing of proton current activation by Zn^{2+} is plotted for several hH_V1 constructs. The constructs listed in the inset are illustrated with His^{140} shown as a circle and His^{193} as a square. Solid symbols indicate His at the indicated position, open symbols indicate replacement with Ala. *Reprinted with permission from Ref. 24, copyright 2010.* (See Color Insert.)

mutant lacked any slowing up to 1 mM Zn^{2+}, suggesting that no other groups on the channel contribute significantly. Control tandem dimers in which both protomers possessed both histidines showed Zn^{2+} sensitivity indistinguishable from WT (Fig. 12.5, blue vs. green symbols) as well as sigmoidal activation kinetics suggestive of cooperative gating. A tandem dimer in which one protomer lacked His^{140} and the other protomer lacked His^{193} ("H193A–H140A tandem") exhibited Zn^{2+} sensitivity similar to single mutants in the WT background (Fig. 12.5). A surprising result was observed for a tandem dimer in which one protomer was WT, with both His present, while the other protomer lacked both His. Just as for the double His mutant (lacking both His), Zn^{2+} did not slow opening in the tandem dimer in which both histidines were present in one protomer, while the other protomer lacked His. In summary, slowing of channel opening by Zn^{2+} was observed only when there was at least one His in each protomer. Taken together, these data indicated that Zn^{2+} is most likely coordinated between protomers, which implies a dimer interface involving S2 and S3 helices (Fig. 12.4B).

The apparent discrepancy between the two suggested dimer interfaces could be explained in at least two ways. In one, the TM domains of the two protomers, although tethered by coiled-coil interactions within the C-termini, may be able to rotate around the tether and adopt different orientations relative to each other. This possibility appears to be less likely in light of the proposed rigidity of the linker between S4 and the coiled-coil region.[25] Nevertheless, a tandem WT–WT dimer, linked with a short (6 AA) connection between the C-terminus of one monomer and the N-terminus of the second, functioned almost identically to the WT channel, with the exception of somewhat slower opening kinetics.[24,52] This result might indicate that the required C-terminal interaction can occur almost normally despite the proximity of an attached N-terminus. Another explanation posits the formation of a tetramer consisting of a dimer-of-dimers; this complex maybe stabilized by Zn^{2+} binding at an interface at the S2/S3 helices. Wherever Zn^{2+} binds, it clearly stabilizes the closed conformation, because it slows channel opening, and shifts the g_H–V relationship in the positive direction.[57]

It might be objected that the molecular model that gave rise to the hypothesis of Zn^{2+} binding between protomers is presumed to be an open-state model, whereas the general interpretation of Zn^{2+} effects presumes that Zn^{2+} binds to the closed channel. The relative positions of His^{140} and His^{193} were estimated in the predicted closed state, based on the positions of the corresponding residues in molecular models of closed $K_V1.2$ VSDs.[61] The estimated closed state positions do not differ greatly from those in the open state—in particular, they still do not appear to approach closely enough to coordinate Zn^{2+}. This is not unexpected, considering that neither His is located on the S4 helix, which is thought to be the main "moving part" of the H_V1. In any case, the results obtained do not depend on perfect accuracy of the molecular model. When the crystal structure of the closed hH_V1 molecule has been obtained, this direct evidence should resolve many questions that at present can only be addressed indirectly.

6. PHYSIOLOGICAL CONSEQUENCES OF DIMERIZATION OF H_V1

The most consistent difference between monomeric and dimeric H_V1 is in gating kinetics. However, this difference, while clear, is not dramatic. In some situations, the three- to sixfold faster monomeric opening rate might seem to be an important distinction, but considering that among species, H_V1 activation kinetics varies over several orders of magnitude,[62] it would appear that a species with a need for quickly opening proton channels would be better served in making a channel that is intrinsically fast. H_V in snail neurons open within a few milliseconds,[63] whereas mammalian H_V1 require seconds.[64] From this perspective, fundamental kinetics would most logically be determined genetically, not by adjusting multimerization. On the other hand, if it were possible for a cell to switch its H_V1 between monomer and dimer status, then a fivefold change in gating kinetics might have important consequences. In fact, Koch *et al.*[3] initially suggested that this kind of mechanism might be responsible for the "enhanced gating mode" of proton channels in phagocytes.

6.1. Does the enhanced gating mode in phagocytes reflect dimer-to-monomer conversion of H_V1?

The clearest function of proton channels in phagocytes (neutrophils, macrophages, and eosinophils) is to facilitate and sustain NADPH oxidase (or "Nox2") activity. Detection of a foreign invader (e.g., a bacterium) by the phagocyte triggers a complex series of biochemical and cellular process culminating in phagocytosis of the detected material. A rapid biochemical consequence of the detection mechanism is the assembly, from its several components, of Nox2 in the phagosome membrane. Nox2 uses intracellular NADPH as a source of electrons, which it translocates across the phagosome membrane to produce reactive oxygen species that are released into the forming or mature phagosome. Electron translocation by Nox2 depolarizes the membrane potential,[65] while NADPH catabolism, which releases $NADP^+$ and H^+,[8,66] decreases the intracellular pH.[65,67–70] Proton efflux through H_V1 compensates for both the membrane depolarization and the pH change.[10,11,64,65,68,69,71–73]

Another consequence of the detection of foreign entities by phagocytes is a radical change in the gating kinetics of H_V1, which turns out to be a significant factor contributing to the activation of H^+ efflux. Four properties of hH_V1 change: opening occurs more quickly, closing slows, the maximum conductance $g_{H,max}$ increases, and the g_H–V relationship shifts negatively by 40 mV.[71,74–77] This transformation of H_V1 properties is termed the "enhanced gating mode." As a result of the four changes in H_V1 behavior, each of which promotes H^+ efflux, during enhanced gating the channel opens sooner and thereby improves the efficiency of NADPH oxidase activity by 15–20%.[78] Is it possible that enhanced gating reflects conversion of the normally dimeric hH_V1 to monomeric status? This suggestion[3] has not yet been resolved beyond doubt, but the bulk of evidence suggests that this is not the case.[79]

Monomeric H_V1 and H_V1 in enhanced gating mode share one property: monomeric H_V1 open several times faster than dimeric channels.[3,5,24,52] Similarly, during enhanced gating, H_V1 opens four to six times faster than in resting cells.[12,71,74–77,80,81] Three other properties evident in the enhanced gating mode differ from those of monomeric constructs. (1) The position of the g_H–V relationship appears to be 10–15 mV more positive in monomeric constructs than WT dimeric channels.[52,53] In enhanced gating mode, the g_H–V relationship shifts in the opposite direction, negatively by ~40 mV.[71,74–77,80,81] (2) During enhanced gating, the Zn^{2+} sensitivity of hH_V1 is not changed,[52] whereas the Zn^{2+} sensitivity of the C-truncated construct was weaker than that of the dimer.[24] (3) During enhanced gating, activation of hH_V1 continues to have a sigmoid time course, in contrast to the exponential activation of the monomer.[24] On balance, the differences in channel properties during enhanced gating mode compared to monomeric constructs appear to outweigh the similarities. In another supporting piece of evidence, Petheő *et al.*[26] did not observe any difference in DSS cross-linking potential when they pretreated human neutrophils with PMA, which effectively induces the enhanced gating mode.

An alternative explanation is that phosphorylation of hH_V1,[82] specifically at Thr[29,83] is responsible for converting the resting channel into enhanced gating mode. The PKC activator PMA produces enhanced gating[12,15,16,74–77,80,82–84] that is prevented or reversed by PKC inhibitors such as staurosporine or GF109203X (GFX).[12,15,76,77,82–84] Other activators of enhanced gating including arachidonic acid and lipopolysaccharide act at least partially through PKC.[81,82,84]

6.2. Might the differences in properties of monomeric and dimeric constructs provide clues to the functional importance of dimerization?

The main differences between WT dimeric H_V1 and monomeric constructs include four properties: (a) activation (channel opening) is approximately five times slower in WT dimeric channels, (b) WT channel opening occurs with a sigmoid rather than exponential time course, (c) the WT g_H–V relationship is 10–15 mV more negative, and (d) the WT voltage dependence is twice steeper. We may for now discount the difference in opening time course (b) and the relatively subtle change in the position of the g_H–V

relationship (c). Although there are certain situations in which these changes could have functional ramifications—in excitable cells like cardiac muscle, for example, a 10 mV shift of a g–V curve has profound physiological significance[85]—in the absence of detailed knowledge about the precise situations in which H_V1 operate in most of the many cells where they exist, it seems pedantic to focus on small differences. The faster opening (a) and steeper voltage dependence (d) do, however, seem important enough to warrant speculation.

As discussed earlier, phagocytes, especially neutrophils and eosinophils, use proton channels during the respiratory burst, when Nox2 is active. The best understood role of H_V1 in this situation is to compensate charge,[65] to prevent excessive depolarization that would otherwise result from the activity of the electrogenic Nox2. Nox2 activity tends to occur with a delay of several seconds or even minutes, depending upon the stimulus, and may continue for hours.[86] Thus, it may be that the ability of H_V1 to open rapidly may not have much value to these cells. The phagocyte "cares" only that enough H_V1 open as soon as Nox2 is active to limit the depolarization. From the vantage point of the phagocyte, there is no rush, because Nox2 turns on slowly and remains active for a long while. In fact, in these nonexcitable cells, proton channel opening is quite slow, compared with the kinetics of gating of most ion channels, and compared with the opening of H_V in other species, such as snail neurons.[63,87,88]

On the other hand, H_V1 opening plays a major role in limiting the extent of the depolarization that occurs when Nox2 is active; membrane depolarization, especially beyond +50 mV, directly inhibits Nox2 activity.[10] To "help" Nox2, H_V1 should open within a relatively negative voltage range. The enhanced gating mode, which shifts Hv1's g_H–V relationship negatively by 40 mV, clearly facilitates this goal. Enhanced gating improves the efficiency of Nox2 by 15–20%, because to open enough channels to compensate Nox2, requires 24–30 mV less depolarization in human neutrophils and eosinophils than would be needed if there were no enhanced gating mode.[78] We have already seen that enhanced gating mode does not seem to represent a change in the dimerization state of H_V1.

The steep voltage dependence of H_V1 also facilitates Nox2 activity. Cooperative gating increases the steepness of the voltage dependence of ion channels.[89] The cooperative gating of the hH_V1 dimer results in a doubling of the voltage dependence compared to the monomer.[22] The result is that substantially less depolarization is required to open a given number of proton channels. One might argue that the cell could enable sufficient activation of g_H simply by shifting the g_H–V relationship negatively, but once enhanced gating is in effect, $V_{threshold}$ is already perilously close to E_H. The cell "wants" H_V1 to turn on rapidly once there is an outward electrochemical gradient for H^+, but it definitely does not want H_V1 to open when the gradient is inward. The main function of H_V1 in most cells is acid extrusion—cells continually produce metabolic acid that they must eliminate; this is the "central problem of pH_i regulation."[90] For this reason, it would be self-destructive to have a constitutively active g_H, because this would act as a proton leak that would flood the cell with unwanted H^+. The most optimal possible design is to poise $V_{threshold}$ near E_H and then to have strong dependence of P_{open} upon depolarization above this point. Nature has produced the most efficient possible mechanism.

It would be interesting to know why H_V in some species appear to be monomeric. However, because we know the functions and the situations in which the channels are active only in the most general terms, it seems premature to attempt to predict why monomeric behavior would be preferable.

7. CONCLUSIONS

In most of the species where it has been identified, H_V1 exists as a dimer. Dimerization is driven and enforced mainly through coiled-coil interactions at the C-terminus; H_V1 sequences from a few single-celled organisms do not have a predicted coiled-coil region and so may exist as monomers. Dimerization in H_V1 provides the opportunity for a cooperative gating mechanism that gives rise to a steeply voltage-dependent conductance. This property appears to be important in the most intensively studied function of H_V1, which exists in human phagocytes, where steep voltage dependence provides optimum compensation for Nox activity and phagocyte function. A number of questions regarding H_V1 dimerization remain open, including the nature of the dimer interface, the mechanism of cooperativity, the impact of multimeric state on trans-acting factors (e.g., kinases) of H_V1, possible effects of transitions between monomeric and multimeric states, and implications of cooperative gating in cell types other than phagocytes.

ACKNOWLEDGMENTS

This work was supported by NSF award MCB-0943362 to T. E. D. and S. M. E. S. and NIH R01-GM087507 to T. E. D. The content is solely the responsibility of the authors and does not necessarily represent the views of the National Institute of General Medical Sciences or the National Institutes of Health.

REFERENCES

1. Ramsey IS, Moran MM, Chong JA, Clapham DE. A voltage-gated proton-selective channel lacking the pore domain. *Nature* 2006;**440**:1213–6.
2. Sasaki M, Takagi M, Okamura Y. A voltage sensor-domain protein is a voltage-gated proton channel. *Science* 2006;**312**:589–92.
3. Koch HP, Kurokawa T, Okochi Y, Sasaki M, Okamura Y, Larsson HP. Multimeric nature of voltage-gated proton channels. *Proc Natl Acad Sci USA* 2008;**105**:9111–6.
4. Lee SY, Letts JA, Mackinnon R. Dimeric subunit stoichiometry of the human voltage-dependent proton channel Hv1. *Proc Natl Acad Sci USA* 2008;**105**:7692–5.
5. Tombola F, Ulbrich MH, Isacoff EY. The voltage-gated proton channel Hv1 has two pores, each controlled by one voltage sensor. *Neuron* 2008;**58**:546–56.
6. Cherny VV, Markin VS, DeCoursey TE. The voltage-activated hydrogen ion conductance in rat alveolar epithelial cells is determined by the pH gradient. *J Gen Physiol* 1995;**105**:861–96.
7. Demaurex N. Functions of proton channels in phagocytes. *Wiley Interdisc Rev Membr Transp Signal* 2012;**1**:3–15.
8. Henderson LM, Chappel JB. NADPH oxidase of neutrophils. *Biochim Biophys Acta* 1996;**1273**:87–107.
9. DeCoursey TE. Voltage-gated proton channels find their dream job managing the respiratory burst in phagocytes. *Physiology (Bethesda)* 2010;**25**:27–40.
10. DeCoursey TE, Morgan D, Cherny VV. The voltage dependence of NADPH oxidase reveals why phagocytes need proton channels. *Nature* 2003;**422**:531–4.
11. Morgan D, Capasso M, Musset B, Cherny VV, Ríos E, Dyer MJS, et al. Voltage-gated proton channels maintain pH in human neutrophils during phagocytosis. *Proc Natl Acad Sci USA* 2009;**106**:18022–7.
12. Musset B, Morgan D, Cherny VV, MacGlashan Jr. DW, Thomas LL, Ríos E, et al. A pH-stabilizing role of voltage-gated proton channels in IgE-mediated activation of human basophils. *Proc Natl Acad Sci USA* 2008;**105**:11020–5.
13. Fischer H. Function of proton channels in lung epithelia. *WIRES Interdiscip Rev Membr Transp Signal* 2012;**1**:247–58.
14. Lishko PV, Botchkina IL, Fedorenko A, Kirichok Y. Acid extrusion from human spermatozoa is mediated by flagellar voltage-gated proton channel. *Cell* 2010;**140**:327–37.
15. Musset B, Clark RA, DeCoursey TE, Petheo GL, Geiszt M, Chen Y, et al. NOX5 in human spermatozoa: expression, function and regulation. *J Biol Chem* 2012;**287**:9376–88.
16. Capasso M, Bhamrah MK, Henley T, Boyd RS, Langlais C, Cain K, et al. HVCN1 modulates BCR signal strength via regulation of BCR-dependent generation of reactive oxygen species. *Nat Immunol* 2010;**11**:265–72.
17. Wang Y, Li SJ, Wu X, Che Y, Li Q. Clinicopathological and biological significance of human voltage-gated proton channel Hv1 over-expression in breast cancer. *J Biol Chem* 2012;**287**:13877–88.
18. Wu LJ, Wu G, Sharif MR, Baker A, Jia Y, Fahey FH, et al. The voltage-gated proton channel Hv1 enhances brain damage from ischemic stroke. *Nat Neurosci* 2012;**15**:565–73.
19. MacKinnon R. Determination of the subunit stoichiometry of a voltage-activated potassium channel. *Nature* 1991;**350**:232–5.
20. Doyle DA, Morais Cabral J, Pfuetzner RA, Kuo A, Gulbis JM, Cohen SL, et al. The structure of the potassium channel: molecular basis of K^+ conduction and selectivity. *Science* 1998;**280**:69–77.
21. Kohout SC, Ulbrich MH, Bell SC, Isacoff EY. Subunit organization and functional transitions in Ci-VSP. *Nat Struct Mol Biol* 2008;**15**:106–8.
22. Gonzalez C, Koch HP, Drum BM, Larsson HP. Strong cooperativity between subunits in voltage-gated proton channels. *Nat Struct Mol Biol* 2010;**17**:51–6.

23. Li SJ, Zhao Q, Zhou Q, Unno H, Zhai Y, Sun F. The role and structure of the carboxyl-terminal domain of the human voltage-gated proton channel Hv1. *J Biol Chem* 2010;**285**:12047–54.
24. Musset B, Smith SM, Rajan S, Cherny VV, Sujai S, Morgan D, et al. Zinc inhibition of monomeric and dimeric proton channels suggests cooperative gating. *J Physiol* 2010;**588**: 1435–49.
25. Fujiwara Y, Kurokawa T, Takeshita K, Kobayashi M, Okochi Y, Nakagawa A, et al. The cytoplasmic coiled-coil mediates cooperative gating temperature sensitivity in the voltage-gated H^+ channel Hv1. *Nat Commun* 2012;**3**:816.
26. Petheő GL, Orient A, Baráth M, Kovács I, Réthi B, Lányi A, et al. Molecular and functional characterization of H_V1 proton channel in human granulocytes. *PLoS One* 2010;**5**: e14081.
27. Smith SME, Morgan D, Musset B, Cherny VV, Place AR, Hastings JW, et al. Voltage-gated proton channel in a dinoflagellate. *Proc Natl Acad Sci USA* 2011;**108**:18162–8.
28. Taylor AR, Brownlee C, Wheeler GL. Proton channels in algae: reasons to be excited. *Trends Plant Sci* 2012;**17**:675–84.
29. Lee SY, Letts JA, MacKinnon R. Functional reconstitution of purified human Hv1 H^+ channels. *J Mol Biol* 2009;**387**:1055–60.
30. Hodgkin AL, Huxley AF. The components of membrane conductance in the giant axon of *Loligo*. *J Physiol* 1952;**116**:473–96.
31. Horn R. Uncooperative voltage sensors. *J Gen Physiol* 2009;**133**:463–6.
32. Hodgkin AL, Huxley AF. A quantitative description of membrane current and its application to conduction and excitation in nerve. *J Physiol* 1952;**117**:500–44.
33. Goldman DE. Potential, impedance, and rectification in membranes. *J Gen Physiol* 1943;**27**:37–60.
34. Ahern CA, Horn R. Focused electric field across the voltage sensor of potassium channels. *Neuron* 2005;**48**:25–9.
35. Asamoah OK, Wuskell JP, Loew LM, Bezanilla F. A fluorometric approach to local electric field measurements in a voltage-gated ion channel. *Neuron* 2003;**37**:85–97.
36. Starace DM, Bezanilla F. A proton pore in a potassium channel voltage sensor reveals a focused electric field. *Nature* 2004;**427**:548–53.
37. Grabe M, Lecar H, Jan YN, Jan LY. A quantitative assessment of models for voltage-dependent gating of ion channels. *Proc Natl Acad Sci USA* 2004;**101**:17640–5.
38. Almers W. Gating currents and charge movements in excitable membranes. *Rev Physiol Biochem Pharmacol* 1978;**82**:96–190.
39. Sigg D, Bezanilla F. Total charge movement per channel. The relation between gating charge displacement and the voltage sensitivity of activation. *J Gen Physiol* 1997;**109**:27–39.
40. Hirschberg B, Rovner A, Lieberman M, Patlak J. Transfer of twelve charges is needed to open skeletal muscle Na^+ channels. *J Gen Physiol* 1995;**106**:1053–68.
41. Islas LD, Sigworth FJ. Voltage sensitivity and gating charge in *Shaker* and *Shab* family potassium channels. *J Gen Physiol* 1999;**114**:723–42.
42. Noceti F, Baldelli P, Wei X, Qin N, Toro L, Birnbaumer L, et al. Effective gating charges per channel in voltage-dependent K^+ and Ca^{2+} channels. *J Gen Physiol* 1996;**108**: 143–55.
43. Schoppa NE, McCormack K, Tanouye MA, Sigworth FJ. The size of gating charge in wild-type and mutant *Shaker* potassium channels. *Science* 1992;**255**:1712–5.
44. Seoh SA, Sigg D, Papazian DM, Bezanilla F. Voltage-sensing residues in the S2 and S4 segments of the Shaker K^+ channel. *Neuron* 1996;**16**:1159–67.
45. Horrigan FT, Cui J, Aldrich RW. Allosteric voltage gating of potassium channels I. Mslo ionic currents in the absence of Ca^{2+}. *J Gen Physiol* 1999;**114**:277–304.

46. Horrigan FT, Aldrich RW. Allosteric voltage gating of potassium channels II. Mslo channel gating charge movement in the absence of Ca^{2+}. *J Gen Physiol* 1999;**114**:305–36.
47. Horn R, Ding S, Gruber HJ. Immobilizing the moving parts of voltage-gated ion channels. *J Gen Physiol* 2000;**116**:461–76.
48. Gagnon DG, Bezanilla F. A single charged voltage sensor is capable of gating the Shaker K^+ channel. *J Gen Physiol* 2009;**133**:467–83.
49. DeCoursey TE, Cherny VV. Effects of buffer concentration on voltage-gated H^+ currents: does diffusion limit the conductance? *Biophys J* 1996;**71**:182–93.
50. DeCoursey TE, Cherny VV. Deuterium isotope effects on permeation and gating of proton channels in rat alveolar epithelium. *J Gen Physiol* 1997;**109**:415–34.
51. Musset B, Cherny VV, Morgan D, Okamura Y, Ramsey IS, Clapham DE, et al. Detailed comparison of expressed and native voltage-gated proton channel currents. *J Physiol* 2008;**586**:2477–86.
52. Musset B, Smith SM, Rajan S, Cherny VV, Morgan D, DeCoursey TE. Oligomerization of the voltage gated proton channel. *Channels (Austin)* 2010;**4**:260–5.
53. Tombola F, Ulbrich MH, Kohout SC, Isacoff EY. The opening of the two pores of the Hv1 voltage-gated proton channel is tuned by cooperativity. *Nat Struct Mol Biol* 2010;**17**:44–50.
54. Levis RA, Rae JL. The use of quartz patch pipettes for low noise single channel recording. *Biophys J* 1993;**65**:1666–77.
55. Cherny VV, Murphy R, Sokolov V, Levis RA, DeCoursey TE. Properties of single voltage-gated proton channels in human eosinophils estimated by noise analysis and by direct measurement. *J Gen Physiol* 2003;**121**:615–28.
56. Hille B. *Ion channels of excitable membranes*. Sunderland, MA: Sinauer Associates Inc.; 2001.
57. Cherny VV, DeCoursey TE. pH-dependent inhibition of voltage-gated H^+ currents in rat alveolar epithelial cells by Zn^{2+} and other divalent cations. *J Gen Physiol* 1999;**114**:819–38.
58. Mahaut-Smith MP. The effect of zinc on calcium and hydrogen ion currents in intact snail neurones. *J Exp Biol* 1989;**145**:455–64.
59. Thomas RC, Meech RW. Hydrogen ion currents and intracellular pH in depolarized voltage-clamped snail neurones. *Nature* 1982;**299**:826–8.
60. DeCoursey TE, Cherny VV. Pharmacology of voltage-gated proton channels. *Curr Pharm Des* 2007;**13**:2400–20.
61. Pathak MM, Yarov-Yarovoy V, Agarwal G, Roux B, Barth P, Kohout S, et al. Closing in on the resting state of the Shaker K^+ channel. *Neuron* 2007;**56**:124–40.
62. DeCoursey TE. Voltage-gated proton channels and other proton transfer pathways. *Physiol Rev* 2003;**83**:475–579.
63. Byerly L, Meech R, Moody Jr. W. Rapidly activating hydrogen ion currents in perfused neurones of the snail, *Lymnaea stagnalis*. *J Physiol* 1984;**351**:199–216.
64. DeCoursey TE, Cherny VV. Potential, pH, and arachidonate gate hydrogen ion currents in human neutrophils. *Biophys J* 1993;**65**:1590–8.
65. Henderson LM, Chappell JB, Jones OTG. The superoxide-generating NADPH oxidase of human neutrophils is electrogenic and associated with an H^+ channel. *Biochem J* 1987;**246**:325–9.
66. Babior BM, Curnutte JT, McMurrich BJ. The particulate superoxide-forming system from human neutrophils. Properties of the system and further evidence supporting its participation in the respiratory burst. *J Clin Invest* 1976;**58**:989–96.
67. Gabig TG, Lefker BA, Ossanna PJ, Weiss SJ. Proton stoichiometry associated with human neutrophil respiratory-burst reactions. *J Biol Chem* 1984;**259**:13166–71.

68. Henderson LM, Chappell JB, Jones OTG. Superoxide generation by the electrogenic NADPH oxidase of human neutrophils is limited by the movement of a compensating charge. *Biochem J* 1988;**255**:285–90.
69. Henderson LM, Chappell JB, Jones OTG. Internal pH changes associated with the activity of NADPH oxidase of human neutrophils. Further evidence for the presence of an H^+ conducting channel. *Biochem J* 1988;**251**:563–7.
70. Molski TF, Naccache PH, Volpi M, Wolpert LM, Sha'afi RI. Specific modulation of the intracellular pH of rabbit neutrophils by chemotactic factors. *Biochem Biophys Res Commun* 1980;**94**:508–14.
71. Bánfi B, Schrenzel J, Nüsse O, Lew DP, Ligeti E, Krause KH, et al. A novel H^+ conductance in eosinophils: unique characteristics and absence in chronic granulomatous disease. *J Exp Med* 1999;**190**:183–94.
72. El Chemaly A, Okochi Y, Sasaki M, Arnaudeau S, Okamura Y, Demaurex N. VSOP/Hv1 proton channels sustain calcium entry, neutrophil migration, and superoxide production by limiting cell depolarization and acidification. *J Exp Med* 2010;**207**:129–39.
73. Ramsey IS, Ruchti E, Kaczmarek JS, Clapham DE. Hv1 proton channels are required for high-level NADPH oxidase-dependent superoxide production during the phagocyte respiratory burst. *Proc Natl Acad Sci USA* 2009;**106**:7642–7.
74. DeCoursey TE, Cherny VV, DeCoursey AG, Xu W, Thomas LL. Interactions between NADPH oxidase-related proton and electron currents in human eosinophils. *J Physiol* 2001;**535**:767–81.
75. DeCoursey TE, Cherny VV, Morgan D, Katz BZ, Dinauer MC. The $gp91^{phox}$ component of NADPH oxidase is not the voltage-gated proton channel in phagocytes, but it helps. *J Biol Chem* 2001;**276**:36063–6.
76. Musset B, Cherny VV, DeCoursey TE. Strong glucose dependence of electron current in human monocytes. *Am J Physiol Cell Physiol* 2012;**302**:C286–C295.
77. Mori H, Sakai H, Morihata H, Kawawaki J, Amano H, Yamano T, et al. Regulatory mechanisms and physiological relevance of a voltage-gated H^+ channel in murine osteoclasts: phorbol myristate acetate induces cell acidosis and the channel activation. *J Bone Miner Res* 2003;**18**:2069–76.
78. Murphy R, DeCoursey TE. Charge compensation during the phagocyte respiratory burst. *Biochim Biophys Acta* 2006;**1757**:996–1011.
79. Musset B, DeCoursey TE. Biophysical properties of the voltage-gated proton channel H_V1. *Wiley Interdiscip Rev Membr Transp Signal* 2012;**1**:605–20.
80. DeCoursey TE, Cherny VV, Zhou W, Thomas LL. Simultaneous activation of NADPH oxidase-related proton and electron currents in human neutrophils. *Proc Natl Acad Sci USA* 2000;**97**:6885–9.
81. Cherny VV, Henderson LM, Xu W, Thomas LL, DeCoursey TE. Activation of NADPH oxidase-related proton and electron currents in human eosinophils by arachidonic acid. *J Physiol* 2001;**535**:783–94.
82. Morgan D, Cherny VV, Finnegan A, Bollinger J, Gelb MH, DeCoursey TE. Sustained activation of proton channels and NADPH oxidase in human eosinophils and murine granulocytes requires PKC but not $cPLA_2\alpha$ activity. *J Physiol* 2007;**579**:327–44.
83. Musset B, Capasso M, Cherny VV, Morgan D, Bhamrah M, Dyer MJS, et al. Identification of Thr^{29} as a critical phosphorylation site that activates the human proton channel *Hvcn1* in leukocytes. *J Biol Chem* 2010;**285**:5117–21.
84. Szteyn K, Yang W, Schmid E, Lang F, Shumilina E. Lipopolysaccharide sensitive H^+ current in dendritic cells. *Am J Physiol Cell Physiol* 2012;**303**:C204–C212.
85. Liao Z, Lockhead D, Larson ED, Proenza C. Phosphorylation and modulation of hyperpolarization-activated HCN4 channels by protein kinase A in the mouse sinoatrial node. *J Gen Physiol* 2010;**136**:247–58.

86. DeCoursey TE, Ligeti E. Regulation and termination of NADPH oxidase activity. *Cell Mol Life Sci* 2005;**62**:2173–93.
87. Doroshenko PA, Kostyuk PG, Martynyuk AE. Transmembrane outward hydrogen current in intracellularly perfused neurones of the snail *Helix pomatia*. *Gen Physiol Biophys* 1986;**5**:337–50.
88. Mahaut-Smith MP. Separation of hydrogen ion currents in intact molluscan neurones. *J Exp Biol* 1989;**145**:439–54.
89. Sigworth FJ. Voltage gating of ion channels. *Q Rev Biophys* 1994;**27**:1–40.
90. Roos A, Boron WF. Intracellular pH. *Physiol Rev* 1981;**61**:296–434.

CHAPTER THIRTEEN

Receptor Heteromeric Assembly—How It Works and Why It Matters: The Case of Ionotropic Glutamate Receptors

Beatriz Herguedas, James Krieger, Ingo H. Greger
Neurobiology Division, MRC Laboratory of Molecular Biology, Cambridge, United Kingdom

Contents

Abstract

The composition and spatial arrangement of subunits in ion channels are essential for their function. Diverse stoichiometries are possible in a multitude of channels. These depend upon cell type-specific subunit expression, which can be tuned in a developmentally regulated manner and in response to activity, on subunit stability in the endoplasmic reticulum, intersubunit affinities, and potentially subunit diffusion within the ER membrane. In concert, these parameters shape channel biogenesis and ultimately tune cellular response properties. The complexity of this assembly process is particularly well illustrated by the ionotropic glutamate receptors, the main mediators of excitatory neurotransmission. These tetrameric cation channels predominantly assemble into heteromers, which is "obligatory" for some iGluR subfamilies but "preferential" for others. Here, we discuss recent insights into the rules underlying these two pathways, the role of individual domains based on an ever increasing list of crystal structures, and how these assembly parameters tune assembly across diverse receptor oligomers.

Progress in Molecular Biology and Translational Science, Volume 117
ISSN 1877-1173
http://dx.doi.org/10.1016/B978-0-12-386931-9.00013-1

1. INTRODUCTION

Signaling receptors operate as oligomers, assembled from either homo- or heteromeric subunits. While homomeric assemblies are common in prokaryotes,[1] genome expansion and gene duplication events in metazoan eukaryotes paved the way for the formation of receptor heteromers.[2] Assembly into heteromeric complexes increases the versatility and facilitates fine-tuning of response properties and of their downstream signaling cascades.

For certain classes of receptors, heteromeric assembly is a prerequisite for function and onward traffic to the cell membrane, with incompletely assembled complexes being selectively retained in the assembly compartment, the endoplasmic reticulum (ER). This group of "obligatory" heteromers includes different families of immune receptors (such as the T-cell receptor),[3,4] some G-protein-coupled receptors (e.g., the gamma-amino-butyric acid $[GABA]_B$ receptor),[5,6] the muscle nicotinic acetylcholine receptor (nAchR),[7] ionotropic glutamate receptors (iGluRs) of the *N*-methyl-D-aspartate (NMDA) subtype (NMDARs)[8], and others. Genes encoding subunits of these complexes can reside in different chromosomes; how their expression and ultimately their spatial proximity for heteromeric assembly in the ER are coordinated is an unresolved question.

Contrasting with obligatory heteromers are ion channels that can form homomeric assemblies under certain conditions but predominantly exist as heteromers. Examples are members of the Cys-loop family ($GABA_A$ receptors[9] and brain nAchRs[7]), certain cation channels (including potassium-, transient receptor potential-,[10] and cyclic-nucleotide gated- [CNG] channels[11]), and non-NMDA iGluRs such as the α-amino-3-hydroxy-5-methyl-4-isoxazole propionic acid (AMPA)- and low-affinity kainate receptors (Fig. 13.1A). These iGluR subtypes heteromerize in most neurons studied (e.g., Refs. 12,13) and can be referred to as "preferential" heteromers. A balance between homo- versus heteromeric assembly routes under certain physiological (and pathological) settings permits further response tuning, which has been highlighted for AMPA receptors (AMPARs) in a number of reports.[14–16]

What are the mechanisms underlying assembly preferences? Parameters likely include (1) availability of a heteromeric partner, which depends upon expression levels in a given cell type and on the dwell time of a given subunit in the ER; (2) translation/translocation kinetics across the ER membrane and

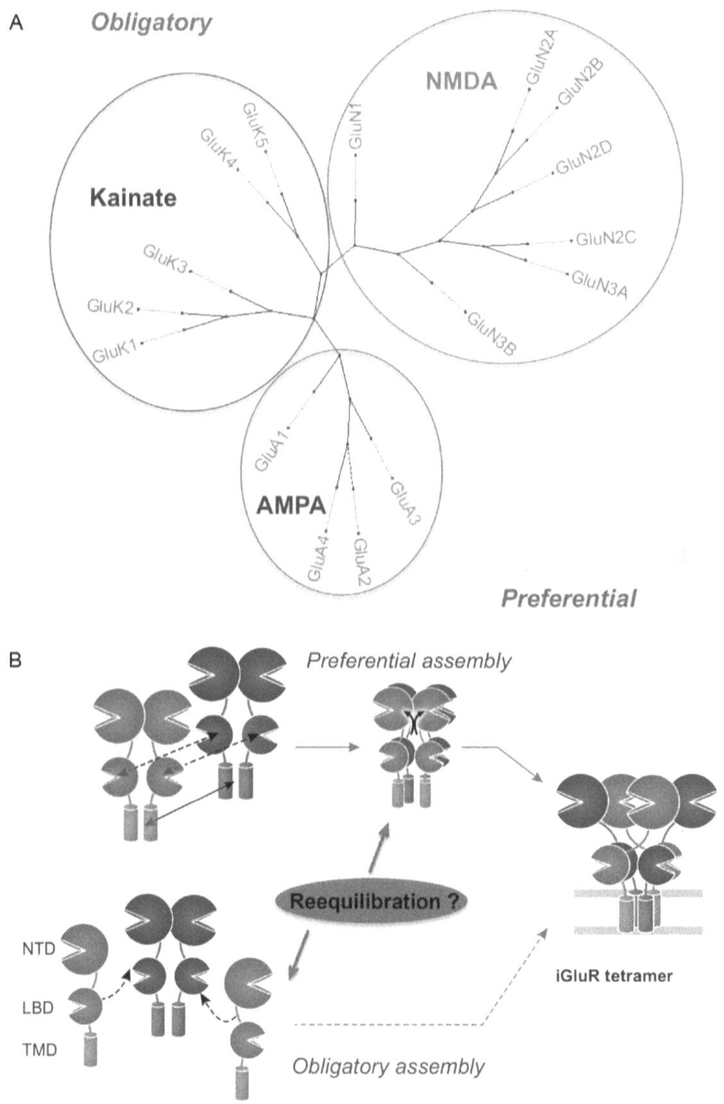

Figure 13.1 Ionotropic glutamate receptor (iGluR) assembly pathways. (A) There are four iGluR subfamilies: AMPA-, kainate-, NMDA-, and delta receptors. AMPA-, kainate-, and NMDA- receptor subunits are shown in a phylogenetic tree and divided into those that form obligatory heteromers that are essential for function (kainate receptor subunits GluK4 and -5, which have high affinity for ligand, and all NMDA receptor subunits) and those that can form functional homomers but preferentially form heteromers (kainate receptor subunits GluK1-3, which have low affinity for ligand, and all AMPA

(Continued)

protein folding rates, a function of topology; this parameter might feature particularly in cases of cotranslational assembly; (3) intersubunit affinity, where both attractive and repelling forces will be decisive—specific domains orchestrating assembly have been identified in a variety of ion channels (see below); and (4) lateral diffusion rates within the ER membrane, where subunits exhibiting higher diffusion coefficients are more likely to encounter heteromeric partners, inserted into the ER membrane at spatially distinct sites. Of relevance is a recent study utilizing fluorescence recovery after photobleaching measurements,[17] which revealed unexpectedly rapid diffusion of ER-retained AMPAR subunits, which closely mimicked an ER-retained version of vesicular stomatitis virus protein G that freely diffuses in the ER membrane. This relatively rapid diffusion seems surprising, given that the large extracellular portion of the receptor faces the crowded environment of the ER lumen, where it is expected to interact with a network of ER-resident chaperones[18] (but see Ref. 19).

The outcome of ER biogenesis is stoichiometry and spatial arrangement of receptor subunits, which ultimately impact signal transmission and integration at the plasma membrane. Whereas assembly pathways are increasingly well described for some receptor classes, such as the T-cell receptor[3] and CNG cation channels,[20] assembly rules are elusive for many others. Here, we discuss these principles for neuronal iGluRs, where a spectrum of sophisticated assembly rules is emerging.[21–24] iGluR biogenesis underlies signal transmission properties at synapses in health and disease and hence matters a great deal for information processing in neuronal networks. Altered subunit stoichiometries are associated with various diseases, which have been covered in a number of recent reviews.[16,24–27]

receptor subunits). Delta receptors are not shown as little is known about their assembly. (B) iGluRs are composed of three main domains that are involved in assembly: a channel-forming transmembrane domain (TMD) and two extracellular clamshells, the membrane-proximal ligand-binding domain (LBD), and the distal N-terminal domain (NTD). Preferential subunits initially form homodimers mediated by NTD contacts. These dimers then assemble into tetramers via weaker *trans*-dimerization of the LBD and tetramerization of the TMD. This is accompanied by reequilibration of interfaces to produce functional heterotetramers, which are likely to contain NTD heterodimers. Obligatory subunits follow a different assembly path (shown below), where some subunits cannot form homodimers but rather invade a homodimer scaffold individually to produce the final heterotetramer. The final tetramer structure shown is a schematic representation of the nearly intact GluA2 structure (PDB 3KG2). (For color version of this figure, the reader is referred to the online version of this chapter.)

2. IONOTROPIC GLUTAMATE RECEPTORS MEDIATE EXCITATORY NEUROTRANSMISSION

L-Glutamate is a signaling currency abundantly used throughout the central- and the peripheral nervous systems, where it activates two receptor systems[16]: G-protein-coupled metabotropic glutamate receptors (mGluRs) and ionotropic cation channels (iGluRs). mGluRs mediate modulatory activity at a relatively slow time scale (0.5–1 s),[28] for example, through shaping transmitter release kinetics and activating ion channels via secondary messengers. iGluRs are responsible for the majority of fast excitatory signal transmission across chemical synapses. Presynaptically released glutamate serves as a ligand to activate (mostly) postsynaptic iGluRs; the resulting cation flux through the channel pore locally depolarizes the postsynaptic neuron. Neuronal dendrites are studded with small protrusions, dendritic spines (tens of thousands/cell), which serve as postsynaptic signaling compartments in most principal neurons. Summation of multiple inputs sufficient to depolarize the postsynaptic cell will trigger an action potential and in turn transmit information. iGluRs are divided into four subfamilies: the AMPA-, kainate-, NMDA-, and delta receptors, all of which fulfill unique roles in neuronal communication.[16,29] In addition to forming core channel tetramers, iGluRs, in particular the non-NMDA subtypes, are associated with an increasing list of auxiliary cofactors,[30–34] in analogy to their distant relatives, the K^+ channels.[35] Subunit stoichiometry is decisive in shaping gating kinetics, ion permeability, and responsiveness to an array of small-molecule channel modulators. This wide functional spectrum is extended further by different types of auxiliary subunits,[33,34,36–39] and ultimately orchestrates synapse-selective neuronal communications that underlie higher-order cognitive processes.

Here, we will contrast subunit assembly between preferential and obligatory iGluRs, discuss the roles of individual receptor domains in the assembly process, highlight the role of the N-terminal domain (NTD) and ion channel sector in detail, and touch upon parallels between iGluRs and the biogenesis of related receptor oligomers.

3. GLUTAMATE-GATED ION CHANNELS ASSEMBLE INTO TETRAMERS

iGluRs assemble into the core tetramer in two steps: subunits first dimerize, which is followed by the assembly of two dimers into tetramers,[40] analogous to K^+ channels.[41] At what stage cofactors join the channel core has not been established firmly, although current evidence suggests that the tetramer serves as the platform for cofactor association.[42,43] iGluRs are modulatory polypeptides and are composed of discrete domains (Figs. 13.1B and 13.4A), which play distinct roles during assembly. The extracellular portion is composed of two globular folds: the distal NTD (Fig. 13.2A) and the membrane-proximal ligand-binding domain (LBD; which binds L-glutamate). The LBD is fused to the membrane-embedded ion channel (Fig. 13.4A) and extends a C-terminus into the cytoplasm. This carboxy-terminal domain varies in size and encompasses 50–80 residues in AMPARs but is up to 10 times longer in NMDAR GluN2 subunits. The tetrameric nature of iGluRs has been revealed at multiple levels including electrophysiology,[44] biochemistry,[43,45] structural elucidation of isolated domains,[21,46] single-particle reconstruction,[47,48] and single-particle counting,[49] and has been concluded more recently with the crystal structure of a complete AMPAR homotetramer.[50] Below we discuss the role of individual domains in the assembly process.

4. THE iGluR ASSEMBLY PATHWAY

4.1. Initiating subunit assembly: The NTD

In stark contrast to the LBD and the channel core, NTDs are sequence diverse between and within subfamilies (e.g., the sequence identity between AMPAR NTD paralogs is ~55%). Lagging ~10 years behind the LBD,[51] crystal structures of the NTD have only been solved recently (reviewed in Ref. 21). This domain encompasses ~50% of a subunit polypeptide and contributes a large symmetrical interface for dimer formation, which initiates receptor biogenesis. As reviewed 5 years ago, the NTD is the first segment to face the unique chemical environment of the ER lumen and may engage nascent "sister" polypeptides, emerging from the same polyribosome, as early as during chain translocation.[52] This scenario is particularly likely for the AMPA-type receptor, which forms high-affinity NTD homodimers in the submicromolar range.[53,54] Homomeric NTD contacts are weaker in kainate receptors (KARs)[21] and appear to be absent in NMDAR NTDs.[55] Hence, a relationship between obligatory (NMDAR) versus preferential (AMPAR) assemblies (Fig. 13.1A) and NTD interface strength is apparent. Below we provide a brief description of the physicochemical features of this strategic interface, for which a series of high-resolution structures across the main subfamilies is now available.

4.1.1 The NTD dimer interface in "preferential" versus "obligatory" assemblies

iGluR NTDs belong to the Type-1 periplasmic-binding protein (PBP) superfamily, which is characterized by two lobes separated by a deep cleft (Figs. 13.1B and 13.2A).[56] In non-NMDARs, both lobes contribute to an extensive dimer interface (~1500 Å^2) (Fig. 13.2A and B).[53,57–61] In AMPARs, the upper lobe (UL) interface is the most highly conserved part of the NTD and is composed primarily of two helices (B and C in GluA2) and a loop of ~20 residues, which wedges into the interface where it is anchored by intramolecular disulfide bridges (Fig. 13.2A and B). The length of this loop differs between iGuR subtypes and is most likely a key assembly determinant.[58] Contact residues as well as UL interface strength are mostly conserved across the four AMPAR paralogs. Interface strength can be quantified and expressed as "local density index," which summarizes interfacing atom contacts and their density.[62,63] The LD metric is shown for iGluR NTD homo- and heterodimers in Fig. 13.3. The lower lobe (LL) interface is generally looser (i.e., has a lower LD value), is less well conserved, and features a greater solvent content (see Ref. 61 for a comparison of all AMPAR NTDs). It is composed mostly of a helix (E in GluA2) and a strand (7 in GluA2) as well as the loops connected to them (Fig. 13.2B). Variable LL contacts likely underlie the wide spectrum of homodimer affinities (ranging from low nanomolar to low micromolar) between AMPAR NTDs[53,54] and may play a role in reequilibrating initial NTD homodimers into heteromers during tetramer formation (Fig. 13.1; "preferential pathway") (see below). Heterodimeric NTDs form with greater affinity in solution,[53] which is particularly evident when comparing the GluA2/1 and the GluA2/3 heterodimers with the GluA1/1 and GluA3/3 homodimers. Accordingly, AMPAR heterotetramers would be composed of heterodimeric NTD assemblies as has been described for NMDARs[55,65,66] and for KARs.[64] In KARs, the preferential heterodimerization of isolated NTDs is substantially more pronounced than in AMPARs.[53,64]

In GluK1-3 KARs, which like AMPARs belong to the "preferential" heteromers (Fig. 13.1A), interface stability seems to be distributed more equally between the UL and LL interfaces,[59,64] resulting in almost equal

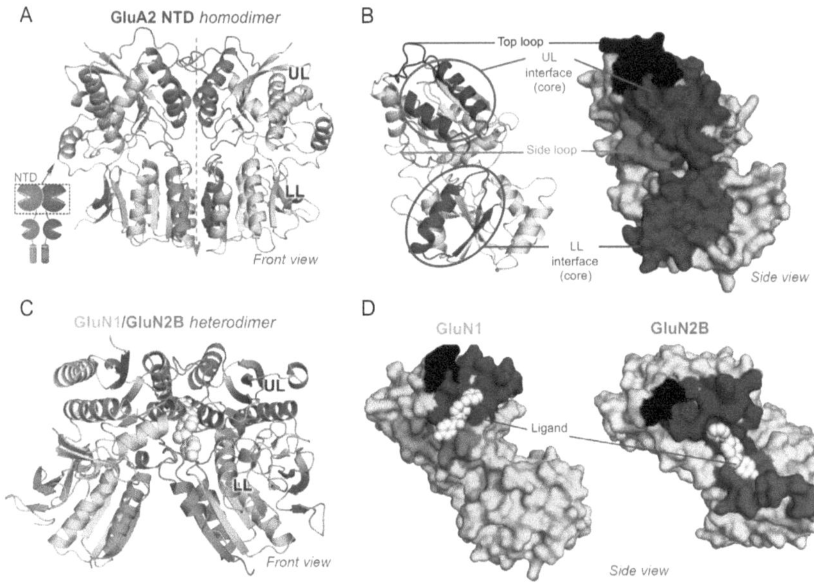

Figure 13.2 Features of the N-terminal domain dimer interface. The NTD has a clamshell structure with two lobes (UL and LL) separated by a deep cleft that is involved in ligand binding in related periplasmic-binding proteins. iGluR family members can be grouped by their heteromerization properties as preferential or obligatory subunits, and NTD dimerization is an early step in iGluR tetramer assembly that is an important determinant of this. An example, NTD dimer is shown for a subunit at each extreme. (A and B) The AMPA receptor subunit GluA2 is capable of forming functional homodimers at the level of the whole receptor and at the level of isolated NTDs (K_d 1.8 nM). (A) GluA2 NTD homodimers (e.g., PDB structure 3H5V shown here) form tight interfaces at both UL and LL. (B) The dimer is opened up and the blue subunit is rotated 90° to highlight the key interface elements. The UL interface is primarily formed by two helices (blue) and a loop at the top (black), which is anchored in place by intramolecular disulfide bridges (sandy colored sulfurs). Kainate receptor UL interfaces also involve a loop at the side (colored in green even though it is shorter in AMPA receptors and is not involved in the interface). The LL interface is primarily formed by a helix and a strand (red). (C and D) NMDA receptors are obligatory heteromers requiring two GluN1 subunits and two GluN2 or GluN3 subunits. Even heterodimeric interactions of NMDA receptor NTDs are weak ($K_d \sim 1$ μM), so a ligand (ifenprodil, yellow) had to be used to help crystallize the GluN1/GluN2B heterodimer shown here (PDB 3QEL). (C) The dimer interface is rather different involving only the UL of GluN1 but both lobes of GluN2B. (D) The interface is broken open and rotated again to show the key elements (colored as in B). There is also an extra helix from GluN1 involved in the interface (colored orange). Both subunits are shown as the interface is asymmetric. (For interpretation of the references to color in this figure legend, the reader is referred to the online version of this chapter.)

LD values for the UL and LLs (Fig. 13.3A vs. C). Similarly, conservation values between the two interface patches are comparable, contrasting with AMPARs. Another key difference between the non-NMDAR NTDs is a longer side loop in the UL of KARs (green in Fig. 13.2B), which forms specific contacts (mostly with the LL) in KAR NTD dimers. GluK2 homodimers associate in the mid-nanomolar range (350 nM),[64] well above GluA2 and GluA4 (both below 10 nM) and GluA1 (~100 nM)[53] and assemble with the obligatory GluK4 and -5 into heterodimers of higher affinity (~77 nM).[64] The obligatory KAR subunits (GluK4, 5) (Fig. 13.1A), however, associate relatively weakly (~400 μM for GluK5) and are not expected to form *in vivo*. Structures of the GluK2 and GluK5 NTD homodimers and the GluK2/5 heterodimer were solved recently, facilitating a comparison of interface rearrangements between homo- and heterodimer (Fig. 13.3C–F).[64] In addition to changes of core interface residues, new contacts were mediated by the large top loop and via a drastic rearrangement of a loop segment at the bottom of the LL. In addition, side loop interactions are different in GluK2/GluK5 versus the GluK2 homodimer.[64] In the GluK5 homodimer, the UL has rotated away from the twofold axis of symmetry, disrupting the UL interface core in two independent structures.[59] Similar to GluK5 homomers, a UL rearrangement is seen for a particular dimer form of GluA3 (PDB: 3O21, chains AB),[63] the weakest AMPAR NTD homodimer.[53] The relevance of this configuration, if any, is currently not known.

In NMDARs, true obligatory heteromers (Fig. 13.1A), NTD LL contacts are mostly absent (Figs. 13.2C and D and 13.3G–I) and homomeric assemblies are too weak to be assessed accurately by analytical ultracentrifugation techniques.[55] As a result, heteromers almost exclusively form via UL–UL contacts and are significantly weaker than their non-NMDAR counterparts (~1 μM for GluN1/2B). Structures of a GluN1 homodimer, a GluN2B monomer, and a GluN1/N2B NTD heterodimer are currently available. The arrangement of the NMDA homo- and heterodimeric NTDs is vastly different.[55,65] In both cases, interfaces observed in these structures have been probed (and confirmed) via disulfide cross-linking in the context of the complete receptor. To what extent sequence-specific contacts, mediated by the GluN2B upper- and lower lobes, the presence of GluN2B ligand, which contributes additional interface contacts (Fig. 13.2D), and differences in crystal packing contributed to this diverse arrangement is open at present.

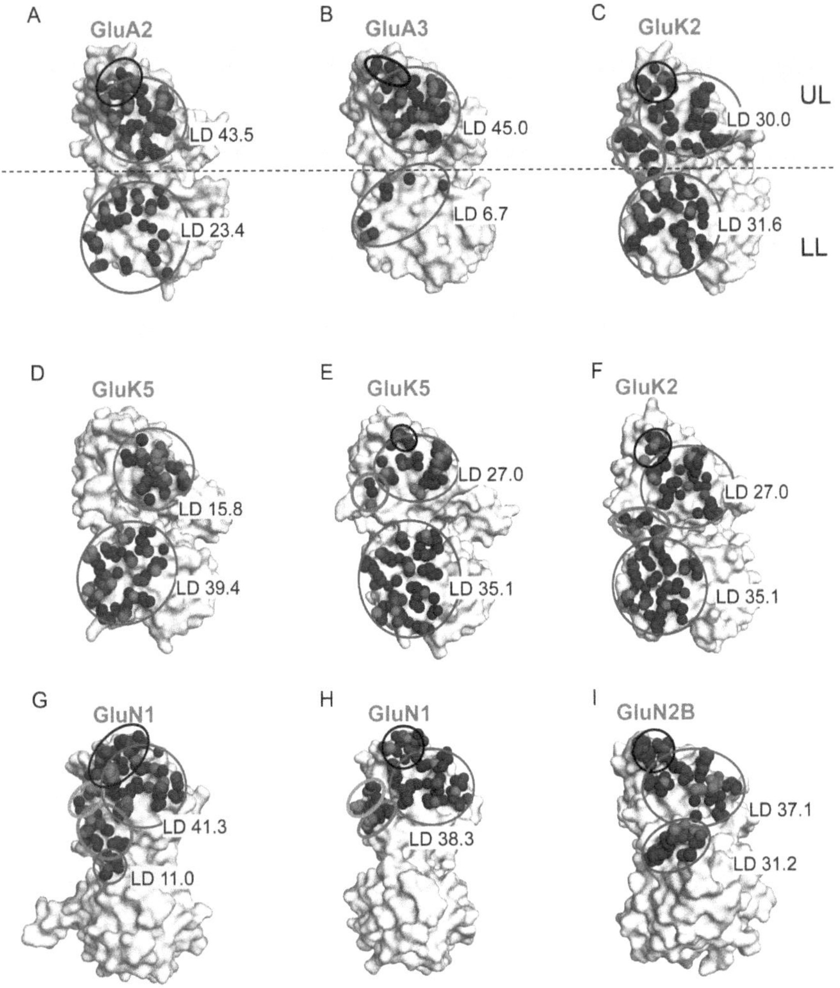

Figure 13.3 N-terminal domain dimer interfaces in ionotropic glutamate receptors. Representative structures are shown for homodimers of preferential subunits (blue background), homodimers of obligatory subunits (orange background), and heterodimers of obligatory subunits (white background). Atoms in one protomer within 4.5 Å of the other protomer are indicated as spheres colored by the number of contacts from blue (1) to red ($\geq$7). Also shown is the local density index (LD)[62] for the upper and lower lobes (UL and LL) of the clamshell using a distance cut-off of 12 Å. This is a measure of interface strength. LDs are averaged over all similar structures of a given dimer. Differences between LD values here and those published previously[61,63] for GluA2 and GluA3 may be due to differences in lobe definition and which structures were analyzed. Key elements are highlighted as in Fig. 13.2: core UL interface, blue; top loop, black; side loop, green; LL interface, red; additional NMDA receptor helix, orange. (A–C) Preferential subunits GluA2, GluA3, and GluK2 show differences in interface strength LD at the upper

4.1.2 Related interfaces in different receptor families

NTD-like Type-1 PBPs exist in prokaryotes where they are mostly monomeric[56] as well as in a variety of other receptor families such as the group C GPCRs[67] and in the natriuretic peptide receptors.[68] It is plausible that some ancestral Type-1 PBP gained a weak dimerization function and that evolution of more specific interactions followed. The dimer interface still remains quite weak at least in some of these receptors allowing movements such as those thought to be involved in signaling in group C GPCRs[69] and natriuretic peptide receptors[68] as well as allosteric modulation in NMDARs.

These interfaces can be modulated by ligands triggering allosteric signaling,[55,68,70] the mechanism of which differs between these receptor families. In the natriuretic peptide receptor, circular peptide ligands bind within the LL dimer interface, which results in interface tightening and triggers opening of the clamshell cleft.[68] A related scenario might take place in NMDARs in response to spermine binding.[70] In group C GPCRs, ligand binding to the cleft induces cleft closure and results in substantial rearrangements of the dimer interface.[69]

In mGluRs and $GABA_B$ receptors (GBRs), both class C GPCRs, the NTD-like ligand-binding cores form similar dimers to those found in NMDARs with only the ULs involved in the dimer interface.[6,69] The dimerization abilities of GBR NTDs are analogous to that of NMDAR NTDs with GBR1 being capable of forming homodimers but GBR2 only

---

and lower lobes, which correlate with their differences in affinity. (A) Isolated NTDs from AMPA receptor subunit GluA2 form strong homodimers (K_d ~1 nM[53]; representative PDB structure 3H5V[58]), which have a high LD for the UL and a fairly high LL LD. (B) Isolated NTDs from AMPA receptor subunit GluA3 form much weaker homodimers (K_d ~1 μM[53]; PDB 3O21[63] dimer CD), which have a similar UL LD to GluA2 but a much lower LL LD. (C) Isolated NTDs from kainate receptor subunit GluK2 form homodimers of intermediate strength (K_d ~350 nM[64]; PDB 3H6H[60]), which have intermediate LD values at both lobes. (D and E) Dimers formed by isolated NTDs from kainate receptor subunit GluK5, which forms obligate heterodimers, have a similar structure to preferential homodimers. (D) GluK5 NTD homodimers (K_d ~400 μM; PDB 3OM1)[64] have disrupted upper lobe interfaces (lower LD). (E) This is recovered in GluK2/GluK5 NTD heterodimers (PDB 3QLU)[64] and the K_d goes down to ~10 nM.[64] (F and G) NTD dimers formed by NMDA receptor subunit GluN1 also show differences between homodimerization and heterodimerization. (F) The homodimer formed by the GluN1 N-terminal domain (PDB 3Q41)[65] is so weak that it cannot be measured in solution and only involves the upper lobe. (G) The GluN1/GluN2B N-terminal domain heterodimer (PDB 3QEL)[55] is stronger (K_d ~1 nM),[55] which is probably due to involvement of the lower lobe from GluN2B interacting with the side loop of GluN1 as well as the presence of the ligand ifenprodil. (See Color Insert.)

being detected as a monomer.[6] mGluRs on the other hand form mostly homodimers, which are linked by intermolecular disulfide bonds,[69] although heterodimeric mGluRs have been reported.[71]

In further analogy to NMDARs, GBRs are obligatory heteromers with functional divergence between the subunits: the GBR1 subunit is involved in ligand recognition and the GBR2 subunit activates G proteins and modulates ligand binding to GBR1; no ligand binding to GBR2 has been identified to date. In NMDARs and in high-affinity KARs, subunits with different ligand specificities combine into obligatory heteromers with a spectrum of pharmacological and functional variability.[72–75] These examples emphasize why heteromers matter.

4.2. Tetramerization of iGluRs: The LBD sector

In a second assembly step, dimers associate into tetramers, which result in the formation of new interfaces and potentially in the rearrangement of existing ones (Fig. 13.1B). Upon tetramerization, the ion conduction pore forms, and as suggested by the AMPAR GluA2 homotetramer ($GluA2_{cryst}$; PDB 3KG2), the LBDs associate into twofold related interdimer assemblies[50] (Fig. 13.1B, "preferential" assembly pathway). This interaction requires a separation of the LBDs at the level of the dimer (Fig. 13.4A), which is observed also in AMPAR structures determined by EM reconstruction.[43] In addition, new and comparatively small interfaces between NTD- and LBD-dimers form in the tetramer (NTD tetramer interface ~300 Å^2 vs. NTD dimer interface ~1500 Å^2; LBD tetramer interface ~150 Å^2 vs. LBD dimer interface ~900 Å^2).

The domain swapping or crossover of the LBD sector in the $GluA2_{cryst}$ structure came as a surprise (Fig. 13.4A); the resulting "*trans*-dimer" arrangement has been supported via disulfide cross-linking of interfaces in both AMPARs and KARs.[50,76] Forcing tight LBD associations at the level of the dimer interferes with tetramerization[43] (but see Refs. 77,78) and secretory onward traffic.[43,77] Hence, formation of the LBD dimer interfaces upon tetramerization appears to be an intermediate step during biogenesis. In AMPARs, these LBD interface contacts can be altered as a result of mRNA processing events, alternative splicing, and RNA editing, which target the LBD dimer interface.[79] These naturally occurring alterations impact subunit assembly by altering subunit ER dwell time (reviewed in Ref. 23), which conceivably targets this dimer-to-tetramer transition. Currently, there is no evidence for changes in interface affinity as a result of LBD alternative splicing and editing. Nevertheless, these processes ultimately shape gating kinetics and thus facilitate conformational changes in the LBD layer, which may be sensed during ER quality control. The possibility that glutamate binding to ER-localized iGluRs could act as a chemical chaperone triggering conformational changes in the ER has been proposed.[29,78,80–84] More recent data suggest that RNA processing in the LBD responds to neuronal activity, which requires Ca^{2+} signaling.[85,86] This mRNA-based regulation thus permits remodeling of AMPAR composition in response to changing inputs, providing homeostatic regulation of excitatory signaling at the level of AMPAR subunit assembly.[85,87] In addition to these alterations of the LBD dimer interface in "*cis*," binding of allosterically active ions in "*trans*" can modulate LBD dimer strength[88–90]; whether this impacts on assembly is currently not known. Another editing site (the Q/R site) locates to the channel pore of the GluA2 subunit, where it destabilizes GluA2 homotetramers and by disfavoring the assembly of edited GluA2 homotetramers increases the dwell time of GluA2 dimers[45] (see below).

4.3. Subunit reequilibration

Based on their synthesis on polyribosomes, and hence a local concentration of multiple copies of a given subunit, and on their relatively tight NTD associations (in particular, in the case of AMPAR subunits), homodimeric iGluR assembly intermediates seem likely. Homodimers have been detected by various means, such as on glycerol gradients, on native PAGE, and by SDS-PAGE for AMPARs, KARs, and for the NR1 subunit of the NMDAR.[45,66,91] As there is evidence for the presence of heterodimers in the receptor tetramer (e.g., Refs. 50,70,92–94), homodimers would have to "reequilibrate" upon tetramerization (Fig. 13.1 "preferential assembly"). Alternatively, subunits that poorly homodimerize such as the high-affinity KAR GluK4, 5 subunits or the glutamate-binding GluN2A and GluN2B subunits may induce reassembly by "invading" more stable homodimers (Fig. 13.1 "obligatory assembly"). The latter has been suggested recently for the NMDAR, where the existence of GluN1 homodimers has been described and where the GluN1 homodimer serves as a scaffold for the formation of tetramers composed of GluN1/2 heterodimers.[65] In either scenario, reequilibration will require the "unzipping" and reassembly of interfaces. In AMPARs, the NTD dimer interface extends across both lobes of the clamshell and is tight; nevertheless, isolated NTDs preferentially assemble into heterodimers.[53] In this iGluR, subclass reequilibration would occur during tetramerization and may involve unzipping of the NTD dimer, which could initiate in the weaker LL interface and proceed toward the ULs (Fig. 13.1B; intermediate in the "preferential" pathway). Evidence for this unzipping of NTD dimers comes from full-atom molecular dynamics simulations, where an initial rupture of the GluA3 LLs propels toward the UL interface.[61] Moreover, ring-like NTD tetramers have been observed in single-particle EM analysis and may represent an intermediate of this process.[48] Clearly, a crystal structure of an AMPAR heteromeric receptor and of NTD heteromers is required to understand this reequilibration and spatial organization better.

4.4. Tetramerization of iGluR: The transmembrane domain

The transmembrane domain (TMD) of iGluRs plays a central role in both dimerization and oligomerization processes.[40] Like the NTD, the TMD forms dimers at early stages of receptor biogenesis, as demonstrated by the structure of GluA2 intermediates obtained by electron microscopy,[43] although the nature of helical packing of the dimeric TMD sector is currently elusive and is expected to be loose. Similarly, the topology of the reentrant loops is likely to be different in dimers; loop equilibration into lipid may only occur upon tetrameric assembly of the TMD. As iGluR subunits lacking the NTD can assemble into functional receptors,[95] TMDs alone or both TMDs and LBDs are sufficient for tetramerization. GluA2$_{cryst}$ provided a unique view of the extensive interfaces that are formed within the TMD subunits. Despite its homomeric nature, the structure revealed two distinct conformations for the four subunits, A/C and B/D, along the entire subunit axis, particularly in the LBD-linker/gate region, which will have functional consequences (Fig. 13.4A). Hence, diverse studies addressed whether different subunits occupy different positions in obligatory heterotetramers.[94,96] How homodimeric intermediates reequilibrate into heterotetramers is therefore of particular interest. Below, we summarize key features of the TMD as seen in the crystal structure of the antagonist-bound GluA2$_{cryst}$.[50]

The TMD of iGluR subunits is composed of three membrane-spanning α-helices (M1, M3, and M4) and a membrane-reentrant pore loop (M2) (Fig. 13.4B–D). Although the M2 segment was initially identified as a transmembrane helix and thus iGluR channels were thought to form pentameric ligand-gated ion channels like GABA$_A$ or acetylcholine receptors, the identification of three rather than four membrane-spanning helices and the discovery of prokaryotic GluR0 channels, which share similarities with potassium channels and iGluRs,[97] suggested a remote structural homology with potassium channels.[98,99] GluA2$_{cryst}$ confirmed a common architecture of the TMD with an inverted potassium channel, with M1, M3, and M2 segments resembling the M1, M2, and P-loop of prokaryotic KscA channels[100] and the S5, S6, and P-loop of the *Shaker* voltage-gated potassium channels.[101]

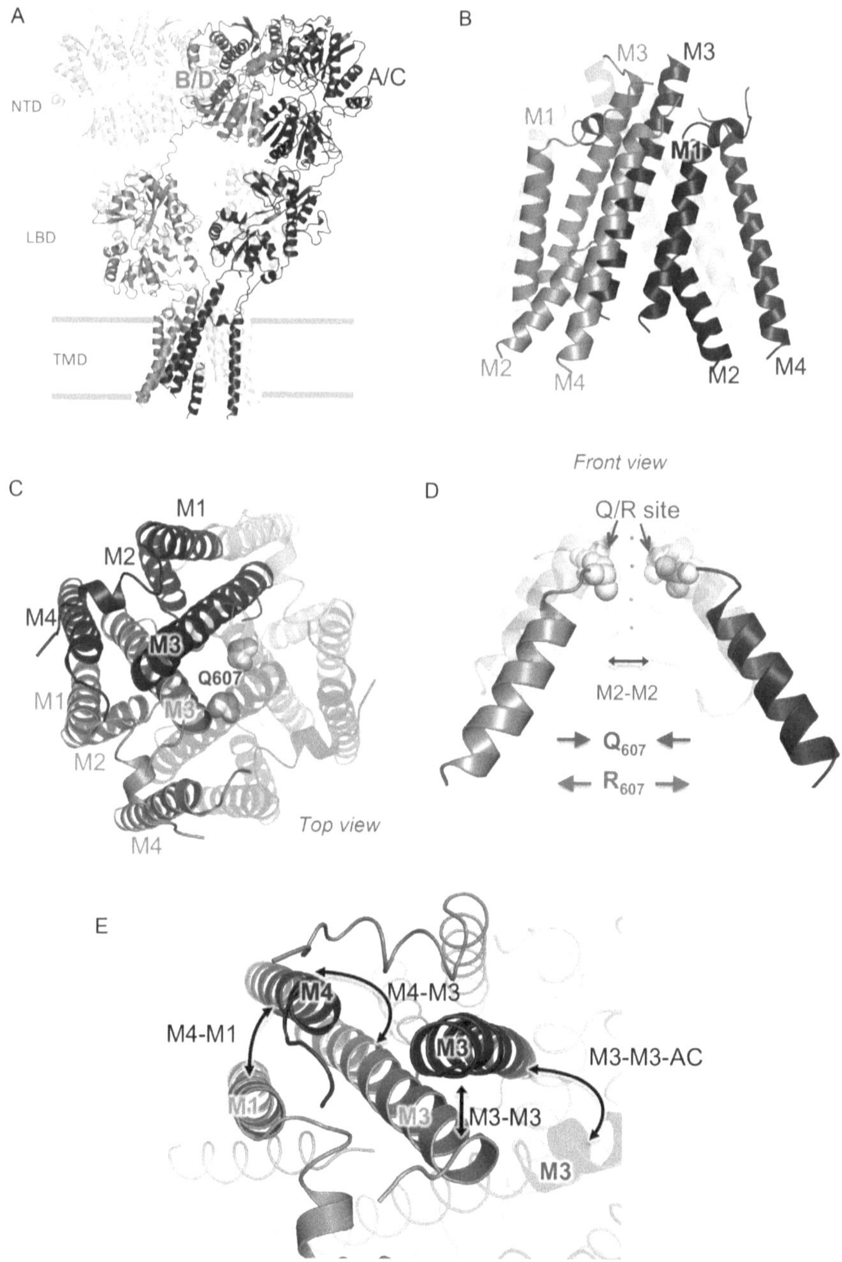

Figure 13.4 Organization of the transmembrane domain plus assembly determinants. (A) $GluA2_{cryst}$ (PDB 3KG2) is composed of two extracellular modules (NTD and LBD) and a TMD. Although it is homotetrameric, it presents two distinct conformations for subunits A/C (colored blue) and subunits B/D (colored orange). (B and C) The TMD of (Continued)

The TMD adopts a fourfold rotational symmetry (Fig. 13.4C), in contrast to the twofold symmetry observed in the extracellular domains (Fig. 13.4A).[50] The four M3 helices, which likely form the gate, line the inner core of the channel and form a narrow constriction in the extracellular vestibule (Fig. 13.4B), closing the ion permeation pathway as a consequence of antagonist binding in the LBD. The external part of the channel is composed of the M1 and M4 helices; the latter is absent in potassium channels and bacterial iGluR homologs. It is noteworthy that packing of the TMD helices leaves some gaps between the membrane and the central cavity of the channel. As noted by Sobolevsky *et al.*,[50] these cavities could be occupied by residues from other auxiliary subunits, such as transmembrane AMPA regulatory proteins (TARPs) and cornichons, which are able to modulate ion channel properties.[32,36,37] The fourfold symmetry of the TMD is disrupted in the C-terminus of M3 helices, a region that constitutes the transition between the fourfold symmetry of the channel pore and the twofold symmetry of the extracellular region. Thus, A/C subunits contain an additional turn in their M3 C-terminus, with residue Met629 pointing toward the pore and closing the ion permeation pathway in A/C subunits and being more than 30 Å apart in B/D subunits.[50] Below this residue, the M3 helix contains the SYTANLAAF motif, which spans the gate and which is highly

$GluA2_{cryst}$ shows a fourfold rotational symmetry around the channel pore. Each subunit contains three membrane-spanning helices (M1, M3, and M4) and a reentrant loop (M2). The M3 helix forms the inner part of the channel and contains the SYTANLAAF motif. The other elements form the external part. Yellow spheres in (C) represent Gln607, which is subject to RNA editing in AMPA and kainate receptors. (D) The TMD of iGluRs shares structural similarity with potassium channels. The M2 reentrant loop is similar to the P-loop of potassium channels, which constitute their selectivity filter. In the KscA crystal structure (PDB 1BL8, colored green and transparent), three K^+ ions and a water molecule line the pore of the channel. In $GluA2_{cryst}$, the residues constituting the inner pore are disordered. The interaction of the four subunits through these residues should form an additional interface similar to the one observed in the KscA ion channel (indicated by M2-M2 arrows). The Q/R site, located at the apex of the loop in $GluA2_{cryst}$ (yellow spheres), determines Ca^{2+} permeability and heteromerization in AMPA receptors. The presence of the edited Arg at this position retains the receptor in the ER and favors heteromerization with unedited subunits. (E) The four subunits of $GluA2_{cryst}$ establish extensive interactions through the TMD, with a total buried surface of ~11,500 Å^2. M1 and M3 helices of one subunit (colored orange) interact with M4 of the adjacent clockwise subunit (colored blue; M4-M1 and M4-M3 arrows). Additional interactions between the C-terminus of the M3 segments are observed in adjacent (M3-M3) and nonadjacent subunits (M3-M3-AC). This latter interaction may only occur in the antagonist-bound (closed) state in subunits (A) and (C). (For interpretation of the references to color in this figure legend, the reader is referred to the online version of this chapter.)

conserved across the iGluR family.[102] The existence of two distinct conformations in the C-terminus of M3 helices (which continue in the M3-S2 linker) is also relevant for obligatory NMDAR heteromers, where each isoform (GluN1 or GluN2) could adopt the same (A/C or B/D) or different conformations, resulting in N1/N1/N2/N2 or N1/N2/N1/N2 arrangements at the TMD level (Fig. 13.4A). Support for the latter model with alternating subunits was provided recently using cross-linking of Cys mutants of residues located in the TMD region, which also indicated that GluN1 subunits always adopt the "inner" A/C conformation whereas GluN2 subunits adopt the distal B/D positions.[94,96] Related questions have been raised for the obligatory GluK2/K5 heteromer, where the GluK2 subunits occupied the B/D positions placing the GluK5 LBDs in proximity analogous to GluN1.[64] For AMPAR heteromers, it remains to be determined whether particular subunits preferentially occupy a given position or whether both conformations (A/C vs. B/D) can be accessed with equal occupancy (Fig. 13.4A). It is worth pointing out that although an alternate arrangement has been suggested for GluA1/GluA2 AMPAR heteromers,[93] subunit stoichiometry and the number of GluA2 subunits within a tetramer have not been clarified.[103] This is an important problem as GluA2 abundance has a key impact on AMPAR properties (e.g., Ref. 104); in neurons expressing low levels of this subunit, channels with less than two GluA2 subunits may exist.

4.4.1 Assembly determinants in the TMD

The M2 segment is disordered in the crystal structure and lacks information for the residues that form the selectivity filter of potassium channels. The first part of M2 forms an α-helix that is oriented toward the central cavity of the ion channel. The short loop descending from this helix contains Gln607 at its apex (Fig. 13.4D). This site is subject to RNA editing from glutamine to arginine (Q/R) in AMPA and KA receptors,[105,106] where it determines Ca^{2+} permeability, channel blocking by polyamines as well as subunit tetramerization.[13,107–110] In the brain, the vast majority of GluA2 subunits are edited to Arg (>99%), whereas GluA1, GluA3, and GluA4 isoforms are unedited.[79] In GluK1 and GluK2, editing at this site is region specific and does not exceed 90% in GluK2.[111] Heteromeric AMPARs containing the edited GluA2 are selective for monovalent ions, whereas receptors lacking this subunit are Ca^{2+} permeable. The effect of the Q/R site on assembly is believed to occur at the dimer-to-tetramer transition, where approximation of two edited GluA2-Arg607 homodimers would result in charge repulsion, resulting in suboptimal packing within the TMD sector.[45] These channels can form but are substantially less stable than the unedited GluA2-Gln607 variety and are not expected to assemble in the presence of Gln607-containing subunit partners (i.e., GluA1, 3, 4), inclusion of which would be energetically more favorable. ER-retained GluA2 assembly intermediates are stable and their concentration in the ER is higher compared to GluA1,[108] the resulting availability of GluA2 for assembly can explain why the majority of AMPARs in the brain are GluA2 containing. Q/R-unedited GluA2 homotetramerizes and integrates into synapses more efficiently than GluA2-lacking AMPARs.[45,112] Underediting at this site and the resulting overload of Ca^{2+}-permeable AMPARs are associated with excitoxicity.[113] Of note, charged residues in transmembrane segments play an orchestrating role in the assembly of the T-cell receptor.[4]

The other TMD helices mediate extensive intersubunit contacts (Fig. 13.4E), with a total buried surface in the TMD of 11,500 Å^2. Viewing the pore from the extracellular side of the membrane, the M1 and M3 helices of one subunit interact clockwise with the M4 helix from a neighboring subunit, forming a large interface of ~1400 Å^2 (Fig. 13.4C). In GluA2, residues 789–817 from the M4 helix form an interfacing surface with helices M1 and M3 from the adjacent subunit, suggesting a role in the stabilization of tetramers (Fig. 13.4B, C, E). In fact, a recent study demonstrated that the interaction of M4 with the adjacent subunits is necessary for AMPAR surface expression, supporting the role of this segment in the biogenesis of functional receptors and oligomerization.[114] Sequence identity in M4 among AMPAR isoforms is very high, with only a few substitutions of residues not located in the interface so it is not very likely that this M4 interaction would determine subunit specificity. In the case of NMDARs, with a lower similarity between subunits, M4 has also been shown to be critical for the surface expression of the receptor, where a single residue in the GluN1 subunit determines surface expression.[115] However, the role of the M4 subunit in NMDAR tetramerization remains unclear, and some studies suggest that the interaction of M4 in GluN1 with the M3 segment of GluN2 and GluN3 subunits would be negating the ER retention signals located at the M3 segment, rather than determining subunit specificity.

4.4.2 Molecular determinants of heteromerization in related tetrameric ion channels

Voltage-gated potassium channels and CNG ion channels are tetramers that share a common topology in the TMD, consisting of six membrane-spanning helices, S1–S6, and a membrane-reentrant loop. S1–S4 segments are the putative voltage sensors, whereas S5–S6 and the pore loop form the pore region, which presents a common architecture with the TMD of iGluRs (see earlier). CNG and potassium channels can form heteromeric receptors; the precise subunit composition and the molecular mechanisms that determine receptor heteromerization have been extensively investigated for both ion channel families.[116,117]

In *Shaker*-related potassium channels, a cytoplasmic N-terminal T1 domain (first tetramerization domain) promotes the interaction of members of the same subfamily and prevents the interaction of members of different subfamilies.[118,119] The presence of a T1 domain is not strictly required for channel tetramerization, as subunits without a T1 domain could also associate into functional tetramers via the TMD, although less efficiently.[120] In Kv7 potassium channels, a C-terminal domain helix forms a tetrameric coiled-coil domain that mediates assembly and specificity.[121] However, in the Kv7.2 subunit, which also contains this C-terminus domain, the pore region of the TMD is the key determinant of channel assembly and surface expression, similar to the Q/R site in AMPARs.[122] Kv7.2 channels are retained in the ER and require the expression of the Kv7.3 subunit to be trafficked to the plasma membrane, and a single mutation in the pore of Kv7.2 restores surface trafficking.[122] Interestingly, Kv7.2/Kv7.3 subunits form heteromers with random stoichiometries and arrangements.[123] In EAG potassium channels, a C-terminal assembly domain was initially implicated in the assembly of α-subunits[124] through a tetramerizing coiled-coil domain. However, a recent study demonstrated that the S6 membrane-spanning helix and the proximal C-terminus are in fact the main assembly determinants.[125]

CNG channels are composed of CNG_A and CNG_B subunits, both contributing to the formation of the pore. These receptors resemble the KAR subclass in their heteromerization propensities—CNG_A subunits 1–3 can form functional homomers when expressed alone in heterologous systems but preferentially form heteromers, while CNG_B subunits cannot form functional homomers and form obligatory heteromers with CNG_A subunits. Stoichiometry and subunit composition differ between the CNG channels from olfactory receptors, rod photoreceptors, and cone receptors,[11,126,127] affecting diverse parameters including Ca^{2+} permeability, nucleotide specificity, and modulation by Ca^{2+}/calmodulin. In rod receptors, which are composed of three CNG_A subunits and one CNG_B subunit, two mechanisms configure the stoichiometry of the receptor. First, a carboxy-terminal leucine zipper (CLZ) located in CNG_A subunits promotes trimerization of CNG_A subunits through a parallel three-helix coiled coil. Second, the high affinity between CNG_A and CNG_B subunits ensures that, if the latter subunit is present, CNG_B is incorporated into the channel.[20]

In summary, the assembly of CNG and K^+ channels is governed by diverse molecular mechanisms, including the oligomerization of specific domains located at the N- and C-termini as well as interactions via the TMDs. Most likely more than one mechanism will contribute to assembly of a given channel, as demonstrated by the fact that both T1 domains and CLZ domains can be deleted and functional receptors still form.

5. CONCLUSION

Precise heteromeric assembly is essential for fine-tuning of oligomer function. We have illustrated the principles of heteromeric assembly using neuronal iGluRs as an example. These receptors form tetrameric ion channels, which mediate fast excitatory synaptic transmission and various forms of synaptic plasticity. The four iGluR subtypes (AMPA-, kainate-, NMDA-, and delta receptors) coassemble only within a subfamily. We have outlined the determinants of subtype-specific heteromeric assembly, which can be either preferential or obligatory, and are encoded in each of the three domains of iGluRs. Residue differences between iGluR subunits as a result of gene differences and cotranscriptional RNA processing fine-tune the relative affinities of different subunit combinations and spatial arrangements. These affinity differences together with relative subunit abundances in the ER where iGluRs are assembled in different brain regions lead to specific receptor populations, which can respond differently to synaptic inputs. This is further modulated by a large number of auxiliary subunits such as TARPs and cornichons for AMPARs, ultimately giving rise to the rich repertoire of postsynaptic response properties in diverse neuronal circuitries.

ACKNOWLEDGMENTS

We would like to thank Ondrej Cais, Andrew Penn, Madhav Sukumaran, Ales Balik, Carol Deutsch, and Teru Nakagawa for critically reading the manuscript. We would also like to thank Jo Westmoreland for help with figure design. We acknowledge the Medical Research Council (MRC) for funding (U105174197 to B. H., J. K., and I. H. G.).

REFERENCES

1. Lynch M. The evolution of multimeric protein assemblages. *Mol Biol Evol* 2012;**29** (5):1353–66.
2. Pereira-Leal JB, Levy ED, Kamp C, Teichmann SA. Evolution of protein complexes by duplication of homomeric interactions. *Genome Biol* 2007;**8**(4):R51.

3. Call ME, Wucherpfennig KW. Common themes in the assembly and architecture of activating immune receptors. *Nat Rev Immunol* 2007;**7**(11):841–50.
4. Call ME, Pyrdol J, Wucherpfennig KW. Stoichiometry of the T-cell receptor-CD3 complex and key intermediates assembled in the endoplasmic reticulum. *EMBO J* 2004;**23**(12):2348–57.
5. Jones KA, Borowsky B, Tamm JA, Craig DA, Durkin MM, Dai M, et al. GABA(B) receptors function as a heteromeric assembly of the subunits GABA(B)R1 and GABA(B)R2. *Nature* 1998;**396**(6712):674–9.
6. Geng Y, Xiong D, Mosyak L, Malito DL, Kniazeff J, Chen Y, et al. Structure and functional interaction of the extracellular domain of human GABA(B) receptor GBR2. *Nat Neurosci* 2012;**15**(7):970–8.
7. Wanamaker CP, Christianson JC, Green WN. Regulation of nicotinic acetylcholine receptor assembly. *Ann N Y Acad Sci* 2003;**998**:66–80.
8. Cull-Candy S, Brickley S, Farrant M. NMDA receptor subunits: diversity, development and disease. *Curr Opin Neurobiol* 2001;**11**:327–35.
9. Tsetlin V, Kuzmin D, Kasheverov I. Assembly of nicotinic and other Cys-loop receptors. *J Neurochem* 2011;**116**(5):734–41.
10. Wu LJ, Sweet TB, Clapham DE. International Union of Basic and Clinical Pharmacology. LXXVI. Current progress in the mammalian TRP ion channel family. *Pharmacol Rev* 2010;**62**(3):381–404.
11. Zheng J, Zagotta WN. Stoichiometry and assembly of olfactory cyclic nucleotide-gated channels. *Neuron* 2004;**42**(3):411–21.
12. Lu W, Shi Y, Jackson AC, Bjorgan K, During MJ, Sprengel R, et al. Subunit composition of synaptic AMPA receptors revealed by a single-cell genetic approach. *Neuron* 2009;**62**(2):254–68.
13. Isaac JT, Ashby M, McBain CJ. The role of the GluR2 subunit in AMPA receptor function and synaptic plasticity. *Neuron* 2007;**54**:859–71.
14. Kauer JA, Malenka RC. Synaptic plasticity and addiction. *Nat Rev Neurosci* 2007; **8**(11):844–58.
15. Sutton MA, Ito HT, Cressy P, Kempf C, Woo JC, Schuman EM. Miniature neurotransmission stabilizes synaptic function via tonic suppression of local dendritic protein synthesis. *Cell* 2006;**125**(4):785–99.
16. Traynelis SF, Wollmuth LP, McBain CJ, Menniti FS, Vance KM, Ogden KK, et al. Glutamate receptor ion channels: structure, regulation, and function. *Pharmacol Rev* 2010; **62**(3):405–96.
17. Cui-Wang T, Hanus C, Cui T, Helton T, Bourne J, Watson D, et al. Local zones of endoplasmic reticulum complexity confine cargo in neuronal dendrites. *Cell* 2012; **148**(1–2):309–21.
18. Ellgaard L, Helenius A. Quality control in the endoplasmic reticulum. *Nat Rev Mol Cell Biol* 2003;**4**:181–91.
19. Nehls S, Snapp EL, Cole NB, Zaal KJ, Kenworthy AK, Roberts TH, et al. Dynamics and retention of misfolded proteins in native ER membranes. *Nat Cell Biol* 2000; **2**(5):288–95.
20. Shuart NG, Haitin Y, Camp SS, Black KD, Zagotta WN. Molecular mechanism for 3:1 subunit stoichiometry of rod cyclic nucleotide-gated ion channels. *Nat Commun* 2012;**2**:457.
21. Kumar J, Mayer ML. Functional insights from glutamate receptor ion channel structures. *Annu Rev Physiol* 2013;**75**:313–37.
22. Mayer ML. Structure and mechanism of glutamate receptor ion channel assembly, activation and modulation. *Curr Opin Neurobiol* 2011;**21**(2):283–90.
23. Penn AC, Greger IH. Sculpting AMPA receptor formation and function by alternative RNA processing. *RNA Biol* 2009;**6**(5):517–22.

24. Sukumaran M, Penn AC, Greger IH. AMPA receptor assembly: atomic determinants and built-in modulators. *Adv Exp Med Biol* 2012;**970**:241–64.
25. Cull-Candy S, Kelly L, Farrant M. Regulation of Ca^{2+}-permeable AMPA receptors: synaptic plasticity and beyond. *Curr Opin Neurobiol* 2006;**16**:288–97.
26. Kwak S, Weiss JH. Calcium-permeable AMPA channels in neurodegenerative disease and ischemia. *Curr Opin Neurobiol* 2006;**16**(3):281–7.
27. Liu SJ, Zukin RS. Ca^{2+}-permeable AMPA receptors in synaptic plasticity and neuronal death. *Trends Neurosci* 2007;**30**:126–34.
28. Batchelor AM, Knopfel T, Gasparini F, Garthwaite J. Pharmacological characterization of synaptic transmission through mGluRs in rat cerebellar slices. *Neuropharmacology* 1997;**36**(3):401–3.
29. Yuzaki M. New (but old) molecules regulating synapse integrity and plasticity: Cbln1 and the delta2 glutamate receptor. *Neuroscience* 2009;**162**(3):633–43.
30. Copits BA, Swanson GT. Dancing partners at the synapse: auxiliary subunits that shape kainate receptor function. *Nat Rev Neurosci* 2012;**13**(10):675–86.
31. Guzman SJ, Jonas P. Beyond TARPs: the growing list of auxiliary AMPAR subunits. *Neuron* 2010;**66**(1):8–10.
32. Jackson AC, Nicoll RA. The expanding social network of ionotropic glutamate receptors: TARPs and other transmembrane auxiliary subunits. *Neuron* 2011;**70**(2):178–99.
33. Schwenk J, Harmel N, Brechet A, Zolles G, Berkefeld H, Müller CS, et al. High-resolution proteomics unravel architecture and molecular diversity of native AMPA receptor complexes. *Neuron* 2012;**74**(4):621–33.
34. Shanks NF, Savas JN, Maruo T, Cais O, Hirao A, Oe S, et al. Differences in AMPA and kainate receptor interactomes facilitate identification of AMPA receptor auxiliary subunit GSG1L. *Cell Rep* 2012;**1**(6):590–8.
35. Norris AJ, Foeger NC, Nerbonne JM. Neuronal voltage-gated K+ (Kv) channels function in macromolecular complexes. *Neurosci Lett* 2010;**486**(2):73–7.
36. Schwenk J, Harmel N, Zolles G, Bildl W, Kulik A, Heimrich B, et al. Functional proteomics identify cornichon proteins as auxiliary subunits of AMPA receptors. *Science* 2009;**323**(5919):1313–9.
37. Tomita S, Adesnik H, Sekiguchi M, Zhang W, Wada K, Howe JR, et al. Stargazin modulates AMPA receptor gating and trafficking by distinct domains. *Nature* 2005;**435**(7045):1052–8.
38. von Engelhardt J, Mack V, Sprengel R, Kavenstock N, Li KW, Stern-Bach Y, et al. CKAMP44: a brain-specific protein attenuating short-term synaptic plasticity in the dentate gyrus. *Science* 2010;**327**(5972):1518–22.
39. Zhang W, Cho Y, Lolis E, Howe JR. Structural and single-channel results indicate that the rates of ligand binding domain closing and opening directly impact AMPA receptor gating. *J Neurosci* 2008;**28**(4):932–43.
40. Ayalon G, Stern-Bach Y. Functional assembly of AMPA and kainate receptors is mediated by several discrete protein-protein interactions. *Neuron* 2001;**31**:103–13.
41. Tu L, Deutsch C. Evidence for dimerization of dimers in K^{+} channel assembly. *Biophys J* 1999;**76**(4):2004–17.
42. Nakagawa T. The biochemistry, ultrastructure, and subunit assembly mechanism of AMPA receptors. *Mol Neurobiol* 2010;**42**(3):161–84.
43. Shanks NF, Maruo T, Farina AN, Ellisman MH, Nakagawa T. Contribution of the global subunit structure and stargazin on the maturation of AMPA receptors. *J Neurosci* 2010;**30**(7):2728–40.
44. Rosenmund C, Stern-Bach Y, Stevens CF. The tetrameric structure of a glutamate receptor channel. *Science* 1998;**280**:1596–9.
45. Greger IH, Khatri L, Kong X, Ziff EB. AMPA receptor tetramerization is mediated by Q/R editing. *Neuron* 2003;**40**:763–74.

46. Mayer ML. Glutamate receptors at atomic resolution. *Nature* 2006;**440**(7083):456–62.
47. Nakagawa T, Cheng Y, Ramm E, Sheng M, Walz T. Structure and different conformational states of native AMPA receptor complexes. *Nature* 2005;**433**(7025):545–9.
48. Tichelaar W, Safferling M, Keinanen K, Stark H, Madden DR. The three-dimensional structure of an ionotropic glutamate receptor reveals a dimer-of-dimers assembly. *J Mol Biol* 2004;**344**(2):435–42.
49. Reiner A, Arant RJ, Isacoff EY. Assembly stoichiometry of the GluK2/GluK5 kainate receptor complex. *Cell Rep* 2012;**1**(3):234–40.
50. Sobolevsky AI, Rosconi MP, Gouaux E. X-ray structure, symmetry and mechanism of an AMPA-subtype glutamate receptor. *Nature* 2009;**462**:745–56.
51. Armstrong N, Sun Y, Chen GQ, Gouaux E. Structure of a glutamate-receptor ligand-binding core in complex with kainate. *Nature* 1998;**395**:913–7.
52. Greger IH, Ziff EB, Penn AC. Molecular determinants of AMPA receptor subunit assembly. *Trends Neurosci* 2007;**30**:407–16.
53. Rossmann M, Sukumaran M, Penn AC, Veprintsev DB, Babu MM, Greger IH. Subunit-selective N-terminal domain associations organize the formation of AMPA receptor heteromers. *EMBO J* 2011;**30**:959–71.
54. Zhao H, Berger AJ, Brown PH, Kumar J, Balbo A, May CA, et al. Analysis of high-affinity assembly for AMPA receptor amino-terminal domains. *J Gen Physiol* 2012;**139**(5):371–88.
55. Karakas E, Simorowski N, Furukawa H. Subunit arrangement and phenylethanolamine binding in GluN1/GluN2B NMDA receptors. *Nature* 2012;**475**(7355):249–53.
56. Quiocho FA, Ledvina PS. Atomic structure and specificity of bacterial periplasmic receptors for active transport and chemotaxis: variation of common themes. *Mol Microbiol* 1996;**20**:17–25.
57. Clayton A, Siebold C, Gilbert RJ, Sutton GC, Harlos K, McIlhinney RA, et al. Crystal structure of the GluR2 amino-terminal domain provides insights into the architecture and assembly of ionotropic glutamate receptors. *J Mol Biol* 2009;**392**(5):1125–32.
58. Jin R, Singh SK, Gu S, Furukawa H, Sobolevsky AI, Zhou J, et al. Crystal structure and association behaviour of the GluR2 amino-terminal domain. *EMBO J* 2009;**28**(12):1812–23.
59. Kumar J, Mayer ML. Crystal structures of the glutamate receptor ion channel GluK3 and GluK5 amino-terminal domains. *J Mol Biol* 2010;**404**(4):680–96.
60. Kumar J, Schuck P, Jin R, Mayer ML. The N-terminal domain of GluR6-subtype glutamate receptor ion channels. *Nat Struct Mol Biol* 2009;**16**(6):631–8.
61. Dutta A, Shrivastava IH, Sukumaran M, Greger IH, Bahar I. Comparative dynamics of NMDA- and AMPA-glutamate receptor N-terminal domains. *Structure* 2012;**20**(11):1838–49.
62. Bahadur RP, Chakrabarti P, Rodier F, Janin J. A dissection of specific and non-specific protein-protein interfaces. *J Mol Biol* 2004;**336**(4):943–55.
63. Sukumaran M, Rossmann M, Shrivastava I, Dutta A, Bahar I, Greger IH. Dynamics and allosteric potential of the AMPA receptor N-terminal domain. *EMBO J* 2011;**30**(5):972–82.
64. Kumar J, Schuck P, Mayer ML. Structure and assembly mechanism for heteromeric kainate receptors. *Neuron* 2011;**71**(2):319–31.
65. Farina AN, Blain KY, Maruo T, Kwiatkowski W, Choe S, Nakagawa T. Separation of domain contacts is required for heterotetrameric assembly of functional NMDA receptors. *J Neurosci* 2011;**31**(10):3565–79.
66. Atlason PT, Garside ML, Meddows E, Whiting P, McIlhinney RA. N-Methyl-D-aspartate (NMDA) receptor subunit NR1 forms the substrate for oligomeric assembly of the NMDA receptor. *J Biol Chem* 2007;**282**(35):25299–307.

67. Pin JP, Galvez T, Prezeau L. Evolution, structure, and activation mechanism of family 3/C G-protein-coupled receptors. *Pharmacol Ther* 2003;**98**(3):325–54.
68. He X, Chow D, Martick MM, Garcia KC. Allosteric activation of a spring-loaded natriuretic peptide receptor dimer by hormone. *Science* 2001;**293**(5535):1657–62.
69. Kunishima N, Shimada Y, Tsuji Y, Sato T, Yamamoto M, Kumasaka T, et al. Structural basis of glutamate recognition by a dimeric metabotropic glutamate receptor. *Nature* 2000;**407**(6807):971–7.
70. Mony L, Zhu S, Carvalho S, Paoletti P. Molecular basis of positive allosteric modulation of GluN2B NMDA receptors by polyamines. *EMBO J* 2011;**30**(15):3134–46.
71. Doumazane E, Scholler P, Zwier JM, Trinquet E, Rondard P, Pin JP. A new approach to analyze cell surface protein complexes reveals specific heterodimeric metabotropic glutamate receptors. *FASEB J* 2011;**25**(1):66–77.
72. Cui C, Mayer ML. Heteromeric kainate receptors formed by the coassembly of GluR5, GluR6, and GluR7. *J Neurosci* 1999;**19**(19):8281–91.
73. Gielen M, Siegler Retchless B, Mony L, Johnson JW, Paoletti P. Mechanism of differential control of NMDA receptor activity by NR2 subunits. *Nature* 2009;**459**(7247):703–7.
74. Perrais D, Veran J, Mulle C. Gating and permeation of kainate receptors: differences unveiled. *Trends Pharmacol Sci* 2010;**31**(11):516–22.
75. Swanson GT, Green T, Sakai R, Contractor A, Che W, Kamiya H, Heinemann SF. Differential activation of individual subunits in heteromeric kainate receptors. *Neuron* 2002;**34**(4):589–98.
76. Das U, Kumar J, Mayer ML, Plested AJ. Domain organization and function in GluK2 subtype kainate receptors. *Proc Natl Acad Sci USA* 2010;**107**(18):8463–8.
77. Greger IH, Akamine P, Khatri L, Ziff EB. Developmentally regulated, combinatorial RNA processing modulates AMPA receptor biogenesis. *Neuron* 2006;**51**:85–97.
78. Penn AC, Williams SR, Greger IH. Gating motions underlie AMPA receptor secretion from the endoplasmic reticulum. *EMBO J* 2008;**27**(22):3056–68.
79. Seeburg PH. The role of RNA editing in controlling glutamate receptor channel properties. *J Neurochem* 1996;**66**:1–5.
80. Coleman SK, Moykkynen T, Jouppila A, Koskelainen S, Rivera C, Korpi ER, Keinänen K. Agonist occupancy is essential for forward trafficking of AMPA receptors. *J Neurosci* 2009;**29**(2):303–12.
81. Gill MB, Vivithanaporn P, Swanson GT. Glutamate binding and conformational flexibility of ligand-binding domains are critical early determinants of efficient kainate receptor biogenesis. *J Biol Chem* 2009;**284**(21):14503–12.
82. Mah SJ, Cornell E, Mitchell NA, Fleck MW. Glutamate receptor trafficking: endoplasmic reticulum quality control involves ligand binding and receptor function. *J Neurosci* 2005;**25**(9):2215–25.
83. Kenny AV, Cousins SL, Pinho L, Stephenson FA. The integrity of the glycine co-agonist binding site of N-methyl-D-aspartate receptors is a functional quality control checkpoint for cell surface delivery. *J Biol Chem* 2009;**284**(1):324–33.
84. She K, Ferreira JS, Carvalho AL, Craig AM. Glutamate binding to the GluN2B subunit controls surface trafficking of N-methyl-D-aspartate (NMDA) receptors. *J Biol Chem* 2012;**287**(33):27432–45.
85. Penn AC, Balik A, Wozny C, Cais O, Greger IH. Activity-mediated AMPA receptor remodeling, driven by alternative splicing in the ligand-binding domain. *Neuron* 2012;**76**(3):503–10.
86. Balik A, Penn AC, Nemoda Z, Greger IH. Activity-regulated RNA editing in select neuronal subfields in hippocampus. *Nucleic Acids Res* 2013;**41**(2):1124–34.
87. Salussolia CL, Wollmuth LP. Flip-flopping to the membrane. *Neuron* 2012;**76**(3):463–5.

88. Chaudhry C, Plested AJ, Schuck P, Mayer ML. Energetics of glutamate receptor ligand binding domain dimer assembly are modulated by allosteric ions. *Proc Natl Acad Sci USA* 2009;**106**(30):12329–34.
89. Hansen KB, Naur P, Kurtkaya NL, Kristensen AS, Gajhede M, Kastrup JS, Traynelis SF. Modulation of the dimer interface at ionotropic glutamate-like receptor delta2 by D-serine and extracellular calcium. *J Neurosci* 2009;**29**(4):907–17.
90. Plested AJ, Mayer ML. Structure and mechanism of kainate receptor modulation by anions. *Neuron* 2007;**53**:829–41.
91. Vivithanaporn P, Lash LL, Marszalec W, Swanson GT. Critical roles for the M3-S2 transduction linker domain in kainate receptor assembly and postassembly trafficking. *J Neurosci* 2007;**27**(39):10423–33.
92. Lee CH, Gouaux E. Amino terminal domains of the NMDA receptor are organized as local heterodimers. *PLoS One* 2011;**6**(4):e19180.
93. Mansour M, Nagarajan N, Nehring RB, Clements JD, Rosenmund C. Heteromeric AMPA receptors assemble with a preferred subunit stoichiometry and spatial arrangement. *Neuron* 2001;**32**(5):841–53.
94. Riou M, Stroebel D, Edwardson JM, Paoletti P. An alternating GluN1-2-1-2 subunit arrangement in mature NMDA receptors. *PLoS One* 2012;**7**(4):e35134.
95. Pasternack A, Coleman SK, Jouppila A, Mottershead DG, Lindfors M, Pasternack M, Keinänzen K. Alpha-amino-3-hydroxy-5-methyl-4-isoxazolepropionic acid (AMPA) receptor channels lacking the N-terminal domain. *J Biol Chem* 2002;**277**(51):49662–7.
96. Salussolia CL, Prodromou ML, Borker P, Wollmuth LP. Arrangement of subunits in functional NMDA receptors. *J Neurosci* 2011;**31**(31):11295–304.
97. Chen GQ, Cui C, Mayer ML, Gouaux E. Functional characterization of a potassium-selective prokaryotic glutamate receptor. *Nature* 1999;**402**(6763):817–21.
98. Kuner T, Seeburg PH, Guy HR. A common architecture for K+ channels and ionotropic glutamate receptors? *Trends Neurosci* 2003;**26**(1):27–32.
99. Wollmuth LP, Sobolevsky AI. Structure and gating of the glutamate receptor ion channel. *Trends Neurosci* 2004;**27**(6):321–8.
100. Doyle DA, Morais Cabral J, Pfuetzner RA, Kuo A, Gulbis JM, Cohen SL, et al. The structure of the potassium channel: molecular basis of K^+ conduction and selectivity. *Science* 1998;**280**(5360):69–77.
101. Long SB, Campbell EB, Mackinnon R. Crystal structure of a mammalian voltage-dependent Shaker family K^+ channel. *Science* 2005;**309**(5736):897–903.
102. Sprengel R, Aronoff R, Volkner M, Schmitt B, Mosbach R, Kuner T. Glutamate receptor channel signatures. *Trends Pharmacol Sci* 2001;**22**(1):7–10.
103. Washburn MS, Numberger M, Zhang S, Dingledine R. Differential dependence on GluR2 expression of three characteristic features of AMPA receptors. *J Neurosci* 1997;**17**(24):9393–406.
104. Bowie D. Redefining the classification of AMPA-selective ionotropic glutamate receptors. *J Physiol* 2012;**590**(Pt 1):49–61.
105. Kohler M, Burnashev N, Sakmann B, Seeburg PH. Determinants of Ca^{2+} permeability in both TM1 and TM2 of high affinity kainate receptor channels: diversity by RNA editing. *Neuron* 1993;**10**(3):491–500.
106. Sommer B, Kohler M, Sprengel R, Seeburg PH. RNA editing in brain controls a determinant of ion flow in glutamate-gated channels. *Cell* 1991;**67**(1):11–9.
107. Burnashev N, Monyer H, Seeburg PH, Sakmann B. Divalent ion permeability of AMPA receptor channels is dominated by the edited form of a single subunit. *Neuron* 1992;**8**(1):189–98.
108. Greger IH, Khatri L, Ziff EB. RNA editing at Arg607 controls AMPA receptor exit from the endoplasmic reticulum. *Neuron* 2002;**34**:759–72.

109. Hume RI, Dingledine R, Heinemann SF. Identification of a site in glutamate receptor subunits that controls calcium permeability. *Science* 1991;**253**(5023):1028–31.
110. Ball SM, Atlason PT, Shittu-Balogun OO, Molnar E. Assembly and intracellular distribution of kainate receptors is determined by RNA editing and subunit composition. *J Neurochem* 2010;**114**(6):1805–18.
111. Bernard A, Ferhat L, Dessi F, Charton G, Represa A, Ben-Ari Y, Khrestchatisky M. Q/R editing of the rat GluR5 and GluR6 kainate receptors in vivo and in vitro: evidence for independent developmental, pathological and cellular regulation. *Eur J Neurosci* 1999;**11**(2):604–16.
112. Sans N, Vissel B, Petralia RS, Wang YX, Chang K, Royle GA, et al. Aberrant formation of glutamate receptor complexes in hippocampal neurons of mice lacking the GluR2 AMPA receptor subunit. *J Neurosci* 2003;**23**(28):9367–73.
113. Brusa R, Zimmermann F, Koh DS, Feldmeyer D, Gass P, Seeburg PH, Sprengel R. Early-onset epilepsy and postnatal lethality associated with an editing-deficient GluR-B allele in mice. *Science* 1995;**270**(5242):1677–80.
114. Salussolia CL, Corrales A, Talukder I, Kazi R, Akgul G, Bowen M, Wollmuth LP. Interaction of the M4 segment with other transmembrane segments is required for surface expression of mammalian alpha-amino-3-hydroxy-5-methyl-4-isoxazolepropionic acid (AMPA) receptors. *J Biol Chem* 2011;**286**(46):40205–18.
115. Kaniakova M, Lichnerova K, Vyklicky L, Horak M. Single amino acid residue in the M4 domain of GluN1 subunit regulates the surface delivery of NMDA receptors. *J Neurochem* 2012;**123**(3):385–95.
116. Clarke OB, Gulbis JM. Oligomerization at the membrane: potassium channel structure and function. *Adv Exp Med Biol* 2012;**747**:122–36.
117. Wang Z, Jiang Y, Lu L, Huang R, Hou Q, Shi F. Molecular mechanisms of cyclic nucleotide-gated ion channel gating. *J Genet Genomics* 2007;**34**(6):477–85.
118. Li M, Jan YN, Jan LY. Specification of subunit assembly by the hydrophilic amino-terminal domain of the Shaker potassium channel. *Science* 1992;**257**(5074):1225–30.
119. Lu J, Robinson JM, Edwards D, Deutsch C. T1-T1 interactions occur in ER membranes while nascent Kv peptides are still attached to ribosomes. *Biochemistry* 2001; **40**(37):10934–46.
120. Deutsch C. Potassium channel ontogeny. *Annu Rev Physiol* 2002;**64**:19–46.
121. Schwake M, Athanasiadu D, Beimgraben C, Blanz J, Beck C, Jentsch TJ, et al. Structural determinants of M-type KCNQ (Kv7) K+ channel assembly. *J Neurosci* 2006;**26** (14):3757–66.
122. Gomez-Posada JC, Etxeberria A, Roura-Ferrer M, Areso P, Masin M, Murrell-Lagnado RD, Villarroel A. A pore residue of the KCNQ3 potassium M-channel subunit controls surface expression. *J Neurosci* 2010;**30**(27):9316–23.
123. Stewart AP, Gomez-Posada JC, McGeorge J, Rouhani MJ, Villarroel A, Murrell-Lagnado RD, Edwardson JM. The Kv7.2/Kv7.3 heterotetramer assembles with a random subunit arrangement. *J Biol Chem* 2012;**287**(15):11870–7.
124. Ludwig J, Owen D, Pongs O. Carboxy-terminal domain mediates assembly of the voltage-gated rat ether-a-go-go potassium channel. *EMBO J* 1997;**16**(21):6337–45.
125. Chen IH, Hu JH, Jow GM, Chuang CC, Lee TT, Liu DJ, Jeng CJ. Distal end of carboxyl terminus is not essential for the assembly of rat Eag1 potassium channels. *J Biol Chem* 2011;**286**(31):27183–96.
126. Peng C, Rich ED, Varnum MD. Subunit configuration of heteromeric cone cyclic nucleotide-gated channels. *Neuron* 2004;**42**(3):401–10.
127. Weitz D, Ficek N, Kremmer E, Bauer PJ, Kaupp UB. Subunit stoichiometry of the CNG channel of rod photoreceptors. *Neuron* 2002;**36**(5):881–9.

SECTION IV

Oligomerization in Enzymes

CHAPTER FOURTEEN

The Structural Basis for the Allosteric Regulation of Ribonucleotide Reductase

Md. Faiz Ahmad*, Chris G. Dealwis*,†

*Department of Pharmacology, School of Medicine, Case Western Reserve University, Cleveland, Ohio, USA

†Center for Proteomics and Bioinformatics, Case Western Reserve University, Cleveland, Ohio, USA

Contents

Abstract

Ribonucleotide reductases (RRs) catalyze a crucial step of *de novo* DNA synthesis by converting ribonucleoside diphosphates to deoxyribonucleoside diphosphates. Tight control of the dNTP pool is essential for cellular homeostasis. The activity of the enzyme is tightly regulated at the S-phase by allosteric regulation. Recent structural studies by our group and others provided the molecular basis for understanding how RR recognizes substrates, how it interacts with chemotherapeutic agents, and how it is regulated by its allosteric regulators ATP and dATP. This review discusses the molecular basis of allosteric regulation and substrate recognition of RR, and particularly the discovery that subunit oligomerization is an important prerequisite step in enzyme inhibition.

Progress in Molecular Biology and Translational Science, Volume 117
ISSN 1877-1173
http://dx.doi.org/10.1016/B978-0-12-386931-9.00014-3

ABBREVIATIONS

5-NITP 5-nitro-indolyl-2′-deoxyribose triphosphate
ADP adenosine diphosphate
AMPPNP adenosine 5-(β,γ-imido) triphosphate tetralithium salt
CDP cytidine diphosphate
dGTP deoxyguanosine triphosphate
dNTP deoxynucleoside triphosphate
hRRM1 human ribonucleotide reductase
ScRR1 *Saccharomyces cerevisiae* RR1

1. INTRODUCTION

The process by which biological macromolecules such as proteins transmit regulatory signals upon ligand binding to a distinct site distant from the functional site is known as allostery.[1] Allostery is important in governing vital functions of cell metabolism and gene regulation.[2,3] Considering that allostery is such an important process, a general understanding of its structural basis has been elusive. In the past four decades, allostery has usually been understood via two classical models: the concerted or symmetrical model also known as the MWC model and the sequential model also known as the KNF model.[4,5] The concerted model proposes that regulatory proteins have a quaternary structure consisting of subunits organized into finite assemblies, or oligomers, and that all subunits in an oligomer switch states at the same time.[3] This model proposes that the conformational coupling occurs between different parts of the molecule as a complete entity.[1] The second model postulates that subunits in the same oligomer can be in different states, but the conversion of one subunit makes the conversion of other subunits in the oligomer more likely. In this model, ligand binding to the allosteric site(s) causes structural changes intrinsic to the protein molecule that affect substrate binding.

Allosteric regulation can occur upon binding of a small molecule at a site distant from the active site. So it is pertinent to ask how the signal is transmitted over such a distance. Ligands that activate and inhibit an enzyme through allostery are called allosteric modulators. A significant amount of information on allosteric mechanisms has come from high-resolution structures determined by X-ray crystallography or NMR spectroscopy. High-resolution structure analysis by these methods has shown how ligand binding triggers the allosteric mechanism in those proteins. Comparison of various structures of apoproteins and the same proteins with bound ligand showed bonds formed or lost on ligand binding that correlated with the functional (T) or relaxed (R) states of the molecule.[6,7] However, in this review, we discuss ribonucleotide reductase (RR); an allosteric enzyme follows neither the concerted nor the sequential model.

2. RIBONUCLEOTIDE REDUCTASE

RR is an allosteric enzyme that provides the balanced pool of deoxyribonucleoside triphosphate (dNTPs) to maintain genomic integrity in all cells.[8,9] RR catalyzes the rate-determining step of dNTP synthesis and hence provides the essential precursors for DNA replication and repair.[10–12]

Historically, RR was first characterized by Peter Reichard and group at the Karolinska Institute in Sweden in the 1960s.[12–16] Four decades after conducting the first enzymatic studies of RR, we still discover novel mechanisms pertaining to its mode of action.

In 1969, Brown and Reichard showed that RR has the amazing ability to catalyze transformation of all four nucleoside diphosphate substrates to their deoxy forms. They also showed that there were two distinct allosteric sites. The first site bound effectors and determined substrate preferences while the second site bound ATP to activate the enzyme or dATP to inhibit it. Mechanistically, reduction of ribonucleoside to deoxyribonucleoside by RR requires the generation of thiyl radical by cysteine residues near the active site of the enzyme. Metal cofactors are involved in the generation of free radicals that ultimately lead to the formation of the important thiyl radical in the active site.[17–19] So far, three classes of RR have been reported based on the metal cofactors that generate their thiyl radicals.[9,17,18,20] In this review, we discuss the Class I enzymes that use a tyrosyl free radical for catalysis.

Class I enzymes are expressed in nearly all eukaryotes, in some prokaryotes, and in certain viruses.[9,20–23] However, in this review, we focus mostly on human, yeast (eukaryotic), and *E. coli* (prokaryotic) RR. The active form of the eukaryotic enzyme consists of two proteins, the larger being RR1 (α) and the smaller RR2 (β). These proteins associate in dimeric or multimeric forms such as $\alpha_2\beta_2$ and $(\alpha)_n(\beta2)_m$ (where $n=4$ or 6 and $m=1$, 2, or 3).[24–30] Substrate catalysis occurs in RR1 (α) which contains two allosteric sites, the specificity site (S-site), and the activity site (A-site) in addition to the catalytic site (C-site)[29,31–34] (Fig. 14.1). The smaller RR2 (β) subunit is associated with the diferric-oxygen center which generates the stable tyrosyl free

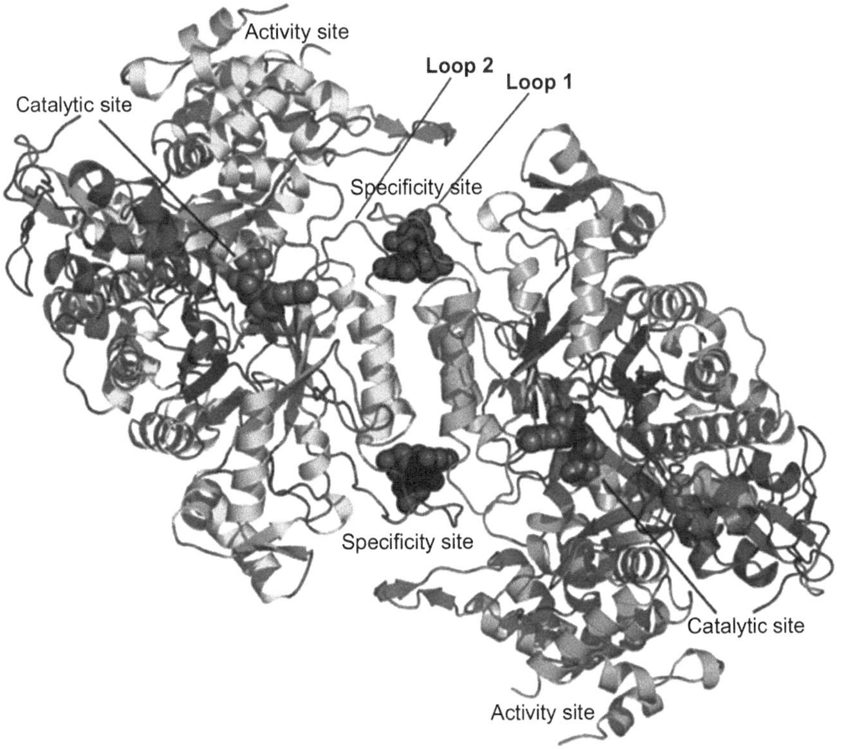

Figure 14.1 Structure of large subunit of ScRR1 in dimeric form. ScRR1 monomers are yellow and green; effector dGTP (violet) and substrate ADP (blue) from the effector substrate pair are shown in the specificity (S-site) and catalytic site (C-site). *Reproduced with permission from Ref. 32. Copyright (2006) National Academy of Sciences, USA.* (For interpretation of the references to color in this figure legend, the reader is referred to the online version of this chapter.)

radical.[35] Class I enzymes are further divided into three subclasses, Class Ia, Ib, and Ic, based on the organization of RR genes, protein topology, and metalo-cluster assembly of the RR2 (β) protein.[9,36,37] Class Ia and Class Ib require a diferric-tyrosyl radical cofactor (Fe^{III} Fe^{III}–$Y^{\bullet}$) associated with their RR2 (β) protein. The Class Ic enzymes do not use $Y^{\bullet}$, but the synthesis of an active metal cluster in the β subunit requires oxygen. Catalysis by Class Ia enzymes involves reduction of cysteine residues near the C-site brought about by transfer of the radical originating at the diferric-tyrosyl 30–35 Å from the C-site.[30,31,38] After catalysis, the cysteines are re-oxidized to regenerate the enzyme.

The early biochemical studies raised a remarkable question concerning RR: by what molecular mechanism did this enzyme catalyze specific reduction of all four nucleoside diphosphate substrates? In other words, what causes this enzyme to switch its specificity when it binds different effectors? Among the two allosteric sites, one, the A-site, regulates the overall activity and the other, the S-site, regulates the substrate specificity. Binding of ATP or dATP at the A-site increases or reduces, respectively, the overall activity of the enzyme.[10,12,33] The S-site has the ability to bind dGTP and TTP in addition to ATP and dATP. RR1 uses allosteric communication to determine C-site substrate preference based on the nucleotide effector bound at the S-site.[9,10,21,32,34] Either ATP or dATP binding at the S-site facilitates both CDP and UDP binding at the C-site; TTP binding at the S-site selects for GDP binding at the C-site, and dGTP binding at the S-site promotes ADP binding at the C-site. The determination of the first three-dimensional structure of RR1 from *E. coli* provided the molecular structure of the two allosteric sites and the C-site.[31] Another important study from Hans Eklund's group in 1997 showed the structure of *E. coli* RR1 in complex with TTP/GDP for the first time.[39] This study illustrated the importance of a polypeptide chain called Loop 2 which was involved in substrate recognition and another called Loop 1 which was important in effector binding. It also laid the groundwork for the definition of the S-site and the C-site. In the same study, they showed how substrate binding at the C-site is structurally regulated by the binding of allosteric effectors.

RR is highly regulated transcriptionally, by subcellular compartmentalization, by small protein inhibitors such as the yeast proteins Sml1 and Spd1 and, during the S-phase of the cell cycle, by allosteric regulation.[9] In this review, we discuss the allosteric regulation of RR by which a single protein provides a balanced supply of all four dNTPs. Understanding RR regulation is important in cancer therapy since many of the clinically approved anticancer drugs such as gemcitabine, clofarabine, cladribine, and fludarabine mimic the substrates and allosteric effectors of this enzyme.[40,41] RR has a highly sophisticated mechanism for synthesis and regulation of dNTPs, and even after more than four decades of RR discovery, research on this enzyme continues to reveal fascinating new aspects of its action. Recent studies from our group and others advance our understanding of the intriguing mechanisms of allosteric regulation in this enzyme.[29,42–46]

2.1. Allosteric regulation

Interestingly, the allosteric regulation occurring between the S- and C-sites is shared by both Class I and Class II enzymes. In contrast, Class III enzymes use a slightly different set of substrate selection rules. The A-site is known as the pyrimidine site and can bind either ATP or dATP. Binding of ATP to the pyrimidine site stimulates reduction of pyrimidine ribonucleotides. The purine site resembles the S-site in Class I and Class II enzymes. dGTP and TTP bind at the purine site and select for ATP or GTP at the C-site, respectively. However, the binding of dATP to either the purine or pyrimidine site is always inhibitory. The molecular basis for substrate selections is described in two studies based on the Class II and Class I structures.[32,47]

2.2. The C- and S-sites of *Saccharomyces cerevisiae* RR and *Homo sapiens* RR (hRRM1)

Our lab has produced hRRM1•TTP•GDP and ScRR1•TTP•GDP ternary complexes by soaking cocrystals of hRRM1•TTP and ScRR1•TTP with the substrate GDP. The structures of these ternary complexes showed that the ribose ring of the substrate adopts a puckered 3′-endo conformation similar to that previously observed in the TmNrd1 substrate–effector complex structures.[29,32,48] Although in these complexes the C-site interactions with the ribose of the GDP are fairly well conserved, there are differences with those of the Class II TmNrd1 complex. For instance, the sulfur of the catalytic cysteine in the hRRM1 and ScRR1 complexes is within hydrogen bonding distance of the 2′ and 3′ OH groups of the ribose, while the corresponding groups are more than 4 Å apart in the TmNrd1 complexes. Comparison of the S-site of the hRRM1•TTP•GDP complex with those of the TTP complexes formed by eukaryotic ScRR1 and prokaryotic EcRR1 and TmNrd1 shows that TTP binds very similarly in all the structures, with the exception of the EcRR1•TTP•GDP complex in which the base is substantially shifted by about 5 Å.

2.3. The role of Loop 1 and Loop 2 in allosteric regulation of RR1

Two very important loops called Loop 1 (residues 245–260) and Loop 2 (residues 286–295) (Fig. 14.1) were identified in the *E. coli* X-ray structure.[39] Nucleoside or dNTP effector binding at the S-site is a prerequisite for the dimerization of the large subunit, which is essential for substrate selection. This is because allosteric communication occurs between subunits by means of Loop 2 (Fig. 14.1) (which connects the S-site on one subunit with the C-site of the adjacent subunit). The mechanism of substrate selection mediated by Loop 2 has been the subject of two crystallographic studies conducted on *T. maritima* Class II RR and *S. cerevisiae* Class I RR. These enzymes share a common allosteric mechanism involving key residues in Loop 2[47] (Fig. 14.2). In the apo form of *S. cerevisiae* RR, this loop sits in a partially closed conformation, leaving little or no room for substrate binding. Upon nucleoside triphosphate binding to the S-site, Loop 2 shifts toward the effector site, concomitantly creating room for substrate entry and creating a face of the binding site that helps select a particular substrate (Figs. 14.1 and 14.2). Once the substrate is bound, Loop 2 yet again changes conformation and moves back toward the substrate.[32] We called this elegant communication between the S-site and C-site "specificity cross talk." Two residues in the Loop 2, R293 and Q288, have been proposed to be critical for substrate recognition.[42,49] A recent study has shown that a number of single point mutations in the ScRR1 Loop 2 have a significant impact on the cellular dNTP pool and the cell growth phenotype.[49,50] Structural studies have shown that Q288 and R293 in ScRR1 form hydrogen bonds with the adenine ring, and the R293 makes an additional stacking interaction with the adenine.[32] Mutation of R293 in ScRR1 to alanine severely compromised ADP binding at the C-site as evident in the crystal structure of the R293A ScRR1-dGTP complex. Furthermore, isothermal titration calorimetry analysis indicated significant loss of affinity for ADP substrate in the case of R293A ScRR1.[42] This is not surprising, as ADP makes a greater number of interactions with residue 293 when bound to wild-type ScRR1 than to the mutant. The Q288A mutation destroys two hydrogen bonds that

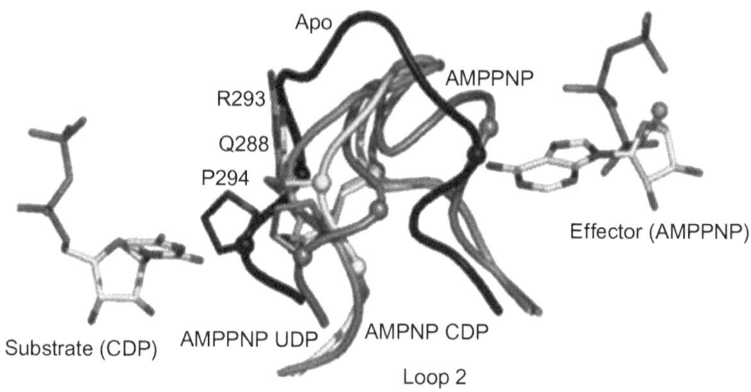

Figure 14.2 Substrate selection mediated via Loop 2: effector AMPPNP and substrate ADP binding at the S and C-site, respectively. The key residues on Loop 2 required for substrate selection are Q288 and R293 as shown. *Reproduced with permission from Ref. 32. Copyright (2006) National Academy of Sciences, USA.* (For color version of this figure, the reader is referred to the online version of this chapter.)

the glutamine side chain forms with ADP and R293. The latter interaction is important for positioning the guanidinium side chain of R293, to form its hydrogen bond and stacking interaction with ADP.

Mutation at R293 in ScRR1 leads to altered CDP substrate binding, such that CDP binds farther from the C-site when compared to wild-type ScRR1. Moreover, this mutation results in a conformational change in Loop 2 accompanied by a shift toward the C-site. The newly acquired loop conformation would cause a steric clash between P294 and the CDP substrate if the CDP bound in the original position as observed in the wild-type ScRR1. This study shows that residues R293 and Q288 are important not only for facilitating substrate selection but also for stabilizing the conformation of Loop 2. It is evident in these structural studies that Loop 2 conformation is crucial for the communication between the S-site and the C-site (Fig. 14.3A–C). The Chabes group has previously shown that expressing R293A and Q288A ScRR1 mutants in yeast skewed the dNTP pool toward higher dATP/dCTP and dGTP/dCTP ratios. Substitution at multiple positions in Loop 2 also gave rise to mutants with imbalanced dNTP pools; however, only R293A and Q288A mutations exhibited synthetic lethality based on meiotic progeny analysis when the alternative RR1 subunit ScRR3 was deleted,[49] indicating the significant role played by Q288 and R293 in ScRR1 (Fig. 14.3D).

2.4. ATP and dATP binding at the A-site (ATP cone)

The A-site in hRRM1 consists of a four-helix bundle. These helices are numbered H1–H4, with H1 comprising residues 15–26, H2 comprising residues 36–46, H3 comprising residues 53–70, and H4 comprising residues 74–90 (Fig. 14.4A and B). This ATP-binding cone is covered at one end by the previously termed "β-cap," a β-hairpin formed by the first 14 N-terminal residues and residues 48–51.[31] Binding of either ATP or dATP requires helix H1 and the β-cap to shift from their positions in the native (TTP-only) form (Fig. 14.4C), involved in some induced fit. However, the two binding modes are markedly different. dATP binds more deeply inside the four-helix bundle and its ribose adopts a half-chair conformation with the C2′ out of plane. In contrast, ATP binds less deeply inside the four-helix bundle and its ribose adopts a 2′-endo conformation (Fig. 14.4). Interestingly, we have observed in our ITC binding experiment that hRRM1 has severalfolds higher affinity for dATP compared to ATP (Ahmad MF, unpublished data). The binding energy of protein–ligand interactions can

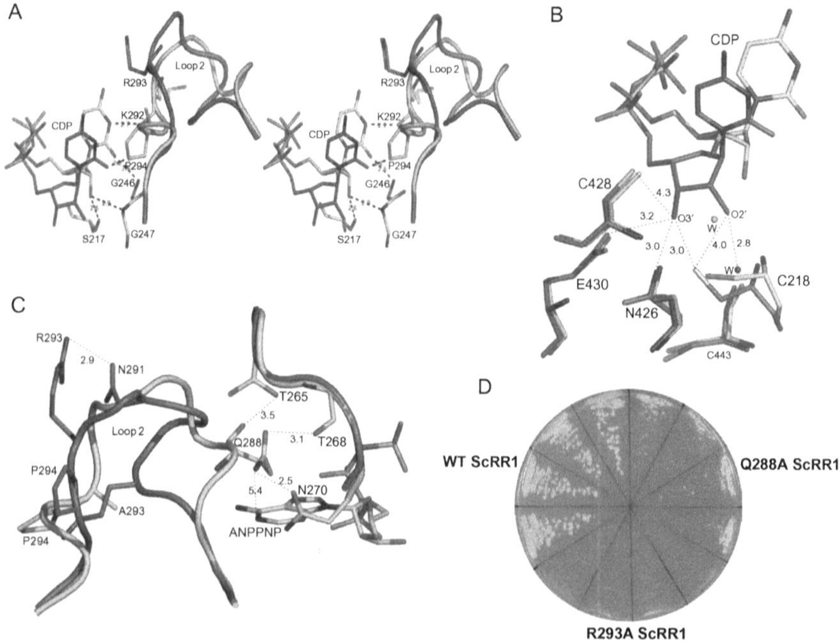

Figure 14.3 Effect of R293A and Q288A mutations on wild-type ScRR1. (A) Structural comparison of AMPPNP- and CDP-bound wild type and R293A ScRR1; stereo view of the substrate CDP binding and its proximity to Loop 2. (B) Comparison of catalytic site of wild type and R293A ScRR1 with CDP bound. The catalytic important residues C218, C428, N426, C428, and E430 are also shown binding to the ribose moiety of wild-type ScRR1 (shown as dashed lines). (C) Effect of R293A and Q288A mutations in the Loop 2 conformation. In these figures, wild-type ScRR1 is colored in magenta and R293A ScRR1 is colored in green. (D) Growth defects in wild-type ScRR1 because of R293A and Q288A mutations; R293A ScRR1 causes lethality whereas Q288A ScRR1 showed slower growth compared to wild-type ScRR1. *Reproduced with permission from Ref. 42. Copyright (2012).* (For interpretation of the references to color in this figure legend, the reader is referred to the online version of this chapter.)

be expressed as the change in solvent-accessible surface area between the bound and unbound states[51]; the deeper binding of dATP and its larger buried surface area compared to ATP must contribute to its higher affinity. The chemical difference at the 2′ position on the ribose moiety between ATP and dATP causes the difference in their binding at the A-site. The 2′-OH of ATP sterically precludes the ribose from binding more deeply inside the four-helix bundle, whereas dATP, lacking the 2′-OH, can penetrate the pocket more deeply. In addition, our structural studies on dATP-bound hRRM1 illustrated that residue I18 acts as a steric gate, as it would clash with the 2′-OH of ATP binding at the dATP position. In the similar study, we

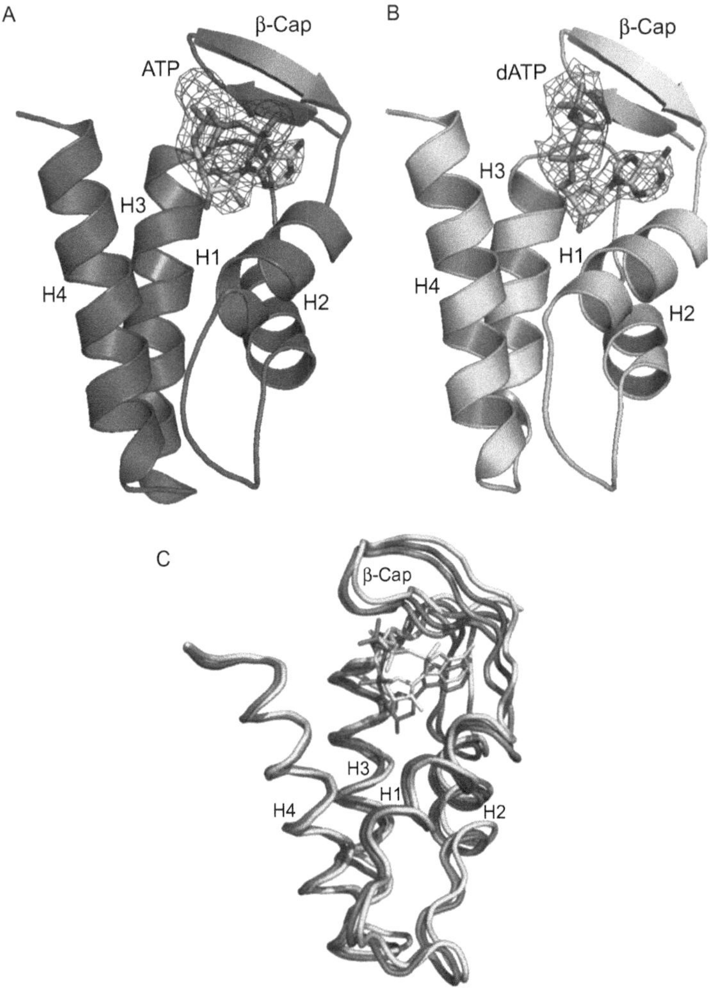

Figure 14.4 Structure of ATP- and dATP-bound hRRM1. (A and B) ATP and dATP binding at the A-site in hRRM1 consist of a four-helix bundle. $2F_O-F_c$ electron density for ATP and dATP is shown which is contoured at 1σ (blue density). (C) Ribbon diagram depicting ATP-binding cones of hRRM1–TTP–ATP ternary complex (blue) and hRRM1–TTP–dATP ternary complex (magenta) were aligned to that of hRRM1–TTP complex (orange). Allosteric modulator ATP and dATP are shown in yellow and green, respectively. *Reproduced with permission from Ref. 29.* (For interpretation of the references to color in this figure legend, the reader is referred to the online version of this chapter.)

proposed the molecular basis for how dATP is able to overcome a severalfold deficit to outcompete ATP at the A-site during the S-phase of the cell cycle.

2.5. Subunit oligomerization and allosteric regulation of RR

Binding of different ligands enables the formation of a biologically functional oligomer in some cases and a nonfunctional oligomer in others. In the case of RR, both the allosteric activator ATP and allosteric inhibitor dATP induce oligomerization upon binding to RR. ATP and dATP bind in a concentration-dependent manner to alter the enzyme's oligomeric state and thereby regulate enzyme activity. Initial studies on mouse RR have shown that dATP binding to the A-site induces formation of a tetramer, which further isomerizes to an inactive form of the enzyme. ATP, in contrast, induces formation of an active hexamer upon binding to a proposed H-site.[25,26] This study also examined mouse D57N RR, which differs from wild-type mouse RR in that it has lost the ability to differentiate between ATP and dATP and hence is activated by both. They argue that the D57N mutation in mouse RR produces this effect by blocking the transformation of active to inactive tetramer.[25,27]

Later studies on mouse RR suggest that despite the fact that ATP and dATP have opposing effects on enzyme activity, both nucleotides facilitate the formation of higher-order oligomers. Anders Hofer's group has shown, using a technique known as gas-phase electrophoretic-mobility macromolecule analysis, that ATP/dATP-bound murine RR1 forms hexamers which can interact with dimers of the small subunit to form the enzymatically active complex $\alpha_6\beta_2$. This $\alpha_6\beta_2$ holoenzyme complex can exist in either an inactivated state or an activated state depending upon whether the bound nucleotide is dATP or ATP.[24] Another study with the drug gemcitabine (F2CDP) which is used clinically in the treatment of various types of cancer showed that F2CDP is a substoichiometric mechanism-based inhibitor of the large subunit of both *E. coli* and human RR in the presence of reductant.[28] In this study, F2CDP-bound RR forms a tight $\alpha_2\beta_2$ complex in *E. coli*. Whereas human RR forms an $\alpha_6\beta_6$ complex upon F2CDP binding that leads to enzyme inactivation.[28] Recently, we demonstrated that dATP-induced oligomerization is concentration dependent.[29] Our studies have shown that in the absence of any effector molecules, human RR is a monomer. At 5 μm dATP, the dimer and hexamer are seen with a reduction of monomer concentration as analyzed by size-exclusion chromatography (SEC) and multiangle light scattering (MALS). Interestingly, at a physiological concentration of 20 μm dATP (the physiological concentration of dATP reported for the S-phase), most of the human RR forms a hexameric complex, with a few dimeric species as revealed by both SEC and MALS. Furthermore, it has been shown that the N-terminus is involved in hexamer formation. Upon deleting the first 74 amino acid residues, commonly known as the ATP cone, from human RR, dATP is no longer able to induce the formation of the hexamer. In addition, in the presence of human RR2, the N-terminally truncated version of hRRM1 forms an $\alpha_2\beta_2$ holocomplex. This result further emphasizes the role of the A-site in oligomerization. In this study, it has been shown that the rate of dCDP and dADP formation by hRRM1 and ScRR1 is severely diminished in the dATP-induced hexamer. However, the dimeric form of eukaryotic RR is not inhibited by dATP. The above results indicate that the dATP-induced hexamerization of eukaryotic RR is a prerequisite for enzyme inhibition. It is important to mention here that these findings deviate from previous studies which reported that the dATP-induced tetramer is the least active RR form and that the hexamer is observed only at very high dATP concentration.[25] The discrepancy regarding dATP-induced oligomerization in the above study is partly due to the difficulty of interpretation of dynamic light scattering data. Since dynamic light scattering cannot resolve differences in the radius of various oligomer species that are less than a factor of 4–5, the reported mammalian RR tetramer may be mixture of dimers and hexamers, which would be consistent with our results.

2.6. The X-ray structure of the dATP-induced hexamer of Class 1 RNR

To understand the structural basis of RR1 oligomerization and enzyme inactivation by dATP, we crystallized the ScRR1 hexamer complex. The complex crystallized in a hexagonal space group and diffracted to 6 Å. Although the low-resolution structure does not provide atomic detail, it is useful for indicating oligomer organization. To analyze the data, two models of ScRR1 hexamer packing were considered. In both models, ScRR1 α_6 is a trimer of dimers, in which the three dimers are related to each other by a threefold axis. However, the models differ in the diameter of the central pore of the hexamer and in how dATP would mediate hexamerization. In one model, only three of the six dATP-bound ATP-binding cones participate in forming the hexamer interface, the other three free to interact with the small subunit. Hence, only three dATP molecules are present at the hexamer interface. In the other model, the interfaces that stabilize the hexamer are entirely formed by six dATP-bound ATP-binding cones. Furthermore, each of the three interfaces is formed by two dATP-bound ATP-binding cones from adjacent RR1 dimers that contact each other in an antiparallel conformation and are related by twofold symmetry[29] (Fig. 14.5A).

Since the hexamer model was built on the low-resolution ScRR1 hexamer structure, it was pertinent to test the hexamer interfaces of both models biochemically. Hence, site-directed mutagenesis was used to disrupt the hexamer interface predicted by each model. Interestingly, only the D16R mutation, at the N-terminus, disrupts hexamer formation in both hRRM1 and ScRR1. The results of this biochemical study validated only one model—the one in which all six ATP-binding cones participate in the hexamer interface (Fig. 14.5A).

It was hypothesized that since the D16R mutation abolishes the ability of hRRM1 to form a dATP-induced hexamer, D16R like the earlier reported

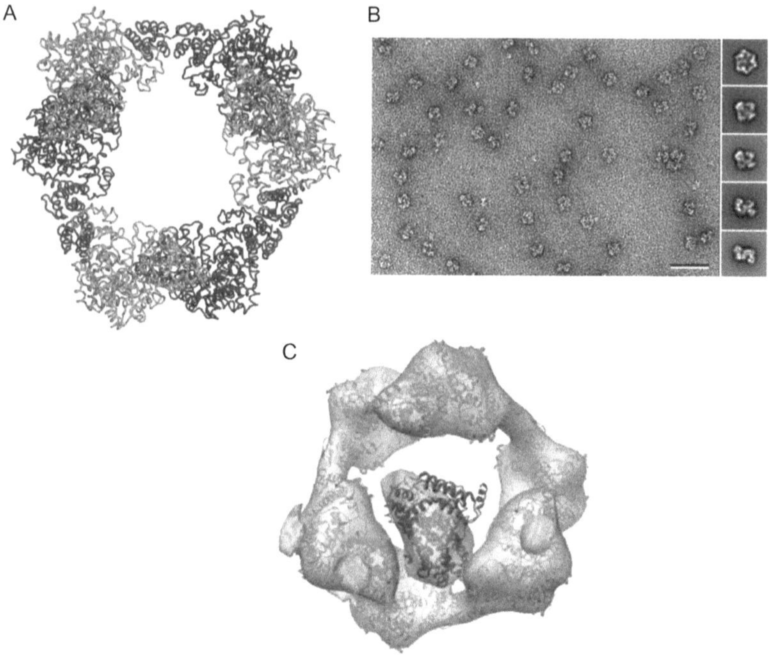

Figure 14.5 Subunit oligomerization packing of RR1 based on the low-resolution X-ray crystal structure of the ScRR1 hexamer. (A) ScRR1 monomers are colored in cyan and blue. ATP-binding cones are colored in red. (B) Electron micrograph of the α_6–ββ′–dATP holocomplex; image showing the negative stain of holocomplex. Scale bar, 50 nm. Model of the α_6•ββ′•dATP holocomplex based on cryo-EM data. *Reproduced with permission from Ref. 29.* (For interpretation of the references to color in this figure legend, the reader is referred to the online version of this chapter.)

mutation D57N which abolishes the ability to discriminate between ATP and dATP would prevent the allosteric inhibition of hRRM1 by dATP at physiological concentration. Interestingly, D16R hRRM1 retains the ability to reduce ADP and CDP substrates and, as expected, its activity is not inhibited by dATP. Unlike D57N, it is not activated by dATP either, however. Furthermore, D16R hRRM1 retains the ability to bind dATP, so this change is not caused by abolishing dATP binding at the A-site. Since D16R hRRM1 will still bind dATP but does not form hexamers as a result and is not inhibited by it, this supports the idea that allosteric inhibition of RR by dATP under physiological conditions requires hRRM1 to be in hexameric form.[29]

Our dATP-induced hexamer structure let us examine subunit packing for the first time. The cryo-EM structure showed that the dATP holoenzyme has a subunit composition of $\alpha_6\beta_2$ (Fig. 14.5B and C). In a similar study, site-directed mutagenesis studies showed that the ATP-induced hexamer adopts an interface different from that of the dATP-induced hexamer. It is quite easy to envisage that there must be structural differences between the ATP-induced hexamer and the dATP-induced hexamer for one to act as a functional entity and the other to be inactive.

Recently, it has been shown by carefully examining the ATP cone of ATP- and dATP-bound Class I RNR that the largest differences between the two complexes are in the loop region that spans residues 45–52.[52] This particular loop is believed to be an important part of the dimer–dimer interface in the eukaryotic hexamer. This observation led to speculation that subtle differences in the conformation of this region upon ATP and dATP binding are responsible for the different oligomeric structures of ATP- versus dATP-bound eukaryotic RR and for the different oligomerization behaviors of eukaryotic and *E. coli* Class 1RR.[52] Rather than form hexamers as in human RR, *E. coli* RR appears to undergo oligomeric interconversion from $\alpha_2\beta_2$ (the active oligomer for *E. coli* RR) to the ring-shaped $\alpha_4\beta_4$ (the inactive oligomer for *E. coli* RR) under the influence of the allosteric effector. Moreover, this interconversion is linked with distinct subunit rearrangements upon ATP and dATP binding, which gives further insight into the molecular mechanism of the activated and inactivated states.[45] Further extending the research on *E. coli* RR ring structure, a later study showed that the clinically approved anticancer drug gemcitabine when added to a solution of both *E. coli* RNR subunits induces an unusual structure known as a concatamer, in which two $\alpha_4\beta_4$ rings interlock to form a unique $(\alpha_4\beta_4)_2$ complex.[46] This study exploited a hybrid approach to structural studies by using small-angle X-ray scattering (SAXS) and electron microscopy (EM) along with X-ray crystallography to verify the formation of this unique complex. The proposed mechanism for the formation of the concatamer is that $\alpha_4\beta_4$ ring formation occurs as described above and the ring subsequently opens to accommodate a second ring and then recloses. The mechanism by which the concatamer forms was further confirmed by SAXS analysis, additionally demonstrating that the structure was not a crystallography artifact. It is important to mention here that despite the structural similarity between ScRR1 and the *E. coli* enzyme, *E. coli* RR does not form a hexamer. The only complexes reported for the *E. coli* enzyme to this point are $\alpha_2\beta_2$, $\alpha_4\beta_4$, and $(\alpha_4\beta_4)$. It is noteworthy that the ability to form different ring structures among Class 1 RR provides to develop new strategies to inhibit the RRs.[46]

2.7. The oligomeric state of hRR in the presence of nonnatural ligand

As described above and in our recent study, we showed that a key to hRRM1 inhibition by the negative regulator dATP is the ability of the natural nucleotide to induce the protein to form an inactive hexamer.[29] In an independent study, the dATP analog, clofarabine, was shown to inhibit hRRM1 by inducing hexamer formation.[44,53] The above study indicates the importance of developing nonnatural ligands that can perturb the quaternary status of protein by inducing the inactive hexamer of RR.[44,53] Recent studies by Ahmad *et al.* have shown that 5-NITP inhibits the activity of hRRM1. This particular nonnatural ligand was identified by *in silico* analysis of possible inhibitors of human RR. This analog's structural similarities to dATP suggested that it would inhibit hRRM1 activity by binding to its allosteric sites. However, 5-NITP binds to the S-site with micromolar affinity and only weakly to the A-site as revealed by ITC studies. Furthermore, 5-NITP binding at the S-site was confirmed by the X-ray crystal structure. Thus, binding of 5-NITP does not induce formation of the inactive hexamer of hRRM1 as does dATP and may account for the moderate inhibitor potency. Regardless of its mechanism, 5-NITP produces cytostatic and cytotoxic effects against human leukemia cells by altering cell-cycle progression, probably through its action against RR.[43] However, it is still important to develop nonnatural ligands that bind the A-site to force formation of an inactive hRRM1 hexamer as another way to exploit RR allosteric control in the service of anticancer and antiviral therapy.

3. TARGETING LARGE SUBUNIT OF RR

Considering the fact that RR plays an essential role in DNA synthesis and repair makes it an attractive target for cancer chemotherapy. In past few years, considerable amount of effort has been made to develop specific inhibitor against this enzyme.[54] In this review, we briefly discuss about the inhibitors for cancer chemotherapy that involves the large subunit of RR. To date, four druggable sites have been identified for the large subunit of RR. As shown in Fig. 14.1, these druggable sites are the A-site, S-site, C-site, and the peptide-binding site (P-site). Among the four sites, three sites are occupied with the nucleoside and deoxynucleoside, whereas the fourth side is the P-site. Mostly nucleotide-based drugs that target RR inhibit its activity by binding at either the A-site or the C-site. Interestingly, most of the analogs that bind the A-site quite often bind the S-site. Now we briefly describe the drugs and their modes of mechanism when binding to the A-site, C-site, and P-site of RR.

Fludarabine, cladribine, and clofarabine[55–58] are nucleoside analogs that target the A-site of the enzyme. These are U.S. FDA-approved clinical drugs that target the A-site of RR1. Clofarabine is the best-characterized drug among the above-mentioned nucleoside analogs and is clinically used in treating childhood leukemia.[58–60] Clofarabine triphosphate is the potent inhibitor of RR (IC50 = 65 nM), apparently binding to allosteric site of large subunit of RR.[40,61] In a recent report by Stubbe and co-workers, it was shown that clofarabine triphosphate inhibits hRRM1 with a Ki equal to 40 nM at the A-site.[53] In addition, they have also shown that the clofarabine is not an irreversible inhibitor. Clofarabine triphosphate induces hexamers by binding at the A-site similar to the allosteric inhibitor dATP. Hence, it is important to mention here that nucleoside analogs that retain the ability to hexamerize the RR1 subunit similar to that of dATP are likely to be categorized as good inhibitors of RR.

Gemcitabine which is a billion dollar drug used in many chemotherapeutic regimes is a mechanism-based inhibitor that targets the C-site of RR1 has been characterized by Stubbe and co-workers.[28,62] It is important to mention here that Gemcitabine is one of the few chemotherapeutic agents that are known to have any efficacy against pancreatic cancer.[63,64] Gemcitabine binds at the C-site of the large subunit of RR and induces the hexamerization, $\alpha_6\beta_6$ holocomplex. We were able to determine the initial interactions of F_2dCDP, the diphosphate metabolite of gemcitabine, at the C-site of ScRR1[41] (yeast RR1). It is interesting to note that while F_2dCDP differs from CDP only by substitution of two fluorines for the hydroxyl group and a hydrogen atom on the 2′ carbon of the ribose, F2CDP adopts a different conformation when binding to RR1.[41] In the similar study, we have shown that differences observed between CDP and F_2dCDP binding were due to unique chemical properties of the germinal fluorine atoms which are more hydrophobic and yet retain the ability to form hydrogen bonds. Our study illustrated the unique interactions that fluorine atoms can make in an instance where chemical space truly invades biological space.

Studies conducted in our lab and Cooperman's group have suggested that P-site is an excellent drug target for anticancer therapy.[32,41,65–72] Drug/molecule that targets the P-site will interfere with the quaternary structure of RR via a specific protein–protein interaction. Exploiting this particular characteristic will enable us to develop the highly specific inhibitors that do not lack the specificity associated with nucleoside analogs that target the A-, S-, and C-site of RR. The P-site comprises two well-formed subsites providing a platform for fragment-based drug design. Currently, there is no clinically used drug available that targets the P-site. However, future drugs are likely to emerge that target the P-site.

4. CONCLUSIONS

RR is a fascinating enzyme that converts ribonucleosides to deoxyribonucleosides and provides raw material for DNA synthesis and repair. Considering its important function, RR is an attractive target for cancer and antiviral therapy. Interestingly, in a recent review, it was noted that hRRM1 is a useful biomarker in cancer therapy similar to BRCA1.[73] A number of clinical trials have shown that patients with tumors expressing low levels of hRRM1 have increased progression-free and overall survival rates when treated with regimens that included gemcitabine.[73,74] Hence, hRR can be used in bench to bedside treatments.[30] The current interest in RR research is increasing our understanding of higher-order oligomerization, enzyme turnover, and the mechanism of allosteric regulation that governs specificity. The molecular basis of activation and deactivation by ATP and dATP was partially clarified by our recent study, which provided an explanation of how dATP achieves a 100-fold higher affinity than ATP for the A-site.[29] dATP achieves its high affinity by binding deeper in the pocket because it does not have an O2′ atom as does ATP. Interestingly, I18 acts as a steric gate for the O2′ atom.[29] In the same study, we showed that dATP regulates RR by forming inactive hexamers, whereas the $\alpha_2\beta_2$ dimer remains active. Moreover, the inactive holoenzyme complex in ScRR1 was shown to be an $\alpha_6\beta_2$ oligomer, which is in agreement with a previously reported higher-order oligomer in mouse RR.[24,75] Recently, using various structural and biophysical techniques, the Drenan group demonstrated that dATP inhibits *E. coli* RR by driving formation of the $\alpha_4\beta_4$ ring-like structure.[45] Furthermore, they have shown formation of the $(\alpha_4\beta_4)_2$ megacomplex of *E. Coli* RR in the presence of gemcitabine and dATP.[46] It is important to mention that the difference in the high-order oligomeric ring structure between *E. coli* and eukaryotic RR is probably due to the difference in the N-terminal ATP-binding cones. Due to the complexity of its functions, understanding the structure, function, and regulation of RR is intellectually very satisfying. As we begin to understand more about how this complicated molecule functions, we will discover novel ways to develop chemotherapies that target hRRM1.

ACKNOWLEDGMENTS

This research was supported by NIH grants 1R01GM100887-01 and R01CA100827-04A1. We would like to thank Dr. Cat Faber for proofreading. We also like to thank the present and past members of Dealwis lab for contributing to the work on ribonucleotide reductase.

REFERENCES

1. Hilser VJ. Biochemistry. An ensemble view of allostery. *Science* 2010;**327**(5966):653–4.
2. Monod J, Changeux JP, Jacob F. Allosteric proteins and cellular control systems. *J Mol Biol* 1963;**6**:306–29.
3. Changeux JP, Edelstein SJ. Allosteric mechanisms of signal transduction. *Science* 2005;**308**(5727):1424–8.
4. Monod J, Wyman J, Changeux JP. On the nature of allosteric transitions: a plausible model. *J Mol Biol* 1965;**12**:88–118.
5. Koshland Jr DE, Nemethy G, Filmer D. Comparison of experimental binding data and theoretical models in proteins containing subunits. *Biochemistry* 1966;**5**(1):365–85.
6. Kantrowitz ER, Lipscomb WN. Escherichia coli aspartate transcarbamylase: the relation between structure and function. *Science* 1988;**241**(4866):669–74.
7. Perutz M. *Cooperativity and allosteric regulation in proteins*. Cambridge, UK: Cambridge University Press; 1990.
8. Kolberg M, Strand KR, Graff P, Andersson KK. Structure, function, and mechanism of ribonucleotide reductases. *Biochim Biophys Acta* 2004;**1699**(1–2):1–34.
9. Nordlund P, Reichard P. Ribonucleotide reductases. *Annu Rev Biochem* 2006;**75**:681–706.
10. Thelander L, Reichard P. Reduction of ribonucleotides. *Annu Rev Biochem* 1979;**48**:133–58.

11. Reichard P, Baldesten A, Rutberg L. Formation of deoxycytidine phosphates from cytidine phosphates in extracts from Escherichia coli. *J Biol Chem* 1961;**236**:1150–7.
12. Brown NC, Canellakis ZN, Lundin B, Reichard P, Thelander L. Ribonucleoside diphosphate reductase. Purification of the two subunits, proteins B1 and B2. *Eur J Biochem* 1969;**9**(4):561–73.
13. Holmgren A, Reichard P, Thelander L. Enzymatic synthesis of deoxyribonucleotides, 8. The effects of ATP and dATP in the CDP reductase system from E. coli. *Proc Natl Acad Sci USA* 1965;**54**(3):830–6.
14. Moore EC, Reichard P. Enzymatic synthesis of deoxyribonucleotides. VI. The cytidine diphosphate reductase system from Novikoff hepatoma. *J Biol Chem* 1964;**239**:3453–6.
15. Blakley RL, Ghambeer RK, Nixon PF, Vitols E. The cobamide-dependent ribonucleoside triphosphate reductase of lactobacilli. *Biochem Biophys Res Commun* 1965;**20**(4):439–45.
16. Berglund O, Karlstrom O, Reichard P. A new ribonucleotide reductase system after infection with phage T4. *Proc Natl Acad Sci USA* 1969;**62**(3):829–35.
17. Reichard P. From RNA, to DNA, why so many ribonucleotide reductases? *Science* 1993;**260**(5115):1773–7.
18. Cotruvo Jr JA, Stubbe J. Class I ribonucleotide reductases: metallocofactor assembly and repair in vitro and in vivo. *Annu Rev Biochem* 2011;**80**:733–67.
19. Licht S, Gerfen GJ, Stubbe J. Thiyl radicals in ribonucleotide reductases. *Science* 1996;**271**(5248):477–81.
20. Jordan A, Reichard P. Ribonucleotide reductases. *Annu Rev Biochem* 1998;**67**:71–98.
21. Reichard P. Ribonucleotide reductases: the evolution of allosteric regulation. *Arch Biochem Biophys* 2002;**397**(2):149–55.
22. Stubbe J. Ribonucleotide reductases. *Adv Enzymol Relat Areas Mol Biol* 1990;**63**: 349–419.
23. Stubbe J, Ge J, Yee CS. The evolution of ribonucleotide reduction revisited. *Trends Biochem Sci* 2001;**26**(2):93–9.
24. Rofougaran R, Vodnala M, Hofer A. Enzymatically active mammalian ribonucleotide reductase exists primarily as an alpha6beta2 octamer. *J Biol Chem* 2006;**281**(38):27705–11.
25. Kashlan OB, Cooperman BS. Comprehensive model for allosteric regulation of mammalian ribonucleotide reductase: refinements and consequences. *Biochemistry* 2003;**42** (6):1696–706.
26. Cooperman BS, Kashlan OB. A comprehensive model for the allosteric regulation of Class Ia ribonucleotide reductases. *Adv Enzyme Regul* 2003;**43**:167–82.
27. Kashlan OB, Scott CP, Lear JD, Cooperman BS. A comprehensive model for the allosteric regulation of mammalian ribonucleotide reductase. Functional consequences of ATP- and dATP-induced oligomerization of the large subunit. *Biochemistry* 2002;**41**:462–74.
28. Wang J, Lohman GJ, Stubbe J. Enhanced subunit interactions with gemcitabine-5'-diphosphate inhibit ribonucleotide reductases. *Proc Natl Acad Sci USA* 2007;**104**(36): 14324–9.
29. Fairman JW, Wijerathna SR, Ahmad MF, et al. Structural basis for allosteric regulation of human ribonucleotide reductase by nucleotide-induced oligomerization. *Nat Struct Mol Biol* 2011;**18**(3):316–22.
30. Wijerathna SR, Ahmad MF, Xu H, Fairman JW, Zhang A, Kaushal PS, et al. Targeting the large subunit of human ribonucleotide reductase for cancer chemotherapy. *Pharmaceuticals* 2011;**4**:1328–54.
31. Uhlin U, Eklund H. Structure of ribonucleotide reductase protein R1. *Nature* 1994;**370** (6490):533–9.
32. Xu H, Faber C, Uchiki T, Fairman JW, Racca J, Dealwis C. Structures of eukaryotic ribonucleotide reductase I provide insights into dNTP regulation. *Proc Natl Acad Sci USA* 2006;**103**(11):4022–7.

33. Brown NC, Reichard P. Role of effector binding in allosteric control of ribonucleoside diphosphate reductase. *J Mol Biol* 1969;**46**(1):39–55.
34. Reichard P, Eliasson R, Ingemarson R, Thelander L. Cross-talk between the allosteric effector-binding sites in mouse ribonucleotide reductase. *J Biol Chem* 2000;**275**(42):33021–6.
35. Stubbe J, Riggs-Gelasco P. Harnessing free radicals: formation and function of the tyrosyl radical in ribonucleotide reductase. *Trends Biochem Sci* 1998;**23**(11):438–43.
36. Jordan A, Pontis E, Atta M, et al. A second class I ribonucleotide reductase in Enterobacteriaceae: characterization of the Salmonella typhimurium enzyme. *Proc Natl Acad Sci USA* 1994;**91**(26):12892–6.
37. Roshick C, Iliffe-Lee ER, McClarty G. Cloning and characterization of ribonucleotide reductase from Chlamydia trachomatis. *J Biol Chem* 2000;**275**(48):38111–9.
38. Nordlund P, Eklund H. Structure and function of the Escherichia coli ribonucleotide reductase protein R2. *J Mol Biol* 1993;**232**(1):123–64.
39. Eriksson M, Uhlin U, Ramaswamy S, et al. Binding of allosteric effectors to ribonucleotide reductase protein R1: reduction of active-site cysteines promotes substrate binding. *Structure* 1997;**5**(8):1077–92.
40. Parker WB, Shaddix SC, Chang CH, et al. Effects of 2-chloro-9-(2-deoxy-2-fluoro-beta-D-arabinofuranosyl)adenine on K562 cellular metabolism and the inhibition of human ribonucleotide reductase and DNA polymerases by its 5'-triphosphate. *Cancer Res* 1991;**51**(9):2386–94.
41. Xu H, Faber C, Uchiki T, Racca J, Dealwis C. Structures of eukaryotic ribonucleotide reductase I define gemcitabine diphosphate binding and subunit assembly. *Proc Natl Acad Sci USA* 2006;**103**(11):4028–33.
42. Ahmad MF, Kaushal PS, Wan Q, et al. Role of arginine 293 and glutamine 288 in communication between catalytic and allosteric sites in yeast ribonucleotide reductase. *J Mol Biol* 2012;**419**(5):315–29.
43. Ahmad MF, Wan Q, Jha S, Motea E, Berdis A, Dealwis C. Evaluating the therapeutic potential of a non-natural nucleotide that inhibits human ribonucleotide reductase. *Mol Cancer Ther* 2012;**11**:2077–86.
44. Aye Y, Brignole EJ, Long MJ, et al. Clofarabine targets the large subunit (alpha) of human ribonucleotide reductase in live cells by assembly into persistent hexamers. *Chem Biol* 2012;**19**(7):799–805.
45. Ando N, Brignole EJ, Zimanyi CM, et al. Structural interconversions modulate activity of Escherichia coli ribonucleotide reductase. *Proc Natl Acad Sci USA* 2012;**108**(52):21046–51.
46. Zimanyi CM, Ando N, Brignole EJ, Asturias FJ, Stubbe J, Drennan CL. Tangled up in knots: structures of inactivated forms of E. coli class Ia ribonucleotide reductase. *Structure* 2012;**20**(8):1374–83.
47. Larsson KM, Jordan A, Eliasson R, Reichard P, Logan DT, Nordlund P. Structural mechanism of allosteric substrate specificity regulation in a ribonucleotide reductase. *Nat Struct Mol Biol* 2004;**11**(11):1142–9.
48. Eliasson R, Pontis E, Sun X, Reichard P. Allosteric control of the substrate specificity of the anaerobic ribonucleotide reductase from Escherichia coli. *J Biol Chem* 1994;**269**(42):26052–7.
49. Kumar D, Viberg J, Nilsson AK, Chabes A. Highly mutagenic and severely imbalanced dNTP pools can escape detection by the S-phase checkpoint. *Nucleic Acids Res* 2010;**38**(12):3975–83.
50. Kumar D, Abdulovic AL, Viberg J, Nilsson AK, Kunkel TA, Chabes A. Mechanisms of mutagenesis in vivo due to imbalanced dNTP pools. *Nucleic Acids Res* 2010;**39**(4):1360–71.

51. Murphy KP, Xie D, Garcia KC, Amzel LM, Freire E. Structural energetics of peptide recognition: angiotensin II/antibody binding. *Proteins* 1993;**15**(2):113–20.
52. Hofer A, Crona M, Logan DT, Sjoberg BM. DNA building blocks: keeping control of manufacture. *Crit Rev Biochem Mol Biol* 2012;**47**(1):50–63.
53. Aye Y, Stubbe J. Clofarabine 5'-di and -triphosphates inhibit human ribonucleotide reductase by altering the quaternary structure of its large subunit. *Proc Natl Acad Sci USA* 2011;**108**(24):9815–20.
54. Cerqueira NM, Fernandes PA, Ramos MJ. Ribonucleotide reductase: a critical enzyme for cancer chemotherapy and antiviral agents. *Recent Pat Anticancer Drug Discov* 2007;**2**(1):11–29.
55. Huang P, Chubb S, Plunkett W. Termination of DNA synthesis by 9-beta-D-arabinofuranosyl-2-fluoroadenine. A mechanism for cytotoxicity. *J Biol Chem* 1990;**265**(27):16617–25.
56. Avery TL, Rehg JE, Lumm WC, Harwood FC, Santana VM, Blakley RL. Biochemical pharmacology of 2-chlorodeoxyadenosine in malignant human hematopoietic cell lines and therapeutic effects of 2-bromodeoxyadenosine in drug combinations in mice. *Cancer Res* 1989;**49**(18):4972–8.
57. Griffig J, Koob R, Blakley RL. Mechanisms of inhibition of DNA synthesis by 2-chlorodeoxyadenosine in human lymphoblastic cells. *Cancer Res* 1989;**49**(24 Pt 1):6923–8.
58. Parker WB, Shaddix SC, Rose LM, et al. Comparison of the mechanism of cytotoxicity of 2-chloro-9-(2-deoxy-2- fluoro-beta-D-arabinofuranosyl)adenine, 2-chloro-9-(2-deoxy-2-fluoro- beta-D-ribofuranosyl)adenine, and 2-chloro-9-(2-deoxy-2,2-difluoro- beta-D-ribofuranosyl)adenine in CEM cells. *Mol Pharmacol* 1999;**55**(3):515–20.
59. Faderl S, Gandhi V, Keating MJ, Jeha S, Plunkett W, Kantarjian HM. The role of clofarabine in hematologic and solid malignancies—development of a next-generation nucleoside analog. *Cancer* 2005;**103**(10):1985–95.
60. Gandhi V. Clofarabine: a viewpoint by Varsha Gandhi. *Paediatr Drugs* 2005;**7**(4):265–6.
61. Xie C, Plunkett W. Metabolism and actions of 2-chloro-9-(2-deoxy-2-fluoro-beta-D-arabinofuranosyl)-adenine in human lymphoblastoid cells. *Cancer Res* 1995;**55**(13):2847–52.
62. Wang J, Lohman GJ, Stubbe J. Mechanism of inactivation of human ribonucleotide reductase with p53R2 by gemcitabine 5'-diphosphate. *Biochemistry* 2009;**48**(49):11622–9.
63. Langer CJ. The role of new agents in advanced non-small-cell lung carcinoma. *Curr Oncol Rep* 2000;**2**(1):76–89.
64. Squadroni M, Fazio N. Chemotherapy in pancreatic adenocarcinoma. *Eur Rev Med Pharmacol Sci* 2010;**14**(4):386–94.
65. Yang FD, Spanevello RA, Celiker I, Hirschmann R, Rubin H, Cooperman BS. The carboxyl terminus heptapeptide of the R2 subunit of mammalian ribonucleotide reductase inhibits enzyme activity and can be used to purify the R1 subunit. *FEBS Lett* 1990;**272**(1–2):61–4.
66. Fisher A, Laub PB, Cooperman BS. NMR structure of an inhibitory R2 C-terminal peptide bound to mouse ribonucleotide reductase R1 subunit. *Nat Struct Biol* 1995;**2**(11):951–5.
67. Pellegrini M, Liehr S, Fisher AL, Laub PB, Cooperman BS, Mierke DF. Structure-based optimization of peptide inhibitors of mammalian ribonucleotide reductase. *Biochemistry* 2000;**39**(40):12210–5.
68. Fisher A, Yang FD, Rubin H, Cooperman BS. R2 C-terminal peptide inhibition of mammalian and yeast ribonucleotide reductase. *J Med Chem* 1993;**36**(24):3859–62.

69. Liehr S, Barbosa J, Smith 3rd AB, Cooperman BS. Synthesis and biological activity of cyclic peptide inhibitors of ribonucleotide reductase. *Org Lett* 1999;**1**(8):1201–4.
70. Pender BA, Wu X, Axelsen PH, Cooperman BS. Toward a rational design of peptide inhibitors of ribonucleotide reductase: structure-function and modeling studies. *J Med Chem* 2001;**44**(1):36–46.
71. Cooperman BS, Gao Y, Tan C, Kashlan OB, Kaur J. Peptide inhibitors of mammalian ribonucleotide reductase. *Adv Enzyme Regul* 2005;**45**:112–25.
72. Sun D, Xu H, Wijerathna SR, Dealwis C, Lee RE. Structure-based design, synthesis, and evaluation of 2'-(2-hydroxyethyl)-2'-deoxyadenosine and the 5'-diphosphate derivative as ribonucleotide reductase inhibitors. *ChemMedChem* 2009;**4**(10):1649–56.
73. Jordheim LP, Guittet O, Lepoivre M, Galmarini CM, Dumontet C. Increased expression of the large subunit of ribonucleotide reductase is involved in resistance to gemcitabine in human mammary adenocarcinoma cells. *Mol Cancer Ther* 2005; **4**(8):1268–76.
74. Jordheim LP, Seve P, Tredan O, Dumontet C. The ribonucleotide reductase large subunit (RRM1) as a predictive factor in patients with cancer. *Lancet Oncol* 2011;**12**: 693–702.
75. Rofougaran R, Crona M, Vodnala M, Sjoberg BM, Hofer A. Oligomerization status directs overall activity regulation of the Escherichia coli class Ia ribonucleotide reductase. *J Biol Chem* 2008;**283**(51):35310–8.

CHAPTER FIFTEEN

Oligomerization of Dynamin Superfamily Proteins in Health and Disease☆

Katja Faelber*, Song Gao*,†, Martin Held‡, York Posor§, Volker Haucke§, Frank Noé‡, Oliver Daumke*

*Max-Delbrück-Centrum for Molecular Medicine, Crystallography, Berlin, Germany
†State Key Laboratory of Oncology in South China, Sun Yat-sen University Cancer Center, Guangzhou, PR China
‡Institute for Mathematics, Freie Universität Berlin, Berlin, Germany
§Leibniz-Institute for Molecular Pharmacology (FMP) & Freie Universität Berlin, Berlin, Germany

Contents

☆Parts of this chapter are based on our recent review (Faelber *et al.*, 2012).[1]

Progress in Molecular Biology and Translational Science, Volume 117
ISSN 1877-1173
http://dx.doi.org/10.1016/B978-0-12-386931-9.00015-5

Abstract

Proteins of the dynamin superfamily are mechanochemical GTPases, which mediate nucleotide-dependent membrane remodeling events. The founding member dynamin is recruited to the neck of clathrin-coated endocytic vesicles where it oligomerizes into helical filaments. Nucleotide-hydrolysis-induced conformational changes in the oligomer catalyze scission of the vesicle neck.

Here, we review recent insights into structure, function, and oligomerization of dynamin superfamily proteins and their roles in human diseases. We describe in detail the molecular mechanisms how dynamin oligomerizes at membranes and introduce a model how oligomerization is linked to membrane fission. Finally, we discuss molecular mechanisms how mutations in dynamin could lead to the congenital diseases, Centronuclear Myopathy and Charcot-Marie Tooth disease.

1. STRUCTURE AND FUNCTION OF DYNAMIN SUPERFAMILY PROTEINS

1.1. Introduction

Members of the dynamin superfamily are mechanochemical GTPases which convert the energy of GTP binding and hydrolysis to a mechanical output (reviewed in Ref. 2). Most of them have a function related to cellular membrane remodeling events such as membrane tubulation, scission, or fusion. Based on theoretical considerations, it has been postulated that in order to deform membranes, peripheral membrane proteins have to be present at high density at the membrane surface.[3] Dynamin superfamily proteins achieve this by binding and assembling into regular oligomers at the membrane surface. Often, this assembly is accompanied by a drastic stimulation of their intrinsic GTPase activity which is a prerequisite for their mechanochemical function.

Dynamin superfamily proteins appear also in bacteria but are particularly abundant in animals and plants with their extensive endomembrane system. All members have similar domain architectures (Figs. 15.1 and 15.2). The most highly conserved region is the amino-terminal GTPase (G) domain. It shows a mixed α–β-fold with a central β-sheet surrounded by α-helices at both sides.[4] Compared to the canonical Ras protein, the G domain is extended by additional β-strands and α-helices, and additional variations are apparent in some members of the superfamily. As in most other GTPases, five highly conserved motifs (G1–G5), including the phosphate binding loop and switch I and switch II, participate in nucleotide binding and

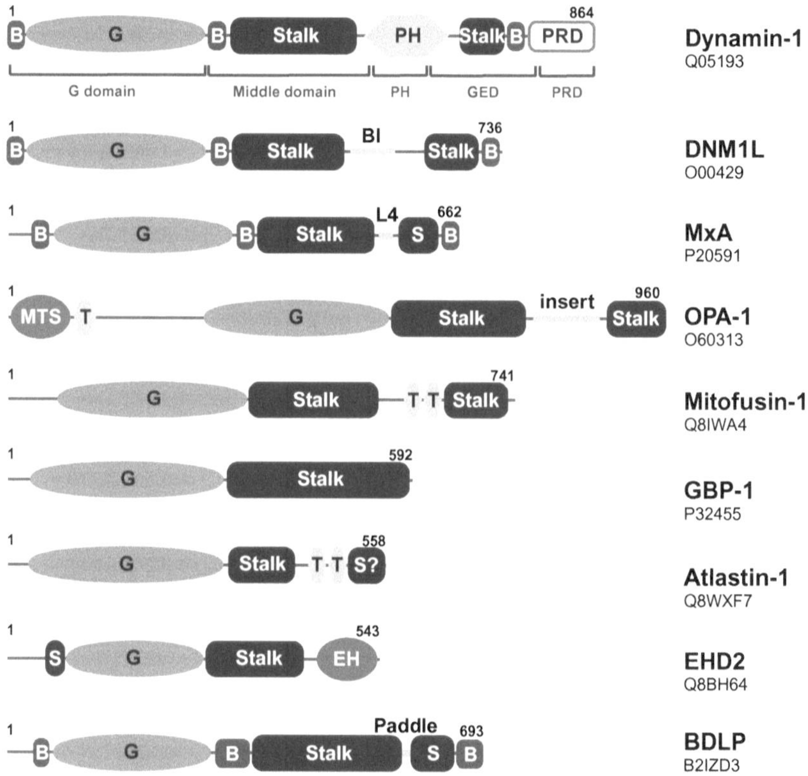

Figure 15.1 Domain architecture of dynamin superfamily proteins. Structure-based domain architecture of dynamin superfamily proteins, with Uniprot accession number indicated (G, G domain; B, BSE; S, stalk; PH, PH domain; PRD - Proline rich domain; BI, B insert; L4, Loop L4; MTS, mitochondrial targeting sequence; T, predicted transmembrane domain; EH, Eps15 homology domain). If no structure was available (OPA-1, Mitofusin-1) the domain architecture was inferred. In dynamin, the original sequence-based domain prediction is indicated below. The C-terminal region of atlastin-1 was not present in the crystallized construct. It forms an alpha helix which is involved in membrane remodeling but it has not been described that this helix associates with the identified part of the stalk. (For color version of this figure, the reader is referred to the online version of this chapter.)

hydrolysis. Dynamin superfamily proteins require magnesium ions to bind nucleotides with low (micromolar) affinity. Consequently, these proteins do not need exchange factors for nucleotide binding and release.

A common characteristic of the G domain in dynamin superfamily proteins is their ability to homo- or heterodimerize via a highly conserved interface across the nucleotide binding site, the G interface. It was demonstrated for some members that dimerization induces rearrangement of catalytic

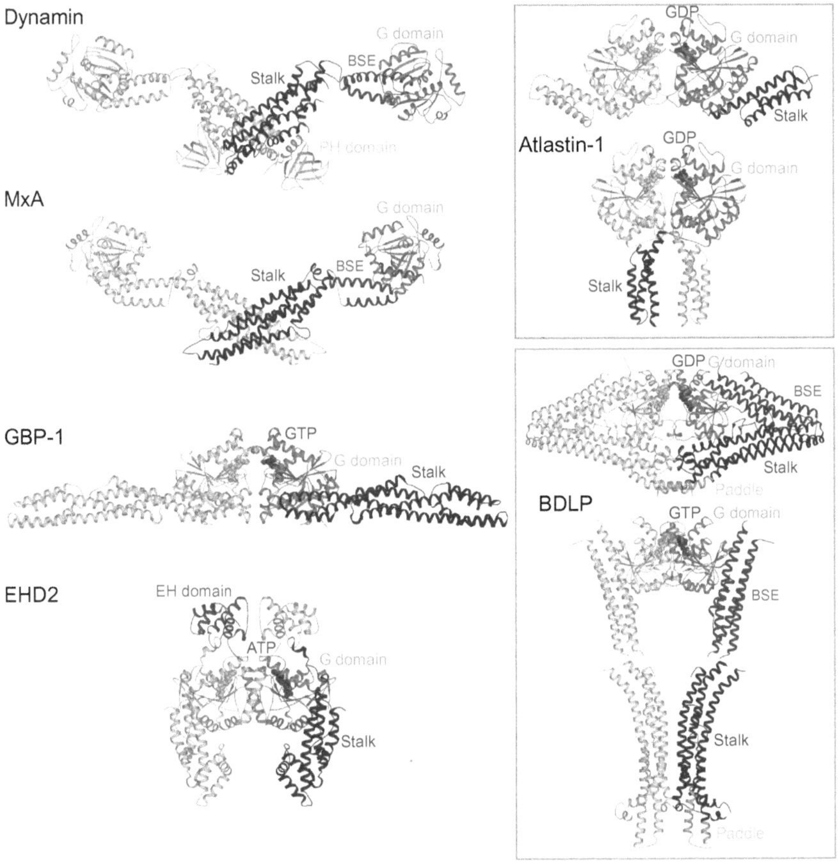

Figure 15.2 Structures of dynamin superfamily proteins. Ribbon type representation of the dynamin dimer (pdb 3SNH), MxA dimer (3SZR), full-length GBP1 (1F5N) assembled via the G domain (modeled using 2B8W), two crystal structures of the GDP-bound atlastin-1 dimer (3Q5D, 3Q5E), the EHD2 dimer in the presence of a nonhydrolyzable ATP analog (2QPT), the BDLP dimer in the GDP-bound closed form (2J68) and the EM reconstruction-based open form in the presence of a nonhydrolyzable GTP analog (2W6D). (For color version of this figure, the reader is referred to the online version of this chapter.)

residues. This often leads to the positioning of a catalytic water molecule followed by a drastic stimulation of the intrinsic GTPases activity. The conformational changes resulting from GTP hydrolysis are thought to be relayed to the other domains of dynamin thereby triggering a mechanical output.

The G domain of dynamin superfamily proteins is followed by one or two helical domains which have been termed the bundle signaling element (BSE) and/or the stalk in eukaryotic proteins. If only one helical domain is present, we will refer to it here as stalk. The length of the helical domains varies considerably, and in some cases, the two domains are fused into a single domain. Consequently, the sequence similarity in the helical domains is low. Nevertheless, a structural comparison shows clear similarities in the architectures of these domains. Both the BSE and stalk are composed of antiparallel helical bundles. The N-terminal helix of these domains leads the polypeptide away from the G domain to the tip of the stalk, whereas an extended long helix at the C-terminus guides the polypeptide back to the G domain. Within the helical domains, structural variations are apparent, for example, insertions of loop regions or domains or deletions of specific helical segments. Further, in some of the members, sequence stretches N-terminal of the G domain participate in the formation of the helical bundles. Stalk and BSE mediate self-assembly of dynamin proteins, and structural differences in these domains might modify self-assembly or assembly with other domains. The variations could also influence the conformational coupling between the G domain and the helical domains.

At the tip of the stalk, at the opposite side of the G domain, the substrate/membrane-binding site is localized. Some of the members have a specialized domain for membrane binding, such as the pleckstrin homology (PH) domain of dynamin. Others, such as atlastin, mitofusins, or bacterial dynamin-like protein (BDLP), have predicted transmembrane helices at this position. Again others, such as the Myxovirus resistance protein A (MxA) and dynamin-1 like protein (DNM1L), bind to membranes via an extended loop at the tip of the stalk, or, as the EH domain containing protein 2 (EHD2), via hydrophobic/basic residues within the stalk. Still other domains are present in some members and mediate specialized functions, such as the recruitment to specific subcellular sites or the regulation of oligomerization activity.

In this chapter, we introduce structure, mechanism, and function of the different dynamin superfamily members. We discuss in detail recent structural insights into oligomerization of the founding member dynamin and propose a structural model how dynamin catalyzes membrane fission. Finally, we discuss the molecular mechanisms how disease-related mutations in dynamin might alter oligomerization and function.

1.2. Dynamins

Dynamin was identified as a microtubule-associated, mechanochemical GTPase displaying sequence similarity to the antiviral Myxovirus resistance protein A (MxA).[5,6] This microtubule binding activity might be related to the regulation of microtubule stability during mitosis.[7] Dynamin's best-examined function, however, is the scission of clathrin-coated vesicles from the plasma membrane, which was discovered by the analysis of the temperature-sensitive dynamin mutant *shibire* in *Drosophila*.[8–10] In mammals, three closely related dynamin isoforms (dynamin 1–3) mediate cleavage of clathrin-coated vesicles for various cellular functions, for example, the recycling of synaptic vesicles in neurons or the uptake of nutrients or signaling factors in almost all cell types.[11] The essential cellular role of dynamin is also reflected in the fact that mutations in the dynamin 2 gene can lead to severe human disease (see Section 3).

Many structural and functional studies have addressed the molecular mechanisms of dynamin-induced membrane remodeling processes. In solution, dynamin and the closely related MxA are predominantly tetrameric.[12,13] The presence of suitable membrane templates promotes the oligomerization of dynamin or MxA into ring-like or helical structures at membranes thereby inducing membrane tubulation.[14] A ring-like dynamin coat was observed at the neck of clathrin-coated vesicles in the presence of a nonhydrolyzable GTP analog.[15] Once oligomerized on a lipid template, the low basal GTPase activity of dynamin is stimulated up to 200-fold.[16,17] Stimulated GTP hydrolysis and the resulting conformational changes in the dynamin oligomer ultimately result in vesicle scission.[18–20]

One dynamin molecule has a size of approximately 100 kDa and is composed of a G domain, a BSE, a stalk, a PH domain, and a proline rich domain (PRD) (Figs. 15.1 and 15.2). Electron microscopy reconstructions yielded first structural insights into the domain organization of the dynamin oligomer assembled on a lipid tubule.[21–23] These revealed three radial layers of electron density, which were arranged in a helical pattern around the lipid tubule. The G domains were assigned to the outer layer, the stalks to the middle layer and the PH domains to the inner layer. These reconstructions also indicated that a multitude of domain interactions is required for the formation of a helical filament as well as for the interconnection of helical turns.

Dynamin's BSE is in direct contact with the G domain and is built of a three helix bundle.[24–28] The three helices α1–3 are derived from different regions of the protein sequence, for example, α1 and α2 are located N- and C-terminally to the G domain and α3 follows the stalk at the C-terminus (Fig. 15.1 and 15.2.). Hydrophobic residues of all three helices participate in an extensive network forming the core of this domain. The BSE is connected via two conserved proline residues (Pro32 and Pro294, hinge 2) to the G domain of the same molecule and via two relatively loose loop regions to the stalk (hinge 1).

The stalk of dynamin is composed of an antiparallel four-helix bundle.[13,27,28] Following helix α2 of the BSE, a region previously known as the middle domain forms helices α1–3. The fourth helix of stalk follows the PH domain of the same molecule and has originally been described as GTPase effector domain (GED). Despite its extended structure, the hydrophobic core of the stalk appears to mediate a high degree of stability.

Between α3 and α4 of the stalk, the globular PH domain is interspersed. It is composed of two orthogonal β-sheets flanked by a C-terminal helix α1.[29] Three variable loop regions at the opposite side bind to negatively charged membranes.[30] The PH domain shows some specificity for phosphatidylinositol-4,5-biphosphate, a phosphoinositide enriched at the plasma membrane[31] that plays a key role in clathrin-mediated endocytosis.

The C-terminal PRD of dynamin is thought to be unstructured. It mediates the recruitment of dynamin to clathrin-coated pits via interaction with Src Homology 3 domains of interaction partners such as the BAR-domain-containing proteins amphiphysin[32] and SNX9.[33]

1.3. Dynamin-like proteins

Dynamin-1-like protein (DNM1L), the closest relative of dynamin, catalyzes the fission of mitochondrial and peroxisomal membranes. Mitochondrial dynamics are crucial for the maintenance of mitochondrial DNA, respiratory activity, and calcium signaling, and play a role in the control of embryonic development, apoptosis, and neuronal plasticity.[34,35] DNM1L is recruited from the cytosol to the mitochondrial outer membrane where it oligomerizes in discrete foci. Some of these foci develop into mitochondrial scission sites.[36,37] In mammals, recruitment to the mitochondrial outer membrane involves the endoplasmic reticulum (ER)[38] and is mediated via adaptor proteins, such as mitochondrial fission factor,[39] mitochondrial elongation factor 1/mitochondrial dynamics proteins of 49 and 51 kDa (MIEF1/MiD49/51),[40,41] and mitochondrial fission protein Fis1.[42,43] For the yeast homologue of DNM1L, Dnm1, recruitment via similar adaptors was shown to nucleate and promote self-assembly of Dnm1 into helical structures, thereby driving membrane scission.[44]

Dysfunction of DNM1L is implicated in a number of mitochondria-based neurologic disorders (reviewed in Ref. 45,46). Impaired mitochondrial function and increased fragmentation of mitochondria were reported in Parkinson's disease patients. Perturbed mitochondrial morphology in some of these patients might be caused by diminished proteasome-dependent degradation and subsequent accumulation of DNM1L.[47] Fibroblasts from Alzheimer's disease patients show significantly elongated mitochondria and a marked decrease in DNM1L expression.[48] Further, S-nitrosylation of DNM1L caused by the expression of the β-amyloid protein was reported to trigger mitochondrial fission and induces synaptic loss and neuronal damage. This implicates DNM1L also in Alzheimer's disease,[49] although these results are controversial.[50] Abnormal mitochondrial dynamics and altered DNM1L expression levels were also reported in Huntington's disease patients.[51,52] Finally, a newborn girl with the A395D mutation in DNM1L showed severe developmental defects such as microcephaly, optic neuropathy, hypoplasia, persistent lactic acidemia, and the elevated plasma levels of long-chain fatty acids.[53]

Based on its sequence, DNM1L is expected to structurally resemble dynamin with a G domain, a BSE, and a stalk. Instead of the PH domain, a 100 amino-acid insertion, the B insert, is present at the tip of the stalk. This insertion might be involved in the recruitment to DNM1L to mitochondria, although others have challenged this.[54,55] Recent electron microscopy-based reconstructions showed that yeast Dnm1 forms broader filaments compared to those of dynamin. This suggests that DNM1L follows an alternative oligomerization mode compared to dynamin, reflecting a potential adaptation to the larger size of the mitochondrial scission sites.[54]

1.4. Myxovirus resistance proteins

MxA was the first member of the dynamin superfamily for which the sequence was reported.[6] It belongs to a family of antiviral proteins in vertebrates that mediate cellular resistance against a wide range of negative-strand RNA viruses and also other pathogens (reviewed in Ref. 56). Two Mx genes are present in humans (encoding MxA and MxB) and mice (encoding Mx1 and Mx2). Under nonpathological conditions, expression levels of MxA in most cell types are low, but MxA is massively upregulated in response to viral infections and interferon signaling. In noninfected cells, MxA localizes to a distinct subcompartment of the ER.[57] Upon infection, MxA redistributes to sites of viral replication and appears to promote missorting of essential viral constituents.[58,59] Most inbred mouse strains lack a functional Mx1 gene,[60] suggesting that under laboratory conditions, Mx1 is not required for survival. However, these mice are susceptible to certain viral infections.

So far, no inactivating mutations in Mx genes have been reported in humans. The molecular basis for the antiviral activity is not completely understood. A structural model has been proposed in which MxA oligomerizes via the stalk around viral ribonucleoprotein complexes leading to their sequestration and thus a block of viral replication.[26] This process was proposed to involve the viral polymerase subunit PB2.[61]

Despite a sequence identity of only 15%, the stalk architecture of MxA is very similar to that of dynamin.[13] It is comprised of a four-helix bundle. A 45 amino-acid loop at the tip of the stalk, at the same sequence position as the PH domain of dynamin, was demonstrated to mediate membrane and likely viral nucleoprotein interaction.[62,63] Dimerization and oligomerization of MxA proceed via three distinct interfaces of the stalk, in a fashion very similar to dynamin (see below).[13]

The overall structure of full-length MxA also shows an architecture akin to dynamin.[26] The G domain of MxA is connected via hinge 2 to the BSE, which connects to the stalk via another hinge region (hinge 1). Mutations in an interface between the BSE and the stalk of the neighboring molecule (interface-5 in dynamin) interfere with oligomerization and its antiviral function, suggesting that both proteins are regulated similarly via stalk–BSE interactions.

1.5. Dynamin-related mitochondrial fusion proteins

The dynamin-like protein fuzzy onion was identified in *Drosophila* as a central player of mitochondrial outer membrane fusion (Ref. 64, reviewed in Ref. 65). The two homologues in mammals, mitofusin 1 and 2, were shown to have similar functions.[66]

Mitofusin 2 is involved in oxidative metabolism, cell cycle, cell death, and mitochondrial axonal transport. Gene knockout mice lacking *Mitofusin 1* or *Mitofusin 2* die during gestation due to a placental defect.[67] Fibroblasts isolated from these mice display a great reduction in mitochondrial fusion activity. In humans, mutations in the *MITOFUSIN2* gene can cause autosomal dominant Charcot-Marie-Tooth (CMT) disease.[68]

No structural information of a mitofusin homologue is available so far. Besides an N-terminal G domain and two helical domains corresponding to the stalk and BSE, two transmembrane helices are predicted at the tip of the stalk that might anchor mitofusins to the outer mitochondrial membrane. The molecular mechanism of membrane fusion is unclear.

Optic atrophy 1 (OPA1), the mammalian ortholog of the yeast protein Mitochondrial Genome Maintenance 1 (Mgm1), is a dynamin-related protein implicated in the fusion of the inner mitochondrial membrane.[69] An additional role of Mgm1 in maintaining cristae morphology has been demonstrated.[70–72] OPA1 has been shown to control apoptosis, cell growth, and oxygen consumption. Mutations in OPA1 are the main causes for autosomal dominant optic atrophy (ADOA), the most common form of inherited optic neuropathy.[73] ADOA patients experience visual impairment starting in early childhood, ranging from reduced visual acuity to blindness. The underlying defect of the condition is the degeneration of retinal ganglion cells whose axons form the optic nerve.

The mitochondrial import sequence of OPA1 is cleaved after the transport across the outer mitochondrial membrane. This long OPA1 form is attached to the inner mitochondrial membrane via a predicted N-terminal transmembrane segment. A shorter soluble form is produced by additional proteolytic cleavage of the N-terminal transmembrane segment. The presence of both forms is crucial for the maintenance of mitochondrial morphology.[65] Instead of the PH domain found in dynamin, OPA1 has a 100 amino-acid domain predicted to be mostly α-helical that might participate in membrane binding.

No high-resolution structural data of OPA1 are available yet. However, OPA1 was shown to oligomerize at liposomes, leading to liposome tubulation and stimulations of its GTPase activity.[74] How oligomerization of OPA1 is linked to its function in mediating membrane fusion is currently not clear.

1.6. Guanylate binding proteins

The guanylate binding proteins (GBPs) are interferon-induced GTPases of the dynamin superfamily (reviewed in Ref. 75). The group comprises 7 members in humans and 11 in mice. Some members mediate protection against intracellular pathogens, such as listeria or mycobacteria.[76] Some GBPs contain a C-terminal CaaX sequence that is isoprenylated *in vivo* and contributes to membrane binding.[77]

The crystal structure of GBP1 was the first full-length structure of a dynamin superfamily member (Fig. 15.2).[78] It showed an extended G domain with two additional β-strands from the N-terminal region. Dimerization of the G domains via the G interface was shown to induce rearrangement of catalytic residues *in cis*.[79] In particular, an arginine residue of the P loop moves toward the phosphates to drive nucleotide hydrolysis. GBPs have the unique property to hydrolyze GTP consecutively to GDP and then further to GMP. This involves a slipping of the β-phosphate in the GDP-bound form to the position where the γ-phosphate is situated in the GTP-bound form.[79] The physiological function of GDP hydrolysis is unknown.

Compared to other dynamin proteins, the helical domains are greatly elongated. Dimerization of GBPs via the G interface in the G domain leads to an extended banana-shaped dimer.[79] In the presence of a transition analog of the GTPase reaction, GDP-AlF_4, the GBP1 dimer was shown to tetramerize.[78] In cell culture experiments, addition of GDP-AlF_4 leads to the recruitment of GBP1 to the Golgi apparatus.[80] Molecular as well as functional details of the oligomer remains to be elucidated.

1.7. Atlastins

Atlastins are a family of dynamin superfamily proteins most closely related to the GBP family. In eukaryotes, a single atlastin homologue resides at the ER and is implicated in the fusion of ER tubules for the creation of an extensive tubular network (reviewed in Ref. 81).

Two crystal structures of atlastin-1 in the presence of GDP were independently determined.[82,83] Although dimerization is dependent on GTP in solution, atlastin-1 in both crystal forms also dimerized in the presence of GDP via the G interface. Compared to GBP1, the stalk of atlastin-1 is shorter, and only three helices were present in the crystallized construct (Fig. 15.2). At the tip of the stalk, two transmembrane helices contribute to oligomerization and membrane remodeling.[84] A fourth helix following the transmembrane segment was recently demonstrated to insert into the membrane bilayer and to induce membrane curvature.[84] In the two crystal forms, the helical domains undergo a dramatic hinge-like movement compared to the G domain.[82,83] It was suggested that this movement mediates membrane fusion by tethering two opposing membranes and actively pulling them together.

Mutations in the *atlastin* gene can cause hereditary spastic paraplegia, a neurodegenerative disease characterized by shortened axons in corticospinal motor neurons.[85] The disease is characterized by progressive spasticity and weakness of the lower limbs. Most of the 28 described missense mutations of atlastin inducing disease cause folding or dimerization deficits.[82] For some of these mutants, the molecular deficits leading to disease are still unclear.

1.8. EH domain containing proteins

The single EH domain containing protein (EHD) in *C. elegans*, named receptor-mediated endocytosis 1 (Rme1), was shown to mediate exit of cargo proteins from the recycling compartment.[86] In mammals, a family of four related proteins, EHD1–4, functions in several membrane trafficking pathways. EHD1 has a similar role as Rme1 at the endocytic recycling compartment,[87,88] whereas EHD2 was recently shown to localize to caveolae and stabilize them at the plasma membrane.[89,90] EHD4/Pincher is implicated in macroendocytosis in neurons for the generation of tropomyosin-receptor-kinase (Trk) containing endosomes with specialized signaling functions.[91]

Despite the presence of an N-terminal GTPase domain, EHDs bind with low affinity to adenine nucleotides.[92,93] For EHD2, the intrinsic ATP hydrolysis rate is very slow (k_{cat} of approximately 1 h^{-1}) but can be 10-fold stimulated in the presence of negatively charged liposomes.[92] EHD2 oligomerizes at the liposome surface into ring-like structures leading to membrane tubulation. The carboxy-terminal Eps15 homology (EH) domain is known to bind with low micromolar affinity to linear peptide motifs containing a central Asn-Pro-Phe (NPF) motif. One to three NPF motifs are present in most of the described interaction partners, such as the BAR-domain proteins PACSIN 1 and 2 (PKC and CK2 substrate in neurons 1 and 2) and *C. elegans* amphiphysin (reviewed in Ref. 94).

The crystal structure of the EHD2 dimer in the presence of a non-hydrolyzable ATP analog showed that dimerization is mediated by a highly conserved interface in the G domain which does not correspond to the G interface of other dynamin superfamily members (Fig. 15.2).[92] At the distal side of the G domain dimer, two disordered loops contain a KPFxxxNPF motif that might bind to the EH domain during oligomerization. The EH domains are positioned on top of the G domain of the opposing monomer. In this position, the C-termini of the EH domains point into the nucleotide binding site of the G domain, thereby shielding the G interface. The known peptide binding groove of the EH domain is occupied by a GPF motif derived from the linker between the helical domain and the EH domain of the opposite monomer.

Sequence stretches at the N- and C-terminus of the G domain form the stalk. Within the EHD2 dimer, a continuous highly curved membrane-binding site is created at the tips of two adjacent stalks. The membrane interaction site includes hydrophobic residues and a polylysine stretch that might mediate binding to negatively charged membrane surfaces. Similar to other dynamin superfamily members, the C-terminus of the stalk contacts the G domain.

An oligomerization model for EHD2 was proposed in which EHD2 dimers assemble into ring-like oligomers via the highly conserved G interface in the G domain.[92] The EH domains might act in an autoinhibitory fashion in solution to restrict EHD2 oligomerization. When recruited by an interaction partner, the EH domains were proposed to switch to the KPFxxxNPF motif in the G domain of the neighboring dimer to stabilize oligomerization. Accordingly, deletion of the EH domain or mutations in the GPF motif interfere with regular assembly of EHD2.[90,92] Interestingly, an EH-domain deletion mutant of EHD2 can still localize to caveolae.[89] Thus, recruitment of EHD2 to caveolae appears to involve the stalk and/or the G domain, and might be mediated by interactions with the caveolar coat component cavin.[89]

1.9. Bacterial dynamins

While many bacteria have genes encoding for dynamin-related proteins, clear sequence similarity is found only in the G domain, not in the helical domains. Further, the cellular functions of these proteins are often not well characterized. Many bacteria express two related dynamin genes within one operon, or the two genes are fused into a single polypeptide chain, suggesting a mechanistic interplay of these heterodimers (reviewed in Ref. 95). The bacterial dynamin-like protein (BDLP) from *Nostoc punctiforme* oligomerizes at the surface of liposomes and induces membrane tubulation. It was suggested that BDLP mediates membrane fusion in this cyanobacterium.[96]

The crystal structure of the BDLP dimer in the nucleotide-free and GDP-bound form showed that the G domains assemble via the G interface.[97] In the extended helical domain, the BSE (here termed the "neck") is separated from the stalk (the "trunk") by a sharp kink around hinge 1. This kink guides the stalk back onto the G domain leading to a compact closed form of GDP-bound BDLP. At the tip of the stalk, a two-helix motif has been termed the "paddle." Interestingly, the paddle is located at a similar site below the stalk as the PH domain in dynamin or the L4 loop in MxA is.

The EM reconstruction of the GTP-bound BDLP oligomer assembled around a lipid tubule revealed dramatic structural rearrangements in the BDLP molecule which now adopts an open conformation.[98] In this conformation, the BSE and stalk align almost in parallel perpendicularly to the liposome surface. The paddle is thought to insert as a pair of transmembrane helices into the membrane. Interestingly, the architecture of the BDLP oligomer differs from that of the dynamin oligomer. Thus, whereas the G domain dimer in dynamin was suggested to interlink neighboring filaments (see below), the G domain dimer in BDLP is 90° rotated relative to the axis of the lipid tubule and appears to stabilize the BDLP dimer. This might be an adaptation to the role of BDLP in mediating membrane fusion versus membrane fission. It was suggested that this function is related to the generation of highly curved membrane tubules that are prone to membrane fusion.[98]

2. STRUCTURAL AND MECHANISTIC INSIGHTS INTO DYNAMIN OLIGOMERIZATION

2.1. The stalk mediates oligomerization of dynamin

Dynamin is the biochemically and functionally best-examined member of the dynamin superfamily. Recent structural studies provided a detailed molecular description of the oligomerization mechanism thus leading to models how dynamin functions at the neck of budding vesicles.

While the structural basis of oligomerization has initially been clarified for MxA,[13] it has subsequently been confirmed for dynamin.[27,28] These studies showed that the stalk is the central assembly hub mediating oligomerization of dynamin and MxA into filaments. Two stalks assemble in a criss-cross fashion via a highly conserved central interface (interface-2, Fig. 15.3). Assembly via this interface results in the formation of stable dynamin and MxA dimers (Fig. 15.3). Accordingly, mutations in interface-2 in MxA yield a monomeric protein that fails to oligomerize,[13] indicating the central importance of this interface for dimer formation. Corresponding mutations in dynamin render the protein insoluble.[27] Based on cross-linking studies, other dimerization models considering domain swapped C-termini have been suggested.[23] However, these are difficult to reconcile with crystal structures of full-length dynamin and mutagenesis data.

In the crystal structures of the isolated MxA stalk,[13] of full-length MxA[26] and dynamin,[27,28] stalk dimers assembled via interface-2 further oligomerized via two additional interfaces, interfaces-1 and 3 (Fig. 15.3B). This raised the idea that tetramerization and oligomerization of MxA and dynamin proceed via these two interfaces. In agreement with this model, mutations in interfaces-1 and 3 in MxA and in interface-3 in dynamin result in stable dimers that are unable to further oligomerize.

Interface-1 mediates interactions with the neighboring stalk dimer above interface-2 (Fig. 15.3B and C). It is mostly hydrophobic and relatively flat. This suggests that it might act as a pivot to control the angle between two stalk dimers and therefore adjust the diameter and rise of the dynamin helix (Fig. 15.3C). Interface-3 has been identified by mutagenesis in dynamin[99] and includes two loop regions of the stalk (L1N and L2). Mutations in interface-3 prevent assembly of dynamin or MxA into ring-like or helical structures and were a prerequisite for their crystallization.[26–28] This interface is not fully resolved in the linear MxA and dynamin filaments in the crystal structures, but it has been modeled for a helical assembly of the dynamin

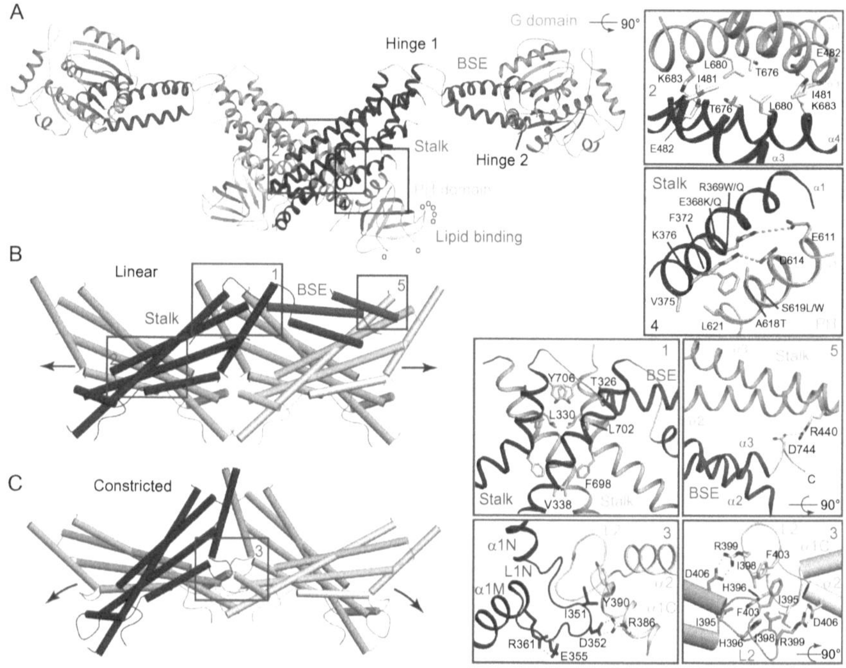

Figure 15.3 Stalk-mediated domain interactions. (A) The dynamin dimer (pdb 3SNH) is the building block of dynamin tetramers and oligomers. Two dynamin molecules interact via the central stalk interface-2. The PH domain folds against another surface of the stalk (interface-4). Insets show structural details of both interfaces, as observed in the crystal structure. Intramolecular interactions are shown in grey boxes and intermolecular interactions in magenta boxes. (B) In the crystal structures, two stalk dimers assemble into a linear filament via interfaces-1 and 3. The red BSE belongs to the blue stalk and interacts with the stalk of the neighboring dimer via interface-5 (shown in magnification at the right). (C) Adjustments of interface-1 and 3 during assembly of dynamin dimers were proposed to adapt the helical dynamin filaments to membrane tubules of various diameters (bottom). Interface-1 and the modelled interface-3 in these rotated stalk dimers are shown at the right. *Modified from Ref. 27.* (For color version of this figure, the reader is referred to the online version of this chapter.)

stalks.[27] Molecular dynamic simulations suggested that rearrangements in the assembly of interfaces-1 and 3 allow dynamin to form helices of different pitch and diameter.[27] This might explain how dynamin oligomerizes around lipid templates of various diameters.[100,101]

2.2. Regulatory interactions of the stalk

The stalk also mediates intra- as well as intermolecular domain interactions, which regulate the activity of dynamin. In the crystal structure of dynamin, the PH domain binds to a highly conserved interface of the stalk at the opposite face of the oligomerization interfaces-1 and 2 (Fig. 15.3A).[27,28] This interaction, involving α1 of the stalk and α1 of the PH domain, is here referred to as interface-4. In this orientation, the lipid-binding loops point toward the G domain, away from the prospective lipid interaction side. Mutations in this interface in the PH domain or the stalk lead to increased oligomerization rates of dynamin in solution.[27,102] Consequently, it was postulated that this interface prevents oligomerization of dynamin in solution. Comparison of the crystal structures with dynamin models fitted into the electron microscopy density indicated that this interface opens during lipid-binding thereby allowing the PH domain to interact with the membrane bilayer.[22,27]

In the crystals of the isolated PH domain, a putative dimerization interface was discovered.[103] In solution, the PH domain is monomeric, and the membrane-binding affinity of this monomeric PH domain was shown to be low.[104] Thus, at least two interconnected PH domains are required for high avidity membrane binding. Interestingly, in the electron microscopy reconstructions of the membrane-bound dynamin filament, two PH domains of neighboring dimers are found in close vicinity to each other below the stalks.[13,21] Future studies need to address whether the PH domains, indeed, dimerize upon lipid binding, for example, for stabilization of the dynamin oligomer in a lipid-bound state. In line with such a hypothesis, the 41 amino-acid substrate binding loop L4 of MxA, which corresponds in sequence position to the PH domain of dynamin, was also shown to be involved in oligomerization.[13]

Based on the influence of PH domain mutations on the GTPase activity of dynamin, a cross-talk between the G and PH domains has been postulated.[12,102] In the linear dynamin oligomers, however, no direct interaction between these two domains was found. Further studies need to address whether a direct contact exists, for example, in the dynamin tetramer, or whether the observed stimulation of GTPase activity in PH domain mutants is indirectly caused by oligomerization.

In the structure of assembled dynamin, the BSE of one dynamin dimer contacts the stalk of the neighboring dimer via a salt bridge between Arg440 and Asp744 (Fig. 15.3B). Mutations in this interface-5 lead to increased rates of dynamin-mediated endocytosis, suggesting a regulatory function.[27] Based on a comparison of the crystal structure with electron microscopy reconstruction-based models of the dynamin oligomer, it was suggested that this interface opens during oligomerization allowing the G domains of neighboring filaments to dimerize (see below). Interface-5 might also be involved in the transduction of nucleotide-induced conformational changes from the G domain to the stalk of the neighboring dimer.

Further, dynamin oligomerization is affected by the PRD at the C-terminus of the protein.[21] Whether the PRD contributes to interface-5 or acts in an alternative way requires clarification.

2.3. G domain dimerization links neighboring helical turns

GTP hydrolysis in the G domains of dynamin is essential for vesicle scission.[18–20] Based on structural studies of the G domain including a minimal BSE, it was demonstrated that GTP triggers dimerization of the G domains via a highly conserved interface across the nucleotide binding site (Fig. 15.4A).[23,25] This interface includes switch I, switch II, and the G4 loop, which contacts the guanine base *in cis and trans*. Dimerization induces rearrangements of the catalytic machinery within the G domain, the positioning of the catalytic water molecule by the conserved threonine in switch I, and the positioning of a catalytic sodium or potassium ion in the active site.[105] The dimerization affinity of the G domains in the GDP-bound state appears to be low, leading to dissociation.[25]

Upon dimerization, the nucleotide loading state of the G domain determines the position of the BSE relative to the G domain, as demonstrated for the G domain-minimal BSE construct of human dynamin 1.[23] Thus, in the GTP-bound state, the BSE was found in an open extended conformation relative to the G domain, whereas it undergoes a 70° rotation to a closed state in the presence of a transition state analog of the GTPase reaction, GDP-AlF_4^- (Fig. 15.4B). In the closed conformation, α1 of the BSE interacts with the G domain via multiple contacts (interface-6, Fig. 15.4B). This closed conformation is also found in the structure of nucleotide-free dynamin. Mutations in this interface lead to unspecific aggregation of dynamin.[27] In a G domain-BSE construct of *Arabidopsis thaliana* dynamin-related protein, an open conformation was found in the presence of GDP-AlF_4^- and a closed conformation in the presence of GDP.[106] Thus, GTP appears to promote an open conformation of the BSE, GDP a closed conformation, whereas the orientation of the BSE during GTP hydrolysis is currently unclear.

Nucleotide-induced movements of the BSE require a conformational coupling across the G domain since the BSE is located at the opposite face of the nucleotide binding site (Fig. 15.4B). It was suggested that this coupling is mediated through the switch regions. However, these undergo only small

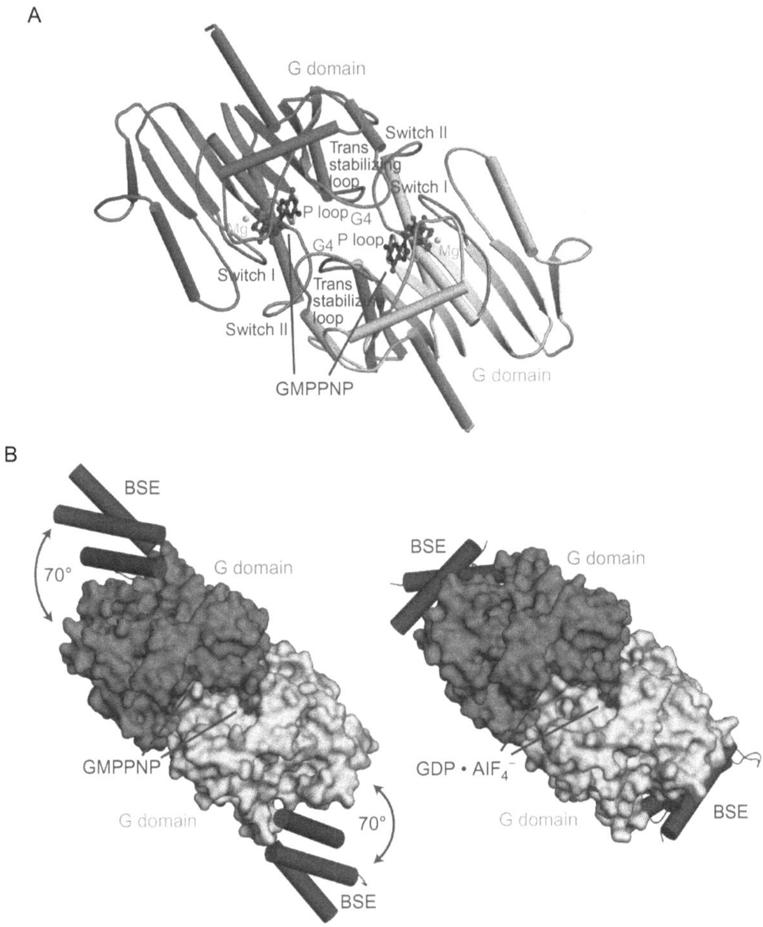

Figure 15.4 Nucleotide-induced movements in the G domains upon dimerization. (A) Structure of the GTP-bound G domain dimer of human dynamin1 (pdb 3ZYC). Structural elements involved in dimerization are indicated. (B) Comparison of the GTP- (3ZYC) and GDP-AlF_4^--bound structures (2X2E) of a G domain-minimal BSE constructs. A 70° rotation of the BSE versus the G domain is apparent during formation of interface-6. (For color version of this figure, the reader is referred to the online version of this chapter.)

structural changes in the open and closed conformations.[23] Interestingly, the temperature-sensitive *shibire(ts2)* phenotype is caused by a mutation in switch II and can be suppressed by a second site mutation in α3 of the BSE[107] supporting the idea of a functional interplay between switch II and the BSE.

Electron microscopy reconstructions-based models of dynamin and MxA indicated that dimerization of the G domain does not occur within a filament assembled via the stalks but rather between neighboring turns of a dynamin helix.[13,22,23] Thus, dimerization of the G domains can only occur after a full turn of the dynamin helix around its membrane template has been established. This has important implications for the assembly of dynamin into a right-handed helix and for its mechanism of action during membrane remodeling.

2.4. Right- versus left-handed dynamin helix

In electron microscopy reconstructions, it was shown that dynamin forms a right-handed helix when assembled on a membrane template.[22] Here, we shortly discuss the architectural principles how a dynamin dimer with a two-fold symmetry can promote the assembly of a right-handed helix on a lipid tubule with rotational symmetry.

The dynamin dimer is an enantiomer, that is, it has a twofold rotational but no mirror symmetry (Fig. 15.5A). During assembly of this dimer, a right-handed helical conformation might be favored. The right-handed assembly of dynamin might be an intrinsic property how the specific enantiomer assembles via the stalk. This asymmetry might also involve the BSE since modification of two cysteines in the stalk and BSE in *Dictyostelium* dynamin induce the formation of dynamin rings rather than helices.[108] Interestingly, in the molecular dynamics simulations used to build our recent dynamical model of dynamin,[27] the isolated stalk dimers of dynamin could in principle assemble as right- or left-handed oligomers (unpublished data). This indicates that the energetic differences between these two assemblies may not be very high.

We suggest that the right-handed dynamin helix is also kinetically preferred. When dynamin is recruited to the neck of a vesicle or to a membrane tubule, oligomerization of the dynamin filament is expected to proceed around the membrane tubule until the two filament ends meet each other (Fig. 15.5B). In a right-handed dynamin helix, the elongated dimer architecture of the dynamin enantiomer allows G domains to assemble before the stalks encounter each other (Fig. 15.5C). In a left-handed dynamin helix, however, additional dynamin dimers would have to be recruited, until the filament reaches a sufficient length before G domains of neighboring filaments could dimerize (Fig. 15.5D). Once the first G domain dimer is formed, the handedness of the helix is fixed.

In vivo, a right-handed dynamin helix might also be favored in cases when dynamin recruitment factors assemble as a right-handed helix, as, for example, shown for the BAR-domain-containing protein endophilin.[109] However,

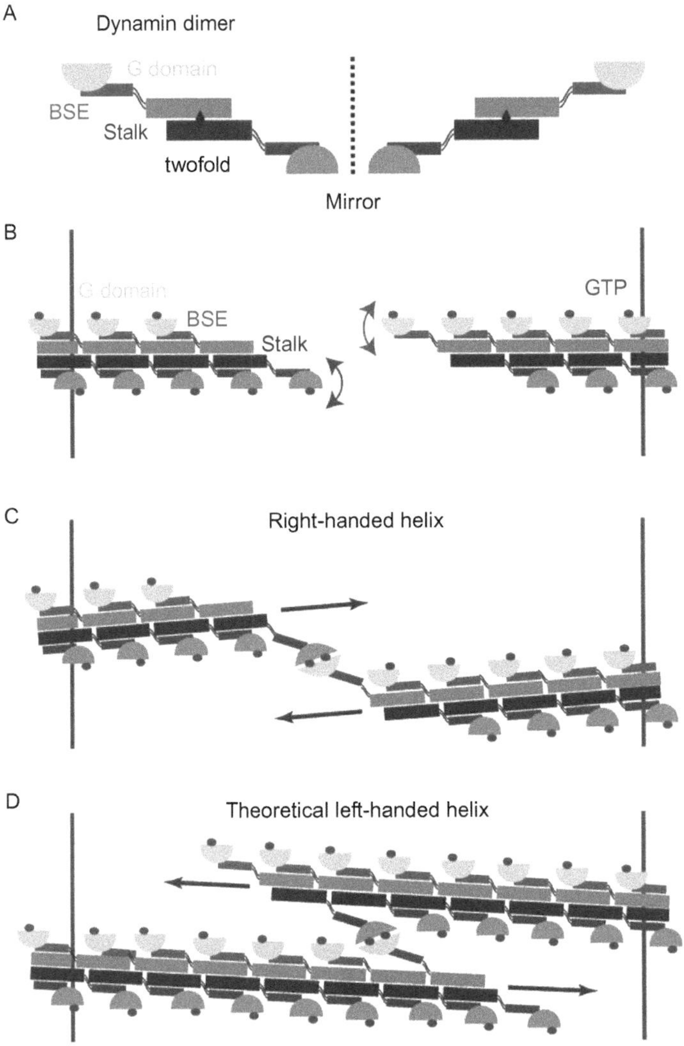

Figure 15.5 Domain rearrangements upon oligomerization in a right-handed dynamin helix. (A) The dynamin dimer is chiral and has a central two-fold axis. Stalk dimers (blue) as well as the G domain—BSE architecture (red–orange) relative to the stalk might contribute to a nonsymmetric assembly of dimers resulting in a right-handed dynamin helix. (B) When the growing dynamin filament has embraced the lipid tubule, two ends of the filament oppose each other. The terminal G domains are not restricted by interface-5 and therefore flexible (indicated as arrows). (C) In a right-handed dynamin helix, the terminal G domains can dimerize before the stalks complete one turn. (D) In a theoretical left-handed dynamin helix, the filaments would need to oligomerize along each other before the G domains could dimerize. Dimerization would involve G domains of dynamin dimers which are already incorporated into the filament. (For interpretation of the references to color in this figure legend, the reader is referred to the online version of this chapter.)

other BAR-domain-containing dynamin interaction partners, such as FBP17, were found to form left-handed helical oligomers.[110] It is currently unclear how such left-handed helical oligomers can recruit dynamin and mediate oligomerization of a right-handed dynamin helix. Further, the chirality of phospholipids was suggested to influence the handedness of the dynamin helix.[111–113] Clearly, further studies are needed to assess the potential contribution of these factors to the handedness of dynamin oligomerization at membranes.

2.5. Scission model

Several models have been put forward to explain how the dynamin oligomer mediates GTPase-dependent membrane remodeling. Electron microscopy reconstructions indicated that nucleotide binding induces constriction of the dynamin oligomer when the underlying membrane template is deformable.[14,114] Consistently, 15 dimers per dynamin helix turn were found in the absence of nucleotide, whereas only 13 dimers per turn were found in the presence of a nonhydrolyzable GTP analog.[22] These findings suggested that nucleotide-free dynamin assembles in a relaxed conformation around tubular membrane templates. Nucleotide-induced conformational changes would then induce constriction of the oligomer followed by membrane scission ("constrictase" model). In the presence of stiff membrane tubules, it was observed that the dynamin helix is closely packed along the helix axis in the GTP-bound form and opens toward a wider rise-per-turn in the presence of GDP.[17] This led to a model where the spring-like opening of the dynamin helix induced by nucleotide-hydrolysis catalyzes vesicle scission ("poppase" model). Finally, microscopy-based live assays monitoring dynamin-mediated membrane remodeling showed that GTP hydrolysis results in twisting of dynamin-coated membrane tubules,[20] implicating dynamin-mediated twisting of the vesicle neck during vesicle scission ("twistase" model).

In light of the recent structural findings and simulation results,[27] we suggest a structural model for the conformational changes in the dynamin helix during membrane scission, which unifies these previous ideas (Fig. 15.6). This model predicts a mechanism that combines aspects of a constrictase, a poppase and a twistase. In the presence of cellular GTP concentrations of >100 μM and considering the low micromolar affinity of dynamin for GTP, dynamin is predicted to be in the GTP-bound state when recruited to the bud neck. Inhibitory domain interactions of the stalk with the PH

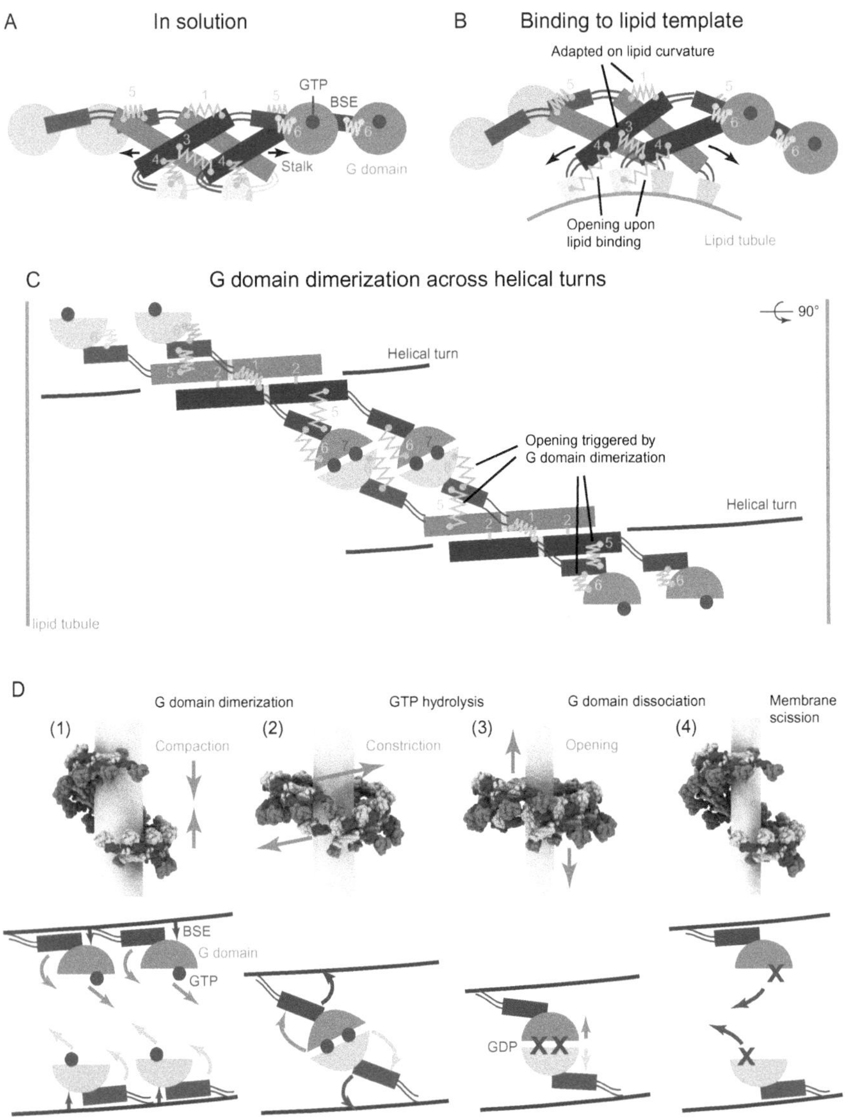

Figure 15.6 Structural model for dynamin-mediated membrane scission. (A) In solution, dynamin is predominantly tetrameric. Domain interactions adapted during assembly are indicated as springs (grey springs for intra- and pink for intermolecular interactions). (B) During membrane binding, interfaces-1 and 3 can adapt to the diameter of the lipid tubule. Membrane binding releases the PH domain from the autoinhibitory interaction with the stalk (opening of interface-4). (C) When G domains of neighboring dynamin filaments dimerize, the BSE might be released from the stalk (e.g., opening of interface-5) and from the G domains (e.g., opening of interface-6). (D) Model for dynamin action during membrane scission. (1) When the growing dynamin filament has

domain and the BSE (interfaces-4 and 5) are released by membrane binding and/or oligomerization, and further dynamin subunits are recruited (Fig. 15.6B). When the dynamin oligomer has formed a single turn around the bud neck, G domains of opposing filament ends dimerize and close the first turn of a right-handed dynamin helix. We assume that G domain dimerization in the presence of GTP opens the G domain-BSE interface-6 (Fig. 15.6C). It is unclear how many dynamin molecules can still be recruited before GTP hydrolysis is triggered, but given the k_{max} of dynamin's stimulated GTPase activity of 4 s^{-1},[27] the time span for further assembly is limited. G domain dimerization induces GTP hydrolysis followed by a conformational movement of the BSE back to the G domain. This movement might act as a power stroke by pulling neighboring filaments along each other.[23,28] Alternatively, G domain dimerization might induce a tightening of the dynamin helix thereby fully or partially compensating the energy required to induce higher curvature on the underlying membrane template. This might lead to a passive sliding of neighboring filaments to adapt the helix to the new diameter of the template.[27] Both scenarios lead to the constriction of the bud neck, similarly to a contracting sling. In order to contract, the right-handed dynamin helix must undergo a right-handed twisting of the dynamin filaments. This was, indeed, observed by light microscopy.[20] Multiple rounds of GTP binding and hydrolysis might lead to the successive tightening of the dynamin helix and bud neck. The resulting increase in membrane curvature will further destabilize the membrane at the bud neck. The dynamin coat itself, however, might act as a scaffold preventing cleavage of the underlying membrane neck by constriction alone.[3] Thus, we suggest that GTP hydrolysis followed by dissociation of the G domains according to the poppase model might separate dynamin filaments and allow the highly curved bud neck to break. In agreement with this, short dynamin collars have been observed to cycle on and off membranes during membrane fission.[100,115]

---

embraced the membrane template once, GTP-loaded G domains of neighboring turns oppose each other. (2) G domains of neighboring filaments dimerize leading to a compaction of the dynamin helix. (3) Nucleotide-hydrolysis-induced conformational changes of the BSE versus the G domain might induce a power stroke actively pulling filaments versus each other or induce higher membrane curvature of the underlying template leading to a passive sliding of neighboring filaments. (4) GDP-bound G domains separate leading to dissociation of neighboring filaments. This might allow the highly strained membrane neck to break. (For color version of this figure, the reader is referred to the online version of this chapter.)

3. STRUCTURAL INSIGHTS INTO DISEASE-INDUCING MUTATIONS IN DYNAMIN

Dynamin 1 and 3 are highly expressed in the brain and are thought to mediate retrieval of synaptic vesicle membranes that occurs concomitant with exocytic fusion.[11] In contrast, dynamin 2 is ubiquitously expressed and thus appears to mediate uptake of clathrin-coated vesicles in all other cell types.

A dynamin 2 gene knockout in mice leads to embryonic lethality[116] whereas dynamin 1 knockout mice are born, but have a reduced life span.[117] The latter mice showed pronounced defects in synaptic vesicle endocytosis during strong exogenous stimulations suggesting an important function of dynamin 1 during high neuronal activity. Dynamin 3 knockout mice have no apparent phenotype. However, additional deletion of dynamin 3 worsens the phenotype of dynamin 1 knockout mice, leading to perinatal lethality.[118] This suggests a cooperative essential role of dynamin 1 and 3 in synaptic vesicle endocytosis.

Mutations in dynamin 2, but not in dynamin 1 or 3, can lead to congenital diseases in humans, for example, to rare forms of autosomal dominant centronuclear myopathy (CNM) and Charcot-Marie-Tooth neuropathy (CMT) neuropathy (reviewed in Ref. 119). CNM is a disease of the peripheral muscles where the cell nuclei of muscle fiber are relocated from the periphery to the center, indicating distortion of the muscle fiber architecture. CMT is a peripheral neuropathy that is characterized by progressive muscle weakness and atrophy, initially apparent in the muscles of the leg.[120] Some dynamin 2 mutations lead to a severe neonatal phenotype of CMT, which progressively improves with age implicating compensatory mechanisms during development.

So far, no disease-inducing mutations have been identified in the most highly conserved G domain, where the majority of inactivating dynamin 2 mutations would be expected. This suggests that disease is not caused by a complete loss of GTPase function of dynamin 2. Further, deficits induced by dynamin 2 mutations appear to be restricted to certain cell types such as peripheral neurons or muscle fibers. This might be related to the specific splice isoforms of dynamin 2 expressed in these tissues.[119] For example, in a dorsal root ganglia culture model, it has been shown that overexpression of dynamin 2 harboring CMT but not CNM mutations cause myelination deficits.[121] In this system, overexpression of dynamin variants with CMT mutations interfered with clathrin-mediated endocytosis whereas CNM mutations did not.

Many dynamin 2 mutations found in CNM patients are located in or in close vicinity of the stalk–PH domain interface (Fig. 15.7). It has been demonstrated that mutations in the C-terminal helix of the PH domain, a hot spot for CNM-related mutations, cause increased oligomerization of recombinant dynamin in solution.[102] Also mutations at the opposing stalk interface lead to increased oligomerization and disease.[27] This suggests that the molecular basis of CNM-inducing dynamin mutations is a disruption of the autoinhibitory interface-4. Accordingly, these dynamin mutants might induce CNM by aberrant oligomerization at cellular sites where dynamin is normally not present. Alternatively, disruption of interface-4 might also prevent the disassembly of the dynamin coat after GTP hydrolysis, by stabilizing the structure of oligomerized dynamin.

Most of the reported mutations causing CMT localize to the PH domain, in particular to the lipid-binding loops (Fig. 15.7). Exceptions are the mutation G358R which localizes to the edge of interface-3 and a microdeletion in the PRD (ΔThr859-Ile860). One of the PH domain mutants, K562E, completely blocks dynamin-dependent endocytosis whereas a microdeletion a few residues upstream, Δ553–555, does not affect clathrin-mediated endocytosis.[122] The latter deletion mutant was found to prominently decorate microtubules and to affect microtubule dynamics. Thus, the molecular basis of CMT disease induced by dynamin mutations might not always be related to dynamin's function in endocytosis but, at least in some cases, to its less well characterized function at microtubules.[122] To understand the deficits in these

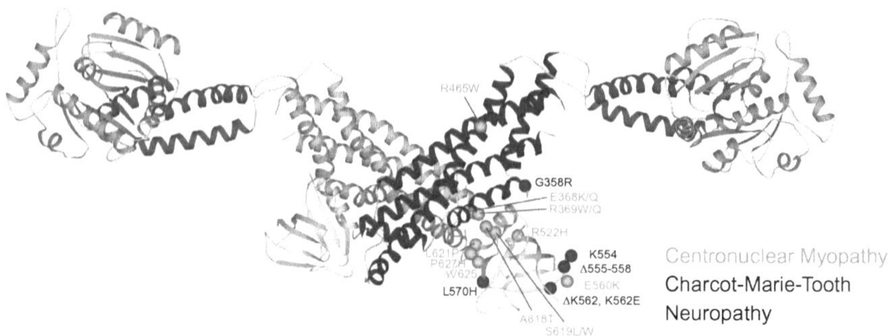

Figure 15.7 Disease mutations in dynamin. The positions of disease mutations leading to CNM (pink) and CMT (black) are mapped as balls on the structure of the dynamin dimer (based on Refs. 119,121). CNM mutations predominantly cluster in the interface between stalk and the PH domain. (For interpretation of the references to color in this figure legend, the reader is referred to the online version of this chapter.)

mutants on the molecular level, the interaction of dynamin with microtubules needs to be investigated by biochemical and structural methods.

Screens to inhibit the function of dynamin yielded small molecular compounds which bind to the G or PH domain and interfere with the GTPase activity and/or endocytosis (reviewed in Ref. 123). So far, the stalk has not been targeted by dynamin inhibitors. The identification of the assembly interfaces in the stalk could boost the structure-based design of compounds which target the oligomerization interfaces. Such compounds could efficiently interfere with the function of dynamin in endocytosis and, possibly, disease.

4. OUTLOOK

In the past years, many studies have provided detailed structural insights into the architectures, functions, and mechanisms of a number of dynamin superfamily proteins. Still, for some members, in particular the mitochondrial fusion proteins, little structural information is available. These gaps need to be filled to mechanistically understand similarities and differences of membrane scission and fusion events catalyzed by different dynamin superfamily members. Further, only for a few members, the structural changes associated with nucleotide binding and, in particular, with membrane binding and oligomerization have been characterized at high resolution. To understand the role of these proteins in membrane remodeling and the exact deficits leading to disease, future work needs to address such membrane-bound oligomers. Moreover, these structural experiments need to be paired with biophysical, theoretical, and functional studies to translate the structural information into novel mechanistic insights.

ACKNOWLEDGMENTS

This work was supported by a grants of the German Research Foundation (SFB 740 and SFB958 to O. D., M. H., F. N., and V. H.), by a Career Development Fellowship of the International Human Frontier Science Program Organization and the EMBO Young Investigator Program YIP to O. D.

REFERENCES

1. Faelber K, Held M, Gao S, Posor Y, Haucke V, Noe F, et al. Structural insights into dynamin-mediated membrane fission. *Structure* 2012;**20**:1621–8.
2. Praefcke GJ, McMahon HT. The dynamin superfamily: universal membrane tubulation and fission molecules? *Nat Rev Mol Cell Biol* 2004;**5**:133–47.

3. Boucrot E, Pick A, Camdere G, Liska N, Evergren E, McMahon HT, et al. Membrane fission is promoted by insertion of amphipathic helices and is restricted by crescent BAR domains. *Cell* 2012;**149**:124–36.
4. Niemann HH, Knetsch ML, Scherer A, Manstein DJ, Kull FJ. Crystal structure of a dynamin GTPase domain in both nucleotide-free and GDP-bound forms. *EMBO J* 2001;**20**:5813–21.
5. Shpetner HS, Vallee RB. Identification of dynamin, a novel mechanochemical enzyme that mediates interactions between microtubules. *Cell* 1989;**59**:421–32.
6. Staeheli P, Haller O, Boll W, Lindenmann J, Weissmann C. Mx protein: constitutive expression in 3T3 cells transformed with cloned Mx cDNA confers selective resistance to influenza virus. *Cell* 1986;**44**:147–58.
7. Ishida N, Nakamura Y, Tanabe K, Li SA, Takei K. Dynamin 2 associates with microtubules at mitosis and regulates cell cycle progression. *Cell Struct Funct* 2011;**36**:145–54.
8. Kosaka T, Ikeda K. Reversible blockage of membrane retrieval and endocytosis in the garland cell of the temperature-sensitive mutant of Drosophila melanogaster, shibirets1. *J Cell Biol* 1983;**97**:499–507.
9. van der Bliek AM, Meyerowitz EM. Dynamin-like protein encoded by the Drosophila shibire gene associated with vesicular traffic. *Nature* 1991;**351**:411–4.
10. Chen MS, Obar RA, Schroeder CC, Austin TW, Poodry CA, Wadsworth SC, et al. Multiple forms of dynamin are encoded by shibire, a Drosophila gene involved in endocytosis. *Nature* 1991;**351**:583–6.
11. Ferguson SM, De Camilli P. Dynamin, a membrane-remodelling GTPase. *Nat Rev Mol Cell Biol* 2012;**13**:75–88.
12. Muhlberg AB, Warnock DE, Schmid SL. Domain structure and intramolecular regulation of dynamin GTPase. *EMBO J* 1997;**16**:6676–83.
13. Gao S, von der Malsburg A, Paeschke S, Behlke J, Haller O, Kochs G, et al. Structural basis of oligomerization in the stalk region of dynamin-like MxA. *Nature* 2010;**465**:502–6.
14. Hinshaw JE, Schmid SL. Dynamin self-assembles into rings suggesting a mechanism for coated vesicle budding. *Nature* 1995;**374**:190–2.
15. Takei K, McPherson PS, Schmid SL, De Camilli P. Tubular membrane invaginations coated by dynamin rings are induced by GTP-gamma S in nerve terminals. *Nature* 1995;**374**:186–90.
16. Tuma PL, Stachniak MC, Collins CA. Activation of dynamin GTPase by acidic phospholipids and endogenous rat brain vesicles. *J Biol Chem* 1993;**268**:17240–6.
17. Stowell MH, Marks B, Wigge P, McMahon HT. Nucleotide-dependent conformational changes in dynamin: evidence for a mechanochemical molecular spring. *Nat Cell Biol* 1999;**1**:27–32.
18. Sweitzer SM, Hinshaw JE. Dynamin undergoes a GTP-dependent conformational change causing vesiculation. *Cell* 1998;**93**:1021–9.
19. Marks B, Stowell MH, Vallis Y, Mills IG, Gibson A, Hopkins CR, et al. GTPase activity of dynamin and resulting conformation change are essential for endocytosis. *Nature* 2001;**410**:231–5.
20. Roux A, Uyhazi K, Frost A, De Camilli P. GTP-dependent twisting of dynamin implicates constriction and tension in membrane fission. *Nature* 2006;**441**:528–31.
21. Zhang P, Hinshaw JE. Three-dimensional reconstruction of dynamin in the constricted state. *Nat Cell Biol* 2001;**3**:922–6.
22. Mears JA, Ray P, Hinshaw JE. A corkscrew model for dynamin constriction. *Structure* 2007;**15**:1190–202.
23. Chappie JS, Mears JA, Fang S, Leonard M, Schmid SL, Milligan RA, et al. A pseudoatomic model of the dynamin polymer identifies a hydrolysis-dependent powerstroke. *Cell* 2011;**147**:209–22.

24. Chappie JS, Acharya S, Liu YW, Leonard M, Pucadyil TJ, Schmid SL. An intramolecular signaling element that modulates dynamin function in vitro and in vivo. *Mol Biol Cell* 2009;**20**:3561–71.
25. Chappie JS, Acharya S, Leonard M, Schmid SL, Dyda F. G domain dimerization controls dynamin's assembly-stimulated GTPase activity. *Nature* 2010;**465**:435–40.
26. Gao S, von der MA, Dick A, Faelber K, Schroder GF, Haller O, et al. Structure of myxovirus resistance protein a reveals intra- and intermolecular domain interactions required for the antiviral function. *Immunity* 2011;**35**:514–25.
27. Faelber K, Posor Y, Gao S, Held M, Roske Y, Schulze D, et al. Crystal structure of nucleotide-free dynamin. *Nature* 2011;**477**:556–60.
28. Ford MG, Jenni S, Nunnari J. The crystal structure of dynamin. *Nature* 2011;**477**:561–6.
29. Ferguson KM, Lemmon MA, Schlessinger J, Sigler PB. Crystal structure at 2.2 A resolution of the pleckstrin homology domain from human dynamin. *Cell* 1994;**79**:199–209.
30. Zheng J, Cahill SM, Lemmon MA, Fushman D, Schlessinger J, Cowburn D. Identification of the binding site for acidic phospholipids on the pH domain of dynamin: implications for stimulation of GTPase activity. *J Mol Biol* 1996;**255**:14–21.
31. Salim K, Bottomley MJ, Querfurth E, Zvelebil MJ, Gout I, Scaife R, et al. Distinct specificity in the recognition of phosphoinositides by the pleckstrin homology domains of dynamin and Bruton's tyrosine kinase. *EMBO J* 1996;**15**:6241–50.
32. Shupliakov O, Low P, Grabs D, Gad H, Chen H, David C, et al. Synaptic vesicle endocytosis impaired by disruption of dynamin-SH3 domain interactions. *Science* 1997;**276**:259–63.
33. Soulet F, Yarar D, Leonard M, Schmid SL. SNX9 regulates dynamin assembly and is required for efficient clathrin-mediated endocytosis. *Mol Biol Cell* 2005;**16**:2058–67.
34. Cho DH, Nakamura T, Lipton SA. Mitochondrial dynamics in cell death and neurodegeneration. *Cell Mol Life Sci* 2010;**67**:3435–47.
35. Nunnari J, Suomalainen A. Mitochondria: in sickness and in health. *Cell* 2012; **148**:1145–59.
36. Labrousse AM, Zappaterra MD, Rube DA, van der Bliek AM. C. elegans dynamin-related protein DRP-1 controls severing of the mitochondrial outer membrane. *Mol Cell* 1999;**4**:815–26.
37. Smirnova E, Griparic L, Shurland DL, van der Bliek AM. Dynamin-related protein Drp1 is required for mitochondrial division in mammalian cells. *Mol Biol Cell* 2001;**12**:2245–56.
38. Friedman JR, Lackner LL, West M, DiBenedetto JR, Nunnari J, Voeltz GK. ER tubules mark sites of mitochondrial division. *Science* 2011;**334**:358–62.
39. Otera H, Wang C, Cleland MM, Setoguchi K, Yokota S, Youle RJ, et al. Mff is an essential factor for mitochondrial recruitment of Drp1 during mitochondrial fission in mammalian cells. *J Cell Biol* 2010;**191**:1141–58.
40. Palmer CS, Osellame LD, Laine D, Koutsopoulos OS, Frazier AE, Ryan MT. MiD49 and MiD51, new components of the mitochondrial fission machinery. *EMBO Rep* 2011;**12**:565–73.
41. Zhao J, Liu T, Jin S, Wang X, Qu M, Uhlen P, et al. Human MIEF1 recruits Drp1 to mitochondrial outer membranes and promotes mitochondrial fusion rather than fission. *EMBO J* 2011;**30**:2762–78.
42. Koch A, Yoon Y, Bonekamp NA, McNiven MA, Schrader M. A role for Fis1 in both mitochondrial and peroxisomal fission in mammalian cells. *Mol Biol Cell* 2005;**16**:5077–86.

43. Yoon Y, Krueger EW, Oswald BJ, McNiven MA. The mitochondrial protein hFis1 regulates mitochondrial fission in mammalian cells through an interaction with the dynamin-like protein DLP1. *Mol Cell Biol* 2003;**23**:5409–20.
44. Lackner LL, Horner JS, Nunnari J. Mechanistic analysis of a dynamin effector. *Science* 2009;**325**:874–7.
45. Chen H, Chan DC. Physiological functions of mitochondrial fusion. *Ann N Y Acad Sci* 2010;**1201**:21–5.
46. Knott AB, Perkins G, Schwarzenbacher R, Bossy-Wetzel E. Mitochondrial fragmentation in neurodegeneration. *Nat Rev Neurosci* 2008;**9**:505–18.
47. Wang H, Song P, Du L, Tian W, Yue W, Liu M, et al. Parkin ubiquitinates Drp1 for proteasome-dependent degradation: implication of dysregulated mitochondrial dynamics in Parkinson disease. *J Biol Chem* 2011;**286**:11649–58.
48. Wang X, Su B, Fujioka H, Zhu X. Dynamin-like protein 1 reduction underlies mitochondrial morphology and distribution abnormalities in fibroblasts from sporadic Alzheimer's disease patients. *Am J Pathol* 2008;**173**:470–82.
49. Cho DH, Nakamura T, Fang J, Cieplak P, Godzik A, Gu Z, et al. S-nitrosylation of Drp1 mediates beta-amyloid-related mitochondrial fission and neuronal injury. *Science* 2009;**324**:102–5.
50. Bossy B, Petrilli A, Klinglmayr E, Chen J, Lutz-Meindl U, Knott AB, et al. S-Nitrosylation of DRP1 does not affect enzymatic activity and is not specific to Alzheimer's disease. *J Alzheimers Dis* 2010;**20**(Suppl. 2):S513–S526.
51. Costa V, Giacomello M, Hudec R, Lopreiato R, Ermak G, Lim D, et al. Mitochondrial fission and cristae disruption increase the response of cell models of Huntington's disease to apoptotic stimuli. *EMBO Mol Med* 2010;**2**:490–503.
52. Shirendeb UP, Calkins MJ, Manczak M, Anekonda V, Dufour B, McBride JL, et al. Mutant huntingtin's interaction with mitochondrial protein Drp1 impairs mitochondrial biogenesis and causes defective axonal transport and synaptic degeneration in Huntington's disease. *Hum Mol Genet* 2012;**21**:406–20.
53. Waterham HR, Koster J, van Roermund CW, Mooyer PA, Wanders RJ, Leonard JV. A lethal defect of mitochondrial and peroxisomal fission. *N Engl J Med* 2007;**356**:1736–41.
54. Mears JA, Lackner LL, Fang S, Ingerman E, Nunnari J, Hinshaw JE. Conformational changes in Dnm1 support a contractile mechanism for mitochondrial fission. *Nat Struct Mol Biol* 2011;**18**:20–6.
55. Strack S, Cribbs JT. Allosteric modulation of Drp1 mechanoenzyme assembly and mitochondrial fission by the variable domain. *J Biol Chem* 2012;**287**:10990–1001.
56. Haller O, Gao S, von der Malsburg A, Daumke O, Kochs G. Dynamin-like MxA GTPase: structural insights into oligomerization and implications for antiviral activity. *J Biol Chem* 2010;**285**:28419–24.
57. Stertz S, Reichelt M, Krijnse-Locker J, Mackenzie J, Simpson JC, Haller O, et al. Interferon-induced, antiviral human MxA protein localizes to a distinct subcompartment of the smooth endoplasmic reticulum. *J Interferon Cytokine Res* 2006;**26**:650–60.
58. Kochs G, Haller O. Interferon-induced human MxA GTPase blocks nuclear import of Thogoto virus nucleocapsids. *Proc Natl Acad Sci U S A* 1999;**96**:2082–6.
59. Reichelt M, Stertz S, Krijnse-Locker J, Haller O, Kochs G. Missorting of LaCrosse virus nucleocapsid protein by the interferon-induced MxA GTPase involves smooth ER membranes. *Traffic* 2004;**5**:772–84.
60. Staeheli P, Grob R, Meier E, Sutcliffe JG, Haller O. Influenza virus-susceptible mice carry Mx genes with a large deletion or a nonsense mutation. *Mol Cell Biol* 1988;**8**:4518–23.

61. Verhelst J, Parthoens E, Schepens B, Fiers W, Saelens X. Interferon-inducible Mx1 protein inhibits influenza virus by interfering with functional viral ribonucleoprotein complex assembly. *J Virol* 2012;**86**:13445–55.
62. von der Malsburg A, Abutbul-Ionita I, Haller O, Kochs G, Danino D. Stalk domain of the dynamin-like MxA GTPase protein mediates membrane binding and liposome tubulation via the unstructured L4 loop. *J Biol Chem* 2011;**286**:37858–65.
63. Zimmermann P, Manz B, Haller O, Schwemmle M, Kochs G. The viral nucleoprotein determines Mx sensitivity of influenza A viruses. *J Virol* 2011;**85**:8133–40.
64. Hales KG, Fuller MT. Developmentally regulated mitochondrial fusion mediated by a conserved, novel, predicted GTPase. *Cell* 1997;**90**:121–9.
65. Zorzano A, Liesa M, Sebastian D, Segales J, Palacin M. Mitochondrial fusion proteins: dual regulators of morphology and metabolism. *Semin Cell Dev Biol* 2010;**21**:566–74.
66. Santel A, Fuller MT. Control of mitochondrial morphology by a human mitofusin. *J Cell Sci* 2001;**114**:867–74.
67. Chen H, Detmer SA, Ewald AJ, Griffin EE, Fraser SE, Chan DC. Mitofusins Mfn1 and Mfn2 coordinately regulate mitochondrial fusion and are essential for embryonic development. *J Cell Biol* 2003;**160**:189–200.
68. Zuchner S, Mersiyanova IV, Muglia M, Bissar-Tadmouri N, Rochelle J, Dadali EL, et al. Mutations in the mitochondrial GTPase mitofusin 2 cause Charcot-Marie-Tooth neuropathy type 2A. *Nat Genet* 2004;**36**:449–51.
69. Wong ED, Wagner JA, Gorsich SW, McCaffery JM, Shaw JM, Nunnari J. The dynamin-related GTPase, Mgm1p, is an intermembrane space protein required for maintenance of fusion competent mitochondria. *J Cell Biol* 2000;**151**:341–52.
70. Meeusen S, DeVay R, Block J, Cassidy-Stone A, Wayson S, McCaffery JM, et al. Mitochondrial inner-membrane fusion and crista maintenance requires the dynamin-related GTPase Mgm1. *Cell* 2006;**127**:383–95.
71. Frezza C, Cipolat S, Martins de BO, Micaroni M, Beznoussenko GV, Rudka T, et al. OPA1 controls apoptotic cristae remodeling independently from mitochondrial fusion. *Cell* 2006;**126**:177–89.
72. Yamaguchi R, Lartigue L, Perkins G, Scott RT, Dixit A, Kushnareva Y, et al. Opa1-mediated cristae opening is Bax/Bak and BH3 dependent, required for apoptosis, and independent of Bak oligomerization. *Mol Cell* 2008;**31**:557–69.
73. Olichon A, Guillou E, Delettre C, Landes T, Arnaune-Pelloquin L, Emorine LJ, et al. Mitochondrial dynamics and disease, OPA1. *Biochim Biophys Acta* 2006;**1763**:500–9.
74. Ban T, Heymann JA, Song Z, Hinshaw JE, Chan DC. OPA1 disease alleles causing dominant optic atrophy have defects in cardiolipin-stimulated GTP hydrolysis and membrane tubulation. *Hum Mol Genet* 2010;**19**:2113–22.
75. Vestal DJ, Jeyaratnam JA. The guanylate-binding proteins: emerging insights into the biochemical properties and functions of this family of large interferon-induced guanosine triphosphatase. *J Interferon Cytokine Res* 2011;**31**:89–97.
76. Kim BH, Shenoy AR, Kumar P, Das R, Tiwari S, MacMicking JD. A family of IFN-gamma-inducible 65-kD GTPases protects against bacterial infection. *Science* 2011;**332**:717–21.
77. Fres JM, Muller S, Praefcke GJ. Purification of the CaaX-modified, dynamin-related large GTPase hGBP1 by coexpression with farnesyltransferase. *J Lipid Res* 2010;**51**:2454–9.
78. Prakash B, Praefcke GJ, Renault L, Wittinghofer A, Herrmann C. Structure of human guanylate-binding protein 1 representing a unique class of GTP-binding proteins. *Nature* 2000;**403**:567–71.
79. Ghosh A, Praefcke GJ, Renault L, Wittinghofer A, Herrmann C. How guanylate-binding proteins achieve assembly-stimulated processive cleavage of GTP to GMP. *Nature* 2006;**440**:101–4.

80. Modiano N, Lu YE, Cresswell P. Golgi targeting of human guanylate-binding protein-1 requires nucleotide binding, isoprenylation, and an IFN-gamma-inducible cofactor. *Proc Natl Acad Sci USA* 2005;**102**:8680–5.
81. Hu J, Prinz WA, Rapoport TA. Weaving the web of ER tubules. *Cell* 2011;**147**:1226–31.
82. Byrnes LJ, Sondermann H. Structural basis for the nucleotide-dependent dimerization of the large G protein atlastin-1/SPG3A. *Proc Natl Acad Sci USA* 2011; **108**:2216–22.
83. Bian X, Klemm RW, Liu TY, Zhang M, Sun S, Sui X, et al. Structures of the atlastin GTPase provide insight into homotypic fusion of endoplasmic reticulum membranes. *Proc Natl Acad Sci U S A* 2011;**108**:3976–81.
84. Liu TY, Bian X, Sun S, Hu X, Klemm RW, Prinz WA, et al. Lipid interaction of the C terminus and association of the transmembrane segments facilitate atlastin-mediated homotypic endoplasmic reticulum fusion. *Proc Natl Acad Sci U S A* 2012;**109**: E2146–E2154.
85. Salinas S, Proukakis C, Crosby A, Warner TT. Hereditary spastic paraplegia: clinical features and pathogenetic mechanisms. *Lancet Neurol* 2008;**7**:1127–38.
86. Grant B, Zhang Y, Paupard MC, Lin SX, Hall DH, Hirsh D. Evidence that RME-1, a conserved C. elegans EH-domain protein, functions in endocytic recycling. *Nat Cell Biol* 2001;**3**:573–9.
87. Lin SX, Grant B, Hirsh D, Maxfield FR. Rme-1 regulates the distribution and function of the endocytic recycling compartment in mammalian cells. *Nat Cell Biol* 2001;**3**:567–72.
88. Caplan S, Naslavsky N, Hartnell LM, Lodge R, Polishchuk RS, Donaldson JG, et al. A tubular EHD1-containing compartment involved in the recycling of major histocompatibility complex class I molecules to the plasma membrane. *EMBO J* 2002;**21**:2557–67.
89. Moren B, Shah C, Howes MT, Schieber NL, McMahon HT, Parton RG, et al. EHD2 regulates caveolar dynamics via ATP-driven targeting and oligomerization. *Mol Biol Cell* 2012;**23**:1316–29.
90. Stoeber M, Stoeck IK, Hanni C, Bleck CK, Balistreri G, Helenius A. Oligomers of the ATPase EHD2 confine caveolae to the plasma membrane through association with actin. *EMBO J* 2012;**31**:2350–64.
91. Shao Y, Akmentin W, Toledo-Aral JJ, Rosenbaum J, Valdez G, Cabot JB, et al. Pincher, a pinocytic chaperone for nerve growth factor/TrkA signaling endosomes. *J Cell Biol* 2002;**157**:679–91.
92. Daumke O, Lundmark R, Vallis Y, Martens S, Butler PJ, McMahon HT. Architectural and mechanistic insights into an EHD ATPase involved in membrane remodelling. *Nature* 2007;**449**:923–7.
93. Lee DW, Zhao X, Scarselletta S, Schweinsberg PJ, Eisenberg E, Grant BD, et al. ATP binding regulates oligomerization and endosome association of RME-1 family proteins. *J Biol Chem* 2005;**280**:17213–20.
94. Naslavsky N, Caplan S. EHD proteins: key conductors of endocytic transport. *Trends Cell Biol* 2011;**21**:122–31.
95. Bramkamp M. Structure and function of bacterial dynamin-like proteins. *Biol Chem* 2012;**393**:1203–14. http://dx.doi.org/10.1515/hsz-2012-0185.
96. Low HH, Lowe J. Dynamin architecture—from monomer to polymer. *Curr Opin Struct Biol* 2010;**20**:791–8.
97. Low HH, Lowe J. A bacterial dynamin-like protein. *Nature* 2006;**444**:766–9.
98. Low HH, Sachse C, Amos LA, Lowe J. Structure of a bacterial dynamin-like protein lipid tube provides a mechanism for assembly and membrane curving. *Cell* 2009;**139**:1342–52.

99. Ramachandran R, Surka M, Chappie JS, Fowler DM, Foss TR, Song BD, et al. The dynamin middle domain is critical for tetramerization and higher-order self-assembly. *EMBO J* 2007;**26**:559–66.
100. Bashkirov PV, Akimov SA, Evseev AI, Schmid SL, Zimmerberg J, Frolov VA. GTPase cycle of dynamin is coupled to membrane squeeze and release, leading to spontaneous fission. *Cell* 2008;**135**:1276–86.
101. Roux A, Koster G, Lenz M, Sorre B, Manneville JB, Nassoy P, et al. Membrane curvature controls dynamin polymerization. *Proc Natl Acad Sci USA* 2010;**107**:4141–6.
102. Kenniston JA, Lemmon MA. Dynamin GTPase regulation is altered by PH domain mutations found in centronuclear myopathy patients. *EMBO J* 2010;**29**:3054–67.
103. Timm D, Salim K, Gout I, Guruprasad L, Waterfield M, Blundell T. Crystal structure of the pleckstrin homology domain from dynamin. *Nat Struct Biol* 1994;**1**:782–8.
104. Klein DE, Lee A, Frank DW, Marks MS, Lemmon MA. The pleckstrin homology domains of dynamin isoforms require oligomerization for high affinity phosphoinositide binding. *J Biol Chem* 1998;**273**:27725–33.
105. Ash MR, Maher MJ, Mitchell GJ, Jormakka M. The cation-dependent G-proteins: in a class of their own. *FEBS Lett* 2012;**586**:2218–24.
106. Yan L, Ma Y, Sun Y, Gao J, Chen X, Liu J, et al. Structural basis for mechanochemical role of Arabidopsis thaliana dynamin-related protein in membrane fission. *J Mol Cell Biol* 2011;**3**:378–81.
107. Ramaswami M, Rao S, van der Bliek A, Kelly RB, Krishnan KS. Genetic studies on dynamin function in Drosophila. *J Neurogenet* 1993;**9**:73–87.
108. Klockow B, Tichelaar W, Madden DR, Niemann HH, Akiba T, Hirose K, et al. The dynamin A ring complex: molecular organization and nucleotide-dependent conformational changes. *EMBO J* 2002;**21**:240–50.
109. Mim C, Cui H, Gawronski-Salerno JA, Frost A, Lyman E, Voth GA, et al. Structural basis of membrane bending by the N-BAR protein endophilin. *Cell* 2012;**149**:137–45.
110. Frost A, Perera R, Roux A, Spasov K, Destaing O, Egelman EH, et al. Structural basis of membrane invagination by F-BAR domains. *Cell* 2008;**132**:807–17.
111. Yager P, Schoen PE, Davies C, Price R, Singh A. Structure of lipid tubules formed from a polymerizable lecithin. *Biophys J* 1985;**48**:899–906.
112. Selinger JV, MacKintosh FC, Schnur JM. Theory of cylindrical tubules and helical ribbons of chiral lipid membranes. *Phys Rev E Stat Phys Plasmas Fluids Relat Interdiscip Topics* 1996;**53**:3804–18.
113. Sarasij RC, Mayor S, Rao M. Chirality-induced budding: a raft-mediated mechanism for endocytosis and morphology of caveolae? *Biophys J* 2007;**92**:3140–58.
114. Chen YJ, Zhang P, Egelman EH, Hinshaw JE. The stalk region of dynamin drives the constriction of dynamin tubes. *Nat Struct Mol Biol* 2004;**11**:574–5.
115. Pucadyil TJ, Schmid SL. Real-time visualization of dynamin-catalyzed membrane fission and vesicle release. *Cell* 2008;**135**:1263–75.
116. Ferguson SM, Raimondi A, Paradise S, Shen H, Mesaki K, Ferguson A, et al. Coordinated actions of actin and BAR proteins upstream of dynamin at endocytic clathrin-coated pits. *Dev Cell* 2009;**17**:811–22.
117. Ferguson SM, Brasnjo G, Hayashi M, Wolfel M, Collesi C, Giovedi S, et al. A selective activity-dependent requirement for dynamin 1 in synaptic vesicle endocytosis. *Science* 2007;**316**:570–4.
118. Raimondi A, Ferguson SM, Lou X, Armbruster M, Paradise S, Giovedi S, et al. Overlapping role of dynamin isoforms in synaptic vesicle endocytosis. *Neuron* 2011;**70**:1100–14.
119. Durieux AC, Prudhon B, Guicheney P, Bitoun M. Dynamin 2 and human diseases. *J Mol Med* 2010;**88**:339–50.

120. Claeys KG, Zuchner S, Kennerson M, Berciano J, Garcia A, Verhoeven K, et al. Phenotypic spectrum of dynamin 2 mutations in Charcot-Marie-Tooth neuropathy. *Brain* 2009;**132**:1741–52.
121. Sidiropoulos PN, Miehe M, Bock T, Tinelli E, Oertli CI, Kuner R, et al. Dynamin 2 mutations in Charcot-Marie-Tooth neuropathy highlight the importance of clathrin-mediated endocytosis in myelination. *Brain* 2012;**135**:1395–411.
122. Tanabe K, Takei K. Dynamic instability of microtubules requires dynamin 2 and is impaired in a Charcot-Marie-Tooth mutant. *J Cell Biol* 2009;**185**:939–48.
123. Harper CB, Popoff MR, McCluskey A, Robinson PJ, Meunier FA. Targeting membrane trafficking in infection prophylaxis: dynamin inhibitors. *Trends Cell Biol* 2013;**23**:90–101.

CHAPTER SIXTEEN

Multimerization of the Dnmt3a DNA Methyltransferase and Its Functional Implications

Albert Jeltsch, Renata Z. Jurkowska
Institute of Biochemistry, Stuttgart University, Stuttgart, Germany

Contents

Abstract

The Dnmt3a DNA cytosine-C5 methyltransferase has been recently shown to exhibit a complex oligomerization and multimerization potential, the structural basis and functional implications of which will be the subject of this contribution. The enzyme forms a linear heterotetramer with Dnmt3L, in which the interaction of Dnmt3a and 3L stimulates the catalytic activity of Dnmt3a. Isolated Dnmt3a forms protein filaments that bind to several DNA molecules oriented in parallel, which plays an essential role in the location of the enzyme to heterochromatin. Dnmt3L disrupts Dnmt3a protein filaments and leads to a redistribution of the enzyme in cells toward euchromatin. Finally, Dnmt3a complexes and Dnmt3a/3L heterotetramers cooperatively multimerize on DNA forming protein–DNA filaments. This leads to a preference of the enzyme for periodic methylation of DNA and supports its heterochromatic localization.

Progress in Molecular Biology and Translational Science, Volume 117
ISSN 1877-1173
http://dx.doi.org/10.1016/B978-0-12-386931-9.00016-7

1. INTRODUCTION TO Dnmt3 ENZYMES AND DNA METHYLATION

All cells of a multicellular organism are derived from a single cell, the zygote, and consequently they all carry the same genetic information. In spite of this, individual cells follow distinct developmental pathways and differentiate into hundreds of different cells types found in the body of vertebrates, for example. The cellular fate and phenotype are determined by epigenetic regulatory mechanisms that are heritable through cell divisions and function without changing the DNA sequence (for general reviews on molecular epigenetics, cf. Refs. 1–3). Epigenetic signals include DNA methylation, covalent modifications of the histone proteins (such as acetylation, phosphorylation, and methylation of histone tails), incorporation of histone variants, as well as generation of noncoding RNAs; which all function in concert to modulate the structure of chromatin and, thereby, determine the transcriptional activity of the genome. Aberrant DNA methylation patterns are associated with several human diseases, including psychiatric diseases and diseases of the immune system, and contribute to both the initiation and the progression of various cancers.[4–8] In mammals, DNA methylation occurs at the C5 position of the cytosine residues, primarily in the CG dinucleotides (for general reviews on DNA methylation, cf. Refs. 9–12). About 60–80% of the CG sites in the human genome are methylated in a tissue and cell type-specific pattern. DNA methylation is introduced into DNA by a group of enzymes called DNA methyltransferases (MTases) (for general reviews on DNA MTases, cf. Refs. 10,11,13,14). These enzymes catalyze the transfer of the methyl group from the cofactor *S*-adenosyl-L-methionine (AdoMet) to the C5 position of cytosine residues in DNA. In mammals, three active DNA MTases, called Dnmt1, Dnmt3a, and Dnmt3b, and one related protein lacking catalytic activity, called Dnmt3L, are present. The initial DNA methylation pattern is set by the so-called *de novo* MTases (which are enzymes from the Dnmt3 family in mammals). This pattern is perpetuated by maintenance MTases (Dnmt1 in mammals), which specifically methylate hemimethylated CpG sites and thereby reestablish the methylation pattern after DNA replication.

Mammalian DNA MTases comprise two parts: a large multidomain N-terminal part of variable size, which has regulatory functions and a C-terminal catalytic part. The N-terminal part guides the nuclear and subnuclear localization of the enzymes, mediates their interactions with other proteins, DNA, and chromatin, and regulates the catalytic activity. In case of the Dnmt3 family, the N-terminal part contains two defined subdomains, called PWWP domain and ADD domain. For Dnmt3a, the PWWP domain has been shown to interact with the H3 histone tails methylated at K36, which is essential for the heterochromatic localization of the enzyme.[15] The ADD domain of Dnmt3a, 3b, and 3L binds to the N-terminal end of the H3 tail if its K4 residue is not di- or trimethylated[16–18] and this interaction has been shown to stimulate the catalytic activity of Dnmt3a.[18,19] The smaller C-terminal part is conserved between eukaryotic and prokaryotic cytosine-C5 DNA MTases. It harbors the active center of the enzyme and contains 10 amino acids motifs diagnostic for all DNA C5 cytosine MTases.[14,20] The catalytic domains of all DNA MTases share a common core structure, called "AdoMet-dependent MTase fold," which consists of a mixed seven-stranded β-sheet, formed by six parallel β-strands, and a seventh strand in an antiparallel orientation, inserted into the sheet between strands five and six. Six helices surround the central β-sheet.[13,14] The catalytic domain is involved in both the cofactor binding (motifs I and X) and the catalysis (motifs IV, VI, and VIII). The nonconserved region between motifs VIII and IX, the so-called target recognition domain (TRD), is involved in DNA recognition and specificity.

The Dnmt3a DNA MTase has been recently shown to exhibit an unexpected and multifaceted oligomerization and multimerization potential, the structural basis and functional implications of which will be the subject of this contribution.

2. INTERACTION OF Dnmt3a AND 3L LEADS TO STIMULATION OF THE CATALYTIC ACTIVITY

The Dnmt3a and 3b enzymes were discovered in 1999 and shown to have de novo DNA methylation activity *in vitro* and *in vivo*.[21–23] The C-terminal catalytic domain of Dnmt3a has been shown to be active in isolated form.[24] Subsequently, this domain had been used as a model system to investigate the catalytic properties of Dnmt3a in several biochemical studies. Genetic studies demonstrated that both Dnmt3a and Dnmt3b are indispensable for the embryonic development in mice.[23] Mouse Dnmt3b$^{-/-}$ embryos die *in utero* (at embryonic day E9.5) and show multiple developmental defects, whereas the Dnmt3a knock-out animals develop to term, but become runted and die shortly after birth. Substantial demethylation of minor satellite repeats was observed in Dnmt3b$^{-/-}$ embryos and ES cells, but not in Dnmt3a$^{-/-}$

embryos, suggesting that these sequences are specific targets of Dnmt3b *in vivo*.[23]

The first indications of the role of protein oligomerization in the activity of these enzymes came with the identification of the third member of the Dnmt3 family, Dnmt3L, which shows clear homology to the Dnmt3a and 3b enzymes.[25] However, its N-terminal part is shorter than that of Dnmt3a and 3b and its C-terminal part only extends up to conserved motif VIII. Strikingly, Dnmt3L carries amino acid exchanges and deletions within all conserved motifs, which contain the catalytic residues of DNA-(cytosine-C5)-methyltransferases, indicating that Dnmt3L adopts the typical AdoMet-dependent MTase fold described above, but it cannot have catalytic activity and is unable to bind the cofactor AdoMet (Fig. 16.1). Experimental studies confirmed these predictions and showed additionally that Dnmt3L binds only very weakly to DNA.[27,28] Dnmt3L colocalizes with both Dnmt3a and Dnmt3b in mammalian cells.[29] It directly interacts with its C-terminal domain with the catalytic domains of Dnmt3a and 3b and stimulates the activity of both enzymes *in vivo* and *in vitro*.[27,28,30–32] Dnmt3L is expressed during gametogenesis and embryonic stages,[29,33,34] showing a similar

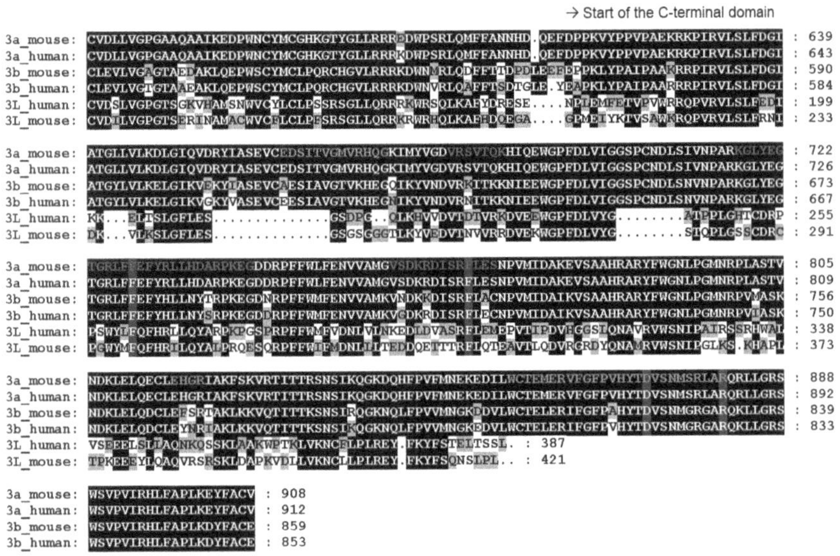

Figure 16.1 Sequence alignment of the C-terminal parts of the human and mouse Dnmt3 enzymes, showing the similarity of the FF and RD interfaces. Residues from the RD and FF interfaces are labeled in orange and red, respectively, the main residues involved in the interaction at both interfaces (F728, F768 at the FF interface, and R881 and D872 at the RD interface) are shaded in blue. *Adapted from Ref. 26.* (See Color Insert.)

expression pattern as the Dnmt3a and 3b enzymes. However, in contrast to the knockout of the other two enzymes, Dnmt3L knock-out mice display a normal phenotype.[29,33,34] Homozygous female mice are fertile, but when crossed with wild-type males are unable to deliver viable pups. Analysis of the DNA methylation pattern showed that the maternal imprint was not properly established in oocytes of Dnmt3L knock-out females.[29,33] Homozygous male knock-out animals are sterile, because of defects in spermatogenesis due to the loss of DNA methylation in spermatogonial stem cells, which lead to male infertility.[29,33,34] While Dnmt3b conditional germ line knock-out animals show no apparent phenotype, the phenotype of a corresponding Dnmt3a knockout is indistinguishable from that of Dnmt3L knock-out mice and is characterized by an altered methylation of imprinted sequences in both male and female germ cells.[29,33–35] These data suggest that Dnmt3L is required for the establishment of DNA methylation during development of the gametes and that it cooperates with Dnmt3a in the generation of imprints in the female germ line. Hence, Dnmt3L interacts with Dnmt3a and 3b, acts as a positive regulator of *de novo* MTases *in vivo* and *in vitro* and this interaction is required for normal development. The interaction of the C-terminal domains of Dnmt3a/3L and Dnmt3b/3L, leading to the stimulation of the enzymes activity is the first example illustrating the important role the oligomerization plays in the function of the Dnmt3 enzymes.

3. Dnmt3a AND 3L FORM A HETEROTETRAMERIC COMPLEX CONTAINING TWO ACTIVE SITES

The structure of the complex of the C-terminal domains of Dnmt3a and Dnmt3L (Dnmt3a/3L) provided the structural basis for the Dnmt3a/3L interaction and revealed an additional and unpredicted dimerization of Dnmt3a.[a;36] It showed that the complex forms a linear heterotetramer, consisting of two Dnmt3L (at the edges of the tetramer) and two Dnmt3a subunits (in the center) (Fig. 16.2). The heterotetrameric state of the Dnmt3a/3L complex was confirmed in solution.[37] In addition, site-directed mutagenesis of interface residues demonstrated that disruption of any of these interfaces leads to the complete loss of catalytic activity,[26,37] indicating that a hypothetical monomeric Dnmt3a would be catalytically inactive.

[a] The following text, "Dnmt3a" or "Dnmt3L" will refer to the C-terminal domains of proteins. Whenever the full length protein is meant, this will be mentioned explicitly.

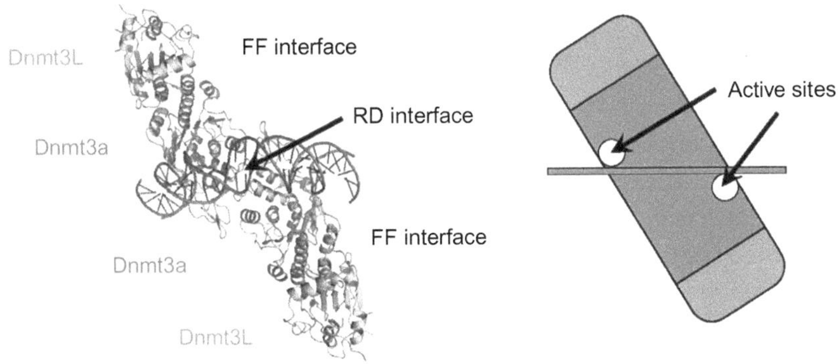

Figure 16.2 Structure of the Dnmt3a/3L heterotetrameric complex and localization of the protein–protein interfaces (adapted from Ref. 36). Dnmt3a is shown in green, Dnmt3L in cyan. The FF and RD interfaces are labeled. The right part shows a schematic picture of the complex. (See Color Insert.)

In the heterotetrameric structure, the Dnmt3a C-terminal domain provides two interfaces for protein/protein contacts: one hydrophobic "FF" Dnmt3a/3L interface (characterized by the stacking interaction of four phenylalanine residues, two from each subunit) and one polar "RD" Dnmt3a/3a interface (characterized by a hydrogen bonding network between arginine and aspartate residues). The interaction of Dnmt3a with Dnmt3L through the FF interface presumably influences the structure of Dnmt3a via the α-helices C, D, and E. Residues from these helices directly interact with the key catalytic or AdoMet binding residues of Dnmt3a, which may explain the stimulatory effect Dnmt3L exerts on Dnmt3a AdoMet binding and catalysis.[36,37]

The RD interface mediates the central Dnmt3a/3a interaction in the Dnmt3a/3L tetramer and arranges the two Dnmt3a subunits, such that one DNA binding cleft is generated (Fig. 16.2).[36,37] In contrast to the FF interface, which is present in Dnmt3a and Dnmt3L, the RD interface is lacking in Dnmt3L (Fig. 16.1). The dimerization of Dnmt3a via the RD interface increases the size of the DNA interface and may compensate for the small TRD of Dnmt3a. It is interesting to note that this arrangement is different in prokaryotic DNA MTases, some of which also dimerize including M.RsrI,[38,39] and M.MboII,[40] which have been structurally characterized. Unlike Dnmt3a, these enzymes form dimers containing two symmetrically related separate DNA binding sites. The presence of two Dnmt3a subunits in the Dnmt3a/3L heterotetramer additionally implies that the complex contains two active sites. Interestingly, although Dnmt3a methylates substrates containing palindromic CpG sites, the arrangement of the two active centers of the two central Dnmt3a subunits does not allow for one complex to methylate one CpG site in both strands directly (Fig. 16.2). In contrast, the active sites are separated by one DNA helical turn, which corresponds to around 10 bp, suggesting that two CpG sites located in this distance in opposite DNA strands could be methylated by the Dnmt3a/3L tetramer simultaneously. Indeed, *in vitro* methylation experiments revealed that there is a correlation of methylation between two sites localized ~10 bp apart in opposite strands,[37,41] clearly demonstrating that the methylation pattern left by the complex directly reflects the structural arrangement of the Dnmt3a dimer. Importantly, an enrichment of CG sites in such distance is observed in the differentially methylated regions (DMRs) of 12 maternally imprinted mouse genes, which are biological substrates of the Dnmt3a/3L complex, suggesting that the favorable CG spacing could make these sequences good substrates for the Dnmt3a/3L complex.[36]

4. Dnmt3a FORMS LONG LINEAR OLIGOMERS AND SHOWS BINDING TO PARALLEL DNA MOLECULES

At the time of discovery of the heterotetrameric Dnmt3a/3L structure, higher levels of oligomerization of Dnmt3a were already known. The tendency of Dnmt3a to self-oligomerize was first reported by Kareta *et al.* who analyzed DNA MTase complexes in solution using size exclusion chromatography and ultracentrifugation, and showed that full length DNMT3A2 forms large structures of heterogeneous sizes.[28] Similar results were later obtained with the catalytic domain of Dnmt3a as well.[26] In addition, it was noticed that the Dnmt3a enzyme tends to aggregate during purification if the protein concentration was too high at the dialysis step when the salt concentration is reduced.[26] Strikingly, the aggregated enzyme preparations could be redissolved after dilution and displayed high catalytic activity, indicating that the aggregation process is reversible and probably occurs through native protein/protein interfaces.[26]

The ability of Dnmt3a to multimerize reversibly could be explained for the first time on the basis of the structure of the Dnmt3a/3L C-terminal domains tetramer, which as mentioned above shows that Dnmt3a has two interfaces for protein/protein interaction, the FF and RD interface. In the structure of the Dnmt3a/3L heterotetramer, the RD interface forms a Dnmt3a/3a contact, while the FF interface forms a Dnmt3a/3L contact. However, biochemical experiments have indicated that the FF interface also supports self-interaction of Dnmt3a, which can be explained by the conservation of key residues of the FF interface between all members of the Dnmt3 family (Fig. 16.1).[26] Hence, the C-terminal domain of Dnmt3a possesses two interfaces for self-interaction, leading to the possibility of forming linear Dnmt3a oligomers, which can explain the experimental result of reversible oligomerization of Dnmt3a full length and C-terminal domain. This finding was further supported by ultracentrifugation studies, which showed that the catalytic domain of Dnmt3a exists as equilibrium of dimers, tetramers, and higher aggregates in solution. Similar experiments with interface mutants of Dnmt3a showed that the FF interface is the primary interface for Dnmt3a homodimerization, because an FF interface mutant was monomeric, while an RD interface mutant retained the ability to form dimers.[26] This finding was in agreement with the structure, because the FF interface is hydrophobic, while the RD interface is hydrophilic in nature, and hydrophobic protein/protein interfaces are known to be more stable.[42,43] Hence, the reversible oligomerization of Dnmt3a-C is mediated by the salt-dependent interaction of Dnmt3a FF dimers via their polar RD interfaces, explaining why Dnmt3a multimer formation is increased at decreasing salt concentration, leading to a reversible precipitation of the protein.

The observation that mixing of separately purified Dnmt3a and Dnmt3L leads to the formation of Dnmt3a/3L heterotetramers[28,36] indicates that the Dnmt3a/3L contact via the FF interface is more stable than the corresponding Dnmt3a/3a contact. Since the Dnmt3a/3L complex shows higher activity than Dnmt3a alone (as described above), the interaction of Dnmt3a and 3L via the FF interface leads to the stimulation of the catalytic activity of Dnmt3a, when compared with a Dnmt3a/3a contact at the FF interface. The existence of Dnmt3a as a stable dimer mediated via the FF interface (and not as monomer) also explains another striking biochemical observation, namely, that separately purified Dnmt3a and Dnmt3L needed a long pre-incubation time after mixing in order to reach quantitative complex formation and maximum stimulation of Dnmt3a.[28] For the formation of Dnmt3a/3L complexes the Dnmt3a homodimer via the FF interface has to first dissociate, which is a slow process.

The formation of Dnmt3a C-terminal domain oligomers via alternating use of the FF and RD interfaces has another interesting consequence. Since each RD interface constitutes a DNA binding site, oligomerization of Dnmt3a creates several potential DNA binding sites (Fig. 16.3). Modeling of a Dnmt3a oligomer predicted that the DNA molecules bound to different DNA binding sites should be oriented next to each other roughly in

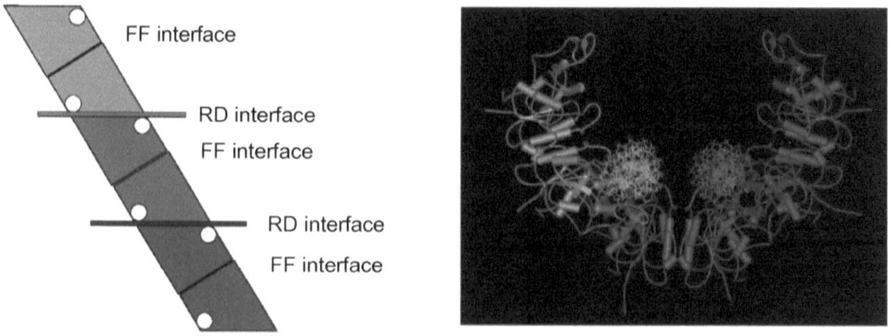

Figure 16.3 Homooligomerization of Dnmt3a and binding to parallel DNA molecules (adapted from Ref. 26). The left part shows a Dnmt3a protein hexamer containing two DNA binding sites in a schematic view. The right image shows a molecular model of the structure suggesting that DNA molecules bound to the two binding sites are roughly oriented in parallel. (See Color Insert.)

parallel.[26] At the same time, it excluded that one Dnmt3a oligomer could bind to DNA in a simple manner, in which the individual DNA binding sites of the Dnmt3a oligomer bind next to each other on the same DNA molecule. This conclusion was confirmed experimentally, by showing that DNA binding of Dnmt3a (which has several DNA binding sites) is not stronger than DNA binding of Dnmt3a/3L heterotetramers, which possess only one DNA binding site (see below), which is in disagreement with the hypothetical multidentate binding mode described above.[26] Instead, the proposed unusual mode of binding of the Dnmt3a oligomers to several DNA molecules oriented in roughly parallel was confirmed experimentally. To study the potential binding of two parallel DNA molecules by the Dnmt3a oligomer in solution, a fluorescence resonance energy transfer (FRET) assay was developed that employed a DNA substrate consisting of two 20 bp double-stranded regions separated by a 10 nucleotide single-stranded region, which is flexible.[26] The ends of the DNA were labeled with fluorophores that constituted a FRET donor/acceptor pair. In the annealed oligonucleotide, the average distance of both ends was too large to permit FRET, but after binding of the two arms of the DNA to adjacent RD interfaces of a Dnmt3a oligomer, the ends approached each other such that FRET was observed, confirming that Dnmt3a oligomers can bind to two DNA molecules oriented in parallel (Fig. 16.4).[26] Binding of long DNA molecules to Dnmt3a oligomers was also studied by scanning force microscopy (SFM), indicating that Dnmt3a and Dnmt3a/3L complexes form DNA-nucleoprotein filaments (Fig. 16.5, see below).[26,37,41] However, with Dnmt3a (in the absence of Dnmt3L) additional structural features, such as

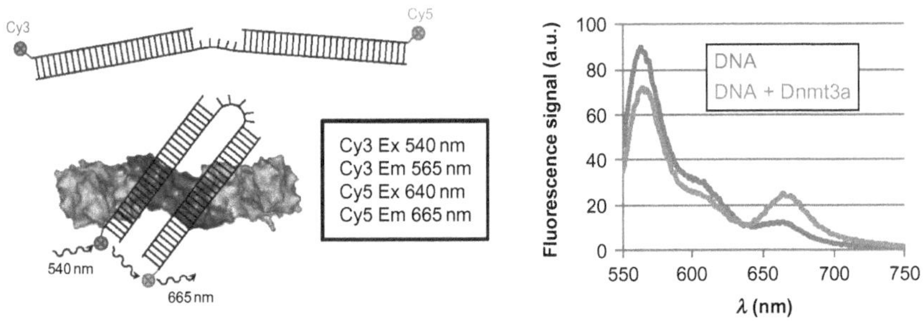

Figure 16.4 Detection of Dnmt3a binding to parallel DNA molecules in solution by a FRET assay, using a DNA substrate that consists of two 20 bp double-stranded regions separated by a 10 nucleotide flexible single-stranded region (adapted from Ref. 26). The ends of the DNA were labeled with Cy3 and Cy5 that constitute an FRET donor/acceptor pair. In the annealed oligonucleotide, no FRET was detected, but after binding of the FRET substrate to a Dnmt3a-C oligomer, the ends approached each other and FRET was observed. (See Color Insert.)

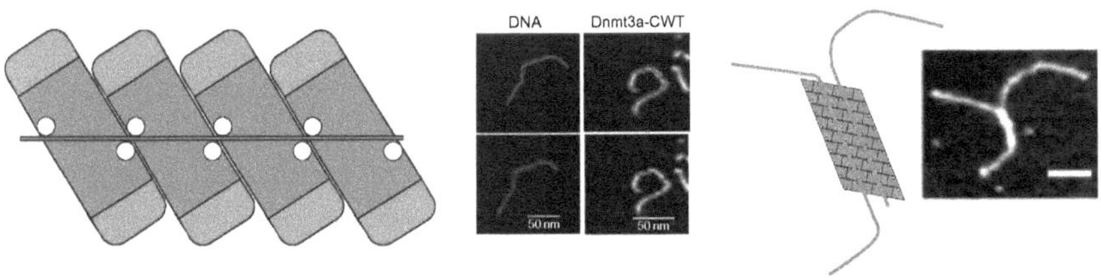

Figure 16.5 Cooperative multimerization of Dnmt3a/3L on DNA. The left image shows a schematic picture of the Dnmt3a/3L–DNA complex together with representative SFM images showing protein–DNA filament formation. On the right, one example of an SFM image showing binding of Dnmt3a fibers to two DNA molecules is given. *Left image: taken from Ref. 41; Right image: taken from Ref. 26.* (See Color Insert.)

sharp curvatures and lariats were also observed, indicating that protein bound DNA stretches from different parts of a long single DNA molecule interacted forming DNA loops. In addition, complexes containing two DNA molecules connected by more or less extended protein–DNA filaments were observed as well (Fig. 16.5), probably formed by a direct interaction between Dnmt3a molecules bound to different DNA molecules oriented in parallel. Hence, direct SFM imaging confirmed the ability of Dnmt3a oligomers to bind simultaneously to two DNA molecules oriented in parallel.

Binding of two DNA molecules in a parallel orientation has been so far observed only with few proteins (one interesting example being RecA[44]) and such DNA binding mode was not anticipated for Dnmt3a. It is likely that this property plays important roles in the cellular function of Dnmt3a, because disruption of Dnmt3a multimerization and its ability to bind to parallel DNA led to strong changes in the subnuclear localization of the enzyme. Dnmt3a has been shown to bind tightly to chromatin and to enrich at heterochromatic spots.[15,26,41,45,46] As described above, its N-terminal part is engaged in two interactions with the H3 tail, mediated by its ADD domain which binds to the end of the H3 tail and reads the methylation state of K4,[17,18] and by its PWWP domain which binds to H3K36me3 and was found to be essential for heterochromatic localization.[15] The observation that heterochromatic localization of Dnmt3a was lost with non-oligomerizing Dnmt3a mutants affected at the FF interface, although both the PWWP and ADD domain were still present in an intact form indicates a central role of Dnmt3a oligomerization in the heterochromatic localization of this enzyme.[26]

To find out, if parallel DNA binding has a direct role in the heterochromatic localization of Dnmt3a, it would be necessary to know the exact molecular structure of heterochromatic DNA. Although this is not available at present, it is clear that heterochromatin contains DNA at high density. Hence, the ability of Dnmt3a oligomers to bind to parallel DNA strands close to each other may contribute to the targeting of the enzyme to such DNA dense regions. One specific model of the 30-nm chromatin fiber, the helical ribbon model,[47] proposes an almost parallel arrangement of the linker DNA regions in a spacing that would fit to the model of the Dnmt3a oligomers binding to parallel DNA molecules. In summary, the unusual DNA binding mode of Dnmt3a likely represents a novel mechanism of heterochromatic targeting of this protein.[26]

Unfortunately, the exact function of the heterochromatic targeting of Dnmt3a for its biological role is currently unknown as well. It may support methylation of these regions (which are enriched in repeats and other highly methylated sequences), which would be in line with the popular model proposing that Dnmt3 and 3b do also contribute to the long-term maintenance of DNA methylation at heterochromatic loci.[10,48] Alternatively, storage of Dnmt3a at heterochromatic sites may help to minimize the risk of aberrant hypermethylation of functionally relevant DNA by restricting the pool of Dnmt3a occurring freely in the cell, because heterochromatic CpG sites are highly methylated anyway and this fraction of the DNA contains only few functional genes.

5. MULTIMERIC COMPLEXES CONTAINING Dnmt3a AND Dnmt3b ARE FORMED AS WELL

Additional studies have provided clear evidence for the formation of mixed heterooligomers of Dnmt3a and 3b as well.[49] FRET experiments with Dnmt3a and 3b coexpressed and carrying different fluorescence tags confirmed the direct interaction of both proteins in cells. DNA methylation experiments revealed a mutual stimulation of both enzymes *in vitro* and also in the methylation of DNA in cells, where both enzymes functioned synergistically to methylate the promoters of the *OCT4* and *NANOG* genes. Based on the conservation of the amino acids sequence of the FF and RD interfaces between Dnmt3a and Dnmt3b (Fig. 16.1), it is very plausible to assume that Dnmt3b/3b and Dnmt3a/3b interactions can be formed via both interfaces, giving the opportunity for the formation of different kinds of heterooligomers. It is currently unknown, whether homo- or heterooligomer formation between Dnmt3a and/or Dnmt3b is preferred through the FF or RD interfaces.

In this context, it is noteworthy that Dnmt3a and Dnmt3b show strong flanking sequence preferences, meaning that CpG sites are methylated with different efficiencies depending on their flanking base pairs.[50–52] The experimental flanking preferences of Dnmt3a and Dnmt3b correlate with the statistical data on the methylation level of CpG sites found in the human genome,[50,53] suggesting that they affect the efficiency of DNA methylation by these enzymes *in vivo*. Since the RD interface forms the DNA binding site of Dnmt3a, the flanking sequence preferences likely are related to processes at the protein/DNA interface in the RD region. Indeed, it was possible to change flanking sequence preferences of Dnmt3a by mutating residues in the RD interface.[54] Based on these observations, it is tempting to speculate that the change from a Dnmt3a/3a interaction to a Dnmt3a/3b interaction at the RD interface will also influence flanking sequence preferences of the corresponding complexes. By this Dnmt3a/3b heteromer formation may fine-tune the selection of target sites modified with a very good or bad efficiency and, therefore, lead to a change in DNA methylation patterns. In this model, homo- and heterooligomerization of Dnmt3a and 3b has a role in the generation of the DNA methylation pattern.

6. Dnmt3L DISRUPTS OLIGOMERIZATION OF Dnmt3a

As mentioned above, both DNMT3A2 full length and Dnmt3a catalytic domain form large structures of heterogeneous sizes in solution and are prone to reversible aggregation and precipitation.[26,28,37] Interestingly, binding of full length DNMT3L to DNMT3A2 led to the resolution of DNMT3A2 multimers and generation of the heterotetrameric structure,[28] and Dnmt3a catalytic domain precipitation was reduced in the presence of the Dnmt3L C-terminal domain.[26] These observations suggest that the oligomerization state of Dnmt3a can be regulated by Dnmt3L. Mechanistically, the influence of Dnmt3L on Dnmt3a oligomer formation can be understood from the amino acid sequence alignment of Dnmt3a and Dnmt3L, which shows that Dnmt3L lacks the RD interface (Fig. 16.1).[26] Therefore, it contains only one interface for the interaction with Dnmt3 enzymes, which explains why it cannot support oligomerization, but instead functions as a cap of linear oligomers, preventing their further growth and restricting them to tetramers (compare Figs. 16.2 and 16.3). This also implies that Dnmt3L would interfere with the simultaneous binding of several DNA molecules to Dnmt3a. Indeed, the binding of Dnmt3a to parallel DNA molecules was lost in the presence of Dnmt3L, as indicated by reversible precipitation, FRET, and SFM experiments.[26] Interestingly, this had a direct effect on the subnuclear localization of Dnmt3a, because its heterochromatic localization was lost in the presence of Dnmt3L.[26] These data indicate that the important biological function of Dnmt3L in the methylation of imprinted regions, as indicated by the knock-out animal studies described above, is not only based on the function of Dnmt3L to stimulate Dnmt3a's activity, but also on its role as a regulator of Dnmt3a multimerization, which indirectly affects the subnuclear localization of the MTase. In this model, the role of Dnmt3L is to release Dnmt3a from dense heterochromatin and make it available for methylation at imprinted loci, which are generally located in euchromatin.[26] Thus, the regulation of Dnmt3a oligomerization appears to play a central role in the biological function of this enzyme.

7. OLIGOMERIZATION OF Dnmt3a AND Dnmt3a/3L COMPLEXES ON DNA

Dnmt3a interacts with DNA, which is a polymeric substrate that provides several binding sites for the DNA MTase. Since both, the heterotetrameric Dnmt3a/3L complex and oligomeric Dnmt3a complexes, bind nonspecifically to DNA, several of them could bind to one DNA molecule. Interestingly, experimental studies showed that binding of several Dnmt3a and Dnmt3a/3L complexes to DNA is not independent, but it occurs in a highly cooperative manner, such that Dnmt3a and Dnmt3a/3L complexes multimerize on the DNA forming large nucleoprotein filaments (Fig. 16.5).[26,36,37]

Multimerization of Dnmt3a along DNA was first observed in electrophoretic mobility shift assays carried out with the catalytic domain of

Dnmt3a. At increasing protein concentrations, protein–DNA complex with very low electrophoretic mobility were observed, the mobility of which corresponded to several Dnmt3a-C complexes being bound to the DNA.[36,37] The cooperativity of the multimerization reaction was indicated by the formation of the multimeric complex without generation of detectable amounts of intermediates. DNA binding by the catalytic domain of Dnmt3a was also investigated in solution using fluorescently labeled double-stranded oligonucleotides. A Hill-analysis of the sigmoidal binding isotherms revealed a cooperativity value of 1.8 for binding to 30mers, indicating that at least two Dnmt3a-C complexes bound to the DNA in a cooperative manner.[55] With longer 60mer substrate, a Hill coefficient of 2.2 was obtained, showing that at least three complexes cooperatively bind to this substrate.[41] In addition, SFM experiments showed the formation of long nucleoprotein filament after incubation of DNA with Dnmt3a/3L heterotetramers or with Dnmt3a oligomers (Fig. 16.5). As described above, with Dnmt3a oligomers, formation of fibers containing two DNA molecules and loops, in which two distal parts of one DNA molecule contact each other, was observed as well.[37,41] These experiments confirmed the multimerization of Dnmt3a complexes on DNA by direct visualization. In addition, DNA methylation studies revealed characteristic patterns of methylation activities along the DNA molecule. Several studies showed that Dnmt3a preferentially methylates CpG sites with a distance of 8–10 bp both on the same and on opposite strands of the DNA.[26,36,37] According to the model of the Dnmt3a/3L heterotetramer in complex with DNA,[36,37] the methylation of two sites in a distance of 8–10 bp in opposite DNA strands could be attributed to the activity of the active sites from the two Dnmt3a subunits that interact via the RD interface. In contrast, the preferential methylation of sites in distances of 8–10 bp located in the same DNA strand indicates that there must be a regular arrangement of Dnmt3a complexes on the DNA, which bind next to each other and thereby generate a periodic methylation pattern. Methylation of both cytosines from one CpG site was observed, indicating that adjacent Dnmt3a (or Dnmt3a/3L) complexes interact with one CpG site and each complex methylates one of the DNA strands (Fig. 16.6).

The cooperative multimerization of proteins on DNA can only happen if a favorable interaction between adjacent protein molecules is present. The putative interface between Dnmt3a complexes bound next to each other on the DNA could be identified by modeling of such complexes, taking into account the periodic methylation of CpG sites in 8–10 bp distance

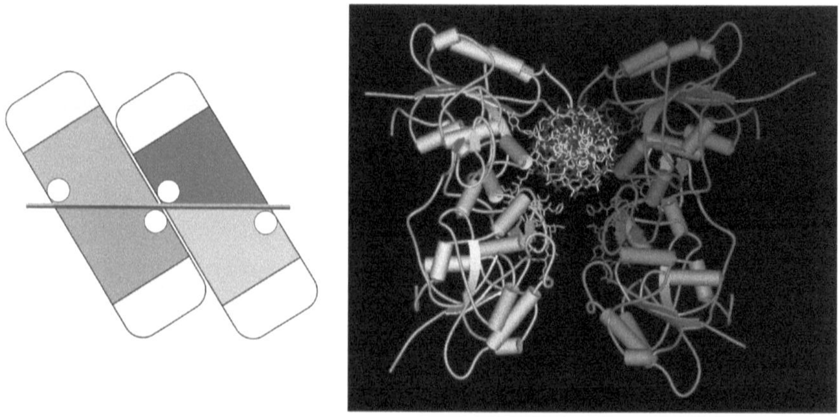

Figure 16.6 Interface of two Dnmt3a/3L complexes bound next to each other on DNA. The contact between two complexes is mediated by two loops originating from the green and red subunits in this illustration, which approach the cytosine from the upper and lower strand of one CpG site (adapted from Ref. 41). The molecular model only shows the inner Dnmt3a molecules of each complex. The DNA is shown in light green and blue for the upper and lower strand, respectively. The green Dnmt3a subunit interacts with the light green DNA strand and methylates the rotated cytosine of the central CpG site, the red subunit interacts with the blue DNA strand and methylates the second cytosine of the CpG site. (See Color Insert.)

(Fig. 16.6).[41] The interface loops are partially disordered in the Dnmt3a-C/3L-C structure,[36] which might be due to the absence of a second enzyme tetramer bound to the DNA. By mutating critical residues in the putative interface and introducing opposite charge, it was possible to disrupt this interaction and prevent multimerization of Dnmt3a complexes on DNA. Noncooperative Dnmt3a variants did not lose DNA binding and retained methylation activity, indicating that cooperative DNA binding and multimerization of Dnmt3a on the DNA is not required for enzyme activity (as opposed to protein oligomer formation using the FF and RD interfaces, which are both needed for activity). However, loss of the cooperative multimerization of Dnmt3a on DNA reduced the heterochromatic localization of the enzyme in NIH3T3 cells, suggesting that it has a role in binding of Dnmt3a to heterochromatic regions.[41] The successful modeling of the interface and its targeted disruption illustrates the good level of understanding we have reached for this important enzyme and its complicated quaternary structure.

Ever since its first observation, the cooperative multimerization of Dnmt3a on the DNA was considered puzzling, because it was difficult to attribute a biological function to this "*in vitro*" property. For example, it appeared questionable whether there is enough naked DNA available in the cell nucleus to allow for the formation of long protein nucleofilaments. The observation that the noncooperative mutants did not lose DNA binding and retained methylation activity indicated that cooperative DNA binding and multimerization of Dnmt3a on the DNA are not required for enzyme activity—a conclusion that further emphasizes the question of the physiological role of this entire process. Two models for the role of cooperative multimerization of Dnmt3a complexes on DNA in cells could be proposed:

i. One possible interpretation is based on the observation that the formation of a defined structured protein complex of Dnmt3a and Dnmt3a/3L on DNA leads to a characteristic methylation pattern of the DNA with peaks in a distance of 8–10 bp.[37] Such patterns have been observed in genome-wide DNA methylation analyses[53,56,57] and a similar periodicity of presentation of CpG sites has been found in imprinted Dnmt3a/3L target sites,[36] suggesting that they may be of importance. Dnmt3a variants which lost cooperative complex formation did not generate such defined patterns.[41] Hence, multimerization of Dnmt3a and stable filament formation is needed for periodic methylation of DNA. One may speculate that cooperative DNA binding and periodic DNA methylation could be possible in linker DNA regions or after chromatin remodeling. However, since the 8–10 bp periodicity coincides with the helical repeat length of DNA, it is still unclear if it contains epigenetic information or just reflects the accessibility of CpG sites when the DNA is bound to nucleosomes.

ii. An alternative model could be based on the finding that Dnmt3a is known to bind very tightly to heterochromatic sites. As described above, the interaction of its PWWP domain with H3K36me3 is essential for this localization,[15] but also the oligomerization of the Dnmt3a protein, together with its ability to bind several DNA molecules oriented in parallel is required.[26] In addition, disruption of the interface of Dnmt3a complexes that is needed for the side by side binding on the DNA, also led to the disruption of heterochromatic targeting of the enzyme. What emerges is a picture of Dnmt3a binding to very DNA dense regions, which have a condensed and regular structure. One may speculate that multimerization of Dnmt3a on DNA stabilizes the binding of the protein to parallel DNA molecules and thereby supports heterochromatic localization.

8. CONCLUSION

Dnmt3a is an interesting example of protein oligomerization in several dimensions. The enzyme's DNA binding site is placed at a subunit interface leading to the positioning of two active sites in a distance suitable for methylation of CpG sites separated by 8–10 bp in the opposite DNA strands. Thereby, characteristic methylation patterns are generated. Dnmt3a further multimerizes and forms linear protein fibers, which contain several DNA binding sites and bind to parallel DNA molecules. The multimerization of Dnmt3a fibers on DNA leads to a two-dimensional meshwork of DNA fibers and protein fibers (Fig. 16.7), which is unique and appears to have an important role for heterochromatic localization of Dnmt3a. The interaction of Dnmt3a with Dnmt3L stimulates the catalytic activity and regulates the oligomerization of Dnmt3a. Reduction of Dnmt3a oligomerization redistributes the enzyme toward euchromatin and stimulates methylation of euchromatic target regions. Formation of mixed oligomers between Dnmt3a and 3b influence catalytic activity of both enzymes and might affect the selection of target sites for methylation.

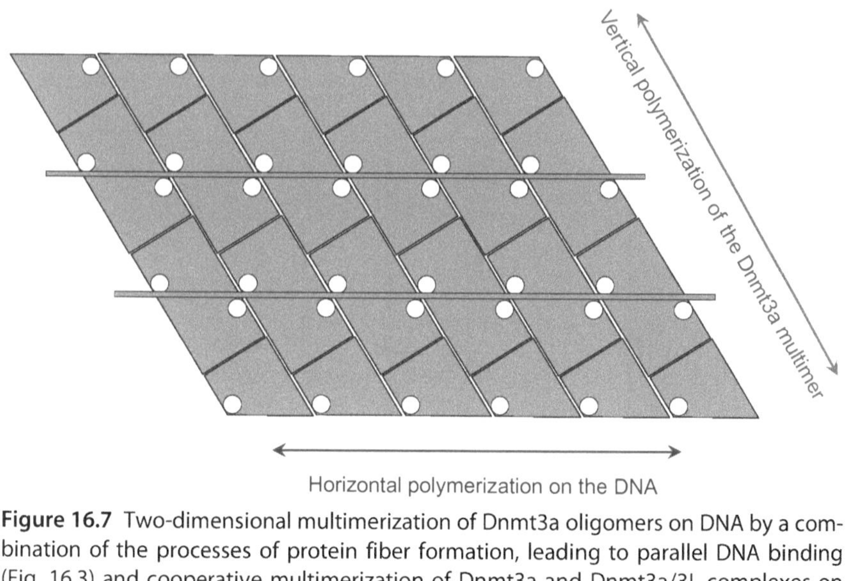

Figure 16.7 Two-dimensional multimerization of Dnmt3a oligomers on DNA by a combination of the processes of protein fiber formation, leading to parallel DNA binding (Fig. 16.3) and cooperative multimerization of Dnmt3a and Dnmt3a/3L complexes on DNA (Fig. 16.5). (For color version of this figure, the reader is referred to the online version of this chapter.)

REFERENCES

1. Bonasio R, Tu S, Reinberg D. Molecular signals of epigenetic states. *Science* 2010;**330**:612–6.
2. Feng S, Jacobsen SE, Reik W. Epigenetic reprogramming in plant and animal development. *Science* 2010;**330**:622–7.
3. Margueron R, Reinberg D. Chromatin structure and the inheritance of epigenetic information. *Nat Rev Genet* 2010;**11**:285–96.
4. Baylin SB, Jones PA. A decade of exploring the cancer epigenome—biological and translational implications. *Nat Rev Cancer* 2011;**11**:726–34.
5. Dawson MA, Kouzarides T. Cancer epigenetics: from mechanism to therapy. *Cell* 2012;**150**:12–27.
6. Feinberg AP. Genome-scale approaches to the epigenetics of common human disease. *Virchows Arch* 2010;**456**:13–21.
7. Kelly TK, De Carvalho DD, Jones PA. Epigenetic modifications as therapeutic targets. *Nat Biotechnol* 2010;**28**:1069–78.
8. Portela A, Esteller M. Epigenetic modifications and human disease. *Nat Biotechnol* 2010;**28**:1057–68.
9. Jones PA. Functions of DNA methylation: islands, start sites, gene bodies and beyond. *Nat Rev Genet* 2012;**13**:484–92.
10. Jurkowska RZ, Jurkowski TP, Jeltsch A. Structure and function of mammalian DNA methyltransferases. *Chembiochem* 2011;**12**:206–22.
11. Klose RJ, Bird AP. Genomic DNA methylation: the mark and its mediators. *Trends Biochem Sci* 2006;**31**:89–97.
12. Suzuki MM, Bird A. DNA methylation landscapes: provocative insights from epigenomics. *Nat Rev Genet* 2008;**9**:465–76.
13. Cheng X, Blumenthal RM. Mammalian DNA methyltransferases: a structural perspective. *Structure* 2008;**16**:341–50.
14. Jeltsch A. Beyond Watson and Crick: DNA methylation and molecular enzymology of DNA methyltransferases. *Chembiochem* 2002;**3**:274–93.
15. Dhayalan A, Rajavelu A, Rathert P, Tamas R, Jurkowska RZ, Ragozin S, et al. The Dnmt3a PWWP domain reads histone 3 lysine 36 trimethylation and guides DNA methylation. *J Biol Chem* 2010;**285**:26114–20.
16. Ooi SK, Qiu C, Bernstein E, Li K, Jia D, Yang Z, et al. DNMT3L connects unmethylated lysine 4 of histone H3 to de novo methylation of DNA. *Nature* 2007;**448**:714–7.
17. Otani J, Nankumo T, Arita K, Inamoto S, Ariyoshi M, Shirakawa M. Structural basis for recognition of H3K4 methylation status by the DNA methyltransferase 3A ATRX-DNMT3-DNMT3L domain. *EMBO Rep* 2009;**10**:1235–41.
18. Zhang Y, Jurkowska R, Soeroes S, Rajavelu A, Dhayalan A, Bock I, et al. Chromatin methylation activity of Dnmt3a and Dnmt3a/3L is guided by interaction of the ADD domain with the histone H3 tail. *Nucleic Acids Res* 2010;**38**:4246–53.
19. Li BZ, Huang Z, Cui QY, Song XH, Du L, Jeltsch A, et al. Histone tails regulate DNA methylation by allosterically activating de novo methyltransferase. *Cell Res* 2011;**21**:1172–81.
20. Cheng X. Structure and function of DNA methyltransferases. *Annu Rev Biophys Biomol Struct* 1995;**24**:293–318.
21. Okano M, Xie S, Li E. Cloning and characterization of a family of novel mammalian DNA (cytosine-5) methyltransferases. *Nat Genet* 1998;**19**:219–20.
22. Gowher H, Jeltsch A. Enzymatic properties of recombinant Dnmt3a DNA methyltransferase from mouse: the enzyme modifies DNA in a non-processive manner and also methylates non-CpG [correction of non-CpA] sites. *J Mol Biol* 2001;**309**:1201–8.

23. Okano M, Bell DW, Haber DA, Li E. DNA methyltransferases Dnmt3a and Dnmt3b are essential for de novo methylation and mammalian development. *Cell* 1999;**99**:247–57.
24. Gowher H, Jeltsch A. Molecular enzymology of the catalytic domains of the Dnmt3a and Dnmt3b DNA methyltransferases. *J Biol Chem* 2002;**277**:20409–14.
25. Aapola U, Kawasaki K, Scott HS, Ollila J, Vihinen M, Heino M, et al. Isolation and initial characterization of a novel zinc finger gene, DNMT3L, on 21q22.3, related to the cytosine-5-methyltransferase 3 gene family. *Genomics* 2000;**65**:293–8.
26. Jurkowska RZ, Rajavelu A, Anspach N, Urbanke C, Jankevicius G, Ragozin S, et al. Oligomerization and binding of the Dnmt3a DNA methyltransferase to parallel DNA molecules: heterochromatic localization and role of Dnmt3L. *J Biol Chem* 2011;**286**:24200–7.
27. Gowher H, Liebert K, Hermann A, Xu G, Jeltsch A. Mechanism of stimulation of catalytic activity of Dnmt3A and Dnmt3B DNA-(cytosine-C5)-methyltransferases by Dnmt3L. *J Biol Chem* 2005;**280**:13341–8.
28. Kareta MS, Botello ZM, Ennis JJ, Chou C, Chedin F. Reconstitution and mechanism of the stimulation of de novo methylation by human DNMT3L. *J Biol Chem* 2006;**281**:25893–902.
29. Hata K, Okano M, Lei H, Li E. Dnmt3L cooperates with the Dnmt3 family of de novo DNA methyltransferases to establish maternal imprints in mice. *Development* 2002;**129**:1983–93.
30. Chedin F, Lieber MR, Hsieh CL. The DNA methyltransferase-like protein DNMT3L stimulates de novo methylation by Dnmt3a. *Proc Natl Acad Sci USA* 2002;**99**:16916–21.
31. Chen ZX, Mann JR, Hsieh CL, Riggs AD, Chedin F. Physical and functional interactions between the human DNMT3L protein and members of the de novo methyltransferase family. *J Cell Biochem* 2005;**95**:902–17.
32. Suetake I, Shinozaki F, Miyagawa J, Takeshima H, Tajima S. DNMT3L stimulates the DNA methylation activity of Dnmt3a and Dnmt3b through a direct interaction. *J Biol Chem* 2004;**279**:27816–23.
33. Bourc'his D, Xu GL, Lin CS, Bollman B, Bestor TH. Dnmt3L and the establishment of maternal genomic imprints. *Science* 2001;**294**:2536–9.
34. Bourc'his D, Bestor TH. Meiotic catastrophe and retrotransposon reactivation in male germ cells lacking Dnmt3L. *Nature* 2004;**431**:96–9.
35. Kaneda M, Okano M, Hata K, Sado T, Tsujimoto N, Li E, et al. Essential role for de novo DNA methyltransferase Dnmt3a in paternal and maternal imprinting. *Nature* 2004;**429**:900–3.
36. Jia D, Jurkowska RZ, Zhang X, Jeltsch A, Cheng X. Structure of Dnmt3a bound to Dnmt3L suggests a model for de novo DNA methylation. *Nature* 2007;**449**:248–51.
37. Jurkowska RZ, Anspach N, Urbanke C, Jia D, Reinhardt R, Nellen W, et al. Formation of nucleoprotein filaments by mammalian DNA methyltransferase Dnmt3a in complex with regulator Dnmt3L. *Nucleic Acids Res* 2008;**36**:6656–63.
38. Scavetta RD, Thomas CB, Walsh MA, Szegedi S, Joachimiak A, Gumport RI, et al. Structure of RsrI methyltransferase, a member of the N6-adenine beta class of DNA methyltransferases. *Nucleic Acids Res* 2000;**28**:3950–61.
39. Kaszubska W, Webb HK, Gumport RI. Purification and characterization of the M.RsrI DNA methyltransferase from Escherichia coli. *Gene* 1992;**118**:5–11.
40. Osipiuk J, Walsh MA, Joachimiak A. Crystal structure of MboIIA methyltransferase. *Nucleic Acids Res* 2003;**31**:5440–8.
41. Rajavelu A, Jurkowska RZ, Fritz J, Jeltsch A. Function and disruption of DNA methyltransferase 3a cooperative DNA binding and nucleoprotein filament formation. *Nucleic Acids Res* 2012;**40**:569–80.
42. Chothia C, Janin J. Principles of protein-protein recognition. *Nature* 1975;**256**:705–8.

43. Bahadur RP, Chakrabarti P, Rodier F, Janin J. Dissecting subunit interfaces in homodimeric proteins. *Proteins* 2003;**53**:708–19.
44. Bell CE. Structure and mechanism of Escherichia coli RecA ATPase. *Mol Microbiol* 2005;**58**:358–66.
45. Ge YZ, Pu MT, Gowher H, Wu HP, Ding JP, Jeltsch A, et al. Chromatin targeting of de novo DNA methyltransferases by the PWWP domain. *J Biol Chem* 2004;**279**:25447–54.
46. Jeong S, Liang G, Sharma S, Lin JC, Choi SH, Han H, et al. Selective anchoring of DNA methyltransferases 3A and 3B to nucleosomes containing methylated DNA. *Mol Cell Biol* 2009;**29**:5366–76.
47. Dorigo B, Schalch T, Kulangara A, Duda S, Schroeder RR, Richmond TJ. Nucleosome arrays reveal the two-start organization of the chromatin fiber. *Science* 2004;**306**:1571–3.
48. Jones PA, Liang G. Rethinking how DNA methylation patterns are maintained. *Nat Rev Genet* 2009;**10**:805–11.
49. Li JY, Pu MT, Hirasawa R, Li BZ, Huang YN, Zeng R, et al. Synergistic function of DNA methyltransferases Dnmt3a and Dnmt3b in the methylation of Oct4 and Nanog. *Mol Cell Biol* 2007;**27**:8748–59.
50. Handa V, Jeltsch A. Profound flanking sequence preference of Dnmt3a and Dnmt3b mammalian DNA methyltransferases shape the human epigenome. *J Mol Biol* 2005;**348**:1103–12.
51. Jurkowska RZ, Siddique AN, Jurkowski TP, Jeltsch A. Approaches to enzyme and substrate design of the murine Dnmt3a DNA methyltransferase. *Chembiochem* 2011;**12**:1589–94.
52. Lin IG, Han L, Taghva A, O'Brien LE, Hsieh CL. Murine de novo methyltransferase Dnmt3a demonstrates strand asymmetry and site preference in the methylation of DNA in vitro. *Mol Cell Biol* 2002;**22**:704–23.
53. Zhang Y, Rohde C, Tierling S, Jurkowski TP, Bock C, Santacruz D, et al. DNA methylation analysis of chromosome 21 gene promoters at single base pair and single allele resolution. *PLoS Genet* 2009;**5**:e1000438.
54. Gowher H, Loutchanwoot P, Vorobjeva O, Handa V, Jurkowska RZ, Jurkowski TP, et al. Mutational analysis of the catalytic domain of the murine Dnmt3a DNA-(cytosine C5)-methyltransferase. *J Mol Biol* 2006;**357**:928–41.
55. Maltseva DV, Baykov AA, Jeltsch A, Gromova ES. Impact of 7,8-dihydro-8-oxoguanine on methylation of the CpG site by Dnmt3a. *Biochemistry* 2009;**48**:1361–8.
56. Zhang X, Yazaki J, Sundaresan A, Cokus S, Chan SW, Chen H, et al. Genome-wide high-resolution mapping and functional analysis of DNA methylation in arabidopsis. *Cell* 2006;**126**:1189–201.
57. Chodavarapu RK, Feng S, Bernatavichute YV, Chen PY, Stroud H, Yu Y, et al. Relationship between nucleosome positioning and DNA methylation. *Nature* 2010;**466**:388–92.

CHAPTER SEVENTEEN

Oligomerization in Endoplasmic Reticulum Stress Signaling

Saïd Taouji[1], Sebastian Wolf[1], Eric Chevet
Inserm U1053, Université Bordeaux Segalen, Bordeaux, France
[1]Equally contributed to this work.

Contents

Abstract

Oligomerization of protein complexes has been involved in various mechanisms that play a major role in endoplasmic reticulum (ER) functions. In particular, in response to the accumulation of misfolded proteins in the ER, an adaptive response is activated and named the unfolded protein response (UPR). To facilitate recovery of ER homeostasis, both the inositol-requiring enzyme-1 (IRE1) and the protein kinase RNA-like ER kinase, two transmembrane ER stress transducers, oligomerize and activate UPR-specific transcription factors to adjust the folding and productive capacity of the ER, to direct misfolded proteins to ER-associated degradation or autophagy. Recent advances in the molecular characterization of how ER protein sensors transduce signals to orchestrate the adaptive cellular response have greatly unlocked the development of tools to dissect their functions in health and disease. Here, we focus on the advances concerning oligomerization of ER stress transducers and, in particular IRE1, describe the oligomerization-dependent mechanisms for modulating UPR signals on and off.

Progress in Molecular Biology and Translational Science, Volume 117
ISSN 1877-1173
http://dx.doi.org/10.1016/B978-0-12-386931-9.00017-9

ABBREVIATIONS

ATF activating transcription factor
CHOP CCAAT/enhancer-binding protein homologous protein (also-called GADD153)
ER endoplasmic reticulum
ERAD endoplasmic reticulum-associated degradation
GRP78 glucose-regulated protein of molecular weight 78 (also-called BiP)
IRE1 inositol-requiring enzyme-1
JNK c-jun N-terminal kinase
PERK protein kinase activated by double-stranded RNA (PKR)-like ER kinase (also-called pancreatic ER kinase)

1. INTRODUCTION

The endoplasmic reticulum (ER) is a large membranous organelle in eukaryotic cells. The rough ER is the main site for the synthesis of secretory and transmembrane proteins, whereas the smooth ER is involved in lipid biogenesis. The ER is a major site for calcium storage. Most secretory and transmembrane proteins fold and mature in the lumen of this intracellular compartment.[1] Proteins enter the ER lumen as nascent unfolded chains through a translocation channel named translocon. These nascent chains have to be folded to reach a correct conformation, mechanism that is assisted and monitored by resident chaperone, foldase, and cofactor proteins. Polypeptide-binding proteins act to slow protein-folding reactions and prevent aberrant aggregation. A large family of disulfide isomerases is also required to ensure proper disulfide bond formation and to prevent the formation of illegitimate disulfide bonds. The ER is the primary Ca^{2+} storage organelle in the cell (1–5 mM) and is therefore involved in many Ca^{2+} signaling pathways[2,3] including ER stress-dependent activation of transcriptional processes, Ca^{2+} signaling and communication to the plasma membrane, Ca^{2+} channels, and endoplasmic reticulum-associated degradation (ERAD). Changes in the ER luminal Ca^{2+} concentration modulate the quality and efficiency of protein folding.[4] The major ER chaperone BiP/GRP78 (binding immunoglobulin protein/glucose-regulated 78 kDa protein) as well as many ER-resident calcium-binding proteins (Calreticulin, Calnexin, etc.) play an important role in the storage of ER Ca^{2+} and key roles in the unfolded protein response (UPR). Mechanisms activated upon perturbation of the cell such as oxidative and inflammatory stresses can also perturb ER homeostasis by affecting ER calcium fluxes.[5]

Correct folding is essential for a protein to be exported from the ER to the Golgi complex and to later compartments. This transport step is conditioned by the supply in ATP to ensure chaperone functions and by the homeostasis of Ca^{2+} stores and of redox potentials. Furthermore, proteins are subjected to posttranslational modifications including N-linked glycosylation. All these processes are highly sensitive to alterations of the homeostasis of the ER.[6]

2. ENDOPLASMIC RETICULUM STRESS SIGNALING: THE UNFOLDED PROTEIN RESPONSE

2.1. General overview

The folding of secretory and transmembrane proteins is fine-tuned by different mechanisms in the ER.[7] The adjustment of folding capacity is required to maintain a high fidelity of the quality of the secreted proteins. Proteins that are misfolded are directed to degradation via a process named ERAD. In such situation, misfolded proteins are retrotranslocated in the cytosol, ubiquitinated, and further degraded by the 26S proteasome.

An imbalance in ER homeostasis provokes a strong adaptive response to restore the homeostasis of this compartment known as UPR.[8] The role of this response has been described in several physiological and pathophysiological states including diabetes,[9,10] inflammation,[11] immunity,[12] neurodegenerative diseases,[13] and cancer.[14] Emerging evidence shows that the tumor microenvironment, in particular, hypoxia and nutrient limitation, can also lead to the perturbation of ER functions. The initial phase of the response is an adaptive component to protect cells. In contrast, sustained UPR activation has been shown to initiate cell death programs through apoptosis-dependent pathways.[6]

UPR signaling has been described in great detail in different reviews.[15–18] It composes a complex signal transduction pathway that is designed to allow the cells to recover from a challenging folding load by attenuating translation and upregulating the expression of chaperone proteins and degradation factors to refold or clear misfolded proteins. If this adaptive program fails to restore ER homeostasis, ER stress induces cell death. Moreover, several studies have shown that the UPR is specifically induced to trigger the activation of autophagy as both adaptive and death mechanisms.[19,20]

The UPR in metazoans is governed by three transmembrane ER stress sensors, namely the inositol-requiring enzyme-1 (IRE1), the protein kinase RNA (PKR)-like ER kinase (PERK), and the activating transcription

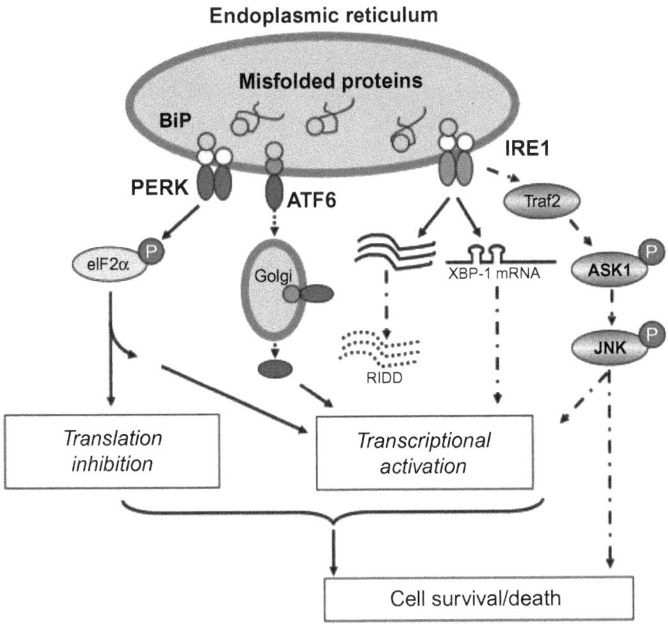

Figure 17.1 The unfolded protein response (UPR). Accumulation of misfolded proteins leads to the activation of the ER stress transducers. PERK and IRE1 activation are dependent on binding immunoglobulin protein (BiP). Activated IRE1 controls three pathways. First the regulated IRE1-dependent mRNA decay (RIDD), second the splicing of *Xbp1* mRNA, and third the activation of JNK (c-jun-N-terminal kinase) pathway through TRAF2 and ASK1. Activated PERK phosphorylates eIF2α, which decreases the general protein synthesis and activates UPR target genes. ATF6 is exported to the Golgi where it is processed and its cytosolic domain is released. It acts as a transcription factor for different UPR target genes. (For color version of this figure, the reader is referred to the online version of this chapter.)

factor-6 (ATF6; Fig. 17.1). As long as binding to the ER chaperone BiP is not altered, these proteins are maintained in an inactive state whereas the accumulation of unfolded or misfolded proteins leads to the activation of the three factors. The signal transduction pathway mediated by these three ER sensors activates a genetic program that regulates the expression of a variety of genes involved in specific secretory pathway-related processes. Those include protein entry to the ER, folding, N-linked glycosylation, redox metabolism, protein quality control, ERAD, and vesicular trafficking.[18]

2.2. ATF6

ATF6 is a type 2 membrane-bound transcription factor. It is composed of a luminal and a cytosolic domain. The activation of ATF6 under ER stress is dependent on BiP (also known as GRP78). Under basal conditions, BiP is bound to the luminal domain of ATF6. Upon activation of UPR, ATF6 is released and displays a Golgi localization sequence.[21] ATF6α disulfide bounds are also remodeled upon ER stress, which allows the molecule to be translocated to the Golgi apparatus.[22] In the latter compartment, it is cleaved by the Site-1 and Site-2 protease (S1P and S2P).[23] ATF6 cytosolic domain ATF6c is then translocated into the nucleus and activates the transcription of UPR target genes via direct binding to the ER stress-response element or acts in collaboration with the general transcription factor nuclear factor-Y.[24] ATF6 target genes mainly encode ER stress transcription factors such as X-box-binding protein 1 (XBP1) or ER chaperones and folding enzymes such as BiP. It is thought that since ATF4 is also a target of ATF6, the overexpression of ATF6 can enhance CHOP (CCAAT/enhancer-binding protein homolgous protein) level, similar to PERK signaling and lead to apoptosis.

2.3. PERK

PERK is a type 1 transmembrane protein with a luminal stress-sensing domain and a cytosolic kinase domain that undergoes trans-autophosphorylation by oligomerization upon ER stress. Activated PERK phosphorylates the eukaryotic translation initiation factor-2 (eIFα), which in turn results in global translation attenuation.[25] PERK activation and eIF2α phosphorylation lead to decreasing the load of total protein entering in the ER lumen and thereby to reducing the misfolding burden in this compartment. A gene regulatory function of PERK has also been described. Upon stress, PERK can phosphorylate the transcription factor nuclear factor (erythroid-derived 2)-like 2 and lead to a modulation of its activity.[26] It controls mainly the regulation of oxidative stress responses[26] but it has also been shown that most of the pathways activated downstream of PERK are eIF2α phosphorylation dependent. The translational activation of ATF4 through the bypass of an inhibitory micro-open reading frame located upstream of ATF4 open reading frame[27] can then be achieved. High levels of eIF2α may also activate nuclear factor kappa B signaling, but the mechanisms involved are still controversially discussed.

Gene expression profiling revealed that genes encoding proteins involved in amino acid biosynthesis and transport functions, antioxidative stress responses, and apoptosis, such as growth arrest and DNA damage 34 (GADD34) and CHOP/GADD153 require PERK, eIF2α phosphorylation, and ATF4.[6] CHOP has a proapoptotic activity with an unclear mechanism of action. It may trigger cell death by downregulation of B-cell-lymphoma-2 and upregulation of BIM, p53-upregulated modulator of apoptosis, and GADD34.[15–17] Many other regulators of apoptosis have been reported to be upregulated upon prolonged ER stress. But there was no threshold or switch found, which decides between survival and death signaling.

The third UPR sensor has been identified as the IRE1, which is the most conserved across evolution and which is fully described below.

3. IRE1 STRUCTURE, SIGNALING, AND INTERACTING PARTNERS

3.1. IRE1 structure and activation in the yeast *Saccharomyces cerevisiae*

In yeast, only one stress sensor mediates the UPR, which is IRE1. Ron and Walter[18] called this pathway the conserved core of the UPR. IRE1 is a type 1 transmembrane protein with three functional domains: a luminal domain acting as the stress sensor, cytosolic kinase, and endoribonuclease domains.[28]

In yeast undergoing ER stress, the dimerization of the core luminal domains (cLD) is able to build a groove that is reminiscent of peptide-binding domains of major histocompatibility complexes (MHCs) in its gross architecture.[29] By this, it is able to directly monitor/bind to unfolded or misfolded proteins.[30] The study furthermore found two contact regions on the opposing sides of the cLD. Mutations of both domains impaired IRE1 activation, thus suggesting that high order linear oligomers are also important for IRE1 activation.

A secondary step of regulation includes the chaperone BiP/Kar2. This chaperone senses the amount of unfolded proteins in the ER lumen leading to its dissociation from IRE1 luminal domain. The dissociation of BiP/Kar2 from the luminal domain of IRE1 enables its activation. BiP acts as a negative regulator of IRE1 activation in yeast.[31] Oligomerization of IRE1 kinase domain results in its trans-autophosphorylation and activation of the endoribonuclease domain that creates an interaction surface for binding the mRNA substrate (Fig. 17.2). It could not be clearly found, if a dimerization of endoribonuclease domains is necessary for its activity. But the appearance of high order homo-oligomers is essential for the endoribonuclease activity.[32]

The kinase domain of IRE1 is not activated by phosphorylation *per se*. It was shown that the autophosphorylation of the kinase domain is needed to induce a conformational change in the molecule. Otherwise, the

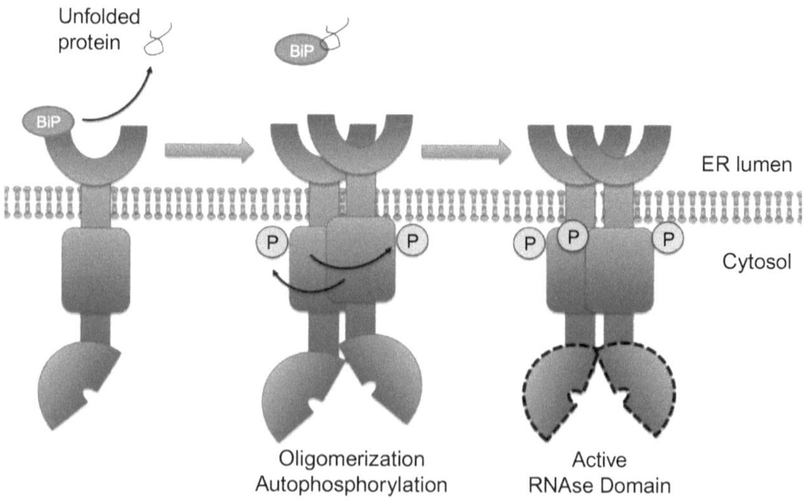

Figure 17.2 IRE1α activation in mammals. IRE1α is bound to BiP under basal conditions. BiP is released under ER stress and IRE1α oligomerizes and autophosphorylates. By a conformal change, the endoribonuclease/RNAse domain of IRE1α is activated. (For color version of this figure, the reader is referred to the online version of this chapter.)

occupancy of the active side with a ligand seems sufficient for RNase activation and could therefore bypass the autophosphorylation by an ATP-competitive drug.[33] ADP was though shown to be a better activator of the kinase domain than ATP. ADP binding changes the conformation of the molecule to an endoribonuclease active form. This might suggest that IRE1 may contribute to a mechanism to monitor the energy balance of the cell. In yeast, it could lately be shown that dephosphorylation of IRE1 in certain residues is required for attenuating IRE1-dependent signals. Furthermore, it was found that sites on IRE1 could contribute to constant activation of the endoribonuclease domain by a mimic conformation of autophosphorylation. This study demonstrated that the kinase domain of IRE1 in yeast is responsible for both activating and attenuating IRE1 endoribonuclease activity.[34]

The most important function of the endoribonuclease domain in yeast is the splicing of Hac1 mRNA.[35] The basic leucine zipper transcription factor HAC1 is required for the UPR and binds to promoters of UPR target genes. The splicing of Hac1 mRNA relieves the translational attenuation mediated by a 3′UTR located structure. The spliced parts are relegated by a tRNA ligase[36] and translation starts. Hac1 mRNA is recruited to IRE1 just by its 3′UTR element, which is sufficient to target other mRNAs to IRE1 foci.[37]

3.2. IRE1 structure and activation in human

Two mammalian isoforms of IRE1 been identified. Whereas IRE1α is ubiquitously expressed, IRE1β primary restricted to intestinal epithelial cells. Human IRE1α shows a similar organization as the yeast molecule. The most important differences are to be found in the luminal domain and its activation. The work of Zhou *et al.*[38] showed that the MHC-like grooves in the luminal domain are different from the ones in yeast IRE1. They are too narrow to bind peptides. In addition, another report showed that the mutation of specific BiP-binding sites significantly activated IRE1α even under normal conditions.[39] The luminal domain of IRE1α is still able to dimerize, and the disruption of specific sides let the molecule fail dimerization and activates the UPR upon ER stress.

Autophosphorylation by the cytoplasmic kinase domain of IREα appears to be crucial to activate its endoribonuclease domain.[40] Additional features from yeast IRE1 cytoplasmic domain crystal show that how dimerization, trans-autophosphorylation, and nucleotide binding were required for IRE1 nuclease activity.[41] The nuclease activity of IRE1α cleaves XBP1 mRNA to remove a 26-nucleotides long intron.[34,42] The cleaved fragments are relegated by a yet unknown ligase although the human tRNA ligase pentameric complex containing HSPC117 could be involved in such mechanism.[43] The XBP1 protein is an important transcription factor for the UPR and regulates the expression many UPR target genes.

In the course of its activation process, IRE1 in mammals has been shown as observed in yeast to be able to form high order oligomers. This was illustrated for human IRE1 but is anticipated in other mammalian species.[44] It was shown that oligomerization and trans-autophosphorylation increased directly the endoribonuclease activity of IRE1α (Fig. 17.2). Moreover, a model of yeast Ire1p activation was proposed following structural study in which Ire1p nuclease activity is controlled via oligomerization Ire1p shift from dimer to polymerization[32] (Fig. 17.3). In a similar manner and as observed in yeast cells, a bypass of autophosphorylation could be used to activate the nuclease. Kinase-inhibited IRE1α showed another activation profile and lead rapidly to apoptosis-related signals.[45] The conformation of the kinase influences directly the nuclease and IRE1α-dependent signaling. The activation profile of the monomers of IRE1α should in theory have a high impact on the formation of oligomers or the activation of the kinase domain.[45]

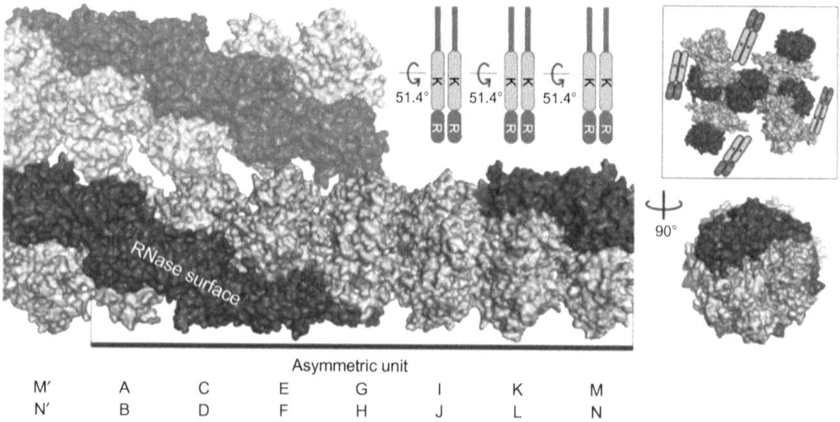

Figure 17.3 High order organization of Ire1p cytosolic domain. Assembly of Ire1KR32⊗28 mutant with the kinase inhibitor APY29. A parallel filament in the crystal packing is shown above the main filament. The kinase domain of Ire1 is colored light brown and the RNase domain is colored purple. The inset shows crystal packing of Ire1 dimers (PDB ID 2RIO). Letters A–N below the structure refers to individual monomers in the asymmetric unit. K, kinase; R, ribonuclease. *Figure from Ref. 32.* (For color version of this figure, the reader is referred to the online version of this chapter.)

4. THE UPROSOME

Beyond the formation of high-order IRE1 oligomers upon ER stress, the formation of hetero-oligomers around IRE1 that could control IRE1 signaling outcomes has been described in many instances and named the UPRosome[15–17] (Fig. 17.4).

4.1. Upstream components influencing the activity of IRE1α

4.1.1 BiP, a master regulator

Binding immunoglobulin protein (BiP) is an Hsp70 chaperone located in the lumen of the ER. The main function of BiP is to enforce protein folding. As described earlier, BiP is bound to IRE1α under basal conditions.[46] Under ER stress, unfolded proteins in the lumen of the ER dramatically increase. After nascent chains enter the lumen, they are directly bound to BiP and the folding process is started. At high levels of unfolded proteins, BiP is tending away from IRE1. The equilibrium of the competitive binding between IRE1α and unfolded proteins lies on the side of the proteins. After BiP is

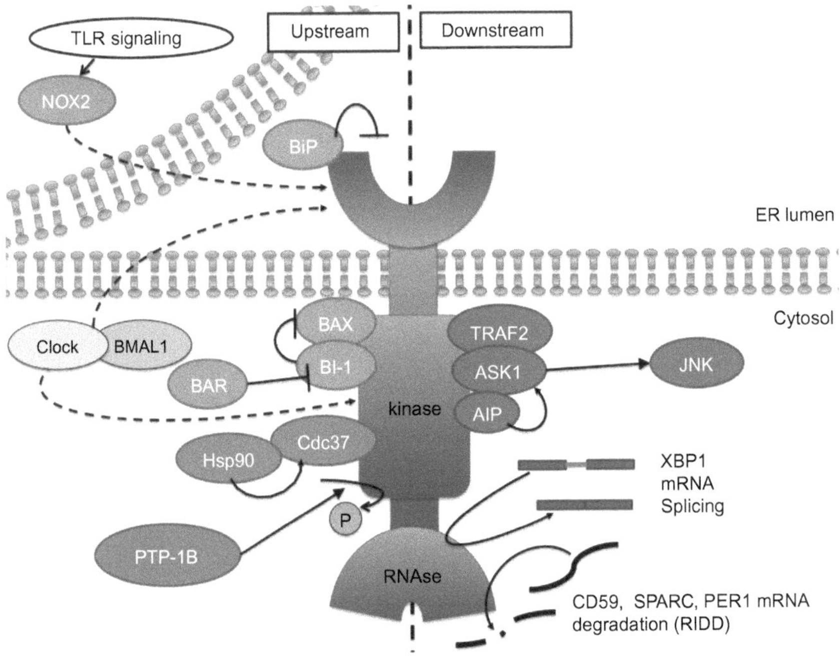

Figure 17.4 IRE1α interaction partners. IRE1α activity is upstream regulated by BiP in the ER lumen. If BiP is bind to the luminal domain IRE1α activity is inhibited. BiP is released from the luminal domain on a high load of unfolded proteins in the ER lumen. TLR signaling and the circadian clock are able to activate IRE1α in specific cell types, while the domain of activation remains unclear. BAX enhances the activity of IRE1α, but is itself inhibited by BI-1, which also binds directly to the IRE1α kinase domain. BAR is able to inhibit BI-1 function and restore BAX mediated IRE1α activation. PTP-1B phosphatase dephosphorylates the kinase domain and is essential for a conformal change that activates kinase and endoribonuclease functions. Hsp90 stabilizes Cdc37 binding to IRE1α. This directly inhibits IRE1α autophosphorylation and oligomerization. IRE1α downstream activities are all located in the cytosol. TRAF2 is binding IRE1α to recruit ASK1 kinase. ASK1 must be activated by AIP to mediate a signal to JNK activation. The endoribonuclease domain of IRE1α has two functions. First and most known is the splicing of Xbp1 mRNA to a more stabile form that is translated to a strong transcription factor that can induce the transcription of UPR target genes. Furthermore, the RNAse domain is able to cleave a subset of mRNAs, which are degraded after. This reaction is called-RIDD. (For color version of this figure, the reader is referred to the online version of this chapter.)

released from IRE1α, dimerization and activation of the downstream signaling follow. The dissociation of BiP from IRE1α is reversible. On reestablished homeostasis in the ER lumen, it can be bound to IRE1α again and the activation is attenuated. It has to be said that the expression of BiP is highly upregulated upon ER stress by all three arms of the UPR.

A high level of BiP in the ER lumen decreases the level of unfolded protein and could lead to a stop of the UPR (Fig. 17.4).

4.1.2 TLR signaling leads to IRE1α activation in macrophages

Martinon *et al.*[47] showed that IRE1α is not only activated upon ER stress. Toll-like receptor (TLR) 2 and 4 specifically activate IRE1α-dependent splicing of Xbp1 mRNA. The reaction was independent of ER stress; moreover, the activation of TLR 2 and 4 inhibited ER stress responses like increased levels of BiP or CHOP. The result of this activation was a high secretion of inflammatory molecules like interleukin 6 (IL6) or tumor necrosis factor (TNF). It was found that XBP1 directly interacts with the IL6 and TNF promoters as a coactivating factor. The authors were able to show that the response is dependent on TNF receptor associated factor 6 (TRAF6), NADPH oxidase 2, and reactive oxygen species presence. Because both pathways, TLR and IRE1α, are known to trigger inflammatory signals through c-jun N-terminale kinasen (JNK), it is interesting to find out strategies that these pathways could have in common. Furthermore, TLR signaling can downregulate ATF4 signaling and by this stop ATF4 to promote CHOP expression. Even low dose of TLR ligands were able to stop CHOP levels and diminish apoptosis of macrophages upon ER stress.[48] The question remains whether pharmacological manipulations in the ER lumen could influence the inflammatory response triggered by TLR activation in macrophages.[12]

4.1.3 PTP-1B potentiates IRE1α signaling upon ER stress

The functions of Protein-tyrosine-phosphatase-1B (PTP-1B) are well established in terms of insulin signaling and metabolism. The molecule is ubiquitously expressed and localized at the ER. Gu *et al.*[49] showed the essential role of PTP-1B in IRE1α-mediated ER stress signaling in mouse embryonic fibroblasts (MEFs). PTP-1B knockout MEFs showed decreased IRE1α kinase and RNase signaling under ER stress. Because there was no PTP-1B influence on BiP and CHOP expression, it was concluded that PTP-1B might directly dephosphorylate IRE1α kinase domain and change the conformation of the molecule. This conformational change is needed to transduce ER stress by IRE1α. Furthermore, the study could show that PTP-1B knockout cells are less sensitive to ER stress-induced apoptosis. Direct or indirect dephosphorylation of IRE1α kinase domain by PTP-1B thus is an important component of the UPR. Further studies showed the role of PTP-1B in mouse livers. PTB-1B was upregulated upon ER stress.

Silencing of PTP-1B changed ER stress responses upon drug treatment or high fat diets.[50] The study pointed out that the UPR and PTP-1B expressions are interlinked (Fig. 17.4).

4.1.4 BAX and BAK positively regulate IRE1α activation

Bcl-2-associated X protein (BAX) and Bcl-2 homologous antagonist/killer (BAK) are both known as essential gateways to cell death via mitochondria. Both molecules interact and signal in concert *in vivo* to modulate IRE1α/XBP1 signaling. Both were found as positive mediators of ER stress.[51] In double knockout cells for BAK/BAX, there was no activation of the IRE1α kinase found after initialization of ER stress. BAX and BAX could be shown to directly bind to IRE1α and interact with the luminal-activated form of IRE1α. The interaction alters the conformation of the whole molecule, because BiP binding was stronger in double knockout cells for BAK/BAX, still both interact with the cytosolic region of the molecule.

BAX Inhibitor-1 (BI-1) is well known as a negative regulator of apoptotic signaling. BI-1 knockout cells showed that Xbp1 mRNA splicing even with low levels of ER stress and overexpression of the molecule down-regulate IRE1α-dependent signals.[52] It was shown that BI-1 acts in two ways to influence apoptosis. It acts in early manners on ER stress at later directly on apoptosis inducer BAX. The molecule directly interacts with IRE1α. It may compete with BAX for a common binding site and so diminish IRE1α activation through BAK/BAX.

The regulation of IRE1α through BI-1 was shown to decrease IRE1α-dependent autophagy signaling by inhibiting the binding of TRAF2. Inhibition of BI-1-dependent autophagy inhibition was able to increase cell survival.[53] Bifunctional apoptosis regulator (BAR) functions as an E3-ligase and promotes proteasomal degradation of BI-1. The molecule seems to have an important function in inhibiting BI-1 function to keep ER stress signaling possible. After prolonged ER stress, BAR is down-regulated, which enables BI-1 to inhibit IRE1α signaling. BAR directly colocalizes with BI-1 and its overexpression decreases BI-1 level significantly. It is a dynamic regulator of IRE1α-dependent UPR response. This chain represents an important upstream influence on IRE1α (Fig. 17.4).

4.1.5 Hsp90 and Cdc37 regulate the kinase domain of IRE1α

The heat shock protein 90 (Hsp90) was known to regulate IRE1α activity in the cytosolic domain.[54] The work of Ota *et al.*[55] showed that the influence is mediated by cell division cycle homolog 37 (Cdc37). Cdc37 is a cochaperone of Hsp90. It directly binds to the kinase domain of IRE1α. The binding of Cdc37 inhibits IRE1α signaling after ER stress. Hsp90 stabilizes the formation of an IRE1–Cdc37 complex. In the absence of Hsp90, Cdc37 dissociates from IRE1α and IRE1α can autophosphorylate and oligomerize. It could be shown that a knockdown of Cdc37 increased XBP1 mRNA splicing, but it had no effect on the IRE1α protein stability. While normally Hsp90 is regulating the stability of many cytosolic kinases with cochaperone Cdc37.

4.1.6 Hsp72 enhances IRE1α/XBP1 signaling

There is another heat shock protein that can build a direct complex together with IRE1α. Hsp72 is a cytosolic chaperon that bound to the cytosolic domain of IRE1α in Rat pheochromocytoma cells. Hsp72 enhances Xbp1 mRNA splicing upon ER stress through this complex. On the other hand, the binding of Hsp72 to IRE1α reduced ER stress-related caspase activation. Consequently, ER stress-induced apoptosis was reduced through Hsp72.[56] As shown multiple inhibitors and activators are bound to IRE1α. Hetero-oligomerization of IRE1α with other proteins can directly influence its formation, status of phosphorylation, and so its downstream signals.

4.2. IRE1α downstream interaction partners and signals

4.2.1 JNK pathway activation through TRAF2/ASK1

The kinase domain of IRE1α is able to initialize a pathway, on which end stands the activation of the JNK. This kinase is known to activate the transcription factors c-jun and ATF2. JNK is activated is a downstream effects of many stress pathway in cells.[57] A first report described that IRE1α directly binds to TRAF2[58] but the binding domain of both proteins is still unclear. Different mutants of IRE1α were not able to bind TRAF2 anymore[58,59] and silencing TRAF2 expression impaired ER stress-dependent activation of JNK. This binding was shown to be ER stress dependent (Fig. 17.4).

The mitogen-activated protein kinase (MAP3K) found to mediate this pathway is Apoptosis signal-regulating kinase 1 (ASK1).[60] It was found in neuronal cells. The knockout of this kinase directly inhibited apoptotic signaling in an ER stress-dependent manner. The kinase builds hetero-oligomers with TRAF2 and IRE1α under ER stress. A complex is build by TRAF2 binding to IRE1α and realizing the conformal conditions to bind ASK1. This interaction is critically influenced by ASK1-interacting protein 1.[61] This protein could be shown to directly interact with IRE1α. AIP is needed to activate ASK1 by release of its inhibitor thioredoxin and 14-3-3.

It also changes IRE1α conformal conditions, and AIP knockout cells show crucial effects on Xbp1 mRNA splicing and JNK activation. The molecule is a positive regulator of IRE1α downstream functions in relation to ASK1.

4.2.2 XBP1 mRNA splicing as a primary response of IRE1α activation

XBP1 is known to control a large number of UPR related target genes as a transcription factor. It is highly expressed in secretory cells. XBP1s (XBP1 spliced protein) influenced genes belonging to ERAD, ER protein load, and lipid biosynthesis. In different cell types, XBP1 can have different interaction partners. In liver cells, it could be shown that XBP1 nuclear localization to the liver is insulin dependent via p85α.[62] In these cells, ER stress alone was not sufficient to activate XBP1s target genes. XBP1 has different regulations and accordingly different gene profiles in different cell types. XBP1 mRNA is spliced by IRE1α in an unconventional way as IRE1a nuclease activity cleaves of 26 nt from uXBP1 mRNA, both fragments are then ligated by a yet unknown ligase in mammals. The resulting RNA, sXBP1 presents an Open Reading Frame shift leading to the translation of a new protein. XBP1 is a target gene of another arm of the UPR. Indeed, the unspliced form or XBP1 mRNA is strongly upregulated upon ATF6 activation.[42]

Furthermore it was shown that the translation product of XBP1u (XBP1 unspliced protein) has a hydrophobic region of about 20 amino acids (aa) in its nascent chain. This region recruits the complex of ribosome-mRNA-and nascent chain to the ER membrane. The nascent chain of XBP1u is normally rapidly degraded. It was shown that a translational stop leaves the nascent chain bound to the ribosome long enough to translocate the complex to the ER membrane. The mRNA is brought to IRE1α in this way and can be spliced by the endoribonuclease of IRE1α.[63,64]

4.2.3 IRE1-dependent decay of mRNA

The splicing of XBP1 mRNA is not the unique function of IRE1 endoribonuclease domain. The molecule is able to degrade a certain number of mRNAs.[65,66] The role of this mechanism was first pointed out in *Drosophila melanogaster* cells. The process was named IRE1-dependent mRNA decay (RIDD). In IRE1α-deficient cells, plenty of mRNAs were upregulated. Under ER stress, these RNAs were downregulated in cells carrying the IRE1 gene in an XBP1-independent manner. The mRNAs have to be localized near the ER membrane to enable this function. One of the first mRNAs substrate for IRE1α nuclease was a secreted protein, acidic, cysteine-rich (SPARC). This molecule modulates the extracellular matrix and was shown to be an important factor in tumor cell migration.[67] In mammalian cells, the RIDD activity was shown to be dependent both on kinase and RNAse domain of IRE1α. The mechanism for activating the RIDD still remains unclear because an activation of only one of the two domains is not sufficient to see the mRNA decay.[65] The mechanism described in mammalian cells was fairly comparable to that observed in drosophila. The difference between the Xbp1 mRNA splicing and the cleavage of the RIDD targets is important. While Xbp1 mRNA is ligated and translated after cleavage, the RIDD target mRNAs are degraded. The degradation leads to a diminishing protein level of the particular target. Since this first report, others RIDD substrates were discovered using tools to identify the cleavage side of RIDD targets.[68] By searching a homolog fragment in the code of further targets, a consensus sequence, which has to appear in a stem-loop structure, was found. The sequence appears in either Xbp1 mRNA splicing side as in CD59 mRNA. This gave an idea of how to find other mRNAs that could be cleaved by IRE1α.

Dealing with human cells, it was possible to generate an IRE1α dominant negative cell line of U87 cancer cells.[69] In IRE1α-deficient tumors, the SPARC mRNA was shown to be upregulated.[70,71] These data led to the supposition that IRE1α has an influence not only to the reaction on ER stress but also to the reaction further on cell migration in human cancer.

5. MANIPULATING PROTEIN OLIGOMERIZATION IN UPR SIGNALING AND NOVEL THERAPEUTIC STRATEGIES

Given the connection of ER stress sensor proteins to diseases, providing compounds able to shift signaling outputs to therapeutic effect might be of great importance. The use of strategies aiming at affecting protein oligomerization status to modulate signaling outcomes and consequently pathophysiological states could in this context represent an interesting avenue.

Chronic ER stress has been for instance associated with damages to tissues with high secretory capacity, for example, the insulin-producing beta cells of the pancreas.[76] Moreover, IRE1 activity has been shown to contribute to some of the deleterious consequences of ER stress, such as insulin resistance.[77] Furthermore, inhibition of XBP1 splicing could be used to impair the IRE1 branch of the UPR, with beneficial effects in fighting cancers, such multiple myeloma, which rely on the secretory pathway for their survival. All these observations are consistent with the identification of IRE1 as a potential therapeutic target.

The endoribonuclease activity of IRE1 has been the target of an intense endeavor to identify inhibitory compounds. As such, flavonols such as quercetin have been shown to enhance IRE1 endoribonuclease activity by direct binding to a specific and newly identified domain of the cytosolic region located at the dimer interface.[72] Moreover, several inhibitors of IRE1 endoribonclease activity have been identified.[73,74] IRE1 RNase inhibitors have also been tested on multiple myeloma with significant success.[75]

The oligomerization equilibrium of IRE1 has also been investigated using peptide-based strategies.[44] Indeed, we used IRE1α cytosolic domain-derived peptides as potential regulators of IRE1 oligomerization state that were tested an activity-based *in vitro* assay. We identified specific peptides that were used to probe IRE1α activity *in vitro* and *in vivo*. We identified a peptide derived from the kinase domain of human IRE1α, which promoted IRE1α oligomerization *in vitro*, enhanced its Xbp1 mRNA cleavage activity and attenuated both ER stress-mediated JNK activation and RIDD. Moreover, cells expressing this peptide were protected from the deleterious effect of ER stress.[44] With the benefit of this insight, chemical modulation of self-association steps in the IRE1 activation process may be a plausible approach to control IRE1 signaling. Inhibiting or preserving proapoptotic and/or prosurvival kinase or nuclease activity from IRE1 may be beneficial in patients with type-II diabetes or in multiple myeloma patients.

6. CONCLUSION

Besides protein productive folding, the ER is the main site of lipid synthesis, membrane biogenesis, xenobiotic detoxification, and cellular calcium storage. Perturbation of ER homeostasis leads to stress and the activation of the UPR. Each arm of the UPR and their downstream transcriptional targets constitute an integrated signaling network that coordinates the overall protein-folding environment in the ER as well as general protein homeostasis (proteostasis) and serves to maintain a balance between cell survival and cell death during ER stress. Chronic activation of ER stress has been shown to have an important role in the development of many diseases. Therefore, identification and targeting of various key components of the UPR for therapeutic purposes could lead to shifting the balance of this coordinated cellular adaptive program toward survival or cell death. It could therefore represent an attractive curative approach for various diseases.

ACKNOWLEDGMENTS

We apologize to all colleagues whose work could not be cited owing to space limitations. This work was funded by grants from INSERM, INCA, and Ligue contre le cancer to E. C.

REFERENCES

1. Campbell NA, Reece JB. *Biologie*. München [u.a.]: Pearson Studium; 2006. pp. 140–141.
2. Coe H, Michalak M. Calcium binding chaperones of the endoplasmic reticulum. *Gen Physiol Biophys* 2009;**28 Spec No Focus**:F96–F103.
3. Michalak M, Groenendyk J, Szabo E, Gold LI, Opas M. Calreticulin, a multi-process calcium-buffering chaperone of the endoplasmic reticulum. *Biochem J* 2009;**417**:651–66.
4. Corbett EF, Michalak M. Calcium, a signaling molecule in the endoplasmic reticulum? *Trends Biochem Sci* 2000;**25**:307–11.
5. Cardozo AK, Ortis F, Storling J, Feng YM, Rasschaert J, Tonnesen M, et al. Cytokines downregulate the sarcoendoplasmic reticulum pump Ca2+ ATPase 2b and deplete endoplasmic reticulum Ca2+, leading to induction of endoplasmic reticulum stress in pancreatic beta-cells. *Diabetes* 2005;**54**:452–61.
6. Malhotra JD, Kaufman RJ. The endoplasmic reticulum and the unfolded protein response. *Semin Cell Dev Biol* 2007;**18**:716–31.
7. Ellgaard L, Helenius A. Quality control in the endoplasmic reticulum. *Nat Rev Mol Cell Biol* 2003;**4**:181–91.
8. Kozutsumi Y, Segal M, Normington K, Gething MJ, Sambrook J. The presence of malfolded proteins in the endoplasmic reticulum signals the induction of glucose-regulated proteins. *Nature* 1988;**332**:462–4.
9. Schenk S, Saberi M, Olefsky JM. Insulin sensitivity: modulation by nutrients and inflammation. *J Clin Invest* 2008;**118**:2992–3002.
10. Scheuner D, Kaufman RJ. The unfolded protein response: a pathway that links insulin demand with beta-cell failure and diabetes. *Endocr Rev* 2008;**29**:317–33.
11. Zhang K, Kaufman RJ. From endoplasmic-reticulum stress to the inflammatory response. *Nature* 2008;**454**:455–62.
12. Kaufman RJ, Cao S. Inositol-requiring 1/X-box-binding protein 1 is a regulatory hub that links endoplasmic reticulum homeostasis with innate immunity and metabolism. *EMBO Mol Med* 2010;**2**:189–92.
13. Yoshida H. ER stress and diseases. *FEBS J* 2007;**274**:630–58.
14. Moenner M, Pluquet O, Bouchecareilh M, Chevet E. Integrated endoplasmic reticulum stress responses in cancer. *Cancer Res* 2007;**67**:10631–4.
15. Hetz C. The unfolded protein response: controlling cell fate decisions under ER stress and beyond. *Nat Rev Mol Cell Biol* 2012;**13**:89–102.
16. Hetz C, Glimcher LH. Fine-tuning of the unfolded protein response: assembling the IRE1alpha interactome. *Mol Cell* 2009;**35**:551–61.
17. Hetz C, Martinon F, Rodriguez D, Glimcher LH. The unfolded protein response: integrating stress signals through the stress sensor IRE1{alpha}. *Physiol Rev* 2011;**91**:1219–43.
18. Ron D, Walter P. Signal integration in the endoplasmic reticulum unfolded protein response. *Nat Rev Mol Cell Biol* 2007;**8**:519–29.
19. Bernales S, McDonald KL, Walter P. Autophagy counterbalances endoplasmic reticulum expansion during the unfolded protein response. *PLoS Biol* 2006;**4**:e423.
20. Kruse KB, Brodsky JL, McCracken AA. Autophagy: an ER protein quality control process. *Autophagy* 2006;**2**:135–7.
21. Shen J, Chen X, Hendershot L, Prywes R. ER stress regulation of ATF6 localization by dissociation of BiP/GRP78 binding and unmasking of Golgi localization signals. *Dev Cell* 2002;**3**:99–111.

22. Nadanaka S, Okada T, Yoshida H, Mori K. Role of disulfide bridges formed in the luminal domain of ATF6 in sensing endoplasmic reticulum stress. *Mol Cell Biol* 2007;**27**:1027–43.
23. Chen X, Shen J, Prywes R. The luminal domain of ATF6 senses endoplasmic reticulum (ER) stress and causes translocation of ATF6 from the ER to the Golgi. *J Biol Chem* 2002;**277**:13045–52.
24. Okada T, Yoshida H, Akazawa R, Negishi M, Mori K. Distinct roles of activating transcription factor 6 (ATF6) and double-stranded RNA-activated protein kinase-like endoplasmic reticulum kinase (PERK) in transcription during the mammalian unfolded protein response. *Biochem J* 2002;**366**:585–94.
25. Schroder M, Kaufman RJ. The mammalian unfolded protein response. *Annu Rev Biochem* 2005;**74**:739–89.
26. Cullinan SB, Zhang D, Hannink M, Arvisais E, Kaufman RJ, Diehl JA. Nrf2 is a direct PERK substrate and effector of PERK-dependent cell survival. *Mol Cell Biol* 2003;**23**:7198–209.
27. Lu PD, Harding HP, Ron D. Translation reinitiation at alternative open reading frames regulates gene expression in an integrated stress response. *J Cell Biol* 2004;**167**:27–33.
28. Shamu CE, Cox JS, Walter P. The unfolded-protein-response pathway in yeast. *Trends Cell Biol* 1994;**4**:56–60.
29. Credle JJ, Finer-Moore JS, Papa FR, Stroud RM, Walter P. On the mechanism of sensing unfolded protein in the endoplasmic reticulum. *Proc Natl Acad Sci USA* 2005;**102**:18773–84.
30. Gardner BM, Walter P. Unfolded proteins are Ire1-activating ligands that directly induce the unfolded protein response. *Science* 2012;**333**:1891–4.
31. Kimata Y, Ishiwata-Kimata Y, Ito T, Hirata A, Suzuki T, Oikawa D, et al. Two regulatory steps of ER-stress sensor Ire1 involving its cluster formation and interaction with unfolded proteins. *J Cell Biol* 2007;**179**:75–86.
32. Korennykh AV, Egea PF, Korostelev AA, Finer-Moore J, Zhang C, Shokat KM, et al. The unfolded protein response signals through high-order assembly of Ire1. *Nature* 2009;**457**:687–93.
33. Papa FR, Zhang C, Shokat K, Walter P. Bypassing a kinase activity with an ATP-competitive drug. *Science* 2003;**302**:1533–7.
34. Chawla A, Chakrabarti S, Ghosh G, Niwa M. Attenuation of yeast UPR is essential for survival and is mediated by IRE1 kinase. *J Cell Biol* 2011;**193**:41–50.
35. Cox JS, Walter P. A novel mechanism for regulating activity of a transcription factor that controls the unfolded protein response. *Cell* 1996;**87**:391–404.
36. Sidrauski C, Cox JS, Walter P. tRNA ligase is required for regulated mRNA splicing in the unfolded protein response. *Cell* 1996;**87**:405–13.
37. Aragon T, van Anken E, Pincus D, Serafimova IM, Korennykh AV, Rubio CA, et al. Messenger RNA targeting to endoplasmic reticulum stress signalling sites. *Nature* 2009;**457**:736–40.
38. Zhou J, Liu CY, Back SH, Clark RL, Peisach D, Xu Z, et al. The crystal structure of human IRE1 luminal domain reveals a conserved dimerization interface required for activation of the unfolded protein response. *Proc Natl Acad Sci USA* 2006;**103**:14343–8.
39. Oikawa D, Kimata Y, Kohno K, Iwawaki T. Activation of mammalian IRE1alpha upon ER stress depends on dissociation of BiP rather than on direct interaction with unfolded proteins. *Exp Cell Res* 2009;**315**:2496–504.
40. Ali MM, Bagratuni T, Davenport EL, Nowak PR, Silva-Santisteban MC, Hardcastle A, et al. Structure of the Ire1 autophosphorylation complex and implications for the unfolded protein response. *EMBO J* 2011;**30**:894–905.
41. Lee KP, Dey M, Neculai D, Cao C, Dever TE, Sicheri F. Structure of the dual enzyme Ire1 reveals the basis for catalysis and regulation in nonconventional RNA splicing. *Cell* 2008;**132**:89–100.

42. Yoshida H, Matsui T, Yamamoto A, Okada T, Mori K. XBP1 mRNA is induced by ATF6 and spliced by IRE1 in response to ER stress to produce a highly active transcription factor. *Cell* 2001;**107**:881–91.
43. Popow J, Englert M, Weitzer S, Schleiffer A, Mierzwa B, Mechtler K, et al. HSPC117 is the essential subunit of a human tRNA splicing ligase complex. *Science* 2010;**331**: 760–4.
44. Bouchecareilh M, Higa A, Fribourg S, Moenner M, Chevet E. Peptides derived from the bifunctional kinase/RNase enzyme IRE1alpha modulate IRE1alpha activity and protect cells from endoplasmic reticulum stress. *FASEB J* 2011;**25**:3115–29.
45. Han D, Lerner AG, Vande Walle L, Upton JP, Xu W, Hagen A, et al. IRE1alpha kinase activation modes control alternate endoribonuclease outputs to determine divergent cell fates. *Cell* 2009;**138**:562–75.
46. Bertolotti A, Wang X, Novoa I, Jungreis R, Schlessinger K, Cho JH, et al. Increased sensitivity to dextran sodium sulfate colitis in IRE1beta-deficient mice. *J Clin Invest* 2001;**107**:585–93.
47. Martinon F, Chen X, Lee AH, Glimcher LH. TLR activation of the transcription factor XBP1 regulates innate immune responses in macrophages. *Nat Immunol* 2010; **11**:411–8.
48. Woo CW, Cui D, Arellano J, Dorweiler B, Harding H, Fitzgerald KA, et al. Adaptive suppression of the ATF4-CHOP branch of the unfolded protein response by toll-like receptor signalling. *Nat Cell Biol* 2009;**11**:1473–80.
49. Gu F, Nguyen DT, Stuible M, Dube N, Tremblay ML, Chevet E. Protein-tyrosine phosphatase 1B potentiates IRE1 signaling during endoplasmic reticulum stress. *J Biol Chem* 2004;**279**:49689–93.
50. Agouni A, Mody N, Owen C, Czopek A, Zimmer D, Bentires-Alj M, et al. Liver-specific deletion of protein tyrosine phosphatase (PTP) 1B improves obesity- and pharmacologically induced endoplasmic reticulum stress. *Biochem J* 2011;**438**:369–78.
51. Hetz C, Bernasconi P, Fisher J, Lee AH, Bassik MC, Antonsson B, et al. Proapoptotic BAX and BAK modulate the unfolded protein response by a direct interaction with IRE1alpha. *Science* 2006;**312**:572–6.
52. Lisbona F, Rojas-Rivera D, Thielen P, Zamorano S, Todd D, Martinon F, et al. BAX inhibitor-1 is a negative regulator of the ER stress sensor IRE1alpha. *Mol Cell* 2009;**33**:679–91.
53. Castillo K, Rojas-Rivera D, Lisbona F, Caballero B, Nassif M, Court FA, et al. BAX inhibitor-1 regulates autophagy by controlling the IRE1alpha branch of the unfolded protein response. *EMBO J* 2011;**30**:4465–78.
54. Marcu MG, Doyle M, Bertolotti A, Ron D, Hendershot L, Neckers L. Heat shock protein 90 modulates the unfolded protein response by stabilizing IRE1alpha. *Mol Cell Biol* 2002;**22**:8506–13.
55. Ota A, Wang Y. Cdc37/Hsp90-mediated regulation of IRE1alpha activity in ER stress response and insulin synthesis in INS-1 cells. *J Biol Chem* 2011;**287**:6266–74.
56. Gupta S, Deepti A, Deegan S, Lisbona F, Hetz C, Samali A. HSP72 protects cells from ER stress-induced apoptosis via enhancement of IRE1alpha-XBP1 signaling through a physical interaction. *PLoS Biol* 2010;**8**:e1000410.
57. Gao YJ, Ji RR. Activation of JNK pathway in persistent pain. *Neurosci Lett* 2008;**437**:180–3.
58. Urano F, Wang X, Bertolotti A, Zhang Y, Chung P, Harding HP, et al. Coupling of stress in the ER to activation of JNK protein kinases by transmembrane protein kinase IRE1. *Science* 2000;**287**:664–6.
59. Kaneko M, Niinuma Y, Nomura Y. Activation signal of nuclear factor-kappa B in response to endoplasmic reticulum stress is transduced via IRE1 and tumor necrosis factor receptor-associated factor 2. *Biol Pharm Bull* 2003;**26**:931–5.

60. Nishitoh H, Matsuzawa A, Tobiume K, Saegusa K, Takeda K, Inoue K, et al. ASK1 is essential for endoplasmic reticulum stress-induced neuronal cell death triggered by expanded polyglutamine repeats. *Genes Dev* 2002;**16**:1345–55.
61. Luo D, He Y, Zhang H, Yu L, Chen H, Xu Z, et al. AIP1 is critical in transducing IRE1-mediated endoplasmic reticulum stress response. *J Biol Chem* 2008;**283**:11905–12.
62. Park SW, Zhou Y, Lee J, Lu A, Sun C, Chung J, et al. The regulatory subunits of PI3K, p85alpha and p85beta, interact with XBP-1 and increase its nuclear translocation. *Nat Med* 2010;**16**:429–37.
63. Yanagitani K, Imagawa Y, Iwawaki T, Hosoda A, Saito M, Kimata Y, et al. Cotranslational targeting of XBP1 protein to the membrane promotes cytoplasmic splicing of its own mRNA. *Mol Cell* 2009;**34**:191–200.
64. Yanagitani K, Kimata Y, Kadokura H, Kohno K. Translational pausing ensures membrane targeting and cytoplasmic splicing of XBP1u mRNA. *Science* 2011;**331**:586–9.
65. Hollien J, Lin JH, Li H, Stevens N, Walter P, Weissman JS. Regulated Ire1-dependent decay of messenger RNAs in mammalian cells. *J Cell Biol* 2009;**186**:323–31.
66. Hollien J, Weissman JS. Decay of endoplasmic reticulum-localized mRNAs during the unfolded protein response. *Science* 2006;**313**:104–7.
67. Brekken RA, Sage EH. SPARC, a matricellular protein: at the crossroads of cell-matrix. *Matrix Biol* 2000;**19**:569–80.
68. Oikawa D, Tokuda M, Hosoda A, Iwawaki T. Identification of a consensus element recognized and cleaved by IRE1 alpha. *Nucleic Acids Res* 2010;**38**:6265–73.
69. Drogat B, Auguste P, Nguyen DT, Bouchecareilh M, Pineau R, Nalbantoglu J, et al. IRE1 signaling is essential for ischemia-induced vascular endothelial growth factor-A expression and contributes to angiogenesis and tumor growth in vivo. *Cancer Res* 2007;**67**:6700–7.
70. Auf G, Jabouille A, Guerit S, Pineau R, Delugin M, Bouchecareilh M, et al. Inositol-requiring enzyme 1alpha is a key regulator of angiogenesis and invasion in malignant glioma. *Proc Natl Acad Sci USA* 2010;**107**:15553–8.
71. Dejeans N, Pluquet O, Lhomond S, Grise F, Bouchecareilh M, Juin A, et al. Autocrine control of glioma cells adhesion/migration through Inositol Requiring enzyme 1alpha (IRE1alpha)-mediated cleavage of Secreted Protein Acidic Rich in Cysteine (SPARC) mRNA. *J Cell Sci* 2012;**125**:4278–87.
72. Wiseman RL, Zhang Y, Lee KP, Harding HP, Haynes CM, Price J, et al. Flavonol activation defines an unanticipated ligand-binding site in the kinase-RNase domain of IRE1. *Mol Cell* 2010;**38**:291–304.
73. Cross BC, Bond PJ, Sadowski PG, Jha BK, Zak J, Goodman JM, et al. The molecular basis for selective inhibition of unconventional mRNA splicing by an IRE1-binding small molecule. *Proc Natl Acad Sci USA* 2012;**109**:E869–E878.
74. Volkmann K, Lucas JL, Vuga D, Wang X, Brumm D, Stiles C, et al. Potent and selective inhibitors of the inositol-requiring enzyme 1 endoribonuclease. *J Biol Chem* 2010;**286**:12743–55.
75. Papandreou I, Denko NC, Olson M, Van Melckebeke H, Lust S, Tam A, et al. Identification of an Ire1alpha endonuclease specific inhibitor with cytotoxic activity against human multiple myeloma. *Blood* 2012;**117**:1311–4.
76. Powers ET, Morimoto RI, Dillin A, Kelly JW, Balch WE. Biological and chemical approaches to diseases of proteostasis deficiency. *Annu Rev Biochem*. 2009;**78**:959–91. http://dx.doi.org/10.1146/annurev.biochem.052308.114844.
77. Ozcan U, Cao Q, Yilmaz E, Lee AH, Iwakoshi NN, Ozdelen E, et al. Endoplasmic reticulum stress links obesity, insulin action, and type 2 diabetes. *Science* 2004;**306** (5695):457–61.

SECTION V

Oligomerization in Regulatory Proteins

CHAPTER EIGHTEEN

Toll-IL-1-Receptor-Containing Adaptor Molecule-1: A Signaling Adaptor Linking Innate Immunity to Adaptive Immunity

Misako Matsumoto, Kenji Funami, Hiroyuki Oshiumi, Tsukasa Seya
Department of Microbiology and Immunology, Hokkaido University Graduate School of Medicine, Sapporo, Japan

Contents

Abstract

The innate immune system senses microbial infections using pattern-recognition receptors and signals to activate adaptive immunity. Type I transmembrane protein Toll-like receptors (TLRs) play important roles in antimicrobial immune responses. Upon the recognition of pathogen-associated molecular patterns, TLRs homo- or heterodimerize and recruit distinct adaptor molecules to the intracellular TIR domains. Toll-IL-1-receptor-containing adaptor

Progress in Molecular Biology and Translational Science, Volume 117
ISSN 1877-1173
http://dx.doi.org/10.1016/B978-0-12-386931-9.00018-0

molecule-1 (TICAM-1) is a signaling adaptor downstream of TLRs 3 and 4 that recognizes virus-derived double-stranded RNA and lipopolysaccharide, respectively. TLR3 is expressed on the endosomal membrane in myeloid DCs, where TLR3-mediated signaling is initiated. Once TICAM-1 is activated, transcription factors, IRF-3, NF-κB, and AP-1, are activated, leading to production of IFN-β and proinflammatory cytokines and maturation of dendritic cells, which are capable of activating NK cells and cytotoxic T cells. Hence, TICAM-1 signaling appears to link innate immunity to adaptive immunity. In this review, we summarize the current knowledge on TICAM-1 and discuss its role in virus infection and antitumor immunity.

1. INTRODUCTION

The innate immune system senses microbial infections using pattern-recognition receptors (PRRs), including membrane-bound Toll-like receptors (TLRs), cytosolic RIG-I-like receptors (RLRs), and NOD-like receptors, which rapidly induce an antimicrobial response.[1,2] PRRs recognize unique microbial components called pathogen-associated molecular patterns (PAMPs) and activate signaling cascades via distinct adaptor proteins, leading to the production of type I IFNs and proinflammatory cytokines in innate immune cells. Moreover, accumulating evidence indicates that PRRs are involved in the recognition of damage-associated molecular patterns (DAMPs), which are closely associated with the development of autoimmune diseases or inflammatory disorders.[3,4]

TLRs are the first PRRs shown to link innate immunity to adaptive immunity through induction of dendritic cell (DC) maturation and cytokine production.[5] In humans, 10 members of the TLR family of proteins (TLR1-10) exist. The TLRs are classified into two groups based on subcellular localization. TLR 1, 2, 4, 5, and 6 are expressed on cell surfaces and detect lipids, carbohydrates, or proteins derived from microbes. TLR 3, 7, 8, and 9 are localized in endosomal compartments, where TLRs recognize microbial or host nucleic acids (Table 18.1). TLRs are glycosylated type I transmembrane proteins that are composed of extracellular, transmembrane, and intracellular domains. The extracellular domain contains 19–25 leucine-rich repeat (LRR) modules sandwiched between N- and C-terminal flanking regions and is responsible for ligand recognition.[6] The cytoplasmic domain contains a linker region composed of 20–40 amino acids and a Toll-IL-1 receptor (TIR) domain which is required for interaction with the adaptor proteins. Upon ligand binding, TLRs form homo- or heterodimers and transmit signals via recruitment of

Table 18.1 TLR recognition of microbial components

TLR	Microbial ligands	Subcellular localization
1/2	Triacyl lipopeptides	Cell surface
2	Peptidoglycan, Porin, Lipoarabinomannan	Cell surface
3	Double-stranded RNA	Cell surface, early endosome
4	Lipopolysaccharide	Cell surface
5	Flagellin	Cell surface
6/2	Diacyl lipopeptides, Lipoteichoic acid	Cell surface
7	Single-stranded RNA	ER, endolysosome
8	Single-stranded RNA	ER, early endosome
9	CpG-DNA	ER, endolysosome
10	Unknown	Cell surface

distinct adaptor proteins to the intrarcellular TIR domains.[7] Five TIR-containing adaptor molecules, including MyD88, TICAM-1 (also known as TIR domain-containing adaptor-inducing interferon-β [TRIF]), TICAM-2 (also known as TRIF-related adaptor molecule [TRAM]), Mal (also called TIRAP), and SARM, have been identified.[8] All TLRs, except for TLR3, use MyD88 as a signaling adaptor and induce NF-κB-dependent cytokine production. TLR7 and TLR9 also induce IRF-7-dependent IFN-α production via MyD88 in plasmacytoid DCs.[9] TICAM-1 is a TLR3 adaptor that activates the transcription factors, IRF-3, NF-κB, and AP-1, leading to the induction of type I IFN (especially IFN-β), cytokine/chemokine production, and DC maturation.[10,11] TICAM-1 is also engaged in TLR4-mediated MyD88-independent signaling cascades downstream of TICAM-2.[12,13] TICAM-2 is directly associated with TLR4 and bridges between TLR4 and TICAM-1. Mal/TIRAP is another bridging adaptor downstream of TLR4 and TLR2[14,15] (Fig. 18.1). The last adaptor to be identified, SARM, is a negative regulator for TLR3–TICAM-1 signaling.[16] Thus, the combinations between TLRs and adaptor proteins determine the properties of PAMP-induced innate immune responses. Among these adaptors, TICAM-1 is unique in inducing a wide range of cellular responses linking innate and adaptive immunity. In this review, we summarize the current knowledge on TICAM-1 function in the innate and adaptive immune systems and discuss the role of TICAM-1 in virus infection and antitumor immunity.

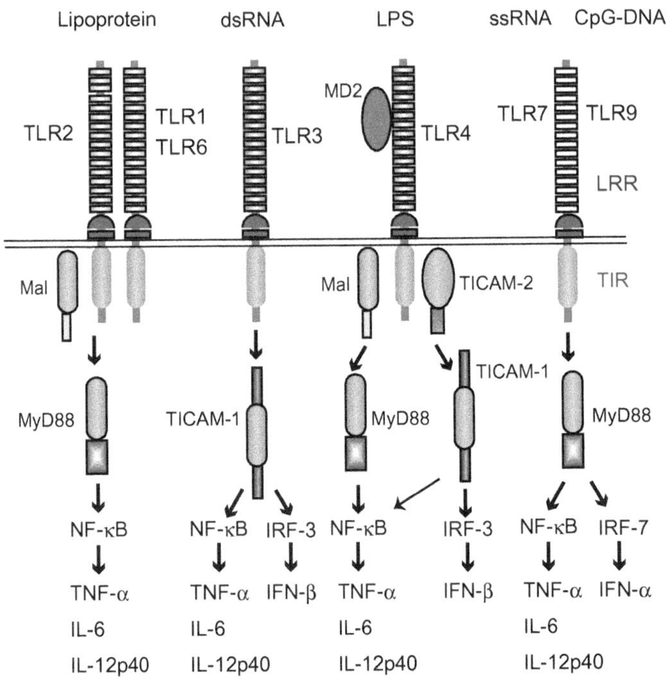

Figure 18.1 TLR transmits signals through distinct adaptor proteins. TLR2/1 and TLR2/6 heterodimers use two adaptor proteins, Mal and MyD88, that activate NF-κB, leading to production of proinflammatory cytokines, including TNF-α, IL-6, and IL-12p40. TLR3 uses TICAM-1, which activates NF-κB and IRF-3 and induces proinflammatory cytokines and IFN-β. TLR4 uses four adaptor proteins, Mal, MyD88, TICAM-2, and TICAM-1. TLR7 and TLR9 use MyD88, which activates NF-κB and also IRF-7 in plasmacytoid DCs and induces IFN-α production. (For color version of this figure, the reader is referred to the online version of this chapter.)

2. TLR3–TICAM-1 PATHWAY

2.1. Expression and localization of TLR3

TLR3 has been functionally identified as a sensor of virus-derived dsRNA and its synthetic analog, polyriboinosinic:polyribocytidylic acid [poly(I:C)].[17,18] Upon dsRNA recognition, TLR3 induces IFN-β and proinflammatory cytokine production from host cells via the adaptor protein, TICAM-1. It was subsequently shown that the cytosolic 5′-triphosphated viral RNA and long dsRNA are recognized by the cytosolic RNA helicases, retinoic-acid inducible gene-I (RIG-I) and melanoma differentiation-associated gene 5 (MDA5), respectively, which induces IFN-α/β production via the adaptor protein, IPS-1 (also referred to as MAVS, Cardif, and VISA) located on the mitochondrial outer membrane.[19–26] More recently, it has been reported that the cytoplasmic RNA helicase complex, DDX1, DDX21, and DHX36, detects dsRNA and induces type I IFNs via TICAM-1, but the precise TICAM-1 activation mechanism is unclear.[27]

TLR3 is expressed in various tissues and cells. In the human central nervous system, TLR3 is expressed constitutively in neurons, astrocytes, and microglia, suggesting a role in the response to viruses causing encephalopathy.[28,29] In immune cells, only myeloid DCs and macrophages express TLR3. Monocytes, polymorphonuclear leukocytes, and T, B, and NK cells do not express TLR3.[30–34] Among DC subsets, TLR3 is highly expressed in the professional antigen-presenting DCs, including human $CD141^+$ DCs and mouse splenic $CD8\alpha^+$ DCs.[35,36] In contrast, the plasmacytoid DCs, which express TLR7 and TLR9 and secrete large amounts of IFN-α in response to viral ssRNA and imidazoquinoline compounds (TLR7 ligands) or CpG-DNA (TLR9 ligand), do not express TLR3.[34] TLR3 is also expressed in fibroblasts and a variety of epithelial cells, including airway, corneal, cervical, biliary, and intestinal cells, which are target sites of virus infection.[37] Notably, TLR3 expression is upregulated by type I IFN, and this positive feedback system is important for TLR3-mediated antiviral responses.[38]

The subcellular localization of TLR3 depends on the type of cell. TLR3 is localized to cell surfaces and endosomes in fibroblasts, macrophages, and some epithelial cell lines.[18] In contrast, myeloid DCs only express TLR3 intracellularly.[34] Immunofluoresence analysis has shown that endogenous TLR3 is localized to the early endosome, but not to late endosomes/lysosomes. The ER resident protein, UNC93B1, is physically associated with TLR3 and facilitates intracellular trafficking of TLR3[39]; however, the linker region between the transmembrane and TIR domains of TLR3 determines the endosomal localization.[40,41] In any cell type, TLR3 signaling arises in the endosomal compartment, requiring endosomal maturation. Hence, based on endosomal localization and signaling of TLR3, it is generally accepted that the TLR3–TICAM-1 pathway is involved in the detection of extrinsic viral dsRNA, while the RLR–IPS-1 pathway participates in cytoplasmic viral RNA sensing.[42]

2.2. Recognition of dsRNA

Human TLR3 consists of an extracellular domain containing 23 LRRs and N- and C-terminal flanking regions (LRR-NT and LRR-CT), a transmembrane domain, and an intracellular linker region and TIR domain. It possesses 15 putative carbohydrate-binding motifs in the extracellular domain. Structural analyses of the human TLR3 ectodomain (ECD) revealed that the LRRs form a large horseshoe-shaped solenoid, one face of which is largely masked by carbohydrate, while the other face is glycosylation-free.[43,44] A subsequent structural analysis of the mouse TLR3–ECD and 46-bp dsRNA oligonucleotide complex demonstrated that dsRNA interacts with N- and C-terminal binding sites on the glycan-free surface of each mTLR3–ECD, which are located on opposite sides of dsRNA.[45] Further, the two LRR-CT domains are brought into proximity and form a series of protein–protein interactions which facilitate dimerization of the cytoplasmic TIR domain. Point mutation analyses revealed that the His539 and Asn541 residues in TLR3–LRR20, which are located on the glycan-free lateral face, and the N-terminal-conserved histidine residues, His39 in LRR-NT, His60 in LRR1, and His108 in LRR3, are critical amino acids for dsRNA binding and signaling.[46–48] Protonation of these imidazole groups under acidic conditions, such as exists in endosomes, appears to generate an ionic interaction between the histidine residues and the negatively charged phosphate backbone of dsRNA. In addition, Asp648, Thr679, and Pro680 in the LRR-CT dimerization site are essential for TLR3 signaling, and this site also influences ligand binding.[49]

Using biochemical techniques, Leonard *et al.*[50] showed that TLR3–ECD binds to a 40–50 bp length of dsRNA as a dimer, and multiple TLR3–ECD dimers bind to long dsRNA strands. The binding affinities increase with both buffer acidity and dsRNA length. Based on these structural and biochemical analyses of the TLR3–dsRNA complex, it has been proposed that the 40–50 bp dsRNA is the minimum signaling unit with two TLR3 molecules. An alternative model for TLR3 dimer formation has also been proposed, in which shorter RNA duplexes of between 21 and 30 bp can form less stable complexes with two TLR3 molecules.[47] Recently, Jelinek *et al.*[51] reported that dsRNAs, >90 bp in length, triggered TLR3 oligomerization and efficiently induced IFN-β production and T cell activation in conventional murine DCs, while 60-bp dsRNA did not. Because the >90-bp length of dsRNA, but not the 40- to 50-bp length of dsRNA, forms a stable complex with TLR3 at the pH within early endosomes (~6.0–6.5),[50] oligomerization of TLR3 at the early endosomes must be required for IFN-β production and efficient DC maturation.

2.3. Mechanism of dsRNA uptake

Because TLR3 is a type I transmembrane protein located on the endosomal membrane, TLR3 is intrinsically involved in the recognition of extracellular viral dsRNA; however, the molecular mechanism by which extracellular dsRNA is delivered to TLR3-positive organelles remains unknown. Itoh *et al.*[52] demonstrated that poly(I:C) is internalized via clathrin-dependent endocytosis. In addition, they reported that poly(I:C) shares its uptake receptor with B- and C-type oligodeoxynucleotides (ODNs). Subsequently, Watanabe *et al.*[53] identified the cytoplasmic lipid raft protein, Raftlin, that is essential for poly(I:C) cellular uptake and TLR3-mediated IFN-β production in human myeloid DCs and epithelial cells. Raftlin induces the internalization of poly(I:C) by cooperating with the clathrin-AP-2 complex. Confocal microscopic analyses showed that upon poly(I:C) stimulation, Raftlin is translocated from the cytoplasm to the plasma membrane, where it colocalizes with poly(I:C), and thereafter moves to TLR3-positive endosomes with internalized poly(I:C). Raftlin is physically associated with clathrin in response to poly(I:C) stimulation in the cell membrane, and dissociates from clathrin before reaching TLR3-positive endosomes. Interestingly, Raftlin participates in the cell entry of B- and C-type ODNs, but not transferrin.[53] Hence, Raftlin appears to modulate cargo sorting and delivery by the clathrin-AP-2 complex.

Several studies have shown that CD14 and the scavenger receptor class A act as an uptake receptor for poly(I:C) in mouse macrophages and human bronchial epithelial cells, respectively.[54,55] However, these receptors do not appear to participate in poly(I:C) cellular uptake in human DCs and epithelial cells.[56] Identification of the poly(I:C)/ODN uptake receptor is important for a full understanding of the mechanism of uptake for extracellular nucleic acids.

3. TLR4–TICAM-1 PATHWAY

TLR4 recognizes LPS cooperatively with MD-2.[5,57] Upon LPS binding, TLR4 undergoes dimerization and associates with a bridging adaptor, Mal/TIRAP, that contains a phosphatidylinositol 4,5-biphosphate [PtdIns(4,5)P_2] binding domain.[58] MyD88 translocates to the cell surface and colocalizes with Mal/TIRAP and TLR4 in PtdIns(4,5)P_2-rich regions, thus allowing for activation of the NF-κB pathway and leading to production of proinflammatory cytokines. TLR4 then moves into the endosome in a dynamin- and clathrin-dependent manner and induces IRF-3 and NF-κB activation, leading to the production of IFN-β.[59] TICAM-2 and TICAM-1 participate in this MyD88-independent signaling pathway.[60]

TICAM-2 was discovered via homology search using the TICAM-1-TIR domain.[12] Human TICAM-2 mRNA is expressed in many tissues and cells, including DCs, macrophages, and natural killer cells, and various B and T cell lines. Human TICAM-2 consists of 235 amino acids with a short N-terminal Ser/Thr-rich domain, a TIR domain, and a C-terminal 20-amino acid stretch (Fig. 18.2A). The TIR motif of TICAM-2 has low similarity to that of Mal and MyD88. TICAM-2 contains a myristoylation motif at the N-terminus, which is essential for localization and function.

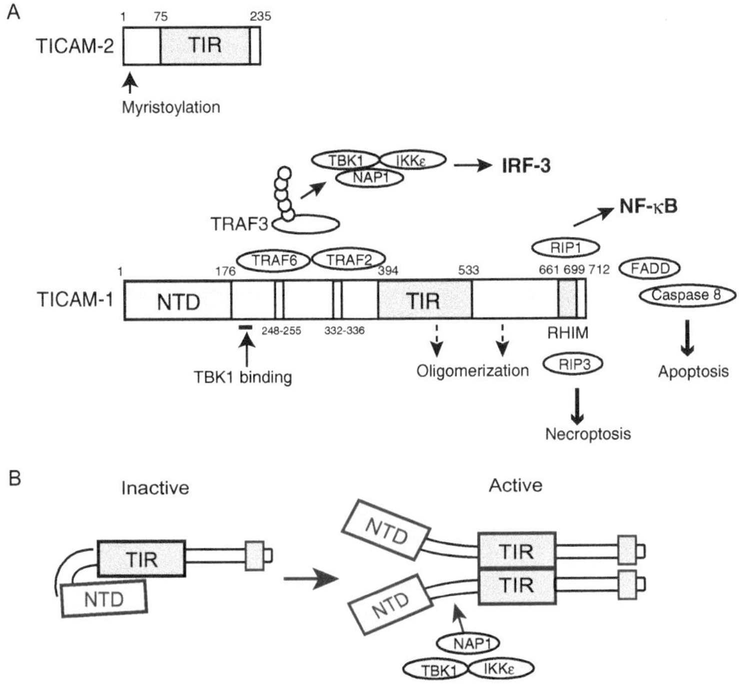

Figure 18.2 Schematic representation of protein domains and motifs in human TICAM-1 and TICAM-2 and the signaling cascades. (A) TICAM-1 and TICAM-2 have a TIR domain. TICAM-2 is myristoylated at the N-terminus. TICAM-1 has an N-terminal domain (NTD) that regulates TICAM-1 activation. TBK1, TRAF2, and TRAF6-binding sites are located between the NTD and TIR domain. TRAF3 and NAP1 participate in the recruitment and activation of IRF-3 kinase, TBK1, and IKKε. The RHIM domain is located at the C-terminus, which is essential for NF-κB activation and induction of apoptosis or necroptosis. (B) Model of TICAM-1 activation. In resting cells, TICAM-1 is inactivated by intramolecular interaction between the NTD and TIR domain (left panel). Upon stimulation of TLR3/4, or TICAM-1 overexpression, TICAM-1 oligomerizes through the TIR domain and the C-terminal region, which breaks the intramolecular association and induces a conformational change that allows access of downstream signaling molecules to their binding sites (right panel). See text for a detailed explanation.

Kagan *et al.*[59] showed that wild-type TICAM-2 is localized to the plasma membrane and early endosome, while a myristration-deficient TICAM-2 mutant is uniformly distributed throughout the cell and fails to mediate signaling. Functional analyses using various TICAM-2 mutants revealed that endosomal but not cell surface localization of TICAM-2 is sufficient to induce TLR4-mediated TICAM-2–TICAM-1 signaling. Upon LPS stimulation, TICAM-2 associates with internalized TLR4 at the early endosome and recruits TICAM-1. Pro714 in the TLR4-TIR domain is critical for interaction with TICAM-2-TIR.[12]

In addition to LPS stimulation, the forced expression of TICAM-2 also activates NF-κB and IRF-3 via endogenous TICAM-1. The TICAM-2-TIR mutant C117H, which has TLR4 binding ability, fails to homodimerize and does not interact with TICAM-1, suggesting that homodimerization of TICAM-2 is critical for TICAM-1 recruitment.[12]

4. TICAM-1 SIGNALING

4.1. Structure of TICAM-1

Human TICAM-1 is composed of 712 amino acids, which consists of an N-terminal region, a TIR domain, and a C-terminal region (Fig. 18.2A). The TIR domain of TICAM-1 is essential for binding to the TIR domain of TLR3 and to the TLR4 adaptor TICAM-2. Pro434 in the TIR domain is essential for homo-oligomerization of TICAM-1, but not for association with TLR3 or TICAM-2.[61] The N-terminal region is crucial for TICAM-1-mediated IRF-3 activation, which contains the N-terminus regulatory domain (NTD) and the TBK1, TRAF6 and TRAF2 binding sites.[62–64] The C-terminal region contains the RIP homotypic-interacting motif (RHIM) domain and is involved in NF-κB activation and apoptosis.[65–67] TICAM-1 is expressed at a low level in most tissues and cells, and is diffusely localized in the cytoplasm of resting cells.[68] When endosomal TLR3 is activated by dsRNA, TICAM-1 transiently colocalizes with TLR3, then dissociates from the receptor, and forms speckled structures that colocalize with downstream signaling molecules. Thus, TICAM-1 alters the distribution profile in the cytosol in response to dsRNA.

4.2. TICAM-1 oligomerization and signaling

TICAM-1 activates the transcription factors, IRF-3, NF-κB, and AP-1, leading to the induction of type I IFN, cytokine/chemokine production, and DC maturation, which then enables the activation of NK cells and CTLs. In contrast to other TLR adaptors, the forced expression of TICAM-1 leads to potent induction of IFN-β promoter activation. Homo-oligomerization of TICAM-1 is critical for TICAM-1-mediated activation of NF-κB and IRF-3.[61] In a yeast two-hybrid system, TICAM-1-TIR mutant P434H failed to interact with the TIR domain of TICAM-1, but bound tightly to the TIR of TLR3 or TICAM-2, suggesting that Pro434 is critical for TICAM-1 dimerization. In addition, the C-terminal region of TICAM-1, with the exception of the RHIM, mediates TICAM-1 oligomerization. Mutation of Pro434 or deletion of the C-terminal region reduces TICAM-1-mediated NF-κB and IRF-3 activation. Thus, Pro434 and the C-terminal region are required for TICAM-1 oligomerization. Once TICAM-1 is oligomerized, the serine–threonine kinases, TANK-binding kinase 1 (TBK1; also called NAK or T2K) and IκB kinase-related kinase ε (IKK-ε; also called IKK-ι), are activated and phosphorylate IRF-3.[69,70] The phosphorylated IRF-3 then translocates into the nuclei, leading to IRF-3-dependent gene expression. The ubiquitin ligase of the TRAF family members, TRAF2, TRAF3, and TRAF6, are downstream signaling molecules of TICAM-1 (Fig. 18.2A). TRAF2 and TRAF6 directly bind to the N-terminal region of TICAM-1 and mediate Lys63-linked TICAM-1 polyubiquitination, which facilitates TICAM-1-mediated IFN-β promoter activation.[63] Lys63-linked autoubiquitination of TRAF3 is also required for IRF-3 activation.[71,72] Further, NF-κB activating kinase (NAK)-associated protein 1 (NAP1) participates in the recruitment of IRF-3 kinases to the N-terminal region of TICAM-1.[73] Although TRAF3 and NAP1 associate with oligomerized TICAM-1 and serve as a critical link between TICAM-1 and downstream IRF-3 kinases, there is no evidence that they bind directly to TICAM-1. Interestingly, Tatematsu *et al.*[64] showed that direct binding of TBK1 to TICAM-1 is necessary for IRF-3 activation. The Leu194 residue in the N-terminal region is critical for TBK1 binding to TICAM-1 (Fig. 18.2A). The L194A mutant of TICAM-1 failed to recruit TBK1, resulting in disabling IRF-3 phosphorylation, although TRAF3 and NAP1 were recruited. In addition, the Ser189, Arg195, and Ser196 residues are involved in TBK1–TICAM-1 binding.

The N-terminal 176 amino acids of TICAM-1 (NTD) form a protease-resistant structural domain. Because the crucial amino acids for TRAF2, TRAF6, and TBK1 binding reside between the NTD and TIR domain, naive TICAM-1 may have a closed conformation that covers these binding sites (Fig. 18.2B). Indeed, protein–protein interaction analysis has revealed that the NTD interacts with the N-terminus of TICAM-1-TIR.[64] Thus, the NTD folds into the TIR domain structure to maintain the naive conformation of TICAM-1. Upon stimulation of TLR3 or TLR4, TICAM-1 oligomerizes through the TIR domain and the C-terminal region, possibly breaking the intramolecular association and inducing a conformational change that allows TBK1 access to TICAM-1 (Fig. 18.2B).

Whereas the N-terminal region is crucial for TICAM-1-mediated IRF-3 activation, the C-terminal region of TICAM-1 is involved in NF-κB activation and apoptosis. Receptor-interacting protein 1 (RIP1), a kinase containing a death domain, associates with TICAM-1 via the RHIM domain in the C-terminal region and acts as an NF-κB inducer and apoptosis mediator in TICAM-1-mediated signaling.[65–67] TRAF6 has also been implicated in NF-κB activation by TICAM-1 in a cell type-dependent manner.[62,74] Notably, the TICAM-1 mutant lacking the RIP1 binding motif failed to recruit NAP1 and TBK1. Thus, full activation and formation of TICAM-1 signalosomes require oligomerization induced at two different sites and RIP1 binding.

TLR3–TICAM-1-mediated type I IFN production is negatively regulated by deubiquitinating enzyme A (DUBA) (Fig. 18.3). DUBA selectively cleaves the Lys63-linked polyubiquitin chains on TRAF3, resulting in its dissociation from the downstream signaling molecules.[75] In addition, the ubiquitin-modifying enzyme A20 inhibits TICAM-1-mediated NF-κB activation by deubiquitinating TRAF6.[76] However, the precise mechanisms by which TRAF3 and TRAF6 are ubiquitinated and interact with downstream signaling molecules are unknown.

4.3. Induction of apoptosis/necroptosis

Overexpressed TICAM-1 potently induces apoptosis. TICAM-1-mediated cell death is inhibited by a dominant-negative form of Fas-associated death domain protein (FADD), a caspase inactive mutant of caspase-8, or the caspase inhibitor CrmA, indicating that TICAM-1 induces apoptosis via the FADD–caspase-8 axis.[66] This signaling event has been shown to be tightly regulated by the cellular FLICE-like inhibitory protein short form (cFLIPs).[77] cFLIPs heterodimerizes with caspase-8 to suppress the self-activation of caspase-8. Mutational analyses have revealed that the RHIM located in the C-terminus of TICAM-1 is essential for TICAM-1-induced apoptosis.[66] TICAM-1 physically interacts with RHIM-containing proteins, RIP1 and RIP3, of which RIP1, but not RIP3, contains the death

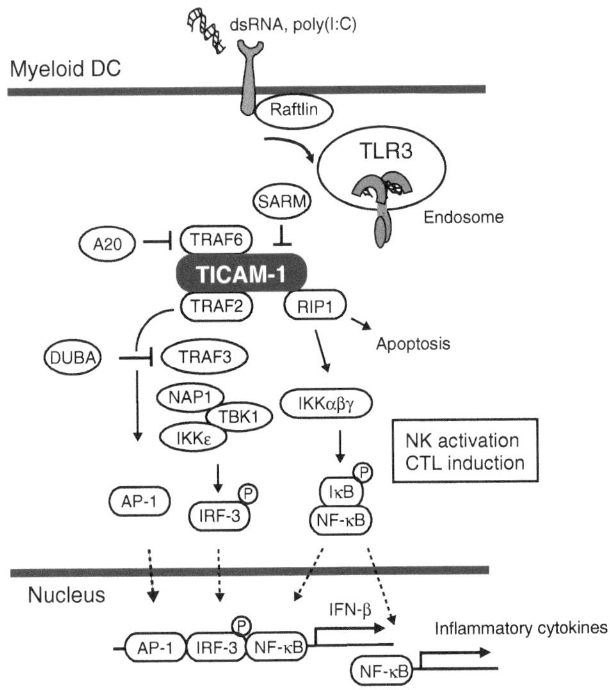

Figure 18.3 TLR3–TICAM-1 signaling pathway. In myeloid DCs, TLR3 is expressed in the early endosome and recognizes viral dsRNA and its synthetic analog, poly(I:C), which are internalized via raftlin- and clathrin-dependent endocytosis. Once TLR3 is dimerized by dsRNA, TLR3 recruits the adaptor protein, TICAM-1. After a transient association between TLR3 and TICAM-1 through the TIR domain, TICAM-1 dissociates from TLR3 to form a speckle-like signalosome with downstream signaling molecules, including RIP1, TRAF2/3/6, NAP1, and TBK1, where TICAM-1-mediated signaling is initiated. RIP1 associates with TICAM-1 via the PHIM domain and acts as a NF-κB activator and apoptosis mediator in TICAM-1-mediated signaling. Phosphorylated IRF-3 translocates into the nucleus, and together with NF-κB and AP-1, induces IFN-β gene transcription. TICAM-1-mediated AP-1 activation pathway is unclear. In addition to IFN-β and inflammatory cytokine production, the TICAM-1 signal induces DC maturation capable of activating NK cells and CTLs. SARM, A20, and DUBA inhibit TICAM-1 signaling via targeting distinct molecules.

domain. Overexpression of TICAM-1 with mutant RHIM cannot recruit RIP1 and failed to induce apoptosis, indicating that TICAM-1 induces RIP1–FADD–caspase 8-dependent apoptosis (Fig. 18.2A). The ability of TICAM-1 to induce apoptosis appears to be segregated from those of the activation of IRF-3 and NF-κB.

Several very recent studies have demonstrated that RIP1 and RIP3 mediate cell death, which is referred to as programmed necrosis or necroptosis.[78–80] In RIP1/3-dependent necrosis, ligating TLR3 or TLR4 activates RIP1 via TICAM-1, which in turn binds to RIP3, thus triggering cell death.[81] These pathways are inhibited by caspase-8, and therefore this type of necrosis appears under some conditions of caspase-8 inhibition. Feoktistova *et al.*[77] reported that the loss of cellular inhibitor of apoptosis proteins (cIAPs) promotes TLR3-induced RIP1–RIP3-mediated necroptosis. Intriguingly, they showed that in the absence of cIAPs, cFLIPs was able to protect from the RIP1-induced apoptosis, but promoted necroptosis, that is caused by the lack of caspase activity. TLR3 is expressed in many tumor cells at a high level, including melanoma, hepatocellular carcinoma, and breast cancer.[82–84] The proapoptotic activity of TLR3/TICAM-1/RIP1/caspase-8 in cancer cells is under the control of inhibitor of apoptosis proteins.

4.4. Induction of type I and III IFNs and proinflammatory cytokines

Upon TLR3 stimulation, TICAM-1 induces gene expression of IFN-β in fibroblasts, epithelial cells, and myeloid DCs, which generate antiviral states in both stimulated and surrounding unstimulated cells by induction of IFN-stimulated genes through IFN-α/β receptor.[85] TICAM-1 activation also induces IL-12p70 production from human, but not mouse myeloid DCs. IL-12p70 is an important cytokine that directs naïve Th0 cells to differentiate into Th1 cells in conjunction with antigen stimulation. Further, mouse $CD8\alpha^+$ DCs and human $CD141^+$ DCs produce type III IFN (IFN-λ) in response to poly(I:C), which depends on TLR3–TICAM-1.[86]

Interestingly, when poly(I:C) is added to cells or injected into mice, both endosomal TLR3 and cytoplasmic MDA5 recognize poly(I:C) and induce gene expression via TICAM-1 and IPS-1, respectively.[87,88] After *in vivo* poly(I:C) injection or *in vitro* bone marrow-derived DC (BMDC) stimulation, production of IFN-α is not impaired in TICAM-$1^{-/-}$ mice compared with wild-type mice, whereas IL-12p40 production is completely dependent on TICAM-1; in contrast, IFN-β and IL-6 productions are partially diminished in TICAM-$1^{-/-}$ mice.

4.5. Induction of adaptive immunity

TICAM-1 signaling is important for DC-mediated activation of NK cells and CTLs (Fig. 18.4). TICAM-1-induced IFN-β and IL-12p70 directly activate NK cells. In addition, the TLR3–TICAM-1 pathway in myeloid DCs facilitates the DC–NK cell interaction following NK cell activation.[89]

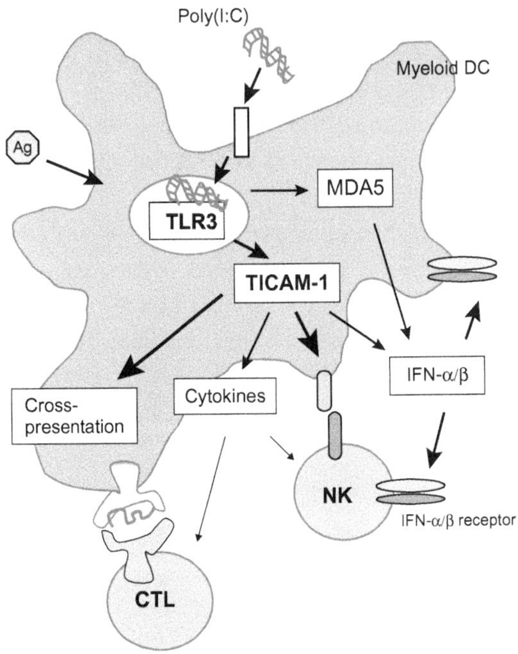

Figure 18.4 Double-stranded RNA-induced TLR3–TICAM-1-mediated cellular responses in myeloid DCs. Myeloid DCs take up extracellular poly(I:C) or apoptotic bodies containing dsRNA from virus-infected cells and induce IFN-α/β and cytokine production, NK cell activation, and CTL induction via the TLR3–TICAM-1 pathway. Extracellular poly(I:C) is further delivered to the cytoplasm and activates MDA5, leading to production of IFN-α/β. IFN-α/β participates in NK cell activation and promotion of cross-priming resulting in CTL induction, which largely depends on the experimental conditions.

Ebihara *et al.*[90] reported that TICAM-1–IRF-3-dependent expression of a novel molecule, namely IRF-3-dependent NK activation molecule (INAM), in myeloid DCs is required for NK cell activation.

Myeloid DCs are the most effective professional antigen-presenting cells, possessing several antigens processing, and transporting machineries.[91,92] One of the most notable features of myeloid DCs is the cross-presentation of exogenous antigens to $CD8^+$ T cells. This pathway is important for effective host CTL induction against viruses that do not directly infect DCs and tumor cells. Among the myeloid DC subsets, the splenic $CD8\alpha^+$ DC subset in mice and the $CD141(BDCA3)^+DNGR\text{-}1(CLEC9A)^+$ DC subset in humans highly express TLR3 and display a superior capacity for cross-presenting apoptotic and necrotic cell antigens after TLR3 stimulation.[36,37] Using TLR3-deficient mice, Schultz *et al.*[93] clearly showed that TLR3 plays an important role in cross-priming. In many cases, virus-infected cells produce IFN-α/β that activates DCs to promote $CD8^+$ T cell cross-priming.[94] Thus, both TLR3- and IFN-α/β-mediated signalings are likely implicated in licensing DCs for the cross-priming of $CD8^+$ T cells. However, the molecules responsible for TICAM-1-mediated cross-priming have not been identified.

5. TICAM-1 AND HOST DEFENSE

5.1. Role of TLR3–TICAM-1 in virus infection

Double-stranded RNA is generated in the cytoplasm during replication of positive-strand RNA and DNA viruses.[95] There is accumulating evidence that the TLR3–TICAM-1 signaling is involved in protection against virus infection. Two independent studies have demonstrated that the TLR3–TICAM-1 pathway is essential for protection against poliovirus infection, a virus belonging to the *Picornaviridae* family.[96,97] Poliovirus receptor (PVR)-transgenic and TICAM-1- or TLR3-deficient mice are more susceptible to PV than PVR-transgenic mice. TLR3-dependent type I IFN production by splenic $CD11c^+$ DCs and macrophages is essential for the protection of PVR-transgenic mice against PV infection. Moreover, Negishi *et al.*[98] showed that $TLR3^{-/-}$ mice are more vulnerable to coxsackievirus group B serotype 3 (a virus belonging to the *Picornaviridae* family) than wild-type mice, in terms of higher mortality and acute myocarditis. In addition to these mouse studies, genetic studies in patients with herpes simplex encephalitis demonstrated that the TLR3–TICAM-1 pathway is involved in protection against herpes simplex virus-1 encephalitis in children.[99–101] TLR3 is expressed in the central nervous system, where it might be required to control HSV-1.

In contrast, a detrimental effect of TLR3 signaling on the outcome of viral infection has been reported in several animal models. Wang *et al.*[102] showed that $TLR3^{-/-}$ mice have impaired cytokine production and enhanced viral loads in the periphery after West Nile virus (a positive-stranded RNA virus) infection, whereas in the brain, the viral load, inflammatory responses, and neuropathology were reduced compared with wild-type mice. TLR3-mediated peripheral inflammatory cytokine production is critical for disruption of the blood–brain barrier, which facilitates viral entry into the brain, thus causing lethal encephalitis. In other RNA viral infections such as respiratory syncytial virus, influenza A virus, and phlebovirus (all negative-stranded RNA viruses), TLR3-dependent inflammatory cytokine, and chemokine production affect virus-induced pathology and host survival.[103–105] Thus, TLR3 plays different roles in virus infections, which depends on the viral genome structure, the TLR3-expressing cell type that encounters viral RNA, and the properties of host antiviral effector functions.

5.2. Antiviral cellular immunity induced by the TLR3–TICAM-1 pathway

CTLs and NK cells are principal effector cells in antiviral immunity. The contribution of TLR3 to antiviral responses has been shown in MCMV infection,[106] during which virus clearance is partly dependent on NK cell activation. TLR3$^{-/-}$ mice are hypersusceptible to MCMV infection. Cytokine production and NK and NKT cell activations are impaired in TLR3$^{-/-}$ mice compared with wild-type mice. Recently, it has been shown that TLR3- and TICAM-1-mediated signaling in the nonhematopoietic compartment contributes to protection from intestinal epithelial rotavirus infection in adult mice.[107] In rotavirus-infected TLR3$^{-/-}$ mice, the expressions of IFN-λ and chemokines are reduced, and the recruitment of CD8^{+} T and NK cells to intestinal tissue is impaired, which might contribute to enhanced virus replication. In humans, Ebihara *et al.*[108] demonstrated the role of TLR3, expressed in myeloid DCs, in the immune response to HCV infection. The phagocytosis of HCV-infected apoptotic cells that contain HCV-derived dsRNA and their interaction with the TLR3 pathway in myeloid DCs, plays a critical role in DC maturation and activation of T and NK cells. In addition, Jongbloed *et al.*[35] reported that CD141^{+} DCs are able to cross-present viral antigens from human cytomegalovirus-infected necrotic fibroblasts. Physiologically, TLR3 in a DC subset specialized for antigen presentation encounters viral dsRNA in the endosome by uptake of apoptotic or necrotic virus-infected cells and signals for cross-presentation of viral antigens.

5.3. TLR3–TICAM-1 pathway in antitumor immunity

Selective expression of TLR3 in professional antigen-presenting DCs is an advantage when using TLR3 ligands as an antitumor vaccine adjuvant (Fig. 18.4). Azuma *et al.*[109] recently demonstrated that the TICAM-1 pathway in mouse CD8α^{+} DCs is crucial for cross-priming for antitumor CTLs induced by tumor-associated antigen and poly(I:C). When tumor cell lysate and poly(I:C) are subcutaneously injected in tumor-bearing mice, tumor growth retardation is observed in wild-type mice, and to a lesser extent IPS-1$^{-/-}$ mice, but not TICAM-1$^{-/-}$ mice. IRF3 and IRF7 are essential, but IPS-1 and type I IFN are minimally involved in poly(I:C)-mediated CTL proliferation. In addition to TICAM-1-dependent CTL activation, DC-mediated NK cell activation is also important for adjuvancy of TLR3 ligands. Akazawa *et al.*[88] showed that the TLR3–TICAM-1 pathway is essential for poly(I:C)-induced NK cell-mediated tumor regression in a syngeneic mouse tumor implant model. NK cell activation requires cell–cell contact with BMDCs preactivated by poly(I:C), but not IFN-α or IL-12. TICAM-1–IRF3-dependent expression of INAM in myeloid DCs is required for NK activation.[90] Further, Shime *et al.*[110] recently reported that TLR3–TICAM-1 signaling converts tumor-supporting myeloid cells to tumoricidal effectors. Thus, the TLR3–TICAM-1 pathway is not only important to mature myeloid DCs for cross-priming and NK cell activation in the induction of tumor immunity, but also critically engaged in tumor suppression by converting tumor-supporting macrophages to those with tumoricidal properties.[111]

Although poly(I:C) has pivotal antitumor activity, the use of poly(I:C) as a vaccine adjuvant is restricted by severe side effects.[112,113] As described above, poly(I:C) activates both TLR3 and MDA5, although the transporting mechanisms are unknown, resulting in the strong induction of type I IFNs, proinflammatory cytokines, and chemokines that are likely responsible for poly(I:C)-induced side effects. Therefore, it is important to devise an effective means for specific delivery of TLR3 ligands to TLR3-positive endosomes or new TLR3 ligands that activate TLR3 alone. Identification of the putative dsRNA uptake receptor is crucial for analyzing the intracellular transport of dsRNA. Further, clarification of the differences between the RLR–IPS-1- and TLR3–TICAM-1-mediated signaling pathways is important for assessment of dsRNA-induced immune responses.

6. CONCLUDING REMARKS

Accumulating evidence indicates that innate immune signaling is compartmentally regulated in PRRs and signaling adaptors, which is important for determination of cellular responses. TLR3- and TLR4-mediated signaling for IFN-β production is initiated from early endosomes through the recruitment of naïve TICAM-1 to endosomal TLR3 and TICAM-2, respectively. TICAM-1 then oligomerizes and forms a speckle-like signalosome with downstream signaling molecules in the cytosol apart from endosomes. Although there is a wide range of signaling cascades downstream from TICAM-1, including type I and type III IFN production, proinflammatory cytokine production, DC-mediated NK and CTL activation, and induction of apoptosis and necroptosis, it is unknown whether or not distinct TICAM-1 signalosomes are formed and how TICAM-1 signaling is regulated according to stimuli. Further, localization sites of TICAM-1 signalosome have not been clearly defined. Proteomic analyses of the TICAM-1 signalosome using different stimuli and responding cells may give new insight into the TICAM-1 function linking innate immunity to adaptive immunity.

ACKNOWLEDGMENTS

We would like to thank Drs. T. Akazawa, M. Sasai, T. Ebihara, H. Shime, M. Azuma, and M. Tatematsu for valuable discussions. This work was supported in part by CREST, JST, and by Grants-in-Aid from the ministry of Education, Science, and Culture, the Ministry of Health and Welfare of Japan, and by the Naito Memorial Foundation, the Uehara Memorial Foundation, and the Akiyama Life Science Foundation.

REFERENCES

1. Janeway Jr CA, Medzhitov R. Innate immune recognition. *Annu Rev Immunol* 2002;**20**:197–216.
2. Akira S, Uematsu S, Takeuchi O. Pathogen recognition and innate immunity. *Cell* 2006;**124**:783–801.
3. Marshak-Rothstein A. Toll-like receptors in systemic autoimmune disease. *Nat Rev Immunol* 2006;**6**:823–35.
4. Kono H, Rock KL. How dying cells alert the immune system to danger. *Nat Rev Immunol* 2008;**8**:279–89.
5. Medzhitov R, Preston-Hurburt P, Janeway Jr CA. A human homologue of the Drosophila Toll protein signals activation of adaptive immunity. *Nature* 1997;**388**:394–7.
6. Bell JK, Mullen GED, Leifer CA, Mazzoni A, Davies DR, Segal DM. Leucine-rich repeats and pathogen recognition in Toll-like receptors. *Trends Immunol* 2003;**24**:528–33.
7. Gay NJ, Gangloff M, Weber ANR. Toll-like receptors as molecular switches. *Nat Rev Immunol* 2006;**6**:693–8.
8. O'Neill LAJ, Bowie AG. The family of five: TIR-domain-containing adaptors in Toll-like receptor signaling. *Nat Rev Immunol* 2007;**7**:353–64.
9. Kawai T, Sato S, Ishii KJ, Coban C, Hemmi H, Yamamoto M, et al. Interferon-α induction through Toll-like receptors involves a direct interaction of IRF7 with MyD88 and TRAF6. *Nat Immunol* 2004;**5**:1061–8.
10. Oshiumi H, Matsumoto M, Funami K, Akazawa T, Seya T. TICAM-1, an adaptor molecule that participates in Toll-like receptor 3-mediated interferon-β induction. *Nat Immunol* 2003;**4**:161–7.
11. Yamamoto M, Sato S, Hemmi H, Hoshino K, Kaisho T, Sanjo H, et al. Role of adaptor TRIF in the MyD88-independent Toll-like receptor signaling pathway. *Science* 2003;**301**:640–3.
12. Oshiumi H, Sasai M, Shida K, Fujita T, Matsumoto M, Seya T. TIR-containing adapter molecule (TICAM)-2, a bridging adapter recruiting to Toll-like receptor 4 TICAM-1 that induces interferon-β. *J Biol Chem* 2003;**278**:49751–62.

13. Fitzgerald KA, Rowe DC, Barnes BJ, Caffrey DR, Visintin A, Latz E, et al. LPS-TLR4 signaling to IRF-3/7 and NF-κB involves the toll adapters TRAM and TRIF. *J Exp Med* 2003;**198**:1043–55.
14. Fitzgerald KA, Palsson-McDermott EM, Bowie AG, Jefferies CA, Mansell AS, Brady G, et al. Mal (Myd88-adapter-like) is required for Toll-like receptor 4 signal transduction. *Nature* 2001;**413**:78–83.
15. Horng T, Barton GM, Medzhitov R. TIRAP: an adaptor molecule in the Toll signaling pathway. *Nat Immunol* 2001;**2**:835–41.
16. Carty M, Goodbody R, Schroder M, Stack J, Moynagh PN, Bowie A. The human afdaptor SARM nrgatively regulates adaptor protein TRIF-dependent Toll-like receptor signaling. *Nat Immunol* 2006;**7**:1074–81.
17. Alexopoulou L, Holt AC, Medzhitov R, Flavell RA. Recognition of double-stranded RNA and activation of NF-κB by Toll-like receptor 3. *Nature* 2001;**413**:732–8.
18. Matsumoto M, Kikkawa S, Kohase M, Miyake K, Seya T. Establishment of a monoclonal antibody against human Toll-like receptor 3 that blocks double-stranded RNA-mediated signaling. *Biochem Biophys Res Commun* 2002;**239**:1364–9.
19. Yoneyama M, Kikuchi M, Natsukawa T, Shinobu N, Imaizumi T, Miyagishi M, et al. The RNA helicase RIG-I has an essential function in double-stranded RNA-induced innate antiviral responses. *Nat Immunol* 2004;**5**:730–7.
20. Hornung V, Ellegast J, Kim S, Brzózka K, Jung A, Kato H, et al. 5'-Triphosphate RNA is ligand for RIG-I. *Science* 2006;**314**:994–7.
21. Pichlmair A, Schulz O, Tan CP, Näslund TI, Liljeström P, Weber F, et al. RIG-I-mediated anti-viral responses to single-stranded RNA bearing 5'-phosphates. *Science* 2006;**314**:997–1001.
22. Kato H, Takeuchi O, Mikamo-Satoh E, Hirai R, Kawai T, Matsushita K, et al. Length-dependent recognition of double-stranded ribonucleic acids by retinoic acid-inducible gene-I and melanoma differentiation-associated gene 5. *J Exp Med* 2008;**205**:1601–10.
23. Kawai T, Takahashi K, Sato S, Coban C, Kumar H, Kato H, et al. IPS-1, an adaptor triggering RIG-I- and Mda5-mediated type I interferon induction. *Nat Immunol* 2005;**6**:981–8.
24. Seth RB, Sun L, Ea CK, Chen ZJ. Identification and characterization of MAVS, a mitochondrial antiviral signaling protein that activates NF-kappaB and IRF 3. *Cell* 2005;**122**:669–82.
25. Meylan E, Curran J, Hofmann K, Moradpour D, Binder M, Bartenschlager R, et al. Cardif is an adaptor protein in the RIG-I antiviral pathway and is targeted by hepatitis C virus. *Nature* 2005;**437**:1167–72.
26. Xu LG, Wang YY, Han KJ, Li LY, Zhai Z, Shu HB. VISA is an adapter protein required for virus-triggered IFN-beta signaling. *Mol Cell* 2005;**19**:727–40.
27. Zhang Z, Kim T, Bao M, Facchinetti V, Jung SY, Ghaffari AA, et al. DDX1, DDX21, and DHX36 helicases form a complex with the adaptor molecule TRIF to sense dsRNA in dendritic cells. *Immunity* 2011;**34**:866–78.
28. Farina C, Krumbholz M, Giese T, Hartmann G, Aloisi F, Meinl E. Preferential expression and function of Toll-like receptor 3 in human astrocytes. *J Neuroimmunol* 2005;**159**:12–9.
29. Town T, Jeng D, Alexopoulou L, Tan J, Flavell RA. Microglia recognize double-stranded RNA via TLR3. *J Immunol* 2006;**176**:3804–12.
30. Muzio M, Bosisio D, Polentarutti N, D'amico G, Stoppacciaro A, Mancinelli R, et al. Differential expression and regulation of Toll-like receptors (TLR) in human leukocytes: selective expression of TLR3 in dendritic cells. *J Immunol* 2000;**64**:5998–6004.
31. Visintin A, Mazzoni A, Spitzer JH, Wyllie DH, Dower SK, Segal DM. Regulation of Toll-like receptors in human monocytes and dendritic cells. *J Immunol* 2001;**166**: 249–54.

32. Kadowaki N, Ho S, Antonenko S, Malefyt RW, Kastelein RA, Bazan F, et al. Subsets of human dendritic cell precursors express different Toll-like receptors and respond to different microbial antigens. *J Exp Med* 2001;**194**:863–70.
33. Hornung V, Rothenfusser S, Britisch S, Krug A, Jahrsdörfer B, Giese T, et al. Quantitative expression of Toll-like receptor 1-10 mRNA in cellular subsets of human peripheral blood mononuclear cells and sensitivity to CpG oligodeoxynucleotides. *J Immunol* 2002;**168**:4531–7.
34. Matsumoto M, Funami K, Tanabe M, Oshiumi H, Shingai M, Seto Y, et al. Subcellular localization of Toll-like receptor 3 in human dendritic cells. *J Immunol* 2003;**171**:3154–62.
35. Jongbloed SL, Kassianos AJ, McDonald KJ, Clark GJ, Ju X, Angel CE, et al. Human CD141 + (BDCA-3)+ dendritic cells (DCs) represent a unique myeloid DC subset that cross-presents necrotic cell antigens. *J Exp Med* 2010;**207**:1247–60.
36. Poulin LF, Salio M, Griessinger E, Anjos-Afonso F, Craciun L, Chen JL, et al. Characterization of human DNGR-1+ BDCA3+ leukocytes as putative equivalents of mouse CD8α+ dendritic cells. *J Exp Med* 2010;**207**:1261–71.
37. Matsumoto M, Seya T. TLR3: interferon induction by double-stranded RNA including poly(I:C). *Adv Drug Deliv Rev* 2008;**60**:805–12.
38. Miettinen M, Sareneva T, Julkunen I, Matikainen S. IFNs activate toll-like receptor gene expression in viral infections. *Genes Immun* 2001;**2**:349–55.
39. Brinkmann MM, Spooner E, Hoebe K, Beutler B, Ploegh HL, et al. The interaction between the ER membrane protein UNC93B and TLR3, 7, and 9 is crucial for TLR signaling. *J Cell Biol* 2007;**177**:265–75.
40. Funami K, Matsumoto M, Oshiumi H, Akazawa T, Yamamoto A, Seya T. The cytoplasmic 'linker region' in Toll-like receptor 3 controls receptor localization and signaling. *Int Immunol* 2004;**16**:1143–54.
41. Nishiya T, Kajita E, Miwa S, DeFranco A. TLR3 and TLR7 are targeted to the same intracellular compartments by distinct regulatory elements. *J Biol Chem* 2005;**280**:37107–17.
42. Seya T, Matsumoto M, Ebihara T, Oshiumi H. Functional evolution of the TICAM-1 pathway for extrinsic RNA sensing. *Immunol Rev* 2009;**227**:44–53.
43. Choe J, Kelker MS, Wilson IA. Crystal structure of human Toll-like receptor 3 (TLR3) ectodomain. *Science* 2005;**309**:581–5.
44. Bell JK, Botos I, Hall PR, Askins J, Shiloach J, Segal DM, et al. The molecular structure of the Toll-like receptor 3 ligand-binding domain. *Proc Natl Acad Sci USA* 2005;**102**:10976–80.
45. Liu L, Botos I, Wang Y, Leonard JN, Shiloach J, Segal DM, et al. Structural basis of Toll-like receptor 3 signaling with double-stranded RNA. *Science* 2008;**320**:379–81.
46. Bell JK, Askins J, Hall PR, Davies DR, Segal DM. The dsRNA binding site of human Toll-like receptor 3. *Proc Natl Acad Sci USA* 2006;**103**:8792–7.
47. Fukuda K, Watanabe T, Tokisue T, Tsujita T, Nishikawa S, Hasegawa T, et al. Modulation of double-stranded RNA recognition by the N-terminal histidine-rich region of the human Toll-like receptor 3. *J Biol Chem* 2008;**283**:22787–94.
48. Pirher N, Ivicak K, Pohar J, Bencina M, Jerala R. A second binding site for double-stranded RNA in TLR3 and consequences for interferon activation. *Nat Struct Mol Biol* 2008;**15**:761–3.
49. Wang Y, Liu L, Davies DR, Segal DM. Dimerization of Till-like receptor 3 (TLR3) is required for ligand binding. *J Biol Chem* 2010;**285**:36836–41.
50. Leonard JN, Ghirlando R, Askins J, Bell JK, Margulies DH, Davies DR, et al. The TLR3 signaling complex forms by cooperative receptor dimerization. *Proc Natl Acad Sci USA* 2008;**105**:258–63.

51. Jelinek I, Leonard JN, Price GE, Brown KN, Meyer-Manlapat A, Goldsmith PK, et al. TLR3-specific double-stranded RNA oligonucleotide adjuvants induce dendritic cell cross-presentation, CTL responses, and antiviral protection. *J Immunol* 2011;**186**:2422–9.
52. Itoh K, Watanabe A, Funami K, Seya T, Matsumoto M. The clathrin-mediated endocytic pathway participates in dsRNA-induced IFN-β production. *J Immunol* 2008;**181**:5522–9.
53. Watanabe A, Tatematsu M, Saeki K, Shibata S, Shime H, Yoshimura A, et al. Raftlin is involved in the nucleocapture complex to induce poly(I:C)-mediated TLR3 activation. *J Biol Chem* 2011;**286**:10702–11.
54. Lee HKS, Dunzendorfer K, Soldau K, Tobias PS. Double-stranded RNA-mediated TLR3 activation is enhanced by CD14. *Immunity* 2006;**24**:153–63.
55. Limmon GV, Arredouani M, McCann KL, Minor RAC, Kobzik L, Imani F. Scavenger receptor class-A is a novel cell surface receptor for double-stranded RNA. *FASEB J* 2008;**22**:159–67.
56. Weber C, Müller C, Podszuweit A, Montino C, Vollmer J, Forsbach A. Toll-like receptor (TLR) 3 immune modulation by unformulated small interfering RNA or DNA and the role of CD14 (in TLR-mediated effects). *Immunology* 2012;**136**:64–77.
57. Akashi S, Saitoh S, Wakabayashi Y, Kikuchi T, Takamura N, Nagai Y, et al. Lipopolysaccharide interaction with cell surface Toll-like receptor 4-MD-2: higher affinity than that with MD-2 or CD14. *J Exp Med* 2003;**198**:1035–42.
58. Kagan JC, Medzhitov R. Phosphoinositide-mediated adaptor recruitment controls Toll-like receptor signaling. *Cell* 2006;**125**:943–55.
59. Kagan JC, Su T, Horng T, Chow A, Akira S, Medzhitov R. TRAM couples endocytosis of Toll-like receptor 4 to the induction of interferon-β. *Nat Immunol* 2008;**9**:361–8.
60. Seya T, Oshiumi H, Sasai M, Akazawa T, Matsumoto M. TICAM-1 and TICAM-2: toll-like receptor adapters that participate in induction of type I interferons. *Int J Biochem Cell Biol* 2005;**37**:524–9.
61. Funami K, Sasai M, Oshiumi H, Seya T, Matsumoto M. Homo-oligomerization is essential for Toll/IL-1 receptor domain-containing adaptor molecule-1 mediated NF-κB and IRF-3 activation. *J Biol Chem* 2008;**283**:18283–91.
62. Sato S, Sugiyama M, Yamamoto M, Watanabe Y, Kawai T, Takeda K, et al. Toll/IL-1 receptor domain-containing adaptor-inducing IFN-β (TRIF) associates with TNFR-associated factor 6 and TANK-binding kinase 1, and activates two distinct transcription factors, NF-κB and IFN-regulatory factor 3, in the Toll-like receptor signaling. *J Immunol* 2003;**171**:4304–10.
63. Sasai M, Tatematsu M, Oshiumi H, Funami K, Matsumoto M, Hatakeyama S, et al. Direct binding of TRAF2 and TRAF6 to TICAM-1/TRIF adaptor participates in activation of the Toll-like receptor 3/4 pathway. *Mol Immunol* 2010;**47**:1283–91.
64. Tatematsu M, Ishii A, Oshiumi H, Horiuchi M, Inagaki F, Seya T, et al. A molecular mechanism for Toll/IL-1 receptor domain-containing adaptor molecule-1-mediated IRF-3 activation. *J Biol Chem* 2010;**285**:20128–36.
65. Meylan E, Burns K, Hofmann K, Blancheteau V, Martinon F, Kelliher M, et al. RIP1 is an essential mediator of Toll-like receptor 3-induced NF-kappa B activation. *Nat Immunol* 2004;**5**:503–7.
66. Han KJ, Su X, Xu LG, Bin LH, Zhang J, Shu HB. Mechanisms of the TRIF-induced interferon-stimulated response element and NF-kappaB activation and apoptosis pathways. *J Biol Chem* 2004;**279**:15652–61.
67. Kaiser WJ, Offermann MK. Apoptosis induced by the Toll-like receptor adaptor TRIF is dependent on its receptor interacting protein homotypic interaction motif. *J Immunol* 2005;**174**:4942–52.

68. Funami K, Sasai M, Ohba Y, Oshiumi H, Seya T, Matsumoto M. Spatiotemporal mobilization of Toll-IL-1 receptor domain-containing adaptor molecule 1 in response to dsRNA. *J Immunol* 2007;**179**:6867–72.
69. Sharma S, tenOever BR, Grandvaux N, Zhou GP, Lin R, Hiscott J. Triggering the interferon antiviral response through an IKK-related pathway. *Science* 2003;**300**:1148–51.
70. Fitzgerald KA, McWhirter SM, Faia KL, Rowe DC, Latz E, Golenbock DT, et al. IKKε and TBK1 are essential components of the IRF3 signaling pathway. *Nat Immunol* 2003;**4**:491–6.
71. Hacker H, Redecke V, Blagoev B, Kratchmarova I, Hsu LC, Wang GG, et al. Specificity in Toll-like receptor signaling through distinct effector functions of TRAF3 and TRAF6. *Nature* 2006;**439**:204–7.
72. Oganesyan G, Saha SK, Guo B, He JQ, Shahangian A, Zarnegar B, et al. Critical role of TRAF3 in the Toll-like recepror-dependent and -independent antiviral response. *Nature* 2006;**439**:208–11.
73. Sasai M, Oshiumi H, Matsumoto M, et al. Cutting Edge: NF-kappaB-activating kinase-associated protein 1 participates in TLR3/Toll-IL-1 homology domain-containing adapter molecule-1-mediated IFN regulatory factor 3 activation. *J Immunol* 2005;**174**:27–30.
74. Gohda J, Matsumura T, Inoue J. Cutting Edge: TNFR-associated factor (TRAF) 6 is essential for MyD88-dependent pathway but not Toll/IL-1 receptor domain-containing adaptor-inducing IFN-β (TRIF)-dependent pathway in TLR signaling. *J Immunol* 2004;**173**:2913–7.
75. Kayagaki N, Phung Q, Chan S, Chaudhari R, Quan C, O'Rourke KM, et al. DUBA: a deubiquitinase that regulates type I interferon production. *Science* 2007;**318**:1628–32.
76. Boone DL, Turer EE, Lee EG, Ahmad RC, Wheeler MT, Tsui C, et al. The ubiquitin-modifying enzyme A20 is required for termination of Toll-like receptor responses. *Nat Immunol* 2004;**5**:1052–60.
77. Feoktistova M, Geserick P, Kellert B, Dimitrova DP, Langlais C, Hupe M, et al. cIAPs block Ripoptosome formation, a RIP1/caspase-8 containing intracellular cell death complex differentially regulated by cFLIP isoforms. *Mol Cell* 2011;**43**:449–63.
78. Galluzzi L, Kepp O, Kroemer G. RIP Kinases initiate programmed necrosis. *J Mol Cell Biol* 2009;**1**:8–10.
79. Cho YS, Challa S, Moquin D, Genga R, Ray TD, Guildford M, et al. Phosphorylation-driven assembly of the RIP1-RIP3 complex regulates programmed necrosis and virus-induced inflammation. *Cell* 2009;**137**:1112–23.
80. Mocarski Edward S, Upton Jason W, Kaiser William J. Viral infection and the evolution of caspase 8-regulated apoptotic and necrotic death pathways. *Nat Rev Immunol* 2012;**12**:79–88.
81. He S, Liang Y, Shao F, Wang X. Toll-like receptors activate programmed necrosis in macrophages through a receptor-interacting kinase-3-mediated pathway. *Proc Natl Acad Sci USA* 2011;**108**:20054–9.
82. Salaun B, Zitvogel L, Asselin-Paturel C, Morel Y, Chemin K, Dubois C, et al. TLR3 as a biomarker for the therapeutic efficacy of double-stranded RNA in breast cancer. *Cancer Res* 2011;**71**:1607–14.
83. Estornes Y, Toscano F, Virard F, Jacquemin G, Pierrot A, Vanbervliet B, et al. dsRNA induces apoptosis through an atypical death complex associating TLR3 to caspase-8. *Cell Death Differ* 2012;**19**:1482–94.
84. Seya T, Shime H, Takaki H, Azuma M, Oshiumi H, Matsumoto M. TLR3/TICAM-1 signaling in tumor cell RIP3-dependent necroptosis. *Oncoimmunology* 2012;**1**:917–23.
85. Samuel CE. Antiviral actions of interferon, Interferon-regulated cellular proteins and their surprisingly selective antiviral activities. *Virology* 1991;**183**:1–11.

86. Lauterbach H, Bathke B, Gilles S, et al. Mouse CD8a+ DCs and human BDCA3+ DCs are major producers of IFN-λ in response to poly(I:C). *J Exp Med* 2010;**207**: 2703–17.
87. Kato H, Takeuchi O, Sato S, Yoneyama M, Yamamoto M, Matsui K, et al. Differential roles of MDA5 and RIG-I helicases in the recognition of RNA viruses. *Nature* 2006;**441**:101–5.
88. Akazawa T, Ebihara T, Okuno M, Okuda Y, Shingai M, Tsujimura K, et al. Antitumor NK activation induced by the TLR3-TICAM-1 (TRIF) pathway in myeloid dendritic cells. *Proc Natl Acad Sci USA* 2007;**104**:252–7.
89. Seya T, Kasamatsu J, Azuma M, Shime H, Matsumoto M. Natural killer cell activation secondary to innate pattern sensing. *J Innate Immun* 2011;**3**:264–73.
90. Ebihara T, Azuma M, Oshiumi H, Kasamatsu J, Iwabuchi K, Matsumoto K, et al. Identification of a polyI:C-inducible membrane protein that participates in dendritic cell-mediated natural killer cell activation. *J Exp Med* 2010;**207**:2675–87.
91. Heath WR, Belz GT, Behrens GM, Smith CM, Forehan SP, Parish IA, et al. Cross-presentation, dendritic cell subsets, and the generation of immunity to cellular antigens. *Immunol Rev* 2004;**199**:9–26.
92. Shen L, Lock KL. Priming of T cells by exogenous antigen cross-presented on MHC class I molecules. *Curr Opin Immunol* 2006;**18**:85–91.
93. Schulz O, Diebold SS, Chen M, Näslund TI, Nolte MA, Alexopoulou L, et al. Toll-like receptor 3 promotes cross-priming to virus-infected cells. *Nature* 2005;**433**: 887–92.
94. Le Bon A, Etchart N, Rossmann C, Ashton M, Hou S, Gewert D, et al. Cross-priming of CD8+ T cells stimulated by virus-induced type I interferon. *Nat Immunol* 2003; **4**:1009–15.
95. Weber F, Wagner V, Rasmussen SB, Hartmann R, Paludan SR. Double-stranded RNA is produced by positive-stranded RNA viruses and DNA viruses but not in detectable amounts by negative-stranded RNA viruses. *J Virol* 2006;**80**:5059–64.
96. Oshiumi H, Okamoto M, Fujii K, Kawanishi T, Matsumoto M, Koike S, et al. The TLR3-TICAM-1 pathway is mandatory for innate immune responses to poliovirus infection. *J Immunol* 2011;**187**:5320–7.
97. Abe Y, Fujii K, Nagata N, Takeuchi O, Akira S, Oshiumi H, et al. The Toll-like receptor 3-mediayed antiviral response is important for protection against poliovirus infection in poliovirus receptor transgenic mice. *J Virol* 2012;**86**:185–94.
98. Negishi H, Osawa T, Ogami K, Ouyang X, Sakaguchi S, Koshiba R, et al. A critical link between Toll-like receptor 3 and type II interferon signaling pathways in antiviral innate immunity. *Proc Natl Acad Sci USA* 2008;**105**:20446–51.
99. Zhang S-Y, Jouanguy E, Ugolini S, Smahi A, Elain G, Romero P, et al. TLR3 deficiency in patients with Herpes Simplex encephalitis. *Science* 2007;**317**:1522–7.
100. Guo Y, Audry M, Ciancanelli M, Alsina L, Azevedo J, Herman M, et al. Herpes simplex virus encephalitis in a patient with complete TLR3 deficiency: TLR3 is otherwise redundant in protective immunity. *J Exp Med* 2011;**208**:2083–98.
101. Sancho-Shimizu V, Pérez de Diego R, Lorenzo L, Halwani R, Alangari A, Israelsson E, et al. Herpes simplex encephalitis in children with autosomal recessive and dominant TRIF deficiency. *J Clin Invest* 2011;**121**:4889–902.
102. Wang T, Town T, Alexopoulou L, Anderson JF, Fikrig E, Flavell RA. Toll-like receptor 3 mediates West Nile virus entry into the brain causing lethal encephalitis. *Nat Med* 2004;**10**:1366–73.
103. Le Goffic R, Balloy V, Lagranderie M, Alexopoulou L, Escriou N, Flavell R, et al. Detrimental contribution of the Toll-like receptor (TLR) 3 to influenza A virus-induced acute pneumonia. *PLoS Pathog* 2006;**2**:526–35.

104. Gowen BB, Hoopes JD, Wong MH, Jung KH, Isakson KC, Alexopoulou L, et al. TLR3 deletion limits mortality and disease severity due to phlebovirus infection. *J Immunol* 2006;**177**:6301–7.
105. Rudd BD, Smit JJ, Flavell RA, Alexopoulou L, Schller MA, Gruber A, et al. Deletion of TLR3 alters the pulmonary immune environment and mucus production during respiratory syncytial virus infection. *J Immunol* 2006;**176**:1937–42.
106. Tabeta K, Georgel P, Janssen E, Du X, Hoebe K, Crozat K, et al. Toll-like receptor 9 and 3 as essential components of innate immune defense against mouse cytomegalovirus infection. *Proc Natl Acad Sci USA* 2004;**101**:3516–21.
107. Pott J, Stockinger S, Torow N, Smoczek A, Lindner C, McInerney G, et al. Age-dependent TLR3 expression of the intestinal epithelium contributes to rotavirus susceptibility. *PLoS Pathog* 2012;**8**:e1002670.
108. Ebihara T, Shingai M, Matsumoto M, Wakita T, Seya T. Hepatitis C virus (HCV)-infected apoptotic cells extrinsically modulate dendritic cell function to activate T cells and NK cells. *Hepatology* 2008;**48**:48–58.
109. Azuma M, Ebihara T, Oshiumi H, Matsumoto M, Seya T. Cross-priming for anti-tumor CTL induced by soluble Ag + polyI:C depends on the TICAM-1 pathway in mouse CD11c+/CD8α+ dendritic cells. *Oncoimmunology* 2012;**1**:581–92.
110. Shime H, Matsumoto M, Oshiumi H, Tanaka S, Nakane A, Iwakura Y, et al. Toll-like receptor 3 signaling converts tumor-supporting myeloid cells to tumoricidal effectors. *Proc Natl Acad Sci USA* 2012;**109**:2066–71.
111. Seya T, Shime H, Matsumoto M. TAMable tumor-associated macrophages in response to innate RNA sensing. *Oncoimmunology* 2012;**1**:1000–1.
112. Absher M, Stinebring WR. Toxic properties of a synthetic double-stranded RNA. Endotoxin-like properties of poly I. poly C, an interferon stimulator. *Nature* 1969; **223**:715–7.
113. Champney KJ, Levine DP, Levy HB, Lerner AM. Modified polyriboinosinic-polyribocytidylic acid complex: sustained interferonemia and its physiological associates in humans. *Infect Immun* 1979;**25**:831–7.

CHAPTER NINETEEN

Assembly of Gamma-Tubulin Ring Complexes: Implications for Cell Biology and Disease

Marie-Hélène Remy*, **Andreas Merdes***, **Lynn Gregory-Pauron**†
*Centre de Biologie du Développement, Université Paul Sabatier, Toulouse, France
†Unité de Recherche Animal et Fonctionnalités des Produits Animaux, Université de Lorraine, INRA, Vandoeuvre-lès-Nancy, France

Contents

Abstract

Microtubules are among the main constituents of the cytoskeleton. They are assembled from dimers of alpha- and beta-tubulin. This assembly occurs preferentially at organizing centers such as the centrosomes, catalyzed by multiprotein complexes of gamma-tubulin. At the beginning of mitosis, the amount of gamma-tubulin complexes at the centrosomes increases sharply, supporting the sudden formation of numerous spindle microtubules. Recent studies on the structure of gamma-tubulin complex proteins have advanced our understanding of the assembly process of gamma-tubulin complexes, and have pointed toward putative mechanisms of microtubule nucleation. Moreover, the discovery of novel proteins associated with gamma-tubulin complexes has illustrated the possibilities of how gamma-tubulin might be recruited and regulated at specific sites of microtubule organization. This chapter highlights recent developments in the field and discusses the potential of the gamma-tubulin complex as a pharmacological target, to control proliferation of cells.

Progress in Molecular Biology and Translational Science, Volume 117
ISSN 1877-1173
http://dx.doi.org/10.1016/B978-0-12-386931-9.00019-2

1. INTRODUCTION

Microtubules are filamentous polymers that constitute an essential part of the cytoskeleton in all eukaryotes. Most prominently, microtubules are visible during cell division, as the main structural elements of the spindle apparatus. They are necessary for the attachment and equal separation of chromosomes. During interphase, microtubules provide a framework for intracellular transport of vesicles, organelles, and other cargoes, by microtubule-dependent motor proteins. Moreover, microtubules are necessary for the polarization and locomotion of cells. Microtubules are dynamic polymers that possess a fast growing plus-end, and a slower growing minus-end.[1] This polarity results from the polar assembly of heterodimers of alpha- and beta-tubulin into the microtubule polymer. The microtubule polymer itself can be described as a hollow cylinder that is constituted of laterally associated protofilaments of alpha/beta-tubulin heterodimers (Fig. 19.1), with beta-tubulin facing toward

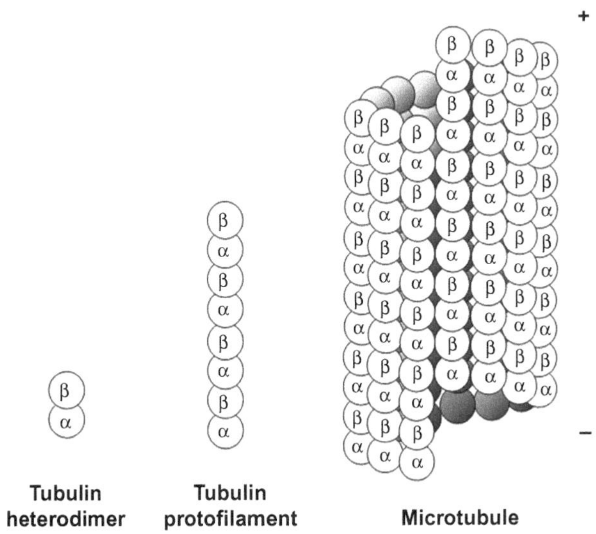

Figure 19.1 Assembly intermediates of microtubules. Left: heterodimer of alpha- and beta-tubulin. Middle: protofilament of longitudinally aligned heterodimers. During assembly, alpha-tubulin subunits are oriented toward the minus-end and beta-tubulin subunits are oriented toward the plus-end. Right: microtubule, containing 13 protofilaments assembled in a typical "A-lattice," corresponding to a 3-start helix. Note the formation of a "seam" at the contact between the 1st and the 13th protofilament, where lateral contacts between alpha- and beta-tubulin occur, in contrast to the remainder of the lattice, where homotypic lateral contacts (alpha–alpha and beta–beta) occur.

the plus-end.[2] Most microtubules in the cell have 13 protofilaments per diameter.[3–6] The protofilaments are nucleated from a template structure that contains a related member of the tubulin family, gamma-tubulin.[7] Gamma-tubulin is found enriched on the centrosome, as a part of the amorphous pericentriolar material on the centrosomal surface.[7–9] The amounts of gamma-tubulin on the centrosome oscillate during the cell cycle, with the highest concentration in mitosis, and the lowest concentration in interphase.[8,10] In mitosis, a large amount of gamma-tubulin is also seen on the surface of spindle microtubules, and at the mid-body during anaphase/telophase.[11–13]

2. COMPOSITION OF GAMMA-TUBULIN COMPLEXES

Gamma-tubulin was originally discovered in a screen for suppressors of a heat-sensitive beta-tubulin mutation in the fungus *Aspergillus nidulans*.[14] Initially thought to encode a nontubulin gene, the suppressor gene was subsequently characterized and sequenced, which revealed its relation to the tubulin family.[7] A role of gamma-tubulin in the nucleation and organization of microtubules from centrosomes and spindle pole bodies was put in evidence.[8,9,15,16] In many organisms, two gamma-tubulin genes exist, encoding two closely related isoforms, gamma-tubulin 1 and 2.[17] In mice, the gamma-tubulin isoform 2 is expressed specifically in the brain.[18] Studies in knockout mice indicate that this isoform may be nonessential, in contrast to gamma-tubulin 1 which is essential and ubiquitously expressed.[18] Gamma-tubulin binds mainly to the minus-ends of microtubules, although a fraction of the protein may also bind to the microtubule lattice.[19–23] In order to form a template for microtubule nucleation, gamma-tubulin associates with a specific set of proteins[24] called "gamma-tubulin complex proteins" or "GCPs."[25] In most eukaryotes, gamma-tubulin complexes contain five distantly related GCPs: GCP2, 3, 4, 5, and 6.[26,27] Together, these GCPs represent a protein family characterized by a "grip 1" motif in the amino-terminal half, and a "grip 2" motif in the carboxy-terminal half of their sequence.[26] Gamma-tubulin itself is sometimes referred to as GCP1, although its sequence has no resemblance to the GCP family.[25] The first member of the GCP family was identified in *Saccharomyces cerevisiae*, as a suppressor of a conditional-lethal gamma-tubulin allele.[28] The protein, named Spc98p, is homologous to the vertebrate GCP3. This led to the subsequent identification of Spc97p, the budding yeast homologue of GCP2, which was shown to interact with Spc98p.[29] The first GCPs in vertebrates and Drosophila were characterized, soon afterwards, by several independent groups.[25–27,30–34]

Two types of gamma-tubulin complexes are found in eukaryotes: a small complex, also named "gamma-TuSC" for "gamma-tubulin small complex," and a large "gamma-tubulin ring complex," also named "gamma-TuRC."[30] Gamma-TuSCs contain a dimer of laterally associated GCP2 and GCP3 that are each bound to one molecule of gamma-tubulin (Fig. 19.2A; Ref. 37). *In vitro*, gamma-TuSCs have the potential to oligomerize into larger structures that bear resemblance to gamma-TuRCs, but that are very sensitive to variation in salt and pH.[35] In contrast to gamma-TuSCs, gamma-TuRCs contain several additional proteins, namely GCP4, 5, 6, and the proteins NEDD1, MOZART 1, MOZART 2, and NME7 (Fig. 19.2B and C; Refs. 26,27,38–42). NEDD1, MOZART 1 and 2, and NME7 are unrelated to the GCP family. NEDD1 possesses WD40 sequence motifs that are involved in protein–protein interactions and that appear to establish the binding to centrosomes and spindles. Because of this sequence feature, NEDD1 is also referred to as GCP-WD in mammalian cells, or Dgp71WD in *Drosophila*.[38,43]

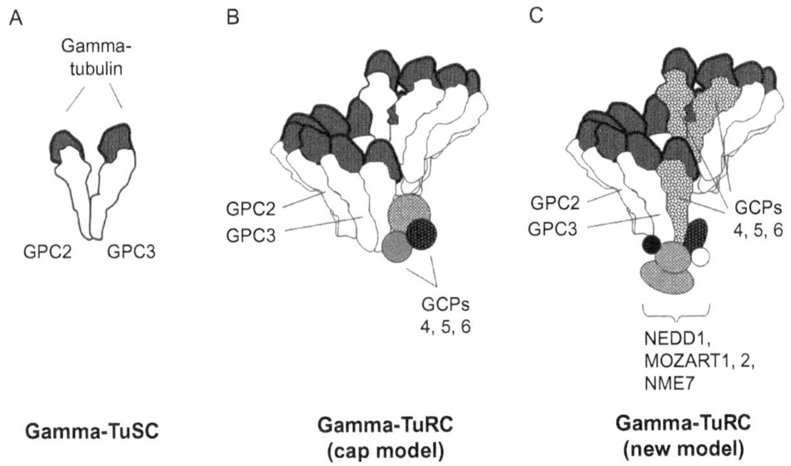

Figure 19.2 Schematic representations of gamma-TuSC and gamma-TuRC complexes, lateral views. (A) Gamma-TuSCs are composed of one molecule of GCP2, one molecule of GCP3, and two molecules of gamma-tubulin that are longitudinally bound to GCPs 2 and 3. (B) Representation of the gamma-TuRC according to Ref. 22. GCPs 4, 5, and 6 are located at the base of the complex, as part of a cap. (C) Revised model of the gamma-TuRC, taking into account structural data by Refs. 35 and 36. GCPs 4, 5, and 6 possess structures similar to GCPs 2 and 3, and are integrated into the wall of the complex. Their low stoichiometric abundance in the complex suggests that they may play specific roles at the start and the end of the gamma-TuRC helix. Recently discovered proteins such as NEDD1, MOZART 1 and 2, and NME7 may bind to the base of the gamma-TuRC and mediate binding of the gamma-TuRC to microtubule organizing centers.

MOZART 1 and 2 are small proteins of 8.5 and 16.2 kDa, respectively.[40] Two related isoforms of MOZART 2 exist, A and B, encoded by two distinct genes in humans.[40] NME7 is a nucleoside diphosphate kinase mainly expressed in tissues with motile cilia and in sperm, and binds to microtubules *in vitro*.[44]

3. STRUCTURE OF GAMMA-TUBULIN COMPLEX COMPONENTS

A combination of recent crystallographic data on GCP4 and cryo-electron microscopic data on gamma-TuSCs enabled structure prediction of all GCP family members[35,36]: GCPs share a common core structure that resembles a flattened "V." The GCP core contains five bundles of alpha-helices, with a kink between bundles 3 and 4 (Fig. 19.3; Ref. 36). The amino-terminal three bundles that form the first diagonal of the "V" correspond to the previously described "grip 1" motif and permit lateral contact with neighboring GCPs in gamma-TuSCs and gamma-TuRCs.[36] The C-terminal two bundles of alpha-helices constitute the second diagonal of the "V"-shaped molecule and correspond to the described "grip 2" motif.[36] Biochemical experiments as well as structural modeling of gamma-tubulin complexes show that this motif enables longitudinal contact of each GCP with gamma-tubulin.[36] Gamma-tubulin itself is structurally similar to

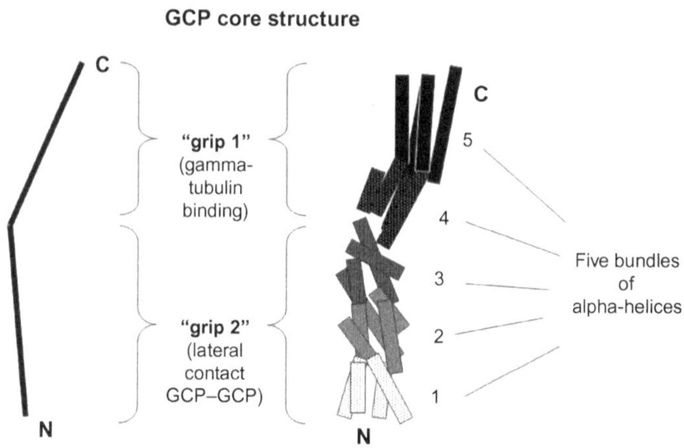

Figure 19.3 Schematic representation of the core structure of GCPs: left, the overall shape can be compared to a shallow "V." GCP-protein domains previously described as "grip 1" and "grip 2" represent stacked bundles of alpha-helices (right), with a kink between bundles 3 and 4, marking the domain border between grip 1 and grip 2.

alpha- and beta-tubulin.[45] Thus, during microtubule nucleation, it is expected that gamma-tubulin forms microtubule-like contacts with alpha/beta-tubulin.[35]

4. ASSEMBLY OF GAMMA-TUBULIN COMPLEXES *IN VITRO* AND *IN VIVO*

The assembly of gamma-tubulin complexes is a highly regulated process. The expression of the three gamma-TuSC proteins gamma-tubulin, GCP2, and GCP3 is interdependent. Knockdown of one component results in downregulation of the other gamma-TuSC constituents.[23,46] This regulation occurs at a posttranscriptional level, although it remains unclear whether it underlies translational or posttranslational control.[23,46] Likewise, an interdependent expression of several gamma-TuRC proteins has been described.[47,48]

Apparently, the presence of gamma-TuSC proteins alone is sufficient to assemble higher-order complexes that serve as microtubule templates. Two observations support this view: first, the genes encoding gamma-TuSC proteins are essential in all eukaryotes,[15,28,29,49–54] whereas the genes encoding the gamma-TuRC-specific proteins GCP4, 5, and 6 are nonessential in a variety of biological systems, such as *Drosophila*, *A. nidulans*, or *Schizosaccharomyces pombe*.[47,55,56] Second, reconstitution of gamma-tubulin complexes that nucleate microtubules can be performed *in vitro* with the purified gamma-TuSCs from *Drosophila* or *S. cerevisiae*.[30,35] Lateral contact between the N-terminal grip 1 domains of GCP2 and GCP3 leads to the formation of various gamma-TuSC oligomers (Fig. 19.4), with equivalent binding at intergamma-TuSC and intragamma-TuSC contact sites.[35] Oligomerization of gamma-TuSCs also produces a small amount of ring complexes that are highly similar to purified gamma-TuRCs.[22,35] The assembly of rings *in vitro* is facilitated by stabilization with a 220 amino acid-long fragment of the spindle pole body protein Spc110p and by the simultaneous addition of glycerol.[35] Closer examination shows that these complexes resemble lock washers rather than rings, with a slight helical pitch. In the presence of the Spc110p fragment, larger helices with multiple turns are produced by oligomerizing gamma-TuSCs, allowing single particle reconstruction of their structure at 8 Å resolution.[35] A helical pitch is also seen in gamma-TuRCs, purified from *Drosophila* embryonic extracts.[22] Considering the similarities between oligomerized gamma-TuSC helices and purified gamma-TuRCs, it seems likely that the two complexes use the same assembly mechanisms. In contrast to oligomerized gamma-TuSCs, gamma-TuRCs

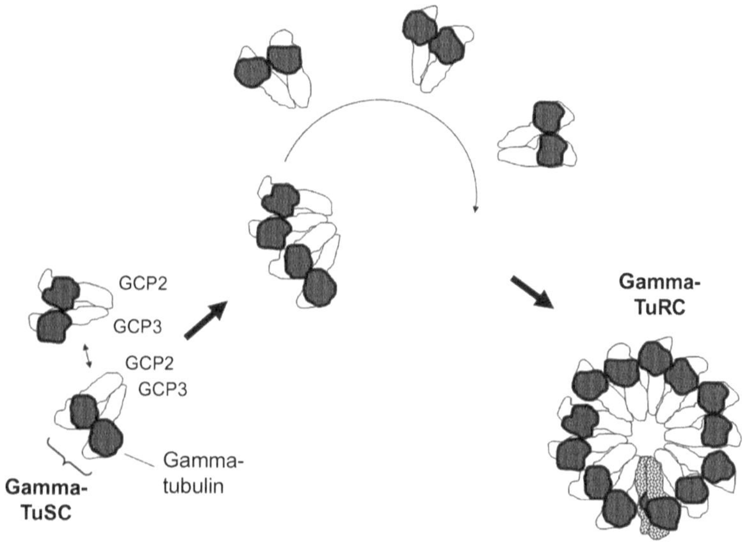

Figure 19.4 Possible assembly mechanism of gamma-TuRCs as viewed from the top. Experiments *in vitro* by Ref. 35 suggest that lateral assembly of gamma-TuSCs leads to the formation of a helix. In a gamma-TuRC, the growth of this helix may be limited by the incorporation of GCPs 4, 5, and 6 at the end-positions (gray), after approximately one full helical turn. These subunits contribute to the stability of the complex.

possess a fixed size, with a relatively well-defined composition of protein subunits, including 12–13 copies of GCP2 and GCP3 combined, in addition to 2–3 copies of GCP4, 1 copy of GCP5, and 0–1 copies of GCP6.[42] Older models of the gamma-TuRC suggest that GCP4, 5, and 6 form a cap at the base of the short helix (Fig. 19.2B; Ref. 22). This model has recently been revised. Based on biochemical and crystallographic data of the protein GCP4, it is now hypothesized that GCP4, 5, and 6 integrate into the helical wall of the gamma-TuRC (Fig. 19.2C; Refs. 35,36). It is, therefore, conceivable that gamma-TuRCs assemble by lateral association of gamma-TuSCs into a helix, and that GCP4, 5, and 6 occupy specific positions at the start and the end of this helix that limit its growth to one full turn.[57] In addition, GCP4, 5, and 6 may stabilize the gamma-TuRC and protect it from disassembly.[57]

5. MECHANISMS OF GAMMA-TUBULIN-DEPENDENT NUCLEATION OF MICROTUBULES

Interestingly, helical gamma-tubulin complexes incorporate 13 molecules of gamma-tubulin per turn, which equals the number of protofilaments in a regular microtubule.[35] Moreover, gamma-tubulin in these complexes is oriented to expose its putative contact site to alpha/beta-tubulin longitudinally, but not laterally.[35] This orientation rules out a previously proposed "protofilament model" of gamma-tubulin complexes.[58] The model had suggested that the helical gamma-tubulin complex would partially unwind into a short filament. This filament would drive nucleation of microtubules by lateral association with alpha/beta-tubulin protofilaments (Fig. 19.5A). Instead, recent structural data support a "template model" of gamma-tubulin complexes attracting the assembly of alpha/beta-tubulin into 13 longitudinally associated protofilaments that close into a hollow cylinder (Fig. 19.5B; Refs. 14,20,57).

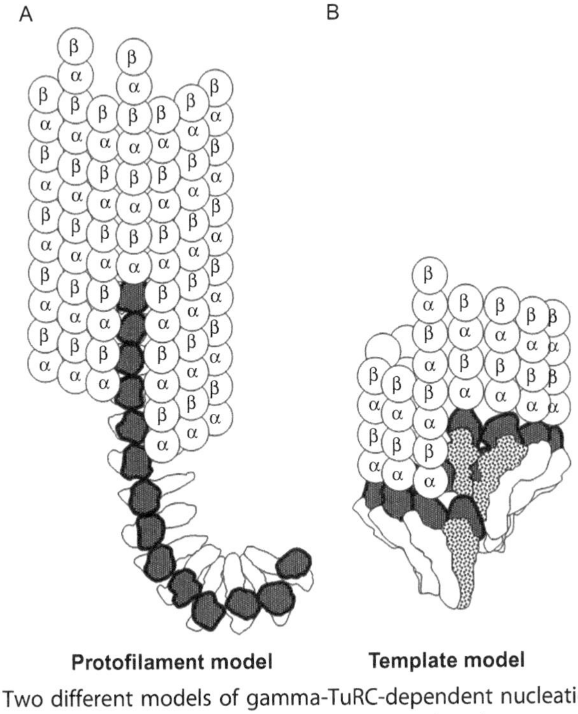

Figure 19.5 Two different models of gamma-TuRC-dependent nucleation of microtubules. (A) The protofilament model by Ref. 58 suggests that the gamma-TuRC uncurls partially into a protofilament-shaped complex that permits nucleation of microtubules by the lateral association of alpha/beta-tubulin into parallel protofilaments. (B) The template model[24] suggests that longitudinal associations of tubulin heterodimers with gamma-tubulin take place, with gamma-tubulin facing uniquely alpha-tubulin. This intermolecular contact is equivalent to the inter-heterodimer contact between alpha and beta-tubulin. Structural work by Ref. 35 supports this model, but rules out the protofilament model presented in (A).

The majority (80%) of gamma-tubulin in a cell[59] is found soluble in the cytoplasm. This soluble protein, albeit in the form of gamma-TuRCs, is unable to nucleate microtubules. Therefore, the recruitment to specific microtubule organizing centers and activation is necessary. One way of activating gamma-TuRCs may involve adaptor proteins that mediate the recruitment of gamma-TuRCs to the centrosome and spindle microtubules. As a possible adaptor protein NEDD1 has been identified.[38,39] Although NEDD1 can be copurified with gamma-TuRCs to various extents, gamma-TuRCs can still assemble in the absence of NEDD1.[38,39,47] Conversely, NEDD1 can bind to centrosomes even in the absence of gamma-TuRCs, in cells depleted of gamma-tubulin.[39] The centrosomal binding domain is located in the amino-terminal half of NEDD1, containing WD40 repeats.[39] At the carboxy-terminal end of NEDD1, the last 62 amino acids mediate binding to the gamma-TuRC.[60] In mitosis, the binding of NEDD1 to both gamma-TuRCs and microtubule nucleation sites is regulated by various kinases, such as Cdk1, Plk1, and Nek9.[38,61–65] Moreover, gamma-tubulin itself as well as GCP5 and GCP6 have been found to be putative substrates of the kinases SADB, GSK3beta, and Plk4, respectively,[66–68], and in biochemical experiments, the kinases Plk1, Wee1, MARK4 (microtubule affinity-regulating kinase 4), Syk tyrosine kinase, phosphoinositide 3-kinase, Lyn-kinase, Src, Fyn, and BubR1 have been found to bind gamma-tubulin or to coimmunoprecipitate with gamma-tubulin complexes.[69–76] The phosphorylation of NEDD1, which is so far the best studied regulation of a gamma-TuRC protein, increases the affinity of NEDD1 to the gamma-TuRC and enables the interaction with proteins at the pericentriolar material and the surface of spindle microtubules.[38,61–65] On the pericentriolar material of mammalian cells, proteins such as pericentrin, AKAP450, ninein, ninein-like protein NLP, CPAP, and Cdk5rap2 may represent potential receptors for gamma-TuRCs.[42,77–82] Related proteins have been found in various eukaryotes, such as the pericentrin-like protein PLP or the Cdk5rap2/myomegalin-related protein centrosomin in *Drosophila*.[83–86] Since a multitude of pericentriolar proteins has been shown to anchor gamma-TuRCs, the question emerges whether all of them recruit gamma-TuRCs randomly, or whether binding to individual pericentriolar proteins is regulated, for example, in a cell cycle-specific manner, or in a cell type-specific manner in tissues. Cell cycle-specific interactions between the gamma-TuRC and the centrosome appear to be mediated by the gamma-TuRC protein MOZART 2, which participates in centrosomal recruitment of gamma-TuRCs via NEDD1 during interphase, but not during mitosis.[41] In addition, there may be cooperative binding of multiple pericentriolar proteins to one gamma-TuRC. Consistent with this latter idea are the findings that pericentrin and AKAP450 can bind directly to GCP2 and GCP3, and may, therefore, interact with gamma-TuRCs independently of NEDD1.[78,79]

During mitosis, a fraction of gamma-TuRCs can be anchored to the lateral surface of spindle microtubules, independent of the centrosome. This is mediated by multiprotein complexes termed "augmin," via NEDD1 as an adaptor.[63,87–89] It is believed that augmin complexes allow for nucleation of spindle microtubules from already existing ones, thereby amplifying the density of microtubules within the spindle, independent of centrosomal nucleation.[90] This mechanism may be particularly important during the early embryonic cell divisions and during female meiosis in various species, since these occur in the absence of centrioles, that is, in the absence of any preformed centrosomal nucleation center.[91]

Although the mechanisms of gamma-TuRC attachment to centrosomes and microtubules are now partly understood, it remains enigmatic whether attachment can directly trigger the activation of microtubule nucleation or whether this requires additional regulatory mechanisms. Moreover, it is not clear what subunits of the gamma-TuRC are involved in the activation process. To induce nucleation of microtubules, conformational changes within the gamma-TuRC must occur. In the helically shaped complex, gamma-tubulins need to be aligned laterally to serve as a template for the polymerization of 13-protofilament microtubules. Cryo-electron microscopy on gamma-TuRC-like helices from gamma-TuSCs has shown that gamma-tubulin molecules bound to GCP3 are out of alignment with GCP2-bound gamma-tubulins (Fig. 19.6; Ref. 35). The initiation of efficient microtubule nucleation requires, therefore, a conformational change in GCP3, to position gamma-tubulin correctly. The grip 2 domain of GCP3 acts as a flexible arm that needs to rotate about 23 degrees around a hinge point, to align its associated gamma-tubulin (Fig. 19.6; Refs. 35,37,92). Comparison between the sequences of GCP3 and GCP4 and structural prediction made it possible to narrow down the region of the putative hinge to the interface between helical bundles 3 and 4 in GCPs.[36] How this conformational change in GCP3 is induced is currently unclear. It is possible that upon binding of soluble gamma-TuRCs to the centrosome or the spindle, allosteric activation occurs. The centrosome protein Cdk5rap2 may be involved in this process, since a conserved domain within its amino-terminal region has been shown to activate the microtubule nucleating capacity of the gamma-TuRC, both *in vitro* and in the cell.[42] This domain, comprising between 31 and 64 amino

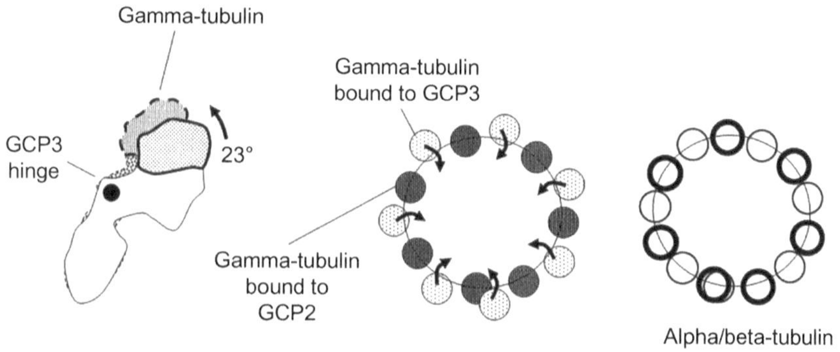

Figure 19.6 GCP3 needs to undergo a conformational change to align its associated gamma-tubulin, to serve as a template for microtubule nucleation. Left, the grip 2 domain of GCP3 needs to rotate by approximately 23° around a pivot point, located between the grip 1 and grip 2 domains. Middle, this rotation permits the lateral alignment of GCP2-bound and GCP3-bound gamma-tubulin subunits in a gamma-TuRC (viewed from the top). After this rotation, all gamma-tubulin molecules are positioned on a circle with a diameter identical to a typical 13-protofilament microtubule (shown on the right).

acids, is found in proteins of the centrosome and of other microtubule organizing centers in a large number of eukaryotes.[42,86] However, the exact mechanism of activation remains to be elucidated.[42] Besides activation by protein binding, phosphorylation of gamma-TuRC components may contribute directly or indirectly to the induction of microtubule nucleation.

Other proteins have been implicated in the process of microtubule nucleation, independently of the gamma-TuRC, such as TPX2 and members of the TOG family.[93–95] The latter is a family of conserved proteins including, for example, Dis1 in fission yeast, Stu2 in budding yeast, Msps in Drosophila, XMAP215 in Xenopus, and chTOG in humans. Proteins of this family contain conserved TOG domains (TOG: tubulin-interacting tumor overexpressed gene) that help catalyzing fast microtubule growth. Unlike the gamma-TuRC, these proteins do not act as structural templates but may support microtubule nucleation by binding to a curved conformation of alpha/beta-tubulin dimers and assist the assembly of protofilaments by controlling assembly dependent conformational changes.[96] The protein TPX2 is a mitosis-specific factor controlled by Ran-GTP and the importin-alpha that stimulates nucleation of microtubules around the mitotic chromosomes.[93,94] Although the exact mechanisms of TPX2 function are still unclear, indirect evidence suggests that TPX2 may promote microtubule elongation or stabilization, rather than acting as a template in microtubule nucleation.[97]

6. GAMMA-TUBULIN COMPLEXES IN DISEASE

Several research groups have shown that cells lacking individual GCPs or carrying mutations in the respective genes fail to assemble functional gamma-TuRCs and show abnormal spindles.[15,28,29,38,39,47–54,64,65,98–110] Whereas loss or mutation of gamma-TuSC components leads to a severe phenotype with monopolar spindles and mitotic arrest, the absence of the gamma-TuRC-specific proteins GCP 4, 5, 6, and NEDD1 often leads to milder spindle defects that slow down mitosis but that can frequently be overcome in model organisms such as *Drosophila*.[47] Spindle defects in cells with functionally impaired gamma-TuRCs can directly lead to errors in mitotic chromosome segregation and hence to raised levels of aneuploidy.[15,28,47,50,52,54,111,112] Aneuploidy is usually a hallmark of cancerous cells, but the induction of aneuploidy can also be used as a strategy to inhibit tumorigenesis.[113] Studying the role of gamma-tubulin complexes should, therefore, be of importance for both basic biological research and biomedical research, related to cancer. Cancerous cells are characterized by their potential to undergo continuous rounds of mitotic cell division. A possible approach of cancer chemotherapy involves induction of mitotic arrest by drugs affecting microtubule dynamics and thus provoking errors in mitotic spindle assembly. Subsequently, cells in which spindle formation is compromised are often undergoing apoptosis,[114–116] a phenomenon that enables to control tumor growth. Several drugs that affect microtubule dynamics are now routinely used in chemotherapy of cancer, such as taxol/paclitaxel and various vinca alkaloids.[117] Although pharmacological inhibitors exist against alpha- and beta-tubulin, so far no inhibitor has been developed against gamma-tubulin or against any of its associated GCPs. Nevertheless, removal of gamma-tubulin or other gamma-TuRC components from the cell induces changes in microtubule dynamics and spindle defects that resemble phenotypes obtained with microtubule drugs.[23,39,118] It will, therefore, be of major value to understand the assembly of gamma-TuRCs, since these are crucial for microtubule assembly and may represent potential pharmacological targets: in a genome-wide screen for chemosensitizers, components of the gamma-TuRC have been described as putative targets whose depletion by siRNA sensitizes a nonsmall-cell lung cancer cell line at 1000-fold reduced doses of paclitaxel.[119] Among various other sensitizer loci in the human genome, nearly all genes encoding gamma-TuRC proteins were identified: gamma-tubulin (both isoforms 1 and 2), GCP2, GCP3, GCP5,

GCP6, NEDD1, and NME7. Besides, gamma-tubulin was found overexpressed in nonsmall-cell lung cancer, as well as in astrocytic gliomas and medulloblastomas.[120–122] Moreover, it has been shown that small amounts of siRNA against the gamma-TuRC component NEDD1 potentiate the antimitotic activity of low doses of a Plk1 inhibitor.[118] These data suggest that if it were possible to suppress the function of the gamma-TuRC during cancer chemotherapy, the sensitizing effect of this treatment may permit lowering the doses of conventional drugs, to avoid side effects.

In a variety of other diseases, including myofibrillar myopathies, "inclusion body myositis," Parkinson's disease, and "dementia with Lewy bodies," gamma-tubulin has been localized to aggresomes, in association with abnormal protein deposits into intracytoplasmic aggregates.[123,124] However, the functional implications of this are still unclear.

7. CONCLUSIONS

Intact gamma-TuRCs are important for fault-free assembly of mitotic spindles. Gamma-TuRCs are formed in the cytoplasm and recruited to microtubule organizing centers by interacting with a variety of structural and regulatory proteins. Future research on microtubule nucleation will have to characterize the conformational changes within the gamma-TuRC that is required for a perfect template, supporting the assembly of 13-protofilament microtubules. Interference with any step during the formation, recruitment, or activation of gamma-TuRCs may lead to defects in mitosis, due to aberrant assembly of spindle microtubules. Because errors in spindle formation are known to activate the mitotic checkpoint and to lead to eventual apoptosis of the cell, the inhibition of gamma-TuRCs may represent an interesting new way of controlling tumor growth. The recently solved atomic structures of gamma-tubulin and GCP4 may help in the rational design of small molecules that bind to protein interfaces within the gamma-TuRC and that inhibit the assembly of this multiprotein complex.

REFERENCES

1. Bergen LG, Borisy GG. Head-to-tail polymerization of microtubules in vitro. Electron microscope analysis of seeded assembly. *J Cell Biol* 1980;**84**:141–50.
2. Nogales E, Whittaker M, Milligan RA, Downing KH. High-resolution model of the microtubule. *Cell* 1999;**96**:79–88.
3. Tilney LG, Bryan J, Bush DJ, Fujiwara K, Mooseker MS, Murphy DB, et al. Microtubules: evidence for 13 protofilaments. *J Cell Biol* 1973;**59**:267–75.

4. Burton PR, Hinkley RE, Pierson GB. Tannic acid-stained microtubules with 12, 13, and 15 protofilaments. *J Cell Biol* 1975;**65**:227–33.
5. Scheele RB, Bergen LG, Borisy GG. Control of the structural fidelity of microtubules by initiation sites. *J Mol Biol* 1982;**154**:485–500.
6. Evans L, Mitchison T, Kirschner M. Influence of the centrosome on the structure of nucleated microtubules. *J Cell Biol* 1985;**100**:1185–91.
7. Oakley CE, Oakley BR. Identification of gamma-tubulin, a new member of the tubulin superfamily encoded by mipA gene of Aspergillus nidulans. *Nature* 1989;**338**:662–4.
8. Zheng Y, Jung MK, Oakley BR. Gamma-tubulin is present in Drosophila melanogaster and Homo sapiens and is associated with the centrosome. *Cell* 1991;**65**:817–23.
9. Stearns T, Evans L, Kirschner M. Gamma-tubulin is a highly conserved component of the centrosome. *Cell* 1991;**65**:825–36.
10. Khodjakov A, Rieder CL. The sudden recruitment of gamma-tubulin to the centrosome at the onset of mitosis and its dynamic exchange throughout the cell cycle, do not require microtubules. *J Cell Biol* 1999;**146**:585–96.
11. Julian M, Tollon Y, Lajoie-Mazenc I, Moisand A, Mazarguil H, Puget A, et al. Gamma-tubulin participates in the formation of the midbody during cytokinesis in mammalian cells. *J Cell Sci* 1993;**105**:145–56.
12. Lajoie-Mazenc I, Tollon Y, Detraves C, Julian M, Moisand A, Gueth-Hallonet C, et al. Recruitment of antigenic gamma-tubulin during mitosis in animal cells: presence of gamma-tubulin in the mitotic spindle. *J Cell Sci* 1994;**107**:2825–37.
13. Shu HB, Li Z, Palacios MJ, Li Q, Joshi HC. A transient association of gamma-tubulin at the midbody is required for the completion of cytokinesis during the mammalian cell division. *J Cell Sci* 1995;**108**:2955–62.
14. Weil CF, Oakley CE, Oakley BR. Isolation of mip (microtubule-interacting protein) mutations of Aspergillus nidulans. *Mol Cell Biol* 1986;**6**:2963–8.
15. Oakley BR, Oakley CE, Yoon Y, Jung MK. Gamma-tubulin is a component of the spindle pole body that is essential for microtubule function in Aspergillus nidulans. *Cell* 1990;**61**:1289–301.
16. Joshi HC, Palacios MJ, McNamara L, Cleveland DW. Gamma-tubulin is a centrosomal protein required for cell cycle-dependent microtubule nucleation. *Nature* 1992;**356**:80–3.
17. Wise DO, Krahe R, Oakley BR. The gamma-tubulin gene family in humans. *Genomics* 2000;**67**:164–70.
18. Yuba-Kubo A, Kubo A, Hata M, Tsukita S. Gene knockout analysis of two gamma-tubulin isoforms in mice. *Dev Biol* 2005;**282**:361–73.
19. Li Q, Joshi HC. Gamma-tubulin is a minus end-specific microtubule binding protein. *J Cell Biol* 1995;**131**:207–14.
20. Keating TJ, Borisy GG. Immunostructural evidence for the template mechanism of microtubule nucleation. *Nat Cell Biol* 2000;**2**:352–7.
21. Wiese C, Zheng Y. A new function for the gamma-tubulin ring complex as a microtubule minus-end cap. *Nat Cell Biol* 2000;**2**:358–64.
22. Moritz M, Braunfeld MB, Guénebaut V, Heuser J, Agard DA. Structure of the gamma-tubulin ring complex: a template for microtubule nucleation. *Nat Cell Biol* 2000;**2**:365–70.
23. Bouissou A, Vérollet C, Sousa A, Sampaio P, Wright M, Sunkel CE, et al. Gamma-tubulin ring complexes regulate microtubule plus end dynamics. *J Cell Biol* 2009;**187**:327–34.
24. Zheng Y, Wong ML, Alberts B, Mitchison T. Nucleation of microtubule assembly by a gamma-tubulin-containing ring complex. *Nature* 1995;**378**:578–83.
25. Murphy SM, Urbani L, Stearns T. The mammalian gamma-tubulin complex contains homologues of the yeast spindle pole body components spc97p and spc98p. *J Cell Biol* 1998;**141**:663–74.

26. Gunawardane RN, Martin OC, Cao K, Zhang L, Dej K, Iwamatsu A, et al. Characterization and reconstitution of Drosophila gamma-tubulin ring complex subunits. *J Cell Biol* 2000;**151**:1513–24.
27. Murphy SM, Preble AM, Patel UK, O'Connell KL, Dias DP, Moritz M, et al. GCP5 and GCP6: two new members of the human gamma-tubulin complex. *Mol Biol Cell* 2001;**12**:3340–52.
28. Geissler S, Pereira G, Spang A, Knop M, Souès S, Kilmartin J, et al. The spindle pole body component Spc98p interacts with the gamma-tubulin-like Tub4p of Saccharomyces cerevisiae at the sites of microtubule attachment. *EMBO J* 1996;**15**:3899–911.
29. Knop M, Pereira G, Geissler S, Grein K, Schiebel E. The spindle pole body component Spc97p interacts with the gamma-tubulin of Saccharomyces cerevisiae and functions in microtubule organization and spindle pole body duplication. *EMBO J* 1997;**16**: 1550–64.
30. Oegema K, Wiese C, Martin OC, Milligan RA, Iwamatsu A, Mitchison TJ, et al. Characterization of two related Drosophila gamma-tubulin complexes that differ in their ability to nucleate microtubules. *J Cell Biol* 1999;**144**:721–33.
31. Martin OC, Gunawardane RN, Iwamatsu A, Zheng Y. Xgrip109: a gamma tubulin-associated protein with an essential role in gamma tubulin ring complex (gammaTuRC) assembly and centrosome function. *J Cell Biol* 1998;**141**:675–87.
32. Tassin AM, Celati C, Moudjou M, Bornens M. Characterization of the human homologue of the yeast spc98p and its association with gamma-tubulin. *J Cell Biol* 1998;**141**:689–701.
33. Fava F, Raynaud-Messina B, Leung-Tack J, Mazzolini L, Li M, Guillemot JC, et al. Human 76p: a new member of the gamma-tubulin-associated protein family. *J Cell Biol* 1999;**147**:857–68.
34. Zhang L, Keating TJ, Wilde A, Borisy GG, Zheng Y. The role of Xgrip210 in gamma-tubulin ring complex assembly and centrosome recruitment. *J Cell Biol* 2000;**151**: 1525–36.
35. Kollman JM, Polka JK, Zelter A, Davis TN, Agard DA. Microtubule nucleating gamma-TuSC assembles structures with 13-fold microtubule-like symmetry. *Nature* 2010;**466**:879–82.
36. Guillet V, Knibiehler M, Gregory-Pauron L, Remy MH, Chemin C, Raynaud-Messina B, et al. Crystal structure of gamma-tubulin complex protein GCP4 provides insight into microtubule nucleation. *Nat Struct Mol Biol* 2011;**18**:915–9.
37. Kollman JM, Zelter A, Muller EG, Fox B, Rice LM, Davis TN, et al. The structure of the gamma-tubulin small complex: implications of its architecture and flexibility for microtubule nucleation. *Mol Biol Cell* 2008;**19**:207–15.
38. Lüders J, Patel UK, Stearns T. GCP-WD is a gamma-tubulin targeting factor required for centrosomal and chromatin-mediated microtubule nucleation. *Nat Cell Biol* 2006;**8**:137–47.
39. Haren L, Remy MH, Bazin I, Callebaut I, Wright M, Merdes A. NEDD1-dependent recruitment of the gamma-tubulin ring complex to the centrosome is necessary for centriole duplication and spindle assembly. *J Cell Biol* 2006;**172**:505–15.
40. Hutchins JR, Toyoda Y, Hegemann B, Poser I, Hériché JK, Sykora MM, et al. Systematic analysis of human protein complexes identifies chromosome segregation proteins. *Science* 2010;**328**:593–9.
41. Teixidó-Travesa N, Villén J, Lacasa C, Bertran MT, Archinti M, Gygi SP, et al. The gammaTuRC revisited: a comparative analysis of interphase and mitotic human gammaTuRC redefines the set of core components and identifies the novel subunit GCP8. *Mol Biol Cell* 2010;**21**:3963–72.
42. Choi YK, Liu P, Sze SK, Dai C, Qi RZ. CDK5RAP2 stimulates microtubule nucleation by the gamma-tubulin ring complex. *J Cell Biol* 2010;**191**:1089–95.

43. Gunawardane RN, Martin OC, Zheng Y. Characterization of a new gamma TuRC subunit with WD repeats. *Mol Biol Cell* 2003;**14**:1017–26.
44. Ikeda T. NDP kinase 7 is a conserved microtubule-binding protein preferentially expressed in ciliated cells. *Cell Struct Funct* 2010;**35**:23–30.
45. Aldaz H, Rice LM, Stearns T, Agard DA. Insights into microtubule nucleation from the crystal structure of human gamma-tubulin. *Nature* 2005;**435**:523–7.
46. Fant X, Gnadt N, Haren L, Merdes A. Stability of the small gamma-tubulin complex requires HCA66, a protein of the centrosome and the nucleolus. *J Cell Sci* 2009;**122**:1134–44.
47. Vérollet C, Colombié N, Daubon T, Bourbon HM, Wright M, Raynaud-Messina B. Drosophila melanogaster gamma-TuRC is dispensable for targeting gamma-tubulin to the centrosome and microtubule nucleation. *J Cell Biol* 2006;**172**:517–28.
48. Vogt N, Koch I, Schwarz H, Schnorrer F, Nüsslein-Volhard C. The gammaTuRC components Grip75 and Grip128 have an essential microtubule-anchoring function in the Drosophila germline. *Development* 2006;**133**:3963–72.
49. Horio T, Uzawa S, Jung MK, Oakley BR, Tanaka K, Yanagida M. The fission yeast gamma-tubulin is essential for mitosis and is localized at microtubule organizing centers. *J Cell Sci* 1991;**99**:693–700.
50. Sunkel CE, Gomes R, Sampaio P, Perdigão J, González C. Gamma-tubulin is required for the structure and function of the microtubule organizing centre in Drosophila neuroblasts. *EMBO J* 1995;**14**:28–36.
51. Vardy L, Toda T. The fission yeast gamma-tubulin complex is required in G(1) phase and is a component of the spindle assembly checkpoint. *EMBO J* 2000;**19**:6098–111.
52. Barbosa V, Yamamoto RR, Henderson DS, Glover DM. Mutation of a Drosophila gamma tubulin ring complex subunit encoded by discs degenerate-4 differentially disrupts centrosomal protein localization. *Genes Dev* 2000;**14**:3126–39.
53. Bobinnec Y, Fukuda M, Nishida E. Identification and characterization of Caenorhabditis elegans gamma-tubulin in dividing cells and differentiated tissues. *J Cell Sci* 2000;**113**:3747–59.
54. Colombié N, Vérollet C, Sampaio P, Moisand A, Sunkel C, Bourbon HM, et al. The Drosophila gamma-tubulin small complex subunit Dgrip84 is required for structural and functional integrity of the spindle apparatus. *Mol Biol Cell* 2006;**17**:272–82.
55. Xiong Y, Oakley BR. In vivo analysis of the functions of gamma-tubulin-complex proteins. *J Cell Sci* 2009;**122**:4218–27.
56. Anders A, Lourenço PC, Sawin KE. Noncore components of the fission yeast gamma-tubulin complex. *Mol Biol Cell* 2006;**17**:5075–93.
57. Kollman JM, Merdes A, Mourey L, Agard DA. Microtubule nucleation by gamma-tubulin complexes. *Nat Rev Mol Cell Biol* 2011;**12**:709–21.
58. Erickson HP, Stoffler D. Protofilaments and rings, two conformations of the tubulin family conserved from bacterial FtsZ to alpha/beta and gamma tubulin. *J Cell Biol* 1996;**135**:5–8.
59. Moudjou M, Bordes N, Paintrand M, Bornens M. Gamma-Tubulin in mammalian cells: the centrosomal and the cytosolic forms. *J Cell Sci* 1996;**109**:875–87.
60. Manning JA, Shalini S, Risk JM, Day CL, Kumar S. A direct interaction with NEDD1 regulates gamma-tubulin recruitment to the centrosome. *PLoS One* 2010;**5**:e9618.
61. Zhang X, Chen Q, Feng J, Hou J, Yang F, Liu J, et al. Sequential phosphorylation of Nedd1 by Cdk1 and Plk1 is required for targeting of the gammaTuRC to the centrosome. *J Cell Sci* 2009;**122**:2240–51.
62. Haren L, Stearns T, Lüders J. Plk1-dependent recruitment of gamma-tubulin complexes to mitotic centrosomes involves multiple PCM components. *PLoS One* 2009;**4**(6):e5976.

63. Zhu H, Fang K, Fang G. FAM29A, a target of Plk1 regulation, controls the partitioning of NEDD1 between the mitotic spindle and the centrosomes. *J Cell Sci* 2009;**122**:2750–9.
64. Gomez-Ferreria M, Bashkurov M, Helbig A, Larsen B, Pawson T, Gingras AC, et al. Novel NEDD1 phosphorylation sites regulate gamma-tubulin binding and mitotic spindle assembly. *J Cell Sci* 2012;**125**:3745–51.
65. Sdelci S, Schütz M, Pinyol R, Bertran MT, Regué L, Caelles C, et al. Nek9 phosphorylation of NEDD1/GCP-WD contributes to Plk1 control of gamma-tubulin recruitment to the mitotic centrosome. *Curr Biol* 2012;**22**:1516–23.
66. Alvarado-Kristensson M, Rodríguez MJ, Silió V, Valpuesta JM, Carrera AC. SADB phosphorylation of gamma-tubulin regulates centrosome duplication. *Nat Cell Biol* 2009;**11**:1081–92.
67. Izumi N, Fumoto K, Izumi S, Kikuchi A. GSK-3beta regulates proper mitotic spindle formation in cooperation with a component of the gamma-tubulin ring complex, GCP5. *J Biol Chem* 2008;**283**:12981–91.
68. Bahtz R, Seidler J, Arnold M, Haselmann-Weiss U, Antony C, Lehmann WD, et al. GCP6 is a substrate of Plk4 and required for centriole duplication. *J Cell Sci* 2012;**125**:486–96.
69. Feng Y, Hodge DR, Palmieri G, Chase DL, Longo DL, Ferris DK. Association of polo-like kinase with alpha-, beta- and gamma-tubulins in a stable complex. *Biochem J* 1999;**339**:435–42.
70. Stumpff J, Kellogg DR, Krohne KA, Su TT. Drosophila Wee1 interacts with members of the gammaTURC and is required for proper mitotic-spindle morphogenesis and positioning. *Curr Biol* 2005;**15**:1525–34.
71. Trinczek B, Brajenovic M, Ebneth A, Drewes G. MARK4 is a novel microtubule-associated proteins/microtubule affinity-regulating kinase that binds to the cellular microtubule network and to centrosomes. *J Biol Chem* 2004;**279**:5915–23.
72. Zyss D, Montcourrier P, Vidal B, Anguille C, Merezegue F, Sahuquet A, et al. The Syk tyrosine kinase localizes to the centrosomes and negatively affects mitotic progression. *Cancer Res* 2005;**65**:10872–80.
73. Kapeller R, Toker A, Cantley LC, Carpenter CL. Phosphoinositide 3-kinase binds constitutively to alpha/beta-tubulin and binds to gamma-tubulin in response to insulin. *J Biol Chem* 1995;**270**:25985–91.
74. Draberova L, Draberova E, Surviladze Z, Draber P, Draber P. Protein tyrosine kinase p53/p56(lyn) forms complexes with gamma-tubulin in rat basophilic leukemia cells. *Int Immunol* 1999;**11**:1829–39.
75. Kukharskyy V, Sulimenko V, Macurek L, Sulimenko T, Draberova E, Draber P. Complexes of gamma-tubulin with nonreceptor protein tyrosine kinases Src and Fyn in differentiating P19 embryonal carcinoma cells. *Exp Cell Res* 2004;**298**:218–28.
76. Müller H, Fogeron ML, Lehmann V, Lehrach H, Lange BM. A centrosome-independent role for gamma-TuRC proteins in the spindle assembly checkpoint. *Science* 2006;**314**:654–7.
77. Dictenberg JB, Zimmerman W, Sparks CA, Young A, Vidair C, Zheng Y, et al. Pericentrin and gamma-tubulin form a protein complex and are organized into a novel lattice at the centrosome. *J Cell Biol* 1998;**141**:163–74.
78. Zimmerman WC, Sillibourne J, Rosa J, Doxsey SJ. Mitosis-specific anchoring of gamma tubulin complexes by pericentrin controls spindle organization and mitotic entry. *Mol Biol Cell* 2004;**15**:3642–57.
79. Takahashi M, Yamagiwa A, Nishimura T, Mukai H, Ono Y. Centrosomal proteins CG-NAP and kendrin provide microtubule nucleation sites by anchoring gamma-tubulin ring complex. *Mol Biol Cell* 2002;**13**:3235–45.

80. Delgehyr N, Sillibourne J, Bornens M. Microtubule nucleation and anchoring at the centrosome are independent processes linked by ninein function. *J Cell Sci* 2005;**118**:1565–75.
81. Casenghi M, Meraldi P, Weinhart U, Duncan PI, Körner R, Nigg EA. Polo-like kinase 1 regulates Nlp, a centrosome protein involved in microtubule nucleation. *Dev Cell* 2003;**5**:113–25.
82. Hung LY, Tang CJ, Tang TK. Protein 4.1 R-135 interacts with a novel centrosomal protein (CPAP) which is associated with the gamma-tubulin complex. *Mol Cell Biol* 2000;**20**:7813–25.
83. Martinez-Campos M, Basto R, Baker J, Kernan M, Raff JW. The Drosophila pericentrin-like protein is essential for cilia/flagella function, but appears to be dispensable for mitosis. *J Cell Biol* 2004;**165**:673–83.
84. Heuer JG, Li K, Kaufman TC. The Drosophila homeotic target gene centrosomin (cnn) encodes a novel centrosomal protein with leucine zippers and maps to a genomic region required for midgut morphogenesis. *Development* 1995;**121**:3861–76.
85. Terada Y, Uetake Y, Kuriyama R. Interaction of Aurora-A and centrosomin at the microtubule-nucleating site in Drosophila and mammalian cells. *J Cell Biol* 2003; **162**:757–63.
86. Sawin KE, Lourenco PC, Snaith HA. Microtubule nucleation at non-spindle pole body microtubule-organizing centers requires fission yeast centrosomin-related protein mod20p. *Curr Biol* 2004;**14**:763–75.
87. Goshima G, Mayer M, Zhang N, Stuurman N, Vale RD. Augmin: a protein complex required for centrosome-independent microtubule generation within the spindle. *J Cell Biol* 2008;**181**:421–9.
88. Zhu H, Coppinger JA, Jang CY, Yates 3rd. JR, Fang G. FAM29A promotes microtubule amplification via recruitment of the NEDD1-gamma-tubulin complex to the mitotic spindle. *J Cell Biol* 2008;**183**:835–48.
89. Lawo S, Bashkurov M, Mullin M, Ferreria MG, Kittler R, Habermann B, et al. HAUS, the 8-subunit human Augmin complex, regulates centrosome and spindle integrity. *Curr Biol* 2009;**19**:816–26.
90. Lüders J, Stearns T. Microtubule-organizing centres: a re-evaluation. *Nat Rev Mol Cell Biol* 2007;**8**:161–7.
91. Dumont J, Desai A. Acentrosomal spindle assembly and chromosome segregation during oocyte meiosis. *Trends Cell Biol* 2012;**22**:241–9.
92. Choy RM, Kollman JM, Zelter A, Davis TN, Agard DA. Localization and orientation of the gamma-tubulin small complex components using protein tags as labels for single particle EM. *J Struct Biol* 2009;**168**:571–4.
93. Schatz CA, Santarella R, Hoenger A, Karsenti E, Mattaj IW, Gruss OJ, et al. Importin alpha-regulated nucleation of microtubules by TPX2. *EMBO J* 2003;**22**: 2060–70.
94. Brunet S, Sardon T, Zimmerman T, Wittmann T, Pepperkok R, Karsenti E, et al. Characterization of the TPX2 domains involved in microtubule nucleation and spindle assembly in Xenopus egg extracts. *Mol Biol Cell* 2004;**15**:5318–28.
95. Popov AV, Severin F, Karsenti E. XMAP215 is required for the microtubule-nucleating activity of centrosomes. *Curr Biol* 2002;**12**:1326–30.
96. Ayaz P, Ye X, Huddleston P, Brautigam CA, Rice LM. A TOG:αβ-tubulin complex structure reveals conformation-based mechanisms for a microtubule polymerase. *Science* 2012;**337**:857–60.
97. Groen AC, Maresca TJ, Gatlin JC, Salmon ED, Mitchison TJ. Functional overlap of microtubule assembly factors in chromatin-promoted spindle assembly. *Mol Biol Cell* 2009;**20**:2766–73.

98. Marschall LG, Jeng RL, Mulholland J, Stearns T. Analysis of Tub4p, a yeast gamma-tubulin-like protein: implications for microtubule-organizing center function. *J Cell Biol* 1996;**134**:443–54.
99. Spang A, Geissler S, Grein K, Schiebel E. Gamma-Tubulin-like Tub4p of Saccharomyces cerevisiae is associated with the spindle pole body substructures that organize microtubules and is required for mitotic spindle formation. *J Cell Biol* 1996;**134**:429–41.
100. Tavosanis G, Llamazares S, Goulielmos G, Gonzalez C. Essential role for gamma-tubulin in the acentriolar female meiotic spindle of Drosophila. *EMBO J* 1997;**16**:1809–19.
101. Nguyen T, Vinh DBN, Crawford DK, Davis TN. A genetic analysis of interactions with Spc110p reveals distinct functions of Spc97p and Spc98p, components of the yeast gamma-tubulin complex. *Mol Biol Cell* 1998;**9**:2201–16.
102. Wilson PG, Borisy GG. Maternally expressed gTub37CD in Drosophila is differentially required for female meiosis and embryonic mitosis. *Dev Biol* 1998;**199**:273–90.
103. Llamazares S, Tavosanis G, Gonzalez C. Cytological characterisation of the mutant phenotypes produced during early embryogenesis by null and loss-of-function alleles of the gammaTub37C gene in Drosophila. *J Cell Sci* 1999;**112**:659–67.
104. Sampaio P, Rebello E, Varmark H, Sunkel CE, Gonzalez C. Organized microtubule arrays in gamma-tubulin-depleted Drosophila spermatocytes. *Curr Biol* 2001;**11**:1788–93.
105. Vardy L, Fujita A, Toda T. The gamma-tubulin complex protein Alp4 provides a link between the metaphase checkpoint and cytokinesis in fission yeast. *Genes Cells* 2002;**7**:365–73.
106. Barbosa V, Gatt M, Rebollo E, Gonzalez C, Glover SM. Drosophila dd4 mutants reveal that gammaTuRC is required to maintain juxtaposed half spindles in spermatocytes. *J Cell Sci* 2003;**116**:929–41.
107. Raynaud-Messina B, Mazzolini L, Moisand A, Cirinesi AM, Wright M. Elongation of centriolar microtubule triplets contributes to the formation of the mitotic spindle in gamma-tubulin-depleted cells. *J Cell Sci* 2004;**117**:5497–507.
108. Zeng CJ, Lee YR, Liu B. The WD40 repeat protein NEDD1 functions in microtubule organization during cell division in Arabidopsis thaliana. *Plant Cell* 2009;**21**:1129–40.
109. Manning JA, Lewis M, Koblar SA, Kumar S. An essential function for the centrosomal protein NEDD1 in zebrafish development. *Cell Death Differ* 2010;**17**:1302–14.
110. Ma W, Baumann C, Viveiros MM. NEDD1 is crucial for meiotic spindle stability and accurate chromosome segregation in mammalian oocytes. *Dev Biol* 2010;**339**:439–50.
111. Hendrickson TW, Yao J, Bhadury S, Corbett AH, Joshi HC. Conditional mutations in gamma-tubulin reveal its involvement in chromosome segregation and cytokinesis. *Mol Biol Cell* 2001;**12**:2469–81.
112. Prigozhina NL, Oakley CE, Lewis AM, Nayak T, Osmani SA, Oakley BR. Gamma-tubulin plays an essential role in the coordination of mitotic events. *Mol Biol Cell* 2004;**15**:1374–86.
113. Weaver BA, Silk AD, Montagna C, Verdier-Pinard P, Cleveland DW. Aneuploidy acts both oncogenically and as a tumor suppressor. *Cancer Cell* 2007;**11**:25–36.
114. Tao W, South VJ, Zhang Y, Davide JP, Farrell L, Kohl NE, et al. Induction of apoptosis by an inhibitor of the mitotic kinesin KSP requires both activation of the spindle assembly checkpoint and mitotic slippage. *Cancer Cell* 2005;**8**:49–59.
115. Matson DR, Stukenberg PT. Spindle poisons and cell fate: a tale of two pathways. *Mol Interv* 2011;**11**:141–50.
116. Orth JD, Loewer A, Lahav G, Mitchison TJ. Prolonged mitotic arrest triggers partial activation of apoptosis, resulting in DNA damage and p53 induction. *Mol Biol Cell* 2012;**23**:567–76.

117. Jordan MA, Wilson L. Microtubules as a target for anticancer drugs. *Nat Rev Cancer* 2004;**4**:253–65.
118. Tillement V, Haren L, Roullet N, Etievant C, Merdes A. The centrosome protein NEDD1 as a potential pharmacological target to induce cell cycle arrest. *Mol Cancer* 2009;**8**:10.
119. Whitehurst AW, Bodemann BO, Cardenas J, Ferguson D, Girard L, Peyton M, et al. Synthetic lethal screen identification of chemosensitizer loci in cancer cells. *Nature* 2007;**446**:815–9.
120. Maounis NF, Dráberová E, Mahera E, Chorti M, Caracciolo V, Sulimenko T, et al. Overexpression of gamma-tubulin in non-small cell lung cancer. *Histol Histopathol* 2012;**27**:1183–94.
121. Katsetos CD, Reddy G, Dráberová E, Smejkalová B, Del Valle L, Ashraf Q, et al. Altered cellular distribution and subcellular sorting of gamma-tubulin in diffuse astrocytic gliomas and human glioblastoma cell lines. *J Neuropathol Exp Neurol* 2006;**65**:465–77.
122. Caracciolo V, D'Agostino L, Dráberová E, Sládková V, Crozier-Fitzgerald C, Agamanolis DP, et al. Differential expression and cellular distribution of gamma-tubulin and betaIII-tubulin in medulloblastomas and human medulloblastoma cell lines. *J Cell Physiol* 2010;**223**:519–29.
123. Ferrer I, Carmona M, Blanco R, Moreno D, Torrejón-Escribano B, Olivé M. Involvement of clusterin and the aggresome in abnormal protein deposits in myofibrillar myopathies and inclusion body myositis. *Brain Pathol* 2005;**15**:101–8.
124. McNaught KS, Shashidharan P, Perl DP, Jenner P, Olanow CW. Aggresome-related biogenesis of Lewy bodies. *Eur J Neurosci* 2002;**16**:2136–48.

CHAPTER TWENTY

Chemokine Oligomerization in Cell Signaling and Migration

Xu Wang*, Joshua S. Sharp†, Tracy M. Handel‡, James H. Prestegard†
*Department of Chemistry and Biochemistry, Arizona State University, Tempe, Arizona, USA
†Complex Carbohydrate Research Center, University of Georgia, Athens, Georgia, USA
‡Skaggs School of Pharmacy & Pharmaceutical Science, University of California, La Jolla, San Diego, California, USA

Contents

Abstract

Chemokines are small proteins best known for their role in controlling the migration of diverse cells, particularly leukocytes. Upon binding to their G-protein-coupled receptors on the leukocytes, chemokines stimulate the signaling events that cause cytoskeletal rearrangements involved in cell movement, and migration of the cells along chemokine gradients. Depending on the cell type, chemokines also induce many other types of cellular responses including those related to defense mechanisms, cell proliferation, survival, and development. Historically, most research efforts have focused on the interaction of chemokines with their receptors, where monomeric forms of the ligands are the functionally relevant state. More recently, however, the importance of chemokine interactions with cell surface glycosaminoglycans has come to light, and in most cases

Progress in Molecular Biology and Translational Science, Volume 117
ISSN 1877-1173
http://dx.doi.org/10.1016/B978-0-12-386931-9.00020-9

appears to involve oligomeric chemokine structures. This review summarizes existing knowledge relating to the structure and function of chemokine oligomers, and emerging methodology for determining structures of complex chemokine assemblies in the future.

1. INTRODUCTION

Oligomerization of proteins is widespread in biological systems, with estimates of homooligomerization occurring with up to 70% of all eukaryotic proteins.[1,2] There are many benefits to this type of association; oligomerization can increase the efficiency of processes involving multiple proteins such as signaling and enzymatic reactions, it can help to stabilize and protect proteins against proteolytic processes, it can amplify affinities of proteins for their ligands by organizing them in polymeric constructs or surface arrays, and it can increase the complexity, diversity, and regulation of signaling events.[1–3] In this review, we highlight the known roles that oligomerization plays in the chemokine-mediated regulation of cell migration and signaling in the immune system. We also highlight the structural features of chemokine oligomerization, and new methodology for the characterization of these oligomeric species.

1.1. The chemokine superfamily

Chemokines, or *chemo*tactic cyto*kines*, are small proteins (8–12 kDa) that provide directional signals for the migration of cells.[4,5] Approximately 45 human chemokines have been identified and are classified into four families based on the sequence of conserved cysteines in their amino termini[6]: XC chemokines contain a single cysteine; CC chemokines have sequential cysteines; in CXC chemokines, the cysteines are separated by one amino acid; and CX3C chemokines have two cysteines separated by three amino acids. In early studies, the ligands were named with a nonsystematic nomenclature (e.g., RANTES, MIP-1α). However, a systematic nomenclature for chemokines was introduced in 2002 and uses the cysteine motif followed by L (for ligand) and a number, to enable assignment of chemokines to their subfamilies (e.g., CCL5, CXCL2, CX3CL1, XCL1).[7] Twenty-two chemokine receptors have been identified and are similarly named by the cysteine motif, R (for receptor) and a number, for example, CCR5. The receptors belong to the G-protein-coupled receptor (GPCR) superfamily and interact predominantly with the Gαi subfamily of G proteins.

Many chemokine receptors bind to more than one ligand and similarly most chemokines bind to more than one receptor, leading to a significant level of potential cross-reactivity. The promiscuous pairing of chemokines and receptors was initially thought to provide redundancy in the system.[8,9] However, the concept of functional selectivity or biased agonism has recently emerged, where different ligands bind the same receptor but activate different sets of signaling pathways, potentially providing for finely tuned signaling responses. It is also possible that the oligomeric properties of chemokines and their specificities for different glycosaminoglycans (GAGs; described later) may break the apparent redundancy by controlling chemokine localization and density, certain aspects of signaling, and ultimately the migration and functional responses of the chemokine receptor expressing cells.[10]

Chemokines are produced by, and act on, a wide variety of cells. In addition to their role in controlling leukocyte migration in the context of immune surveillance and inflammation, chemokines and their receptors are involved in the migration and signaling responses of cells during other processes such as development, angiogenesis, wound repair, and neural regeneration.[11,12] Further, many diseases are associated with deregulated expression or utilization of chemokines and their receptors: For example, unresolved chemokine-mediated cell migration and activation contribute to many chronic inflammatory diseases such as multiple sclerosis and rheumatoid arthritis.[13–15] Specific chemokines and receptors are also key mediators in the metastasis of cancer cells from primary tumors to secondary sites.[16] And HIV utilizes primarily two chemokine receptors as portals for entry into cells.[17,18] Accordingly, chemokines and their receptors have been high-priority therapeutic targets for many years.[6,19]

1.2. Chemokine-mediated cell migration

Despite the central role of chemokines and their receptors in cell migration, it is important to note that it is a complex multistep process, involving many proteins and interactions.[20,21] For example, in the context of an inflammatory response, mucin-like selectin ligands on leukocytes (e.g., P-selectin glycoprotein ligand 1, PSGL-1, and L-selectin) interact with counter receptors (e.g., P- and E-selectins) on endothelial cells to capture the leukocytes from rapid diffusion through the vasculature and to promote slow rolling of the leukocytes along the blood vessel endothelium through labile interactions between the cell surfaces. The slow rolling of the leukocytes enables them to sample the endothelial surface where they encounter chemokines produced by or deposited on the endothelium. Interaction of receptors on the leukocytes with chemokines then triggers the activation of integrins into their high-affinity state, binding of the integrins to their ligands (e.g., ICAMs) on the endothelial cells, and rapid leukocyte arrest. The end result is the migration of the leukocytes through the vascular endothelial layer following a gradient of chemokines and adhesive molecules. While the selectins, integrins, their ligands, and the chemokine receptors are all membrane tethered, the immobilization of chemokines on the endothelial surface and the formation of chemokine gradients involve a different mechanism: interactions with GAGs. Thus, chemokines have two types of receptors: seven transmembrane GPCRs, and the long linear sulfated carbohydrate chains of proteoglycans. The ability of chemokines to interact with both of these macromolecules relies on the ability of chemokines to reversibly oligomerize as described in the next sections.

2. FUNCTIONAL EFFECTS OF CHEMOKINE OLIGOMERIZATION

As implied above and detailed in Section 3, most chemokines have a propensity to oligomerize; however, it is widely accepted that monomeric chemokines are the forms with the highest affinity for their receptors and the capacity to promote cell migration. This finding was demonstrated with mutant chemokines that were impaired in their ability to oligomerize, but were still capable of binding and activating their receptors with equivalent potencies as the WT proteins.[22–24] Further, in some CC chemokines, the residues that are critical for receptor activation are buried in the dimers, and therefore it is predictable that CC dimers would not be capable of inducing cell migration.[24] Beyond the knowledge that it is the monomeric form that activates the receptor, and in the absence of a crystal structure, the prevailing view of chemokine:receptor complexes is that of the so-called two-site model of receptor interaction in which the N-terminus of a chemokine monomer interacts with the helical bundle of the receptor (site 2) and the rest of the chemokine (the core domain), binds to the N-terminus of the receptor (site 1) (see figures in Refs. 25,26). The two-site model is supported by mutagenesis studies which indicate that the chemokine N-terminus is the key signaling domain, as modifications to the N-terminus can dramatically change the signaling response of the receptor to which it binds, often with only a small impact on binding affinity (see Section 3). Secondly, structural studies by solution NMR have demonstrated that peptides from the N-termini of chemokine receptors bind to the chemokine core domain.[27,28]

So the question is, if receptors are activated with chemokine monomers then what, if any, is the purpose of chemokine oligomerization? To address the relevance of oligomerization, monomeric variants of several chemokines (RANTES (CCL5), MCP-1 (CCL2), and MIP-1β (CCL4)) were used in *in vivo* models of cell migration where they were shown to be defective in promoting migration, in contrast to their WT counterparts and in contrast to their ability to induce migration *in vitro*.[22] These data clearly demonstrated that oligomerization was required for some aspects of chemokine function. Similarly, it was shown that binding to GAGs was required *in vivo*, using GAG-binding deficient mutants.[22] These two observations were then linked by *in vitro* biochemical studies, which demonstrated that chemokines oligomerize on GAG chains.[29] In the context of cell migration, oligomerization of chemokines on GAGs is thought to concentrate them in localized areas, thereby forming a gradient, providing the affinity needed for retaining chemokines on cell surfaces particularly in the presence of shear forces from blood flow.[30] Oligomerization and interactions with GAGs also protect chemokines from proteolysis and thus contribute to the duration of the "chemokine signal."[31]

However, chemokine oligomerization has also been associated with other aspects of cell migration and chemokine biology. Oligomerization of RANTES (CCL5) has been shown to be necessary for monocyte arrest through CCR1-mediated activation of integrins; furthermore, the oligomers were proposed to bridge surface bound RANTES through GAGs on the endothelium with CCR1 on the monocytes.[32] Oligomerization of IP-10 (CXCL10) was also shown to be required for transendothelial migration of T-cells and attributed to the fact that only the WT form, but not a monomeric form, could be retained on endothelial cell surfaces.[30] GAG interactions, and in fact specific sulfation patterns of GAGs, were shown to be important for the pericellular transport and transcytosis of IL-8 (CXCL8) across endothelial cells, processes needed for the presentation of chemokines on the requisite cell surface.[33] Similarly, heparin sulfate was shown to be required for binding of CCL21 to its receptor CCR7 and the induction of migration.[34] While both of these examples were not specifically attributed to oligomerization, there is overwhelming evidence that chemokines oligomerize on GAGs and thus are likely required for these processes. Finally, an artificially stabilized "locked dimer" of SDF-1 (CXCL12) has been shown to bind CXCR4, but does not cause cell migration, perhaps suggesting that it could act as a "stop signal" for migrating cells.[27] Indeed, it inhibited the metastasis of melanoma cells to the lung, as well as the growth of established pulmonary tumors.[35] In addition to its antagonism of cell migration, part of the mechanism was shown to be the superior serum stability of the locked dimer compared to the WT protein, due to protection from proteolysis. These data also suggest the possibility that, at least in the case of CXC chemokines, different oligomeric forms of the same chemokine can show biased signaling.

Oligomeric chemokines have also been shown to be involved directly in alternative signaling processes. WT RANTES (CCL5) but not oligomerization-deficient variants induce apoptosis and T-cell activation through G-protein-independent protein tyrosine kinase activation and interactions with CD44.[36,37] Additionally, early studies showed that MIP-1α (CCL3), MIP-1β (CCL4), and RANTES (CCL5) are secreted from the activated cytotoxic T cells as large 400–600 K complexes with proteoglycans, and importantly, that these macromolecular proteoglycan-associated complexes blocked HIV entry into monocytes and lysed virus producing cells.[38] Many chemokine receptors have been demonstrated to form homodimers and possibly higher-order oligomers. However, whether oligomerized forms of chemokines bind receptor oligomers is currently unknown; it is not even clear whether chemokine monomers bind receptors in a 1:1 or 1:2 stoichiometry.[39] Nevertheless, one could envision a great deal of regulated signaling that could arise from reversible oligomerization of both the chemokine and chemokine receptor as suggested above with the apparent-biased signaling of SDF-1 (CXCL12).[27]

Much less is known about chemokine hetero-oligomers in terms of their functional affects. However, many chemokines have been shown to hetero-oligomerize not only with other subfamily members but between subfamilies (e.g., CC and CXC).[40] One of the best-studied examples is PF-4 (CXCL4), which associates with RANTES (CCL5).[41] These heteromers are produced by activated platelets and deposited on endothelial cells, where the presence of PF-4 (CXCL4) has been shown to significantly enhance monocyte arrest induced by RANTES (CCL5). Further, peptides that block formation of the heteromers proved effective in inhibiting atherosclerosis in hyperlipidemic mice, underscoring the importance of these interactions.[42]

Despite the above examples, there is still limited information on the functional significance of chemokine oligomerization as well as their interactions with GAGs. Nevertheless, many studies have demonstrated that interfering with either oligomerization or GAG binding of chemokines can provide therapeutic benefit in a number of disease models.[43–45] To pursue such strategies, a detailed understanding of the structures in question is required, as described in the next sections.

3. CHEMOKINE STRUCTURE

The structures of chemokine oligomers are key to the mechanistic understanding of their functions. There are now numerous structures of chemokines in monomeric and dimeric forms. However, structure elucidation of oligomeric forms, and particularly forms in complex with GAG chains, has been much more challenging. In fact, a number of the early structure determinations, as well as functional studies, were conducted on mutants that did not form structures of higher complexity than dimers (the E66S mutant in the case of RANTES (CCL5)) or were based on data collected at pH conditions that inhibited oligomer formation (pH ~3 in the case of RANTES (CCL5)). As the importance of chemokine oligomerization to function has become more clear, a few structures of higher oligomers have emerged. We review these structures here, and subsequently highlight the evolving techniques that promise to provide more structural information on these highly relevant macromolecular complexes in the future.

3.1. Tertiary structure

Despite high-sequence variability among chemokines, their tertiary structures are remarkably similar.[46,47] More than 30 chemokine structures have been solved by high-resolution techniques so far (Table 20.1), and all adopt the "chemokine" fold (Fig. 20.1). The fold consists of an extended N-terminus followed by a core structure made up of a three-stranded beta sheet and a C-terminal helix arranged in a "Greek key" motif. The structure is usually stabilized by the presence of two disulfide bonds, connecting the N-terminus to the core structure. However, chemokines containing only one (lymphotactin/XCL1) as well as three (I-309/CCL1, HCC-2/CCL15, and Ckβ8/CCL23) disulfide bonds also exist. Besides these common features, chemokines also possess an irregular loop referred to as the "N-loop" which connects the N-terminus to the first beta strand, and a long linker connecting the first and second beta strands in the sheet, often referred to as the 30s loop. The 30s loop contains one of the cysteines involved in forming a disulfide bond and also plays a role in the formation of higher-order oligomers.

The affinity of most chemokines for their receptors is quite high, with dissociation constants (K_ds) in the low nanomolar range. Numerous

Table 20.1 Chemokines with known structures and oligomerization states

Subfamily	Name	Oligomeric state
CXC	MGSA (CXCL1)	Dimer (pH 5.5)[48]
	Gro-β (CXCL2)	Dimer (pH 5.5)[49]
	PF4 (CXCL4)	Tetramer (pH 5.9)[50,51]
	NAP-2 (CXCL7)	Tetramer (X-ray).[52] Monomer(NMR, pH 3.6).[53]
	IL-8 (CXCL8)	Dimer (pH 5.2)[54]
	IP-10 (CXCL10)	Tetramer (pH 4.4)[55]
	I-TAC (CXCL11)	Monomer (pH 5.0)[56]
	SDF-1α (CXCL12)	Dimer (X-ray).[57] Monomer (NMR, pH 4.9).[58]
CC	I-309 (CCL1)	Monomer (pH 5.0)[59]
	MCP-1 (CCL2)	Dimer (pH 5.4)[60,61]
	MIP-1α (CCL3)	Dimer (pH 4.5)[62]
	MIP-1β (CCL4)	Dimer (pH 2.5)[63]
	RANTES (CCL5)	Dimer (pH 3.2)[64]
	MCP-3 (CCL7)	Monomer (pH 5.1)[65]
	MCP-2 (CCL8)	Monomer (pH 7.5)[66]
	Eotaxin-1 (CCL11)	Monomer (pH 5.0)[67]
	MCP-4 (CCL13)	Dimer (pH 6.5)[68]
	HCC-1 (CCL14)	Dimer (pH 6.2)[69]
	HCC-2 (CCL15)	Monomer (pH 3.0)[70]
	TARC (CCL17)	Dimer(pH 5.6)[71]
	Ckβ9 (CCL21)	Monomer (pH 5.9)[72]
	MPIF-1 (CCL23)	Monomer (pH 5.2)[73]
	Eotaxin-4 (CCL24)	Monomer (pH 4.0)[74]
	Eotaxin-3 (CCL26)	Monomer (pH 5.0)[75]
	CTACK (CCL27)	Monomer (pH 5.6)[76]
C	Lymphotactin (XCL1)	Monomer–dimer equilibrium (pH 6.0)[77]
CX_3C	Fractalkine (CX_3CL1)	Monomer (NMR, pH 5.0)[78]/dimer (X-ray, pH 3.0)[79]

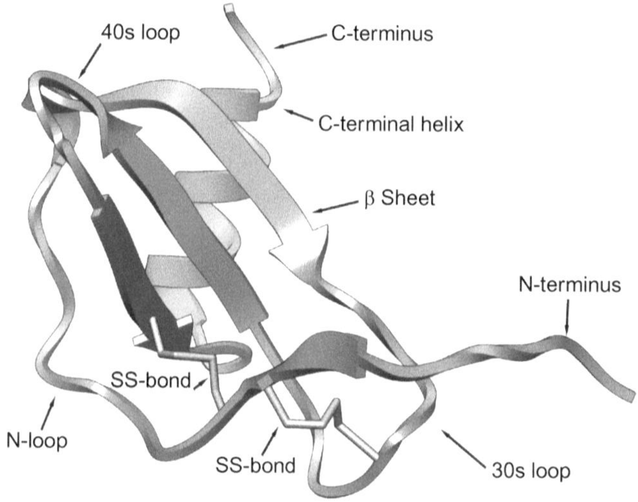

Figure 20.1 Monomeric structure of the chemokine RANTES (CCL5) (PDB ID 1U4L).

mutagenesis studies have shown that the N-terminus (preceding the first cysteine) is the key segment for causing receptor activation. For example, many chemokines lacking all or a few amino acids from the N-terminus can still bind to the receptor, often with high affinity, but they no longer induce the conformational changes required for activation, thereby functioning as antagonists.[80] Indeed, proteolytic processing of the N-terminus is a natural regulatory mechanism for turning chemokine signaling off and in some cases has been shown to alter receptor specificity.[81] Other minor modifications such as extensions of the N-terminus with a single amino acid or single point mutations can convert chemokines from agonists into antagonists, or change the trafficking behavior of the receptors. For example, addition of a single methionine to the N-terminus of RANTES (CCL5) confers antagonist activity.[82] Several groups have taken advantage of this general feature to produce high-affinity chemokine antagonists for proof of concept studies of the role of chemokine:receptor pairs in specific diseases.[83] Further, variants of RANTES (CCL5) have been discovered that cause its receptor, CCR5, to internalize for prolonged periods of time, making for powerful agents that block HIV infection.[84]

While the chemokine N-terminus is important for receptor activation, several studies have confirmed that other segments are involved in controlling binding to receptors and likely the orientation of the N-terminal domain with respect to the receptor-binding pocket. The presence of the chemokine core domain (everything but the N-terminus) seems to facilitate a tighter interaction as peptides derived from only the N-termini of chemokines bind to receptors with affinities that can be many orders of magnitude weaker than wild-type proteins.[85,86] Specific regions of the core domain involved in receptor interactions have been identified: The N-loop (residues 10–20) is a region that has been shown to be critical for receptor binding.[87] In some cases, mutation of one residue in the N-loop can result in a three orders of magnitude decrease in the affinity of a chemokine for its receptor. Several studies involving engineered chimeric chemokines with core residues (beta strands and alpha helix) from one chemokine and N-terminal/N-loop residues from another chemokine also implicated the first 20 residues as the primary determinants of receptor binding and activation.[88,89] The 30s and 40s loops, often rich in basic residues, have also been shown to be crucial to receptor interactions.[90–93] One of the few structural studies done with whole receptor, as opposed to peptides, has produced evidence for a remote interaction with residues 59 and 63 of MIP1-α (CCL3),[94] but generally the C-termini of chemokines are not thought to directly interact with the receptors. Despite the rich mutational data, theoretical prediction of chemokine-receptor specificity remains difficult. The only exception is the ELR sequence motif in the N-termini of CXC chemokines, which usually predestines the chemokines as ligands for CXCR1 and CXCR2, and confers them with the proangiogenic properties of stimulating endothelial cell proliferation and migration.[95]

3.2. Dimer structures

Many if not most chemokines exhibit tendencies to form dimers and higher-order oligomers. Although monomeric chemokines are believed to be the dominant high-affinity form capable of interacting and activating receptors, oligomeric forms are increasingly recognized as being important to function. Some chemokines such as RANTES (CCL5) readily oligomerize, forming high-order polymers under physiological conditions. Other chemokines, such as MCP-3 (CCL7) and fractalkine (CX_3CL1), do not oligomerize, or oligomerize only at very high concentrations, respectively. For most dimer-forming chemokines, the K_d for self-association is in the low micromolar range; however, environmental conditions such as pH and ionic strength can strongly influence dimer stability. Although the physiological concentration of chemokine is believed to be in the nanomolar range, such that soluble oligomers of chemokines should be a small fraction of the total population *in vivo*, local concentrations of chemokines on the endothelium are likely high enough to induce formation of dimers and other higher-order oligomers. Moreover, the interactions of chemokines with GAGs, which cover the endothelial surface, are known to facilitate oligomer formation, even for chemokines that form weak dimers alone in solution. That these interactions are important has been convincingly demonstrated by several studies which showed that chemokine mutants devoid of their ability to oligomerize or interact with GAGS are functionally deficient *in vivo*.[22,45,96]

There is slightly more diversity in dimer structures compared to tertiary structures, and the mode of dimerization seems to be largely dictated by the chemokine subfamily. CC-type chemokines dimerize through residues near the N-terminus, encompassing the first cysteine. These residues in one monomer form an intermolecular antiparallel two-stranded beta-sheet with the same residues in another monomer, resulting in an elongated dumbbell-shaped structure (Fig. 20.2). The subunits are stabilized by a combination of hydrogen-bonding, hydrophobic, and electrostatic interactions. CXC-type chemokines form a distinctly different dimer structure. The dimerization of CXC-type chemokines relies on both the pairing of residues

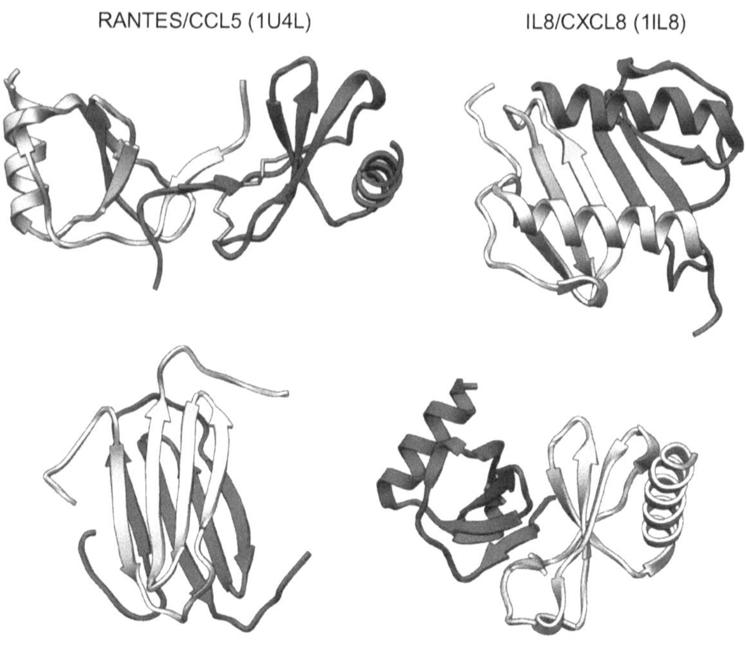

Figure 20.2 Dimer structures of chemokines. Monomeric units in each dimer are colored light or dark gray. The PDB ID of each structure is shown in parentheses.

in the first strand of the core beta-sheet in each monomer, and the positioning of the C-terminal helices across beta-sheets from both monomers (Fig. 20.2). The net result is the formation of a large intermolecular beta-sheet of six strands with more contacts between the monomers than in CC-type dimers. The overall complex is also more globular and resembles major histocompatibility proteins.

CX3C and XC chemokines form unique dimer structures. The XC lymphotactins, which have only a single cystine at their N-terminus, adopt two conformations that are in slow exchange with one another. One conformation is monomeric and has the conventional chemokine fold. The other conformation contains a four-stranded beta-sheet, formed when the usually extended N-terminus pairs with the third strand in the sheet (Fig. 20.2). The expanded beta-sheet is also the dimerization domain for the dimer. The sheets are packed against one another in a beta-sandwich configuration and are held together by hydrophobic interactions. Interestingly, only the canonical chemokine conformation of lymphotactin is capable of binding the receptor and only the alternative oligomer is capable of binding heparin, suggesting that both conformations maybe functional.[77] However, the conformational flexibility of the lymphotactins is likely the exception rather than the rule, since the lack of the second disulfide bond connecting the N-terminus to the 30s loop probably allows the alternative dimer to form.

The second unusual chemokine is fractalkine. In its intact form, fractalkine is a mucin-like membrane protein with an extracellular chemokine domain that overall can function as a tight adhesion molecule, independent of G-protein coupling. However, the chemokine domain can be liberated from the membrane anchor by proteases and then behaves like a traditional chemokine. Although it only dimerizes at high concentrations, the existence of dimers similar to CC-type dimers was observed in a crystal structure. One notable difference does exist between fractalkine and regular CC-type dimers: the unique positions of the disulfide bonds in fractalkine force the core of each monomer to be much closer to one another, thus adopting a much more compact shape.

3.3. Higher-order quaternary structures

For many chemokines, higher-order oligomer formation is common, and the variety of oligomer configurations appears to be extensive and unpredictable. Many of the oligomeric forms have been demonstrated to exist in solution by techniques such as size-exclusion chromatography, analytical ultracentrifugation, and small angle X-ray scattering (SAXS), either alone or induced by binding to GAGs. Others have been captured in crystal structures, suggesting a propensity to form under certain conditions such as bound to GAGs (described below). Among the best-characterized structures are those for MCP-1 (CCL2) and PF4 (CXCL4). MCP-1 (CCL2) is a dimer in solution but forms a tetramer in the presence of GAGs as small as octasaccharides. In the MCP1 (CCL2) tetramers, each monomer contacts one neighbor with a CC-type interface and the other neighbor with a CXC-type interface, giving the tetramer two CC-type dimers (the primary, more stable dimers), two CXC-type dimers, and an overall globular shape[97] (Fig. 20.3). PF4 (CXCL4) also forms similar tetramers; however, it is a stable tetramer, even in the absence of GAGs.

IP-10 (CXCL10) also forms tetramers. However, three types of tetramers were observed in crystals from different space groups. While one of the tetramers resembles the MCP-1 (CCL2) tetramer (M form), the other two (T and H forms) are more extended and involve a network of intermolecular contacts among beta-sheet residues and the formation of a large curved beta-sheet.[55] The structure of murine IP-10 (CXCL10), determined by the same authors, adopts an even more extended form with a 12-stranded beta-sheet.[98] Crystal structures of SDF-1α (CXCL12) also revealed an extended oligomer that contains CC-type dimer contacts as well as unique intermolecular contacts involving both the C-terminal helix and the first beta strand[99] (Fig. 20.3). Of the chemokines that have been characterized, RANTES (CCL5) and MIP-1α/β (CCL3/4) are unique in their ability to form very large oligomers in solution at physiological pH. The oligomer distribution of RANTES (CCL5) has recently been characterized by native MS; even at the deaggregating pH of 4.5, octamers and hexamers have been observed (see Section 5.4).[100] Oligomers of MIP-1 α/β (CCL3/4) have been crystallized and show that chemokines are oligomerized by stacking CC-type dimeric subunits in a helical fashion[101] (Fig. 20.3). The oligomers are held together by several key interdimer electrostatic contacts. The tetramer structure of RANTES (CCL5) in solution has also been characterized using an integrated multitechnique approach combining solution NMR, SAXS, and mass spectrometry (MS). The oligomerization interface of the extensible model involves the same residues as the MIP-1α/β (CCL3/4) oligomerization interface, but the oligomer is much more extended because the interdimer interactions take place between only a single monomer in each neighboring dimeric unit, whereas the MIP-1 α/β (CCL3/4)

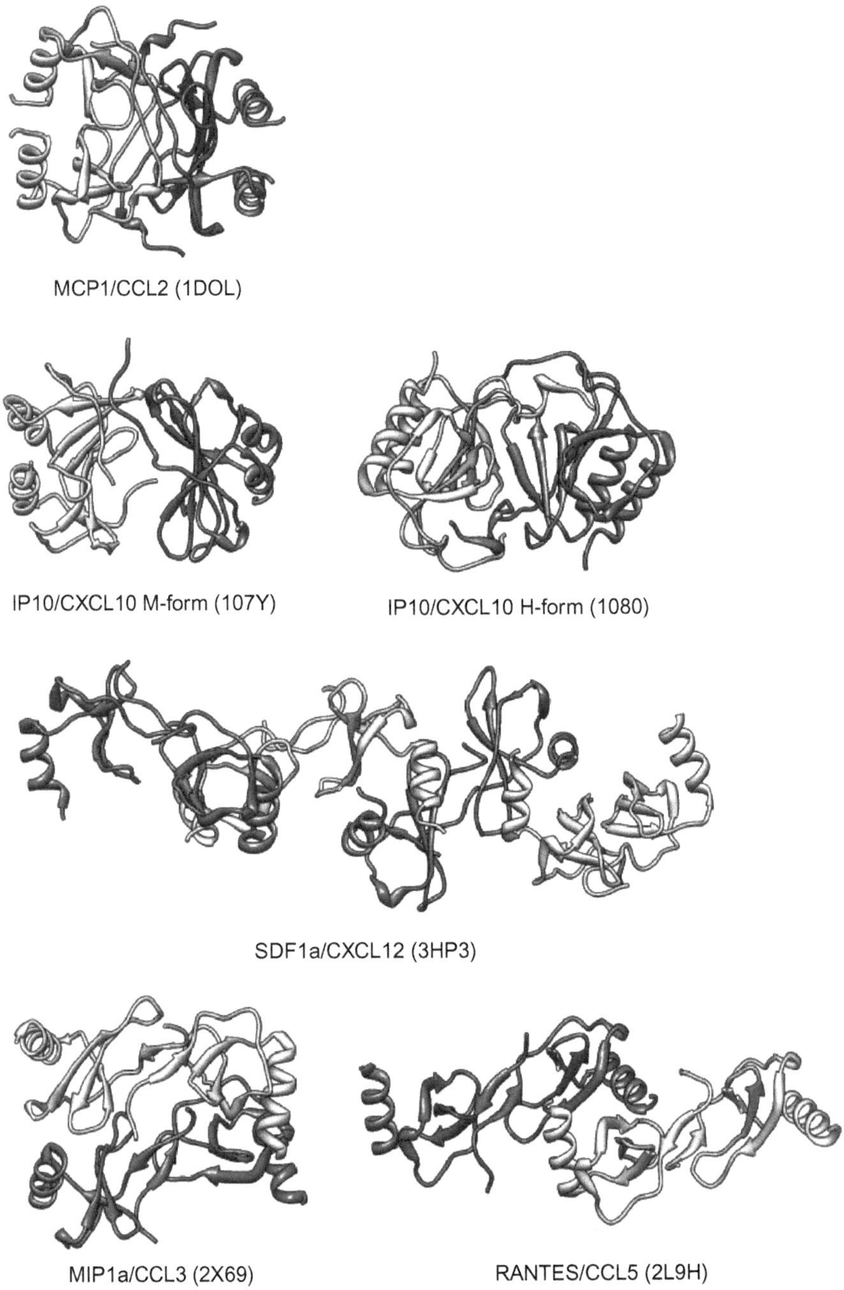

Figure 20.3 Known quaternary structures of chemokines. Each dimeric unit is colored in light and dark gray alternately. The PDB ID of each structure is shown in parentheses.

oligomers utilize both monomers of each dimeric unit to contact adjacent dimers[100] (Fig. 20.3).

Oligomers with exposed oligomerization surfaces at the ends of extended structures, as in the case of MIP-1α/β (CCL3/4), naturally lend themselves to formation of oligomers of ever increasing sizes. In fact, a simple model in which the free energy of association would be the same at each successive addition of a dimer would lead to an extremely broad distribution of oligomer sizes. For an association constant of 10^4 and monomer concentration of 10^{-4} M, concentrations of tetramer, hexamer, octamer, etc., would be 10^{-4}, 10^{-4}, 10^{-4}, etc. That unrestricted oligomerization does not happen in nature very likely is the result of the high-charge density these molecules normally carry. RANTES (CCL5) at neutral pH would have a net charge of +12 for the dimer. Even if counter ion association reduced the effective charge to +3, the attenuation of high oligomer forms by charge repulsion would be substantial. With a simple charge repulsion model that assumes (i) the charge is effectively concentrated at the dimer center, (ii) that the oligomers are linear, and (iii) that addition of each subunit is characterized by a noncharge compensated association constant of 10^5, one can calculate that the teramer, hexamer, and octamer formation constants would be 1.67×10^4, 1.14×10^4, and 0.42×10^4. Studies characterizing the oligomerization interface of RANTES (CCL5) were done at pH of 4.5 where tetramers dominate. Partial protonation of acidic amino acids would artificially increase the effective positive charge. An effective charge of +4 on each dimer and an uncompensated association constant of 10^6 predicts that more than 90% of the species would be tetramers using this simple model. The actual *in vivo* situation is likely to be more complex for many reasons, including the presence of GAG interactions as discussed below. The highly negatively charged GAGs would further reduce the effective charge of a positively charged chemokine and shift the equilibrium to higher-order oligomers—as is observed.

4. CHEMOKINE OLIGOMERIZATION AND GAG BINDING

GAGs are linear polysaccharides that are found on cell surfaces or in the extracellular matrix. They consist of repeating disaccharide units with one saccharide being either galactosamine (chondroitin sulfate or dermatan sulfate) or glucosamine (hyaluronan, heparin, heparan sulfate, or keratin sulfate) and the other being either a uronic acid (all types except keratan sulfate) or galactose (keratan sulfate). GAGs are extremely anionic due to a high content of uronic acids and sulfation. The presence of the high charge density allows GAGs to interact strongly with many signaling molecules. The existence of chemokine:GAG interactions is not surprising given the highly basic nature of most chemokines and the need for some mechanism of cell surface immobilization; thus the importance of these interactions was recognized early on.[102] Further, the interactions are by no means nonspecific. Studies comparing salt elution concentrations of different chemokines on Mono-S and heparin columns showed affinities for Mono-S resin and heparin resin do not correlate strictly with charge.[103] Competitive adhesion assays with immobilized heparin or cells not expressing chemokine receptors also demonstrated the ability of chemokines to discriminate between different types of GAGs.[103,104] Functionally, GAG–chemokine interactions are believed to be important in several aspects of chemotaxis, including prevention of chemokine proteolysis, and stabilization of chemokine gradients on the endothelial surfaces under blood flow, so that they can provide directional cues for migrating cells. The fact that models of leukocyte activation stipulate that leukocytes should only be activated by chemokines close to the endothelium surface also implies that chemokines bound to GAGs may in fact be active *in vivo*, making understanding the structures of GAG-bound chemokines a high priority.

GAG-binding proteins often oligomerize upon binding GAGs. This phenomenon has been shown to be crucial in the role of GAGs as coreceptors for a number of growth factors and as facilitators of amyloid fibril formation *in vivo*. The role of GAG-induced chemokine oligomerization in promoting localization of chemokines was revealed through *in vitro* studies conducted by Hoogewerf *et al.* They showed that binding of GAGs to chemokines induced oligomerization of a number of chemokines and that removal of GAGs from cell surfaces reduced the binding of chemokines to the cell surface significantly. Whether chemokines can bind simultaneously to GAGs and receptors is currently unknown. However, as described above, the functional requirement for the oligomerization of RANTES in inducing monocyte arrest was proposed to provide a mechanism for bridging surface bound RANTES through GAGs on the endothelium with CCR1 on the monocytes.[32]

Mutagenesis studies on a handful of chemokines have identified some motifs important for binding GAGs. *In vivo* assays designed by Proudfoot and coworkers were especially instrumental in establishing the fact that although GAG-binding deficient mutants were able to induce cell migration *in vitro*, they failed to do so *in vivo*.[22] In keeping with early studies on the amino preferences for heparin, these studies identified arginine and lysines in chemokines as the residues most responsible for GAG binding, but histidines were also found in GAG-binding motifs of some chemokines. The BBXB motif known to bind GAGs has been found in MIP-1α/β (CCL3/4) (40s loop), RANTES (CCL5) (40s loop), and SDF1α (CXCL12) (20s loop). All are confirmed to be crucial to GAG binding.[93,105–107] MCP-1(CCL2) and IL-8 (CXCL8) also contain clusters of basic amino acid that have been shown to be important for GAG interactions.[91,104,108]

Although no experimental structures of chemokines bound to anything longer than a disaccharide GAG exist, chemokine oligomer structures without GAGs do offer us some clues as to how GAGs may stabilize oligomerization. The MCP-1 (CCL2), PF-4 (CCL4), and IP-10 (CXCL10) M tetramer structures show basic residues in each monomer to be arranged on the surface to form a binding strip across the entire tetramer structure, prompting the proposal of a "beads-on-a-string" model for these GAG–chemokine interactions (Fig. 20.4). The overall acidic nature of GAGs should also neutralize charge repulsion among the basic chemokine monomers, thus stabilizing the oligomer structure. However, not all chemokines oligomerize in a globular fashion. MIP-1α/β (CCL3/4) oligomerizes in a linear fashion, forming a basic channel on the oligomer surface in which GAGs can bind. Surprisingly, SAX analysis of MIP-1α/β (CCL3/4) showed oligomers formed in the presence of heparin are smaller than without heparin,[101] contrary to a study of the chemokine carried out with other techniques.[29] This complex response may be the result of the rather unique charge distribution in MIP-1α/β (CCL3/4). In contrast to the high-positive charge of RANTES (CCL5), MIP-1α/β (CCL3/4) actually have a net negative charge (−1).

In the oligomer structure of RANTES (CCL5), the BBXB motifs form two regular arrays along the length of the oligomer, providing a possible GAG-binding site on the oligomer. However, RANTES (CCL5) is sufficiently basic that the formation of oligomers also creates large basic patches elsewhere in the structure due to close packing of other basic amino acids on the molecule. The potential of these basic patches as GAG-binding sites means RANTES (CCL5) may possess multiple GAG-binding sites as an oligomer.

The discovery of these linear chemokine oligomers prompts the question of their *in vivo* relevance. Although no chemokine oligomer structure has been solved in the presence of GAGs, the presentations of basic amino acid clusters on these linear oligomers are usually compatible with simultaneous interactions with both GAGs and receptors. Their existence produces

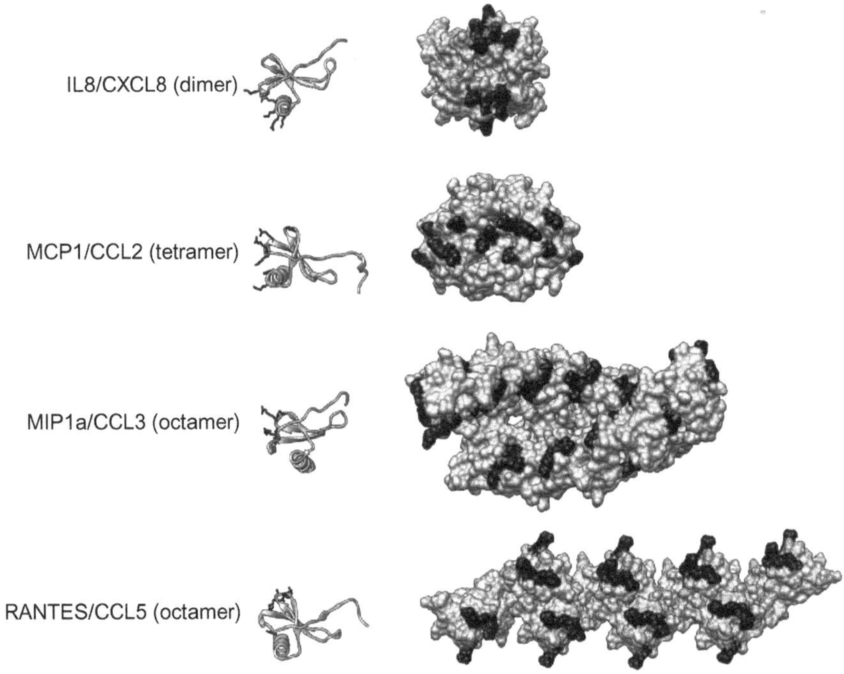

Figure 20.4 GAG-binding motifs of chemokines. Amino acids known to bind GAGs are shown as black sticks. The same residues are also colored black on the oligomer surface depiction. For IL-8 (CXCL8), K20, R60, K64, K67, and R68 are shown. For MCP-1 (CCL2), R18, K19, R24, K49, K58, and H66 are shown. For MIP1α (CCL3), R18, K45, R46, and K48 are shown. For RANTES (CCL5), R17, K44, R45, and K47 are shown.

interesting hypotheses for the chemotaxis process. The primary role of a chemokine concentration gradient is to provide directional cues for the migration of leukocytes and their activation. The random distribution of chemokines around the inflammatory site with concentration gradients controlled by surface diffusion still presents leukocytes with substantial translational migration freedom, which may translate to slower targeting efficiency. The formation of large linear oligomers could favor a more efficient leukocyte targeting process by limiting the movement of leukocytes to a one-dimensional process.

It is worth noting that residues identified as GAG-binding epitopes in several chemokines are also involved in receptor binding. R17 of RANTES (CCL5) has been identified as crucial to activation of CCR1, but strong evidence implicates the same residue in GAG-binding. Similarly, mutations in the BBXB motif of RANTES (CCL5) and MIP-1β (CCL4) reduced its affinity for CCR1 by 200-fold and identical changes in MIP-1β (CCL4) reduced its affinity for CCR5 by 77-fold.[22] The recent crystal structure of CXCR4 showed a large acidic binding site for its ligand SDF-1 (CXCL12), providing a rationale for overlap in GAG and receptor-binding sites in chemokines. The functional consequence of competition between the two interactions for the shared binding motif has yet to be investigated, but it is reasonable to hypothesize that higher-order oligomerization may have evolved to allow simultaneous GAG- and receptor-interaction by displaying multiple binding sites in the same entity.[109] The linear oligomerization states of chemokines are especially suited for this purpose.

Given the known tissue-specific variations in GAG species and sulfation patterns, it is tempting to speculate as to whether chemokines are capable of recognizing specific GAGs. Because of the difficulties in purifying structurally homogeneous GAGs of sufficient length, characterizing specific GAG–chemokine interactions remains difficult. However, current knowledge of chemokine oligomer structures already reveals much complexity in GAG–chemokine interactions: lymphotactin only interacts strongly with heparin as a noncanonical dimer while the monomer is potent in binding its receptor, thus excluding the possibility of simultaneous interactions with both GAGs and receptors. IP-10 (CXCL10) can adopt different tetramer forms with different displays of basic amino acid patterns on their surface and consequently different GAG specificities. SDF-1γ (CXCL12) possesses an extended basic C-terminal tail, which may endow it with considerable promiscuity, allowing it to access GAGs of different species without oligomerizing. These observations suggest flexibility in GAG–chemokine interactions, which may influence their specificity for GAGs under different conditions.[93,105–107] Some attempts have been made to characterize the GAG length dependence of chemokine:GAG interactions, and a wide variety of affinities have been demonstrated. Spillmann *et al.* showed that IL-8 (CXCL8) dimers only bind to HS weakly and that a HS fragment 18 saccharides long was required to achieve measureable affinity.[110] On the other end of the spectrum, RANTES (CCL5) binds to HS hexasaccharides with a K_d of less than 100 nM.[111] In fact, it has been observed that GAG fragments as short as hexasaccharides are capable of inducing formation of higher-order oligomers in RANTES (CCL5), even though the size of the dimer is capable of accommodating a much longer fragment. While no K_d has been reported for MIP1α (CCL3) binding to GAGs, a heparinase protection study showed that a HS fragment with 14 saccharides seems to be the optimum size.[112] Although the original study speculated that dimers were the dominant oligomeric forms in solution, in view of the work by Ren *et al.*,[101] it is highly likely that HS fragments were binding a much larger oligomer.

5. STRUCTURAL METHODS FOR OLIGOMERS AND GAG INTERACTIONS

Given the importance of oligomerization and GAG-binding interactions, it is worth considering how data on these interactions will be acquired in the future. Much of the data presented above was obtained by X-ray crystallography. The resolution in most cases is at a level seldom obtained by other methodology. However, there are limitations. The oligomerization interactions, at least in the case of linear oligomers, have no definitive limits on oligomer size, increasing the likelihood that crystal packing will impose restrictions that do not exist in solution or *in vivo*. Also, the GAGs to which chemokines bind are extraordinarily heterogeneous. This makes homogeneous samples of any significant length difficult to obtain, and the use of heterogeneous samples inhibits crystallization of complexes. Furthermore, it is very likely that hetero-oligomers may be functionally relevant, as are complexes involving multiple receptor molecules, and receptors anchored in their native membrane environment. All of this demands a fresh look at options for structural characterization.

5.1. NMR of protein–protein complexes

NMR offers some advantages when it comes to examining interactions between multiple proteins (chemokine oligomers or chemokine–receptor interactions) or interactions between proteins and ligands (GAGs). It does not require crystallization and is reasonably tolerant of heterogeneity, at least in the case of rapidly exchanging heterogeneous ligands. It is also a solution-based technique applicable in aqueous buffers that approach physiological conditions or media that mimic membrane environments. However, it does have limitations in that the sensitivity of the technique requires solutions that are several hundred micromolar in the protein/complex of interest, enrichment with NMR observable isotopes (^{13}C, ^{15}N) is required, and investigations of large assemblies are limited by resonance broadening (this increases linearly with the molecular weight of the system and causes loss of both resolution and sensitivity). Nevertheless, NMR has made a substantial contribution to investigation of chemokines and their oligomerization properties.

NMR structures of monomeric and dimeric chemokines are abundant.[35,60,64,67,113–117] The small size of monomeric chemokines, less than 10 kDa in most cases, has for years put chemokines well within resolution and sensitivity limits of NMR. Early structures were produced without isotopic labeling or labeling with ^{15}N only. All relied on distance constraints from short-range ^{1}H–^{1}H nuclear Overhauser effects (NOEs). Later, structures, and particularly larger dimeric structures, relied on triple resonance methods for resonance assignment and resolution of larger numbers of NOE cross-peaks (^{13}C and ^{15}N labeling). All have produced useful atomic resolution structures.

The utility of NMR in detecting homotypic chemokine–chemokine interactions, as well as heterotypic interactions with other chemokines or receptors exists at many levels. The most widely practiced approach utilizes chemical shift perturbation. Chemical shifts of resonances, or more commonly the positions of cross-peaks in ^{15}N–^{1}H heteronuclear single quantum coherence (HSQC) spectra, change on contact with ligands or other proteins due to the magnetic susceptibility anisotropy of nearby groups such as phenyl rings and electric field effects of nearby charged groups. They also change (especially for heteronuclei such as ^{15}N or ^{13}C) because of induced distortions of bond angles in the peptide backbone. These multiple contributions make it difficult to quantitatively convert perturbations to distances of approach. Even occasional remote changes can occur due to allosteric effects on protein geometry, but when perturbations to resonances from several clustered residues are observed, identification of regions of contact can be done quite reliably. There are examples where likely dimer interfaces have been identified in this way. A good example is in the characterization of a dimeric variant of the viral analog of human MIP1-β (CCL4), vMIP-II,[118] which is a broad-spectrum antagonist, produced to subvert the human immune system. Shifting the dominant monomer form toward a dimer through a single mutation alters both GAG affinity and binding geometry making identification of the dimer interaction surface particularly relevant. Chemical shift perturbations have also been used to identify chemokine sites that interact with receptors. Most notable is the early work on identification of the importance of N-terminal residues of RANTES (CCL5) on interaction of peptides from the CCR5 receptor.[28] More recently perturbations of ^{15}N–^{1}H cross-peaks from the N-terminal portion of the CXCR4 receptor (1–38) have been used to infer different modes of interaction for monomeric and dimeric forms of the CXCL12 chemokine.[35]

There are also more quantitative methods for identifying interface residues that involve diffusion of perturbed magnetization across an interface (an NMR cross-saturation experiment).[119] Here, perdeuterated proteins with amide protons at ^{15}N-labeled sites (reintroduced by back-exchange) or having only ^{13}C methyl groups protonated are mixed with fully protonated and nonisotope-labeled protein. Proton resonance characteristics of the non-labeled protonated species are saturated, saturation spreads across the protein–protein interface in a distance-dependent fashion and saturation is detected through the isotopically labeled sites at the interface. The very steep distance dependence ($1/r^6$) ensures that sites at the interface are selectively detected. An example of this type of experiment can be found in the determination of a tetramer structure for CCL5 where both monomer–monomer and dimer–dimer interfaces were identified.[100] There is also a very interesting application to the interaction of the CCR5 receptor with MIP1-α (CCL3).[94] Chemokine receptors are integral membrane proteins and CCR5, in particular, seems to require more bilayer-like membrane mimetics to maintain a functional configuration. Nanodiscs are assemblies that fulfill this purpose, but they are very large.[120] The Shimada group has used a modified version of the cross-saturation experiment (transferred cross-saturation), and the higher sensitivity protonated methyl labeling of the chemokine CCL3, to facilitate investigations of this very large complex. Unexpectedly, valines 59 and 63 were identified as components of the interaction site and add to the chemokine N-terminal regions normally implicated in interaction with receptors. The method does require rapid exchange on and off the receptor, so that detection of an excess of free chemokine population can be used to report on the complex; therefore, it is possible that the interactions arise from a secondary (weaker) binding event. Nevertheless, the application does illustrate the potential for applying NMR to some very large systems.

Actual structural characterization of oligomers and complexes is more of a challenge. Traditionally, this would be approached by seeking short-range NOEs connecting residues on different proteins of a complex. Acquiring these data is particularly challenging for symmetric homo-oligomers because resonances from partners in the complex are degenerate. However, there are strategies referred to as X-filtered experiments in which partners are labeled with ^{13}C and ^{12}C, respectively, and signals filtered for ones arising from ^{13}C-labeled sites and detected on ^{12}C sites or vice-versa.[121] These are effective, but labor intensive in terms of the need for preparation of multiple samples and also sensitivity limited in the case of homo-oligomers due to detection of only half the complexes formed. As most interface NOEs involve interaction of protons on side chains, they also require complete side chain assignments, something that becomes more difficult as the size of complexes increases.

There are additional long range distance-dependent interactions that can be useful in the structural characterization of chemokine oligomers. In particular, paramagnetic relaxation enhancement (PRE) experiments, in which line broadening or spin relaxation times show a $1/r^6$ dependence, can be used. They do require the introduction of a paramagnetic site in a protein or in a GAG ligand. Most commonly a nitroxide-carrying TEMPO adduct is used. Because of the high-magnetic moment of the unpaired electron on the nitroxide (650 times larger than that of a proton) and the square dependence of relaxation enhancement on this moment, these effects extend over many tens of angstroms and are potentially effective in positioning one protein relative to another or positioning a ligand in its binding site. While there are few applications to positioning of chemokines in protein complexes at this point, there are applications to positioning GAGs in chemokine binding sites. These will be discussed in the following section.

Long-range NMR constraints can also come from orientation-dependent NMR measurements, the most common of which are residual dipolar coupling (RDC) measurements.[84,122,123] RDCs are measured in orientable media (liquid crystals or stretched polyacrylamide gels) that can induce partial alignment of molecules of interest. They are usually measured as a change in couplings between directly bonded pairs of nuclei (e.g., $^{15}N-^{1}H$ amide bonds) when through space dipole–dipole contributions fail to average to zero. Their dependence on the average of the function, $(3\cos^2\theta - 1)$, where θ is the angle between the spectrometer's magnetic field and the internuclear (bond) vector, provides a powerful determinant of molecular geometry. When sufficient numbers of RDCs are measured for well-defined structures, alignment axes can be determined, and when two different proteins are involved superposition of alignment axes greatly restricts allowed structures.[124,125] Typically, these RDC-based restrictions are combined with data on interfacial contacts and translational restraints to produce unique structures. For symmetric homo-oligomers, there are symmetry-based rules that allow identification of the symmetry axis in the coordinate frame of the monomer and generation of symmetry-related elements by rotation about this axis. Applications involving the determination of structures for symmetric dimers have appeared.[124,126] RDCs are particularly advantageous when working with higher-order oligomers as only backbone resonance assignments are needed and RDCs can be measured with high precision in relatively large systems. This gives them an advantage over more traditional X-filtered experiments for determination of structures, particularly when structures of constituent monomers are known from other sources. A good example of an RDC application to a higher-order chemokine oligomer is the structure determination of a CCL5 tetramer.[100] Previously available structures for CCL5 existed from both crystallography[127] and solution NMR studies.[64,115] These were produced at low pH (~3) where the structure is predominantly a dimer of the canonical CC form. At higher pH (4.5), the system is predominantly a tetramer as confirmed by light scattering and NMR diffusion studies. Comparing experimental RDCs to calculated RDCs based on a monomer from the crystal structure confirmed that the monomer structure was preserved in the tetramer. The appearance of just a single set of ^{15}N–^{1}H HSQC cross-peaks suggested a symmetric structure (at least as averaged over the timescale, ~1 ms, needed to resolve NMR resonances). More surprising was that the symmetry axis of the tetramer, as identified from RDCs, coincided with the symmetry axis of the dimer in the crystal structure. There are only a few possible ways to construct a tetramer while preserving this symmetry axis. Adding interface residue data and SAXS data as discussed below, an elongated structure was deduced. This is one in which higher concentrations and GAG interactions can promote the addition of other dimer units to make progressively longer hexamers, octamers, etc. These structures are in fact observed in subsequent mass spectrometric analysis of low-concentration samples prepared at higher pH.

In the future, residual chemical shift anisotropies (RCSAs) are likely to be added to RDCs for the determination of monomer orientations in oligomeric structures.[128] These display identical angle dependencies to RDCs, but the vectors are principal axes of shift tensors and an average over the three axes are observed. Their advantage is that they depend only on measuring chemical shift changes for highly anisotropic sites such as a ^{13}C carbonyl in a peptide or acetyl group, and can likely be measured in even larger systems. Again applications to chemokines are few at present, but there are applications to the determination of preferred conformations for GAG oligomers.[129]

5.2. NMR of GAG–protein complexes

Interest in structural aspects of GAG–protein complexes remains high both because of scientific curiosity and the potential application of GAGs in treatment of a range of diseases from cancer to atherosclerosis. However, structural characterization of GAG–protein complexes presents significant challenges. Among them is the tendency of GAGs to induce oligomerization of chemokines. Many of the NMR methods discussed above can be effective in the investigation of such complexes. However, significant technical challenges also arise from the lack of a library of structurally homogeneous GAG fragments large enough to reproduce conditions experienced *in vivo*, and the relatively small number of routes to the production of isotopically labeled forms of these fragments.

NMR methods: NMR has always been instrumental in the unambiguous determination of GAG structures, but its ability to solve high-resolution protein structures combined with its sensitivity to protein–ligand interactions and tolerances of heterogeneous GAG fragment mixtures has made it an excellent tool for characterizing GAG–chemokine interactions. The most popular use of NMR involves identifying the GAG-binding site of the protein using ligand-induced chemical shift changes for isotopically labeled sites in chemokine, much like the identification of interfaces in chemokine oligomers discussed above. Binding of GAGs to proteins often changes the magnetic environment in nearby atoms by electrostatic or structure perturbation mechanisms, thus generating a change in a labeled atom's chemical shift. If the assignments of the resonances are known, residues most perturbed by ligand binding can be determined easily. Since both mechanisms tend to be short range, those perturbed most by the ligand are often deemed to be residues close to the GAG-binding site. Interactions between GAG fragments and proteins often fall in the fast exchange regime on the NMR time scale, resulting in a gradual shift of resonances with increasing ligand concentration (Fig. 20.5). This also allows the K_d of binding to be extracted easily and precisely. The most popular atoms monitored are the backbone amide nitrogen and protons, whose chemical shifts are sensitive even to minute perturbations in the magnetic environment. The two-dimensional NMR experiment, ^{1}H–^{15}N HSQC spectroscopy, allows simultaneous measurement of both chemical shifts easily.

If the changes in chemical shifts of the protein atoms are too small, or if the protein undergoes significant conformational changes upon binding, GAGs functionalized with paramagnetic ligands can be used to probe the protein–GAG interactions. Gemma *et al.* coupled amino TEMPO, a stable nitroxide radical, to the carboxyl group of uronic acid in a heparin disaccharide and successfully probed the GAG-binding sites on CCP module 7 of complement factor H using the adduct.[130] Although this is the only reported case of a protein–GAG interaction study utilizing paramagnetic GAG ligands to date, the ease by which GAG fragments can be functionalized on the reducing end should encourage more examples in the future.

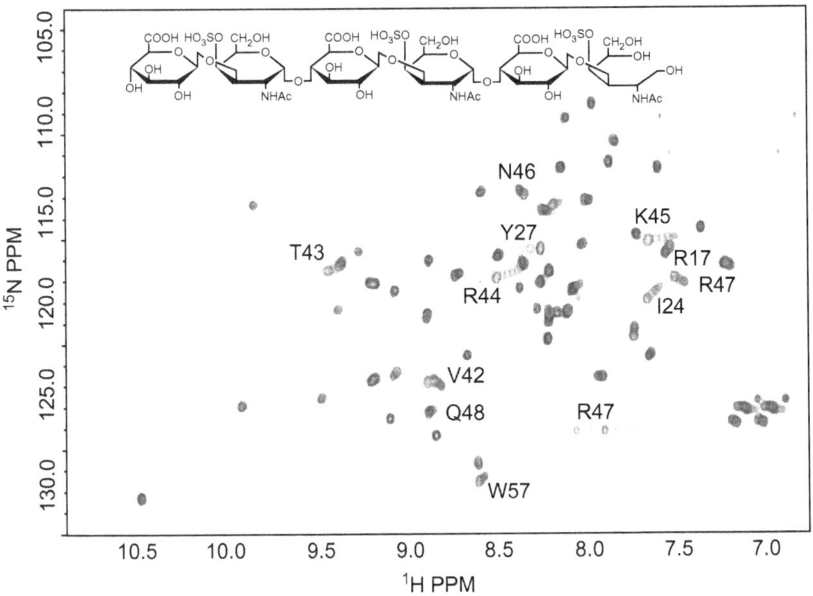

Figure 20.5 Chemical shift perturbation on addition of a four-sufated chondroitin sulfate hexamer to ^{15}N-labeled RANTES (CCL5). A series of ^{15}N–^{1}H HSQC spectra are superimposed to show chemical shift changes as increasing amounts of the hexamer are added. Among cross-peaks shifting are those assigned to residues from the BBXB motif, R44, K45, N46, R47. Spectra are taken at 600 MHz for ^{1}H on a ~400 μM, pH 3.5, sample of wild-type RANTES (CCL5).

Despite the popularity of chemical shift mapping in GAG–protein studies and emerging paramagnetic methods, solving high-resolution structures of GAG–protein complexes remains a challenge for NMR. The dominant contributions to protein–GAG interactions are electrostatic interactions between sulfates on the GAG fragment and side chains of basic amino acids such as lysine, arginine, and histidine. The lack of intermolecular proton contacts in these interactions makes their detection through traditional homonuclear NOESY, which is sensitive to only proton–proton contacts, difficult. So far, the traditional approach of solving protein–ligand interactions has not produced a single structure of GAG–protein complexes. However, some innovative approaches are being developed. One method identifies the basic amino acid involved by combining cross-linking with solution NMR. By slowing down the exchange of protons between lysine side chains and solvent using low pH (3.0) and low temperature (10 °C), the proton signal from side chain amino groups of lysines can be observed. Cross-linking of the activated carboxyls in the GAG fragments to the lysines during the binding event produced significant changes to the spectrum, therefore allowing unambiguous identification of lysine residues involved in GAG binding.[131] This covalent modification technique forms an excellent complement to the chemical shift mapping approach, which is usually carried out using only backbone amide protons that may or may not respond to GAG-binding.

Besides observing perturbation of protein by GAG ligands, NMR is also well suited for studying GAG conformation in the presence of protein. Interactions of chemokines and small GAG fragments usually occur on the fast exchange regime on the NMR time scale. This allows well-established ligand techniques such as saturation transfer difference (STD) and transferred NOE (trNOE) to be applied to the study of GAG conformations in the presence of proteins.[103,132,133] The only limitation on these techniques is the insolubility of GAG–chemokine complexes at high chemokine concentrations (>100 μM). However, STD only requires low concentrations of protein (~10 μM), which can be a great advantage. The only published example where STD is utilized to investigate GAG–chemokine interactions is a study of CXCL12α with a ^{13}C-labeled HS-like octasaccharide derived from *E. coli* K5 capsular polysaccharide.[134] The labeling allowed the use of 2D methods to improve resolution, and the research identified many protons on both glucosamine and glucuronic acid as being in the interface. The information allowed the construction of a model of GAGs bound to the CXCL12α dimer. Another potential source of structural information obtainable from ^{13}C-enriched GAG fragments is RCSA from carbonyl carbons. The amide bond in most *N*-acetyl hexosamines is quite rigid,[135] and therefore orientation of the acetate group can serve as a good indicator of hexosamine ring orientation. In a study of chondroitin sulfate conformation, Yu *et al.* measured RCSA from ^{13}C acetate-labeled CS hexasaccharides and used the information to supplement existing RDC data for polysaccharide conformation determination.[129] Although not yet applied to GAG–chemokine interactions, the technique will be valuable in defining bound GAG conformation and orientation in GAG–chemokine complexes. So far, there is no reported example of trNOE applied to the study of GAG–chemokine interactions. But, the technique has been utilized to determine the bound conformations of GAG fragments in other systems, including the well-characterized GAG interactions of antithrombin[136] and cobra cardiotoxin.[137]

Other than obtaining high-resolution structures for both chemokine and GAGs, NMR is also capable of studying conformational dynamics in these molecules, especially if appropriate isotopic labels are present. NMR signals are sensitive to motions of atoms at a wide range of time scales, allowing frequencies of rotational motions to be characterized in terms of the rotational correlation times. Protein dynamics can also be measured on a residue-specific basis if assignments of resonances from backbone amide nitrogens and protons are available. Such analysis has been applied to both FGF-2 and hepatocyte growth factor (HGF). In the cases of HGF, residues in the GAG-binding regions of the protein experienced significant increases in motions on micro-to-millisecond time scale.[138] FGF-2 was shown to have an overall increase in rotational correlation time, indicative of GAG-induced oligomerization.[139] However, no dynamic studies have been conducted on protein-induced changes in GAG dynamics. The ligand dynamics information is potentially valuable in understanding the kinetics and thermodynamics of protein–GAG interactions possessing multiple binding modes. The main technical barrier preventing this type of study lies in difficulties in obtaining isotopically labeled GAG fragments.

Sources of isotopically labeled GAGs: Traditionally, all homogeneous GAG fragments have been derived from natural GAG polysaccharides. Because of the complexity in GAG structures and challenges of the underlying carbohydrate chemistry, organic syntheses of GAG fragments have remained the privilege of a few synthetic chemists.[140] The isolation of GAG fragments from natural sources usually follows a three step procedure[141]: Natural GAG polymers are partially fragmented using appropriate endohydrolases or endolyases. For instance, heparin is most efficiently depolymerized with heparinase I from *Flavobacterium heparinum* and chondroitin sulfate can be depolymerized with either chondroitinases ABC, or AC, or hyaluronidase. The digested fragments are then separated based on size using size-exclusion chromatography. Finally, fragments of similar sizes are further separated using strong-anion exchange (SAX) HPLC. The success of the last step relies on the fact that SAX HPLC columns are capable of distinguishing fragments possessing different sulfation patterns despite identical total charge on each fragment. Because of sulfation pattern uniformity in some GAGs, SAX HPLC is sufficient to purify short chondroitin sulfate A or DS fragments to homogeneity.[142] However, rigorous separation of heparin beyond the size of octasaccharides continues to be challenging. Although many alternative analytical techniques for GAGs have been developed for GAG analysis, the above procedure has remained the only viable protocol for obtaining GAG fragments in preparative quantities.

Besides vertebrate sources of GAGs, GAG mimics produced using bacterial capusular polysaccharide (CPS), either as the final product or as the precursor for further chemical sulfation, are also becoming more prevalent. Many bacteria generate CPS that are similar to GAGs in carbohydrate composition. Streptococcus group C produces CPS that are identical to hyaluronan while CPS from *E. coli* strain K5 have the same carbohydrate structure as heparin and heparan sulfate.[143] They can be used as GAG mimics either without modification or after N-deacetylation followed by N- and O-sulfation.[144] So far, the sulfation methods used are limited to sulfur trioxide complexed with either trimethyl amine or pyridine, both of which have limited specificity.[144,145] Although N-sulfation and 6-O-sulfation have proven to be quite easy, specific 2-O-sulfation on uronic acids or 4-O-sulfation on GalNAc has proven to be challenging[145]; furthermore, no chemical method to convert glucuronic acid to iduronic acid exists. Difficulties in specifically modifying GAGs can be partially lessened through chemoenzymatic approaches to GAG synthesis. *In vitro* production of GAGs using isolated bacterial enzymes are already feasible,[146] and the technology is quite mature for production of hyaluronan molecular weight standards.[147,148] A recent study produced the potent antithrombin-binding heparin pentasaccharide using a chemoenzymatic approach *in vitro* starting from disaccharide units.[149] Such methods have the promise of producing size- and structure-defined fragments, therefore greatly simplifying purification procedures.

Despite its disadvantages, production of GAG mimics from bacterial CPS can provide new opportunities for the study of GAG–protein interactions using NMR. Lack of GAG fragments enriched in NMR active nuclei such as ^{13}C and ^{15}N has prevented unambiguous identification of protein-contacting GAG protons in cases of long fragments and made the study of GAG dynamics limited to proton relaxation measurements, which can be difficult to interpret because of the numerous factors involved in proton relaxation. The main reason for the deficiency is because vertebrate cells, which are often the sole source of GAGs, cannot yet be grown in isotopically enriched media in an economically feasible fashion. To overcome this deficiency, bacterial CPS grown in isotope-enriched media can be used as a substitute. This was the strategy used by Laguri *et al.* to obtain ^{13}C-labeled heparin octasaccharide for studying the interaction of CXCL12 with GAGs,[134] the only study of chemokine–GAG interaction using ^{13}C-labeled GAG mimics so far. However, the procedure remains a complicated one and limited to heparin/HS mimics only. ^{15}N enrichment of the same polysaccharide has also been carried out[150] and was demonstrated to be of value in the study of GAG dynamics as interpretation of amide nitrogen relaxation is well established.

An alternative to metabolic incorporation of isotopic labels is through the use of chemical modification.[129] *N*-acetyl groups can be chemically deacetylated and then reacetylated using ^{13}C acetic anhydride. In the cited work on chondroitin sulfate, the identity of each carbonyl signal in the ^{13}C NMR spectrum was determined based on correspondence between ^{13}C NMR signal intensity and percentage isotopic enrichment as indicated by MS/MS analysis. The labeling allowed additional structural information to be gained using RCSAs and is potentially applicable to any GAG fragment containing *N*-acetyl hexosamine.

Although the production of isotopically enriched GAGs using mammalian systems is still difficult, one feature of mammalian carbohydrate metabolism has been explored to create ^{15}N-labeled GAGs using mammalian cell culture. In vertebrates, the nitrogen in *N*-acetyl-hexosamine is derived mainly from the side chain amide of glutamine. Supplementing the media with ^{15}N glutamine therefore offers an elegant and economical way to produce ^{15}N-labeled GAGs in vertebrate cell cultures. Pomin *et al.* carried out such a study and was able to produce ^{15}N-labeled CS and HS using both mouse lung endothelial and Chinese hamster ovary cells.[151] As the ^{15}N chemical shift of GAGs is sensitive to GAG types, labeling allowed accurate quantification of ratios of different GAGs existing on the cell surface. The application of such techniques can make quantification of cell surface GAGs possible.

5.3. Global structure of oligomers from SAXS

SAXS is an important technique in elucidating the structural characteristics of macromolecular complexes in solution.[152,153] It utilizes interferences between X-ray photons coherently scattered by atoms to measure geometrical arrangements of atoms in a molecule. Unlike X-ray diffraction using crystals, SAXS samples are solutions of macromolecules with no orientational homogeneity, and thus scattering intensity is dependent on the scattering angle alone. The plot of the scattering intensity versus diffraction angle offers the main experimental data in SAX experiments, and intensities at smaller scattering angles are dependent on the global shape and size of the molecular complex in solution. Although the resolution of the structural information obtained is quite low, it is often valuable in determining the solution states of biomolecule complexes. The ease with which SAXS samples can be prepared and the advent of commercially available benchtop SAXS instruments also made the technique more accessible to researchers eager to combine SAXS data with structural information obtained from other sources, especially NMR spectroscopy. The theory of molecular scattering is well understood, and accurate prediction of the scattering curve of known molecular structures offers a powerful way to determine the solution state of many macromolecules. For samples containing a single species, extraction of the radius of gyration as well as inter-atomic distance distribution is routine. If the description of the geometrical shape of the molecule is simple, existing models can be also used to estimate a number of other important parameters, such as the lengths of rods or radii of spheres. Even a medium resolution shape envelope of the molecule can be obtained through sophisticated combinatorial search algorithms that generate the shape envelope best reproducing experimental results. Since intensity at zero angle is directly proportional to the molecular mass of the complex in solution, the total mass of a complex can be estimated using SAXS data, provided data acquired on samples of known molecular composition are also available. For samples containing a heterogeneous mixture of oligomerization states, which is often the case for many chemokine oligomers, the data interpretation is more challenging. However, if shapes of species in existence are available, scattering contribution from each species can be calculated and summed to produce the best fit to experimental data.

At least two studies on chemokine oligomerization have incorporated SAXS data. In the study by Ren *et al.*, SAXS was used to estimate the size distributions of MIP1α and MIP1β (CCL3/4) in the presence and absence of heparin.[101] The scattering curve of the two chemokines showed characteristics of rod-shaped molecules, which agreed with the crystal structures of the oligomeric chemokines. Fitting of the scattering curve to theoretical prediction of polymer size distributions using a reversible polymerization model allowed the distribution of polymers to be extracted. Interestingly, it was discovered that heparin reduced the size of wild-type MIP1α (CCL3) polymers, but promoted the oligomerization of a nonaggregating mutant of MIP1α (CCL3). The results were also confirmed using size-exclusion chromatography. In the study of the RANTES (CCL5) polymer, SAXS data also played a crucial role.[100] In this study, the major oligomerization state of the chemokine was characterized using data from NMR relaxation and pulse field gradient diffusion experiments, and RDCs were used to determine the orientation of the dimeric units relative to one another. This information greatly reduced the possible structural configurations of a RANTES (CCL5) oligomer. Further analysis by SAXS showed the oligomer to adopt an irregular and extended shape. Using a simple two-dimensional search algorithm, a set of models satisfying the NMR data was generated. The final structural model was then selected based on the agreement between the theoretically predicted scattering curves of each model and the experimental scattering curve. A configuration with each monomeric unit of the dimer making contacts with adjacent dimers was found to fit both NMR and SAXS data well.

5.4. MS analysis of chemokine oligomerization

MS is a powerful and versatile tool for the analysis of protein oligomerization. Depending on the methods used, MS is capable of determining the stoichiometry of oligomerization (even for polydisperse oligomer mixtures) and analysis of the structure of the oligomer. MS is only capable of analyzing ions in the gas phase. Therefore, in order to use MS to study solution-phase oligomerization, it is either necessary to transfer the oligomer intact into the gas phase (referred to as "native spray") or else to modify the protein in solution in a manner that transfers information regarding oligomerization into the gas phase. Both strategies can give valuable information on chemokine oligomerization, but each requires some thought regarding their strengths and limitations.

5.4.1 "Native spray" analysis

MS is capable of measuring the mass-to-charge ratio (m/z) of chemokine oligomers in the gas phase. Based on either high-resolution analysis of ^{13}C isotopomers or analysis of the spacing between neighboring charge states of the same oligomeric state, the actual charge state (and therefore the actual mass) of each oligomer can be determined, allowing for the direct measurement of the size and dispersion of oligomers. However, in order for the information generated to be biologically relevant, the oligomer(s) that exist in solution must be transferred intact into the gas phase.

Fortunately, the development of electrospray allows for the intact transfer of noncovalent complexes from solution to the gas phase in many situations.[154] In order to achieve these "native spray" conditions, several changes to the standard electrospray conditions most widely used for protein and peptide analysis must be allowed. The samples must be sprayed in solutions amenable to complex formation, which often eliminates the use of standard modifiers such as organic solvents and mild organic acids. These conditions often lead to decreased sensitivity compared to standard electrospray conditions. Additionally, the parameters of the ionization source must be tuned to make the transfer from solution to gas phase as gentle as possible. Typically, ions are allowed to undergo energetic collisions with neutral gas molecules to aid in the desolvation of the ion; however, such collisions will also disassemble noncovalent complexes in many cases. Lowering the energy of these collisions helps to maintain the native complex during the transfer to the gas phase, but decreases the efficiency of desolvation and generally reduces sensitivity. Lowering these desolvation potentials also can result in the gas phase transfer of partially desolvated protein-solvent "clusters" that retain various numbers of solvent molecules and/or buffer ions noncovalently bound to the protein. This added dispersity causes one individual protein complex to have multiple, closely-related m/z values consisting of the complex and different numbers and compositions of solvent clusters, leading to broad and poorly resolved peaks in the mass spectrum. Additionally, properly folded proteins and protein complexes usually retain less charge than unfolded proteins, resulting in signals at much higher m/z values than are typically seen in denatured protein analysis. These m/z values are above the scan range of many common commercially available mass spectrometers, and often require specialized or customized hardware to detect.[155] In order to minimize the difficulties of desolvation and declustering, complexes are often ionized in very low-flow nanospray experiments. Nanospray tends to form smaller solvent droplets than pneumatically assisted electrospray processes, requiring less energy for desolvation even under aqueous conditions.[156]

Even with these added difficulties, MS can still be an invaluable tool for the analysis of chemokine oligomerization. In 2005, Yu *et al.* performed native spray analysis of various chemokine ligands of CCR2 in the presence and absence of heparin sulfate octasaccharides. In native spray analysis, MCP-1 (CCL2) and MCP-2 (CCL8) were found to exist in solution in an equilibrium between homodimer and monomer. However, using native spray MS, it was shown that only the dimer bound to the heparan sulfate octasaccharide. Similar studies of other CCR2 chemokines did not reveal any oligomers by native spray, and additionally showed heparan sulfate octasaccharide binding to the chemokine monomer. In all cases, the native spray MS results confirmed previous solution-phase analysis results.[157] Further work by Crown *et al.* examined heterodimerization of these CCR2 chemokine ligands. Using native spray MS, the authors were able to measure apparent equilibriums between homodimer and heterodimer complexes of these CCR2 ligand chemokines. Heterodimerization dominated in mixtures of CCL2 and CCL8 in the absence of Arixtra, a heparan sulfate pentasaccharide derivative. All other combinations of CCL2, CCL8, and other CCR2 chemokine ligands examined revealed only homodimerization of CCL2 or CCL8, with the other paired chemokines existing only as monomers. However, upon addition of Arixtra, all pairs tested were observed to generate significant amounts of heterodimer, with the exception of CCL2–CCL13.[158] More recently, Carlson *et al.* have examined the heterodimerization of CXCL4 and CXCL7 with other platelet-derived chemokines. Using native spray MS, the authors were able to screen 24 different combinations of CC and CXC chemokines and their ability to form heterodimers with CXCL4 and CXCL7.[159] This screening process is made possible by the relative speed and sensitivity of native spray MS when compared to other analytical techniques used for oligomerization analysis, such as NMR and analytical ultracentrifugation.

The previous examples reveal how native spray MS can be used to study hetero- and homodimerization of chemokines; however, the analysis of polydisperse mixtures of oligomers can also be successfully analyzed by native spray MS. An example is shown in Fig. 20.6, where native spray MS was performed on CCL5, examining the oligomerization that occurred spontaneously in a 10 μM buffered aqueous solution of RANTES (CCL5) at pH 4.5. The peaks in the spectrum are broad, representing poor desolvation and declustering. The poor resolution makes it impossible to assign the charge state of any individual peak in isolation. However, as the homo-oligomer is made up of some unknown number of subunits of known size, and as incomplete desolvation and declustering will only *add* mass to the protein complex, we can algebraically determine which combinations of oligomer number and charge state would be consistent with each peak,

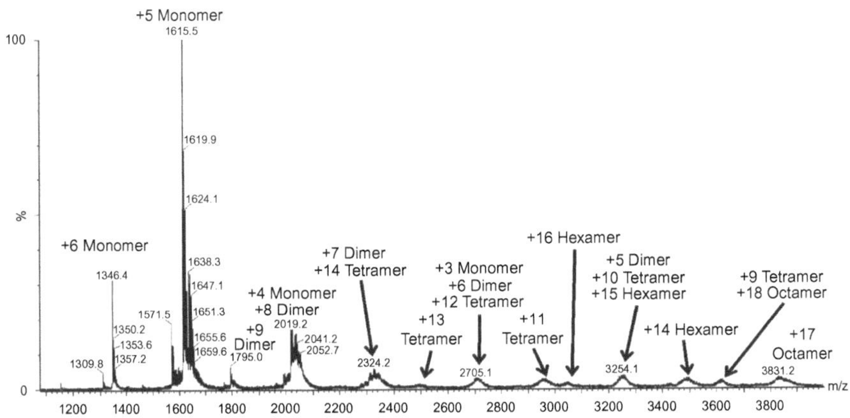

Figure 20.6 Native-spray analysis of 10 μM wild-type RANTES (CCL5) at pH 4.5. Only even numbered oligomers are observed from dimer to octamer, indicating that the RANTES (CCL5) oligomer is assembled via concatenation of the dimer subunit.

by considering the low-mass edge of the peak as being closest to the fully desolvated mass of the complex. Some peaks, such as the peak at *m/z* 2019, can be fit to multiple combinations of oligomeric state and charge state; a monomer with four charges yields the same *m/z* as a homodimer with 8 charges, a homotetramer with 16 charges, etc. However, proteins and protein complexes in electrospray do not exist as a single charge state; they exist as a distribution of multiple charge states. Therefore, if we are detecting a homodimer with eight charges, we should also detect a homodimer with seven charges and/or a homodimer with nine charges, neither of which share an *m/z* ratio with any charge state of a monomer. The peaks at *m/z* 1795 and *m/z* 2334 clearly reveal that the dimer exists in the gas phase. Similar characteristic charge states can be identified in the mass spectrum in Fig. 20.6 for tetramer, hexamer, and octamer, but not for any odd-numbered oligomer. These data from this polydisperse collection of oligomerization states clearly demonstrate that RANTES (CCL5) oligomerizes by the sequential addition of dimers.[100]

5.4.2 Hydroxyl radical protein footprinting analysis

In addition to transferring intact chemokine oligomers to the gas phase for analysis, it is also possible to use solution-phase protein chemistry to modify proteins in a structurally sensitive manner, and view how the chemistry changes upon oligomerization of the chemokine. A technique known as hydroxyl radical protein footprinting (HRPF) has previously been used by Wang *et al.* to study the structure of the RANTES (CCL5) tetramer. In HRPF, proteins in dilute aqueous solution are exposed to hydroxyl radicals formed in solution. These hydroxyl radicals can be formed by a wide variety of methods: chemical catalysis,[160] photochemistry,[161,162] electrochemistry,[163,164] and radiolysis.[165–167] The hydroxyl radicals diffuse to the surface of the protein, where they react with the amino acid side chains present on the surface of the folded protein or protein complex. The modifications are stable, so after oxidation the protein is denatured, proteolytically digested, and the sites of oxidation are identified and quantified by LC-MS/MS. The LC-MS/MS is done either using an isotopically labeled internal standard to measure the loss of unoxidized peptide, or without the use of a standard by comparison of ion intensity of the unoxidized peptide to the intensity of all of the oxidized versions of that peptide.

The apparent rate of reaction has been shown to be a function of primarily two factors: the inherent chemical reactivity of the amino acid side chain[168,169] and the solvent accessibility of that side chain.[170,171] By comparing proteins of identical or very similar sequences but different conformations, changes in the apparent rate of oxidation of a side chain can be directly correlated with changes in the average solvent accessibility of that side chain. One confounding factor in HRPF is the fact that protein oxidation can trigger conformational changes, causing further oxidation to probe nonnative states. Therefore, in order to ensure that HRPF is probing native states, either the protein oxidation must be followed kinetically to ensure that oxidation follows a single rate constant,[172] oxidation must be limited to a single oxidation event per protein,[160] or oxidation must be completed on a timescale shorter than the protein can undergo large-scale oxidation-induced conformational changes.[173] If these procedures are followed, along with proper quenching of secondary oxidants, HRPF data can accurately and reliably detect and describe changes in protein conformation.[174]

Wang *et al.* used HRPF, combined with NMR and SAXS, to compare the structure of the RANTES (CCL5) tetramer with that of the dimer in an attempt to determine the dimer–dimer interface by the protection upon oligomerization. At MS compatible concentrations (~20 μM), wild-type RANTES (CCL5) at pH 4.5 exists primarily as a tetramer, with a small amount of dimer and a small amount of higher-order oligomer also present. The E66S mutant of RANTES (CCL5) at pH 4.5 exists primarily as dimer, with some monomer present. Wang *et al.* used HRPF to compare the structures of wild-type RANTES (CCL5) to the E66S mutant at pH 4.5. The results of this analysis are shown in Fig. 20.7, mapped upon the final structure

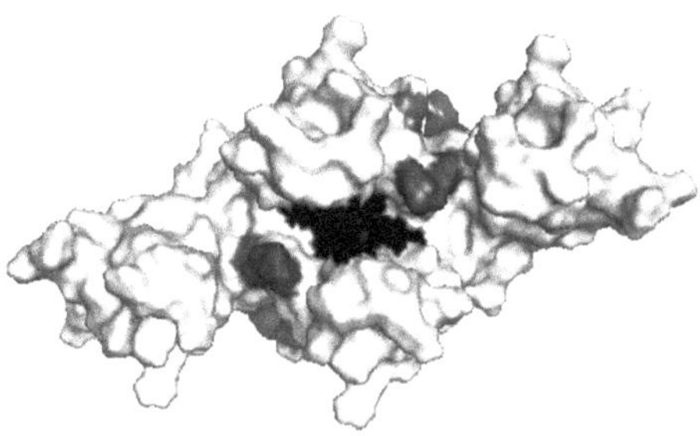

Figure 20.7 Hydroxyl radical protein footprinting of wild-type RANTES (CCL5) at pH 4.5 (tetramer) compared to the E66S mutant at pH 4.5 (dimer) mapped on a model of the RANTES (CCL5) tetramer. Regions of weak protection (<1.5× reduction in oxidation) from oxidation in the wild-type compared to the E66S mutant (gray) are found in the monomer–monomer interface and can be explained by decrease in free monomer in the wild-type sample. Regions of strong protection (>2× reduction in oxidation) from oxidation in the wild-type compared to the E66S mutant (black) are not in the monomer–monomer interface, and are therefore protected by the dimer–dimer interaction.

proposed for the RANTES (CCL5) tetramer. Protected sites can be divided into two broad quantitative categories: amino acids that exhibited strong protection (>2 × decrease in oxidation) and sites that exhibited weak protection (<1.5 × decrease in oxidation). Regions that exhibited weak protection were found to be present in the monomer–monomer interface, and could be attributed to the decrease in the concentration of monomer in the WT RANTES (CCL5) compared to the E66S mutant. The region of strong protection was determined to be the core of the tetramer interface, which was verified by NMR cross-saturation data.[100]

Wild-type RANTES (CCL5) at pH 7 spontaneously forms a polydisperse mixture of very large oligomers. These oligomers are far too large to probe by NMR and exist as a complex mixture of different sizes of oligomer. However, they can be probed by HRPF by comparing wild-type RANTES (CCL5) at pH 4.5 (mostly tetramer) to wild-type RANTES (CCL5) at pH 7 (large oligomers). The inherent rate of reaction of hydroxyl radicals is very pH-insensitive, allowing for the ready comparison of pH-induced changes.[168,175] The results of this analysis are shown in Fig. 20.8, mapped upon the structure of an octomer. The protection resulting from the transition from wild-type at pH 4.5–7 mirror the protection resulting from the transition from dimer to tetramer, only greater in scale. The region of "strong protection" in the dimer–dimer interface is actually completely protected from oxidation in the pH 7 sample, while the regions of "weak protection" at the monomer–monomer interface are still oxidized in the pH 7 sample, albeit at lower amounts than in the

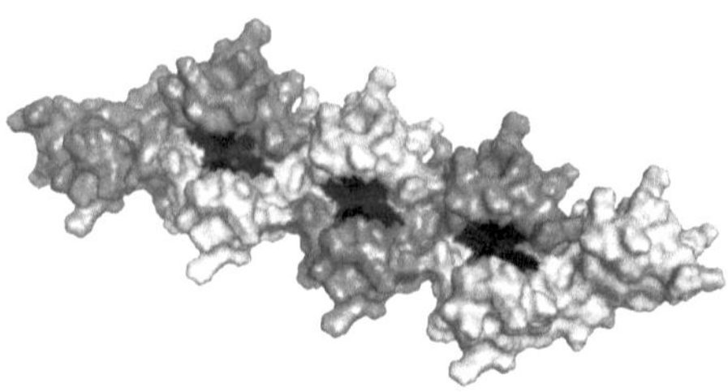

Figure 20.8 Hydroxyl radical protein footprinting of wild-type RANTES (CCL5) at pH 4.5 (tetramer) compared to pH 7 (large oligomers) mapped on a model of the RANTES (CCL5) octamer. Regions of complete protection from oxidation at pH 7 but found heavily oxidized at pH 4.5 (black) are found at the dimer–dimer interface, while regions of partial protection at pH 7 compared to pH 4.5 (gray) are found at the monomer–monomer interface. These pattern of protection mirrors the results found for the comparison of wild-type RANTES (CCL5) at pH 4.5 (tetramer) compared to the E66S mutant at pH 4.5 (dimer), with the extent of protection increasing dramatically.

pH 4.5 sample. These results are consistent with the larger oligomers existing as linear expansions of the tetramer, as shown in Fig. 20.8. HRPF is one of the few techniques that can actually generate structural data on these polydisperse mixtures of chemokine oligomers, and could be quite valuable in the future for the analysis of these biologically critical structures.

6. CONCLUSIONS

It is clear that the complexities of chemokine oligomerization, interaction with GAGs, and interaction with receptors are just beginning to be appreciated. Nevertheless, there is substantial evidence of the importance of these interactions. For example, some chemokine–receptor pairs are believed involved in metastasis of malignant cells,[16] and in the case of the CXCL12–CXCR4 pair, a constitutive dimeric form of the chemokine, as opposed to the monomeric form inhibited metastasis.[35] Also, in the case of inhibition of HIV entry, some of the most effective designed RANTES (CCL5) mutants are know to have reduced GAG binding and oligomerization tendencies.[176] We expect our understanding of oligomerization, GAG interaction, and receptor interaction, to have further impact on strategies to intervene in disease processes in the future.

Developing this understanding will be challenging. The systems of interest will be of increasing size and complexity. In the case of chemokine oligomerization, we have just begun to examine heteromeric complexes, but there are cases such as PF4 (CCL4)/RANTES (CCL5) complexes that are clearly biologically significant. In the case of GAG interactions, there are major issues with the availability of homogeneous GAG oligomers with which to carry out biophysical characterization. In the case of receptor interactions, there are additional challenges associated with the need to examine complexes in their native membrane associated environment.

There are, however, reasons to be optimistic about progress in this area. Technology suitable for the characterization of chemokine oligomers and chemokine complexes is rapidly evolving. NMR methods based on orientational and long-range paramagnetic constraints are well suited to the study of larger complexes assembled from monomers of known structure. MS methods for interface identification are now validated and applicable to complexes inaccessible to NMR because of size or concentration. Synthetic and enzymatic methods for the preparation of pure GAG oligomers are quickly advancing. We have not reviewed methods for macroscale structural characterization such as cryo-electron microscopy and atomic force microscopy, but these will clearly play a role in the future. The resulting improved view of the structure of complex assemblies involving chemokines, and the interactions that control their formation, will clearly have an impact on our ability to combat disease.

ACKNOWLEDGMENTS

Financial support for this work was provided by NIH grants from the National Center for Research Resources (5 P41 RR005351-23) and the National Institute of General Medical Sciences (8 P41 GM103390-23), in support of the Resource for Integrated Glycotechnology at the University of Georgia and from the National Institute of General Medical Sciences (U01 GM094612) and the National Institute of Allergy and Infectious Diseases (R01 AI37113-14) in support of research at the University of California San Diego.

REFERENCES

1. Goodsell DS, Olson AJ. Structural symmetry and protein function. *Annu Rev Biophys Biomol Struct* 2000;**29**:105–53.
2. Levy ED, Pereira-Leal JB, Chothia C, Teichmann SA. 3D complex: a structural classification of protein complexes. *PLoS Comput Biol* 2006;**2**(11):1395–406.
3. Ali MH, Imperiali B. Protein oligomerization: how and why. *Bioorg Med Chem* 2005;**13**(17):5013–20.
4. Sallusto F, Baggiolini M. Chemokines and leukocyte traffic. *Nat Immunol* 2008;**9**(9):949–52.
5. Baggiolini M. Chemokines and leukocyte traffic. *Nature* 1998;**392**(6676):565–8.
6. Scholten DJ, Canals M, Maussang D, et al. Pharmacological modulation of chemokine receptor function. *Br J Pharmacol* 2012;**165**(6):1617–43.
7. Bacon K, Baggiolini M, Broxmeyer H, et al. Chemokine/chemokine receptor nomenclature. *J Interferon Cytokine Res* 2002;**22**(10):1067–8.
8. Mantovani A. The chemokine system: redundancy for robust outputs. *Immunol Today* 1999;**20**(6):254–7.
9. Devalaraja MN, Richmond A. Multiple chemotactic factors: fine control or redundancy? *Trends Pharmacol Sci* 1999;**20**(4):151–6.
10. Johnson Z, Proudfoot AE, Handel TM. Interaction of chemokines and glycosaminoglycans: a new twist in the regulation of chemokine function with opportunities for therapeutic intervention. *Cytokine Growth Factor Rev* 2005;**16**(6):625–36.
11. Raman D, Sobolik-Delmaire T, Richmond A. Chemokines in health and disease. *Exp Cell Res* 2011;**317**(5):575–89.
12. Li MZ, Hale JS, Rich JN, Ransohoff RM, Lathia JD. Chemokine CXCL12 in neurodegenerative diseases: an SOS signal for stem cell-based repair. *Trends Neurosci* 2012;**35**(10):619–28.
13. Tuttolomondo A, Di Raimondo D, Pecoraro R, Arnao V, Pinto A, Licata G. Atherosclerosis as an inflammatory disease. *Curr Pharm Des* 2012;**18**(28):4266–88.
14. Gerard C, Rollins BJ. Chemokines and disease. *Nat Immunol* 2001;**2**(2):108–15.
15. Christopherson 2nd KW, Hromas RA. Endothelial chemokines in autoimmune disease. *Curr Pharm Des* 2004;**10**(2):145–54.
16. O'Hayre M, Salanga CL, Handel TM, Allen SJ. Chemokines and cancer: migration, intracellular signalling and intercellular communication in the microenvironment. *Biochem J* 2008;**409**:635–49.

17. Wilen CB, Tilton JC, Doms RW. HIV: cell binding and entry. *Cold Spring Harb Perspect Med* 2012;**2**(8):1–13.
18. Wilen CB, Tilton JC, Doms RW. Molecular mechanisms of HIV entry. *Adv Exp Med Biol* 2012;**726**:223–42.
19. Schall TJ, Proudfoot AE. Overcoming hurdles in developing successful drugs targeting chemokine receptors. *Nat Rev Immunol* 2011;**11**(5):355–63.
20. Springer TA. Traffic signals for lymphocyte recirculation and leukocyte emigration: the multistep paradigm. *Cell* 1994;**76**(2):301–14.
21. Springer TA. Traffic signals on endothelium for lymphocyte recirculation and leukocyte emigration. *Annu Rev Physiol* 1995;**57**:827–72.
22. Proudfoot AEI, Handel TM, Johnson Z, et al. Glycosaminoglycan binding and oligomerization are essential for the in vivo activity of certain chemokines. *Proc Natl Acad Sci USA* 2003;**100**(4):1885–90.
23. Rajarathnam K, Sykes BD, Kay CM, et al. Neutrophil activation by monomeric interleukin-8. *Science* 1994;**264**(5155):90–2.
24. Paavola CD, Hemmerich S, Grunberger D, et al. Monomeric monocyte chemoattractant protein-1 (MCP-1) binds and activates the MCP-1 receptor CCR2B. *J Biol Chem* 1998;**273**(50):33157–65.
25. Blanpain C, Doranz BJ, Bondue A, et al. The core domain of chemokines binds CCR5 extracellular domains while their amino terminus interacts with the transmembrane helix bundle. *J Biol Chem* 2003;**278**(7):5179–87.
26. Stephens B, Handel TM. Chemokine receptor oligomerization and allostery. *Prog in Mol Biol Transl Sci* 2013;**115**:375–420.
27. Veldkamp CT, Seibert C, Peterson FC, et al. Structural basis of CXCR4 sulfotyrosine recognition by the chemokine SDF-1/CXCL12. *Sci Signal* 2008;**1**(37):ra4.
28. Duma L, Haussinger D, Rogowski M, Lusso P, Grzesiek S. Recognition of RANTES by extracellular parts of the CCR5 receptor. *J Mol Biol* 2007;**365**(4):1063–75.
29. Hoogewerf AJ, Kuschert GS, Proudfoot AE, et al. Glycosaminoglycans mediate cell surface oligomerization of chemokines. *Biochemistry* 1997;**36**(44):13570–8.
30. Campanella GS, Grimm J, Manice LA, et al. Oligomerization of CXCL10 is necessary for endothelial cell presentation and in vivo activity. *J Immunol* 2006;**177**(10):6991–8.
31. Ellyard JI, Simson L, Bezos A, Johnston K, Freeman C, Parish CR. Eotaxin selectively binds heparin. An interaction that protects eotaxin from proteolysis and potentiates chemotactic activity in vivo. *J Biol Chem* 2007;**282**(20):15238–47.
32. Baltus T, Weber KS, Johnson Z, Proudfoot AE, Weber C. Oligomerization of RANTES is required for CCR1-mediated arrest but not CCR5-mediated transmigration of leukocytes on inflamed endothelium. *Blood* 2003;**102**(6):1985–8.
33. Wang L, Fuster M, Sriramarao P, Esko JD. Endothelial heparan sulfate deficiency impairs L-selectin- and chemokine-mediated neutrophil trafficking during inflammatory responses. *Nat Immunol* 2005;**6**(9):902–10.
34. Yin X, Truty J, Lawrence R, et al. A critical role for lymphatic endothelial heparan sulfate in lymph node metastasis. *Mol Cancer* 2010;**9**:316.
35. Drury LJ, Ziarek JJ, Gravel S, et al. Monomeric and dimeric CXCL12 inhibit metastasis through distinct CXCR4 interactions and signaling pathways. *Proc Natl Acad Sci USA* 2011;**108**(43):17655–60.
36. Appay V, Dunbar PR, Cerundolo V, McMichael A, Czaplewski L, Rowland-Jones S. RANTES activates antigen-specific cytotoxic T lymphocytes in a mitogen-like manner through cell surface aggregation. *Int Immunol* 2000;**12**(8):1173–82.
37. Roscic-Mrkic B, Fischer M, Leemann C, et al. RANTES (CCL5) uses the proteoglycan CD44 as an auxiliary receptor to mediate cellular activation signals and HIV-1 enhancement. *Blood* 2003;**102**(4):1169–77.

38. Wagner L, Yang OO, Garcia-Zepeda EA, et al. Beta-chemokines are released from HIV-1-specific cytolytic T-cell granules complexed to proteoglycans. *Nature* 1998;**391**(6670):908–11.
39. Wu BL, Chien EYT, Mol CD, et al. Structures of the CXCR4 chemokine GPCR with small-molecule and cyclic peptide antagonists. *Science* 2010;**330**(6007):1066–71.
40. Kramp BK, Sarabi A, Koenen RR, Weber C. Heterophilic chemokine receptor interactions in chemokine signaling and biology. *Exp Cell Res* 2011;**317**(5):655–63.
41. von Hundelshausen P, Koenen RR, Sack M, et al. Heterophilic interactions of platelet factor 4 and RANTES promote monocyte arrest on endothelium. *Blood* 2005;**105**(3):924–30.
42. Koenen RR, von Hundelshausen P, Nesmelova IV, et al. Disrupting functional interactions between platelet chemokines inhibits atherosclerosis in hyperlipidemic mice. *Nat Med* 2009;**15**(1):97–103.
43. Shahrara S, Proudfoot AE, Park CC, et al. Inhibition of monocyte chemoattractant protein-1 ameliorates rat adjuvant-induced arthritis. *J Immunol* 2008;**180**(5):3447–56.
44. Proudfoot AE, Power CA, Rommel C, Wells TN. Strategies for chemokine antagonists as therapeutics. *Semin Immunol* 2003;**15**(1):57–65.
45. Johnson Z, Kosco-Vilbois MH, Herren S, et al. Interference with heparin binding and oligomerization creates a novel anti-inflammatory strategy targeting the chemokine system. *J Immunol* 2004;**173**(9):5776–85.
46. Allen SJ, Crown SE, Handel TM. Chemokine: receptor structure, interactions, and antagonism. *Annu Rev Immunol* 2007;**25**:787–820.
47. Fernandez EJ, Lolis E. Structure junction, and inhibition of chemokines. *Annu Rev Pharmacol* 2002;**42**:469–99.
48. Fairbrother WJ, Reilly D, Colby TJ, Hesselgesser J, Horuk R. The solution structure of melanoma growth-stimulating activity. *J Mol Biol* 1994;**242**(3):252–70.
49. Qian YQ, Johanson KO, McDevitt P. Nuclear magnetic resonance solution structure of truncated human GRO beta [5-73] and its structural comparison with CXC chemokine family members GRO alpha and IL-8. *J Mol Biol* 1999;**294**(5):1065–72.
50. Mayo KH, Roongta V, Ilyina E, et al. NMR solution structure of the 32-Kda platelet factor-4 Elr-motif N-terminal chimera—a symmetrical tetramer. *Biochemistry* 1995;**34**(36):11399–409.
51. Zhang XH, Chen LQ, Bancroft DP, Lai CK, Maione TE. Crystal-structure of recombinant human platelet factor-4. *Biochemistry* 1994;**33**(27):8361–6.
52. Malkowski MG, Wu JY, Lazar JB, Johnson PH, Edwards BFP. The crystal-structure of recombinant human neutrophil-activating peptide-2 (M6l) at 1.9-Angstrom resolution. *J Biol Chem* 1995;**270**(13):7077–87.
53. Young H, Roongta V, Daly TJ, Mayo KH. NMR structure and dynamics of monomeric neutrophil-activating peptide 2. *Biochem J* 1999;**338**:591–8.
54. Clore GM, Appella E, Yamada M, Matsushima K, Gronenborn AM. 3-Dimensional structure of Interleukin-8 in solution. *Biochemistry* 1990;**29**(7):1689–96.
55. Swaminathan GJ, Holloway DE, Colvin RA, et al. Crystal structures of oligomeric forms of the IP-10/CXCL10 chemokine. *Structure* 2003;**11**(5):521–32.
56. Booth V, Clark-Lewis I, Sykes BD. NMR structure of CXCR3 binding chemokine CXCL11 (ITAC). *Protein Sci* 2004;**13**(8):2022–8.
57. Dealwis C, Fernandez EJ, Thompson DA, Simon RJ, Siani MA, Lolis E. Crystal structure of chemically synthesized [N33A] stromal cell-derived factor 1 alpha, a potent ligand for the HIV-1 "fusin" coreceptor. *Proc Natl Acad Sci USA* 1998;**95**(12):6941–6.
58. Crump MP, Gong JH, Loetscher P, et al. Solution structure and basis for functional activity of stromal cell-derived factor-1; dissociation of CXCR4 activation from binding and inhibition of HIV-1. *EMBO J* 1997;**16**(23):6996–7007.

59. Keizer DW, Crump MP, Lee TW, Slupsky CM, Clark-Lewis I, Sykes BD. Human CC chemokine I-309, structural consequences of the additional disulfide bond. *Biochemistry* 2000;**39**(20):6053–9.
60. Handel TM, Domaille PJ. Heteronuclear (H-1, C-13, N-15) NMR assignments and solution structure of the monocyte chemoattractant protein-1 (MCP-1) dimer. *Biochemistry* 1996;**35**(21):6569–84.
61. Lubkowski J, Bujacz G, Boque L, Domaille PJ, Handel TM, Wlodawer A. The structure of MCP-1 in two crystal forms provides a rare example of variable quaternary interactions. *Nat Struct Biol* 1997;**4**(1):64–9.
62. Czaplewski LG, McKeating J, Craven CJ, et al. Identification of amino acid residues critical for aggregation of human CC chemokines macrophage inflammatory protein (MIP)-1 alpha, MIP-1 beta, and RANTES - Characterization of active disaggregated chemokine variants. *J Biol Chem* 1999;**274**(23):16077–84.
63. Lodi PJ, Garrett DS, Kuszewski J, et al. High-resolution solution structure of the beta-chemokine hMIP-1-beta by multidimensional NMR. *Science* 1994;**263**(5154):1762–7.
64. Chung CW, Cooke RM, Proudfoot AEI, Wells TNC. The 3-dimensional solution structure of RANTES. *Biochemistry* 1995;**34**(29):9307–14.
65. Kim KS, Rajarathnam K, ClarkLewis I, Sykes BD. Structural characterization of a monomeric chemokine: monocyte chemoattractant protein-3. *FEBS Lett* 1996;**395** (2–3):277–82.
66. Blaszczyk J, Van Coillie E, Proost P, et al. Complete crystal structure of monocyte chemotactic protein-2, a CC chemokine that interacts with multiple receptors. *Biochemistry* 2000;**39**(46):14075–81.
67. Crump MP, Rajarathnam K, Kim KS, Clark-Lewis I, Sykes BD. Solution structure of eotaxin, a chemokine that selectively recruits eosinophils in allergic inflammation. *J Biol Chem* 1998;**273**(35):22471–9.
68. Barinka C, Prahl A, Lubkowski J. Structure of human monocyte chemoattractant protein 4 (MCP-4/CCL13). *Acta Crystallogr D Biol Crystallogr* 2008;**64**:273–8.
69. Blain KY, Kwiatkowski W, Zhao Q, et al. Structural and functional characterization of CC chemokine CCL14. *Biochemistry* 2007;**46**(35):10008–15.
70. Sticht H, Escher SE, Schweimer K, Forssmann WG, Rosch P, Adermann K. Solution structure of the human CC chemokine 2: a monomeric representative of the CC chemokine subtype. *Biochemistry* 1999;**38**(19):5995–6002.
71. Asojo OA, Boulegue C, Hoover DM, Lu WY, Lubkowski J. Structures of thymus and activation-regulated chemokine (TARC). *Acta Crystallogr D Biol Crystallogr* 2003;**59**:1165–73.
72. Love M, Sandberg JL, Ziarek JJ, et al. Solution Structure of CCL21 and Identification of a Putative CCR7 Binding Site. *Biochemistry* 2012;**51**(3):733–5.
73. Rajarathnam K, Li YL, Rohrer T, Gentz R. Solution structure and dynamics of myeloid progenitor inhibitory factor-1 (MPIF-1), a novel monomeric CC chemokine. *J Biol Chem* 2001;**276**(7):4909–16.
74. Mayer KL, Stone MJ. NMR solution structure and receptor peptide binding of the CC chemokine eotaxin-2. *Biochemistry* 2000;**39**(29):8382–95.
75. Ye J, Mayer KL, Mayer MR, Stone MJ. NMR solution structure and backbone dynamics of the CC chemokine eotaxin-3. *Biochemistry* 2001;**40**(26):7820–31.
76. Jansma AL, Kirkpatrick JP, Hsu AR, Handel TM, Nietlispach D. NMR analysis of the structure, dynamics, and unique oligomerization properties of the chemokine CCL27. *J Biol Chem* 2010;**285**(19):14424–37.
77. Tuinstra RL, Peterson FC, Kutlesa S, Elgin ES, Kron MA, Volkman BF. Interconversion between two unrelated protein folds in the lymphotactin native state. *Proc Natl Acad Sci USA* 2008;**105**(13):5057–62.

78. Mizoue LS, Bazan JF, Johnson EC, Handel TM. Solution structure and dynamics of the CX3C chemokine domain of fractalkine and its interaction with an N-terminal fragment of CX(3)CR1. *Biochemistry* 1999;**38**(5):1402–14.
79. Hoover DM, Mizoue LS, Handel TM, Lubkowski J. The crystal structure of the chemokine domain of fractalkine shows a novel quaternary arrangement. *J Biol Chem* 2000;**275**(30):23187–93.
80. Gong JH, Clarklewis I. Antagonists of monocyte chemoattractant protein-1 identified by modification of functionally critical NH2-terminal residues. *J Exp Med* 1995;**181**(2):631–40.
81. Guan E, Wang JH, Roderiquez G, Norcross MA. Natural truncation of the chemokine MIP-1 beta/CCL4 affects receptor specificity but not Anti-HIV-1 activity. *J Biol Chem* 2002;**277**(35):32348–52.
82. Proudfoot AEI, Power CA, Hoogewerf AJ, et al. Extension of recombinant human RANTES by the retention of the initiating methionine produces a potent antagonist. *J Biol Chem* 1996;**271**(5):2599–603.
83. Williams SA, Harata-Lee Y, Comerford I, Anderson RL, Smyth MJ, McColl SR. Multiple functions of CXCL12 in a syngeneic model of breast cancer. *Mol Cancer* 2010; **9**(250):1–10.
84. Gaertner H, Cerini F, Escola JM, et al. Highly potent, fully recombinant anti-HIV chemokines: reengineering a low-cost microbicide. *Proc Natl Acad Sci USA* 2008;**105**(46):17706–11.
85. Loetscher P, Gong JH, Dewald B, Baggiolini M, Clark-Lewis I. N-terminal peptides of stromal cell-derived factor-1 with CXC chemokine receptor 4 agonist and antagonist activities. *J Biol Chem* 1998;**273**(35):22279–83.
86. Luo JS, Luo ZW, Zhou NM, Hall JW, Huang ZW. Attachment of C-terminus of SDF-1 enhances the biological activity of its N-terminal peptide. *Biochem Biophys Res Commun* 1999;**264**(1):42–7.
87. Pakianathan DR, Kuta EG, Artis DR, Skelton NJ, Hebert CA. Distinct but overlapping epitopes for the interaction of a CC-chemokine with CCR1, CCR3, and CCR5. *Biochemistry* 1997;**36**(32):9642–8.
88. Clarklewis I, Dewald B, Loetscher M, Moser B, Baggiolini M. Structural requirements for interleukin-8 function identified by design of analogs and CXC chemokine hybrids. *J Biol Chem* 1994;**269**(23):16075–81.
89. Lowman HB, Slagle PH, DeForge LE, et al. Exchanging interleukin-8 and melanoma growth-stimulating activity receptor binding specificities. *J Biol Chem* 1996;**271**(24):14344–52.
90. Beall CJ, Mahajan S, Kuhn DE, Kolattukudy PE. Site-directed mutagenesis of monocyte chemoattractant protein-1 identifies two regions of the polypeptide essential for biological activity. *Biochem J* 1996;**313**:633–40.
91. Hemmerich S, Paavola C, Bloom A, et al. Identification of residues in the monocyte chemotactic protein-1 that contact the MCP-1 receptor, CCR2. *Biochemistry* 1999;**38**(40):13013–25.
92. Martin L, Blanpain C, Garnier P, Wittamer V, Parmentier M, Vita C. Structural and functional analysis of the RANTES-glycosaminoglycans interactions. *Biochemistry* 2001;**40**(21):6303–18.
93. Proudfoot AEI, Fritchley S, Borlat F, et al. The BBXB motif of RANTES is the principal site for heparin binding and controls receptor selectivity. *J Biol Chem* 2001;**276**(14):10620–6.
94. Yoshiura C, Kofuku Y, Ueda T, et al. NMR Analyses of the Interaction between CCR5 and Its Ligand Using Functional Reconstitution of CCR5 in Lipid Bilayers. *J Am Chem Soc* 2010;**132**(19):6768–77.

95. Strieter RM, Burdick MD, Gomperts BN, Belperio JA, Keane MP. CXC chemokines in angiogenesis. *Cytokine Growth Factor Rev* 2005;**16**(6):593–609.
96. Ali S, O'Boyle G, Hepplewhite P, Tyler JR, Robertson H, Kirby JA. Therapy with nonglycosaminoglycan-binding mutant CCL7: a novel strategy to limit allograft inflammation. *Am J Transplant* 2010;**10**(1):47–58.
97. Lau EK, Paavola CD, Johnson Z, et al. Identification of the glycosaminoglycan binding site of the CC chemokine, MCP-1 - Implications for structure and function in vivo. *J Biol Chem* 2004;**279**(21):22294–305.
98. Jabeen T, Leonard P, Jamaluddin H, Acharya KR. Structure of mouse IP-10, a chemokine. *Acta Crystallogr D Biol Crystallogr* 2008;**64**:611–9.
99. Murphy JW, Yuan H, Kong Y, Xiong Y, Lolis EJ. Heterologous quaternary structure of CXCL12 and its relationship to the CC chemokine family. *Proteins* 2010;**78**(5):1331–7.
100. Wang X, Watson C, Sharp JS, Handel TM, Prestegard JH. Oligomeric structure of the chemokine CCL5/RANTES from NMR, MS, and SAXS data. *Structure* 2011;**19**(8):1138–48.
101. Ren M, Guo Q, Guo LA, et al. Polymerization of MIP-1 chemokine (CCL3 and CCL4) and clearance of MIP-1 by insulin-degrading enzyme. *EMBO J* 2010;**29**(23):3952–66.
102. Handin RI, Cohen HJ. Purification and binding properties of human platelet factor 4. *J Biol Chem* 1976;**251**(14):4273–82.
103. Handel TM, Johnson Z, Crown SE, Lau EK, Sweeney M, Proudfoot AE. Regulation of protein function by glycosaminoglycans—as exemplified by chemokines. *Annu Rev Biochem* 2005;**74**:385–410.
104. Kuschert GSV, Coulin F, Power CA, et al. Glycosaminoglycans interact selectively with chemokines and modulate receptor binding and cellular responses. *Biochemistry* 1999;**38**(39):12959–68.
105. Amara A, Lorthioir O, Valenzuela A, et al. Stromal cell-derived factor-1 alpha associates with heparan sulfates through the first beta-strand of the chemokine. *J Biol Chem* 1999;**274**(34):23916–25.
106. Koopmann W, Krangel MS. Identification of a glycosaminoglycan-binding site in chemokine macrophage inflammatory protein-1 alpha. *J Biol Chem* 1997;**272**(15):10103–9.
107. Sadir R, Baleux F, Grosdidier A, Imberty A, Lortat-Jacob H. Characterization of the stromal cell-derived factor-l alpha-heparin complex. *J Biol Chem* 2001;**276**(11):8288–96.
108. Jarnagin K, Grunberger D, Mulkins M, et al. Identification of surface residues of the monocyte chemotactic protein 1 that affect signaling through the receptor CCR2. *Biochemistry* 1999;**38**(49):16167–77.
109. Salanga CL, Handel TM. Chemokine oligomerization and interactions with receptors and glycosaminoglycans: the role of structural dynamics in function. *Exp Cell Res* 2011;**317**(5):590–601.
110. Spillmann D, Witt D, Lindahl U. Defining the interleukin-8-binding domain of heparan sulfate. *J Biol Chem* 1998;**273**(25):15487–93.
111. Rek A, Brandner B, Geretti E, Kungl AJ. A biophysical insight into the RANTES-glycosaminoglycan interaction. *Biochim Biophys Acta* 2009;**1794**(4):577–82.
112. Stringer SE, Forster MJ, Mulloy B, Bishop CR, Graham GJ, Gallagher JT. Characterization of the binding site on heparan sulfate for macrophage inflammatory protein 1 alpha. *Blood* 2002;**100**(5):1543–50.
113. Kim S, Jao S, Laurence JS, LiWang PJ. Structural comparison of monomeric variants of the chemokine MIP-1 beta having differing ability to bind the receptor CCR5. *Biochemistry* 2001;**40**(36):10782–91.

114. Meunier S, Bernassau JM, Guillemot JC, Ferrara P, Darbon H. Determination of the three-dimensional structure of CC chemokine monocyte chemoattractant protein 3 by H-1 two-dimensional NMR spectroscopy. *Biochemistry* 1997;**36**(15):4412–22.
115. Skelton NJ, Aspiras F, Ogez J, Schall TJ. Proton NMR assignments and solution conformation of RANTES, a chemokine of the C-C type. *Biochemistry* 1995;**34**(16):5329–42.
116. Veldkamp CT, Ziarek JJ, Su JD, et al. Monomeric structure of the cardioprotective chemokine SDF-1/CXCL12. *Protein Sci* 2009;**18**(7):1359–69.
117. Volkman BF, Liu TY, Peterson FC. Lymphotactin structural dynamics. In: Handel TM, Hamel DJ, editors. *Chemokines, Part B. Methods in enzymology*, vol. 461. Elsevier; 2009. p. 51–70.
118. Zhao B, LiWang PJ. Characterization of the Interactions of vMIP-II, and a Dimeric Variant of vMIP-II, with Glycosaminoglycans. *Biochemistry* 2010;**49**(33):7012–22.
119. Shimada I, Ueda T, Matsumoto M, et al. Cross-saturation and transferred cross-saturation experiments. *Prog Nucl Magn Reson Spectrosc* 2009;**54**(2):123–40.
120. Bayburt TH, Sligar SG. Membrane protein assembly into Nanodiscs. *FEBS Lett* 2010;**584**(9):1721–7.
121. Greenfield NJ, Huang YJ, Palm T, et al. Solution NMR structure and folding dynamics of the N terminus of a rat non-muscle alpha-tropomyosin in an engineered chimeric protein. *J Mol Biol* 2001;**312**(4):833–47.
122. Bax A, Kontaxis G, Tjandra N. Dipolar couplings in macromolecular structure determination. Nuclear Magnetic Resonance of Biological Macromolecules, Part B. *Meth Enzymol* 2001;**339**:127–74.
123. Prestegard JH, Bougault CM, Kishore AI. Residual dipolar couplings in structure determination of biomolecules. *Chem Rev* 2004;**104**(8):3519–40.
124. Lee HW, Wylie G, Bansal S, et al. Three-dimensional structure of the weakly associated protein homodimer SeR13 using RDCs and paramagnetic surface mapping. *Protein Sci* 2010;**19**(9):1673–85.
125. Wang X, Lee HW, Liu YZ, Prestegard JH. Structural NMR of protein oligomers using hybrid methods. *J Struct Biol* 2011;**173**(3):515–29.
126. Wang X, Bansal S, Jiang M, Prestegard JH. RDC-assisted modeling of symmetric protein homo-oligomers. *Protein Sci* 2008;**17**(5):899–907.
127. Shaw JP, Kryger G, Cleasby A, et al. Crystallization and preliminary-X-ray diffraction studies of human RANTES. *J Mol Biol* 1994;**242**(4):589–90.
128. Liu YZ, Prestegard JH. A device for the measurement of residual chemical shift anisotropy and residual dipolar coupling in soluble and membrane-associated proteins. *J Biomol NMR* 2010;**47**(4):249–58.
129. Yu F, Wolff JJ, Amster IJ, Prestegard JH. Conformational preferences of chondroitin sulfate oligomers using partially oriented NMR spectroscopy of C-13-labeled acetyl groups. *J Am Chem Soc* 2007;**129**(43):13288–97.
130. Gemma E, Hulme AN, Jahnke A, et al. DMT-MM mediated functionalisation of the non-reducing end of glycosaminoglycans. *Chem Commun* 2007;**26**:2686–8.
131. Blaum BS, Deakin JA, Johansson CM, et al. Lysine and arginine side chains in glycosaminoglycan-protein complexes Investigated by NMR, cross-linking, and mass spectrometry: a case study of the factor H-heparin Interaction. *J Am Chem Soc* 2010;**132**(18):6374–81.
132. Roldos V, Canada FJ, Jimenez-Barbero J. Carbohydrate-protein interactions: a 3D View by NMR. *Chembiochem* 2011;**12**:990–1005 12(9):CP27-CP27.
133. Meyer B, Peters T. NMR Spectroscopy techniques for screening and identifying ligand binding to protein receptors. *Angew Chem Int Ed* 2003;**42**(8):864–90.
134. Laguri C, Sapay N, Simorre JP, et al. C-13-labeled heparan sulfate analogue as a tool to study protein/heparan sulfate interactions by NMR spectroscopy: application to the CXCL12 alpha chemokine. *J Am Chem Soc* 2011;**133**(25):9642–5.

135. Davis JT, Hirani S, Bartlett C, Reid BR. H-1-NMR studies on an Asn-linked glycopeptide—Glcnac-1 C2-N2 bond is rigid in H_2O. *J Biol Chem* 1994;**269**(5):3331–8.
136. Hricovini M, Guerrini M, Bisio A, Torri G, Petitou M, Casu B. Conformation of heparin pentasaccharide bound to antithrombin III. *Biochem J* 2001;**359**:265–72.
137. Tjong SC, Chen TS, Huang WN, Wu WG. Structures of heparin-derived tetrasaccharide bound to cobra cardiotoxins: heparin binding at a single protein site with diverse side chain interactions. *Biochemistry* 2007;**46**(35):9941–52.
138. Zhou HJ, Casas-Finet JR, Coats RH, et al. Identification and dynamics of a heparin-binding site in hepatocyte growth factor. *Biochemistry* 1999;**38**(45):14793–802.
139. Moy FJ, Safran M, Seddon AP, et al. Properly oriented heparin-decasaccharide-induced dimers are the biologically active form of basic fibroblast growth factor. *Biochemistry* 1997;**36**(16):4782–91.
140. Boons GJ. Modular synthesis of heparan sulfate oligosaccharides for array development. Abstracts of Papers of the American Chemical Society. Aug 2010. p. 240.
141. Deepa SS, Yamada S, Fukui S, Sugahara K. Structural determination of novel sulfated octasaccharides isolated from chondroitin sulfate of shark cartilage and their application for characterizing monoclonal antibody epitopes. *Glycobiology* 2007;**17**(6):631–45.
142. Pomin VH, Park Y, Huang RR, et al. Exploiting enzyme specificities in digestions of chondroitin sulfates A and C: production of well-defined hexasaccharides. *Glycobiology* 2012;**22**(6):826–38.
143. DeAngelis PL. Glycosaminoglycan polysaccharide biosynthesis and production: today and tomorrow. *Appl Microbiol Biotechnol* 2012;**94**(2):295–305.
144. Casu B, Grazioli G, Razi N, et al. Heparin-like compounds prepared by chemical modification of capsular polysaccharide from Escherichia-coli K5. *Carbohydr Res* 1994;**263**(2):271–84.
145. Nagasawa K, Uchiyama H, Wajima N. Chemical sulfation of preparations of chondroitin 4-sulfate and 6-sulfate, and dermatan sulfate—preparation of chondroitin sulfate E-Like materials from chondroitin 4-sulfate. *Carbohydr Res* 1986;**158**:183–90.
146. DeAngelis PL, Jing W, Tracy BS, Oatman LC, Gay DF. Chemoenzymatic synthesis of glycosaminoglycans with pasteurella synthases. *Glycobiology* 2003;**13**(11):838–9.
147. DeAngelis PL, Oatman LC, Gay DF. Rapid chemoenzymatic synthesis of monodisperse hyaluronan oligosaccharides with immobilized enzyme reactors. *J Biol Chem* 2003;**278**(37):35199–203.
148. Jing W, Haller FM, Almond A, DeAngelis PL. Defined megadalton hyaluronan polymer standards. *Anal Biochem* 2006;**355**(2):183–8.
149. Xu YM, Masuko S, Takieddin M, et al. Chemoenzymatic synthesis of homogeneous ultralow molecular weight heparins. *Science* 2011;**334**(6055):498–501.
150. Mobli M, Nilsson M, Almond A. The structural plasticity of heparan sulfate NA-domains and hence their role in mediating multivalent interactions is confirmed by high-accuracy (15)N-NMR relaxation studies. *Glycoconj J* 2008;**25**(5):401–14.
151. Pomin VH, Sharp JS, Li XY, Wang LC, Prestegard JH. Characterization of glycosaminoglycans by N-15 NMR spectroscopy and in vivo isotopic labeling. *Anal Chem* 2010;**82**(10):4078–88.
152. Putnam CD, Hammel M, Hura GL, Tainer JA. X-ray solution scattering (SAXS) combined with crystallography and computation: defining accurate macromolecular structures, conformations and assemblies in solution. *Q Rev Biophys* 2007;**40**(3):191–285.
153. Svergun DI. Small-angle X-ray and neutron scattering as a tool for structural systems biology. *Biol Chem* 2010;**391**(7):737–43.
154. Rostom AA, Robinson CV. Disassembly of intact multiprotein complexes in the gas phase. *Curr Opin Struct Biol* 1999;**9**(1):135–41.

155. Sobott F, Hernandez H, McCammon MG, Tito MA, Robinson CV. A tandem mass spectrometer for improved transmission and analysis of large macromolecular assemblies. *Anal Chem* 2002;**74**(6):1402–7.
156. Chung EW, Henriques DA, Renzoni D, et al. Probing the nature of interactions in SH2 binding interfaces—evidence from electrospray ionization mass spectrometry. *Protein Sci* 1999;**8**(10):1962–70.
157. Yu Y, Sweeney MD, Saad OM, et al. Chemokine-glycosaminoglycan binding: specificity for CCR2 ligand binding to highly sulfated oligosaccharides using FTICR mass spectrometry. *J Biol Chem* 2005;**280**(37):32200–8.
158. Crown SE, Yu Y, Sweeney MD, Leary JA, Handel TM. Heterodimerization of CCR2 chemokines and regulation by glycosaminoglycan binding. *J Biol Chem* 2006;**281**(35):25438–46.
159. Carlson J, Baxter SA, Dreau D, Nesmelova IV. The heterodimerization of platelet-derived chemokines. *Biochim Biophys Acta* 2012;**1834**(1):158–68.
160. Sharp JS, Becker JM, Hettich RL. Protein surface mapping by chemical oxidation: structural analysis by mass spectrometry. *Anal Biochem* 2003;**313**(2):216–25.
161. Sharp JS, Becker JM, Hettich RL. Topology characterization of model proteins by photochemical side chain oxidation and mass spectrometry. *FASEB J* 2003;**17**(5):A988.
162. Hambly DM, Gross ML. Laser flash photolysis of hydrogen peroxide to oxidize protein solvent-accessible residues on the microsecond timescale. *J Am Soc Mass Spectrom* 2005;**16**(12):2057–63.
163. McClintock C, Kertesz V, Hettich RL. Development of an electrochemical oxidation method for probing higher order protein structure with mass spectrometry. *Anal Chem* 2008;**80**(9):3304–17.
164. Maleknia SD, Chance MR, Downard KM. Electrospray-assisted modification of proteins: a radical probe of protein structure. *Rapid Commun Mass Spectrom* 1999;**13**(23):2352–8.
165. Maleknia SD, Brenowitz M, Chance MR. Millisecond radiolytic modification of peptides by synchrotron X-rays identified by mass spectrometry. *Anal Chem* 1999;**71**(18):3965–73.
166. Sharp JS, Sullivan DM, Cavanagh J, Tomer KB. Measurement of multisite oxidation kinetics reveals an active site conformational change in Spo0F as a result of protein oxidation. *Biochemistry* 2006;**45**(20):6260–6.
167. Watson C, Janik I, Zhuang T, Charvatova O, Woods RJ, Sharp JS. Pulsed electron beam water radiolysis for submicrosecond hydroxyl radical protein footprinting. *Anal Chem* 2009;**81**(7):2496–505.
168. Buxton GV, Greenstock CL, Helman WP, Ross AB. Critical-review of rate constants for reactions of hydrated electrons, hydrogen-atoms and hydroxyl radicals (.Oh/.O-) in aqueous-solution. *J Phys Chem Ref Data* 1988;**17**(2):513–886.
169. Xu G, Chance MR. Radiolytic modification and reactivity of amino acid residues serving as structural probes for protein footprinting. *Anal Chem* 2005;**77**(14):4549–55.
170. Chance MR. Unfolding of apomyoglobin examined by synchrotron footprinting. *Biochem Biophys Res Commun* 2001;**287**(3):614–21.
171. Charvatova O, Foley BL, Bern MW, Sharp JS, Orlando R, Woods RJ. Quantifying protein interface footprinting by hydroxyl radical oxidation and molecular dynamics simulation: application to galectin-1. *J Am Soc Mass Spectrom* 2008;**19**(11):1692–705.
172. Guan JQ, Vorobiev S, Almo SC, Chance MR. Mapping the G-actin binding surface of cofilin using synchrotron protein footprinting. *Biochemistry* 2002;**41**(18):5765–75.
173. Gau BC, Sharp JS, Rempel DL, Gross ML. Fast photochemical oxidation of protein footprints faster than protein unfolding. *Anal Chem* 2009;**81**(16):6563–71.
174. Watson C, Sharp JS. Conformational analysis of therapeutic proteins by hydroxyl radical protein footprinting. *AAPS J* 2012;**14**(2):206–17.

175. Vahidi S, Stocks BB, Liaghati-Mobarhan Y, Konermann L. Mapping pH-induced protein structural changes under equilibrium conditions by pulsed oxidative labeling and mass spectrometry. *Anal Chem* 2012;**84**(21):9124–30.
176. Jin HJ, Kagiampakis I, Li PW, LiWang PJ. Structural and functional studies of the potent anti-HIV chemokine variant P2-RANTES. *Proteins* 2010;**78**(2):295–308.

CHAPTER TWENTY-ONE

Oligomerization of Rab/Effector Complexes in the Regulation of Vesicle Trafficking

Amir R. Khan
School of Biochemistry and Immunology, Trinity Biomedical Sciences Institute, Trinity College Dublin, Dublin 2, Ireland

Contents

Abstract

Rabs comprise the largest member of the Ras superfamily of small GTPases with over 60 proteins in mammals and 11 proteins in yeast. Like all small GTPases, Rabs oscillate

Progress in Molecular Biology and Translational Science, Volume 117
ISSN 1877-1173
http://dx.doi.org/10.1016/B978-0-12-386931-9.00021-0

between an inactive GDP-bound conformation and an active GTP-bound state that is tethered to lipid membranes via a C-terminal prenylation site on conserved cysteine residues. In their active state, Rabs regulate various aspects of membrane trafficking, including vesicle formation, transport, docking, and fusion. The critical element of biological activity is the recruitment of cytosolic effector proteins to specific endomembranes by active Rabs. The importance of Rabs in cellular processes is apparent from their links to genetic disorders, immunodeficiency, cancer, and pathogen invasion. During the last decade, numerous structures of complexes have shed light on the molecular basis for Rab/effector specificity and their topological organization on subcellular membranes. Here, I review the known structures of Rab/effector complexes and their modes of oligomerization. This is followed by a brief discussion on the thermodynamics of effector recruitment, which has not been documented sufficiently in previous reviews. A summary of diseases associated with Rab/effector trafficking pathways concludes this chapter.

1. INTRODUCTION

Spatial and temporal regulation of biological processes in eukaryotic cells requires the Ras superfamily of small GTPases. Ras-like proteins, such as Ras and Rab, are molecular switches that are involved in growth, differentiation, apoptosis, cytoskeletal dynamics, and membrane trafficking.[1] The founding member of this superfamily was Ras, which was originally discovered by its ability to transform human NIH/3T3 cells upon DNA transfection.[2] Point mutations in the three Ras genes in humans—H-ras, K-ras, and N-ras—are strongly associated with tumors and Ras signaling pathways are a target for therapeutics.[3] Additional members of the superfamily include Ran GTPases (nuclear import[4]), Arf GTPases (vesicle trafficking and cytoskeletal dynamics[5]), and Rho GTPases (cell motility and cytoskeletal dynamics[6]).

Ras and Ras-like proteins are generally 150–230 amino acids long, and in their active state, they are noncovalently associated with GTP. In addition, active Ras and Rab GTPases are localized to a specific subcellular compartment via prenylation of their C-termini, which is essential for downstream activation.[7,8] Upon activation, small GTPases recruit "effector" proteins which specifically recognize the GTP-bound conformation with high affinity, thus mediating a downstream signaling pathway that elicits the biological activity. For example, Ras is posttranslationally prenylated at a C-terminal cysteine residue with a 15-carbon lipid group by farnesyltransferase.[8] Upon GTP binding, prenylated Ras stably associates with the inner leaflet of the plasma membrane through its lipid moiety and forms a signaling platform. Ras(GTP) then recruits Raf, a kinase that phosphorylates other proteins downstream of the Ras(GTP)/Raf complex, thus mediating a signaling cascade that leads to altered gene expression and cell transformation (reviewed in Ref. 9).

Canonical small GTPases adopt a conserved α/β fold consisting of a six-stranded (mixed) β-sheet flanked by five α-helices. Conserved sequence motifs in the superfamily are a consequence of GTP binding and hydrolysis, and include the highly conserved P-loop (Walker motif A[10]) that stabilizes the phosphate arm of GTP. In addition, a pair of motifs adjacent to the γ-phosphate is critical for the biological activity of small GTPases, termed switch I and switch II. By definition, the conformations of these two regions are influenced by the presence of GTP versus GDP. Generally, the switch regions are more flexible in the presence of GDP. Binding to GTP leads to direct interactions between switch I/II and the γ-phosphate, resulting in a more rigid conformation that is recognized by effector proteins. The conformational switch has been aptly termed a "spring-loaded" mechanism,[11] which refers to the tenuous nature of the switch/γ-phosphate interactions that are unraveled upon hydrolysis of GTP to GDP.

The hallmark of an active (GTP-bound) small GTPase is the recruitment of cytosolic effector proteins to distinct subcellular compartments and subsequent activation of downstream signaling pathways. However, the slow intrinsic hydrolysis rate of small GTPases lends itself to regulation of the GTP/GDP cycle by accessory proteins. GTP hydrolysis is accelerated by GTPase activating proteins (GAPs), and small GTPases are reactivated by the action of GDP/GTP exchange factors (GEFs), which have been the topic of a recent review.[12] Thus, the overall biological activity of Ras-like GTPases is a complex function encompassing active (GTP) residence time on subcellular compartments, dictated by accessory proteins, and the nature of effector recruitment that mediates dynamic cellular responses to intracellular and extracellular signals. Here, the focus is on the molecular basis for the recruitment of effectors by Rab GTPases and subsequent control of vesicle trafficking pathways.

2. MEMBRANE TRAFFICKING BY Rab GTPases

Rabs comprise the most diverse members of the Ras superfamily with nearly 70 proteins in mammals and 11 proteins in budding yeast.[13,14] Following biosynthesis, nascent Rabs bind to Rab escort protein (REP) which shuttles them to geranylgeranyl transferase (GGTase).[15,16] At this stage, Rabs are prenylated at one or two C-terminal cysteine residues and subsequently escorted to a distinct subcellular compartment, where they recruit effector proteins.[17] Rabs contain flexible 30–40 residue tails preceding the prenylated cysteine residues at their extreme C-termini, and it is the N-terminal GTP-bound globular domain that facilitates oligomerization with effectors. In contrast to Rabs, Arf small GTPases have an additional amphipathic N-terminal helix that undergoes a conformational rearrangement following GTP binding, and it inserts into the lipid bilayer, thus placing Arf (and related Arl GTPases) much closer to the lipid bilayer. In many trafficking events, Rab and Arf GTPases cooperate in recruiting effectors, and constraints imposed by their membrane localization influences the nature of these regulatory interactions.

An additional regulatory protein is GTP/GDP dissociation inhibitor (GDI), which is structurally related to REP and extracts inactive Rab(GDP) from membranes into the cytosolic fraction following hydrolysis of GTP to GDP (aided by GAPs). Thus, the Rab life cycle is completed by the action of GEFs, which catalyze the exchange of GDP for GTP and facilitate reinsertion of Rabs into membrane bilayers. The subcellular distribution of Rabs is a complex balance of Rab binding to GAPs, GEFs, GDI, and effectors.[18]

Despite a common α/β fold that consists of a central six-stranded β-sheet flanked by five α-helices, active Rabs recruit cellular effectors to their globular (cytosolic) face with exquisite specificity. GTP specificity is imparted by a pair of regions proximal to the γ-phosphate, termed switch I and switch II. By definition, these regions undergo conformational changes during the GTP/GDP cycle. The GTP conformation is generally highly conserved in Rabs as the γ-phosphate forms hydrogen bonds with the switches and bridges them together to form a contiguous effector-binding site.[19,20] In contrast, upon hydrolysis of GTP to GDP, these interactions are lost and the resulting switch I/II conformations are highly heterogenous and often disordered.

Cellular effectors are typically modular, with a Rab-binding domain (RBD) and additional domains that exert biological effects such as organelle structure, vesicle formation, transport, docking, and fusion.[21] Identification of effectors is typically performed by yeast two-hybrid studies and/or pulldowns, using a GTP-locked form of the Rab as bait. Mutation of the Gln $\rightarrow$ Leu in switch II of Rabs, originally identified as a GTP-locked transforming mutant of Ras,[22,23] will constitutively interact with genuine effectors. As a negative control for effector recruitment, a GDP-locked variant of Rabs can be prepared by mutation of the conserved Ser/Thr residue in the P-loop to glutamine, thus, confirming a bona fide effector of Rab GTPases.

The classic phenotype of Rab-mediated effector recruitment is the distinct subcellular localization of effectors upon expression of active (GTP) Rabs in cells. Most known effectors are cytosolic proteins that contain a ~20- to 150-residue RBD, which enables binding to Rab(GTP), tethered to the lipid bilayer. The biological effects elicited by Rabs are mediated through the additional effector domains and are reliant upon oligomerization with Rabs and other regulatory proteins and enzymes. The strict definition of an effector involves several properties: (i) recognition of the GTP-bound conformation of a Rab with higher affinity than the GDP-bound conformation, (ii) regulation of biological activity through its binding to an active Rab, and (iii) loss of Rab recruitment and subsequent biological activity following GAP-mediated hydrolysis of GTP to GDP. *In vitro*, a higher affinity for the GTP state has been confirmed for numerous effectors using various binding assays, including fluorescence polarization, surface plasmon resonance, and isothermal titration calorimetry.[24–27]

This review will focus on the nature of Rab/effector assembly and thermodynamics of effector recruitment, which has not been sufficiently addressed in literature reviews of the topic. The diversity of oligomeric complexes, their overall modes of binding, and the links to subsequent biological functions and human diseases will also be addressed briefly.

3. STRUCTURAL BASIS FOR Rab/Effector RECOGNITION

There are effectively nine Rab/effector complexes that have been determined by X-ray crystallography (Fig. 21.1). Structural homologues are counted as a single complex—a total of 14 independent structures have been deposited in the Protein Data Bank. The structure of a complex between Rab8 and LidA from *Legionella pneumophila* has also been determined.[28] Protein LidA from *L. pneumophila*[29] binds to several Rabs (Rab1, Rab6, and Rab8) and interferes with host secretory pathways.[30] LidA may enable trafficking of ER-derived vesicles to recplicative organelles termed Legionella-containing vacuoles in infected cells.[31] However, LidA binds to Rabs with extremely high affinity, estimated at 30 pM for Rab1, and has been termed a "supereffector."[28] Since the interaction is extremely tight for both the GTP and GDP forms of Rabs, the complex may not lend itself to conventional regulation by GAPs and GEFs. Since it does not fit the general criteria for a Rab effector, LidA will not be discussed in detail here.

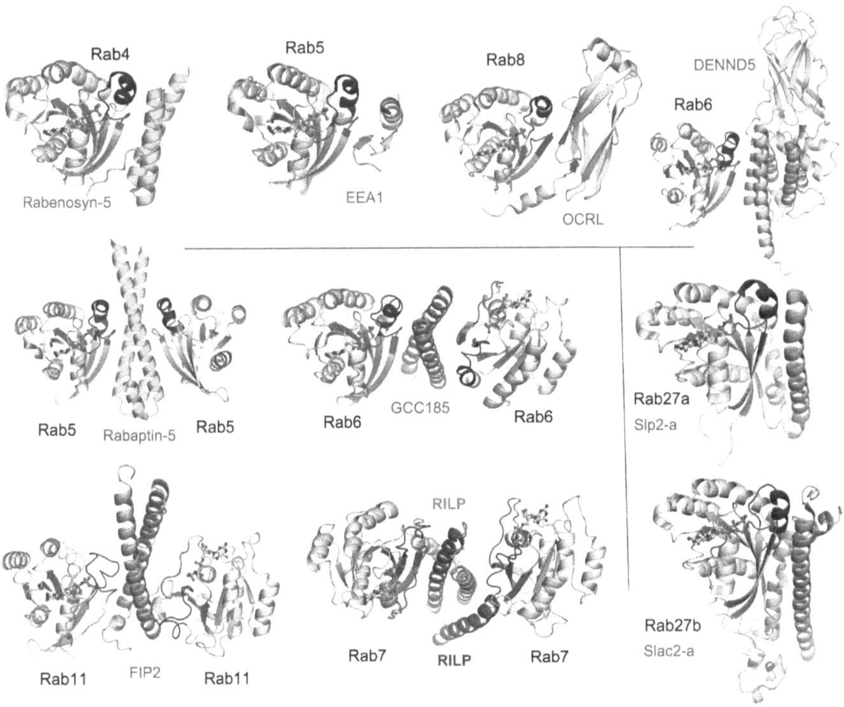

Figure 21.1 Crystal structures of Rab/effector complexes. Rab GTPases are gray, switch I is red, and switch II is blue. The nucleotide GTP (or its analog, GppNHp) is shown as a ball-and-stick model. Effectors are green ribbons, and the magenta spheres are Zn^{2+} ions. The view is approximately identical in all of the complexes, with helix $\alpha5$ of Rabs running parallel to the line of view. Following $\alpha5$ and the C-terminal tails, which are considered to be flexible, Rabs are tethered to the bilayer via prenylated cysteine residues. Thus, the view is roughly facing the plane of the lipid bilayer from the cytosolic face of Rab/effector complexes. Lines denote the organization of structures into heterotetrameric assemblies (2:2 complexes) and monovalent (1:1) complexes. A subset of the 1:1 assemblies adopt nonhelical conformations (Rab5/EEA1, Rab8/OCRL). (See Color Insert.)

3.1. Structures of Rab3 and Rab27 complexes

Rab3 and Rab27 mediate exocytosis in a variety of cell types via recruitment of a family of at least 11 effector proteins that contain a conserved RBD27 at their N-termini[32,33] Rab3 is implicated in neurotransmitter release via its interactions with RIM (Rab3-interacting molecule),[34] while both Rab3 and Rab27 appear to cooperate in sequential exocytic steps of dense-core granules in human sperm.[35] In melanocytes, Rab27 mediates exocytosis of pigment-bearing melanosomes via the recruitment of myosin V. However, the binding of actin-based myosin V is indirect and involves an adaptor protein termed Slac2-a (melanophilin),[33] which contains a conserved N-terminal RBD27 and a downstream myosin-binding region. A similar scenario is achieved in granulocytes, where Rab27 recruits Slp4-a (granuphilin),[36] which in turn binds to actin-based myosin motors.

The first Rab/effector structure was a complex between Rab3 and Exophilin-1 (also called Rabphilin-3).[37] Exophilin-1 forms a pistol-shaped molecule in which the barrel of the gun consists of an N-terminal long α-helix (α1) that interacts with switch I and II. In addition to these GTP-dependent interactions, the handle of the pistol adopts a globular β-sheet fold that is stabilized by two Zn^{2+} ions, followed by a second short helix (α2) at the C-terminus which also interacts with Rab3.

The structure of Rab3/Rabphilin-3 demonstrated some of the general principles of GTP-dependent engagement of Rab GTPases by effectors. GTP specificity arises from the interactions between α1 of Rabphilin-3 and switch I/II—in the GDP state, these contacts would be disrupted. The Zn^{2+}-binding globular domain would reside proximal to the membrane, possibly mediating lipid contacts;[37] thus, the topology of the Rab/effector complex is consistent with constraints imposed by the lipid bilayer. Helix α2 also packs against Rab3 (N-terminus, α3–β5 loop, and the C-terminus). These segments of Rab3 come from the complementarity-determining regions (CDRs 1–3) whose conformations are independent of the nucleotide state of Rab3. Thus, Rab/effector interactions at CDRs (N-terminus, α3–β5 loop, and C-terminal helix α5) were thought to play a general role in Rab/effector specificity, given their sequence diversities.[37] However, subsequent structural analyses of other Rab/effector complexes have revealed that effector recruitment is primarily mediated by switch I, switch II, and the interswitch region (see below). Finally, a key hydrophobic interface is formed by an aromatic triad of residues from switch I (Phe59), interswitch (Trp76), and switch II of Rab3 (Tyr91). The conformation of these residues is a key determinant of effector specificity,[38] and they are observed to mediate effector recruitment in all complexes save one (Rab8/OCRL, see below). Although Exophilin-1 binds to Rab3 *in vitro*, the physiologically relevant partner may be Rab27, with whom it forms a complex both *in vitro* and in PC12 cells.[39]

In addition to Rab3/Rabphilin-3, the structures of Rab27a/Slac2-a (also called melanophilin) and Rab27b/Slp2-a have been determined.[37,40,41] The isoforms of Rab27—Rab27a and Rab27b—both bind to all known effectors containing the conserved RBD.[33,42,43] However, the two isoforms demonstrate some differences in tissue-specific expression, particularly in neurons where Rab27b is the predominant species.[44,45] Termed the synaptotagmin-like (Slp) proteins and the synaptotagmin-like proteins lacking C2 domains (Slac), these Rab27 effectors contain the same conserved N-terminal RBD27 that resembles Rabphilin-3. However, some members of the RBD27 family such as Slp2-a lack the Zn^{2+} domain, and therefore the two α-helices are connected together directly as a helical hairpin. The structures of these additional complexes have shed some light into the specificity of RBD27 for Rab3 versus Rab27. Although Exophilin-1 can recognize both Rab3 and Rab27, Slp2-a and Slac2-a are exclusive effectors of Rab27. The structures reveal a complex contribution of small differences at the Rab/effector interface, particularly at the α3/β5 loop (CDR3), that may account for specificity among the family of RBD27 effectors.[40,41,46] Furthermore, there are additional interactions between Slac2-a and the β2/β3 loop of Rab27b,[41] which are not observed in the Rab3/Rabphilin-3 complex.

3.2. Structures of Rab4 and Rab22 in complex with Rabenosyn-5

Rabenosyn-5 is implicated in the regulation of endocytic trafficking pathways.[47] The 784-residue polypeptide contains a Zn^{2+}-associated FYVE-finger domain, situated near its N-terminus,[47] which mediates lipid binding. Rab-binding regions of coiled coil are localized toward the middle and C-terminus of the protein, and the effector mediates the transport of clathrin-coated vesicles from the plasma membrane into early endosomes.[48] In *Drosphila melanogaster*, a role for Rabenosyn-5 has been linked to polarized endocytosis in wing epithelial cells.[49]

Rab4 and Rab22 bind to discrete motifs of Rabenosyn-5, although the two motifs are structurally identical.[24] Crystal structures of the complexes Rab4/Rabenosyn-5 (457–503) and Rab22/Rabenosyn-5 (728–784) have been determined.[24] This was a landmark study in understanding the molecular basis for effector discrimination using a genome-wide survey of effector binding to human Rab GTPases. The structure of the effector consists of two helices connected by a loop (an α-helical hairpin) which interacts with the switch I, switch II, and interswitch regions of Rab4 and Rab22. Extensive structural, mutational, and biophysical analyses revealed that conformational diversity of conserved switch and interswitch regions combine with subtle sequence differences to contribute, in a global way, to positive and negative determinants of effector recognition.[24]

3.3. Structure of Rab5 in complex with EEA1

The Rab5 effector early endosomal autoantigen 1 (EEA1) enhances endosomal fusion through its interactions with Rab5[50] and several other soluble factors, including Rabenosyn-5, hVPS45, and Rabex-5, an exchange factor for Rab5.[47,51–53] EEA1 is composed of an N-terminal zinc-finger, followed by coiled coil regions and a C-terminal FYVE-finger domain.[54] Rab5 regulates a complex network of phosphorylation and dephosphorylation through its direct interactions with both phosphatases and kinases,[55,56] resulting in the net production of phosphatidyl-inositol(PI)-3-phosphate. Rabenosyn-5 (a phosphoinositide-binding protein), which has already been described as an effector of Rab4, also binds to Rab5 via a C_2H_2 zinc-finger located at its N-terminus. Thus, endocytosis is intricately regulated through the recruitment of numerous effectors, such as EEA1 and Rabenosyn-5, via small GTPases Rab4 and Rab5.

Early endosomal antigen 1 (EEA1) consists of an N-terminal Zn^{2+}-finger, long segments of α-helical coiled coils, and a C-terminal FYVE-finger domain. The N-terminal Zn^{2+}-finger has high affinity for Rab5, and the crystal structure of Rab5 with this domain has recently been determined.[57] The Zn^{2+}-finger motif (C_2H_2) adopts a ββα topology that resembles DNA-binding motifs, but the overall charge of the Rab5 RBD is negative.[57] The N-terminus of the motif, β1, and the α-helix engage the switch and interswitch region of Rab5. The complex was the first example of an RBD that is partly composed of a β-sheet and provided insight into the structural basis for Rab5 specificity. Sequence and structural comparisons, along with mutational analyses, revealed that the active conformations of Rab5 and Rab4 switch regions are variable. In particular, the switch II conformation is subtly affected by α3, which in turn can be influenced by multiple substitutions in the protein core.[57] These and other structural studies of Rab/effector complexes reveal that specificity is a complex phenomenon that is a function of the global fold of active switch regions, and it is difficult to predict from sequence differences alone.

3.4. Structure of Rab5 in complex with Rabaptin-5

Rabaptin-5 is an 862-residue multidomain effector that regulates endocytic trafficking.[58,59] In addition to Rab5, Rabaptin-5 contains a separate binding site for Rab4;[60] thus, it may provide a molecular link between the two Rab trafficking pathways. Rabaptin-5 also interacts with Rabex5,[61] tuberin,[62] and GAP-43,[63] which are all linked to endosomal trafficking functions.

Rab5 GTPase forms a heterotetrameric complex with its effector, Rabaptin-5.[64] The oligomeric assembly is organized with the effector forming a central symmetric coiled coil, which binds to two Rab5 molecules on either side. The two Rabs do not contact each other, and this is a recurring theme in Rab/effector oligomers mediated by symmetric coiled coils. All known active Rabs are monomers *in vitro*, and oligomerization into multimeric assemblies is facilitated by homodimers of effectors.

3.5. Structure of Rab6 in complex with GCC185

The structure of Rab6 in complex with a short α-helical motif from GCC185 (GRIP and *c*oiled *c*oil domain-containing protein) has been determined. Rab6 mediates trafficking into and out of Golgi, while GCC185 belongs to a family of proteins termed golgins,[65–67] which mediate tethering in various Golgi compartments. The complex of Rab6/GCC185 is a heterotetramer in which the effector forms a central symmetric coiled coil.[68] Full-length GCC185 is a large 1684-residue protein that contains long stretches of coiled coils that mediate trafficking from late endosomes to the *trans*-Golgi network.[69,70] In addition to Rab6, GCC185 apparently contains a nonoverlapping site for Arl1 binding, termed the GRIP domain. Structural studies and *in vitro* assays suggest that the Rab6/GCC185 complex enables more efficient loading of the GRIP domain onto Arl1. Active Arl1 GTPase resides proximal to the lipid bilayer via an amphipathic N-terminal helix, since it is closely related to the Arf family. Meanwhile, Rab6 may be more distantly situated due to a 30-residue tail following α5 that leads to its prenylation sites at the C-terminus. From their *in vitro* binding data and the binary structures of Arl1/GCC185 and Rab6/GCC185, the authors propose a model in which the long coiled coils of GCC185 can be simultaneously engaged by Rab6 and Arl1.

The model has come under some scrutiny from cellular studies suggesting that the localization of GCC185 in cells is independent of Arl1 and Rab6.[71] More recently, Arl4a has been identified as an effector of GCC185, and the complex apparently maintains the structural integrity of Golgi compartments.[72] Arl4a binds to the effector in a GTP-dependent manner and colocalizes with GCC185 in HeLa cells.[72]

3.6. Structure of Rab6 in complex with DENND5

The structure of Rab6 in complex with DENND5 (alternatively, Rab6-interacting Protein 1) has been determined.[73] DENND5 is a 1263-residue multidomain protein that contains a series of DENN domains (differentially expressed in normal vs. neoplastic cells) at its N-terminus. The C-terminal half of DENND5 consists of two RUN domains (named for RPIP8, UNC-14, and NESCA) separated by a PLAT domain (named for polycystin-1, lipoxygenase, and alpha-toxin). RUN domains are ~150-residue interaction modules that are found in numerous cytosolic proteins.[74] Genome-wide analyses of RUN domains reveal that they are associated with a limited number of Rabs (Rab1, Rab6, Rab35),[75] and are more frequently involved in other signal transduction pathways that are poorly characterized. The isolated crystal structure of the RUN domain of Rap2-interacting protein X (RPIPX) has also been determined,[76] and like DENND5, the structure adopts an eight-helix bundle that resembles a death domain fold, albeit with a different topology. However, the structure of the Rap2/RPIPX complex remains unknown.

The complex Rab6/DENND5 reveals that an all α-helical RUN domain in the middle of the 1263-residue DENND5 mediates binding to the switch and interswitch region of Rab6. The first and last α-helices of the RUN domain (a1 and a8) stack in a parallel fashion and interact with Rab6. Strikingly, the orientation and composition of the RBD interface (hydrophobic, hydrophilic) are identical to the Rab6/GCC185 complex, which is a short α-helical coiled coil. In addition to the RUN domain, which encompasses the RBD for Rab6, the crystal structure includes a subsequent PLAT domain.[77] PLAT domains adopt a β-sandwich fold and are frequently observed in proteins and enzymes that lie proximal to lipid bilayers. Although their roles are ill defined, they are generally considered to have an affinity for phospholipids.[78]

In full-length DENND5, the PLAT domain is followed by a second RUN domain (RUN2) at the extreme C-terminus of the protein. The two-domain (RUN1-PLAT) module in the complex of Rab6/DENND5 provides a glimpse of the possible topology of the complex on Golgi membranes. The putative lipid-binding region of PLAT is located distal from the Rab6/RUN interface.[73] Furthermore, the downstream RUN2 domain is reported to interact with a sorting nexin (SNX1),[79,80] which is part of a family of concave-shaped α-helical proteins that bind to convex phospholipids and participate in membrane trafficking.[81,82] As mentioned earlier, the N-terminus of DENND5 contains the eponymous DENN domains, which may function generally as Rab GEFs (reviewed in Ref. 83). The structure of a DENN domain from DENN1B, a GEF for Rab35,[84] has recently been determined in complex with its substrate.[85] However, the function of the

DENN domain of DENND5 and whether it acts as a GEF for Rabs has not been clarified. Molecular details of the mechanisms of vesicle trafficking in Golgi await further structures of binary and higher-order protein complexes.

3.7. Structure of Rab7 in complex with RILP

The structure of Rab7 in complex with the RBD from Rab-interacting lysosomal protein (RILP) has been determined.[86] Rab7 links vesicular traffic between endosmal and lysosomal compartments via several unrelated effector proteins. RILP is a 401-residue protein in which the RBD comprises residues 244–308 and, apart from brief segments of coiled coil, contains no other significant homologies to domains with known structure and function.[87] RILP recruitment by Rab7 mediates fusion of phagosytic vesicles with late endosomes and/or lysosomes.[88] More recently, RILP and RILP-related proteins have been identified as effectors for Rab36 in mediating retrograde transport of melanosomes.[89]

In the complex Rab7/RILP, a helical hairpin motif dimerizes and binds to two Rab7 molecules on equivalent and opposite sides of the effector, organized as a Rab7-RILP$_2$-Rab7 complex. The RILP dimer (chains A/B) is enabled by both α-helices in the helical hairpin, such that the two α-helices interdigitate with each other (Fig. 21.1). Each interface with Rab7 involves an N-terminal helix from one monomer (e.g., A) and a C-terminal helix from the partner (e.g., chain B). Mutagenesis of RILP and colocalization studies support the notion that dimerization is required for binding to Rab7. Mutations at the homodimeric interface reduce or abolish binding, and *in vitro* studies of RILP show that it is a dimer.[86] In addition to the switch and interswitch regions including the hydrophobic triad, the C-terminal helix of RILP engages CDR3 (C-terminus) and the N-terminal helix engages CDR1 (N-terminus) of Rab7, thus forming an extensive interface.

3.8. Structure of Rab8 in complex with OCRL1

The structure of Rab8 in complex with the RBD from OCRL1 (oculocerebrorenal syndrome of Lowe protein 1) has been determined.[90] OCRL1 is a 901-residue phosphatase that preferentially dephosphorylates PI-4,5-bisphosphate and PI-3,4,5-triphosphate at the fifth position.[91] In addition to a centrally located phosphatase domain, OCRL1 also contains an N-terminal pleckstrin-homology domain and a C-terminal RhoGAP domain that is catalytically inactive.[92] The Rab-binding region of OCRL1 was localized to a segment of the protein termed the ASPM-SPD-2-Hydin (ASH) domain (residues 540–725), located between the phosphatase domain and the RhoGAP domain.[93] Proteins containing these domains are associated with inherited disorders, including Lowe oculocerebrorenal syndrome that is linked to OCRL1.[94] The syndrome is X-linked and manifests in renal dysfunction, cataracts in the eye, and disorders of the nervous system.

OCRL1 regulates vesicle trafficking from early endosomes to Golgi,[95] endocytic recycling,[96] as well as cell migration.[97] The OCRL1 fragment (residues 540–678) forms an N-terminal α-helix followed by the ASH domain, which adopts a β-sandwich that is similar to immunoglobulin domains.[90] Binding is mediated by the α-helix and β9 of the ASH domain, which interacts with the switch, interswitch, and CDR1 (N-terminus) of Rab8, distributed into regions termed Site I and Site II. One particularly tight interaction is the side chain of Phe668 of OCRL1 (Site II), which packs into a hydrophobic pocket formed between switch I (Ile41) and switch II (Tyr77). The structure of the complex between Rab8/OCRL1 has provided some insight into the molecular basis for disease. Missense mutations in OCRL1 that are associated with disease either cause a global problem of protein folding (L634P in the ASH domain) or weaken the Rab/effector interface (F668V).[90] The Rab/effector complex is unusual in the extent of interactions between OCRL and α1 (CDR1) and β2 (interswitch). Rab8 contacts in this region involve α1 of OCRL1 and a linker to the ASH domain, and are mainly polar in nature (Site 1[90]). Another peculiarity of the Rab8/OCRL1 structure is the absence of contacts with the conserved aromatic triad—only a peripheral interaction between Phe45 (switch I) of Rab8 and OCRL1 is observed.

3.9. Structures of Rab11 in complex with FIPs

Three related structures of Rab11 in complex with members of the Rab11-family interacting proteins (FIPs) have been published.[98–100] Rab11/FIP complexes regulate various aspects of endosomal trafficking. FIPs are multidomain effectors that comprise several varieties of lipid-binding and protein-binding domains at their N-termini, but all FIPs have a highly conserved RBD11 at their extreme C-termini. The butterfly-shaped complex is assembled as a heterotetramer with FIPs forming a central coiled coil that forms equivalent Rab11-binding sites on opposite sides, related by two-fold symmetry down the long axis of the coiled coil. Each of the two α-helices of FIPs interact with a single Rab interface, which involves the switch and interswitch regions. Toward the C-terminus, the central FIP helices are interrupted by a proline that nucleates a short 3_{10} helix perpendicular to the central coiled coil. This short helix rubs against β2 of Rab11 and contributes to Rab11 binding and specificity. Thus, rather than a canonical coiled coil, dimeric FIPs adopt a T-shaped structure. Unusually, the switch II conformation of Rab11 is nonhelical and undergoes a conformational rearrangement upon binding to FIPs.[98–100] The nonhelical and flexible nature of switch II has been presumed to be a determinant of Rab11/FIP specificity. Active Rabs have a generally well-defined α-helical conformation in their active (GTP) state, and all other known structures of Rab/effector complexes involve little or no conformation rearrangements upon effector recruitment.

Classically, the function of Rab11 has been linked to the recycling of transferrin receptor from endosomes back to the plasma membrane.[101] Rab11 is localized to recycling endosomes and the *trans*-Golgi network,[102,103] and several independent groups initially identified a groups of effectors (FIPs) that contain a conserved RBD11 at their C-termini.[104–106] In addition to recycling, Rab11 function has also been linked to delivery of cargo to the growing cleavage furrow in cell division.[107,108] Furthermore, Rab11 has been linked to higher-order complexes with other small GTPases via interactions with dual-specificity effectors. FIP3 is a dual effector of Rab11 and Arf6 and mediates links to the exocyst.[109] Rab11 links to Golgi compartments via Rab6 have also been identified through the dual-specificity effector DENND5.[110] Simultaneous or stepwise engagement of effectors with multiple Rabs from distinct compartments provides a physical association of trafficking pathways in cells.

4. MODES OF Rab/Effector OLIGOMERIZATION

The mode of Rab/effector oligomerization can be organized into several distinct groups. Such an organization of complexes provides a conceptual framework for understanding how the complexes regulate trafficking.

4.1. Heterotetrameric assemblies

Effectors that form parallel coiled coils can effectively form two equivalent RBDs by consequence of the twofold symmetry axis. Given that all known active Rabs are monomeric (at least *in vitro*), these effectors would enable the formation of heterotetramers of Rab/effectors at the membrane. Currently, four Rab/effector complexes fit into this category: Rab5/Rabaptin-5, Rab11/FIPs, Rab6/GCC185 and Rab7/RILP (Fig. 21.1). Rabaptin-5 and FIPs have an additional helix at their C-termini, and are roughly oriented the same way through their respective Rab interfaces, although the long axes of their coiled coils are offset by ~60°. In contrast, GCC185 is oriented (topologically) in the opposite manner in terms of N-/C-termini, relative to Rab11/FIP complexes. The orientation of the N- and C-termini of effectors, relative to the contraints imposed by the lipid bilayer, is critical for subsequent biological activity.

The Rab7/RILP complex is a variant of this theme in which the coiled coil is formed by interdigitation of a helical hairpin. In all of these four cases, interactions with the coiled coil involve the switch and interswitch regions, and both α-helices contribute to interactions with each Rab. It is worth noting that, currently, a single isolated α-helix is insufficient to form a complete RBD. Although Rabphilin-3 mediates contacts with the switch and interswitch of Rab3 via a single α-helix, there are additional stabilizing contacts between Rab CDRs and the "barrel" of the gun (Fig. 21.1).

The topologies of heterotetrameric assemblies of Rab/effector complexes on lipid bilayers, extrapolated from crystal structures of Rab/RBDs, may provide clues to their biological roles. In both Rab5 and Rab11 complexes, the RBD is localized to the C-terminus and the orientation of the Rab/effector complexes would project the N-termini away from the membrane plane. The N-termini of FIPs consist of lipid-binding C2 domains[111,112] (Class I FIPs), a myosin-binding region (FIP2), and Ca^{2+}-binding EF-hand motifs which may also mediate protein/protein interactions in neuronal cells.[113,114] The RBD of RILP (244–308) is situated toward the middle of the 401-residue polypeptide, with additional predicted coiled coils toward the N-terminus, although the significance of a central RBD in trafficking is unknown.[86]

The putative RBD of GCC185 is also localized toward the C-terminus of the 1584-residue protein, although the physiological relevancy of the complex is disputed.[71] In principle, long coiled coils are ideally suited for long-range tethering functions and for linking separate Rab pathways (or Arfs) via multiple recognition motifs. Rab11 effectors FIP3 and FIP4 belong to this category of coiled coils and also bind to Arf5[105,107,115]. Heterotetrameric assemblies, by virtue of a two-fold symmetric RBD, may also enable more stable association of Rab/effector complexes during trafficking (see Section 5). Interestingly, the 3269-residue protein Giantin has been shown to interact with both Rab1 and Rab6.[116,117] Giantin cooperates with numerous other proteins to regulate trafficking in Golgi,[118–120] and it is also predicted to contain segments of coiled coil throughout its primary sequence. Thus, bivalent effectors such as Giantin may link separate transport pathways or subcellular compartments through their non-overlapping binding sites for multiple Rab GTPases.

4.2. Monovalent α-helical RBDs

Rab/effector complexes in this category are formed primarily by α-helical RBDs that interact as a 1:1 complex. The complexes are Rab4/Rabenosyn-5 (and Rab22/Rabenosyn-5), Rab6/DENND5, and the various Rab3 and Rab27 complexes with effectors (Fig. 21.1). The folds of RBDs vary from α-helical hairpins (Rabenosyn-5, Rabphilin-3) to globular α-helical domains (RUN domain of DENND5). The complexes are asymmetric and are therefore unable to hetero-oligomerize through their RBDs. However, higher-order complexes may nevertheless be facilitated by coiled coil regions of effectors, independent of the RBDs. Indeed, experimental evidence supports a dimeric form of DENND5 in cells,[110] while EEA1 contains regions of coiled coil following the N-terminal RBD.

4.3. Nonhelical RBDs

More recently, the crystal structures of nonhelical effectors have been determined. The Zn^{2+}-finger motif of EEA1 and the ASH domain of OCRL1 exploit β-sheets to bind to the switch regions of their cognate Rabs. Both complexes are 1:1 stoichiometry, and it is worth noting that despite the Rab interactions with β-sheets, both effectors also utilize α-helical segments to mediate Rab contacts.

A second example was the recent crystal structure of Rab8 in complex with OCRL,[90] an inositol 5-phosphatase that is linked to an X-linked hereditary disease termed Lowe syndrome.[94] The complex Rab8/OCRL reveals that an immunoglobin-like β-sandwich (ASH domain) mediates the majority of contacts with the Rab,[90] although an α-helix and loop ahead of the ASH domain also packs against strand β2 of the interswitch region of Rab8. In addition to β2, the other key regions of contacts in Rab8 are switch I and switch II. The complex Rab8/OCRL is the only example of a Rab/effector interaction that does not exploit the invariant tryptophan (Trp62 of Rab8) to mediate hydrophobic contacts.

5. THERMODYNAMICS OF Rab/Effector BINDING

For a complete understanding of the biological roles of effectors, it is important understand the thermodynamics and kinetics of complex formation. Measurement of affinities has posed a significant challenge due to the large size and relative insolubility of full-length effectors. In some cases, the RBDs could only be obtained in a soluble form upon coexpression with the cognate Rab.[37,73] In recent years, the thermodynamic parameters of many diverse Rab/effector complexes have been determined using various techniques such as fluorescence polarization, isothermal titration calorimetry, and surface plasmon resonance. A summary of Rab/effector complexes and their corresponding dissociation constants (K_d) is shown (Table 21.1).

The measured affinities of effectors for the active form of cognate Rabs span three orders of magnitude (Table 21.1). In general, the size of the interface correlates with binding affinity, with the highest affinity involving the Rab3/Rab27 effector family (10–100 nM), followed by the Rab11 family (200–500 nM). The remaining Rab/effector complexes interact with relatively weak, low micromolar affinity (1–10 μM) and bury 600–900 Å^2 of surface. Although a single Rab11/effector interface buries the same surface, the complexes are heterotetramers. It is likely that upon binding of one Rab11 molecule to a pre-formed FIP homodimer, the second-site gains additional affinity due to a rigidification of the complex and a reduction in the cost of conformational entropy.[99,100] Indeed, a 2:2 complex of Rab11:FIP2 is observed *in vitro* even in the case of excess effector (A.R. Khan, unpublished data). There is one significant outlier among these general trends—the complex Rab27a/Slp2-a buries far less surface and involves fewer hydrogen bonds and salt bridges than Rab27b/Slac2-a (data not shown), but the affinity is stronger. The bacterial protein LidA has been excluded from this analysis due to its unusually strong affinity for several Rabs in both GDP and GTP states.[28]

A few general comments can be noted from the various parameters. In those complexes for which ΔH and ΔS have been measured by isothermal titration calorimetry (Rab11/effectors[26] and Rab6/effectors[27]), complex formation is mediated by a favorable enthalpic component. Also, relatively weak affinities and fast on/off rates may relate to a role in transient tethering functions, as has been noted for the Rab6 and its effectors.[27] Effectors that act as tethers (such as Giantin, golgins) adopt extended coiled coils to facilitate stretching over long distances to enable the specificity that underpins vesicle trafficking. However, the majority of Rab/effector complexes are likely multimeric, built up directly from a symmetric coiled coil, or through effector oligomerization motifs elsewhere. Given the generally weak affinities of Rab/RBDs, multimeric assemblies may be important for increasing the effective affinities *in vivo*.

Table 21.1 Structural details and affinities of Rab/effector complexes

	PDB	Res. (Å)	Effector region	Effect or source	Stoichiometry	Shared interface ($Å^2$)	K_d (μM)	Effector specificity	Role	References
Rab3/ Rabphilin-3	1zbd	2.6	40–170	Rat	1:1	1425	–	Rab3, Rab27	Exocytosis in neurons	37,121
Rab27/Slp2-a	3bc1	1.8	1–59	Human	1:1	988	0.013	Rab27	Exocytosis	25,40
Rab27/Slac2-a	2zet	3.0	1–153	Mouse	1:1	1871	0.112	Rab27	Exocytosis	25,41,122
Rab4/ Rabenosyin5	1z0k	1.92	440–503	Human	1:1	893	6.2	Rab4, Rab14	Early endosome traffic	24
Rab22/ Rabenosyn5	1z0j	1.32	728–784	Human	1:1	671	9.7	Rab5, Rab22	Early endosome traffic	24
Rab5/ Rabaptin5	1tu3	2.31	789–862	Human	2:2	707	–	Rab5	Endocytic recycling	60,64
Rab5/EEA1	3mjh	2.03	36–69	Human	1:1	586	2.4	Rab5	Endosomal trafficking	57
Rab6/GCC185	3bbp	3.0	1446–1511	Human	2:2	626	2.3	Rab6	Golgi trafficking	68,123
Rab6/ DENND5	3cwz	3.2	707–1145	Mouse	1:1	741	0.5[a]	Rab6	Golgi trafficking	73,110

Rab7/RILP	1yhn	3.0	244–308	Human	2:2	673	–	Rab7, Rab34	Lysosome trafficking	86,124
Rab11/FIP2	2gzh	2.47	447–512	Human	2:2	705	0.25[a]	Rab11, Rab14, Rab25	Endocytic recycling	100,125
Rab11/FIP3	2hv8	1.86	695–756	Human	2:2	785	0.32	Rab11	Endocytic recycling	99,126
Rab11/FIP3	2d7c	1.75	715–756	Human	2:2	900	–	Rab11	Endocytic recycling	98,126
Rab8/OCRL1	3qbt	2.0	540–678	Human	1:1	959	0.9	Rab1, Rab5, Rab6, Rab8	Endosome-Golgi trafficking	90

[a]Values for these Rab/effector complexes were obtained by isothermal titration calorimetry (laboratory of A.R. Khan, unpublished data).
Details of the structures of Rab/effector complexes and their corresponding affinities (K_d, dissociation constant).

In contrast, the monovalent Rab27/RBD complex is relatively strong and is enabled by an unusually large area of complementarity. Rab27-driven motility via myosin V may necessitate a more stable interaction, relative to tethering functions of other complexes. The relationship between Rab trafficking and the biophysics of the Rab/RBD interaction is further dependent on the number and distribution of Rab GTPases on the surface of endomembrane compartments. Such detailed information necessitates a proteomics approach to the nature of vesicular compartments, as performed for synaptic vesicles.[127,128]

Finally, it is instructive to compare Rab/effector binding affinities to Ras and its effectors. All Ras effectors share a conserved ubiquitin-like fold[129–134] and complex formation is driven by electrostatics with fast association constants (k_{on}).[134,135] Dissociation constants (K_d) for Ras/effector complexes are in the low nanomolar range, comparable to Rab27/effector complexes.[25] The kinetics of Ras/effector interactions appear to be evolutionarily conserved,[136] suggesting a correlation with their signaling roles in cells. Thus, the thermodynamics and kinetics of Rab/effector complexes may also be linked to their biological role. Relatively weak affinities may enable promiscuity in recognizing diverse motifs, and may also enable the transient interactions (fast on/off rates) that underpin vesicle trafficking in eukaryotic cells.[27]

6. Rabs AND DISEASE

Defects in membrane trafficking are associated with numerous inherited, acquired, and transmitted diseases (Table 21.2; reviewed in Ref. 164). Many diseases are linked to Rab regulatory proteins, and may have broad effects on posttranslational modifications, membrane targeting and regulation of their active states. Perhaps the most well known example is choroideremia,[165] an inherited X-linked disorder that causes blindness due to degeneration of the choroid and retina. The disease is associated with REP-1 loss-of-function which has a general effect on the prenylation of Rabs, although Rab27 is particularly underprenylated in cells of choroideremia patients.[165,166] Here, I will briefly summarize human diseases that are associated with trafficking by Rab/effector complexes. A more complete catalog of Rabs and human disease has recently been published elsewhere.[167]

6.1. Rab27 and Griscelli disease

The first Rab-associated genetic disorder to be identified was Griscelli Disease, which was originally associated with albinism (skin/hair

Table 21.2 Rabs and associated human diseases

	Disease/pathogen	Effector	Mechanism	References
Rab1	Parkinson's disease	Specific effector links unknown	Relieves ER-to-Golgi block that is induced by α-synuclein inclusions (yeast model)	137,138
Rab1, Rab5, Rab6, Rab8	Oculocerebrorenal syndrome of Lowe (X-linked)	OCRL1	Possible disruption of phosphatase activity/membrane trafficking/cell migration functions of OCRL1, mechanism unclear	90,93,97,139
Rab27	Griscelli disease	Slac2-a and myosinV	Defect in delivery of exocytic vesicles in melanocytes and granulocytes	140–142
Rab25, Rab31	Cancer	FIP1 (RCP), Rabaptin-5	Aberrant endocytic recycling of adhesion molecules and EGF receptor	143–149
Rab14, Rab4	*Chlamydia*	CT229	Lipid delivery, subversion of slow endocytic recycling, sphingomyelin/nutrient delivery	150,151
Rab1, Rab6, Rab10, Rab11	*Chlamydia*	Cpn0585	GTP-dependent recruitment of multiple Rabs to inclusion, chlamydial development, Golgi fragmentation	152–154
Rab7	*Salmonella*	SifA	Direct binding to Rab7, disengagement of SCV from RILP/kinesin	155,156
Rab9	*Salmonella*	SifA	Direct binding to Rab9 (weak affinity), bridging of SifA/SKIP interactions	156
Rab1, Rab35	*Legionella*	AnkX	Phosphocholination, vacuole remodeling	157,158
Rab1, Rab35	*Legionella*	DrrA	Dual GEF activity and adenylylation, vacuole remodeling	159–162
Rab1, Rab35	*Legionella*	SidD	Removal of AMP, vacuole remodeling	130,162,163

discolouration) and immunodeficiency.[168] Patients with this rare autosomal disorder exhibit dilution of skin and hair pigment from accumulation of pigment-bearing melanosomes in melanocytes and neurological disorders.[169] Many patients also develop virus-associated hemophagocytic syndrome (HS) that involves uncontrolled T-cell and macrophage activation, which leads to death in the absence of a bone marrow transplant.[169,170] Initially, the genetic mutation was mapped to the gene corresponding to myosin Va (*MYOVA*)[170] and an associated gene, *RAB27A*.[140] Further studies of the analagous condition in mice revealed that Slac2-a (melanophilin), which was mutated in the *leaden* phenotype,[171] was an effector of Rab27 and provided a physical bridge to myosin Va.[172,173]

Although both myosin and Rab27 mutations are associated with defects in vesicle transport, deleterious Rab27 mutants are specifically linked to the most severe phenotype, HS.[141] The inability of Rab27-defective T lymphocytes to deliver cytotoxic granules in HS patients and a subsequent sustained immune response has been linked to this condition.[141] At the molecular and cellular level, null *RAB27A* mutants result in the inability to deliver perforin-containing granules by cytotoxic T cells and Natural Killer cells.[174] A third phenotype in which patients are restricted to hypopigmentation has been linked to missense mutations in Slac2-a (melanophilin), thus resembling the *leaden* phenotype in mice. The severity of the phenotype is associated with the extent of loss-of-function of Rab27a-mediated delivery of exocytic vesicles in various tissues, such as skin cells, hair cells, lymphocytes and neurons. Missense mutations of Slac2-a in skin and hair cells disrupt the tri-partite complex (Rab27a/Slac2-a/myosin Va), but Slac2-a and myosin Va are not expressed in cytotoxic cells.[172] Indeed a missense mutation in the hydrophobic triad of Rab27 (Trp73 → Gly) in Griscelli Disease patients results in the inability to recruit Slac2-a and leads to a breakdown of melanosome motility.[142] Trp73 forms a key hydrophobic interface with the long α-helix from the effector,[41] and although the mutant protein is folded and competent for GTP binding,[142] a decrease in affinity with Slac2-a presumably disrupts normal trafficking *in vivo*.

6.2. Subversion of Rab GTPases by pathogens

Intracellular bacterial pathogens such as *Chlamydia trachomatis*, *Salmonella enterica*, and *L. pneumophila* invade host cells and create a protective niche termed a vacuole that is actively remodeled during replication.[175,176] Recent cellular studies have revealed that vesicle trafficking pathways associated with Rab small GTPases are exploited by pathogens to generate the replicative niche. Although the detailed mechanisms and subcellular locations of parasites are diverse, several principles of the infection cycle are similar. Following phagocytosis into early endosome-like compartments in host cells, pathogens remodel the endomembrane compartment to create an environment suitable for replication. For some pathogens, diverting the conventional endocytic pathway toward fusion with lysosomes is critical for survival. Lysosomal compartments are acidified and contain proteolytic enzymes which would not be ideal for pathogen survival.[177,178] The early phagosytosed *Salmonella*-containing vacuole (SCV) is trafficked toward the microtubule organizing center (MTOC). However, *Salmonella* tubules incorporate lysosomal markers, and in macrophages models of infection, the SCV survives after it has apparently fused fully with lysosomes.[179]

Modulation of the type and levels of phosposphatidylinositol phosphates (PIPs) is also a common trait among several pathogens. In particular, PI (3,4,5)-phosphate (PIP3) is a key regulator of signal transduction.[180] In addition, the host membrane trafficking machinery is subverted by the bacteria to deliver nutrients to replicating vacuoles. Subversion of trafficking is dependent on bacterial effector proteins (or virulence factors) that are injected directly into the cytosol via a Type III secretion system in *Chlamydia* and *Salmonella*.[181,182] Small GTPases such as Rab and Rho are directly targeted by bacterial effectors to subvert host trafficking and enable bacterial replication.

Endosomal trafficking and the retrograde endosome–Golgi pathways are a particularly striking target for subversion by bacterial pathogens.[176] In *Salmonella* infections, recruitment of Rab7 and its cellular effector RILP enables SCV migration from the cell periphery to a juxtanuclear location, adjacent to the MTOC.[155,183] An intriguing aspect of chlamydial infection is the subversion of host Golgi trafficking pathways.[184,185] Secretory vesicles are hijacked and their phospholipids and proteins are delivered to chlamydial vacuoles as part of their nutrient requirements, a mechanism first characterized by the group of Marci Scidmore (Cornell University). In *Chlamydia trachomatis* infections, Golgi-derived sphingolipids are delivered to the growing bacteria via Rab14-associated pathways.[186] Subversion of intracellular trafficking leads to fragmentation of the Golgi compartment in infected cells, and this striking property of infection has been linked to improvement of the growth capabilities of the pathogen.[152] Small GTPases Rab6 and Rab11 are also intimately involved in this process of Golgi breakdown.[153]

Details on the nature and function of specific Rab/effector complexes in the life cycle of most intracellular bacteria remain poorly defined. However,

Rab subversion by *L. pneumophila* has been well documented in terms of structure, function and biology, and there already exist numerous reviews on the subject.[159,187,188] A novel and exciting theme is the covalent modification of switch regions by bacterial effector proteins, as demonstrated by DrrA targeting of Rab1. The bacterial protein covalently modifies Tyr77 of switch II with adenosine monophophosrylation, thus rendering it constitutively active by virtue of the inability of GAPs to bind Rab1 and hydrolyze GTP.[160] Another Legionella protein, AnkX, covalently modifies an adjacent serine residue (Ser79) of Rab1 with phosphocholine.[157] This reversible modification modulates Rab1 interactions with various cellular and bacterial proteins,[189] presumably playing into the hands of the growing bacterium. Bacterial subversion of Rab activity by covalent modification represents a new frontier in membrane trafficking, and may provide a valuable source of novel therapeutics.

6.3. Rabs associated with cancer and neurodegeneration

Rab GTPases are associated with the progression of several human diseases. Expression of endosomal protein Rab25 is strongly associated with aggressiveness of ovarian, breast, and prostate cancer cells,[143,190,191] as well as mammary cell lines.[191] Rab25 was shown to drive metastasis via its control of the recycling of adhesion molecules (integrins) from endosomes back to the pseudopodial tip of migrating cells.[144] Conflicting data regarding the function of Rab25 as a tumor suppressor has also been published,[145–147,192] and a model for Rab25 effects as a function of the expression of the FIP1 (RCP) has been proposed to resolve the discrepancy.[193,194] Rab31 has also recently been linked to progression of breast cancers.[195] Interestingly, hypoxia inducible factor, which is linked to cell survival and tumor progression, significantly downregulates Rabaptin-5.[148] Reduction of Rabaptin-5 transcripts apparently results in slower endocytic recycling pathways and delayed deactivation of receptors such as EGF receptor, thus linking oxygen metabolism to endosomal traffic and tumor progression.[148]

Membrane trafficking pathways have also recently become associated with the progression of neurodegenerative diseases. Misfolding of a small protein of unknown function—α-synuclein—is associated with several disorders including Parkinson's disease.[196] In a yeast model of the disease, α-synuclein inclusions block ER-to-Golgi transport, but Rab1 overexpression rescues the cytotoxic effects of α-synuclein and restores the trafficking pathway.[137] Additional Rab GTPases including Rab3 and Rab8 also contribute to reducing disease progression in neurons.[138] High-throughput screening of small compounds was successful in identifying molecules that, among other things, reversed the ER-to-Golgi trafficking defect in a yeast model of Parkinson's disease.[197] These overall studies demonstrate that, in principle, Rab trafficking pathways are feasible targets of novel therapeutics.

6.4. Rab/effector oligomers and drug design

The critical role of small GTPases and their regulatory pathways in health and disease makes them obvious targets for novel therapeutics. Ras signaling pathways have long been a target of anticancer drugs since the observation that peptide inhibitors of farnesyltransferase (CAAX peptidomimetics) prevented Ras targeting to membranes,[198] and the early success of inhibitors of transformed cells and blocking of mouse tumors.[199–201] However, these successful cellular and animal models have not been translated to human clinical trials for various reasons, including side effects and incomplete Ras inactivation.[202,203]

Despite these challenges in development of therapeutics to target small GTPases, there is growing potential for targeting of Ras and Rab signaling pathways for treatment of communicable and noncommunicable human diseases. Novel targets and pathways have been identified through an understanding of the molecular basis for signaling networks, and the role of GEFs, GAPs, and other accessory proteins and enzymes involved in regulation of Rabs. Targeting specific Rab/effector oligomeric complexes would be a sophisticated means of developing novel and specific drugs. Methodology for targeting protein–protein interfaces with small molecules is evolving,[204] and combined with new high-throughput interaction screens,[205] provides avenues for future drug discovery. Various endocytic pathways are particularly intriguing as targets, given their pivotal role in viral and bacterial invasion of cells, GPCR signaling, and immunity.[206] Furthermore, covalent modification of Rab1 by *L. pneumophila* to subvert membrane trafficking opens a new frontier in the development of novel therapeutics.[157,160] Bacterial enzymes such as DrrA and AnkX are an obvious target for structure-guided drug design. However, the idea that Rab interfaces can be covalently modified, thus affecting their oligomerization with binding partners, suggests that novel or existing enzymes could be developed and/or exploited to modulate membrane trafficking pathways controlled by other Rab GTPases.

7. CONCLUSIONS

The structures of numerous Rabs in complex with their cognate RBDs have now been determined, thus providing insight into the molecular basis for specificity. However, the field of Rab/effector trafficking is hampered by the lack of a full-length structure of an effector. Effectors are generally large, multidomain proteins that may contain hinge regions and significant interdomain flexibility to enable various steps in vesicle formation, transport, tethering, and fusion. Several effectors also contain a regulatory domain (GEF or GAP), thus adding further complexity to the regulation of effector recruitment. For example, Sec2 contains extended regions of coiled coil that function as a GEF for Sec4 and an effector for Ypt32 to mediate exocytosis in yeast.[207–210] Similarly, the Rab5 effector Rabaptin-5 also binds to Rabex-5 (a GEF for Rab5).[211] However, apart from RBDs, the structures and functions of the various regions in multidomain effectors remain largely obscure. Conceptual advances in the precise role of effectors await a full structural description of Rab/effector and higher-order oligomeric complexes to understand the molecular basis for biological activity.

REFERENCES

1. Takai Y, Sasaki T, Matozaki T. Small GTP-binding proteins. *Physiol Rev* 2001;**81**: 153–208.
2. Bos JL. ras oncogenes in human cancer: a review. *Cancer Res* 1989;**49**:4682–9.
3. Kloog Y, Cox AD. RAS inhibitors: potential for cancer therapeutics. *Mol Med Today* 2000;**6**:398–402.
4. Kalab P, Pu RT, Dasso M. The ran GTPase regulates mitotic spindle assembly. *Curr Biol* 1999;**9**:481–4.
5. Nie Z, Hirsch DS, Randazzo PA. Arf and its many interactors. *Curr Opin Cell Biol* 2003;**15**:396–404.
6. Spiering D, Hodgson L. Dynamics of the Rho-family small GTPases in actin regulation and motility. *Cell Adh Migr* 2011;**5**:170–80.
7. Kikuchi A, Williams LT. The post-translational modification of ras p21 is important for Raf-1 activation. *J Biol Chem* 1994;**269**:20054–9.
8. Hancock JF, Magee AI, Childs JE, Marshall CJ. All ras proteins are polyisoprenylated but only some are palmitoylated. *Cell* 1989;**57**:1167–77.
9. McCubrey JA, Steelman LS, Chappell WH, Abrams SL, Wong EW, Chang F, et al. Roles of the Raf/MEK/ERK pathway in cell growth, malignant transformation and drug resistance. *Biochim Biophys Acta* 2007;**1773**:1263–84.
10. Walker JE, Saraste M, Runswick MJ, Gay NJ. Distantly related sequences in the alpha- and beta-subunits of ATP synthase, myosin, kinases and other ATP-requiring enzymes and a common nucleotide binding fold. *EMBO J* 1982;**1**:945–51.
11. Vetter IR, Wittinghofer A. The guanine nucleotide-binding switch in three dimensions. *Science* 2001;**294**:1299–304.
12. Barr F, Lambright DG. Rab GEFs and GAPs. *Curr Opin Cell Biol* 2010;**22**:461–70.
13. Pereira-Leal JB, Seabra MC. Evolution of the Rab family of small GTP-binding proteins. *J Mol Biol* 2001;**313**:889–901.
14. Pereira-Leal JB. The Ypt/Rab family and the evolution of trafficking in fungi. *Traffic* 2008;**9**:27–38.

15. Andres DA, Seabra MC, Brown MS, Armstrong SA, Smeland TE, Cremers FP, et al. cDNA cloning of component A of Rab geranylgeranyl transferase and demonstration of its role as a Rab escort protein. *Cell* 1993;**73**:1091–9.
16. Leung KF, Baron R, Seabra MC. Thematic review series: lipid posttranslational modifications. Geranylgeranylation of Rab GTPases. *J Lipid Res* 2006;**47**:467–75.
17. Zerial M, McBride H. Rab proteins as membrane organizers. *Nat Rev Mol Cell Biol* 2001;**2**:107–17.
18. Pfeffer S. A model for Rab GTPase localization. *Biochem Soc Trans* 2005;**33**:627–30.
19. Stroupe C, Brunger AT. Crystal structures of a Rab protein in its inactive and active conformations. *J Mol Biol* 2000;**304**:585–98.
20. Lee MT, Mishra A, Lambright DG. Structural mechanisms for regulation of membrane traffic by rab GTPases. *Traffic* 2009;**10**:1377–89.
21. Itzen A, Goody RS. GTPases involved in vesicular trafficking: structures and mechanisms. *Semin Cell Dev Biol* 2011;**22**:48–56.
22. Ahmadian MR, Zor T, Vogt D, Kabsch W, Selinger Z, Wittinghofer A, et al. Guanosine triphosphatase stimulation of oncogenic Ras mutants. *Proc Natl Acad Sci USA* 1999;**96**:7065–70.
23. Der CJ, Finkel T, Cooper GM. Biological and biochemical properties of human rasH genes mutated at codon 61. *Cell* 1986;**44**:167–76.
24. Eathiraj S, Pan X, Ritacco C, Lambright DG. Structural basis of family-wide Rab GTPase recognition by rabenosyn-5. *Nature* 2005;**436**:415–9.
25. Fukuda M. Distinct Rab27 binding affinities of Slp2-a and Slac2-a/melanophilin: hierarchy of Rab27A effectors. *Biochem Biophys Res Commun* 2006;**343**:666–74.
26. Junutula JR, Schonteich E, Wilson GM, Peden AA, Scheller RH, Prekeris R. Molecular characterization of Rab11 interactions with members of the family of Rab11-interacting proteins. *J Biol Chem* 2004;**279**:33430–7.
27. Bergbrede T, Chuky N, Schoebel S, Blankenfeldt W, Geyer M, Fuchs E, et al. Biophysical analysis of the interaction of Rab6a GTPase with its effector domains. *J Biol Chem* 2009;**284**:2628–35.
28. Schoebel S, Cichy AL, Goody RS, Itzen A. Protein LidA from Legionella is a Rab GTPase supereffector. *Proc Natl Acad Sci USA* 2011;**108**:17945–50.
29. Conover GM, Derre I, Vogel JP, Isberg RR. The Legionella pneumophila LidA protein: a translocated substrate of the Dot/Icm system associated with maintenance of bacterial integrity. *Mol Microbiol* 2003;**48**:305–21.
30. Derre I, Isberg RR. LidA, a translocated substrate of the Legionella pneumophila type IV secretion system, interferes with the early secretory pathway. *Infect Immun* 2005;**73**:4370–80.
31. Machner MP, Isberg RR. Targeting of host Rab GTPase function by the intravacuolar pathogen Legionella pneumophila. *Dev Cell* 2006;**11**:47–56.
32. Fukuda M. Rab27 and its effectors in secretory granule exocytosis: a novel docking machinery composed of a Rab27.effector complex. *Biochem Soc Trans* 2006;**34**:691–5.
33. Fukuda M. Synaptotagmin-like protein (Slp) homology domain 1 of Slac2-a/melanophilin is a critical determinant of GTP-dependent specific binding to Rab27A. *J Biol Chem* 2002;**277**:40118–24.
34. Wang Y, Sugita S, Sudhof TC. The RIM/NIM family of neuronal C2 domain proteins. Interactions with Rab3 and a new class of Src homology 3 domain proteins. *J Biol Chem* 2000;**275**:20033–44.
35. Bustos MA, Lucchesi O, Ruete MC, Mayorga LS, Tomes CN. Rab27 and Rab3 sequentially regulate human sperm dense-core granule exocytosis. *Proc Natl Acad Sci USA* 2012;**109**:E2057–E2066.
36. Fukuda M. Slp4-a/granuphilin-a inhibits dense-core vesicle exocytosis through interaction with the GDP-bound form of Rab27A in PC12 cells. *J Biol Chem* 2003;**278**:15390–6.

37. Ostermeier C, Brunger AT. Structural basis of Rab effector specificity: crystal structure of the small G protein Rab3A complexed with the effector domain of rabphilin-3A. *Cell* 1999;**96**:363–74.
38. Merithew E, Hatherly S, Dumas JJ, Lawe DC, Heller-Harrison R, Lambright DG. Structural plasticity of an invariant hydrophobic triad in the switch regions of Rab GTPases is a determinant of effector recognition. *J Biol Chem* 2001;**276**:13982–8.
39. Fukuda M, Kanno E, Yamamoto A. Rabphilin and Noc2 are recruited to dense-core vesicles through specific interaction with Rab27A in PC12 cells. *J Biol Chem* 2004;**279**:13065–75.
40. Chavas LM, Ihara K, Kawasaki M, Torii S, Uejima T, Kato R, et al. Elucidation of Rab27 recruitment by its effectors: structure of Rab27a bound to Exophilin4/Slp2-a. *Structure* 2008;**16**:1468–77.
41. Kukimoto-Niino M, Sakamoto A, Kanno E, Hanawa-Suetsugu K, Terada T, Shirouzu M, et al. Structural basis for the exclusive specificity of Slac2-a/melanophilin for the Rab27 GTPases. *Structure* 2008;**16**:1478–90.
42. Fukuda M. Distinct Rab binding specificity of Rim1, Rim2, rabphilin, and Noc2. Identification of a critical determinant of Rab3A/Rab27A recognition by Rim2. *J Biol Chem* 2003;**278**:15373–80.
43. Shirakawa R, Higashi T, Tabuchi A, Yoshioka A, Nishioka H, Fukuda M, et al. Munc13-4 is a GTP-Rab27-binding protein regulating dense core granule secretion in platelets. *J Biol Chem* 2004;**279**:10730–7.
44. Yi Z, Yokota H, Torii S, Aoki T, Hosaka M, Zhao S, et al. The Rab27a/granuphilin complex regulates the exocytosis of insulin-containing dense-core granules. *Mol Cell Biol* 2002;**22**:1858–67.
45. Gomi H, Mori K, Itohara S, Izumi T. Rab27b is expressed in a wide range of exocytic cells and involved in the delivery of secretory granules near the plasma membrane. *Mol Biol Cell* 2007;**18**:4377–86.
46. Itzen A, Goody RS. Key determinants of Rab specificity. *Structure* 2008;**16**:1437–9.
47. Nielsen E, Christoforidis S, Uttenweiler-Joseph S, Miaczynska M, Dewitte F, Wilm M, et al. Rabenosyn-5, a novel Rab5 effector, is complexed with hVPS45 and recruited to endosomes through a FYVE finger domain. *J Cell Biol* 2000;**151**:601–12.
48. Navaroli DM, Bellve KD, Standley C, Lifshitz LM, Cardia J, Lambright D, et al. Rabenosyn-5 defines the fate of the transferrin receptor following clathrin-mediated endocytosis. *Proc Natl Acad Sci USA* 2012;**109**:E471–E480.
49. Mottola G, Classen AK, Gonzalez-Gaitan M, Eaton S, Zerial M. A novel function for the Rab5 effector Rabenosyn-5 in planar cell polarity. *Development* 2010;**137**:2353–64.
50. Callaghan J, Nixon S, Bucci C, Toh BH, Stenmark H. Direct interaction of EEA1 with Rab5b. *Eur J Biochem: FEBS* 1999;**265**:361–6.
51. Simonsen A, Lippe R, Christoforidis S, Gaullier JM, Brech A, Callaghan J, et al. EEA1 links PI(3)K function to Rab5 regulation of endosome fusion. *Nature* 1998;**394**:494–8.
52. Christoforidis S, McBride HM, Burgoyne RD, Zerial M. The Rab5 effector EEA1 is a core component of endosome docking. *Nature* 1999;**397**:621–5.
53. Ohya T, Miaczynska M, Coskun U, Lommer B, Runge A, Drechsel D, et al. Reconstitution of Rab- and SNARE-dependent membrane fusion by synthetic endosomes. *Nature* 2009;**459**:1091–7.
54. Callaghan J, Simonsen A, Gaullier JM, Toh BH, Stenmark H. The endosome fusion regulator early-endosomal autoantigen 1 (EEA1) is a dimer. *Biochem J* 1999;**338**(Pt 2):539–43.
55. Shin HW, Hayashi M, Christoforidis S, Lacas-Gervais S, Hoepfner S, Wenk MR, et al. An enzymatic cascade of Rab5 effectors regulates phosphoinositide turnover in the endocytic pathway. *J Cell Biol* 2005;**170**:607–18.
56. Christoforidis S, Miaczynska M, Ashman K, Wilm M, Zhao L, Yip SC, et al. Phosphatidylinositol-3-OH kinases are Rab5 effectors. *Nat Cell Biol* 1999;**1**:249–52.

57. Mishra A, Eathiraj S, Corvera S, Lambright DG. Structural basis for Rab GTPase recognition and endosome tethering by the C2H2 zinc finger of Early Endosomal Autoantigen 1 (EEA1). *Proc Natl Acad Sci USA* 2010;**107**:10866–71.
58. Stenmark H, Vitale G, Ullrich O, Zerial M. Rabaptin-5 is a direct effector of the small GTPase Rab5 in endocytic membrane fusion. *Cell* 1995;**83**:423–32.
59. Gournier H, Stenmark H, Rybin V, Lippe R, Zerial M. Two distinct effectors of the small GTPase Rab5 cooperate in endocytic membrane fusion. *EMBO J* 1998;**17**:1930–40.
60. Vitale G, Rybin V, Christoforidis S, Thornqvist P, McCaffrey M, Stenmark H, et al. Distinct Rab-binding domains mediate the interaction of Rabaptin-5 with GTP-bound Rab4 and Rab5. *EMBO J* 1998;**17**:1941–51.
61. Lippe R, Miaczynska M, Rybin V, Runge A, Zerial M. Functional synergy between Rab5 effector Rabaptin-5 and exchange factor Rabex-5 when physically associated in a complex. *Mol Biol Cell* 2001;**12**:2219–28.
62. Xiao GH, Shoarinejad F, Jin F, Golemis EA, Yeung RS. The tuberous sclerosis 2 gene product, tuberin, functions as a Rab5 GTPase activating protein (GAP) in modulating endocytosis. *J Biol Chem* 1997;**272**:6097–100.
63. Neve RL, Coopersmith R, McPhie DL, Santeufemio C, Pratt KG, Murphy CJ, et al. The neuronal growth-associated protein GAP-43 interacts with rabaptin-5 and participates in endocytosis. *J Neurosci* 1998;**18**:7757–67.
64. Zhu G, Zhai P, Liu J, Terzyan S, Li G, Zhang XC. Structural basis of Rab5–Rabaptin5 interaction in endocytosis. *Nat Struct Mol Biol* 2004;**11**:975–83.
65. Barr FA, Short B. Golgins in the structure and dynamics of the Golgi apparatus. *Curr Opin Cell Biol* 2003;**15**:405–13.
66. Kooy J, Toh BH, Pettitt JM, Erlich R, Gleeson PA. Human autoantibodies as reagents to conserved Golgi components. Characterization of a peripheral, 230-kDa compartment-specific Golgi protein. *J Biol Chem* 1992;**267**:20255–63.
67. Fritzler MJ, Hamel JC, Ochs RL, Chan EK. Molecular characterization of two human autoantigens: unique cDNAs encoding 95- and 160-kD proteins of a putative family in the Golgi complex. *J Exp Med* 1993;**178**:49–62.
68. Burguete AS, Fenn TD, Brunger AT, Pfeffer SR. Rab and Arl GTPase family members cooperate in the localization of the golgin GCC185. *Cell* 2008;**132**:286–98.
69. Derby MC, Lieu ZZ, Brown D, Stow JL, Goud B, Gleeson PA. The trans-Golgi network golgin, GCC185, is required for endosome-to-Golgi transport and maintenance of Golgi structure. *Traffic* 2007;**8**:758–73.
70. Reddy JV, Burguete AS, Sridevi K, Ganley IG, Nottingham RM, Pfeffer SR. A functional role for the GCC185 golgin in mannose 6-phosphate receptor recycling. *Mol Biol Cell* 2006;**17**:4353–63.
71. Houghton FJ, Chew PL, Lodeho S, Goud B, Gleeson PA. The localization of the Golgin GCC185 is independent of Rab6A/A' and Arl1. *Cell* 2009;**138**:787–94.
72. Lin YC, Chiang TC, Liu YT, Tsai YT, Jang LT, Lee FJ. ARL4A acts with GCC185 to modulate Golgi complex organization. *J Cell Sci* 2011;**124**:4014–26.
73. Recacha R, Boulet A, Jollivet F, Monier S, Houdusse A, Goud B, et al. Structural basis for recruitment of Rab6-interacting protein 1 to Golgi via a RUN domain. *Structure* 2009;**17**:21–30.
74. Callebaut I, de Gunzburg J, Goud B, Mornon JP. RUN domains: a new family of domains involved in Ras-like GTPase signaling. *Trends Biochem Sci* 2001;**26**:79–83.
75. Fukuda M, Kobayashi H, Ishibashi K, Ohbayashi N. Genome-wide investigation of the Rab binding activity of RUN domains: development of a novel tool that specifically traps GTP-Rab35. *Cell Struct Funct* 2011;**36**:155–70.
76. Kukimoto-Niino M, Takagi T, Akasaka R, Murayama K, Uchikubo-Kamo T, Terada T, et al. Crystal structure of the RUN domain of the RAP2-interacting protein x. *J Biol Chem* 2006;**281**:31843–53.

77. Bateman A, Sandford R. The PLAT domain: a new piece in the PKD1 puzzle. *Curr Biol* 1999;**9**:R588–R590.
78. Aleem AM, Jankun J, Dignam JD, Walther M, Kuhn H, Svergun DI, et al. Human platelet 12-lipoxygenase, new findings about its activity, membrane binding and low-resolution structure. *J Mol Biol* 2008;**376**:193–209.
79. Wassmer T, Attar N, Harterink M, van Weering JR, Traer CJ, Oakley J, et al. The retromer coat complex coordinates endosomal sorting and dynein-mediated transport, with carrier recognition by the trans-Golgi network. *Dev Cell* 2009;**17**:110–22.
80. Fernandes H, Franklin E, Jollivet F, Bliedtner K, Khan AR. Mapping the interactions between a RUN domain from DENND5/Rab6IP1 and sorting nexin 1. *PLoS One* 2012;**7**:e35637.
81. Bujny MV, Popoff V, Johannes L, Cullen PJ. The retromer component sorting nexin-1 is required for efficient retrograde transport of Shiga toxin from early endosome to the trans Golgi network. *J Cell Sci* 2007;**120**:2010–21.
82. Bonifacino JS, Hurley JH. Retromer. *Curr Opin Cell Biol* 2008;**20**:427–36.
83. Marat AL, Dokainish H, McPherson PS. DENN domain proteins: regulators of Rab GTPases. *J Biol Chem* 2011;**286**:13791–800.
84. Allaire PD, Marat AL, Dall'Armi C, Di Paolo G, McPherson PS, Ritter B. The Connecdenn DENN domain: a GEF for Rab35 mediating cargo-specific exit from early endosomes. *Mol Cell* 2010;**37**:370–82.
85. Wu X, Bradley MJ, Cai Y, Kummel D, De La Cruz EM, Barr FA, et al. Insights regarding guanine nucleotide exchange from the structure of a DENN-domain protein complexed with its Rab GTPase substrate. *Proc Natl Acad Sci USA* 2011;**108**:18672–7.
86. Wu M, Wang T, Loh E, Hong W, Song H. Structural basis for recruitment of RILP by small GTPase Rab7. *EMBO J* 2005;**24**:1491–501.
87. Cantalupo G, Alifano P, Roberti V, Bruni CB, Bucci C. Rab-interacting lysosomal protein (RILP): the Rab7 effector required for transport to lysosomes. *EMBO J* 2001;**20**:683–93.
88. Harrison RE, Bucci C, Vieira OV, Schroer TA, Grinstein S. Phagosomes fuse with late endosomes and/or lysosomes by extension of membrane protrusions along microtubules: role of Rab7 and RILP. *Mol Cell Biol* 2003;**23**:6494–506.
89. Matsui T, Ohbayashi N, Fukuda M. The Rab interacting lysosomal protein (RILP) homology domain functions as a novel effector domain for small GTPase Rab36: Rab36 regulates retrograde melanosome transport in melanocytes. *J Biol Chem* 2012;**287**:28619–31.
90. Hou X, Hagemann N, Schoebel S, Blankenfeldt W, Goody RS, Erdmann KS, et al. A structural basis for Lowe syndrome caused by mutations in the Rab-binding domain of OCRL1. *EMBO J* 2011;**30**:1659–70.
91. Lowe M. Structure and function of the Lowe syndrome protein OCRL1. *Traffic* 2005;**6**:711–9.
92. Erdmann KS, Mao Y, McCrea HJ, Zoncu R, Lee S, Paradise S, et al. A role of the Lowe syndrome protein OCRL in early steps of the endocytic pathway. *Dev Cell* 2007;**13**:377–90.
93. Hyvola N, Diao A, McKenzie E, Skippen A, Cockcroft S, Lowe M. Membrane targeting and activation of the Lowe syndrome protein OCRL1 by rab GTPases. *EMBO J* 2006;**25**:3750–61.
94. Attree O, Olivos IM, Okabe I, Bailey LC, Nelson DL, Lewis RA, et al. The Lowe's oculocerebrorenal syndrome gene encodes a protein highly homologous to inositol polyphosphate-5-phosphatase. *Nature* 1992;**358**:239–42.
95. Choudhury R, Diao A, Zhang F, Eisenberg E, Saint-Pol A, Williams C, et al. Lowe syndrome protein OCRL1 interacts with clathrin and regulates protein trafficking between endosomes and the trans-Golgi network. *Mol Biol Cell* 2005;**16**:3467–79.

96. Noakes CJ, Lee G, Lowe M. The PH domain proteins IPIP27A and B link OCRL1 to receptor recycling in the endocytic pathway. *Mol Biol Cell* 2011;**22**:606–23.
97. Coon BG, Mukherjee D, Hanna CB, Riese 2nd DJ, Lowe M, Aguilar RC. Lowe syndrome patient fibroblasts display Ocrl1-specific cell migration defects that cannot be rescued by the homologous Inpp5b phosphatase. *Hum Mol Genet* 2009;**18**:4478–91.
98. Shiba T, Koga H, Shin HW, Kawasaki M, Kato R, Nakayama K, et al. Structural basis for Rab11-dependent membrane recruitment of a family of Rab11-interacting protein 3 (FIP3)/Arfophilin-1. *Proc Natl Acad Sci USA* 2006;**103**:15416–21.
99. Eathiraj S, Mishra A, Prekeris R, Lambright DG. Structural basis for Rab11-mediated recruitment of FIP3 to recycling endosomes. *J Mol Biol* 2006;**364**:121–35.
100. Jagoe WN, Lindsay AJ, Read RJ, McCoy AJ, McCaffrey MW, Khan AR. Crystal structure of rab11 in complex with rab11 family interacting protein 2. *Structure* 2006;**14**:1273–83.
101. Ren M, Xu G, Zeng J, De Lemos-Chiarandini C, Adesnik M, Sabatini DD. Hydrolysis of GTP on rab11 is required for the direct delivery of transferrin from the pericentriolar recycling compartment to the cell surface but not from sorting endosomes. *Proc Natl Acad Sci USA* 1998;**95**:6187–92.
102. Chen W, Feng Y, Chen D, Wandinger-Ness A. Rab11 is required for trans-golgi network-to-plasma membrane transport and a preferential target for GDP dissociation inhibitor. *Mol Biol Cell* 1998;**9**:3241–57.
103. Ullrich O, Reinsch S, Urbe S, Zerial M, Parton RG. Rab11 regulates recycling through the pericentriolar recycling endosome. *J Cell Biol* 1996;**135**:913–24.
104. Prekeris R, Klumperman J, Scheller RH. A Rab11/Rip11 protein complex regulates apical membrane trafficking via recycling endosomes. *Mol Cell* 2000;**6**:1437–48.
105. Hales CM, Griner R, Hobdy-Henderson KC, Dorn MC, Hardy D, Kumar R, et al. Identification and characterization of a family of Rab11-interacting proteins. *J Biol Chem* 2001;**276**:39067–75.
106. Lindsay AJ, Hendrick AG, Cantalupo G, Senic-Matuglia F, Goud B, Bucci C, et al. Rab coupling protein (RCP), a novel Rab4 and Rab11 effector protein. *J Biol Chem* 2002;**277**:12190–9.
107. Hickson GR, Matheson J, Riggs B, Maier VH, Fielding AB, Prekeris R, et al. Arfophilins are dual Arf/Rab 11 binding proteins that regulate recycling endosome distribution and are related to Drosophila nuclear fallout. *Mol Biol Cell* 2003;**14**:2908–20.
108. Neto H, Collins LL, Gould GW. Vesicle trafficking and membrane remodelling in cytokinesis. *Biochem J* 2011;**437**:13–24.
109. Fielding AB, Schonteich E, Matheson J, Wilson G, Yu X, Hickson GR, et al. Rab11-FIP3 and FIP4 interact with Arf6 and the exocyst to control membrane traffic in cytokinesis. *EMBO J* 2005;**24**:3389–99.
110. Miserey-Lenkei S, Waharte F, Boulet A, Cuif MH, Tenza D, El Marjou A, et al. Rab6-interacting protein 1 links Rab6 and Rab11 function. *Traffic* 2007;**8**:1385–403.
111. Cho W. Membrane targeting by C1 and C2 domains. *J Biol Chem* 2001;**276**:32407–10.
112. Lindsay AJ, McCaffrey MW. The C2 domains of the class I Rab11 family of interacting proteins target recycling vesicles to the plasma membrane. *J Cell Sci* 2004;**117**:4365–75.
113. Muto A, Arai K, Watanabe S. Rab11-FIP4 is predominantly expressed in neural tissues and involved in proliferation as well as in differentiation during zebrafish retinal development. *Dev Biol* 2006;**292**:90–102.
114. Muto A, Aoki Y, Watanabe S. Mouse Rab11-FIP4 regulates proliferation and differentiation of retinal progenitors in a Rab11-independent manner. *Dev Dyn* 2007;**236**:214–25.
115. Shin OH, Couvillon AD, Exton JH. Arfophilin is a common target of both class II and class III ADP-ribosylation factors. *Biochemistry* 2001;**40**:10846–52.

116. Rosing M, Ossendorf E, Rak A, Barnekow A. Giantin interacts with both the small GTPase Rab6 and Rab1. *Exp Cell Res* 2007;**313**:2318–25.
117. Beard M, Satoh A, Shorter J, Warren G. A cryptic Rab1-binding site in the p115 tethering protein. *J Biol Chem* 2005;**280**:25840–8.
118. Allan BB, Moyer BD, Balch WE. Rab1 recruitment of p115 into a cis-SNARE complex: programming budding COPII vesicles for fusion. *Science* 2000;**289**:444–8.
119. Weide T, Teuber J, Bayer M, Barnekow A. MICAL-1 isoforms, novel rab1 interacting proteins. *Biochem Biophys Res Commun* 2003;**306**:79–86.
120. Weide T, Bayer M, Koster M, Siebrasse JP, Peters R, Barnekow A. The Golgi matrix protein GM130: a specific interacting partner of the small GTPase rab1b. *EMBO Rep* 2001;**2**:336–41.
121. Tsuboi T. Molecular mechanism of attachment process of dense-core vesicles to the plasma membrane in neuroendocrine cells. *Neurosci Res* 2009;**63**:83–8.
122. Hume AN, Tarafder AK, Ramalho JS, Sviderskaya EV, Seabra MC. A coiled-coil domain of melanophilin is essential for Myosin Va recruitment and melanosome transport in melanocytes. *Mol Biol Cell* 2006;**17**:4720–35.
123. Brown FC, Schindelhaim CH, Pfeffer SR. GCC185 plays independent roles in Golgi structure maintenance and AP-1-mediated vesicle tethering. *J Cell Biol* 2011;**194**:779–87.
124. Wang T, Wong KK, Hong W. A unique region of RILP distinguishes it from its related proteins in its regulation of lysosomal morphology and interaction with Rab7 and Rab34. *Mol Biol Cell* 2004;**15**:815–26.
125. Gidon A, Bardin S, Cinquin B, Boulanger J, Waharte F, Heliot L, et al. A Rab11A/MyosinVb/Rab11-FIP2 complex frames two late recycling steps of langerin from the ERC to the plasma membrane. *Traffic* 2012;**13**(6):815–33.
126. Wilson GM, Fielding AB, Simon GC, Yu X, Andrews PD, Hames RS, et al. The FIP3-Rab11 protein complex regulates recycling endosome targeting to the cleavage furrow during late cytokinesis. *Mol Biol Cell* 2005;**16**:849–60.
127. Takamori S, Holt M, Stenius K, Lemke EA, Gronborg M, Riedel D, et al. Molecular anatomy of a trafficking organelle. *Cell* 2006;**127**:831–46.
128. Pavlos NJ, Jahn R. Distinct yet overlapping roles of Rab GTPases on synaptic vesicles. *Small GTPases* 2011;**2**:77–81.
129. Nassar N, Horn G, Herrmann C, Scherer A, McCormick F, Wittinghofer A. The 2.2 A crystal structure of the Ras-binding domain of the serine/threonine kinase c-Raf1 in complex with Rap1A and a GTP analogue. *Nature* 1995;**375**:554–60.
130. Pacold ME, Suire S, Perisic O, Lara-Gonzalez S, Davis CT, Walker EH, et al. Crystal structure and functional analysis of Ras binding to its effector phosphoinositide 3-kinase gamma. *Cell* 2000;**103**:931–43.
131. Geyer M, Herrmann C, Wohlgemuth S, Wittinghofer A, Kalbitzer HR. Structure of the Ras-binding domain of RalGEF and implications for Ras binding and signalling. *Nat Struct Biol* 1997;**4**:694–9.
132. Vavvas D, Li X, Avruch J, Zhang XF. Identification of Nore1 as a potential Ras effector. *J Biol Chem* 1998;**273**:5439–42.
133. Wohlgemuth S, Kiel C, Kramer A, Serrano L, Wittinghofer F, Herrmann C. Recognizing and defining true Ras binding domains I: biochemical analysis. *J Mol Biol* 2005;**348**:741–58.
134. Kiel C, Selzer T, Shaul Y, Schreiber G, Herrmann C. Electrostatically optimized Ras-binding Ral guanine dissociation stimulator mutants increase the rate of association by stabilizing the encounter complex. *Proc Natl Acad Sci USA* 2004;**101**:9223–8.
135. Sydor JR, Engelhard M, Wittinghofer A, Goody RS, Herrmann C. Transient kinetic studies on the interaction of Ras and the Ras-binding domain of c-Raf-1 reveal rapid equilibration of the complex. *Biochemistry* 1998;**37**:14292–9.

136. Kiel C, Aydin D, Serrano L. Association rate constants of ras-effector interactions are evolutionarily conserved. *PLoS Comput Biol* 2008;**4**:e1000245.
137. Cooper AA, Gitler AD, Cashikar A, Haynes CM, Hill KJ, Bhullar B, et al. Alpha-synuclein blocks ER-Golgi traffic and Rab1 rescues neuron loss in Parkinson's models. *Science* 2006;**313**:324–8.
138. Gitler AD, Bevis BJ, Shorter J, Strathearn KE, Hamamichi S, Su LJ, et al. The Parkinson's disease protein alpha-synuclein disrupts cellular Rab homeostasis. *Proc Natl Acad Sci USA* 2008;**105**:145–50.
139. Fukuda M, Kanno E, Ishibashi K, Itoh T. Large scale screening for novel rab effectors reveals unexpected broad Rab binding specificity. *Mol Cell Proteomics* 2008;**7**:1031–42.
140. Menasche G, Pastural E, Feldmann J, Certain S, Ersoy F, Dupuis S, et al. Mutations in RAB27A cause Griscelli syndrome associated with haemophagocytic syndrome. *Nat Genet* 2000;**25**:173–6.
141. Menasche G, Ho CH, Sanal O, Feldmann J, Tezcan I, Ersoy F, et al. Griscelli syndrome restricted to hypopigmentation results from a melanophilin defect (GS3) or a MYO5A F-exon deletion (GS1). *J Clin Invest* 2003;**112**:450–6.
142. Menasche G, Feldmann J, Houdusse A, Desaymard C, Fischer A, Goud B, et al. Biochemical and functional characterization of Rab27a mutations occurring in Griscelli syndrome patients. *Blood* 2003;**101**:2736–42.
143. Cheng KW, Lahad JP, Kuo WL, Lapuk A, Yamada K, Auersperg N, et al. The RAB25 small GTPase determines aggressiveness of ovarian and breast cancers. *Nat Med* 2004;**10**:1251–6.
144. Caswell PT, Spence HJ, Parsons M, White DP, Clark K, Cheng KW, et al. Rab25 associates with alpha5beta1 integrin to promote invasive migration in 3D microenvironments. *Dev Cell* 2007;**13**:496–510.
145. Cheng JM, Volk L, Janaki DK, Vyakaranam S, Ran S, Rao KA. Tumor suppressor function of Rab25 in triple-negative breast cancer. *Int J Cancer* 2010;**126**:2799–812.
146. Goldenring JR, Nam KT. Rab25 as a tumour suppressor in colon carcinogenesis. *Br J Cancer* 2011;**104**:33–6.
147. Nam KT, Lee HJ, Smith JJ, Lapierre LA, Kamath VP, Chen X, et al. Loss of Rab25 promotes the development of intestinal neoplasia in mice and is associated with human colorectal adenocarcinomas. *J Clin Invest* 2010;**120**:840–9.
148. Wang Y, Roche O, Yan MS, Finak G, Evans AJ, Metcalf JL, et al. Regulation of endocytosis via the oxygen-sensing pathway. *Nat Med* 2009;**15**:319–24.
149. Agarwal R, Jurisica I, Mills GB, Cheng KW. The emerging role of the RAB25 small GTPase in cancer. *Traffic* 2009;**10**:1561–8.
150. Rzomp KA, Moorhead AR, Scidmore MA. The GTPase Rab4 interacts with Chlamydia trachomatis inclusion membrane protein CT229. *Infect Immun* 2006;**74**:5362–73.
151. Scidmore MA. Recent advances in Chlamydia subversion of host cytoskeletal and membrane trafficking pathways. *Microbes Infect* 2011;**13**:527–35.
152. Heuer D, Rejman Lipinski A, Machuy N, Karlas A, Wehrens A, Siedler F, et al. Chlamydia causes fragmentation of the Golgi compartment to ensure reproduction. *Nature* 2009;**457**:731–5.
153. Rejman Lipinski A, Heymann J, Meissner C, Karlas A, Brinkmann V, Meyer TF, et al. Rab6 and Rab11 regulate Chlamydia trachomatis development and golgin-84-dependent Golgi fragmentation. *PLoS Pathog* 2009;**5**:e1000615.
154. Cortes C, Rzomp KA, Tvinnereim A, Scidmore MA, Wizel B. Chlamydia pneumoniae inclusion membrane protein Cpn0585 interacts with multiple Rab GTPases. *Infect Immun* 2007;**75**:5586–96.
155. Harrison RE, Brumell JH, Khandani A, Bucci C, Scott CC, Jiang X, et al. Salmonella impairs RILP recruitment to Rab7 during maturation of invasion vacuoles. *Mol Biol Cell* 2004;**15**:3146–54.

156. Jackson LK, Nawabi P, Hentea C, Roark EA, Haldar K. The Salmonella virulence protein SifA is a G protein antagonist. *Proc Natl Acad Sci USA* 2008;**105**:14141–6.
157. Mukherjee S, Liu X, Arasaki K, McDonough J, Galan JE, Roy CR. Modulation of Rab GTPase function by a protein phosphocholine transferase. *Nature* 2011;**477**:103–6.
158. Goody PR, Heller K, Oesterlin LK, Muller MP, Itzen A, Goody RS. Reversible phosphocholination of Rab proteins by Legionella pneumophila effector proteins. *The EMBO J* 2012;**31**(7):1774–84.
159. Goody RS, Muller MP, Schoebel S, Oesterlin LK, Blumer J, Peters H, et al. The versatile Legionella effector protein DrrA. *Commun Integr Biol* 2011;**4**:72–4.
160. Muller MP, Peters H, Blumer J, Blankenfeldt W, Goody RS, Itzen A. The Legionella effector protein DrrA AMPylates the membrane traffic regulator Rab1b. *Science* 2010;**329**:946–9.
161. Schoebel S, Oesterlin LK, Blankenfeldt W, Goody RS, Itzen A. RabGDI displacement by DrrA from Legionella is a consequence of its guanine nucleotide exchange activity. *Mol Cell* 2009;**36**:1060–72.
162. Muller MP, Shkumatov AV, Oesterlin LK, Schoebel S, Goody PR, Goody RS, et al. Characterization of enzymes from Legionella pneumophila involved in reversible adenylylation of Rab1 protein. *J Biol Chem* 2012;**287**:35036–46.
163. Neunuebel MR, Chen Y, Gaspar AH, Backlund Jr PS, Yergey A, Machner MP. De-AMPylation of the small GTPase Rab1 by the pathogen Legionella pneumophila. *Science* 2011;**333**:453–6.
164. Seabra MC, Mules EH, Hume AN. Rab GTPases, intracellular traffic and disease. *Trends Mol Med* 2002;**8**:23–30.
165. Seabra MC. New insights into the pathogenesis of choroideremia: a tale of two REPs. *Ophthalmic Genet* 1996;**17**:43–6.
166. Larijani B, Hume AN, Tarafder AK, Seabra MC. Multiple factors contribute to inefficient prenylation of Rab27a in Rab prenylation diseases. *J Biol Chem* 2003;**278**: 46798–804.
167. Hutagalung AH, Novick PJ. Role of Rab GTPases in membrane traffic and cell physiology. *Physiol Rev* 2011;**91**:119–49.
168. Griscelli C, Prunieras M. Pigment dilution and immunodeficiency: a new syndrome. *Int J Dermatol* 1978;**17**:788–91.
169. Klein C, Philippe N, Le Deist F, Fraitag S, Prost C, Durandy A, et al. Partial albinism with immunodeficiency (Griscelli syndrome). *J Pediatr* 1994;**125**:886–95.
170. Pastural E, Ersoy F, Yalman N, Wulffraat N, Grillo E, Ozkinay F, et al. Two genes are responsible for Griscelli syndrome at the same 15q21 locus. *Genomics* 2000;**63**:299–306.
171. Matesic LE, Yip R, Reuss AE, Swing DA, O'Sullivan TN, Fletcher CF, et al. Mutations in Mlph, encoding a member of the Rab effector family, cause the melanosome transport defects observed in leaden mice. *Proc Natl Acad Sci USA* 2001;**98**:10238–43.
172. Hume AN, Collinson LM, Hopkins CR, Strom M, Barral DC, Bossi G, et al. The leaden gene product is required with Rab27a to recruit myosin Va to melanosomes in melanocytes. *Traffic* 2002;**3**:193–202.
173. Wu XS, Rao K, Zhang H, Wang F, Sellers JR, Matesic LE, et al. Identification of an organelle receptor for myosin-Va. *Nat Cell Biol* 2002;**4**:271–8.
174. Stepp SE, Dufourcq-Lagelouse R, Le Deist F, Bhawan S, Certain S, Mathew PA, et al. Perforin gene defects in familial hemophagocytic lymphohistiocytosis. *Science* 1999;**286**:1957–9.
175. Hilbi H, Haas A. Secretive bacterial pathogens and the secretory pathway. *Traffic* 2012;**13**(9):1187–97.
176. Alix E, Mukherjee S, Roy CR. Host–pathogen interactions: subversion of membrane transport pathways by vacuolar pathogens. *J Cell Biol* 2011;**195**:943–52.
177. Fields KA, Hackstadt T. The chlamydial inclusion: escape from the endocytic pathway. *Annu Rev Cell Dev Biol* 2002;**18**:221–45.

178. Luzio JP, Pryor PR, Bright NA. Lysosomes: fusion and function. *Nat Rev Mol Cell Biol* 2007;**8**:622–32.
179. Oh YK, Alpuche-Aranda C, Berthiaume E, Jinks T, Miller SI, Swanson JA. Rapid and complete fusion of macrophage lysosomes with phagosomes containing Salmonella typhimurium. *Infect Immun* 1996;**64**:3877–83.
180. Di Paolo G, De Camilli P. Phosphoinositides in cell regulation and membrane dynamics. *Nature* 2006;**443**:651–7.
181. Dehoux P, Flores R, Dauga C, Zhong G, Subtil A. Multi-genome identification and characterization of chlamydiae-specific type III secretion substrates: the Inc proteins. *BMC Genomics* 2011;**12**:109.
182. Schlumberger MC, Hardt WD. Salmonella type III secretion effectors: pulling the host cell's strings. *Curr Opin Microbiol* 2006;**9**:46–54.
183. Ramsden AE, Holden DW, Mota LJ. Membrane dynamics and spatial distribution of Salmonella-containing vacuoles. *Trends Microbiol* 2007;**15**:516–24.
184. Hackstadt T, Scidmore MA, Rockey DD. Lipid metabolism in Chlamydia trachomatis-infected cells: directed trafficking of Golgi-derived sphingolipids to the chlamydial inclusion. *Proc Natl Acad Sci USA* 1995;**92**:4877–81.
185. Hackstadt T, Rockey DD, Heinzen RA, Scidmore MA. Chlamydia trachomatis interrupts an exocytic pathway to acquire endogenously synthesized sphingomyelin in transit from the Golgi apparatus to the plasma membrane. *EMBO J* 1996;**15**:964–77.
186. Capmany A, Damiani MT. Chlamydia trachomatis intercepts Golgi-derived sphingolipids through a Rab14-mediated transport required for bacterial development and replication. *PLoS One* 2010;**5**:e14084.
187. Itzen A, Blankenfeldt W, Goody RS. Adenylylation: renaissance of a forgotten post-translational modification. *Trends Biochem Sci* 2011;**36**:221–8.
188. Ham H, Orth K. De-AMPylation unmasked: modulation of host membrane trafficking. *Sci Signal* 2011;**4**:pe42.
189. Goody PR, Heller K, Oesterlin LK, Muller MP, Itzen A, Goody RS. Reversible phosphocholination of Rab proteins by Legionella pneumophila effector proteins. *EMBO J* 2012;**31**:1774–84.
190. Calvo A, Xiao N, Kang J, Best CJ, Leiva I, Emmert-Buck MR, et al. Alterations in gene expression profiles during prostate cancer progression: functional correlations to tumorigenicity and down-regulation of selenoprotein-P in mouse and human tumors. *Cancer Res* 2002;**62**:5325–35.
191. Wang W, Goswami S, Lapidus K, Wells AL, Wyckoff JB, Sahai E, et al. Identification and testing of a gene expression signature of invasive carcinoma cells within primary mammary tumors. *Cancer Res* 2004;**64**:8585–94.
192. Cheng JM, Ding M, Aribi A, Shah P, Rao K. Loss of RAB25 expression in breast cancer. *Int J Cancer* 2006;**118**:2957–64.
193. Tang BL. Is Rab25 a tumor promoter or suppressor—context dependency on RCP status? *Tumour Biol* 2010;**31**:359–61.
194. Kessler D, Gruen GC, Heider D, Morgner J, Reis H, Schmid KW, et al. The action of small GTPases Rab11 and Rab25 in vesicle trafficking during cell migration. *Cell Physiol Biochem Int J Exp Cell Physiol, Biochem Pharmacol* 2012;**29**:647–56.
195. Grismayer B, Soelch S, Seubert B, Kirchner T, Schafer S, Baretton G, et al. Rab31 expression levels modulate tumor-relevant characteristics of breast cancer cells. *Mol Cancer* 2012;**11**:62.
196. Lee VM, Trojanowski JQ. Mechanisms of Parkinson's disease linked to pathological alpha-synuclein: new targets for drug discovery. *Neuron* 2006;**52**:33–8.
197. Su LJ, Auluck PK, Outeiro TF, Yeger-Lotem E, Kritzer JA, Tardiff DF, et al. Compounds from an unbiased chemical screen reverse both ER-to-Golgi trafficking defects and mitochondrial dysfunction in Parkinson's disease models. *Dis Model Mech* 2010;**3**:194–208.

198. Reiss Y, Goldstein JL, Seabra MC, Casey PJ, Brown MS. Inhibition of purified p21ras farnesyl:protein transferase by Cys-AAX tetrapeptides. *Cell* 1990;**62**:81–8.
199. Bishop WR, Bond R, Petrin J, Wang L, Patton R, Doll R, et al. Novel tricyclic inhibitors of farnesyl protein transferase. Biochemical characterization and inhibition of Ras modification in transfected Cos cells. *J Biol Chem* 1995;**270**:30611–8.
200. Liu M, Bryant MS, Chen J, Lee S, Yaremko B, Lipari P, et al. Antitumor activity of SCH 66336, an orally bioavailable tricyclic inhibitor of farnesyl protein transferase, in human tumor xenograft models and wap-ras transgenic mice. *Cancer Res* 1998;**58**:4947–56.
201. Sun J, Qian Y, Hamilton AD, Sebti SM. Ras CAAX peptidomimetic FTI 276 selectively blocks tumor growth in nude mice of a human lung carcinoma with K-Ras mutation and p53 deletion. *Cancer Res* 1995;**55**:4243–7.
202. Basso AD, Kirschmeier P, Bishop WR. Lipid posttranslational modifications. Farnesyl transferase inhibitors. *J Lipid Res* 2006;**47**:15–31.
203. Van Cutsem E, van de Velde H, Karasek P, Oettle H, Vervenne WL, Szawlowski A, et al. Phase III trial of gemcitabine plus tipifarnib compared with gemcitabine plus placebo in advanced pancreatic cancer. *J Clin Oncol: Official Journal of the American Society of Clinical Oncology* 2004;**22**:1430–8.
204. Jubb H, Higueruelo AP, Winter A, Blundell TL. Structural biology and drug discovery for protein–protein interactions. *Trends Pharmacol Sci* 2012;**33**:241–8.
205. Gogolin L, Schroeder H, Itzen A, Goody RS, Niemeyer CM, Becker CF. Protein–DNA arrays as tools for detection of protein–protein interactions by mass spectrometry. *ChemBioChem: Eur J Chem Biol* 2013;**14**:92–9.
206. Rajendran L, Knolker HJ, Simons K. Subcellular targeting strategies for drug design and delivery. *Nat Rev Drug Discov* 2010;**9**:29–42.
207. Ortiz D, Medkova M, Walch-Solimena C, Novick P. Ypt32 recruits the Sec4p guanine nucleotide exchange factor, Sec2p, to secretory vesicles; evidence for a Rab cascade in yeast. *J Cell Biol* 2002;**157**:1005–15.
208. Walch-Solimena C, Collins RN, Novick PJ. Sec2p mediates nucleotide exchange on Sec4p and is involved in polarized delivery of post-Golgi vesicles. *J Cell Biol* 1997;**137**:1495–509.
209. Itzen A, Rak A, Goody RS. Sec2 is a highly efficient exchange factor for the Rab protein Sec4. *J Mol Biol* 2007;**365**:1359–67.
210. Dong G, Medkova M, Novick P, Reinisch KM. A catalytic coiled coil: structural insights into the activation of the Rab GTPase Sec4p by Sec2p. *Mol Cell* 2007;**25**:455–62.
211. Horiuchi H, Lippe R, McBride HM, Rubino M, Woodman P, Stenmark H, et al. A novel Rab5 GDP/GTP exchange factor complexed to Rabaptin-5 links nucleotide exchange to effector recruitment and function. *Cell* 1997;**90**:1149–59.

INDEX

Note: Page numbers followed by "*f*" indicate figures, and "*t*" indicate tables.

B

C

K

L

M

N

U

W

X

Y

Z

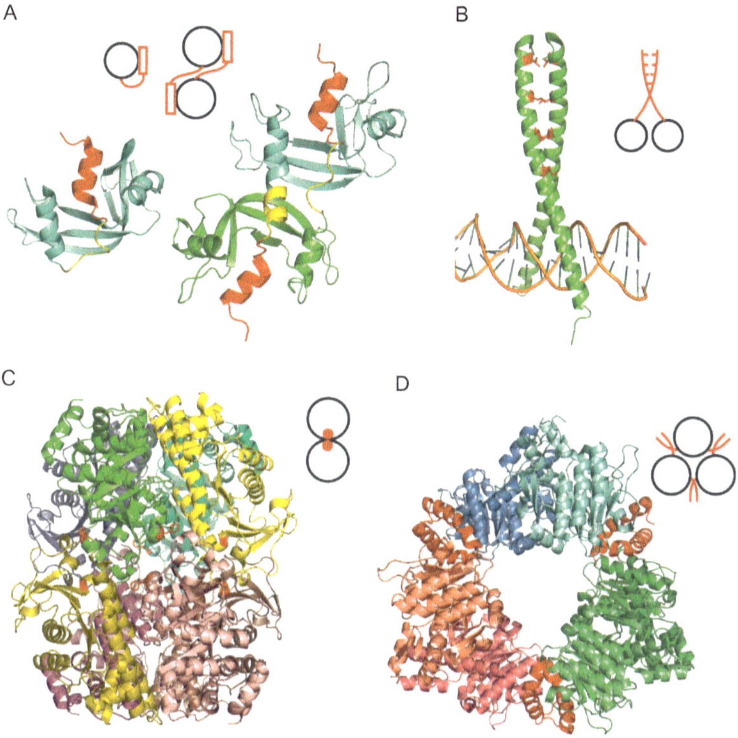

Nishi *et al.*, Figure 1.3

Levy and Teichmann, Figure 2.4

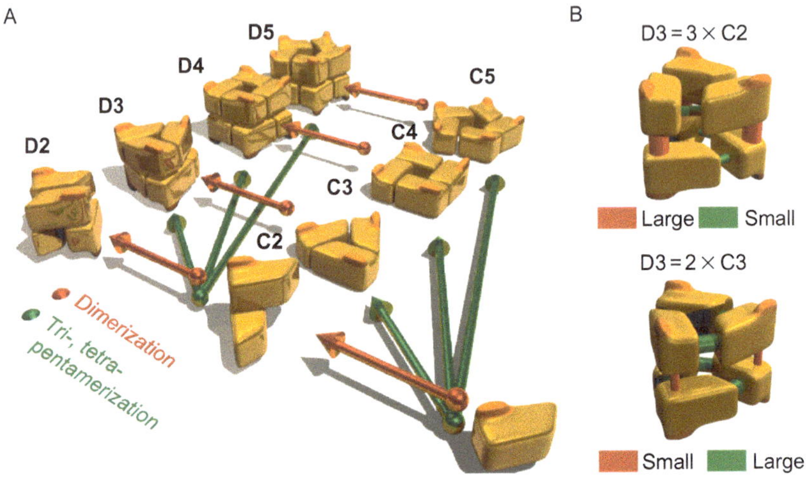

LEVY AND TEICHMANN, FIGURE 2.9

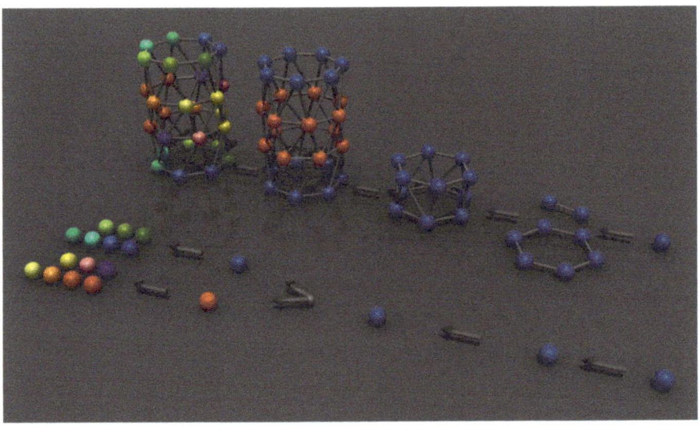

LEVY AND TEICHMANN, FIGURE 2.10

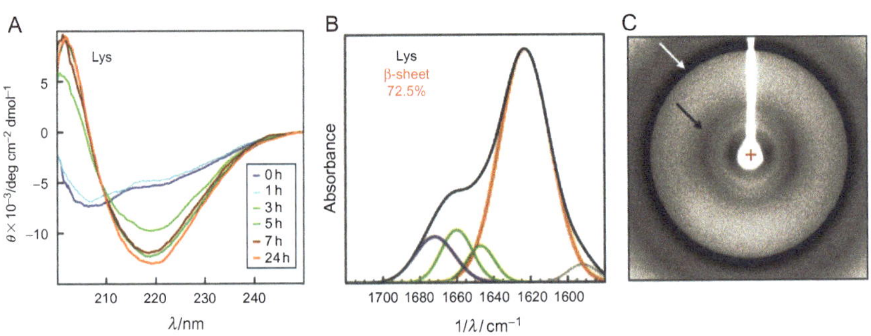

SALVATELLA, FIGURE 4.7

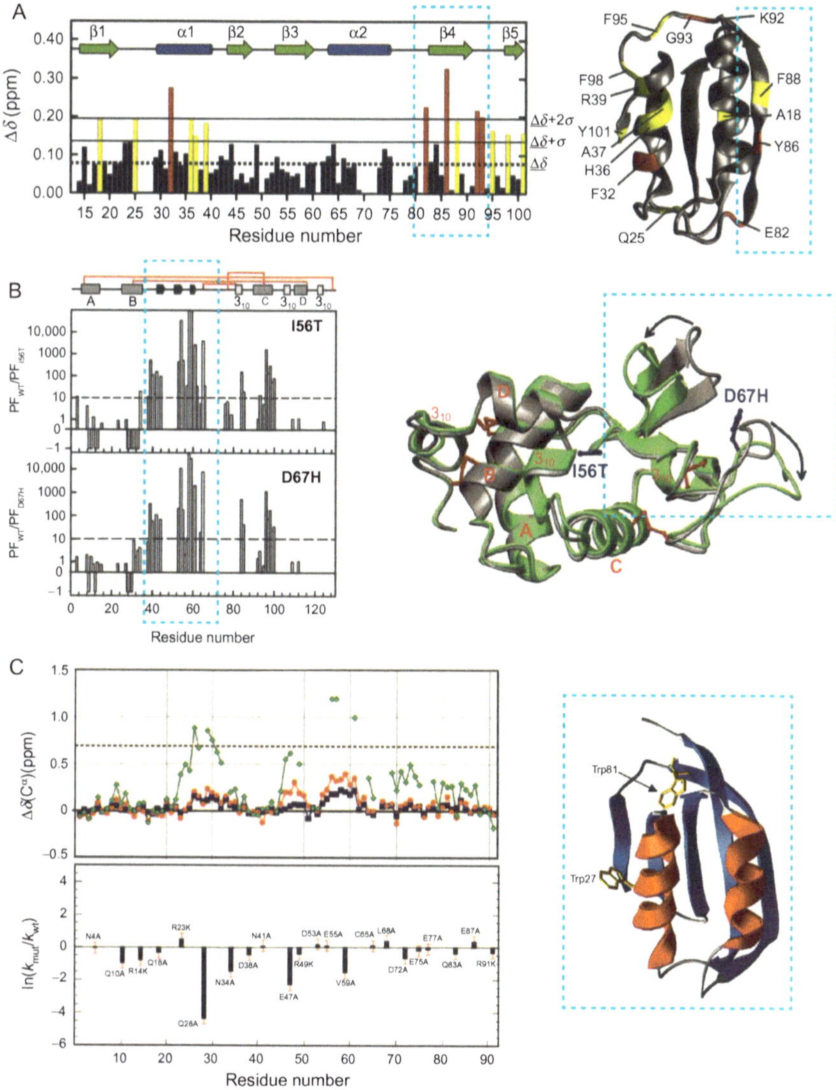

Salvatella, Figure 4.8

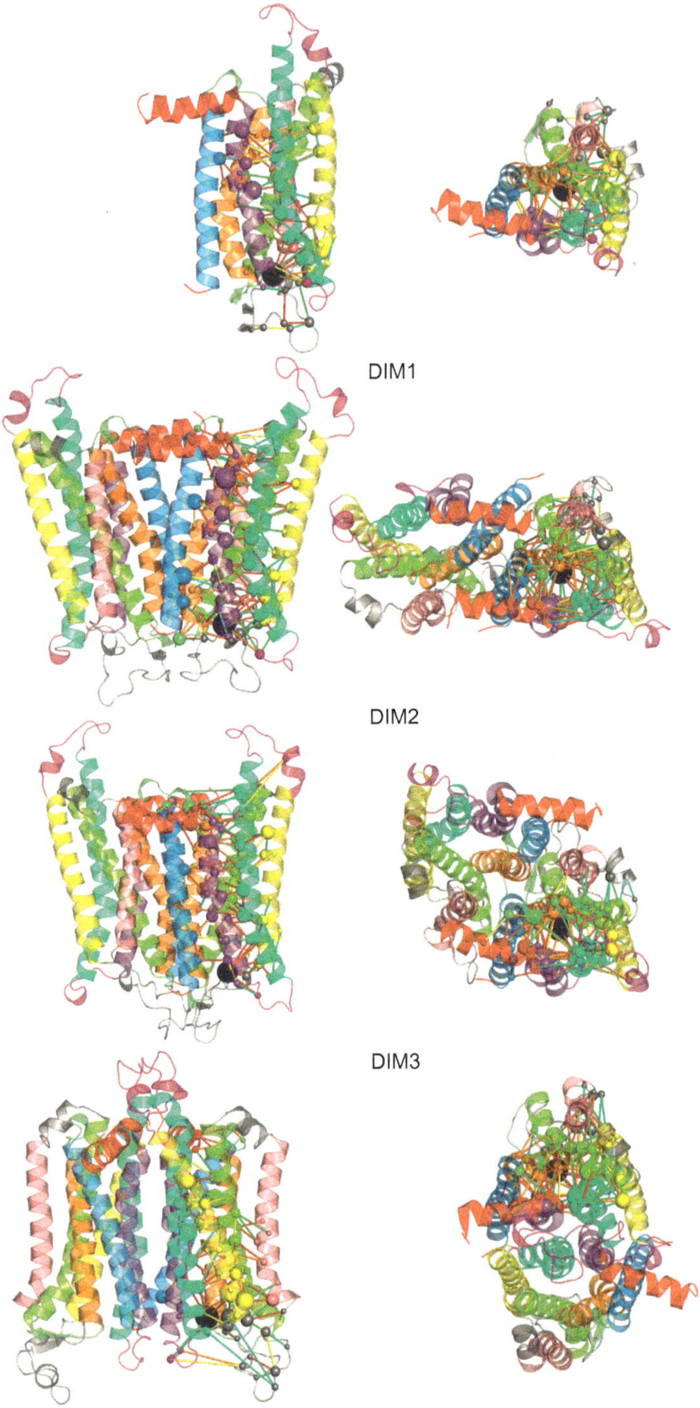

Fanelli *et al.*, Figure 5.2

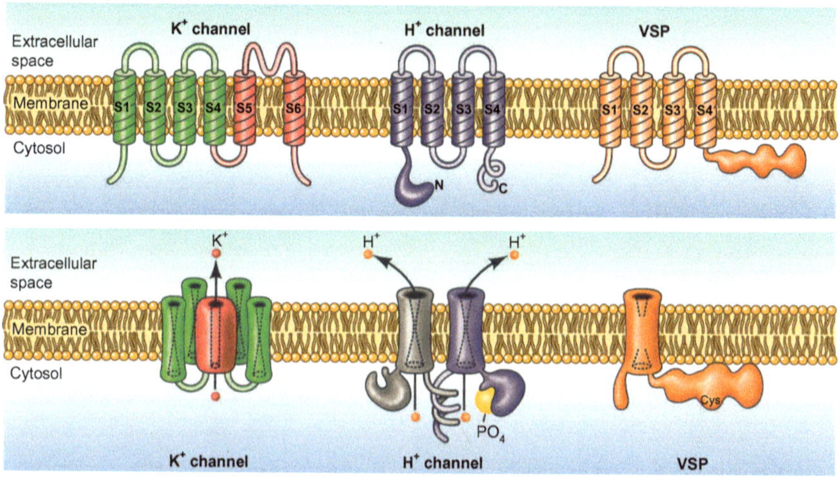

Smith and DeCoursey, Figure 12.1

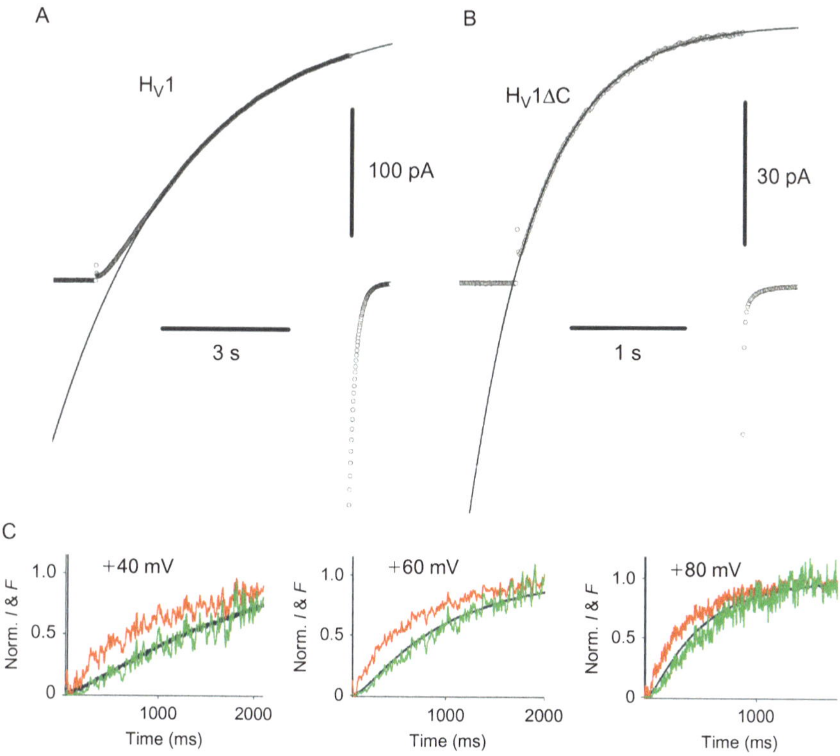

Smith and DeCoursey, Figure 12.3

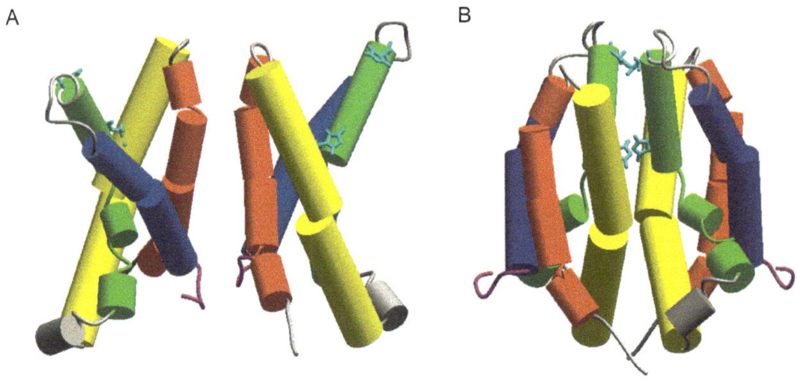

Smith and DeCoursey, Figure 12.4

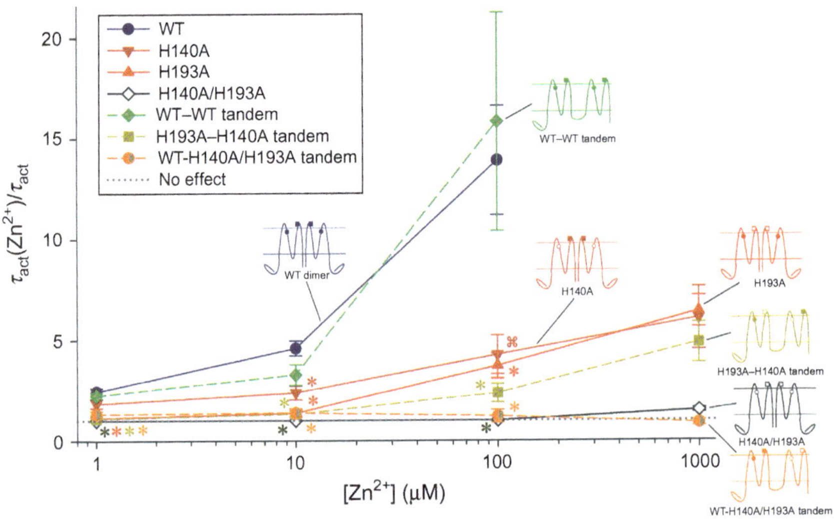

Smith and DeCoursey, Figure 12.5

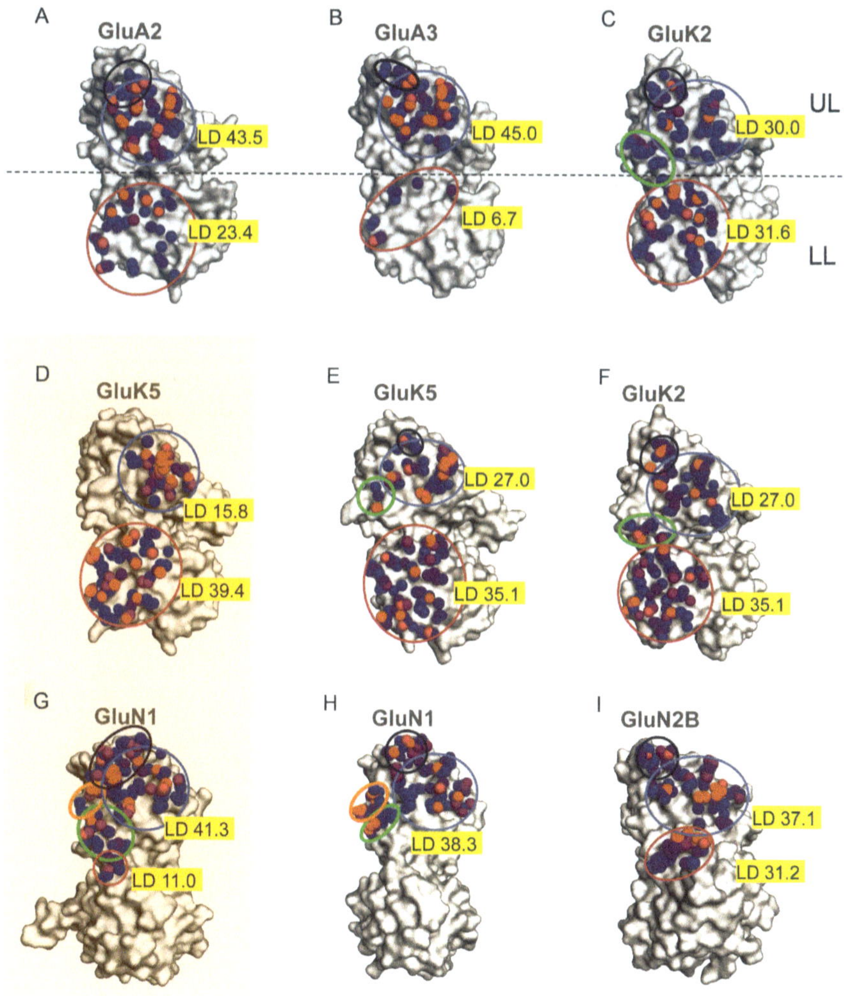

HERGUEDAS *ET AL.*, FIGURE 13.3

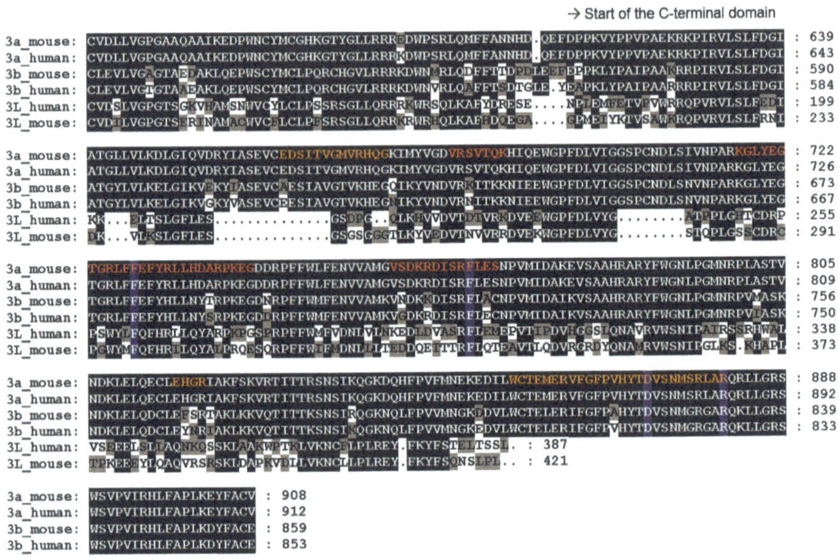

JELTSCH AND JURKOWSKA, FIGURE 16.1

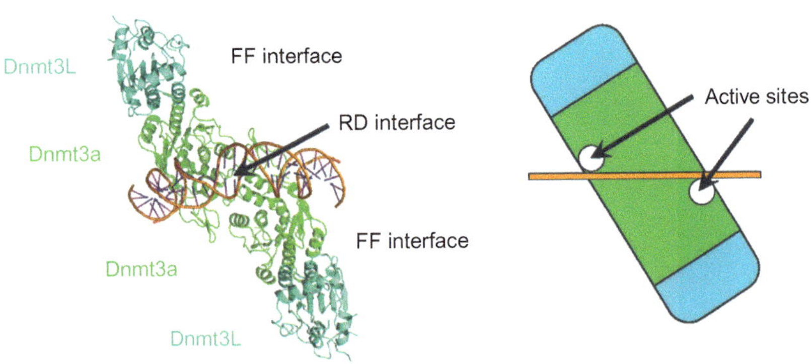

JELTSCH AND JURKOWSKA, FIGURE 16.2

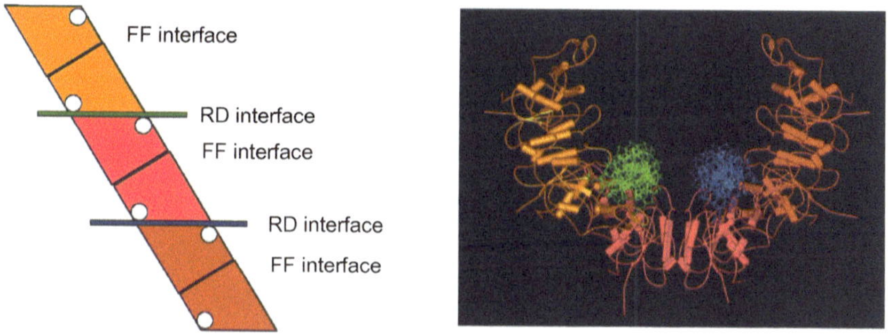

Jeltsch and Jurkowska, Figure 16.3

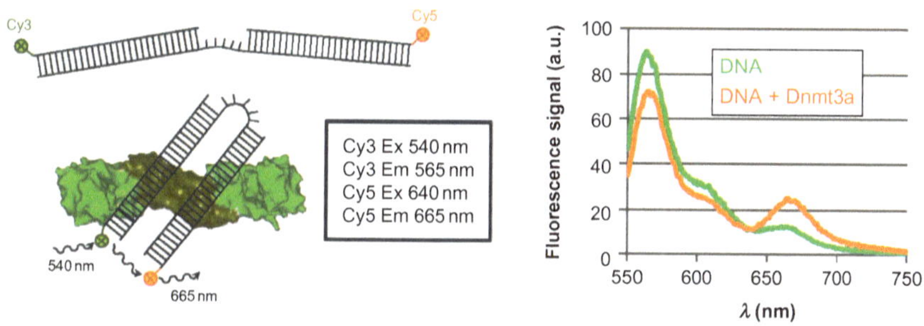

Jeltsch and Jurkowska, Figure 16.4

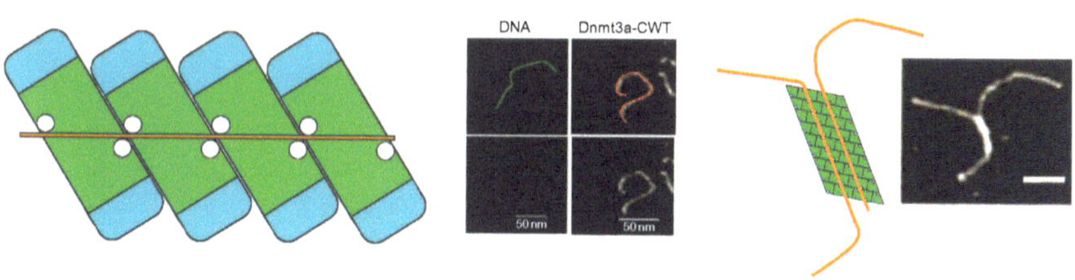

Jeltsch and Jurkowska, Figure 16.5

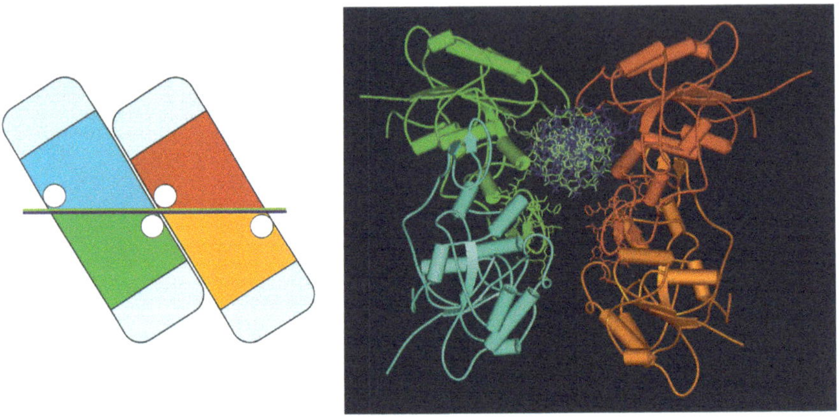

JELTSCH AND JURKOWSKA, FIGURE 16.6

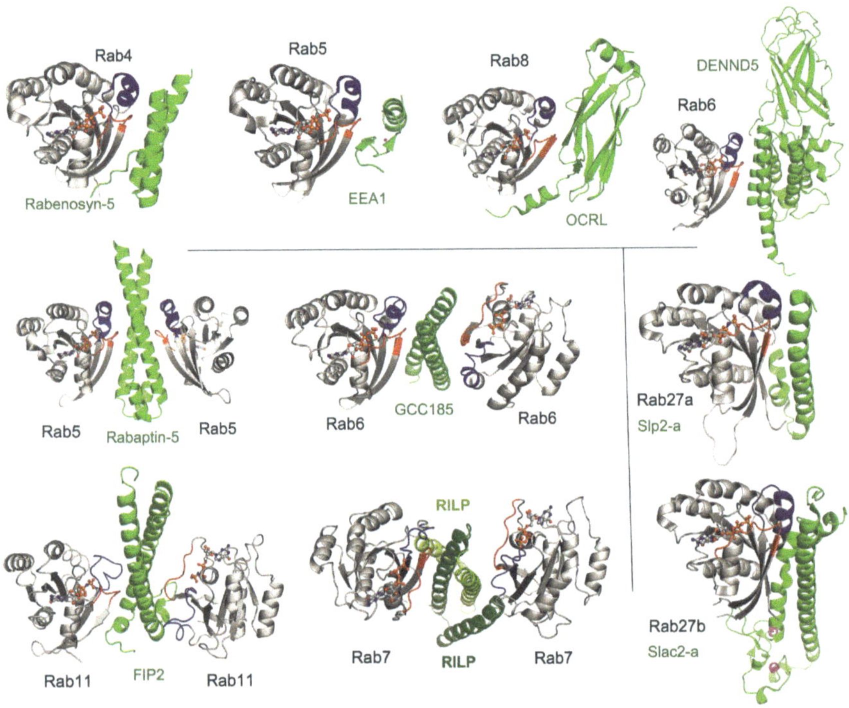

KHAN, FIGURE 21.1

FSC
www.fsc.org
MIX
Papier | Fördert
gute Waldnutzung
FSC® C083411

Kolibri 360 GmbH
Ferdinand-Jühlke-Straße 7
99095 Erfurt, Deutschland
produktsicherheit@kolibri360.de